Electronic Circuits and Applications

Muhammad H Rashid, Ph.D.
Fellow IET (UK), Life Fellow IEEE (USA)
Electrical and Computer Engineering
University of West Florida

Cengage

Australia • Brazil • Canada • India • Mexico • Singapore • United Kingdom • United States

Cengage

Electronic Circuits and Applications
Muhammad H Rashid

ISBN-13: 978-81-315-2284-4
ISBN-10: 81-315-2284-9

Cengage Learning India Pvt. Ltd.
T-303, 3rd Floor
DAV Complex, Plot 3
Mayur Vihar Phase 1
Delhi 110091

Printed by:
Saurabh Printers Pvt. Ltd.
67A-68, Ecotech Ext-I,
Greater Noida,-201308

Printed in India
Print Number: 03 Print Year: 2025

To my parents,
my wife, Fatema,
my family, Faeza, Farzana, Hasan,
Hannah, Laith, Laila, and Nora

CONTENTS

Chapter 2 Feedback Amplifiers

Chapter 3 Power Amplifiers

Chapter 4 Active Filters

Chapter 5 Oscillators

PREFACE

Semiconductor diodes and transistors are the fundamental devices for electronic circuits and applications. Op-amps can be used as the building blocks to design circuits for practical applications. An op-amp is also made of semiconductor devices. The electronic devices are covered in the first course. The second electronic course usually covers applications of electronic devices and op-amps such as power amplifier, feedback, filters, and oscillators. One or more of these applications are often integrated to build an electronic system. Feedback techniques are often applied to electronic circuits to maintain the desired bandwidth and the amplifier gain while the electronic devices and circuit elements are subjected to variations. A differential amplifier is the input stage of op-amps, and it is often covered to introduce internal parameters and characteristics of op-amps. This book focuses on the applications and detailed analysis of electronic circuits. The objectives of this book are as follows:

- To develop an understanding of the principles and techniques in the applications of electronic circuits.
- To develop analytical skills and design techniques of feedback amplifiers, oscillators, power amplifiers, and electronic filters.
- To design electronic circuits to meet desired specifications.
- To introduce students to the various elements of the engineering design process, including formulation of specifications, analysis of alternative solutions, synthesis, decision making, iterations, consideration of cost factors, simulation, and tolerance issues.

Approach

The National Board of Accreditation (NBA) criteria and other engineering accreditation criteria that satisfy the international engineering educational standards under the Washington Accord graduate attributes (*I*) require the integration of design and computer usage throughout the curriculum. The lack of opportunities for design credits in engineering curricula is a common concern. Electronics is generally the first electrical engineering course well suited to the integration of design components and computer usage. This book is structured to permit design content to constitute at least 50% of the course, and it integrates computer usage through PSpice. Many design examples use PSpice to verify the design requirements, and the numerous computer-aided design examples illustrate the usefulness of personal computers as design tools, especially in cases in which design variables are subjected to component tolerances.

Student Course Learning Outcomes (CLOs or COs): Course Outcomes are narrower statements that describe what students are expected to know, and be able to do at the end of each course. These relate to the skills, knowledge, and behavior that students acquire in their matriculation through the course.

After successfully completing the course with a grade of C (2.0/4.0) or better, the student should be able to do the following:

- Outline basic analog electronic circuit design techniques and analytical skills using multiple transistor circuits and Op-Amps.
- Apply basic engineering science to the design, analysis, and operation of active filters, circuits with feedback, oscillators, and power amplifiers.

- Develop problem-solving skills for design of electronic circuits in order to meet desired specifications.
- Apply modern simulation tools such as PSICE or PSpice for the design, analyses, and performance evaluations of electronic circuits.

Content and Organization

Since a differential amplifier is a combination of two single-stage amplifiers, it follows immediately after the first electronic course or at the beginning of the second electronic course. Since the feedback technique is applied to many applications, it is covered next. Active filters are built with op-amps and the oscillators can be built with either op-amps or transistors. The chapters are organized in order of differential amplifiers, feedback amplifiers, power amplifiers, active filters, and oscillators.

As a guide to the instructors for course planning, the table of contents of the book on *Electronic Devices and Circuits* by M.H. Rashid, Cengage Learning India (1/e, 2014) includes the following chapters:

Chapter I: Introduction to Electronics and Design
Chapter 2: Introduction to Amplifiers and Frequency Response
Chapter 3: Introduction to Operational Amplifiers and Applications
Chapter 4: Semiconductor Diodes
Chapter 5: Applications of Diodes
Chapter 6: Semiconductors and *pn* Junction Characteristics
Chapter 7: Metal Oxide Semiconductor Field-Effect Transistors
Chapter 8: Bipolar Junction Transistors and Amplifiers
Chapter 9: Differential Amplifiers

The table of contents of the book on *Linear Integrated Circuits* by M.H. Rashid, Cengage Learning India (1/e, 2014) includes the following chapters:

Chapter 1: Differential Amplifiers
Chapter 2: Operational Amplifiers
Chapter 3: Introduction to Digital Electronics
Chapter 4: Integrated Analog Circuits and Applications

Pedagogy and Supplements

Mathematical derivations are kept to a minimum by using approximate circuit models of operational amplifiers, transistors, and diodes. The significance of these approximations is established by computer-aided analysis using PSpice. Important circuits are analyzed in worked-out examples in order to introduce the basic techniques and emphasize the effects of parameter variations. At the end of each chapter, review questions and problems test students' learning of the concepts developed in the chapter. The student learning outcomes (SLOs) are listed at the beginning of each chapter. Symbols and their meanings have been uniquely identified at the beginning of each chapter to serve as a quick reference to the students. Every chapter opens with an introduction that puts the content of the chapter in perspective of the field of microelectronics. Solved examples carry captions that identify the objective of the example. Notes interspersed through the text provide a link to other chapters and serve to guide students against common misconceptions and mistakes. Key points of most of the sections are summarized in a box in addition to an end-of-chapter summary. A list of references is included at the end of each chapter for those interested in further reading. End-of-chapter exercises are divided into Review Questions and Problems. Design problems and PSpice problems are identified by relevant symbols.

Support for Instructors

The solutions manual and slides of the figures in this book are available on request from Cengage Learning.
Teaching plans and suggested course outlines using this book are included just after this preface.

Acknowledgments

Thanks are due to the editorial team at Cengage Learning for their guidance and support.
I would also like to thank the following reviewers for their comments and suggestions:

Dr. Ezzat G. Bakhoum
University of West Florida

Dr. William T. Baumann
Virginia Polytechnic Institute and State University

Dr. Paul J. Benkeser
Georgia Institute of Technology

Dr. Alok K. Berry
George Mason University

Dr. Michael A. Bridgwood
Clemson University

Dr. Nadeem N. Bunni
Clarkson University

Dr. Wai-Kai Chen
University of Illinois at Chicago

Dr. Shirshak K. Dhali
Southern Illinois University

Dr. Constantine Hatziadoniu
Southern Illinois University

Dr. Bruce P. Johnson
University of Nevada-Reno

Dr. Frank Kornbaum
South Dakota State University

Dr. Oguz Kucur
Gebze Institute of Technology, Turkey

Dr. John A. McNeill
Worcester Polytechnic Institute

Dr. Bahram Nabet
Drexel University

Dr. Hemanshu R. Pota
Australian Defense Force Academy

Dr. Jack R. Smith
University of Florida

Dr. Robert D. Strattan
University of Tulsa

Finally, thanks to my family for their patience while I was occupied with this and other projects.
Any comments and suggestions regarding this book are welcome. They should be sent to the author at mrashidfl@gmail.com.

Muhammad H. Rashid
Web: http://uwf.edu/mrashid

Support for Instructors

Acknowledgements

TEACHING PLANS AND SUGGESTED COURSE OUTLINES

This text has more material than can be covered in a single-semester course. Instructors are advised to select topics that are appropriate to the course depending on the students' background preparations. A course on electronic devices and circuits is a pre-requisite for this course on electronic circuits and applications.

Second Electronics Course

This course usually covers (a) differential amplifiers (b) feedback amplifiers, (c) power amplifiers, (d) active filters. The differential amplifiers maybe covered in a third electronic course on Linear Integrated Circuits. The suggested course topics are shown in Table 1.

TABLE 1 Suggested topics for Second electronics course

Number of Weeks	Topics	Chapters	Sections
3	Differential Amplifiers	1	1.1–1.10
3	Feedback Amplifiers	2	2.1–2.12
3	Power Amplifiers	3	3.1–3.14
3	Active Filters	4	4.1–4.19
2	Oscillators	5	5.1–5.7
1	Exam		

Types of Design Projects

Engineering design involves open-ended problems whose objectives are only partially defined. Problem definition and identification of constraints are needed to achieve a satisfactory solution. Often considerable ingenuity is required to find an acceptable solution from alternative pathways, and generally several iterations are needed to reach the solution. Also, it is necessary to verify the solution by simulation and/or testing to ensure that the design objectives have been satisfied. The design projects that appear in this text vary in complexity; the time required to complete them varies from 1 h to 1 month. The design projects can be classified into four categories, depending on the time required: short design projects, mini design projects, medium design projects, and large design projects.

Short Design Projects

Short design projects can be completed in 1 or 2 h. Some of the problems at the end of each chapter fall into this category, including the following:

1. Design of simple current source biasing circuits (Chapter 1)
2. MOS Differential Amplifiers (1)
3. BJT Differential Amplifiers (1)
4. Delay Filters (4)

Mini Design Projects

Mini design projects can be completed in approximately 1 week. Following are some of the end-of-chapter problems in the text that fall into this category:

1. Design of power amplifiers (Chapter 11)
2. BiCMOS Amplifiers (1)
3. Tuned Amplifiers (3)
4. Class D Amplifiers (3)
5. All-Pass Filters (4)
6. Stagger-Tuned Band-Pass Filter (4)
7. Chebyshev Filters (4)

Medium Design Projects

Medium design projects can be completed in approximately 2 to 3 weeks. Following are some of the end-of-chapter problems in the text that fall into this category:

1. Design of active filters (Chapter 4)
2. Design of single-stage feedback amplifiers (Chapter 2)
3. Design of oscillators (Chapter 5)
4. Design of active current sources (Chapter 1)
5. Design of differential amplifiers with current source biasing (Chapter 1)

Large Design Projects

Large design projects can be completed in approximately 4 to 5 weeks. Following are some of the end-of-chapter problems in the text that fall into this category:

1. Design of higher-order active filters (Chapter 4)
2. Design of power amplifiers with current source biasing (Chapter 3)
3. Design of multistage feedback amplifiers (Chapter 1)
4. Cascode BiCMOS Amplifiers (1)
5. Double-Cascode BiCMOS Amplifiers (1)

Design Report

It is recommended that in your design reports you do the following:

- Give the complete design, including the ratings and values of each component.

- Justify the use of a particular circuit topology.
- Verify your design objectives by simulating your circuit using PSpice/SPICE/OrCAD or Electronics Workbench. Include a worst-case analysis (with 10% tolerances for all passive components, unless specified).
- Give a cost estimate. The project should be as inexpensive as possible.

A suggested format for design reports is as follows:

1. Title page (including your name, the course number, and the date)
   Include the following statement on the cover sheet:
   I/we certify that this assignment is the result of my/our own efforts.
   Signature(s): Date:
2. Design objectives and specifications
3. Design steps (including the circuit topology)
4. Design modifications
5. Computer simulation and design verification
6. Components and costs
7. Flowchart of the design process
8. Costs versus reliability and safety considerations
9. Conclusions

ABOUT THE AUTHOR

Muhammad H. Rashid is currently employed by the University of West Florida as a *Professor of Electrical and Computer Engineering*. Previously, he was employed by the University of Florida as *Professor and Director* of UF/UWF Joint Program. Rashid received B.Sc. degree in Electrical Engineering from the Bangladesh University of Engineering and Technology, and M.Sc. and Ph.D. degrees from the University of Birmingham in UK. Previously, he worked as Professor of Electrical Engineering and the Chair of the Engineering Department at Indiana University-Purdue University at Fort Wayne. He also worked as Visiting Assistant Professor of Electrical Engineering at the University of Connecticut, Associate Professor of Electrical Engineering at Concordia University (Montreal, Canada), Professor of Electrical Engineering at Purdue University Calumet, and Visiting Professor of Electrical Engineering at King Fahd University of Petroleum and Minerals (Saudi Arabia), as a design and development engineer with Brush Electrical Machines Ltd. (England, UK), a Research Engineer with Lucas Group Research Centre (England, UK), a Lecturer and Head of Control Engineering Department at the Higher Institute of Electronics (in Libya and Malta).

Dr. Rashid is actively involved in teaching, researching, and lecturing in electronics, power electronics, and professional ethics. He has published 18 books listed in the US Library of Congress and more than 160 technical papers. His books are adopted as textbooks all over the world. His book, *Power electronics* has translations in Spanish, Portuguese, Indonesian, Korean, Italian, Chinese, Persian, and Indian edition. His book *Microelectronics* has translations in Spanish in Mexico and in Spain, Italian, and Chinese.

He has received many invitations from foreign governments and agencies to give keynote lectures and consult, by foreign universities to serve as an external examiner for undergraduate, master's, and Ph.D. examinations, by funding agencies to review research proposals, and by US and foreign universities to evaluate promotion cases for professorship. Dr. Rashid has worked as a regular employee or consultant in Canada, Korea, the United Kingdom, Singapore, Malta, Libya, Malaysia, Saudi Arabia, Pakistan, and Bangladesh. Dr. Rashid has traveled to almost all States in the USA and many countries to lecture and present papers (Japan, China, Hong Kong, Indonesia, Taiwan, Malaysia, Thailand, Singapore, India, Pakistan, Turkey, Saudi Arabia, United Arab Emirates, Qatar, Libya, Jordan, Egypt, Morocco, Malta, Italy, Greece, United Kingdom, Brazil, and Mexico).

He is a *Fellow* of the Institution of Engineering and Technology (IET, UK) and a *Life Fellow* of the Institute of Electrical and Electronics Engineers (IEEE, USA). He was elected as an IEEE Fellow with the citation "*Leadership in power electronics education and contributions to the analysis and design methodologies of solid-state power converters*." Dr. Rashid is the recipient of the *1991 Outstanding Engineer Award* from The Institute of Electrical and Electronics Engineers (IEEE). He received the 2002 IEEE Educational Activity Award (EAB) Meritorious Achievement Award in Continuing Education with the following citation *"for contributions to the design and delivery of continuing education in power electronics and computer-aided simulation*." He is the recipient of the 2008 IEEE Undergraduate Teaching Award with citation "*For his distinguished leadership and dedication to quality undergraduate electrical engineering education, motivating students and publication of outstanding textbooks*." He is also the recipient of the IEEE 2013 Industry Applications Society *Outstanding Achievement Award.*

Dr. Rashid is an ABET program evaluator for electrical and computer engineering (also from 1995 to 2000) and was an engineering evaluator for the Southern Association of Colleges and Schools (SACS, USA). He is also an ABET program evaluator for (general) engineering program. He is the Series Editors of *Power Electronics and Applications*, and *Nanotechnology and Applications* with the CRC Press. He serves as the Editorial Advisor of *Electric Power and Energy* with Elsevier Publishing. He lectures and conducts workshops on Outcome-Based Education (OBE) and its implementations including assessments.

Dr. Rashid is a Distinguished Lecturer for the IEEE Education Society and a Regional Speaker (previously Distinguished Lecture) for the IEEE Industrial Applications Society. He also authored a book on *The Process of Outcome-Based EducationImplementation, Assessment and Evaluations,* 2012 UiTM Press, Malaysia.

CHAPTER 1

DIFFERENTIAL AMPLIFIERS

Learning Outcomes

After completing this chapter, students should be able to do the following:

- Design, analyze, and evaluate different types of MOSFET constant-current sources for biasing MOSFET amplifiers.
- Design, analyze, and evaluate different types of BJT constant-current sources for biasing BJT amplifiers.
- Describe and analyze the characteristics of differential amplifiers and their DC and small-signal characteristics.
- Identify the parameters influencing the differential and the common-mode gains of differential amplifiers.
- Analyze and evaluate cascode-connected transistors in order to obtain higher differential voltage gains.

Symbols and Their Meanings

Symbol	Meaning
A_d, A_c	Small-signal differential and common-mode voltage gains
CMRR	Common-mode rejection ratio of an amplifier
v_{id}, v_{ic}	Small-signal differential and common-mode signal
v_{o1}, v_{o2}	Small-signal output voltages due to input voltages at inverting and noninverting terminals

Symbol	**Meaning**
r_{o1}, r_{o2}	Small-signal output resistances of transistors
V_M, V_A	MOSFET channel modulation and BJT Early voltages
V_t, V_p	Threshold voltage of enhancement MOSFET and pinch-off voltage of depletion MOSFET

1.1 Introduction

Differential amplifiers are commonly used as an input stage in various types of analog ICs, such as operational amplifiers, voltage comparators, voltage regulators, video amplifiers, power amplifiers, and balanced modulators and demodulators [1–3]. A differential amplifier is a very important transistor stage and determines many of the performance characteristics of an IC. In ICs, including differential amplifiers, it is unnecessary to bias transistors by setting the values of biasing resistors. Because of variations in resistor values, power supply, and temperature, the quiescent point of transistors changes. Transistors can be used to generate the characteristics of DC constant-current sources. Transistors can also be used to produce an output voltage source that is independent of its load or, equivalently, of the output current. This chapter covers the operation, analysis, and characteristics of differential amplifiers using BJTs and MOSFETs. It also covers active current sources and voltage sources.

1.2 Internal Structure of Differential Amplifiers

A differential amplifier acts as an input stage; its output voltage is proportional to the difference between its two input voltages v_1 and v_2. It has a high voltage gain and is directly DC coupled to the input voltages and the load. As we will see later in this chapter, the voltage gain of a differential amplifier depends directly on the output resistance of the current source acting as an active load.

In amplifiers with discrete components, passive components such as resistors and capacitors are less expensive than active devices such as transistors (i.e., MOSFETs and BJTs); thus, in multistage amplifiers, interstage coupling is accomplished with capacitors. However, in monolithic circuits, the die area is the principal determining cost factor. Capacitors of the values and sizes used in amplifiers made with discrete components cannot be included in ICs and must be external to the chip. But using external capacitors increases the pin count of the package and the cost of the IC. A DC-coupled circuit is used to eliminate capacitors. The cheapest component in an IC is the one that can be fabricated within the least area, usually the transistor. The optimal IC has as little resistance as possible and more transistors.

1.2.1 Characteristics of Differential Amplifiers

The differential stage can be represented by an equivalent amplifier, as shown in Fig. 1.1(a). If the two input voltages are equal, a differential amplifier gives an output voltage of almost zero. Its voltage gain is very large, so the input voltage is low, typically less than 50 mV. Thus, we can consider the input voltages as small signals with zero DC components. That is, $v_{G1} = v_{g1}$ and $v_{G2} = v_{g2}$.

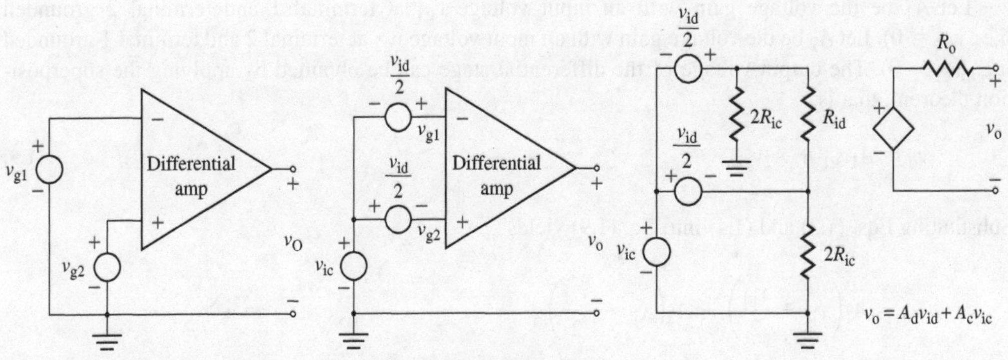

(a) **Equivalent differential amp**

(b) **Small-signal differential and common-mode inputs**

(c) **Equivalent circuit**

FIGURE 1.1 Small-signal equivalent circuit with differential and common-mode inputs

Let us define a differential voltage v_{id} as

$$v_{id} = v_{g1} - v_{g2} \tag{1.1}$$

and a common-mode voltage v_{ic} as

$$v_{ic} = \frac{v_{g1} + v_{g2}}{2} \tag{1.2}$$

From Eqs. (1.1) and (1.2), the two input voltages can be expressed as

$$v_{g1} = v_{ic} + \frac{v_{id}}{2} \tag{1.3}$$

and $$v_{g2} = v_{ic} - \frac{v_{id}}{2} \tag{1.4}$$

By replacing the input signals with the equivalent differential and common-mode signals, we can represent the differential stage by an equivalent amplifier, as shown in Fig. 1.1(b). Let v_{o1} be the output voltage due to v_{g1} only, and let v_{o2} be the output voltage due to v_{g2} only. Then we can define a differential output voltage v_{od} as

$$v_{od} = v_{o1} - v_{o2} \tag{1.5}$$

and a common-mode output voltage as

$$v_{oc} = \frac{v_{o1} + v_{o2}}{2} \tag{1.6}$$

From Eqs. (1.5) and (1.6), the two output voltages can be expressed as

$$v_{o1} = v_{oc} + \frac{v_{od}}{2} \tag{1.7}$$

and $$v_{o2} = v_{oc} - \frac{v_{od}}{2} \tag{1.8}$$

Let A_1 be the voltage gain with an input voltage v_{g1} at terminal 1 and terminal 2 grounded (i.e., $v_{g2} = 0$). Let A_2 be the voltage gain with an input voltage v_{g2} at terminal 2 and terminal 1 grounded (i.e., $v_{g1} = 0$). The output voltage of the differential stage can be obtained by applying the superposition theorem; that is,

$$v_o = A_1 v_{g1} + A_2 v_{g2} \tag{1.9}$$

Substituting Eqs. (1.3) and (1.4) into Eq. (1.9) yields

$$v_o = A_1\left(v_{ic} + \frac{v_{id}}{2}\right) + A_2\left(v_{ic} - \frac{v_{id}}{2}\right)$$

$$= \left(\frac{A_1 - A_2}{2}\right)v_{id} + (A_1 + A_2)v_{ic}$$

$$= A_d v_{id} + A_c v_{ic} \tag{1.10}$$

$$= A_d\left(v_{id} + \frac{A_c}{A_d}v_{ic}\right) \tag{1.11}$$

where $A_d = (A_1 - A_2)/2$ = differential voltage gain and $A_c = A_1 + A_2$ = common-mode voltage gain.

The output voltage v_o in Eq. (1.11) is due to a common-mode input voltage v_{ic} and a differential input voltage v_{id}. If A_d is much greater than A_c, the output voltage will be almost independent of the common-mode signal v_{ic}. A differential amplifier is expected to amplify the differential voltage as much as possible while rejecting (not amplifying) common-mode signals such as noise or other unwanted signals, which will be present in both terminals.

The ability of an amplifier to reject common-mode signals is defined by a performance criterion called the *common-mode rejection ratio* (CMRR), which is defined by

$$\text{CMRR} = \left|\frac{A_d}{A_c}\right| \tag{1.12}$$

$$= 20 \log\left|\frac{A_d}{A_c}\right| \quad \text{(in dB)} \tag{1.13}$$

Substituting Eq. (1.12) into Eq. (1.11) gives the output voltage

$$v_o = A_d\left(v_{id} + \frac{1}{\text{CMRR}}v_{ic}\right) \tag{1.14}$$

which shows that, to reduce the effect of v_{ic} on the output voltage v_o—that is, to get v_{oc} to approach zero—the value of CMRR must be very large, tending to infinity for an ideal amplifier. Thus, a differential amplifier should behave differently for common-mode and differential signals. The small-signal equivalent circuit is shown in Fig. 1.1(c). R_{id} and R_{ic} are the input resistances due to the differential and common-mode signals, respectively. The parameters of a differential amplifier are A_d (ideally ∞), A_c (0), CMRR (∞), R_{id} (∞), and R_{ic} (∞). In the following sections, we will determine the circuit elements affecting these parameters.

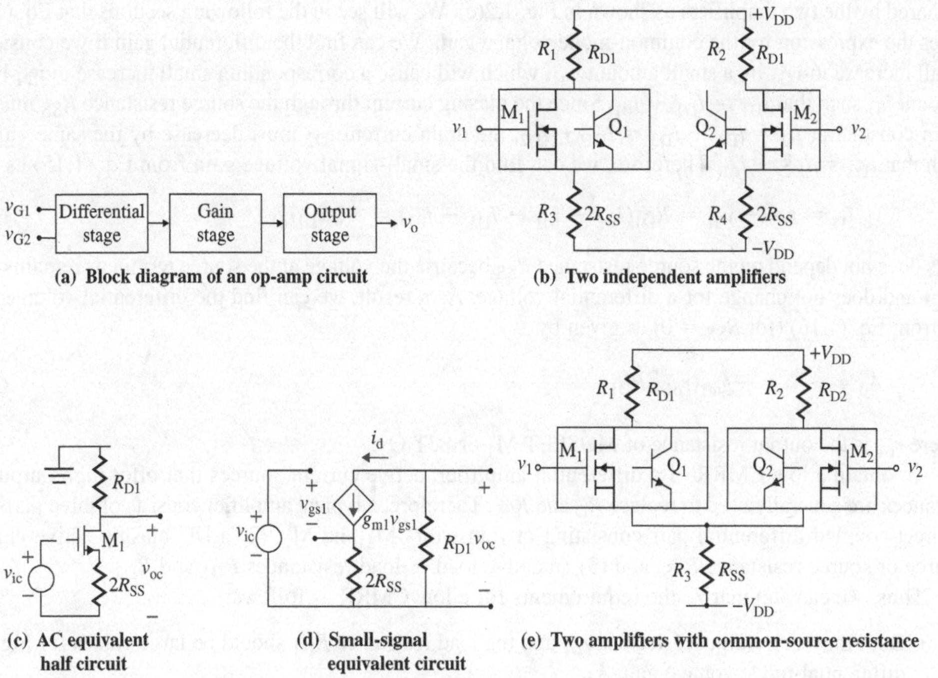

FIGURE 1.2 Basic structure of a differential amplifier

1.2.2 Internal Structure of Differential Amplifiers

A differential amplifier [4] acts as an input stage of an op-amp circuit as shown in Fig. 1.2(a). It serves as a direct DC-coupled differential stage. It may be viewed as consisting of two common-source (CS) amplifiers as shown in Fig. 1.2(b). If the two amplifiers are identical—that is, $R_{D1} = R_{D2}$—their DC biasing drain currents will be the same, and $i_{D1} = i_{D2}$. The DC output voltages are given by

$$v_{o1} = V_{DD} - R_{D1}i_{D1} \quad \text{and} \quad v_{o2} = V_{DD} - R_{D2}i_{D2}$$

The output voltage between the two output terminals of M_2 and M_1 is

$$v_{o(21)} = v_{o2} - v_{o1} = V_{DD} - R_{D2}i_{D2} - V_{DD} + R_{D1}i_{D1} = R_{D1}(i_{D2} - i_{D1}) \tag{1.15}$$

Therefore, we get $v_{o(21)} = v_{o(12)} = 0$ for identical amplifiers of the condition $i_{D1} = i_{D2}$. The small-signal AC equivalent half-circuit is shown in Fig. 1.2(c), which, after replacing the transistor by it transconductance model, becomes as shown in Fig. 1.2(d). Assume that the output of the transistor M_1 is $r_{o1} >> R_{D1}$. From Fig. 1.2(d), we get $v_{gs1} = v_{ic} - g_{m1}v_{gs1}(2R_{SS})$, $v_{gs1}(1 + g_{m1}\,2R_{SS}) = v_{ic}$, $v_{oc} = g_{m1}v_{gs1}R_{D1}$. Now we can find the small-signal common-mode voltage gain as given by

$$A_c = \frac{v_{oc}}{v_{ic}} = -\frac{g_{m1}R_{D1}}{1 + 2g_{m1}R_{SS}} \tag{1.16}$$

The derivation of Eq. (1.16) assumes that no interaction exists between the two amplifiers, but there is an interaction between them. The source (or emitter) resistance, whose value is $R_3 \parallel R_4 (= 2R_{SS} \parallel 2R_{SS} = R_{SS})$, is shared by the two amplifiers as shown in Fig. 1.2(e). We will see in the following sections that Eq. (1.16) gives the expression for the common-mode voltage gain. We can find the differential gain if we consider a small increase in v_{G1} by a small amount v_{g1}, which will cause a corresponding small increase in i_{D1} by an amount i_{d1} such that $i_{D1} = I_{D1} + i_{d1}$. Since the biasing current through the source resistance R_{SS} must remain constant at $I_O = i_{D1} + i_{D2} = I_{D1} + I_{D2}$, the drain current i_{D2} must decrease by the same amount such that $i_{D2} = I_{D2} - i_{d1}$. Therefore, we can find the small-signal voltage gain from Eq. (1.15) as

$$v_{o(21)} = v_{o2} - v_{o1} = R_{D1}(I_{D2} - i_{d1} - I_{D1} - i_{d1}) = -2R_{D1}i_{d1} \tag{1.17}$$

This does not depend on the source resistance R_{SS} because the voltage at the source terminals remains constant and does not change for a differential voltage. As a result, we can find the differential voltage gain A_d from Eq. (1.16) (for $R_{SS} = 0$) as given by

$$A_d = \frac{v_{od}}{v_{id}} = -g_{m1}(r_{o1} \| R_{D1}) \tag{1.18}$$

where r_{o1} is the output resistance of MOSFET M_1 or BJT Q_1.

To obtain a low CMRR of a differential amplifier, active current sources that offer high output resistances are generally used to replace R_D and R_{SS}. Therefore, a typical amplifier consists of three parts: (1) a direct-coupled differential pair consisting of transistors M_1 and M_2, (2) a DC biasing active current source or source resistance R_{SS}, and (3) an active load or load resistances R_{D1} and R_{D2}.

Thus, we can summarize the requirements for a low CMRR as follows:

- Both transistor output resistance r_{o1} and the load resistance R_{D1} should be large values for high differential-mode voltage gain A_d.
- The source resistance R should be large values for a low common-mode voltage gain A_c.

KEY POINTS OF SECTION 1.2

- A differential amplifier consists of an active biasing circuit, an active load, and a differential transistor pair.
- The performance of a differential amplifier is measured by a differential gain A_d that occurs in response to a differential voltage between two input terminals, a common-mode gain A_c that occurs in response to a voltage common to both input terminals, and a common-mode rejection ratio CMRR.
- The CMRR is the ratio of the differential gain to the common-mode gain, and it is a measure of the ability of an amplifier to amplify the differential signal and reject common-mode signals.

1.3 MOSFET Current Sources

An active current source has an inherent high output resistance. Thus, replacing the load resistance of a transistor with a current source increases the voltage gain significantly. Transistor current sources are widely used in analog ICs both as biasing elements and as loads for amplifying stages. Current sources are less sensitive to variations in DC power supply and temperature. Especially, for a small value of bias current, the current sources are more economical than resistors in terms of die area required for resistors. A current source can be designed by using either MOSFETs or BJTs [5].

MOSFET current sources are analogous to BJT current sources. We can convert a BJT current source to an equivalent MOSFET current source by assuming that the β_F of the BJTs is infinite. Since MOSFETs

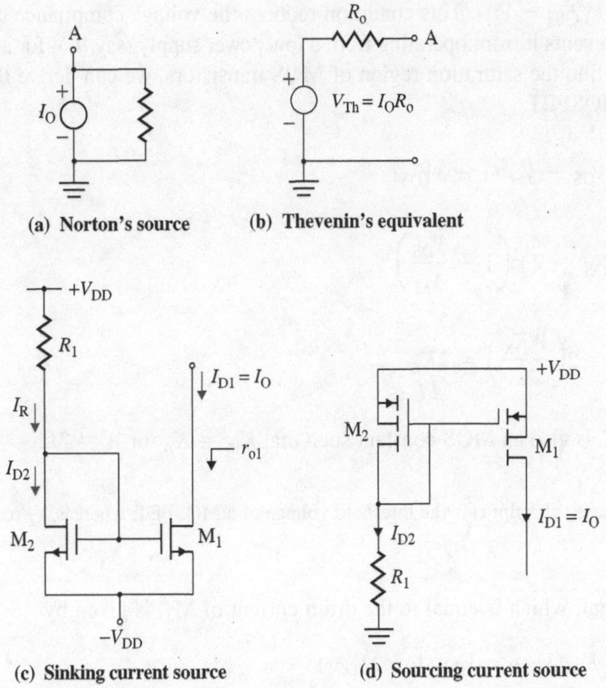

FIGURE 1.3 Basic MOSFET current source

do not draw any gate current, there is no need for gate current compensation as there is with BJTs. The choice of a BJT source or a MOSFET source generally depends on the type of integrated circuit involved (e.g., bipolar or MOS). BJT sources have some advantages over MOSFET sources, such as a wider compliance range and a higher output resistance. However, a higher output resistance can be obtained by cascode-like connections of MOSFETs.

Norton's equivalent of a current source I_O having an output resistance R_o is shown in Fig. 1.3(a), and its Thevenin's equivalent is in Fig. 1.3(b). An ideal current source should maintain a constant current at an infinite output resistance under all operating conditions. To achieve this goal, a number of current sources are developed. The commonly used current sources are the basic current source the modified basic source, the cascoded current source, and the Wilson current source.

1.3.1 Basic Current Source

A basic MOSFET current source is shown in Fig. 1.3(c). The direction of the current flow in Fig. 1.3(c) is *into* the current source circuit; this type of constant-current source is often referred to as a *current sink.* Let us assume that the two transistors M_1 and M_2 are identical. Since their gate–source voltages are equal, their drain currents will be the same; that is, $I_{D1} = I_{D2}$. Thus, the output current I_O $(=I_{D1})$ will be the mirror of I_{D2}. Since $V_{DS2} = V_{GS2}$, M_2 will be in saturation. Let V_{t1} and V_{t2} be the threshold voltages of M_1 and M_2, respectively. For M_1 also to be in saturation, V_{DS1}, which is greater than or equal to $(V_{GS2} - V_{t2})$, must be greater than $(V_{GS1} - V_{t1})$. This condition reduces the voltage compliance range of the MOSFET current source and prevents it from operating from a low power supply (say, 1 V for a battery source). Using the equations that define the saturation region of MOS transistors, we can derive the drain current of an enhancement-type MOSFET:

$$I_D = \frac{WK_x}{2L}(V_{GS} - V_t)^2(1 + \lambda V_{DS}) \tag{1.19}$$

$$= \frac{K_m}{2}(V_{GS} - V_t)^2\left(1 + \frac{V_{DS}}{V_M}\right) \tag{1.20}$$

$$K_m = \frac{WK_x}{L} = 2\left(\frac{WK_x}{2L}\right) = 2K_n \tag{1.21}$$

where $K_n = WK_x/2L$ is also an MOS constant such that $K_m = K_n$ for $W = 2L$.

▶ **NOTE** V_t (lowercase subscript t) is the threshold voltage of a MOSFET, whereas V_T (capital subscript T) is the thermal voltage.

The output current, which is equal to the drain current of M_1, is given by

$$I_{D1} = I_O = K_{n1}(V_{GS1} - V_{t1})^2(1 + \lambda V_{DS1}) \tag{1.22}$$

Drain current I_{D2}, which is equal to the reference current I_R, is given by

$$I_{D2} = I_R = K_{n2}(V_{GS2} - V_{t2})^2(1 + \lambda V_{DS2}) \tag{1.23}$$

In practice, all the components of the current source are processed on the same IC, and hence all physical parameters such as K_x and V_t are identical for both devices. Thus, the ratio of I_O to I_R is given by

$$\frac{I_O}{I_R} = \frac{K_{n1}(1 + \lambda V_{DS1})}{K_{n2}(1 + \lambda V_{DS2})} = \frac{(W/L)_1}{(W/L)_2} \times \frac{(1 + \lambda V_{DS1})}{(1 + \lambda V_{DS2})} \tag{1.24}$$

In practice, $\lambda V_{DS} << 1$. Thus, Eq. (1.24) can be approximated by

$$\frac{I_O}{I_R} = \frac{(W/L)_1}{(W/L)_2} \tag{1.25}$$

By controlling the ratio W/L, therefore, we can change the output current. The gate length L is usually held fixed, and the gate width W is varied from device to device to give the desired current ratio I_O/I_R. By choosing identical transistors with $W_1 = W_2$ and $L_1 = L_2$, a designer can ensure that the output current I_O is almost equal to the reference current I_R.

Since $V_{GS2} = V_{DD} - R_1 I_R$ and $V_{DS2} = V_{GS2}$, the reference current I_R can be found approximately from Eq. (1.23); that is,

$$I_R = I_{D2} = K_{n2}(V_{DD} - R_1 I_R - V_{t2})^2 \tag{1.26}$$

can be solved for known values of V_{t2}, K_{n2}, V_{DD}, and R_1. The current source in Fig. 1.3(c) behaves as a current sink rather than a source. A current source equivalent to this current sink can be obtained by using *p*-channel MOS (PMOS) transistors as shown in Fig. 1.3(d).

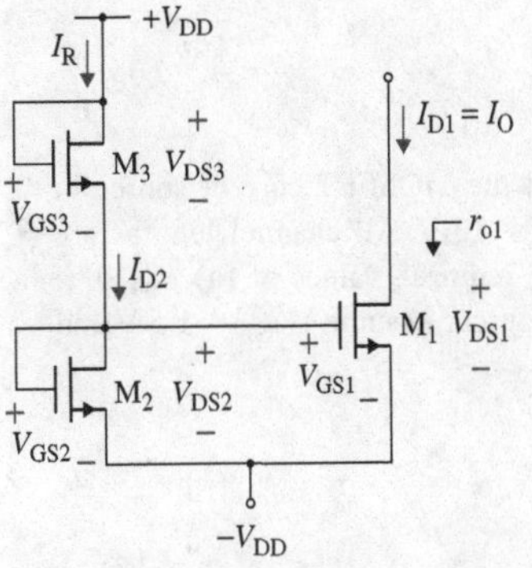

FIGURE 1.4 Basic MOSFET current source without resistance

1.3.2 Modified Basic Current Source

The reference resistance R_1 can be replaced by another MOSFET M_3, as shown in Fig. 1.4. Transistors M_2 and M_3 are used as voltage dividers to control the gate–source voltage of transistor M_1. If M_1 and M_2 are identical, the output current I_O exactly mirrors the drain current through M_2 and M_3. The value of V_{GS1} should be made as low as possible without taking M_1 out of the saturation region.

Since $V_{GS1} = V_{GS2}$, the drain current I_{D1} ($= I_O$) is equal to the drain current I_{D2} and is given by

$$I_{D1} = I_{D2} = I_R = K_{n1}(V_{GS1} - V_{t1})^2(1 + \lambda V_{GS1}) \tag{1.27}$$

Since $V_{DS2} = V_{GS2}$, the drain current I_{D2} is equal to the reference current I_R and is given by

$$I_{D2} = I_R = K_{n2}(V_{GS2} - V_{t2})^2(1 + \lambda V_{GS2}) \tag{1.28}$$

Since $V_{GS3} = V_{DD} - V_{GS2}$, the drain current of M_3 is given by

$$\begin{aligned} I_{D3} = I_R &= K_{n3}(V_{GS3} - V_{t3})^2(1 + \lambda V_{DS3}) \\ &= K_{n3}(V_{DD} - V_{GS2} - V_{t3})^2[1 + \lambda(V_{DD} - V_{GS2})] \end{aligned} \tag{1.29}$$

Since $I_{D2} = I_{D3} = I_R$, from Eqs. (1.28) and (1.29) we get

$$\frac{K_{n2}(V_{GS2} - V_{t2})^2(1 + \lambda V_{GS2})}{K_{n3}(V_{DD} - V_{GS2} - V_{t3})^2[1 + \lambda(V_{DD} - V_{GS2})]} = 1 \tag{1.30}$$

Thus, by controlling the constants K_{n2} and K_{n3}, we can obtain the desired value of $V_{GS2} = V_{GS1}$, which will give the desired output current.

Output Resistance R_o

The small-signal drain-source resistance r_{ds1} can be derived from Eq. (1.20):

$$\frac{1}{r_{ds1}} = \frac{\delta i_{D1}}{\delta v_{DS1}} = \frac{K_{n1}}{V_M}(V_{GS} - V_t)^2 \approx \frac{I_{D1}}{V_M} \tag{1.31}$$

Thus, the small-signal output resistance of the current source becomes

$$R_o = r_{ds1} = \frac{V_M}{I_{D1}} = \frac{1}{\lambda I_{D1}} = r_{o1} \tag{1.32}$$

which is relatively small. This small output resistance is a disadvantage of having only one MOSFET M_1 at the output side of a current source.

EXAMPLE 1.1

D

Designing a simple MOSFET current source The parameters of the MOSFET current source in Fig. 1.4 are $V_t = 1$ V, $I_O = 50$ μA, $I_R = 40$ μA, $V_{DD} = 10$ V, and $V_M = 10$ V. All channel lengths are equal, $L_1 = L_2 = L_3 = L = 10$ μm, and $K_x = 20$ μA/V^2. Calculate the required values of **(a)** K_{n1}, W_1, **(b)** K_{n2}, W_2, **(c)** K_{n3}, W_3, and **(d)** the output resistance R_o of the current source. Assume $V_{GS1} = 1.5$ V and $V_{DS1} = 5$ V.

SOLUTION

(a) From Eq. (1.22),

$$50 \times 10^{-6} = K_{n1}(1.5 - 1)^2\left(1 + \frac{5}{10}\right)$$

which gives $K_{n1} = 133.3$ μA/V^2. From Eq. (1.21),

$$133.3 \times 10^{-6} = \frac{W_1 \times 20 \times 10^{-6}}{10 \times 10^{-6}}$$

which gives $W_1 = 66.65$ μm.

(b) $V_{DS2} = V_{GS2} = V_{GS1} = 1.5$ V. From Eq. (1.28),

$$40 \times 10^{-6} = K_{n2}(1.5 - 1)^2\left(1 + \frac{1.5}{10}\right)$$

which gives $K_{n2} = 139.1$ μA/V^2. From Eq. (1.21),

$$139.1 \times 10^{-6} = \frac{W_2 \times 20 \times 10^{-6}}{10 \times 10^{-6}}$$

which gives $W_2 = 69.55$ μm.

(c) $V_{GS3} = V_{DS3} = V_{DD} - V_{DS2} = V_{DD} - V_{GS1} = 10 - 1.5 = 8.5$ V. From Eq. (1.29),

$$40 \times 10^{-6} = K_{n3}(8.5 - 1)^2\left(1 + \frac{8.5}{10}\right)$$

which gives $K_{n3} = 0.384$ μA/V^2. From Eq. (1.21),

$$0.384 \times 10^{-6} = \frac{W_3 \times 20 \times 10^{-6}}{10 \times 10^{-6}}$$

which gives $W_3 = 0.192$ μm. In practice, because of manufacturing limitations, the minimum value of L or W is 10 μm. Since the value of W_3 is smaller than 10 μm, V_M must be increased to make W_3 at least 10 μm.

(d) From Eq. (1.32), the output resistance R_o is

$$R_o = r_{ds1} = \frac{V_M}{K_{n1}(V_{GS1} - V_t)^2} = \frac{10}{133.3 \times 10^{-6} \times (1.5 - 1)^2} = 300\ \text{k}\Omega$$

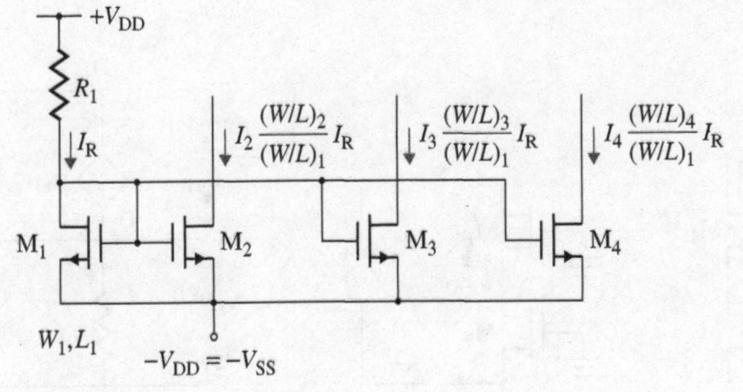

FIGURE 1.5 Multiple MOSFET current sources

1.3.3 Multiple Current Sources

Since there is no gate current in a MOSFET, a number of MOSFETs can be connected to a single reference MOSFET M_1, as shown in Fig. 1.5. Different output currents can be obtained by suitably adjusting the width-to-length ratios of MOSFETs (i.e., M_2, M_3, and M_4). In practice, the gate length L is normally kept constant, and the gate widths (W) of M_2, M_3, and M_4 are varied, to give the desired output currents. Thus, for equal L, Eq. (1.25) gives the relationship of the output currents I_2, I_3, and I_4 to I_R as

$$I_2 = \left(\frac{W_2}{W_1}\right) I_R$$

$$I_3 = \left(\frac{W_3}{W_1}\right) I_R$$

$$I_4 = \left(\frac{W_4}{W_1}\right) I_R$$

Output Resistance R_o

The small-signal output resistance of the current source is

$$R_o = r_{ds1} = \frac{V_M}{I_{D1}} = \frac{1}{\lambda I_{D1}} = r_{o1}$$

1.3.4 Cascode Current Source

The output resistance of the basic current source in Fig. 1.3(c) can be increased by adding two more MOSFETs in a cascodelike connection, as shown in Fig. 1.6(a). The analysis of the circuit is straightforward. The small-signal circuit for finding the output resistance is shown in Fig. 1.6(b), and its small-signal equivalent is shown in Fig. 1.6(c); r_{o2} is the output resistance of transistor M_2. Using KVL and the relation $v_{gs1} = -r_{o2} i_x$, we get

$$\begin{aligned} v_x &= r_{o1} i_1 + r_{o2} i_x \\ &= r_{o1}(i_x - g_{m1} v_{gs1}) + r_{o2} i_x \\ &= r_{o1}(i_x + g_{m1} r_{o2} i_x) + r_{o2} i_x \end{aligned}$$

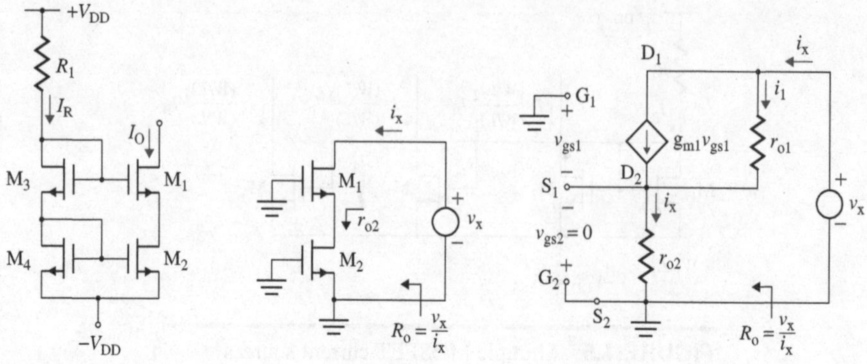

FIGURE 1.6 Cascode current source

which gives the output resistance R_o of the current source as

$$R_o = r_{o1}(1 + g_{m1}r_{o2}) + r_{o2} \tag{1.33}$$

For identical transistors, $r_{o1} = r_{o2} = r_o$, and R_o becomes

$$R_o = r_o(2 + g_{m1}r_o) \tag{1.34}$$

$$\approx g_{m1}r_o^2 \tag{1.35}$$

Thus, the output resistance can be significantly increased, to a level comparable to that of a BJT source. However, the voltage compliance range will be reduced because of the two drain-source voltages in series (i.e., $V_{DS1} + V_{DS2}$).

1.3.5 Wilson Current Source

The MOSFET version of the Wilson current source is shown in Fig. 1.7(a). The equivalent circuit for finding the output resistance R_o is shown in Fig. 1.7(b). We have

$$v_{gs3} = \frac{i_x}{g_{m2}}$$

and $\quad v_{gs1} + v_{gs3} = -g_{m3}v_{gs3}r_{o3}$

which can be simplified to relate v_{gs1} to v_{gs3} and v_{gs2} by

$$v_{gs1} = -(1 + g_{m3}r_{o3})v_{gs3} = -\frac{(1 + g_{m3}r_{o3})i_x}{g_{m2}}$$

Applying KVL to Fig. 1.7(b), we get

$$v_x = (i_x - g_{m1}v_{gs1})r_{o1} + \frac{i_x}{g_{m2}}$$

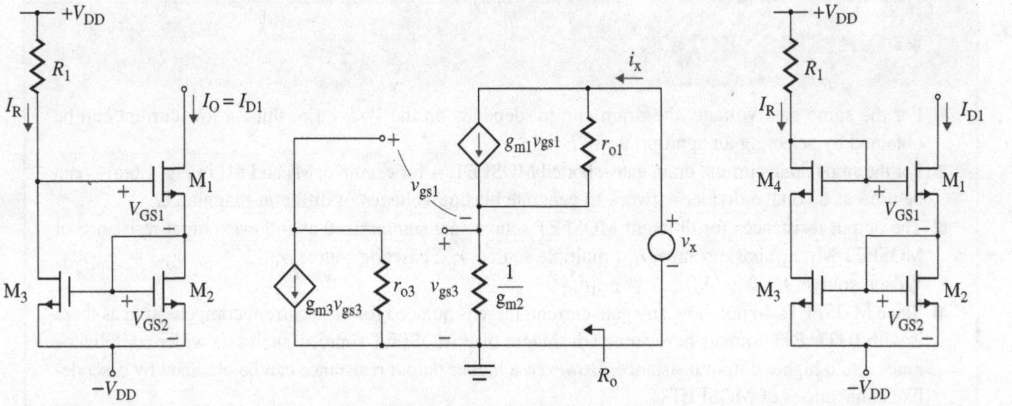

(a) Wilson current source **(b) Equivalent circuit for finding R_o** **(c) Modified Wilson current source**

FIGURE 1.7 Wilson current source

which, after substituting for v_{gs1} and simplifying, gives the output resistance R_o as

$$R_o = \frac{v_x}{i_x} = r_{o1} + \frac{1}{g_{m2}} + \frac{g_{m1}}{g_{m2}} r_{o1}(1 + g_{m3}r_{o3})$$

$$\approx r_{o1} + r_{o1}(1 + g_{m3}r_{o3}) \quad (\text{for } g_{m1} = g_{m2} = g_{m3}) \tag{1.36}$$

where r_{o2} and r_{o1} are the output resistances of transistors M_2 and M_1, respectively.

The problem with this circuit is that the drain voltages V_{D1} and V_{D3} of M_1 and M_3 are unequal. As a result, their drain currents I_{D1} and I_{D3} are also unequal. This problem can be solved by adding one diode-connected MOSFET, as shown in Fig. 1.7(c). This modification ensures that M_1 and M_3 have equal drain voltages and thus equal drain currents.

1.3.6 Design of Active Current Sources

The specifications for designing a current source will include the output current I_Q, the output resistance R_o, and the DC supply voltage V_{DD}. The design sequence is as follows:

Step 1. Determine the design specifications: output current and output resistance.

Step 2. Decide on the type of device to use—either BJTs or MOSFETs.

Step 3. Choose the circuit topology best suited to the specifications. Use simple transistor models for hand analysis to find the circuit-level solution, including component values and specifications of BJTs or MOSFETs.

Step 4. Use the standard values of components—for example, $R_1 = 5.6\ \text{M}\Omega \pm 5\%$ instead of $5.72\ \text{M}\Omega$, $R_2 = 30\ \text{k}\Omega \pm 5\%$ instead of $29.3\ \text{k}\Omega$, and $R_3 = 27\ \text{k}\Omega \pm 5\%$ instead of $27.5\ \text{k}\Omega$. Evaluate your design and modify the values, if necessary.

Step 5. Use PSpice/SPICE verification, employing complex circuit models to calculate the worst-case results due to component and parameter variations. Modify your design, if necessary.

KEY POINTS OF SECTION 1.3

- For the same gate voltage, the drain current depends on the W/L ratio; thus, a low current can be obtained by selecting an appropriate W/L ratio.
- For the same drain current, drain gate–shorted MOSFETs—for example, M_3 and M_4 in Fig. 1.6(a)—can be used as a voltage divider network to generate biasing voltages of different magnitudes.
- The output resistances for different MOSFET sources are summarized as follows: output resistance of MOSFET M_1, r_{o1}; basic source, r_{o1}; multiple source, r_{o1}; cascode source, $r_{o1}(2 + g_{m1}r_{o1}) \simeq g_{m1}r_{o1}^2$; Wilson source, $r_{o1}(2 + g_{m1}r_{o1}) \simeq g_{m1}r_{o1}^2$.
- Since MOSFETs do not draw any gate current, there is no need for base current compensation as there is with BJTs. BJT sources have some advantages over MOSFET sources, such as a wider compliance range and a higher output resistance. However, a higher output resistance can be obtained by cascode-like connections of MOSFETs.

1.4 MOS Differential Amplifiers

During the past few years, MOS technology has developed considerably. MOS transistors are being used increasingly in analog integrated circuits. It is relatively easy to connect MOS transistors in cascode form to control the drain current and give high output resistance. MOS differential pairs are the building blocks in MOS ICs.

1.4.1 NMOS Differential Pair

An n-channel NMOS pair is shown in Fig. 1.8. The DC biasing is normally done by a MOS current source. Although a resistor R_D is shown as the load in Fig. 1.8, a MOS active current mirror is normally used as the load.

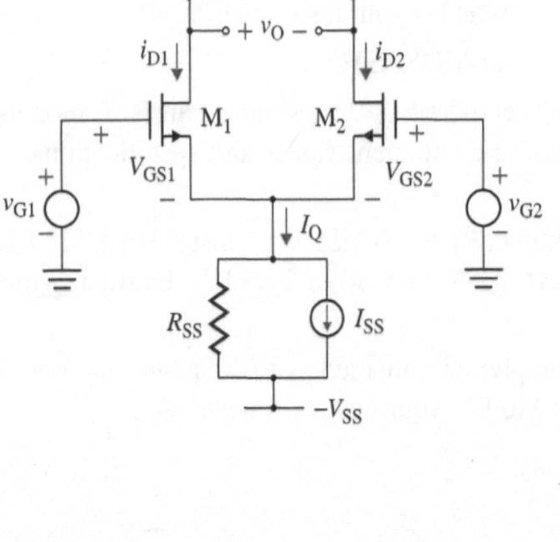

FIGURE 1.8 MOS differential pair

DC Transfer Characteristics

The following analysis is performed for an n-channel MOSFET pair, but it is equally applicable to a p-channel pair with appropriate sign changes. The analysis can be simplified by making the following assumptions:

1. The output resistances of the MOSFETs are infinite: $r_o = \infty$.
2. The MOSFETs are identical and operate in the saturation region. The threshold voltages are the same, $V_{t1} = V_{t2} = V_t$, and the constants are equal, $K_{n1} = K_{n2} = K_n$.
3. The output resistance of the transistor current source is infinite: $R_{SS} = \infty$.

Assuming that the drain current is related to v_{GS} by the approximate square law relationship in Eq. (G.14) in Appendix G, we can write

$$i_D = K_n(v_{GS} - V_t)^2 \tag{1.37}$$

Taking the square root of both sides of Eq. (1.37), we can write the square root of the drain currents as

$$\sqrt{i_{D1}} = \sqrt{K_n}(v_{GS1} - V_t) \tag{1.38}$$

$$\sqrt{i_{D2}} = \sqrt{K_n}(v_{GS2} - V_t) \tag{1.39}$$

We can subtract $\sqrt{i_{D2}}$ from $\sqrt{i_{D1}}$ to find a relation for the differential voltage $v_{id} = v_{GS1} - v_{GS2}$:

$$\begin{aligned}\sqrt{i_{D1}} - \sqrt{i_{D2}} &= \sqrt{K_n}(v_{GS1} - V_t) - \sqrt{K_n}(v_{GS2} - V_t) = \sqrt{K_n}(v_{GS1} - v_{GS2}) \\ &= \sqrt{K_n}v_{id}\end{aligned} \tag{1.40}$$

The sum of i_{D1} and i_{D2} must equal I_Q; that is,

$$I_Q = i_{D1} + i_{D2} \tag{1.41}$$

Substituting Eq. (1.41) into Eq. (1.40) and solving the resultant quadratic, we find the drain currents:

$$i_{D1} = \frac{I_Q}{2} + \sqrt{2K_nI_Q}\left(\frac{v_{id}}{2}\right)\left[1 - \frac{(v_{id}/2)^2}{(I_Q/2K_n)}\right]^{1/2} \tag{1.42}$$

$$i_{D2} = \frac{I_Q}{2} - \sqrt{2K_nI_Q}\left(\frac{v_{id}}{2}\right)\left[1 - \frac{(v_{id}/2)^2}{(I_Q/2K_n)}\right]^{1/2} \tag{1.43}$$

At the quiescent point $v_{id} = 0$, we get

$$i_{D1} = i_{D2} = \frac{I_Q}{2}$$

$$v_{GS1} = v_{GS2} = V_{GS}$$

$$I_Q = 2I_D = 2K_n(V_{GS} - V_t)^2$$

$$\sqrt{2K_n I_Q} = 2K_n \frac{(V_{GS} - V_t)^2}{(V_{GS} - V_t)} = \frac{I_Q}{(V_{GS} - V_t)}$$

$$\left(\frac{I_Q}{2K_n}\right) = (V_{GS} - V_t)^2$$

Substituting these relations into Eqs. (1.42) and (1.43), we can rewrite $i_{D1} = i_{D2}$ as

$$i_{D1} = \frac{I_Q}{2} + \left(\frac{I_Q}{V_{GS} - V_t}\right)\left(\frac{v_{id}}{2}\right)\left[1 - \left(\frac{v_{id}/2}{V_{GS} - V_t}\right)^2\right]^{1/2} \tag{1.44}$$

$$i_{D2} = \frac{I_Q}{2} - \left(\frac{I_Q}{V_{GS} - V_t}\right)\left(\frac{v_{id}}{2}\right)\left[1 - \left(\frac{v_{id}/2}{V_{GS} - V_t}\right)^2\right]^{1/2} \tag{1.45}$$

For $v_{id}/2 << (V_{GS} - V_t)$, i_{D1} and i_{D2} can be approximated by

$$i_{D1} = \frac{I_Q}{2} + \left(\frac{I_Q}{V_{GS} - V_t}\right)\left(\frac{v_{id}}{2}\right) \tag{1.46}$$

$$i_{D2} = \frac{I_Q}{2} - \left(\frac{I_Q}{V_{GS} - V_t}\right)\left(\frac{v_{id}}{2}\right) \tag{1.47}$$

Thus, the change in drain current from the quiescent value of $I_Q/2$ is given by

$$\Delta I_D = \left(\frac{I_Q}{V_{GS} - V_t}\right)\left(\frac{v_{id}}{2}\right) \tag{1.48}$$

which can be normalized with respect to the maximum value $I_Q/2$ as

$$\frac{\Delta I_D}{I_Q/2} = \frac{v_{id}}{V_{GS} - V_t} = \frac{v_{id}}{\sqrt{I_Q/2K_n}} \tag{1.49}$$

If v_{id} is sufficiently large, all of the biasing current I_D must flow through only one of the MOSFETs. The range of v_{id} for which both transistors conduct can be found from Eq. (1.49) under the condition $\Delta I_D = I_Q/2$. That is,

$$v_{id} \leq \left(\frac{I_Q}{2K_n}\right)^{1/2} \tag{1.50}$$

This equation gives the value of v_{id} for which the current I_Q is carried by one of the two transistors. Thus, outside the range defined by Eq. (1.50), currents i_{D1} and i_{D2} will be either zero or I_Q. The plots of the normalized currents i_{D1} and i_{D2} against the differential voltage v_{id}/v_n are shown in Fig. 1.9, where $v_n = \sqrt{I_Q/2K_n}$.

The output voltages of a MOSFET pair are as follows:

$$v_{O1} = V_{DD} - i_{D1}R_D \tag{1.51}$$

$$v_{O2} = V_{DD} - i_{D2}R_D \tag{1.52}$$

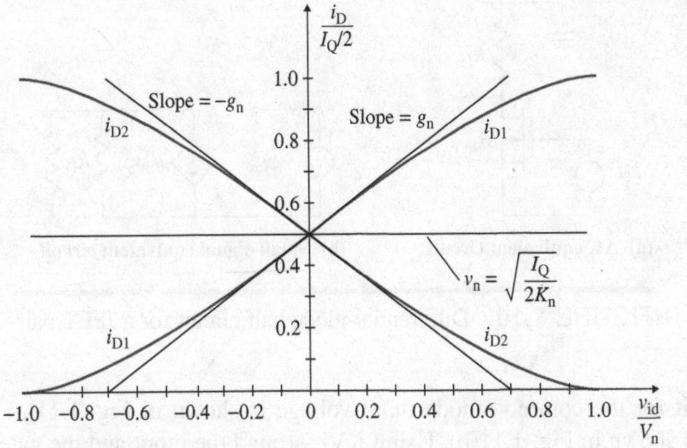

FIGURE 1.9 Normalized DC transfer characteristic of MOSFET pair

The differential DC output voltage is

$$v_{od} = v_{O1} - v_{O2} = R_D(i_{D2} - i_{D1}) = -R_D \Delta I_D$$

Substituting for ΔI_D from Eq. (1.47), we get

$$\begin{aligned} v_{od} &= -R_D\left(\frac{I_Q}{V_{GS} - V_t}\right)\left(\frac{v_{id}}{2}\right) \\ &= -R_D\left(\frac{K_n I_Q}{2}\right)^{1/2} v_{id} \end{aligned} \tag{1.53}$$

which shows the relation between the output voltage v_{od} and the differential voltage v_{id}. If v_{id} is zero, v_{od} is also zero.

Small-Signal Analysis

The transconductance of the small-signal model is given by

$$\begin{aligned} g_m &= 2K_n(V_{GS} - V_t) \\ &= \sqrt{2K_n I_Q} \end{aligned} \tag{1.54}$$

which shows that for a MOSFET a higher value of g_m requires a higher value of the biasing current I_Q.

The half circuit for differential input voltage is shown in Fig. 1.10(a), and its small-signal equivalent circuit is shown in Fig. 1.10(b). The $(r_{o1} \| R_D)$ differential voltage gain A_d for a single-ended output can easily be derived as

$$A_d = \frac{v_{od}/2}{v_{id}/2} = -g_m(r_{o1} \| R_D) \tag{1.55}$$

which, for $r_{o1} >> R_D$, can be approximated to $A_d = -g_m R_D$.

▶ **NOTE** If we define $v_{od} = v_{o2} - v_{o1}$ and $v_{gd} = v_{G1} - v_{G2}$, then $A_d = g_m(R_D \| r_{o1})$.

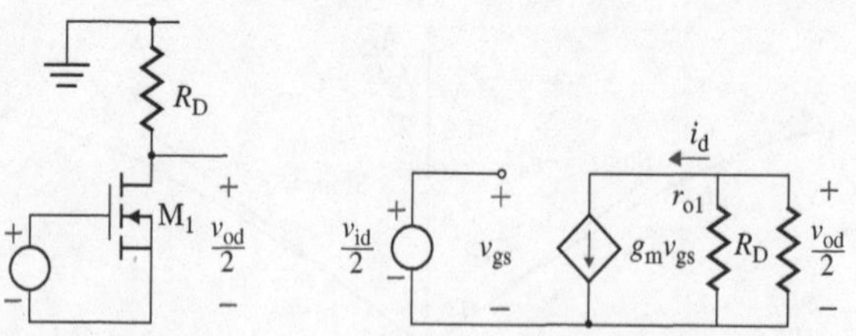

(a) **AC equivalent circuit** (b) **Small-signal equivalent circuit**

FIGURE 1.10 Differential-mode half circuit for a JFET pair

The half circuit for the common-mode input voltage is shown in Fig. 1.11(a), and its small-signal equivalent circuit is shown in Fig. 1.11(b). Using KVL around the input and the gate–source loop, we get

$$v_{gs} = v_{ic} - v_{gs} g_m 2R_{SS}$$

which leads to the following relationship between v_{gs} and v_{ic}:

$$v_{gs}(1 + g_m 2R_{SS}) = v_{ic} \tag{1.56}$$

The common-mode output voltage is given by

$$v_{oc} = -R_D i_d = -R_D g_m v_{gs}$$

Substituting v_{gs} from Eq. (1.56) into the above equation, we get

$$v_{oc} = v_{ic}\left[\frac{-g_m R_D}{1 + g_m 2R_{SS}}\right]$$

which gives the common-mode voltage gain A_c (for a single-ended output) as

$$A_c = \frac{v_{oc}}{v_{ic}} = \frac{-g_m R_D}{1 + g_m 2R_{SS}} \tag{1.57}$$

▶ **NOTE** To simplify the analysis, the effect of transistor output resistance r_{o1} is not included in Eq. (1.57).

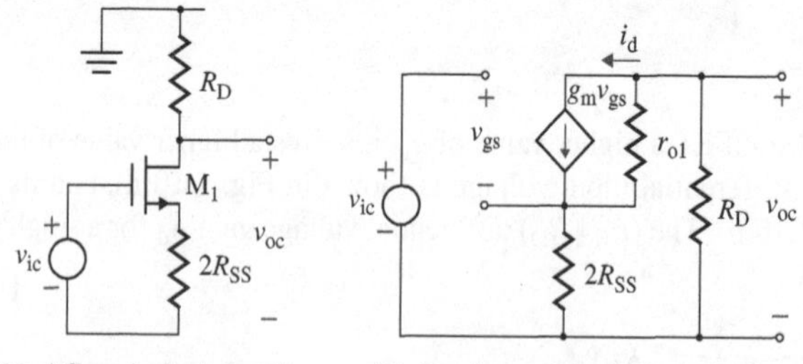

(a) **AC equivalent circuit** (b) **Small-signal equivalent circuit**

FIGURE 1.11 Common-mode half circuit for depletion MOSFET pair

From Eqs. (1.55) and (1.57) for $r_{o1} >> R_D$, we can find CMRR as

$$\text{CMRR} = \frac{A_d}{A_c} = 1 + 2g_m R_{SS} \tag{1.58}$$

which is valid only for single-ended output; that is, the output is taken out from one of the output terminals.

The common-mode and differential input resistances are given by

$$R_{ic} = R_{id} = \infty \tag{1.59}$$

▶ **NOTE** To simplify the analysis, the effect of transistor output resistance r_o is not included in Eqs. (1.57) and (1.58).

EXAMPLE 1.2

Analyzing a MOS differential pair with an active current source The parameters of the MOS differential pair in Fig. 1.8 are $R_{SS} = 50\ \text{k}\Omega$, $I_Q = 10$ mA, $V_{DD} = 30$ V, and $R_D = 5\ \text{k}\Omega$. The NMOSs are identical and have $K_n = 1.25\ \text{mA/V}^2$ and $V_t = 1.0$ V. Assume $V_M = 100$ V.

(a) Calculate the DC drain currents through the MOSFETs if $v_{id} = 10$ mV.

(b) Assuming $I_{D1} = I_{D2}$, calculate A_d, A_c, and CMRR; R_{id} and R_{ic}; and the small-signal output voltage if $v_{g1} = 10$ mV and $v_{g2} = 20$ mV.

(c) Find the drain voltage V_D.

SOLUTION

(a) For $v_{id} = 10$ mV, Eq. (1.42) gives the DC drain current i_{D1} for transistor M_1 as

$$i_{D1} = \frac{10\ \text{m}}{2} + \sqrt{2 \times 1.25\ \text{m} \times 10\ \text{m}} \times \left(\frac{10\ \text{m}}{2}\right)\left[1 - \frac{(10\ \text{m}/2)^2}{10\ \text{m}/(2 \times 1.25\ \text{m})}\right]^{1/2} = 5.25\ \text{mA}$$

$$i_{D2} = I_D - i_{D1} = 10\ \text{mA} - 5.25\ \text{mA} = 4.75\ \text{mA}$$

(b) We know that $I_{D1} = I_{D2} = I_Q/2 = 10\ \text{mA}/2 = 5$ mA. From Eq. (1.54),

$$g_m = \sqrt{2 \times K_n I_Q} = \sqrt{2 \times 1.25\ \text{m} \times 10\ \text{m}} = 5\ \text{mA/V}$$

$$r_{o1} = \frac{V_M}{I_D} = \frac{100}{5\ \text{m}} = 20\ \text{k}\Omega$$

From Eq. (1.55), the single-ended differential voltage gain A_d is

$$A_d = -g_m(R_D \| r_{o1}) = -5\ \text{m} \times (5\ \text{k} \| 20\ \text{k}) = -20\ \text{V/V}$$

From Eq. (1.57), the single-ended common-mode voltage gain A_c is

$$A_c = \frac{-g_m R_D}{1 + g_m 2R_{SS}} = \frac{-5\ \text{m} \times 5\ \text{k}}{1 + 5\ \text{m} \times 2 \times 50\ \text{k}} = -0.0399\ \text{V/V}$$

Thus, CMRR = $|A_d/A_c| = 20/0.0399 = 501$ (or 54 dB).

From Eq. (1.59),

$$R_{id} = R_{ic} = \infty$$

We know that

$$v_{id} = v_{g2} - v_{g1} = 20\text{ mV} - 10\text{ mV} = 10\text{ mV}$$

and $$v_{ic} = \frac{v_{g1} + v_{g2}}{2} = \frac{10\text{ mV} + 20\text{ mV}}{2} = 15\text{ mV}$$

Using Eq. (1.10), we have

$$v_o = A_d v_{id} + A_c v_{ic} = -20 \times 10\text{ mV} - 0.0399 \times 15\text{ mV} = -201\text{ mV}$$

(c) The DC drain voltage at the drain terminal of a transistor is

$$V_D = V_{DD} - I_D R_D = 30\text{ V} - 5\text{ mA} \times 5\text{ k}\Omega = 5\text{ V}$$

Thus, for $A_d = -20$, the maximum differential voltage will be $v_{id} = 5/20 = 250$ mV. Therefore, V_{DD} must be greater than $I_D R_D$ in order to allow output voltage swing due to the input voltages.

EXAMPLE 1.3

Analyzing an NMOS differential pair with an active current source Repeat Example 1.2 if the transistor current source is replaced by the resistance $R_{SS} = 50$ kΩ; that is, $I_{SS} = 0$. Assume $V_M = 100$ V.

SOLUTION

(a) The DC drain current and the gate–source voltage of the MOSFETs can be determined from the DC common-mode half circuit for $v_{g1} = v_{g2} = 0$, shown in Fig. 1.12:

$$V_{GS} + 2I_D R_{SS} = V_{SS}$$

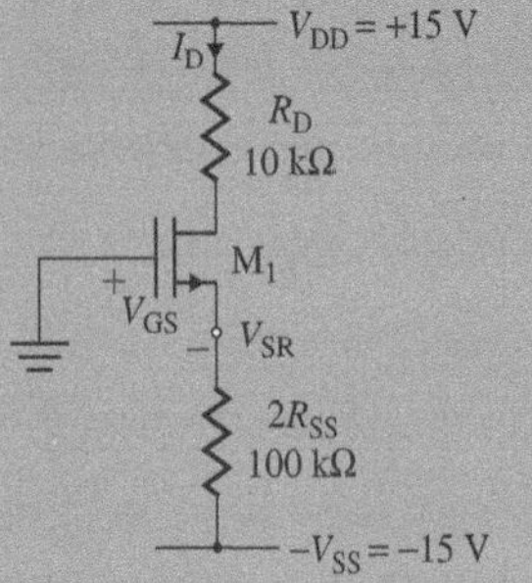

FIGURE 1.12 Common-mode half circuit for DC biasing of MOS pair

Substituting $i_D = I_D$ and $v_{GS} = V_{GS}$ from Eq. (1.37) into the preceding equation, we get

$$V_{GS} + 2R_{SS}K_n(V_{GS} - V_t)^2 = V_{SS}$$

Substituting the values, we get

$$V_{GS} + 2 \times 50\text{ k} \times (V_{GS} - 1)^2 = 15$$

This quadratic equation yields the solution for the gate–source voltage $V_{GS} = 1.478$ V, or 0.514 V. Since $V_{GS} > V_t$, the acceptable value is $V_{GS} = 1.478$ V, which gives the drain current as

$$I_{D1} = K_n(V_{GS} - V_t)^2 = 1.25\text{ m} \times (1.478 - 1)^2 = 285.6\ \mu\text{A}$$

Therefore, the voltage at the source with respect to the ground is

$$V_{SR} = -V_{GS} = -1.478\text{ V} \quad \text{and} \quad I_Q = 2I_D = 2 \times 285.6\ \mu = 571.2\ \mu\text{A}$$

(b) From Eq. (1.54),

$$g_m = \sqrt{2K_nI_Q} = \sqrt{2 \times 1.25\text{ m} \times 571.2\ \mu} = 1.194\text{ mA/V}$$

$$r_{o1} = \frac{V_M}{I_D} = \frac{100}{285.6\ \mu} = 350\text{ k}\Omega$$

From Eq. (1.55),

$$A_d = -g_m(R_D \| r_{o1}) = -1.194\text{ m} \times (5\text{ k} \| 350\text{ k}) = -5.89\text{ V/V}$$

From Eq. (1.57),

$$A_c = \frac{-g_mR_D}{1 + g_m \times 2R_{SS}} = \frac{-1.194\text{ m} \times 5\text{ k}}{1 + 1.194\text{ m} \times 2 \times 50\text{ k}} = -0.0489$$

Thus, CMRR $= |A_d/A_c| = 5.89/0.0489 = 120.4$, or 41.61 dB. From Eq. (1.59),

$$R_{id} = R_{ic} = \infty$$

We know that

$$v_{id} = v_{g2} - v_{g1} = 20\text{ mV} - 10\text{ mV} = 10\text{ mV}$$

$$v_{ic} = \frac{v_{g1} + v_{g2}}{2} = \frac{10\text{ mV} + 20\text{ mV}}{2} = 15\text{ mV}$$

Using Eq. (1.10), we have

$$v_o = A_dv_{id} + A_cv_{ic} = -5.887 \times 10\text{ mV} - 0.489 \times 15\text{ mV} = -59.6\text{ mV}$$

Thus, the output voltage and the voltage gain are much lower than with current-source biasing because the biasing current is low.

EXAMPLE 1.4

Designing a MOS differential pair with an active current source

(a) Design a MOS differential pair as shown in Fig. 1.13, in which one input terminal is grounded. The output is taken from the drain of transistor M_1. The DC biasing current is $I_Q = 10$ mA, and $V_{DD} = V_{SS} = 15$ V. The MOSFETs are identical and have $K_n = 1.25\ \text{mA/V}^2$ and $V_t = 1$ V. A small-signal voltage gain of $A_1 - 10$ V/V is required. Assume $V_M = \infty$.

(b) Calculate the design values of A_d, A_c, and CMRR.

(c) Find the maximum permissible value of R_D and the corresponding voltage gain A_d.

SOLUTION

(a) $I_Q = 10$ mA, and $I_{D1} = I_{D2} = I_Q/2 = 10\ \text{mA}/2 = 5$ mA. From Eq. (1.54),

$$g_m = \sqrt{2 \times K_n I_Q} = \sqrt{2 \times 1.25\ \text{m} \times 10\ \text{m}} = 5\ \text{mA/V}$$

From Eq. (1.37), the DC gate–source voltage is

$$V_{GS1} = -\left(\sqrt{\frac{I_{D1}}{K_n}} + V_t\right) = -\left(\sqrt{\frac{5\ \text{mA}}{1.25\ \text{m}}} + 1\right) = -2.236\ \text{V}$$

Therefore, the voltage at the source terminal with respect to the ground is $V_{SR} = -V_{GS} = -2.236$ V, and

$$R_{SS} = \frac{V_{SR} - V_{SS}}{I_{SS}} = \frac{(-2.236 + 15)\ \text{V}}{10\ \text{mA}} = 1.3\ \text{k}\Omega$$

(b) $v_{G1} = V_{G1} + v_{g1}$, $V_{G1} = 0$, and $v_{G2} = v_{g2} = 0$. Then

$$v_{id} = v_{g1} - v_{g2} = v_{g1}$$

$$v_{ic} = \frac{v_{g1} + v_{g2}}{2} = \frac{v_{g1}}{2}$$

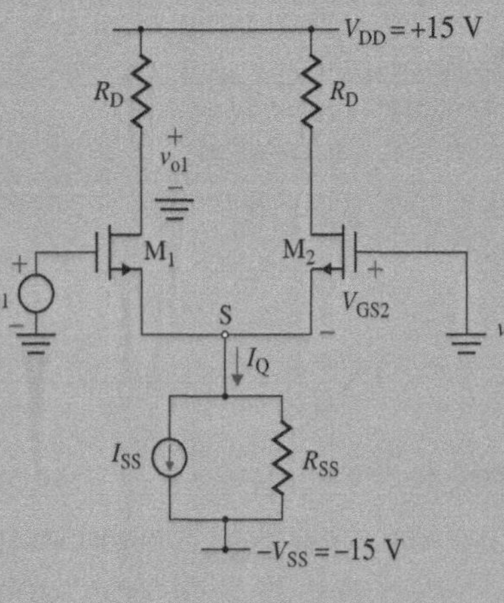

FIGURE 1.13 MOS differential pair with a single input

From Eq. (1.7),

$$v_{o1} = v_{oc} + \frac{v_{od}}{2} = A_c v_{ic} + A_d v_{id} = A_c \frac{v_{g1}}{2} + A_d v_{g1} = \frac{v_{g1}}{2}(A_c + 2A_d)$$

Substituting for A_d and A_c from Eqs. (1.55) and (1.57) gives the voltage gain A_1 as

$$A_1 = \frac{v_{o1}}{v_{g1}} = \frac{1}{2}(A_c + 2A_d) = -\frac{1}{2}\left[\frac{g_m R_D}{1 + 2g_m R_{SS}} + 2g_m R_D\right]$$

Substituting $A_1 = -10$, $g_m = 5$ mA/V, and $R_{SS} = 1.3$ kΩ into the above equation gives $R_D = 1.93$ kΩ.

$$r_{o1} = \frac{V_M}{I_{D1}} = \infty$$

From Eq. (1.55),

$$A_d = -g_m(R_D \| r_{o1}) = -5\text{ m} \times (1.93\text{ k} \| 20\text{ k}) = -8.8\text{ V/V}$$

From Eq. (1.57),

$$A_c = \frac{-g_m R_D}{1 + g_m \times 2R_{SS}} = \frac{-5\text{ m} \times 1.93\text{ k}}{1 + 5\text{ m} \times 2 \times 1.7\text{ k}} = -0.64\text{ V/V}$$

Thus, CMRR = $|A_d/A_c| = 8.8/0.64 = 13.76$, or 22.78 dB.

(c) For $I_D = I_Q/2 = 5$ mA and $V_{DD} = 15$ V, the maximum value of R_D is

$$R_{D(max)} = \frac{V_{DD}}{I_D} = \frac{15\text{ V}}{5\text{ mA}} = 3\text{ k}\Omega$$

which gives the maximum value of

$$A_{d(max)} = -g_m R_{D(max)} = -5\text{ mA} \times 3\text{ k}\Omega = -15\text{ V/V}$$

1.4.2 MOS Differential Pair with Active Load

MOS differential amplifiers are normally used with current mirror active loads. A commonly used configuration is shown in Fig. 1.14. The current mirror consists of transistors M_3 and M_4. Increasing the input voltage to M_1 by $v_{id}/2$ will cause the drain current of M_1 to increase by an amount $g_m v_{id}/2$. This increase will cause a similar increase in the drain current of M_4 due to the current mirror effect and also a decrease in the drain current of M_2. Since M_1 and M_2 are NMOS transistors and their complements (M_3 and M_4) are PMOS types, this configuration is known as a *CMOS amplifier*; the manufacturing process by which the amplifiers are produced is known as *CMOS technology*.

The small-signal equivalent of the output side of the CMOS amplifier is shown in Fig. 1.15. The output resistance R_o is the parallel combination of r_{o2} and r_{o4}. That is,

$$R_o = r_{o2} \| r_{o4} \tag{1.60}$$

where

$$r_{o2} = \frac{2V_M}{I_Q} = \text{output resistance of transistor } M_2$$

$$r_{o4} = \frac{2V_M}{I_Q} = \text{output resistance of transistor } M_4$$

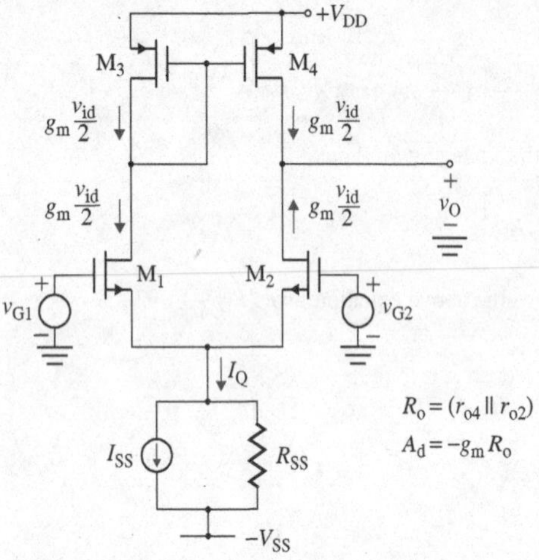

FIGURE 1.14 CMOS differential amplifier

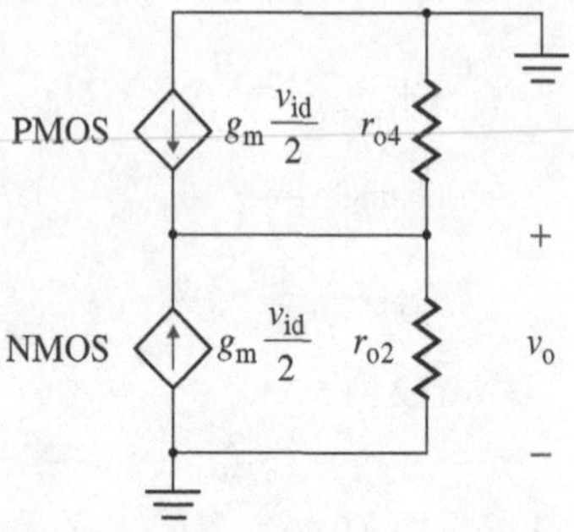

FIGURE 1.15 Small-signal equivalent of the CMOS amplifier

Using Eq. (1.55), we find the differential voltage gain A_d:

$$A_d = \frac{v_o}{v_{id}} = -g_m(r_{o2} \| r_{o4}) \quad (1.61)$$

where g_m, which is the transconductance of MOSFET M_2, is given by Eq. (1.54). Substituting $r_{o4} = r_{o2} = 2V_M/I_Q$ and $g_m = \sqrt{2K_n I_Q}$ and simplifying, we get

$$A_d = -\sqrt{\frac{2K_n}{I_Q}}V_M \quad (1.62)$$

which gives a higher voltage gain for a lower value of I_Q.

EXAMPLE 1.5

D

Designing a CMOS amplifier

(a) Design the CMOS amplifier shown in Fig. 1.16 by determining the *W/L* ratios of the MOSFETs and the threshold voltage V_t. The differential voltage gain should be $A_d = 60$ V/V at biasing current $I_Q = 10$ μA. Assume identical transistors whose channel modulation voltage is $V_M = 20$ V, channel constant is $K_x = 20$ μA/V^2, and channel length is $L = 10$ μm. The *W/L* ratio of the current source is 2, and $K_x = 10$ μA/V^2. Assume $V_{DD} = V_{SS} = 5$ V.

(b) Use PSpice/SPICE to find the small-signal differential voltage gain A_d for $v_{G1} = 1$ mV and $v_{G2} = 0$.

SOLUTION

(a) $A_d = 60$, $I_Q = 10$ μA, $V_M = 20$ V, $K_x = 20$ μA/V^2, and $L = 10$ μm. From Eq. (1.62), we find the MOS constant K_n as

$$K_{n2} = \left(\frac{I_Q}{2}\right)\left(\frac{A_d}{V_M}\right)^2 = \left(\frac{10\ \mu A}{2}\right)\left(\frac{60}{20}\right)^2 = 45\ \mu A/V^2$$

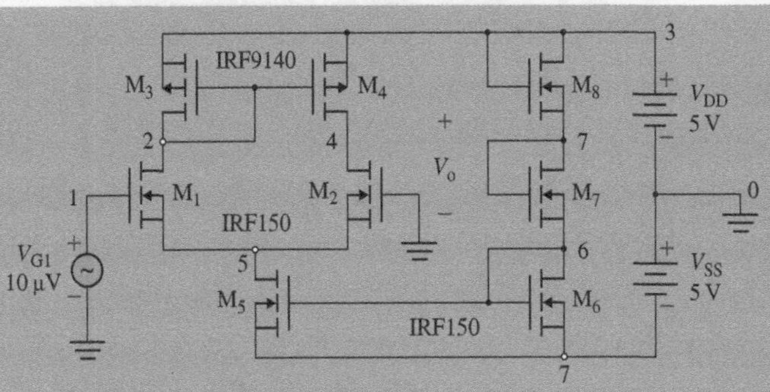

FIGURE 1.16 CMOS amplifier

From Eq. (1.21), we find the W/L ratio as

$$\frac{W}{L} = \frac{K_n}{K_x} = \frac{45\ \mu}{20\ \mu} = 2.25$$

Since $L = 10\ \mu\text{m}$, the desired value is $W = 2.25 \times 10\ \mu\text{m} = 22.5\ \mu\text{m}$. For identical transistors for the current source,

$$V_{GS6} = \frac{V_{DD} + V_{SS}}{3} = \frac{(5 + 5)\ \text{V}}{3} = 3.33\ \text{V}$$

Since $W = 2L$ for the MOSFETs of the current source, the MOS constant is

$$K_{n6} = \frac{W_6 K_x}{2L_6} = \frac{2 \times 10}{2}\ \mu\text{A/V}^2 = 10\ \mu\text{A/V}^2$$

The biasing current is given by

$$I_Q = K_{n6}(V_{GS6} - V_t)^2$$

which, for $V_{GS6} = 3.33$ V, $I_Q = 10\ \mu$A, and $K_{n6} = 10\ \mu\text{A/V}^2$, gives $V_t = 2.33$ V. Then

$$r_{o2} = r_{o4} = \frac{2V_M}{I_Q} = \frac{2 \times 20}{10\ \mu} = 4\ \text{M}\Omega$$

$$R_o = r_{o2} \parallel r_{o4} = 2\ \text{M}\Omega$$

$$r_{o5} = \frac{V_M}{I_Q} = \frac{20}{10\ \mu} = 2\ \text{M}\Omega$$

(b) The CMOS amplifier for PSpice simulation is shown in Fig. 1.17. The plot of the output voltage is shown in Fig. 1.18, which gives a voltage gain 81.72 V/V (expected value 60). Note there is no phase shift of 180° because the output is taken from the drain terminal of the pair.

The results of simulation (.TF analysis) are as follows. (The expected values are listed on the right.)

`****  SMALL-SIGNAL CHARACTERISTICS`	
`V(4)/VID1=8.189E+01=81.89`	($A_{id} = 60$ V/V)
`INPUT RESISTANCE AT VG1=1.000E+20`	($R_{id} = \infty$)
`OUTPUT RESISTANCE AT V(4)=2.221E+06=2.221 MΩ`	($R_{od} = 2$ MΩ)
`ID1=5.44E-06=5.44 μA`	($I_{D1} = 5.44\ \mu$A)
`ID=ID5=1.09E-05=10.9 μA`	($I_Q = 10\ \mu$A)

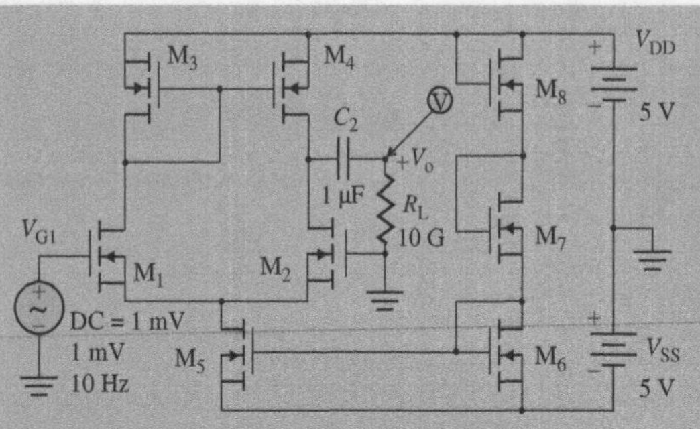

FIGURE 1.17 CMOS amplifier for a PSpice simulation for Example 1.5

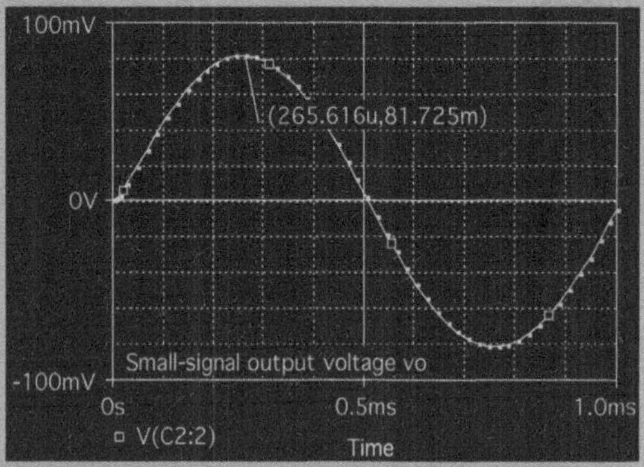

FIGURE 1.18 PSpice plot of small-signal output voltage for Example 1.5

NOTE: As expected, the PSpice results depend on the values of the *W/L* ratio for the MOSFETs. If you run PSpice from the schematic, you will need to change the model parameters of the MOSFETs; otherwise, the results will be different from those shown above.

1.4.3 Cascoded MOS Differential Amplifier

We can notice from Eq. (1.61) that the differential gain increases with the output resistance R_o of the active amplifier load. MOSFETs can be connected in cascode configuration as shown in Fig. 1.6(a) to increase the output resistance and to improve the frequency response. A common modification to Fig. 1.14(a) is shown in Fig. 1.19(a). Transistors M_1 and M_2 are connected in a common-source configuration and form a common-source differential pair, whereas M_3 and M_4 are connected in a common-gate configuration and form a common-gate differential stage. M_6 and M_8 act as the load. For identical MOSFETs, the DC biasing currents are equal, $I_{D1} = I_{D2} = I_{D4} = I_{D6} = I_{D8} = I_Q/2$. All MOSFETs have equal output resistances, $r_{o1} = r_{o2} = r_{o4} = r_{o6} = r_{o8} = V_M/I_{D1}$, and equal transconductances, $g_{m1} = g_{m2} = g_{m4} = g_{m6} = g_{m8} = 2K_n(V_{GS1} - V_t)$. The small-signal half-circuit is shown in Fig. 1.19(b). Using Eq. (1.34), we can find the effective load resistance offered by M_6 and M_8 as given by

$$r'_{o6} = r_{o6}(2 + g_{m1}r_{o6}) \tag{1.63}$$

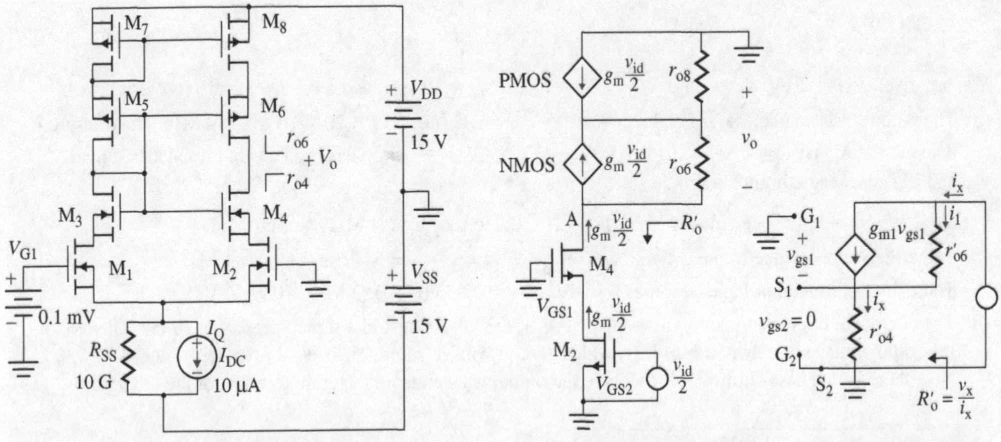

(a) Cascoded MOS amplifier (b) Small-signal half-circuit (c) Equivalent circuit

FIGURE 1.19 Cascoded MOS differential amplifier

Similarly, we can find the effective resistance offered by M_2 and M_4 as given by

$$r'_{o4} = r_{o4}(2 + g_{m1}r_{o4}) \tag{1.64}$$

Therefore, the effective output resistance at the output terminal as shown in Fig. 1.19(c) is given by

$$R_o = r'_{o4} \| r'_{o6} \tag{1.65}$$

Applying Eq. (1.61) we can find the small-signal differential output voltage as given by

$$A_d = \frac{v_{od}}{v_{id}} = -g_{m1}R_o = -g_{m1}(r'_{o4} \| r'_{o6}) \tag{1.66}$$

This should give a much larger differential gain compared to that of Fig. 1.14(a). For example, if $|V_M| = 20$ V, $I_{D1} = 5\ \mu$A, $g_{m1} = 31.5\ \mu$A/V, $r_{o4} = r_{o6} = 5$ MΩ, $r'_{o6} = r'_{o8} = 514$ MΩ, and $R_o = 257$ MΩ, we get $A_d = 8126$ V/V.

EXAMPLE 1.6

Analyzing cascoded MOS amplifiers The DC biasing current of the MOS amplifier shown in Fig. 1.19(a) is kept constant at $I_Q = 10\ \mu$A. All MOS transistors are identical: $V_M = 20$ V, $K_n = 25\ \mu\text{A/V}^2$, $W = 30\ \mu$m, and $L = 10\ \mu$m.

(a) Determine the differential voltage gain A_d for single-ended output at the drain terminal of M_4.
(b) Use PSpice to verify the result.

SOLUTION

You are given $V_M = 20$ V, $K_n = 25\ \mu A/V^2$, $W = 30\ \mu m$, $L = 10\ \mu m$, and $I_D = I_Q/2 = 10\ \mu A/2 = 5\ \mu A$.

(a) The output resistance of M_2 is $r_{o2} = r_{o4} = r_{o6} = r_{o8} = 2V_M/I_Q = 2 \times 20\ V/10\ \mu A = 4\ M\Omega$. The transconductance of M_2 is $g_{m2} = \sqrt{2K_n I_Q} = \sqrt{2 \times 25\ \mu \times 10\ \mu} = 22.36\ \mu A/V$. The load resistance due to M_2 and M_4 cascoded combination is

$$r'_{o4} = r'_{o6} = r_{o4}(2 + g_{m2} r_{o4}) = 4\ M\Omega \times (2 + 22.36\ \mu A/V \times 4\ M\Omega) = 366\ M\Omega.$$

The effective resistance of the active load is $R_o = r'_{o4} \| r'_{o6} = 366\ M \| 366\ M = 183\ M\Omega$.

Thus, the differential voltage becomes $A_d = g_{m2} R_o = 22.36\ \mu \times 183\ M = 4089$ V/V.

(b) The circuit for PSpice simulation is shown in Fig. 1.20 with a sinusoidal signal of 1 μV at 10 Hz. The plot of the output voltage is shown in Fig. 1.21, which gives a voltage gain of 5370 V/V (expected value 4089 V/V). Note there is no phase shift of 180° because the output is taken from the drain terminal of the other pair.

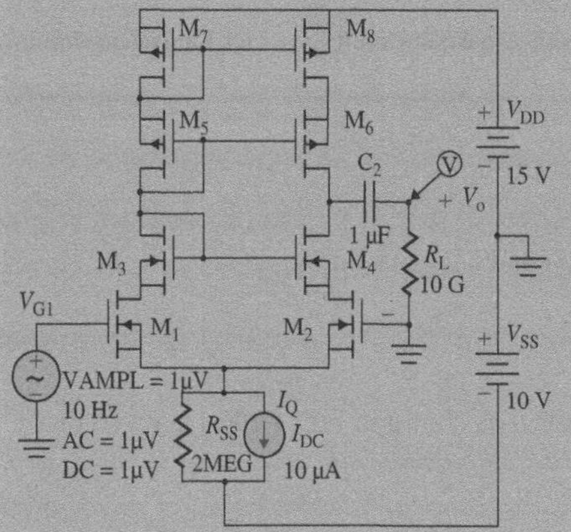

FIGURE 1.20 PSpice schematic for Example 1.6

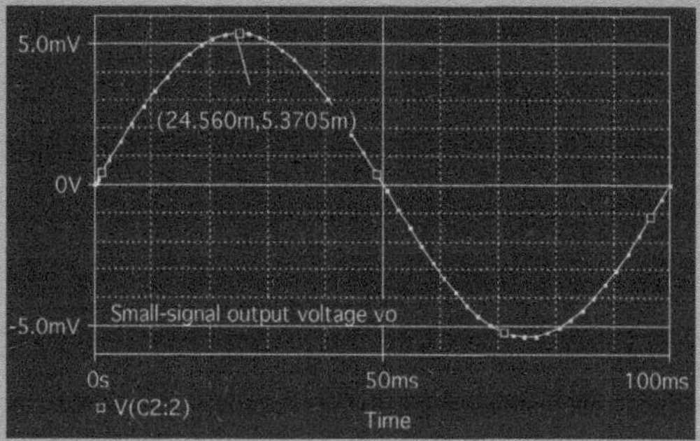

FIGURE 1.21 PSpice plot of small-signal output voltage for Example 1.6

KEY POINTS OF SECTION 1.4

- A MOS amplifier exhibits a linear DC characteristic and has a very high input resistance, tending to infinity.
- It is relatively easy to connect MOS transistors in cascode form in order to control the drain current and give high output resistance. A high-voltage gain can be obtained with a cascode connection.

1.5 Depletion MOS Differential Amplifiers

The depletion MOS amplifiers are similar to the enhancement MOSs, except that the gate–source voltage of a depletion MOS can range from a negative value to a positive value. The drain current becomes the maximum $i_{D(max)} = I_{DSS}$ when $v_{GS} = 0$.

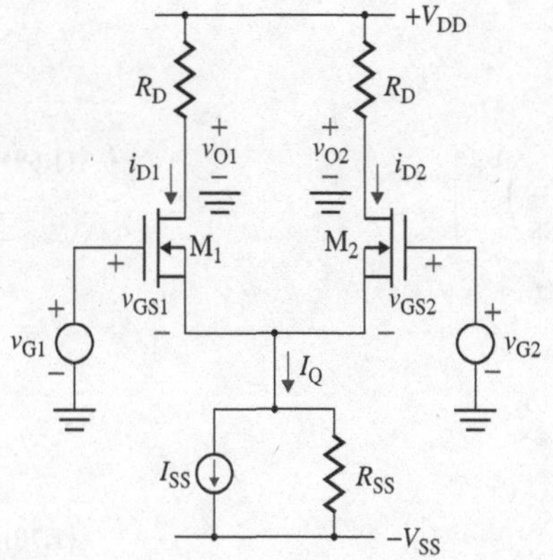

FIGURE 1.22 Depletion MOSFET differential pair

1.5.1 Depletion MOS Differential Pair with Resistive Load

A depletion NMOS source-coupled pair is shown in Fig. 1.22. Although the DC biasing circuit can be either a simple resistor (in which case the equivalent current generator will be zero) or a transistor current source, a current source is generally used.

DC Transfer Characteristics

The following analysis is performed for an *n*-channel MOS pair, but it is equally applicable to a *p*-channel pair with appropriate sign changes. The analysis can be simplified by making the following assumptions:

1. The output resistances of the MOSFETs are infinite: $r_d = \infty$.
2. The MOSFETs are identical and operate in the saturation region. The pinch-down voltages are the same, $V_{p1} = V_{p2} = V_p$, and the drain currents (with drain $V_{GS} = 0$) are equal, $I_{DSS1} = I_{DSS2} = I_{DSS}$.
3. The output resistance of the transistor current source is infinite: $R_{SS} = \infty$.

Using KVL around the loop formed by the two input voltages and two gate–source junctions, we get

$$v_{G1} - v_{GS1} + v_{GS2} - v_{G2} = 0 \tag{1.67}$$

Assuming that the drain current is related to v_{GS} by the approximate square law relationship in Eq. (G.16) in Appendix G, we get

$$i_D = I_{DSS}\left(1 - \frac{v_{GS}}{V_p}\right)^2$$

from which we can find

$$\frac{v_{GS}}{V_p} = 1 - \left(\frac{i_D}{I_{DSS}}\right)^{1/2} \tag{1.68}$$

Substituting Eq. (1.68) into Eq. (1.67) yields

$$-v_{G1} + v_{G2} = -v_{GS1} + v_{GS2}$$

$$= -V_p\left[1 - \left(\frac{i_{D1}}{I_{DSS}}\right)^{1/2}\right] + V_p\left[1 - \left(\frac{i_{D2}}{I_{DSS}}\right)^{1/2}\right] \quad (1.69)$$

$$= V_p\left(\frac{i_{D1}}{I_{DSS}}\right)^{1/2} - V_p\left(\frac{i_{D2}}{I_{DSS}}\right)^{1/2}$$

which gives

$$-\frac{v_{G1} + v_{G2}}{V_p} = \frac{v_{id}}{V_p} = \left(\frac{i_{D1}}{I_{DSS}}\right)^{1/2} - \left(\frac{i_{D2}}{I_{DSS}}\right)^{1/2} \quad (1.70)$$

where $v_{id} = v_{G2} - v_{G1}$ is the differential DC voltage.

Using KCL at the source nodes of the transistors gives

$$I_Q = i_{D1} + i_{D2} \quad (1.71)$$

Substituting Eq. (1.71) into Eq. (1.70) and solving the resultant quadratic, we get

$$i_{D1} = \frac{I_Q}{2} + \frac{I_Q}{2}\left(\frac{v_{id}}{V_p}\right)\left[2\left(\frac{I_{DSS}}{I_Q}\right) - \left(\frac{v_{id}}{V_p}\right)^2\left(\frac{I_{DSS}}{I_Q}\right)^2\right]^{1/2} \quad (1.72)$$

$$i_{D2} = \frac{I_Q}{2} - \frac{I_Q}{2}\left(\frac{v_{id}}{V_p}\right)\left[2\left(\frac{I_{DSS}}{I_Q}\right) - \left(\frac{v_{id}}{V_p}\right)^2\left(\frac{I_{DSS}}{I_Q}\right)^2\right]^{1/2} \quad (1.73)$$

If v_{id} is sufficiently large, all of the biasing current I_Q must flow through only one of the MOSFETs. The range of v_{id} for which both transistors conduct can be found from Eq. (1.72) with the condition $i_{D2} = 0$; that is,

$$\frac{I_Q}{2} - \frac{I_Q}{2}\left(\frac{v_{id}}{V_p}\right)\left[2\left(\frac{I_{DSS}}{I_Q}\right) - \left(\frac{v_{id}}{V_p}\right)^2\left(\frac{I_{DSS}}{I_Q}\right)^2\right]^{1/2} = 0$$

which gives

$$\left|\frac{v_{id}}{V_p}\right| \le \sqrt{\frac{I_Q}{I_{DSS}}} \quad (1.74)$$

This equation gives the value of v_{id} for which the current I_Q is carried by one of the two transistors. Typical drain currents for various values of I_Q are shown in Fig. 1.23. The range of v_{id} for which the circuit exhibits a linear characteristic is approximately equal to the pinch-down voltage $|V_p|$ (typically 2 V to 5 V), compared to V_T (26 mV) for BJTs.

The output voltages for a MOS differential pair are

$$v_{O1} = V_{DD} - i_{D1}R_D \quad (1.75)$$

$$v_{O2} = V_{DD} - i_{D2}R_D \quad (1.76)$$

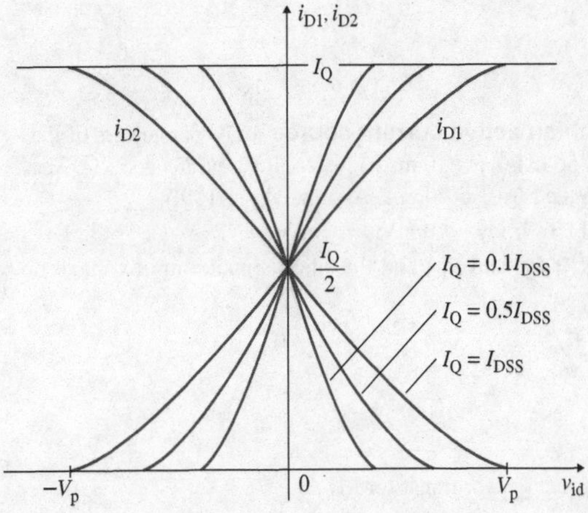

FIGURE 1.23 DC transfer characteristic of depletion NMOS pairs

The differential DC output voltage is

$$v_{od} = v_{O1} - v_{O2} = R_D(i_{D2} - i_{D1})$$

Substituting i_{D1} and i_{D2} from Eqs. (1.72) and (1.73) into the above equation and simplifying, we get

$$v_{od} = -\frac{I_Q R_D}{V_p} v_{id}\left[2\left(\frac{I_{DSS}}{I_Q}\right) - \left(\frac{v_{id}}{V_p}\right)^2\left(\frac{I_{DSS}}{I_Q}\right)^2\right]^{1/2} \quad (1.77)$$

As with the any differential circuit, if v_{id} is zero, v_{od} is also zero. A MOS-coupled pair allows direct coupling of cascoded stages without introducting DC offsets.

Small-Signal Analysis

The drain current is $i_D = \frac{I_{DSS}}{V_p^2}(V_{GS} - V_p)^2$ and the transconductance is given by

$$g_m = \frac{di_D}{dv_{GS}}$$

$$= \frac{2}{|V_p|}[|I_D I_{DSS}|]^{1/2} \quad (1.78)$$

The small-signal analyses for differential and common-mode gains are the same as the enhancement MOS pair. From Eqs. (1.55) and (1.57), we get

$$A_d = -g_m(R_D \parallel r_{o1}) \quad (1.79)$$

$$A_c = \frac{-g_m R_D}{1 + g_m 2R_{SS}} \quad (1.80)$$

▶ **NOTE** To simplify the analysis, the effect of transistor output resistance r_o is not included in Eq. (1.80).

EXAMPLE 1.7

Analyzing a depletion MOS differential pair with an active current source The parameters of the depletion MOS differential pair in Fig. 1.22 are $R_{SS} = 50\ \text{k}\Omega$, $I_Q = 10$ mA, $V_{DD} = 30$ V, and $R_D = 5\ \text{k}\Omega$. The depletion MOSFETs are identical and have $V_p = -4$ V and $I_{DSS} = 20$ mA. Assume $V_M = 100$ V.

(a) Calculate the DC drain currents through the MOSFETs if $v_{id} = 10$ mV.

(b) Assuming $I_{D1} = I_{D2}$, calculate A_d, A_c, and CMRR; R_{id} and R_{ic}; and the small-signal output voltage if $v_{g1} = 10$ mV and $v_{g2} = 20$ mV.

(c) Find the drain voltage V_D.

SOLUTION

(a) For $v_{id} = 10$ mV, Eq. (1.72) gives the DC drain current i_{D1} for transistor M_1 as

$$i_{D1} = \frac{10\ \text{m}}{2}\left\{1 + \frac{100\ \text{m}}{-4}\left[2\left(\frac{20\ \text{m}}{10\ \text{m}}\right) - \left(\frac{100\ \text{m}}{-4}\right)^2\left(\frac{20\ \text{m}}{10\ \text{m}}\right)^2\right]^{1/2}\right\} = 5.25\ \text{mA}$$

$$i_{D2} = I_Q - i_{D1} = 10\ \text{mA} - 5.25\ \text{mA} = 4.75\ \text{mA}$$

(b) We know that $I_{D1} = I_{D2} = I_Q/2 = 10\ \text{mA}/2 = 5$ mA. From Eq. (1.78),

$$g_m = \frac{2[|I_{D1}I_{DSS}|]^{1/2}}{|V_p|} = \left(\frac{2}{4}\right) \times \sqrt{5\ \text{mA} \times 20\ \text{mA}} = 5\ \text{mA/V}$$

$$r_{o1} = \frac{V_M}{I_D} = \frac{100}{5\ \text{m}} = 20\ \text{k}\Omega$$

From Eq. (1.79), the single-ended differential voltage gain A_d is

$$A_d = -g_m(R_D \| r_{o1}) = -5\ \text{m} \times (5\ \text{k} \| 20\ \text{k}) = -20\ \text{V/V}$$

From Eq. (1.80), the single-ended common-mode voltage gain A_c is

$$A_c = \frac{-g_m R_D}{1 + g_m 2R_{SS}} = \frac{-5\ \text{m} \times 5\ \text{k}}{1 + (5\ \text{m} \times 2 \times 50\ \text{k})} = -0.0399\ \text{V/V}$$

Thus, CMRR $= |A_d/A_c| = 20/0.0399 = 501$ (or 54 dB). From Eq. (1.59),

$$R_{id} = R_{ic} = \infty$$

We know that

$$v_{id} = v_{g2} - v_{g1} = 20\ \text{mV} - 10\ \text{mV} = 10\ \text{mV}$$

and $$v_{ic} = \frac{v_{g1} + v_{g2}}{2} = \frac{10\ \text{mV} + 20\ \text{mV}}{2} = 15\ \text{mV}$$

Using Eq. (1.10), we have

$$v_o = A_d v_{id} + A_c v_{ic} = -20 \times 10\ \text{mV} - 0.0399 \times 15\ \text{mV} = -250.6\ \text{mV}$$

(c) The DC drain voltage at the drain terminal of a transistor is

$$V_D = V_{DD} - I_D R_D = 30\ \text{V} - 5\ \text{mA} \times 5\ \text{k}\Omega = 5\ \text{V}$$

Thus, for $A_d = -20$, the maximum differential voltage will be $v_{id} = 5/20 = 250$ mV. Therefore, V_{DD} must be greater than $I_D R_D$ in order to allow output voltage swing due to the input voltages.

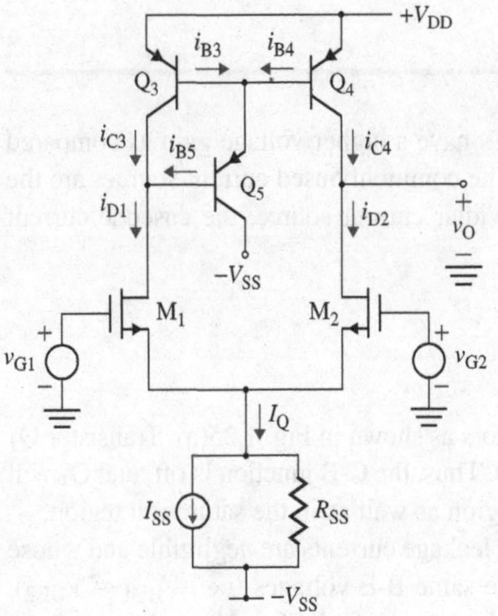

FIGURE 1.24 Depletion MOS differential amplifier with current mirror active load

1.5.2 Depletion MOS Differential Pair with Active Load

Like MOS differential amplifiers, depletion MOS differential amplifiers use a current mirror active load to achieve a large voltage gain. The active load can be made either with MOSFETs similar to that in Fig. 1.24 or with BJTs. A depletion MOS differential amplifier with a basic current source as the BJT active load is shown in Fig. 1.24. From Eq. (1.60), we know that the output resistance R_o is the parallel combination of r_{o2} and r_{o4}; that is,

$$R_o = r_{o2} \| r_{o4} \tag{1.81}$$

where $r_{o2} = 2V_M/I_Q$ = output resistance of transistor M_2 and $r_{o4} = 2V_A/I_Q$ = output resistance of transistor Q_4.

Using Eq. (1.61), we find the differential voltage gain A_d:

$$A_d = \frac{v_o}{v_{id}} = -g_m(r_{o2} \| r_{o4}) \tag{1.82}$$

where g_m, which is the transconductance of MOSFET M_2, is given by Eq. (1.78).

KEY POINTS OF SECTION 1.5

- A MOSFET amplifier has a very high input resistance, in the range of $10^9\ \Omega$ to $10^{12}\ \Omega$.
- An active load considerably increases the differential gain of a MOSFET amplifier.
- The range of v_{id} for which the circuit exhibits a linear DC characteristic is much higher for MOSFETs than for BJTs. For MOSFETs, this range is approximately equal to the pinch-down voltage $|V_p|$ (typically 2 V to 5 V), compared to V_T (26 mV) for BJTs.

1.6 BJT Current Sources

The BJT current sources are similar to MOS sources. Since BJTs have a higher voltage gain as compared to MOSFETs, BJT sources can offer higher output resistances. The commonly used current sources are the basic current sources, the modified basic current source, the Widlar current source, the cascode current source, and the Wilson current source.

1.6.1 Basic Current Source

The simplest current source consists of a resistor and two transistors as shown in Fig. 1.25(a). Transistor Q_1 is diode connected, and its C-B voltage is forced to zero: $v_{CB} = 0$. Thus, the C-B junction is off, and Q_1 will operate in the active region. Transistor Q_2 can be in the active region as well as in the saturation region.

Let us assume that Q_1 and Q_2 are identical transistors whose leakage currents are negligible and whose output resistances are infinite. Since the two transistors have the same B-E voltages (i.e., $V_{BE1} = V_{BE2}$), the collector and base currents are equal: $I_{C1} = I_{C2}$ and $I_{B1} = I_{B2}$. Applying KCL at the collector of Q_1 gives the reference current:

$$I_R = I_{C1} + I_{B1} + I_{B2} = I_{C1} + 2I_{B1}$$

Since $I_{C1} = \beta_F I_{B1}$,

$$I_R = I_{C1} + 2I_{B1} = I_{C1} + \frac{2I_{C1}}{\beta_F}$$

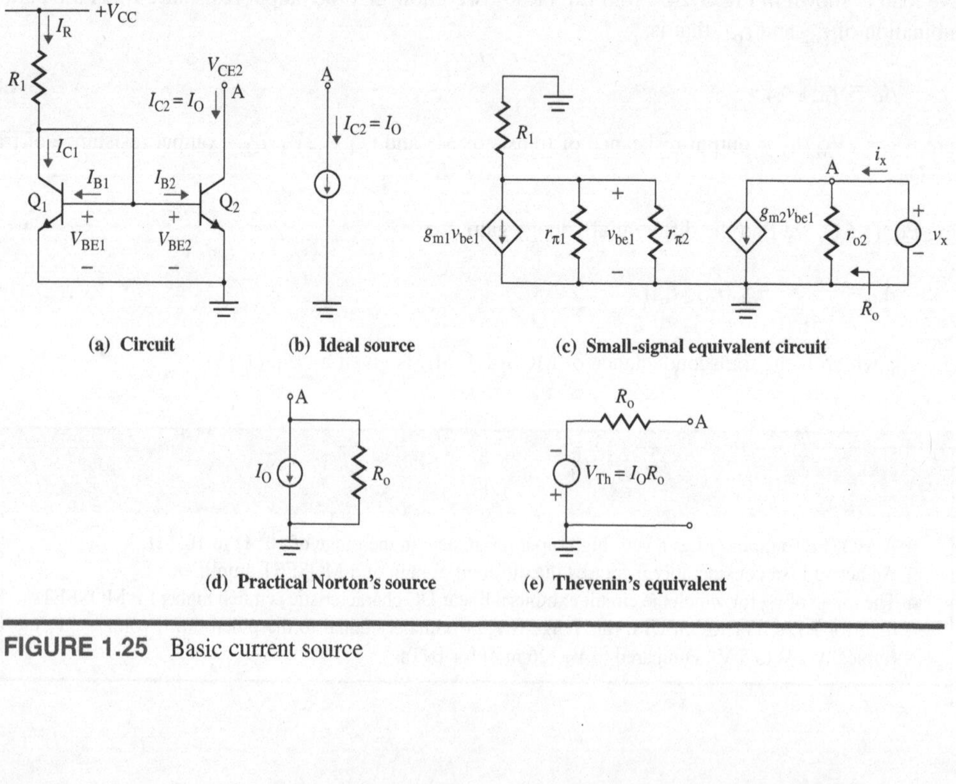

FIGURE 1.25 Basic current source

which gives the collector current I_{C1} as

$$I_{C1} = I_{C2} = \frac{I_R}{1 + 2/\beta_F} \tag{1.83}$$

$$= \frac{V_{CC} - V_{BE1}}{R_1} \times \frac{1}{1 + 2/\beta_F} \tag{1.84}$$

If the DC current gain $\beta_F >> 2$, which is usually the case, Eq. (1.84) is reduced to

$$I_{C1} = I_{C2} \approx I_R$$

Thus, for two identical transistors, the reference and output currents are almost equal. I_{C2}, which is the mirror image of I_{C1}, is known as the *mirror current* of I_{C1}. For transistors with small values of β_F, the current ratio will not be unity. In practice, however, the transistors may not be identical, and the two collector currents will have a constant ratio. The equivalent circuit for the ideal current source is shown in Fig. 1.25(b).

For a transistor with finite output resistance, the effect of the Early voltage V_A should be taken into account, and the collector current in Eq. (G.19) in Appendix G can be modified to

$$I_C = I_S\left[\exp\left(\frac{V_{BE}}{V_T}\right) - 1\right]\left(1 + \frac{V_{CE}}{V_A}\right) \tag{1.85}$$

If we take into account the variation in the collector current due to the C-E voltage, the ratio of the two collector currents can be found from

$$\frac{I_{C2}}{I_{C1}} = \frac{1 + V_{CE2}/V_A}{1 + V_{CE1}/V_A} \tag{1.86}$$

Output Resistance R_o

The small-signal AC equivalent circuit for determining the output resistance is shown in Fig. 1.25(c). Output resistance R_o is the same as r_{o2}; that is,

$$R_o = \frac{v_x}{i_x} = r_{o2} = \frac{V_A}{I_{C2}} \tag{1.87}$$

Thevenin's equivalent voltage is given by

$$V_{Th} = I_O R_o = I_{C2} R_o = I_{C2}\frac{V_A}{I_{C2}} = V_A \tag{1.88}$$

Norton's and Thevenin's equivalents of the current source are shown in Fig. 1.25[(d) and (e)]. If the output of the current source is open-circuited, a voltage of $-V_{Th}$ is expected to appear across transistor Q_1. However, this will not actually happen because transistor Q_1 will saturate when the voltage across the current source (i.e., the C-E voltage of Q_1) reaches zero.

The current source in Fig. 1.25(a) behaves as a current sink rather than a source. A current source equivalent to this current sink can be obtained by using *pnp* transistors and a negative power supply. This arrangement is shown in Fig. 1.26.

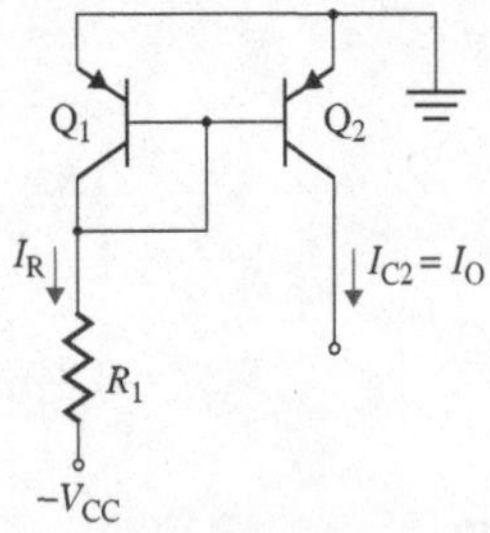

FIGURE 1.26 Current source using *pnp* transistors

EXAMPLE 1.8

D

Designing a simple basic current source

(a) Design the basic current source in Fig. 1.25(a) to give an output current of $I_O = 5\ \mu A$. The transistor parameters are $\beta_F = 100$, $V_{CC} = 30$ V, $V_{BE1} = V_{BE2} = V_{CE1} = 0.7$ V, and $V_A = 150$ V.

(b) Calculate the output resistance R_o, Thevenin's equivalent voltage V_{Th}, and the collector current ratio if $V_{CE2} = 20$ V.

SOLUTION

$V_{BE1} = V_{BE2} = V_{CE1} = 0.7$ V, and $V_A = 150$ V.

(a) From Eq. (1.83),

$$I_O = I_{C2} = I_{C1} = \frac{I_R}{1 + 2/\beta_F}$$

which, for $I_O = 5\ \mu A$, gives $I_R = (5\ \mu A)(1 + 2/\beta_F) = 5.1\ \mu A$. From Eq. (1.84),

$$R_1 = \frac{V_{CC} - V_{BE1}}{I_R} = \frac{30\text{ V} - 0.7\text{ V}}{5.1\ \mu\text{A}} = 5.75\text{ M}\Omega$$

(b) From Eq. (1.87),

$$R_o = \frac{V_A}{I_{C2}} = \frac{150\text{ V}}{5\ \mu\text{A}} = 30\text{ M}\Omega$$

From Eq. (1.88),

$$V_{Th} = V_A = 150\text{ V}$$

From Eq. (1.86),

$$\frac{I_{C2}}{I_{C1}} = \frac{1 + V_{CE2}/V_A}{1 + V_{CE1}/V_A} = \frac{1 + 20/150}{1 + 0.7/150} = 1.128$$

Thus, $I_{C2} = 1.128 \times I_{C1} = 1.128 \times 5\ \mu A = 5.64\ \mu A$, which agrees, with a degree of error, with the desired value of $I_O = 5\ \mu A$.

▶ **NOTES**

1. For a current source of 5 μA, a resistor of 5.75 MΩ would be required. Resistors of such high value are very costly in terms of the die area. Resistors over 50 kΩ are generally avoided for IC applications. Thus, this current source is not suitable for generating a current of less than about 0.6 mA at $V_{CC} = 30$ V and 0.3 mA at $V_{CC} = 15$ V.
2. If the output of the current source is open-circuited, a voltage of $-V_{Th} = -150$ V will not appear across transistor Q_1. Rather, the voltage will be $V_{CE1(sat)} \approx 0.2$ V.

1.6.2 Modified Basic Current Source

Notice from Eq. (1.83) that the collector current I_{C2} ($=I_{C1}$) differs from the reference current I_R by a factor of $(1 + 2/\beta_F)$. For low-gain transistors (especially *pnp* types), I_{C2} can differ significantly from I_R. The error can be reduced by adding another transistor so that I_{C2} becomes less dependent on the transistor parameter β_F. This type of circuit is shown in Fig. 1.27(a). Applying KCL at the emitter of transistor Q_3 gives

$$I_{E3} = I_{B1} + I_{B2} = \frac{I_{C1}}{\beta_F} + \frac{I_{C2}}{\beta_F} \tag{1.89}$$

Since $V_{BE1} = V_{BE2}$, it follows that $I_{C1} = I_{C2}$. Thus, Eq. (1.89) becomes

$$I_{E3} = \frac{2}{\beta_F} I_{C2}$$

The base current of Q_3 is related to I_{E3} by

$$I_{B3} = \frac{I_{E3}}{1 + \beta_F} = \frac{2}{\beta_F(1 + \beta_F)} I_{C2} \tag{1.90}$$

Using KCL at the collector of Q_1 gives

$$I_R = I_{C1} + I_{B3} = I_{C1} + \frac{2}{\beta_F(1 + \beta_F)} I_{C2} \tag{1.91}$$

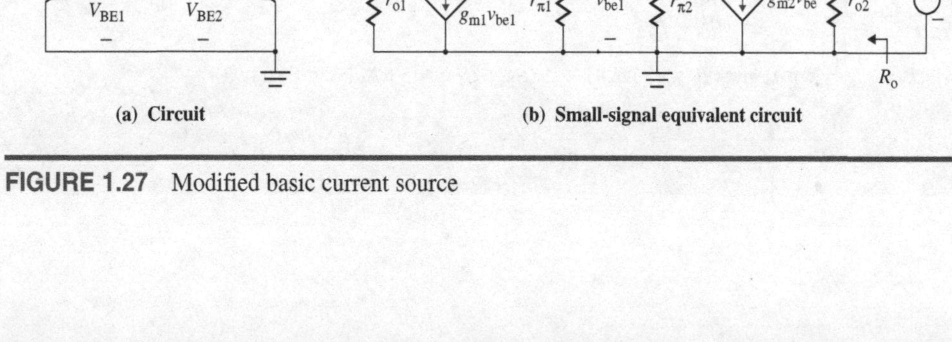

(a) Circuit (b) Small-signal equivalent circuit

FIGURE 1.27 Modified basic current source

Since $I_{C1} = I_{C2}$, Eq. (1.91) gives the output current I_O as

$$I_O = I_{C2} = \frac{I_R}{1 + 2/(\beta_F^2 + \beta_F)} \tag{1.92}$$

which indicates that the reference current I_R is related to the output current I_O by a factor of only $[1 + 2/(\beta_F^2 + \beta_F)]$. The reference current can be found from

$$I_R = \frac{V_{CC} - V_{BE1} - V_{BE3}}{R_1} \tag{1.93}$$

In the derivation of Eq. (1.92), the output resistances of the transistors are ignored. However, for a finite transistor output resistance, the collector current ratio in Eq. (1.86) is applicable.

Output Resistance R_o

The small-signal equivalent circuit for determining the output resistance is shown in Fig. 1.27(b). The output resistance R_o is the same as r_{o2}; that is,

$$R_o = \frac{v_x}{i_x} = r_{o2} = \frac{V_A}{I_{C2}} \tag{1.94}$$

Thevenin's equivalent voltage is given by

$$V_{Th} = I_O R_o = I_{C2} R_o = I_{C2} \frac{V_A}{I_{C2}} = V_A \tag{1.95}$$

EXAMPLE 1.9

D

Designing a simple modified basic current source

(a) Design the modified basic current source in Fig. 1.27(a) to give an output current of $I_O = 5\ \mu A$. The transistor parameters are $\beta_F = 100$, $V_{CC} = 30$ V, $V_{BE1} = V_{BE2} = V_{BE3} = 0.7$ V, and $V_A = 150$ V.

(b) Calculate the output resistance R_o, Thevenin's equivalent voltage V_{Th}, and the collector current ratio if $V_{CE2} = 20$ V.

SOLUTION

$I_O = I_{C2} = 5\ \mu A$, $V_{BE1} = V_{BE2} = V_{BE3} = V_{CE1} = 0.7$ V, and $V_A = 150$ V.

(a) From Eq. (1.92),

$$I_O = I_{C2} = \frac{I_R}{1 + 2/(\beta_F^2 + \beta_F)}$$

which, for $I_O = 5\ \mu A$, gives $I_R = (5\ \mu A)[1 + 2/(\beta_F^2 + \beta_F)] = 5\ \mu A$. From Eq. (1.93),

$$R_1 = \frac{V_{CC} - V_{BE1} - V_{BE3}}{I_R} = \frac{30 - 0.7 - 0.7}{5\ \mu A}$$

which gives the required value of the resistor as $R_1 = 5.72\ \text{M}\Omega$.

(b) From Eq. (1.94),

$$R_o = \frac{V_A}{I_{C2}} = \frac{150}{5\ \mu\text{A}} = 30\ \text{M}\Omega$$

From Eq. (1.95),

$$V_{Th} = V_A = 150\ \text{V}$$

From Eq. (1.86),

$$\frac{I_{C2}}{I_{C1}} = \frac{1 + V_{CE2}/V_A}{1 + (V_{BE1} + V_{BE3})/V_A} = \frac{1 + 20/150}{1 + (0.7 + 0.7)/150} = 1.123$$

Neglecting I_{B3}, we have $I_{C1} \approx I_O = 5\ \mu\text{A}$. Thus, $I_{C2} = 1.123 \times I_{C1} = 1.123 \times 5\ \mu\text{A} = 5.62\ \mu\text{A}$, which agrees, with a degree of error, with the desired value of $I_O = 5\ \mu\text{A}$.

1.6.3 Widlar Current Source

Biasing currents of low magnitudes, typically on the order of 5 μA, are required in a variety of applications. Currents of low magnitude can be obtained by inserting a resistance of moderate value in series with the emitter Q_2 in Fig. 1.25(a). A circuit with this modification, as shown in Fig. 1.28(a), is known as a *Widlar current source*. As a result of the addition of R_2 into the circuit, I_{C2} is no longer equal to I_R, and the value of I_{C2} can be made much smaller than that of I_{C1}. This circuit can give currents in the microampere range, with acceptable circuit resistance values of less than 50 kΩ.

Using KVL around the B-E loop in Fig. 1.28(a) gives

$$V_{BE1} - V_{BE2} - (I_{C2} + I_{B2})R_2 = 0$$

Since $I_{C2} >> I_{B2}$,

$$V_{BE1} - V_{BE2} - I_{C2}R_2\left(1 + \frac{1}{\beta_F}\right) = 0 \tag{1.96}$$

Assuming that the transistors have infinite output resistances, $V_A = \infty$, Eq. (1.85) becomes

$$I_C = I_S \exp\left(\frac{V_{BE}}{V_T}\right)$$

which gives

$$V_{BE} = V_T \ln\left(\frac{I_C}{I_S}\right) \tag{1.97}$$

Applying V_{BE} from Eq. (1.97) in Eq. (1.96), we get

$$V_T \ln\left(\frac{I_{C1}}{I_{S1}}\right) - V_T \ln\left(\frac{I_{C2}}{I_{S2}}\right) - I_{C2}R_2 = 0 \quad (\text{for } \beta_F >> 1)$$

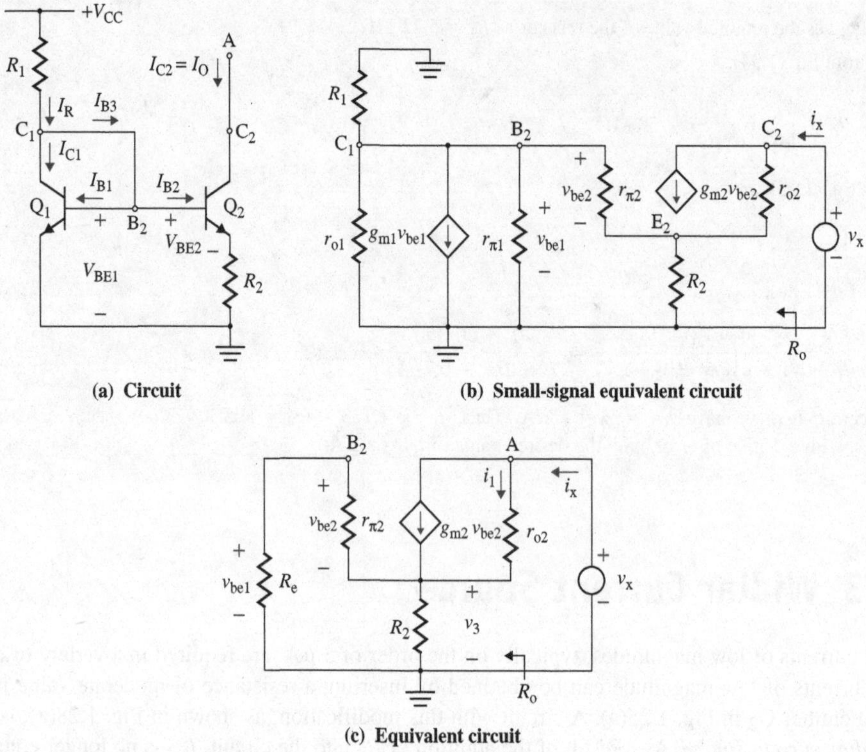

FIGURE 1.28 Widlar current source

which, for identical transistors with $I_{S1} = I_{S2}$, can be simplified to

$$V_T \ln\left(\frac{I_{C1}}{I_{S1}} \times \frac{I_{S2}}{I_{C2}}\right) - I_{C2}R_2 = 0$$

This equation gives the relation of I_{C2} to I_{C1} and R_2 as

$$V_T \ln\left(\frac{I_{C1}}{I_{C2}}\right) = I_{C2}R_2 \tag{1.98}$$

The reference current I_R can be found from

$$I_R = \frac{V_{CC} - V_{BE1}}{R_1} \tag{1.99}$$

which is also related to the base and collector currents; that is,

$$I_R = I_{C1} + I_{B1} + I_{B2}$$

$$= I_{C1}\left(1 + \frac{1}{\beta_F}\right) + \frac{I_{C2}}{\beta_F} \tag{1.100}$$

Substituting I_{C1} from Eq. (1.98) into the preceding equation, we get I_R in terms of I_{C2} and R_2 as

$$I_R = \frac{1+\beta_F}{\beta_F} I_{C2} \exp\left(\frac{I_{C2}R_2}{V_T}\right) + \frac{I_{C2}}{\beta_F} \tag{1.101}$$

Thus, I_{C2} is a nonlinear function of I_R and R_2. To find I_{C2}, we must solve this transcendental equation by trial and error, using known values of I_R and R_2. However, for design purposes, I_{C2} and I_R are known, and it is necessary only to find the value of R_2.

Output Resistance R_o

The small-signal equivalent circuit for determining the output resistance is shown in Fig. 1.28(b). This circuit can be reduced to Fig. 1.28(c), where R_e is the parallel equivalent of $r_{\pi1}$, $1/g_{m1}$, r_{o1}, and R_1; that is,

$$R_e = r_{\pi1} \left\| \frac{1}{g_{m1}} \right\| R_1 \| r_{o1} \tag{1.102}$$

Since transconductance is $g_m = \dfrac{I_C}{V_T} = \dfrac{\beta_E}{r_\pi}$, we can relate the input resistance $r_{\pi1}$ to the transconductance g_{m1} and the small-signal current gain $\beta_f(\approx\beta_F)$:

$$r_{\pi1} = \frac{\beta_f(\approx\beta_F)}{g_{m1}}$$

Since $\beta_f >> 1$, $r_{\pi1} >> 1/g_{m1}$, and in general $R_1 >> 1/g_{m1}$, Eq. (1.102) can be approximated by

$$R_e \approx \frac{1}{g_{m1}} \tag{1.103}$$

The collector current flows through the parallel combination of R_2 and $(r_{\pi2} + R_e)$, so

$$v_3 = i_x[R_2 \| (r_{\pi2} + R_e)] \tag{1.104}$$

The voltage v_3 is related to v_2 by

$$v_2 = -\frac{v_3 r_{\pi2}}{r_{\pi2} + R_e} = -\frac{i_x r_{\pi2}[R_2 \| (r_{\pi2} + R_e)]}{r_{\pi2} + R_e} \tag{1.105}$$

The current i_1 thus becomes

$$i_1 = i_x - g_{m2}v_2$$

The test voltage v_x is

$$v_x = v_3 + r_{o2}i_1 = v_3 + r_{o2}i_x - r_{o2}g_{m2}v_2$$

$$= v_3 + i_x r_{o2} + \frac{r_{o2}g_{m2}v_3 r_{\pi2}}{r_{\pi2} + R_e} \tag{1.106}$$

Substituting v_3 from Eq. (1.104) into Eq. (1.106) and simplifying, we get the resistance at the collector of the transistor:

$$R_o = \frac{v_x}{i_x} = R_2 \| (r_{\pi2} + R_e) + r_{o2}\left[1 + \frac{g_{m2}r_{\pi2}}{r_{\pi2} + R_e} R_2 \| (r_{\pi2} + R_e)\right] \tag{1.107}$$

Since r_{o2} is large, the first term is much smaller than the second one. If the first term is neglected, Eq. (1.107) can be reduced to

$$R_o \approx r_{o2}\left[1 + \frac{g_{m2}r_{\pi2}}{r_{\pi2} + R_e} R_2 \,\|\, (r_{\pi2} + R_e)\right] \tag{1.108}$$

Since transconductance is $g_m = \dfrac{I_C}{V_T} = \dfrac{\beta_E}{r_\pi}$, we get

$$\frac{1}{g_{m1}} = \frac{1}{I_{C1}/V_T} = \frac{V_T}{I_{C1}} \tag{1.109}$$

$$r_{\pi2} = \frac{\beta_F}{g_{m2}} = \frac{\beta_F}{I_{C2}/V_T} = V_T\frac{\beta_F}{I_{C2}} = \beta_F\frac{V_T}{I_{C1}} \times \frac{I_{C1}}{I_{C2}} = \frac{\beta_F}{g_{m1}} \times \frac{I_{C1}}{I_{C2}} \tag{1.110}$$

Since $I_{C1} >> I_{C2}$, $r_{\pi2} >> 1/g_{m1}$ and $r_{\pi2} >> R_e$; that is,

$$R_e + r_{\pi2} \approx r_{\pi2} \tag{1.111}$$

Substituting R_e from Eq. (1.103) into Eq. (1.107) gives

$$R_o \approx r_{o2}[1 + g_{m2}(R_2 \,\|\, r_{\pi2})] \tag{1.112}$$

Since $\beta_F >> 1$ and $r_{\pi2} = \beta_F/g_{m2}$, Eq. (1.112) can be rewritten as

$$R_o = r_{o2}\frac{g_{m2}R_2(1 + \beta_F) + \beta_F}{g_{m2}R_2 + \beta_F} = r_{o2}\left[\frac{1 + g_{m2}R_2}{1 + g_{m2}R_2/\beta_F}\right] \tag{1.113}$$

Generally, $\beta_F >> g_{m2}R_2$, so Eq. (1.113) can be approximated by

$$R_o \approx r_{o2}(1 + g_{m2}R_2) \tag{1.114}$$

$$\approx r_{o2}\left(1 + \frac{I_{C2}R_2}{V_T}\right) \tag{1.115}$$

Thus, R_o depends on $I_{C2}R_2$, which is the DC voltage drop across R_2. The larger this drop is made, the higher the output resistance becomes. For the Widlar source, $I_{C2}R_2$ is limited to several hundred millivolts for practical current ratios, and the corresponding value of V_{Th} is limited to about $10V_A$.

EXAMPLE 1.10

D

Designing a Widlar current source

(a) Design the Widlar current source in Fig. 1.28(a) to give $I_O = 5$ μA and $I_R = 1$ mA. The parameters are $V_{CC} = 30$ V, $V_{BE1} = 0.7$ V, $V_T = 26$ mV, $V_A = 150$ V, and $\beta_F = 100$.

(b) Calculate the output resistance R_o and Thevenin's voltage V_{Th}. Assume $V_T = 26$ mV.

SOLUTION

$I_{C2} = 5\ \mu A$, and $V_T = 26$ mV.

(a) From Eq. (1.99),

$$1\text{ mA} = \frac{30 - 0.7}{R_1}$$

so $R_1 = 29.3$ kΩ. From Eq. (1.100),

$$1\text{ mA} = I_{C1}\left(1 + \frac{1}{100}\right) + \frac{5\ \mu A}{100}$$

which gives $I_{C1} \approx 990\ \mu A$. From Eq. (1.98),

$$26\text{ mV} \times \ln\left(\frac{990\ \mu A}{5\ \mu A}\right) = 5\ \mu A \times R_2$$

so $R_2 = 27.5$ kΩ.

(b) $r_{o2} = V_A/I_{C2} = 150/(5\ \mu A) = 30$ MΩ. From Eq. (1.110),

$$r_{\pi 2} = \frac{V_T \beta_F}{I_{C2}} = 26\text{ m} \times \frac{100}{5\ \mu} = 520\text{ k}\Omega$$

Also from Eq. (1.109),

$$g_{m2} = \frac{I_{C2}}{V_T} = \frac{5\ \mu A}{26\text{ mV}} = 192.3\ \mu A/V$$

From Eq. (1.112),

$$R_o \approx 30\text{ M}\Omega \times [1 + 192.3\ \mu A/V \times (27.5\text{ k}\Omega \parallel 520\text{ k}\Omega)] = 180.68\text{ M}\Omega$$

Using the approximation in Eq. (1.115), we have

$$R_o \approx 30\text{ M}\Omega \times \left(1 + \frac{5\ \mu A \times 27.5\text{ k}\Omega}{26\text{ mV}}\right) = 188.66\text{ M}\Omega$$

and $V_{Th} = R_o I_{C2} = 188.65\text{ M}\Omega \times 5\ \mu A = 943.3$ V

NOTE: The Widlar current source gives a low output current at high output resistance, and Thevenin's equivalent voltage is very high.

1.6.4 Cascode Current Source

The emitter resistance R_2 in the Widlar current source in Fig. 1.28(a) can be replaced by a basic current source consisting of two transistors Q_3 and Q_4. This arrangement, shown in Fig. 1.29, will give a larger output resistance. In a cascodelike connection, two or more transistors are connected in series so that their collector biasing currents are almost identical (e.g., transistors Q_1 and Q_3), whereas in a cascodelike connection, the transistors operate in parallel fashion so that one transistor drives the other (e.g., Q_3 and Q_4). According to Eq. (1.55), the larger the output resistance, the greater the voltage gain of an amplifier

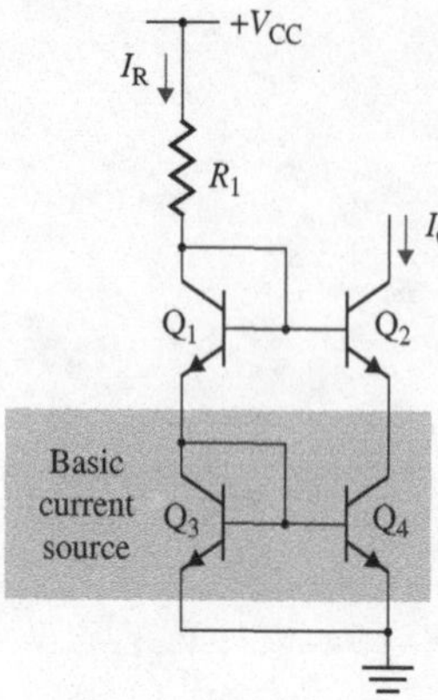

FIGURE 1.29 Cascode current source

becomes. Substituting the output resistance r_{o4} of transistor Q_4 for R_2 in Eq. (1.112) gives the output resistance of this cascode source:

$$R_o \approx r_{o2}[1 + g_{m2}(r_{o4} \| r_{\pi 2})] = r_{o2}(1 + g_{m2}r_{\pi 2}) = r_{o2}(1 + \beta_F) \tag{1.116}$$

1.6.5 Wilson Current Source

A Wilson current source, shown in Fig. 1.30(a), also gives a high output resistance. However, the output current is approximately equal to the reference current. Base current I_{B2}, which is the difference between the reference current I_R and the collector current I_{C1}, is multiplied by $(1 + \beta_F)$ to give I_{E2}, which flows in the diode-connected transistor Q_3 and causes a collector current of the same magnitude to flow in Q_1. There is a feedback path that regulates I_{C1}, which is approximately equal to I_{E2} and I_{C2}. Thus, I_{C2} remains very nearly equal to I_{C1} and nearly constant, giving a high output resistance:

$$I_{E2} = (1 + \beta_F)I_{B2} = (I_R - I_{C1})(1 + \beta_F)$$

Assuming that the transistors have infinite output resistances, $V_A = \infty$, and that the transistors are identical so that $I_{C1} = I_{C3}$, we can write

$$\begin{aligned} I_{E2} &= I_{C3} + I_{B3} + I_{B1} = I_{C3}\left(1 + \frac{1}{\beta_F}\right) + \frac{I_{C1}}{\beta_F} \\ &= I_{C3}\left(1 + \frac{2}{\beta_F}\right) \end{aligned} \tag{1.117}$$

Using Eq. (1.117), we can relate I_{C2} to I_{E2} and I_{C3}:

$$I_{C2} = I_{E2}\frac{\beta_F}{1 + \beta_F} = I_{C3}\left(1 + \frac{2}{\beta_F}\right)\frac{\beta_F}{1 + \beta_F} = I_{C3}\frac{2 + \beta_F}{1 + \beta_F}$$

which gives

$$I_{C3} = I_{C2}\frac{1 + \beta_F}{2 + \beta_F} \tag{1.118}$$

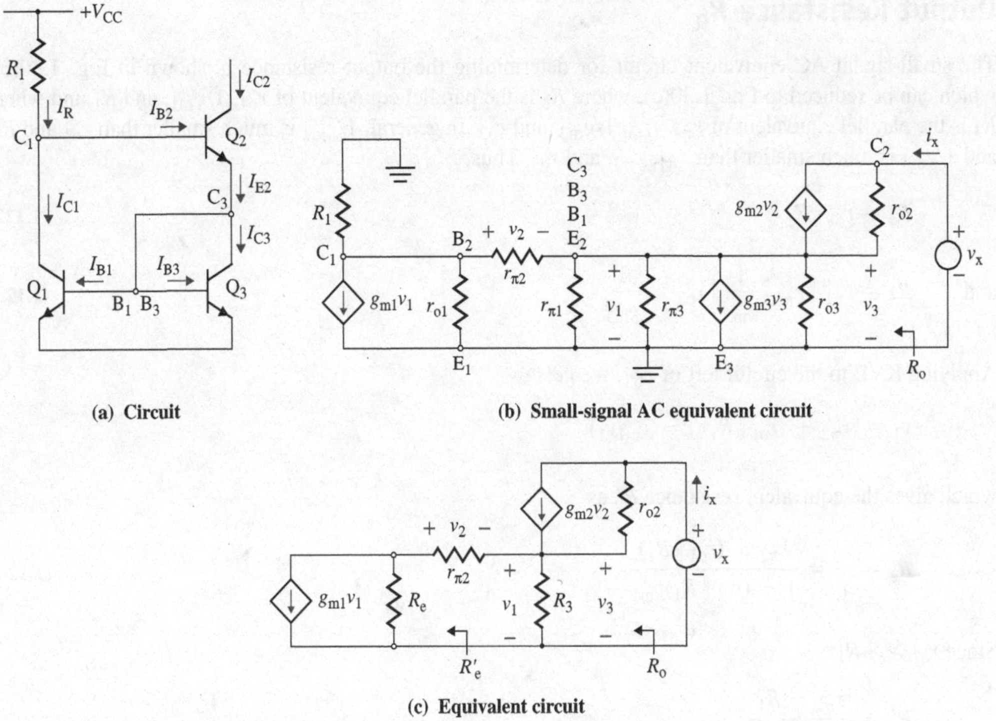

FIGURE 1.30 Wilson current source

I_{C1} is related to the reference current I_R by

$$I_{C1} = I_R - I_{B2} = I_R - \frac{I_{C2}}{\beta_F} \tag{1.119}$$

Since $I_{C3} = I_{C1}$, equating Eq. (1.118) to Eq. (1.119) yields

$$I_{C1} = I_{C3} = I_{C2}\frac{1+\beta_F}{2+\beta_F} = I_R - \frac{I_{C2}}{\beta_F}$$

which gives the output current $I_O = I_{C2}$ as

$$I_O = I_{C2} = I_R\left[\frac{(2+\beta_F)\beta_F}{\beta_F^2 + 2\beta_F + 2}\right] = I_R\left[1 - \frac{2}{\beta_F^2 + 2\beta_F + 2}\right] \tag{1.120}$$

Since $\beta_F >> 1$, Eq. (1.120) can be approximated by

$$I_O \approx I_R \tag{1.121}$$

Thus, the output current almost equals the reference current and is less sensitive to the current gain β_F, which varies in response to temperature changes.

Output Resistance R_o

The small-signal AC equivalent circuit for determining the output resistance is shown in Fig. 1.30(b), which can be reduced to Fig. 1.30(c), where R_e is the parallel equivalent of r_{o1}, $1/g_{m1}$, and R_1 and where R_3 is the parallel equivalent of $r_{\pi 1}$, $r_{\pi 3}$, $1/g_{m3}$, and r_{o3}. In general, $1/g_{m1}$ is much smaller than r_{o1} and R_1, and $1/g_{m3}$ is much smaller than $r_{\pi 1}$, $r_{\pi 3}$, and r_{o3}. Thus,

$$R_e = r_{o1} \| R_1 \tag{1.122}$$

and
$$R_3 = r_{\pi 1} \| r_{\pi 3} \| \frac{1}{g_{m3}} \| r_{o3} \approx \frac{1}{g_{m3}} \tag{1.123}$$

Applying KVL to the circuit left of $r_{\pi 1}$, we get

$$v_1 = i r_{\pi 2} + (r_{o1} \| R_1)(i - g_{m1} v_1)$$

which gives the equivalent resistance R'_e as

$$R'_e = \frac{v_1}{i} = \frac{r_{\pi 2} + (r_{o1} \| R_1)}{1 + (r_{o1} \| R_1) g_{m1}} \tag{1.124}$$

Since $r_{o1} >> R_1$,

$$R'_e = \frac{r_{\pi 2} + R_1}{1 + g_{m1} R_1}$$

which, for $R_1 >> r_{\pi 2}$ and $g_{m1} R_1 >> 1$, can be simplified to

$$R'_e \approx \frac{1}{g_{m1}}$$

The output resistance can be found from Eq. (1.107) by substituting R_3 for R_2 and R'_e for $(r_{\pi 2} + R_e)$. That is,

$$R_o = \frac{v_x}{i_x} = R_3 \| R'_e + r_{o2}\left[1 + \frac{g_{m2} r_{\pi 2}}{R'_e}(R_2 \| R'_e)\right] \tag{1.125}$$

Since r_{o2} is large, the first term is much smaller than the second one and can be ignored. Equation (1.124) can be approximated by

$$R_o \approx r_{o2}\left[1 + \frac{g_{m2} r_{\pi 2}}{R'_e}(R_3 \| R'_e)\right] \tag{1.126}$$

Assuming $(R_3 \| R'_e)/R'_e = 1/2$, $R'_e = 1/g_{m1}$, and $g_{m1} = g_{m3}$, Eq. (1.126) becomes

$$R_o \approx r_{o2}\left(1 + \frac{g_{m2} r_{\pi 2}}{2}\right)$$

$$\approx r_{o2}\left(1 + \frac{\beta_F}{2}\right) \tag{1.127}$$

EXAMPLE 1.11

D

Designing a Wilson current source

(a) Design the Wilson current source in Fig. 1.30(a) to give $I_O = 5\ \mu A$. The parameters are $V_{CC} = 30$ V, $V_{BE1} = V_{BE2} = V_{BE3} = 0.7$ V, $V_T = 26$ mV, $V_A = 150$ V, and $\beta_F = 100$. Assume that all transistors are identical.

(b) Calculate the output resistance R_o and Thevenin's equivalent voltage V_{Th}.

(c) Use PSpice/SPICE to calculate the output current, the output resistance, and the reference current for $\beta_F = 100$ and 400. Assume that all transistors are identical, and $V_{CE} = 10$ V.

SOLUTION

$I_{C2} = 5\ \mu A$, and $V_T = 26$ mV.

(a) From Eq. (1.120),

$$5\ \mu A = I_R\left[1 - \frac{2}{(100^2 + 2 \times 100 + 2)}\right]$$

so $I_R \approx 5\ \mu A$. The reference current is

$$I_R = \frac{V_{CC} - V_{BE1} - V_{BE2}}{R_1}$$

That is,

$$5\ \mu A = \frac{30 - 0.7 - 0.7}{R_1}$$

which gives $R_1 = 5.72\ M\Omega$.

(b) From Eq. (1.118),

$$I_{C3} = I_{C1} = \frac{5\ \mu A \times (1 + 100)}{2 + 100} = 4.95\ \mu A$$

Then $r_{o1} = r_{o3} = \dfrac{V_A}{I_{C1}} = \dfrac{150}{4.95\ \mu} = 30.3\ M\Omega$

$$r_{o2} = \frac{V_A}{I_{C2}} = \frac{150}{5\ \mu} = 30\ M\Omega$$

Since $1/g_{m1} = 1/g_{m3} = V_T/I_{C1} = 26\ mV/4.95\ \mu A = 5.25\ k\Omega$,

$$g_{m1} = g_{m3} = 190.4\ \mu A/V$$

Since $1/g_{m2} = 26\ mV/5\ \mu A = 5.2\ k\Omega$,

$$g_{m2} = 192.3\ \mu A/V$$

Then $r_{\pi 1} = r_{\pi 3} = \dfrac{\beta_F}{g_{m1}} = 100 \times 5.25\text{ k} = 525\text{ k}\Omega$

$$r_{\pi 2} = 100 \times 5.2\text{ k} = 520\text{ k}\Omega$$

From Eq. (1.124),

$$R'_e = \frac{520\text{ k}\Omega + (30.3\text{ M}\Omega \parallel 5.72\text{ M}\Omega)}{[1 + (30.3\text{ M}\Omega \parallel 5.72\text{ M}\Omega) \times 190.4\ \mu\text{A/V}]}$$

$$= 5.81\text{ k}\Omega$$

From Eq. (1.123),

$$R_3 = 525\text{ k} \parallel 525\text{ k} \parallel 5.25\text{ k} \parallel 30\text{ M} \approx 5.15\text{ k}\Omega$$

From Eq. (1.126),

$$R_o = (30\text{ M})\left[1 + \frac{192.3\ \mu \times 520\text{ k}}{5.81\text{ k}} \times 5.15\text{ k} \parallel 5.81\text{ k}\right] = 1.44\text{ G}\Omega$$

and $V_{Th} = 1.44\text{ G} \times 5\ \mu = 7.2\text{ kV}$

NOTE: If we use Eq. (1.127),

$$R_o \approx (30\text{ M})\left(1 + \frac{100}{2}\right) = 1.53\text{ G}\Omega$$

(c) The Wilson current source for PSpice simulation is shown in Fig. 1.31. We will use parametric sweep for the model parameter β_F of the transistors. The voltage source V_y acts as an ammeter for the output current.

The results of simulation (.TF analysis) are as follows. (The hand calculations are shown in parentheses.) For $\beta_F = 100$, the simulation gives

```
VOLTAGE SOURCE CURRENTS
NAME     CURRENT
VCC      -5.022E-06
```

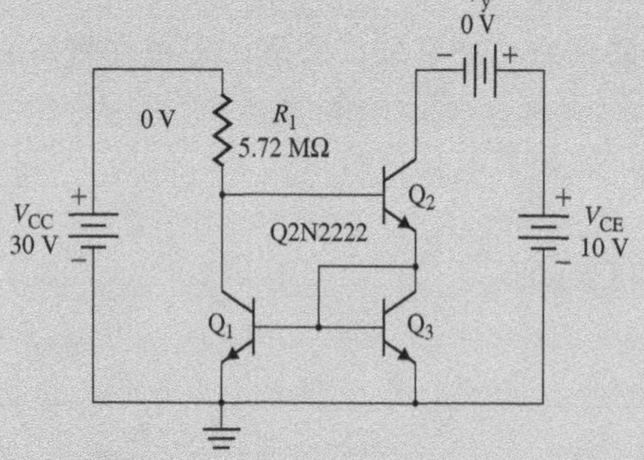

FIGURE 1.31 Wilson current source for PSpice simulation

VX	5.022E-06	($I_R = 5\ \mu A$)		
VY	5.006E-06	($I_R = 5\ \mu A$)		
VCE	-5.006E-06			

**** BIPOLAR JUNCTION TRANSISTORS

NAME	Q1	Q2	Q3	
IB	4.95E-08	4.73E-08	4.95E-08	
IC	4.98E-06	5.01E-06	4.95E-06	
VBE	6.37E-01	6.36E-01	6.37E-01	
VBC	-6.36E-01	-8.73E+00	0.00E+00	
VCE	1.27E+00	9.36E+00	6.37E-01	
BETADC	1.00E+02	1.06E+02	1.00E+02	
GM	1.92E-04	1.94E-04	1.92E-04	
RPI	5.22E+05	5.47E+05	5.22E+05	($r_{\pi 1} = 525\ k\Omega$, $r_{\pi 2} = 520\ k\Omega$, $r_{\pi 3} = 525\ k\Omega$)
RO	3.03E+07	3.17E+07	3.03E+07	($r_{o1} = 30.3\ M\Omega$, $r_{o2} = 30\ M\Omega$, $r_{o3} = 30.3\ M\Omega$)

**** SMALL-SIGNAL CHARACTERISTICS

I(VY)/VCC=1.739E-07

INPUT RESISTANCE AT VCC=5.730E+06 ($R_1 = 5.72\ M\Omega$)

OUTPUT RESISTANCE AT I(VY)=1.602E+09 ($R_o = 1.44\ G\Omega$)

For $\beta_F = 400$, the simulation gives

I(VY)/VCC=1.738E-07

INPUT RESISTANCE AT VCC=5.730E+06

OUTPUT RESISTANCE AT I(VY)=5.441E+09 ($R_o = 6.03\ G\Omega$)

NOTE: R_o changes from 1.602E+09 Ω (for $\beta_F = 100$) to 5.441E+09 Ω (for $\beta_F = 400$ Ω) and depends on β_F, as expected.

1.6.6 Multiple Current Sources

A DC reference current can be generated in one location and reproduced in another location for biasing amplifier circuits in ICs. A group of current sources with only one reference current is shown in Fig. 1.32. This is an extension of the modified basic current source in Fig. 1.25. Transistor Q_1 and resistor R_1 serve as the reference for current-sink transistors Q_3 through Q_6. Transistor Q_2 supplies the total base currents for the transistors and makes the collector current of Q_1 almost equal to the reference current I_R; that is,

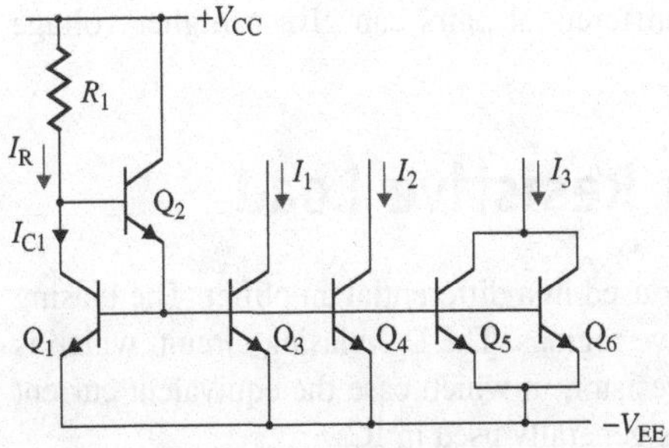

FIGURE 1.32 Multiple current sources

TABLE 1.1 Comparison of BJT current sources

Current Source Type	Output Resistance R_o	Beta (β_F) Dependency of I_o	Comments
Basic source	r_o	$-2/\beta_F$	Not suitable for low currents (i.e., typically less than 0.3 mA)
Modified source	r_o	$-2/(\beta_F^2 + \beta_F)$	Not suitable for low currents (i.e., typically less than 0.3 mA)
Widlar source	$r_{o2}(1 + g_{m2}R_2)$	Nonlinear	Current as low as 5 μA; higher output resistance
Cascode source	$r_{o2}(1 + \beta_F)$	$-2/\beta_F$	Not suitable for low currents; higher output resistance
Wilson source	$r_{o2}(1 + \beta_F/2)$	$-2/(\beta_F^2 + 2\beta_F + 2)$	Not suitable for low currents; higher output resistance

$I_R \approx I_{C1}$. The collector currents I_1 and I_2 will be mirrors of current I_R. Since two transistors Q_5 and Q_6 are connected in parallel, I_3 will be two times I_R (i.e., $I_3 = 2I_R$). The parallel combination of Q_5 and Q_6 should be equivalent to a single transistor whose E-B junction has double the area of Q_1. Therefore, the emitter areas of transistors can be scaled in ICs so as to provide multiples of the reference current simply by designing the transistors so that they have an area ratio equal to the desired multiple.

KEY POINTS OF SECTION 1.6

- Transistors can generate the characteristics of a constant-current source. An ideal current source has a very high output resistance, and its output current is not sensitive to the transistor parameter β_F.
- Table 1.1 compares various BJT current sources. The output resistances for different BJT sources are also summarized in the table.

1.7 BJT Differential Amplifiers

The BJT differential pairs are analogous to MOS differential pairs. Since BJTs, in general, have a higher transconductance compared to identical MOSFETs, BJT differential pairs can give a higher voltage gain, but they have limitations of a lower input resistance.

1.7.1 BJT Differential Pair with Resistive Load

An emitter-coupled pair, as shown in Fig. 1.33, is commonly used in a differential amplifier. The biasing current should be such that the transistors operate in the active regions. The DC biasing circuit, which is shown as a constant-current source, can be either a simple resistor, in which case the equivalent current generator will be zero, or a transistor current source, which is generally used in ICs.

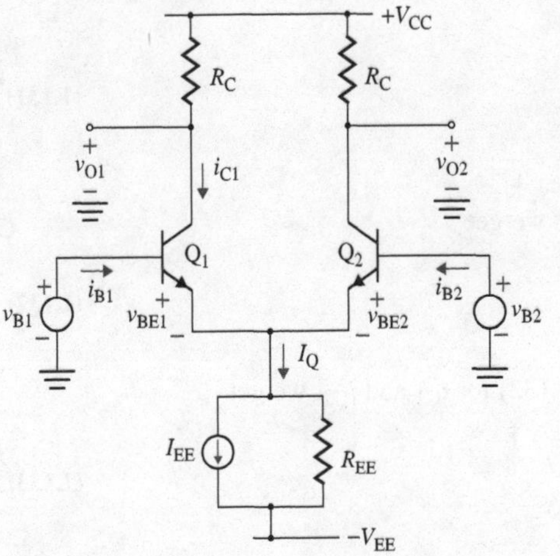

FIGURE 1.33 Emitter-coupled differential pair

DC Transfer Characteristics

The DC transfer characteristic, which gives the relation between the input and output voltages, can be determined from the large-signal analysis, and it should be linear over a wide range. The analysis can be simplified by making the following assumptions:

1. The output resistances of the transistors are infinite: $r_o = \infty$.
2. The output resistance of the transistor current source is infinite: $R_{EE} = \infty$.

Using KVL around the loop formed by the two input voltages and the two B-E junctions, we get

$$v_{B1} - v_{BE1} + v_{BE2} - v_{B2} = 0 \tag{1.128}$$

Assuming $v_{BE1}, v_{BE2} >> V_T$ and equal leakage currents $I_{S1} = I_{S2} = I_S$ and using the transistor current equations, we get

$$v_{BE1} = V_T \ln\left(\frac{i_{C1}}{I_{S1}}\right) = V_T \ln\left(\frac{i_{C1}}{I_S}\right) \tag{1.129}$$

$$v_{BE2} = V_T \ln\left(\frac{i_{C2}}{I_{S2}}\right) = V_T \ln\left(\frac{i_{C2}}{I_S}\right) \tag{1.130}$$

Substituting Eqs. (1.129) and (1.130) into Eq. (1.128) gives

$$v_{B1} - V_T \ln\left(\frac{i_{C1}}{I_S}\right) + V_T \ln\left(\frac{i_{C2}}{I_S}\right) - v_{B2} = 0$$

That is,

$$v_{B1} - v_{B2} = V_T\left[\ln\left(\frac{i_{C1}}{I_S}\right) - \ln\left(\frac{i_{C2}}{I_S}\right)\right]$$

which can be simplified to

$$\frac{i_{C1}}{i_{C2}} = \exp\left(\frac{v_{B1} - v_{B2}}{V_T}\right) = \exp\left(\frac{v_{id}}{V_T}\right) \tag{1.131}$$

where $v_{id} = v_{B1} - v_{B2}$ is the differential DC input voltage.

Using KCL at the emitter terminal of the transistors, we get

$$I_Q = \frac{1}{\alpha}(i_{C1} + i_{C2}) \tag{1.132}$$

where $\alpha = \beta_F/(1 + \beta_F) \approx 1$. Solving Eqs. (1.131) and (1.132) for i_{C1} and i_{C2}, we get

$$i_{C1} = \frac{\alpha I_Q}{1 + \exp(-v_{id}/V_T)} \tag{1.133}$$

$$\approx \alpha I_Q \quad \text{for } v_{id} >> V_T \tag{1.134}$$

$$i_{C2} = \frac{\alpha I_Q}{1 + \exp(v_{id}/V_T)} \tag{1.135}$$

$$\approx 0 \quad \text{for } v_{id} >> V_T \tag{1.136}$$

Thus, if i_{C1} increases, i_{C2} decreases such that $i_{C1} + i_{C2} = \alpha I_Q = \alpha I_{EE}$ remains constant.

The plots of the two collector currents are shown as a function of v_{id} in Fig. 1.34(a). Notice that, for $v_{id} >> V_T$, i_{C1} and i_{C2} become independent of v_{id} and all the currents flow through one of the transistors. For $v_{id} \leq V_T$, i_{C1} and i_{C2} have an approximately linear relation. The differential voltage change Δv_{id} required to shift the current distribution from $i_{C1} = 0.9I_Q$ and $i_{C2} = 0.1I_Q$ to the opposite case, $i_{C1} = 0.1I_Q$ and $i_{C2} = 0.9I_Q$, is called the *transition voltage*; it will have a value of approximately $2V_T = 52.6$ mV.

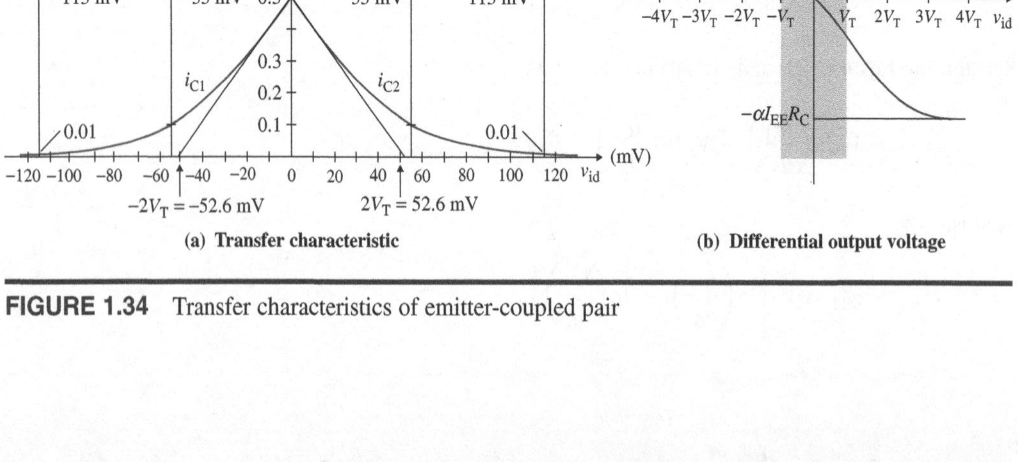

FIGURE 1.34 Transfer characteristics of emitter-coupled pair

The DC output voltages are

$$v_{O1} = V_{CC} - i_{C1}R_C \tag{1.137}$$

$$v_{O2} = V_{CC} - i_{C2}R_C \tag{1.138}$$

The differential DC output voltage is

$$v_{od} = v_{O1} - v_{O2} = R_C(i_{C2} - i_{C1})$$

Substituting Eqs. (1.133) and (1.135) into the above equation and simplifying yields

$$v_{od} = \alpha I_{EE} R_C \tanh\left(-\frac{v_{id}}{2V_T}\right) \tag{1.139}$$

Since tanh $x \approx x$ for a small value of x, Eq. (1.139) can be approximated by

$$v_{od} \approx -\alpha I_{EE} R_C \left(\frac{v_{id}}{2V_T}\right) \tag{1.140}$$

The plot of v_{od} as a function of v_{id} is shown in Fig. 1.34(b). If v_{id} is zero, v_{od} is also zero, and this feature allows direct coupling of cascoded stages without introducing DC offsets. Thus, the amplifier is a true differential, or difference, amplifier, responding only to the difference in the voltages applied to the two input terminals. If $v_{B1} = -v_{B2}$, then $v_{ic} = (v_{B1} + v_{B2})/2$ is zero and there will be only a differential voltage. If, on the other hand, $v_{B1} = v_{B2}$, then v_{id} is zero and there will be a pure common-mode voltage (i.e., no output voltage). However, the range of the differential input voltage v_{id} over which the emitter-coupled pair exhibits a linear characteristic is very small, typically two or three times V_T. This range can be extended by inserting emitter degeneration resistors, as shown in Fig. 1.35(a); in this case, the range over which the characteristic is linear is approximately equal to $I_Q R_E$. The factor by which the voltage gain is reduced is approximately the same as the factor by which the input range is increased. The variations of v_{od} with several values of R_E are shown in Fig. 1.35(b).

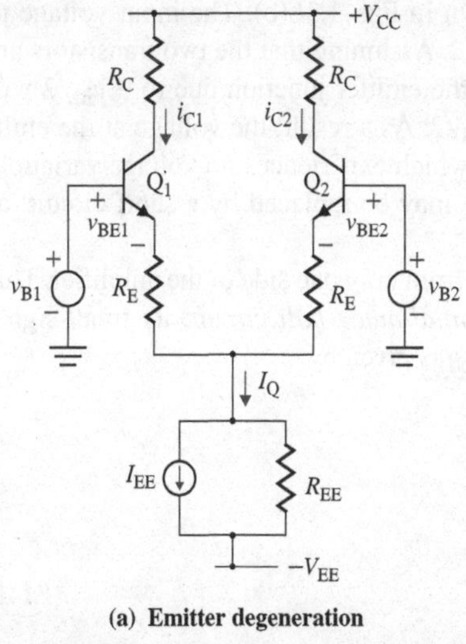

(a) Emitter degeneration

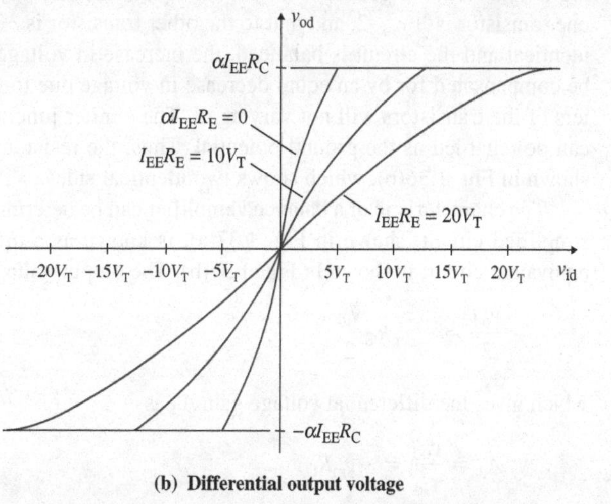

(b) Differential output voltage

FIGURE 1.35 Emitter-coupled pair with degeneration resistors

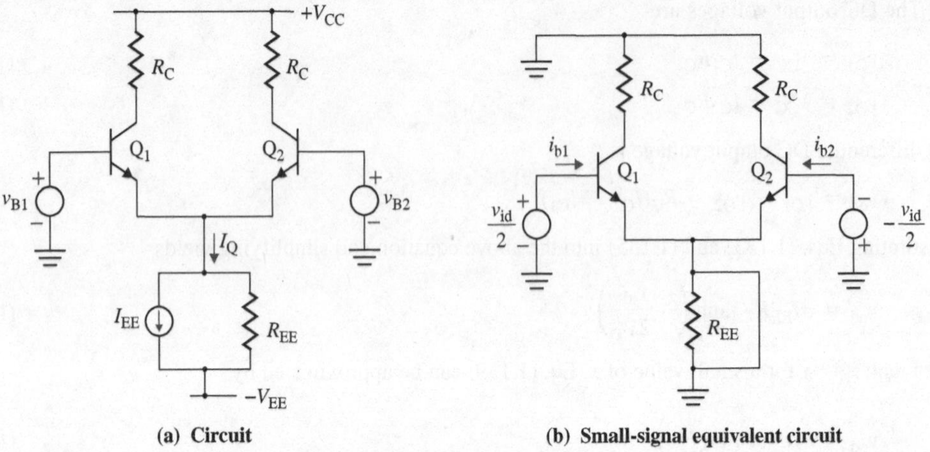

FIGURE 1.36 Emitter-coupled pair with differential input

Small-Signal Analysis

In studying a BJT differential amplifier, it is often of interest to examine the small-signal behavior for small DC differential voltages near zero, when the amplifier operates in the linear portion of the transfer characteristic. Circuit properties such as differential voltage gain A_d, common-mode gain A_c, common-mode rejection ratio (CMMR), common-mode input resistance R_{ic}, and differential mode input resistance R_{id} can be determined from the small-signal analysis.

Small Differential Signal Let us assume that the common-mode signal is zero, $v_{ic} = 0$, and only the differential input voltage v_{id} is applied. This situation is shown in Fig. 1.36(a). If the DC differential biasing is removed, we get the small-signal equivalent circuit shown in Fig. 1.36(b). The input voltage to one transistor is $+v_{id}/2$, and that to the other transistor is $-v_{id}/2$. Assuming that the two transistors are identical and the circuit is balanced, the increase in voltage at the emitter junction due to $+v_{id}/2$ will be compensated for by an equal decrease in voltage due to $-v_{id}/2$. As a result, the voltage at the emitters of the transistors will not vary at all. The emitter junction, which experiences no voltage variation, can be regarded as the ground potential. Thus, the resistor R_{EE} may be replaced by a short circuit, as shown in Fig. 1.36(b), which shows two identical sides.

The characteristic of a balanced amplifier can be determined from only one side of the amplifier. This simplified circuit, shown in Fig. 1.37(a), is known as a *differential-mode half circuit*; its small-signal equivalent circuit is shown in Fig. 1.37(b). The output voltage v_{od} is given by

$$\frac{v_{od}}{2} = -g_m R_C \frac{v_{id}}{2}$$

which gives the differential voltage gain A_d as

$$A_d = \frac{v_{od}}{v_{id}} = -g_m R_C \tag{1.141}$$

▶ **NOTE** If we include the output resistance r_{o1} of Q_1, Eq. (1.141) becomes $A_d = -g_m(R_C \parallel r_{o1})$.

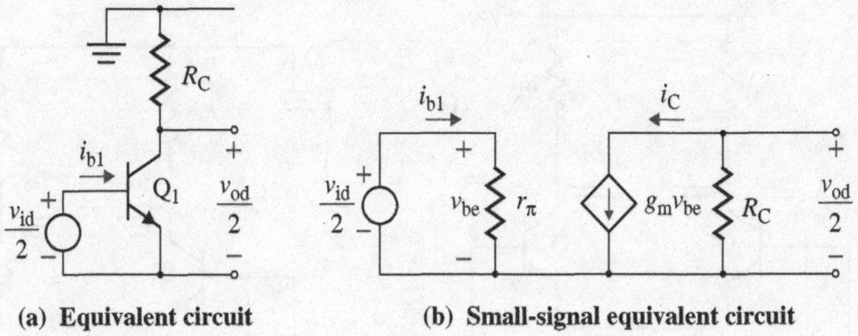

FIGURE 1.37 Differential-mode half circuit

High values of R_C and g_m are required to give a high value of A_d. An active load (discussed in Sec. 1.6) rather than a discrete resistance ensures a high value of R_C and hence of A_d. From Fig. 1.37(b), we can write

$$\frac{v_{id}}{2} = i_{b1} r_\pi$$

which gives the differential input resistance R_{id} as

$$R_{id} = \frac{v_{id}}{i_{b1}} = 2r_\pi \tag{1.142}$$

A high value of r_π $(=V_T/I_C)$—that is, a low collector biasing current $(I_{C1} = I_{C2} = I_Q/2)$—is required to achieve a higher value of R_{id}.

Small Common-Mode Signal An emitter-coupled pair with only common-mode input v_{ic} is shown in Fig. 1.38(a). If the DC common-mode biasing is removed, the result is the small-signal equivalent circuit shown in Fig. 1.38(b). Assuming that the two transistors are identical, the collector currents must be identical, and the voltage at the emitter junction will increase by the same amount in response to inputs at both transistors. Since the voltage across R_{EE} will be the same for both inputs, the resistor R_{EE} can be split into two parallel resistors, each of value $2R_{EE}$, as shown in Fig. 1.38(c). As a result of symmetry, no current will flow through the lead that connects the two sides, and $i_x = 0$. Thus, this lead can be disconnected without affecting the circuit behavior; as shown in Fig. 1.38(d), the two half circuits may be considered to be completely independent. The common-mode behavior can be determined from only one side, as shown in Fig. 1.39(a). The small-signal equivalent circuit is shown in Fig. 1.39(b), from which we get

$$v_{ic} = i_{b1} r_\pi + i_b(1 + \beta_F)2R_{EE} \tag{1.143}$$

which gives the common-mode input resistance R_{ic} as

$$R_{ic} = \frac{v_{ic}}{i_{b1}} = r_\pi + (1 + \beta_F)2R_{EE} \tag{1.144}$$

Equation (1.144) will give a high value for R_{ic}, so R_{ic} should be calculated using r_μ in the BJT model. The current through r_μ can be found from

$$i_\mu = \frac{v_{ic} - v_o}{r_\mu} = \frac{v_{ic} + \beta_F R_C i_{b1}}{r_\mu}$$

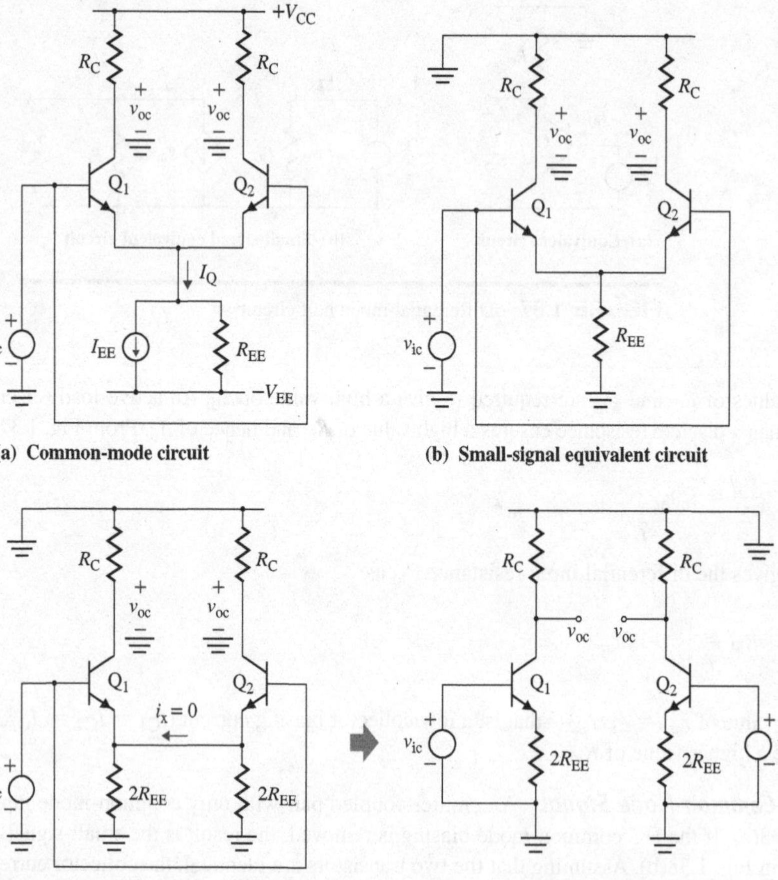

(a) Common-mode circuit (b) Small-signal equivalent circuit

(c) Small-signal equivalent circuit with split R_{EE} (d) Independent half circuits

FIGURE 1.38 Emitter-coupled pair with common-mode input

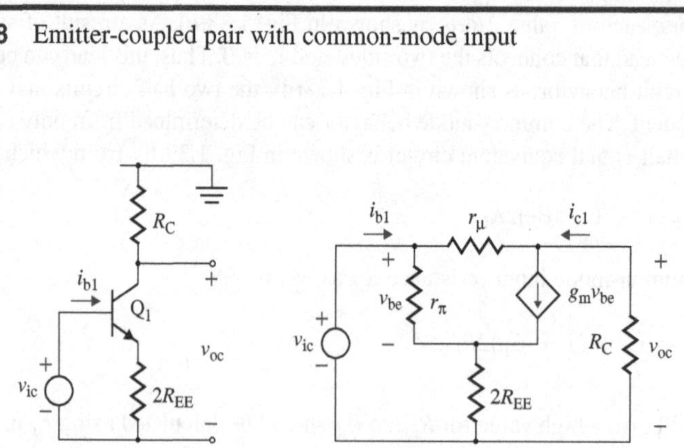

(a) Common-mode half circuit (b) Small-signal equivalent circuit

FIGURE 1.39 Common-mode half circuit

which, after substitution of $i_{b1} = v_{ic}/[r_\pi + (1 + \beta_F)2R_{EE}]$, gives the effective resistance as

$$R_{i\mu} = \frac{r_\mu}{1 + \beta_F R_C/[r_\pi + (1 + \beta_F)2R_{EE}]}$$

Thus, the common-mode resistance R_{ic} becomes

$$R_{ic} = [r_\pi + 2R_{EE}(1 + \beta_F)] \parallel \frac{r_\mu}{r_\pi + (1 + \beta_F)2R_{EE}}$$

The common-mode output voltage v_{oc} is

$$v_{oc} = -R_C i_{c1} = -R_C \beta_F i_{b1} = v_{ic}\left[\frac{-\beta_F R_C}{r_\pi + (1 + \beta_F)2R_{EE}}\right]$$

which gives the common-mode voltage gain A_c as

$$A_c = \frac{v_{oc}}{v_{ic}} = \frac{-\beta_F R_C}{r_\pi + (1 + \beta_F)2R_{EE}} \tag{1.145}$$

$$= \frac{-g_m R_C}{1 + 2g_m R_{EE}(1 + 1/\beta_F)} \tag{1.146}$$

Small-Signal CMRR From Eqs. (1.141) and (1.146), we can find the CMRR as

$$\text{CMRR} = \left|\frac{A_d}{A_c}\right| = 1 + 2g_m R_{EE}\left(1 + \frac{1}{\beta_F}\right) \tag{1.147}$$

$$\approx 1 + 2g_m R_{EE} \quad \text{for } \beta_F >> 1 \tag{1.148}$$

which indicates that a high output resistance R_{EE} on the biasing current source will improve the CMRR; that is, the value of R_{EE} should be as large as possible. To obtain a high value of g_m $(=I_C/V_T)$, the collector biasing current $(I_{C1} = I_{C2} = I_Q/2 = I_{EE}/2)$ should be made large.

The small-signal input currents, which will flow when both v_{id} and v_{ic} are applied, can be found by superposition. Since R_{ic} is common to both i_{b1} and i_{b2}, we get

$$i_{b1} = \frac{v_{ic}}{2R_{ic}} + \frac{v_{id}}{R_{id}} \tag{1.149}$$

$$i_{b2} = \frac{v_{ic}}{2R_{ic}} + \frac{v_{id}}{R_{id}} \tag{1.150}$$

The input resistance can thus be represented by the π-equivalent circuit of Fig. 1.40(a), where R_{ic} is assumed to be much larger than R_{id}. The T-equivalent circuit is shown in Fig. 1.40(b), and its values are shown in Fig. 1.40(c).

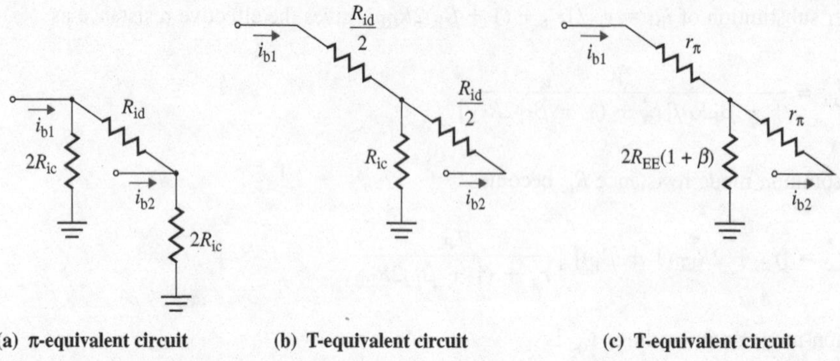

FIGURE 1.40 π- and T-equivalent circuits

EXAMPLE 1.12

Finding the performance parameters of an emitter-coupled pair The parameters of the emitter-coupled pair in Fig. 1.36(a) are $\beta_F = 100$, $R_{EE} = 50\ \text{k}\Omega$, $I_Q = 1$ mA, $V_{CC} = 15$ V, and $R_C = 10\ \text{k}\Omega$.

(a) Calculate the DC collector currents through the transistors if $v_{id} = 5$ mV.

(b) Assuming $i_{C1} = i_{C2}$, calculate A_d, A_c, and CMRR; R_{id} and R_{ic}; and the small-signal output voltage if $v_{B1} = 20$ mV and $v_{B2} = 10$ mV. Assume $V_T \approx 26$ mV.

SOLUTION

$\alpha = \beta_F/(1 + \beta_F) = 100/(1 + 100) = 0.99$.

(a) From Eq. (1.133),

$$i_{C1} = \frac{\alpha I_Q}{1 + \exp(-v_{id}/V_T)} = \frac{0.99 \times 1\,\text{mA}}{1 + \exp(-5\text{ m}/26\text{ m})} = 0.543\text{ mA}$$

and $i_{C2} = 1\text{ m} - 0.543\text{ m} = 0.457$ mA

(b) We know that $i_{C1} = i_{C2} = 1\text{ mA}/2 = 0.5$ mA. Thus,

$$g_m = \frac{i_{C1}}{V_T} = \frac{0.5\text{ mA}}{26\text{ mV}} = 19.23\text{ mA/V}$$

From Eq. (1.141),

$$A_d = -g_m R_C = -19.23\text{ m} \times 10\text{ k} = -192.3\text{ V/V}$$

From Eq. (1.146),

$$A_c = \frac{-g_m R_C}{1 + 2g_m R_{EE}(1 + 1/\beta_F)}$$

$$= \frac{-19.23\text{ mA/V} \times 10\text{ k}\Omega}{1 + 2 \times 19.23\text{ mA/V} \times 50\text{ k}\Omega \times (1 + 1/100)} = -0.099\text{ V/V}$$

Thus, CMRR = $|A_d/A_c|$ = 192.3/0.099 = 1942.4 (or 65.77 dB). From Eq. (1.148),

$$\text{CMRR} \approx 1 + 2g_m R_{EE} = 1 + 2 \times 19.23\text{ m} \times 50\text{ k} = 1924$$

We know that $r_\pi = \beta_F/g_m = 100/19.23\text{ m} = 5.2\text{ k}\Omega$. From Eq. (1.142),

$$R_{id} = 2r_\pi = 2 \times 5.2\text{ k} = 10.4\text{ k}\Omega$$

From Eq. (1.144),

$$R_{ic} = r_\pi + (1 + \beta_F)2R_{EE} = 5.2\text{ k} + (1 + 100) \times 2 \times 50\text{ k} = 10.1\text{ M}\Omega$$

We know that $v_{id} = 20 - 10 = 10$ mV, and $v_{ic} = (20 + 10)/2 = 15$ mV. From Eq. (1.10),

$$v_o = A_d v_{id} + A_c v_{ic} = -192.3 \times 10\text{ mV} - 0.099 \times 15\text{ mV} = -1924.5\text{ mV}$$

▶ **NOTE** To apply Eq. (1.10) and other equations in Sec. 1.2, we must have $|v_{id}| \leq V_T$. The DC collector voltage of a transistor is

$$V_C = V_{CC} - I_C R_C = 15\text{ V} - 0.5\text{ mA} \times 10\text{ k}\Omega = 10\text{ V}$$

Thus, for $|A_d| = 192.3$, the maximum differential voltage will be $v_{id} = 10/192.3 = 52$ mV. Therefore, to allow output voltage swing due to the input voltages, V_{CC} must be greater than $I_C R_C$.

EXAMPLE 1.13

D

Designing an emitter-coupled pair

(a) Design an emitter-coupled pair as shown in Fig. 1.41 in which one input terminal is grounded. The output is taken from the collector of transistor Q_1. The biasing current is $I_{EE} = 1$ mA, and $V_{CC} = V_{EE} = 15$ V. The transistors are identical. Assume $V_{BE} = 0.7$ V, $V_T = 26$ mV, $V_A = \infty$, and $\beta_F = 100$. A small-signal voltage gain of $A_1 = -250$ V/V is required.

(b) Calculate the design values of A_d, A_c, and CMRR.

SOLUTION

(a) $I_{EE} = 1$ mA. Since $v_{B2} = 0$, we can write

$$R_{EE} I_{EE} + V_{BE} = V_{EE} = 15\text{ V}$$

Thus,

$$R_{EE} = \frac{15 - 0.7}{1\text{ m}} = 14.3\text{ k}\Omega$$

We have

$$i_{C1} = i_{C2} = \frac{I_{EE}}{2} = \frac{1\text{ mA}}{2} = 0.5\text{ mA}$$

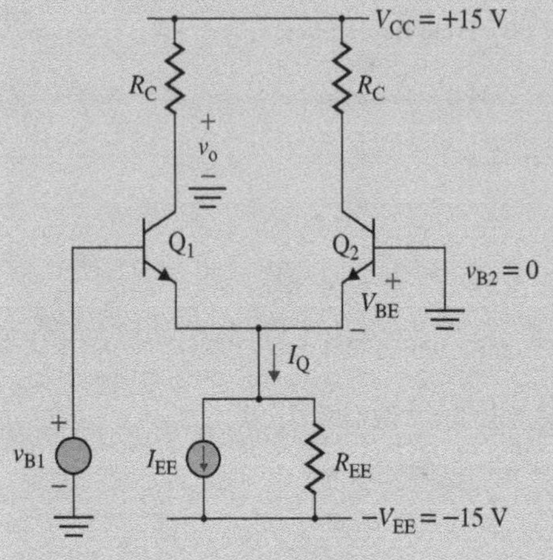

FIGURE 1.41 Emitter-coupled pair with single input

and $g_m = \dfrac{i_{C1}}{V_T} = \dfrac{0.5 \text{ mA}}{26 \text{ mV}} = 19.23 \text{ mA/V}$

Since $v_{B2} = V_{B2} + v_{b2} = 0$, $v_{b2} = 0$. Then

$$v_{id} = v_{b1} - v_{b2} = v_{b1}$$

and $v_{ic} = \dfrac{v_{b1} + v_{b2}}{2} = \dfrac{v_{b1}}{2}$

From Eq. (1.10), we get

$$v_{o1} = A_c v_{ic} + \frac{A_d v_{id}}{2} = A_c \frac{v_{b1}}{2} + \frac{A_d v_{b1}}{2} = \frac{v_{b1}}{2}(A_c + A_d) \qquad \textbf{(1.151)}$$

Substituting for A_d and A_c from Eqs. (1.141) and (1.146) gives the voltage gain A_1 as

$$A_1 = \frac{v_{o1}}{v_{b1}} = \frac{1}{2}(A_c + A_d) = -\frac{1}{2}\left[\frac{g_m R_C}{1 + 2g_m R_{EE}(1 + 1/\beta_F)} + g_m R_C\right] \qquad \textbf{(1.152)}$$

Substituting $A_1 = -250$, $g_m = 19.23$ mA/V, $R_{EE} = 14.3$ kΩ, and $\beta_F = 100$ into Eq. (1.152) gives $R_C = 25.95$ kΩ.

(b) From Eq. (1.141),

$$A_d = -g_m R_C = -19.23 \text{ m} \times 25.95 \text{ k} = -499.01$$

From Eq. (1.146),

$$A_c = \frac{-19.23 \text{ mA/V} \times 25.95 \text{ k}\Omega}{1 + 2 \times 19.23 \text{ mA/V} \times 14.3 \text{ k}\Omega \times (1 + 1/100)} = -0.897$$

Thus, CMRR $= |A_d/A_c| = 499.01/0.897 = 556.3$ (or 54.91 dB).

1.7.2 BJT Differential Amplifiers with Basic Current Mirror Active Load

We saw in Sec. 1.7.1 that the differential gain of a differential pair with a resistive load R_C is $-g_m R_C = -R_C I_C/V_T = -R_C I_Q/2V_T$. In differential amplifiers, a very small value of biasing current I_Q, in the microampere range, is often used. As a result, a very large value of R_C, on the order of megaohms, will be required to give a substantial voltage gain. However, a large value of R_C will cause a large DC voltage drop, reducing the collector voltage to $V_{CC} - R_C I_Q/2$, which will be substantially less than V_{CC}. This low collector voltage will reduce the allowable input voltage range of the amplifier.

Active devices such as transistors occupy much less silicon area than medium-size or large resistors. In practical amplifiers, the load resistor R_C is normally replaced by a constant-current source, which offers a very high load resistance to the amplifier and hence can give a high voltage gain. This type of load, known as an *active load*, has a small voltage drop, typically 0.7 V, and hence allows a wider input voltage range.

A differential amplifier with a basic current source as the active load is shown in Fig. 1.42. The active load consists of transistors Q_3 and Q_4. Since their B-E voltages are the same, their collector currents will be equal. That is, $i_{C3} \approx i_{C4}$. Thus, the current through Q_4 will be the mirror of the current through Q_3. Under quiescent conditions, the differential amplifiers will be balanced such that $I_{C1} = I_{C2}$. Since $I_{C1} = I_{C2}$ and $I_{C3} = I_{C4}$, we can find the quiescent load current:

$$I_O = I_{B3} + I_{B4} = \frac{I_{C3} + I_{C4}}{\beta_F} \approx \frac{I_{C1} + I_{C2}}{\beta_F} = \frac{\alpha I_Q}{\beta_F} = \frac{I_Q}{1 + \beta_F}$$

Since $\beta_F >> 1$, I_O will be very small.

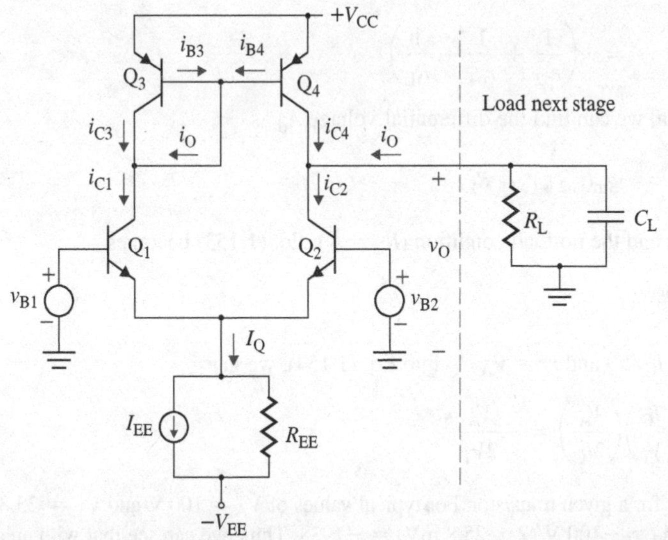

FIGURE 1.42 Differential amplifier with a basic current mirror active load

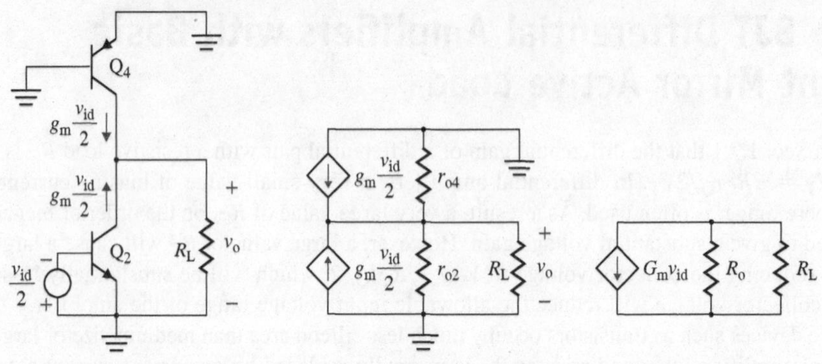

FIGURE 1.43 Equivalent basic current mirror circuit

Small-Signal Analysis

Small Differential Signal If the input voltages change by a small differential amount v_{id}, the collector current of Q_1 will change by a small amount $g_m v_{id}/2$, and the collector current of Q_3 will change by the same amount. Since i_{C4} is a mirror of i_{C3}, the collector current of Q_4 will change by $g_m v_{id}/2$. The B-E voltage of Q_2 will decrease by $v_{id}/2$, causing its collector current to change by an amount $-g_m v_{id}/2$. The equivalent half circuit is shown in Fig. 1.43(a), and its small-signal equivalent appears in Fig. 1.43(b).

Using KCL at the collector of transistors Q_2 and Q_4, we can find the output voltage v_o in terms of v_{id}:

$$\frac{-2 \times g_m v_{id}}{2} = \frac{v_o}{r_{o2}} + \frac{v_o}{r_{o4}} + \frac{v_o}{R_L}$$

$$= v_o\left(\frac{1}{r_{o2}} + \frac{1}{r_{o4}} + \frac{1}{R_L}\right)$$

After simplification, we can find the differential voltage A_d as

$$A_d = \frac{v_o}{v_{id}} = -g_m(r_{o2} \parallel r_{o4} \parallel R_L) \tag{1.153}$$

For $r_{o2} = r_{o4} = r_o$ and the no-load condition ($R_L = \infty$), Eq. (1.153) becomes

$$A_d = -\frac{g_m r_o}{2} \tag{1.154}$$

Substituting $g_m = I_C/V_T$ and $r_o = V_A/I_C$ into Eq. (1.154), we get

$$A_d = -\left(\frac{I_C}{V_T}\right)\left(\frac{V_A}{2I_C}\right) = -\frac{V_A}{2V_T} \tag{1.155}$$

which is a constant for a given transistor. For typical values of $V_A = 100$ V and $V_T = 25.8$ mV, the differential gain becomes $A_d = -100$ V/(2 × 25.8 mV) = −1938. Thus, we can see that with an active load, a very large voltage gain can be obtained with only a single amplifier stage. Also, the gain A_d depends only on the physical parameters V_A and V_T. Since V_T is temperature dependent, A_d will be temperature dependent too.

The differential input resistance R_{id} is given by

$$R_{id} = \frac{v_{id}}{i_d} = 2r_\pi \quad (\text{for } I_C > 0) \tag{1.156}$$

Equation (1.156) is not valid for the no-load condition—that is, when $I_C = 0$. Substituting $r_\pi = \beta_F V_T / I_C$ into Eq. (1.156) gives

$$R_{id} = 2\frac{\beta_F V_T}{I_C} = 2 \times \frac{\beta_F V_T}{I_Q/2} = \frac{4\beta_F V_T}{I_Q} \tag{1.157}$$

which indicates that a lower biasing current I_Q will give a higher value of R_{id} while still maintaining a high voltage gain A_d. A very low value of I_Q, however, will affect the frequency and transient responses of the amplifier, which is undesirable. If a low biasing current is desired to achieve a high input resistance, then a MOSFET differential amplifier is preferable because the amplifier can be operated at a relatively higher value of biasing current without affecting the frequency and transient responses.

The output resistance R_o is the parallel combination of r_{o2} and r_{o4}; that is,

$$R_o = r_{o2} \parallel r_{o4} = \frac{r_o}{2} \tag{1.158}$$

(since $r_{o2} = r_{o4}$).

A differential amplifier is normally followed by other stages. The input resistance R_L of the next stage acts as a load of the amplifier and hence influences the overall voltage gain. The amplifier is often represented as a transconductance amplifier so that the short-circuit current and the effect of load resistance on the output voltage can be determined easily. This arrangement is shown in Fig. 1.43(c). The total current is $2g_m v_{id}/2 = g_m v_{id}$, which gives the effective G_m as

$$G_m = g_m = \frac{I_C}{V_T} = \frac{I_Q}{2V_T} \tag{1.159}$$

Small Common-Mode Signal The common-mode input resistance R_{ic} can be found from Eq. (1.144). The approximate value of the common-mode gain A_c can be found from Eq. (1.145) by replacing R_C by R_{o4}. For practical amplifiers, A_c is generally very small and can be ignored in finding the output voltage.

1.7.3 Differential Amplifier with Modified Current Mirror

Let us use the modified basic current shown in Fig. 1.27(a) as the active load. This arrangement is shown in Fig. 1.44. The active load consists of transistors Q_3, Q_4, and Q_5. The addition of Q_5 makes the ratio i_{C4}/i_{C3} independent of current gain β_F, and i_{C4} approximates a true mirror of i_{C3}. Transistors Q_6 and Q_7 belong to the second stage and act as the load of the differential amplifier. Q_6 and Q_7 form a compound transistor (Darlington pair), in which the emitter current of Q_6 becomes the base current of Q_7. As a result, the effective current gain becomes β_F^2 such that $i_{C7} = \beta_F^2 i_{B6}$.

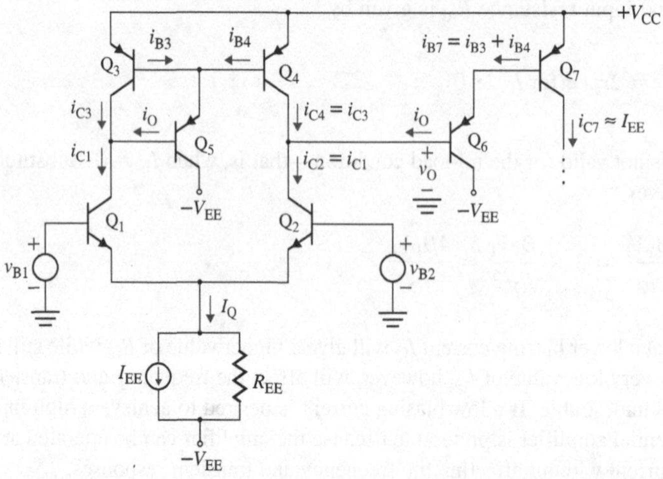

FIGURE 1.44 Differential amplifier with a modified current mirror active load

Under quiescent conditions with identical transistors, $I_{C1} + I_{C2} = I_Q$, $I_{C1} = I_{C2}$, and $I_{C3} = I_{C4}$. Thus, $I_{B5} = I_{B6}$, so $I_{E5} = I_{E6}$; that is,

$$I_{B7} = I_{E6} = I_{E5} = I_{B3} + I_{B4} = \frac{I_{C3} + I_{C4}}{\beta_F} = \frac{I_Q}{\beta_F}$$

Thus, the collector current of Q_7 becomes $I_{C7} = \beta_F I_{B7} = I_Q$. This current mirror causes the quiescent current of the next stage to have the same value as the differential amplifier while having the same performance as the amplifier of Fig. 1.42.

EXAMPLE 1.14

Analyzing a BJT differential amplifier with a current mirror active load The parameters of the differential amplifier in Fig. 1.44 are $\beta_F = 100$, $I_Q = 20\ \mu A$, and $V_{CC} = 15$ V. Calculate A_d, R_{id}, R_o, and the overall voltage gain with load $A_{d(load)}$. Assume $V_T = 26$ mV and $V_A = 100$ V.

SOLUTION

We have

$$I_{C1} = I_{C2} = I_{C4} = \frac{20\ \mu A}{2} = 10\ \mu A$$

$$I_{B6} = \frac{I_Q}{\beta_{F7}\beta_{F8}} = \frac{20\ \mu A}{100 \times 100} = 2\ nA$$

$$g_m = \frac{I_{C2}}{V_T} = \frac{10\ \mu A}{26\ mV} = 387.6\ \mu A/V$$

$$r_{o2} = r_{o3} = r_o = \frac{V_A}{I_C} = \frac{100\ V}{10\ \mu A} = 10\ M\Omega$$

$$r_\pi = \frac{\beta_F V_T}{I_C} = \frac{100 \times 26\ mV}{10\ \mu A} = 260\ k\Omega$$

From Eq. (1.155),

$$A_d = -\frac{V_A}{2V_T} = -\frac{100\ V}{2 \times 26\ mV} = -1923\ V/V$$

From Eq. (1.156),

$$R_{id} = 2r_\pi = 2 \times 260\ k\Omega = 520\ k\Omega$$

and $$R_o = r_{o2} \| r_{o4} = \frac{r_o}{2} = \frac{10\ M\Omega}{2} = 5\ M\Omega$$

Since Q_6 and Q_7 form a compound transistor, its effective B-E voltage is that of two B-E junctions in series. Thus,

$$R_L = 2r_{\pi 6} = \frac{2V_T}{I_{B6}} = \frac{2 \times 26\ mV}{2\ nA} = 26\ M\Omega$$

The effective transconductance is $G_m = g_m = 387.6\ \mu A/V$. Thus, the overall voltage gain with load $A_{d(load)}$ is

$$A_{d(load)} = -G_m(R_o \| R_L) = -387.6\ \mu A/V \times (5\ M\Omega \| 26\ M\Omega) = -1625\ V/V$$

1.7.4 Cascode Differential Amplifier

Notice from Eq. (1.154) that the differential gain increases with the output resistance R_o of the differential amplifier. Transistors are often connected in cascode configurations to increase the output resistance and also to improve the frequency response. A common modification to Fig. 1.44 is shown in Fig. 1.45. Transistors Q_5 and Q_6 are connected in a common-base configuration and form a common-base differential stage. Note that Q_5 and Q_6 must be biased by connecting the base terminals of Q_5 and Q_6 to the DC supply V_{CC} through a potential divider.

The equivalent half circuit is shown in Fig. 1.46(a), which can be simplified to Fig. 1.46(b). Q_4 is replaced by its equivalent circuit. R'_o, which is the equivalent output resistance of the Q_2 and Q_6 combination, can be determined from the test circuit shown in Fig. 1.46(c). The emitter resistance of Q_6 is the parallel combination of $r_{\pi 6}$ and r_{o2}; that is,

$$R'_E = r_{\pi 6} \| r_{o2} \approx r_{\pi 6} \tag{1.160}$$

(since $r_{o2} >> r_{\pi 6}$). Using Eq. (1.114), we get R'_o:

$$\begin{aligned} R'_o &= r_{o6}(1 + g_{m6}R'_E) = r_{o6}(1 + g_{m6}r_{\pi 6}) \\ &= r_{o6}(1 + \beta_{F6}) \approx \beta_{F6}r_{o6} \end{aligned} \tag{1.161}$$

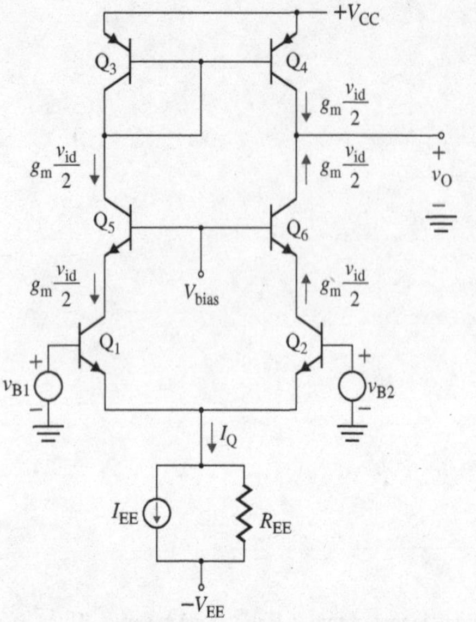

FIGURE 1.45 Cascode differential amplifier

Since all devices are biased at the same current $r_{o6} = r_{o3} = r_{o2} = r_o$, all β_F are the same. Thus,

$$R_o' = \beta_F r_o \tag{1.162}$$

The output resistance of the amplifier becomes

$$R_o = r_o \,\|\, R_o' = r_o \,\|\, \beta_F r_o \approx r_o \tag{1.163}$$

The differential gain becomes

$$A_d = -g_m R_o = -g_m r_o \tag{1.164}$$

FIGURE 1.46 Circuit for determination of output resistance R_o'

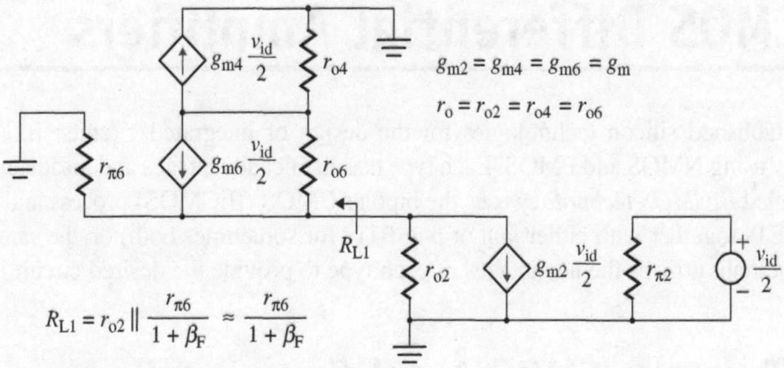

FIGURE 1.47 Circuit for determination of load resistance R_{L1} at the collector of Q_2

Substituting $g_m = I_C/V_T$ and $r_o = V_A/I_C$ into Eq. (1.164), we get

$$A_d = -\frac{V_A}{V_T} \tag{1.165}$$

Thus, the differential gain is twice that of the modified current mirror amplifier in Fig. 1.44. Transistor Q_6 offers a very high resistance seen from the output side.

Transistor Q_6 offers a load resistance R_{L1} seen from the collector of Q_2. R_{L1} can be determined from the test circuit shown in Fig. 1.47; it is the parallel combination of r_{o2}, r_{o6}, and $r_{\pi 6}$ (referred to as the emitter of transistor Q_6). That is,

$$R_{L1} = r_{o2} \| \frac{r_{\pi 6}}{1+\beta_F} \| r_{o6} \equiv \frac{r_{\pi 6}}{1+\beta_F} \tag{1.166}$$

which will have a low value. Thus, transistor Q_5 (or Q_6) acts as the *current buffer*, accepting the signal current ($g_m v_{id}/2$) from the collector of Q_1 at a low resistance R_{L1} and delivering an almost equal current ($g_m v_{id}/2$) to the load at a very high resistance R_o'. The low resistance at the collector of Q_2 improves the frequency response. Additionally, the internal capacitances of Q_6 (from collector to base and emitter to base) are connected to the ground, and there is no Miller multiplication effect. As a result, the high cutoff frequency is increased.

KEY POINTS OF SECTION 1.7

- The DC transfer characteristic of a BJT differential pair is nonlinear. However, a BJT differential pair is normally operated in the linear region where $v_{id} < V_T$.
- The values of the load resistance R_C and the current source resistance R_{EE} should be large for large values of differential gain and CMRR, respectively. A discrete resistor limits the maximum differential input voltage range.
- The voltage gain is $g_m R_C$ for the difference output and $g_m R_C/2$ for the single-sided output. A small biasing current I_Q increases the transconductance and the voltage gain; however, the differential input resistance is reduced.
- An active load increases the differential gain considerably. The gain is directly proportional to the ratio V_A/V_T. Also, a cascodelike connection increases the gain.

1.8 BiCMOS Differential Amplifiers

There are two established silicon technologies for the design of integrated circuits: BJT technology and CMOS technology using NMOS and PMOS. Each type has distinct advantages and disadvantages. An emerging technology called *BiCMOS technology* uses the bipolar-CMOS (BiCMOS) process and combines *n*- and *p*-channel MOSFETs together with either *npn* or *pnp* BJTs (or sometimes both) on the same semiconductor chip. A BiCMOS circuit utilizes the advantages of each type to provide the desired circuit functions.

1.8.1 BJT versus CMOS Amplifiers

We will begin by considering the basic BJT and CMOS amplifiers. Figure 1.48(a) shows a BJT amplifier with active load that is similar to the half circuit of a BJT differential pair. The output resistance is given by

$$R_o = r_o = \frac{V_A}{I_C} \tag{1.167}$$

where V_A is the Early voltage. Typically, $V_A = 50$ V and $I_C = 5$ μA, so $r_o = 50$ V/(5 μA) = 10 MΩ. Assuming that the current source load has infinite resistance, the voltage gain is given by

$$\begin{aligned} A_d &= -g_m R_o = -g_m r_o \\ &= -\frac{I_C}{V_T} \times \frac{V_A}{I_C} = -\frac{V_A}{V_T} \end{aligned} \tag{1.168}$$

which is independent of the biasing current I_C. Since typically $V_A = 50$ V and $V_T = 25.8$ mV at room temperature, the intrinsic gain of a BJT amplifier is $A_d = -1938$ V/V. The input resistance is

$$R_i = r_\pi = \beta_F \frac{V_T}{I_C} \tag{1.169}$$

which is generally low. Typically, $\beta_F = 60$ and $I_C = 5$ μA, so $R_i = 60 \times 25$ mV/5 μA = 300 kΩ. Although lowering the value of I_C will increase R_i, it will lower the g_m of the transistor and hence the upper frequency limit of the amplifier f_T, which is directly proportional to g_m.

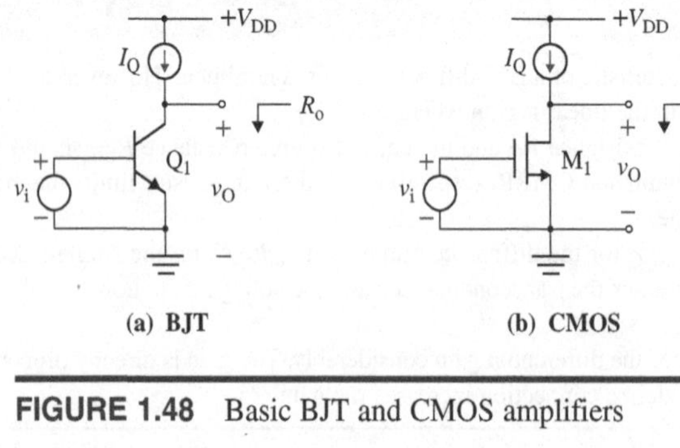

FIGURE 1.48 Basic BJT and CMOS amplifiers

Now consider the MOSFET amplifier shown in Fig. 1.48(b). The output resistance is given by

$$R_o = r_o = \frac{V_M}{I_D} \tag{1.170}$$

where V_M is the channel modulation voltage. Typically, $V_A = 20$ V and $I_D = 5$ μA, so $r_o = 20$ V/(5 μA) = 4 MΩ. Assuming that the current source load has infinite resistance, the voltage gain is given by

$$A_d = -g_m R_o = -g_m r_o$$

$$= -2\sqrt{I_D K_n}\left(\frac{V_M}{I_D}\right) = -2\left(\frac{K_n}{I_D}\right)^{1/2} V_M \tag{1.171}$$

Thus, the gain is inversely proportional to $\sqrt{I_D}$ and will increase as the biasing current is lowered. Decreasing the DC biasing current, however, reduces the amplifier bandwidth. For example, if $I_D = 5$ μA, $V_M = 20$ V, and $K_n = 25$ μA/V^2, $A_d = -89$ V/V.

In summary, for the same value of biasing current I_D, the values of g_m and r_o are much larger for a BJT amplifier than for a MOSFET amplifier—typically 2.5 times as large. The V_A of a BJT amplifier (typically 50 V) is greater than the V_M of a MOSFET (typically 20 V). The voltage gain of a BJT amplifier is greater than that of a MOSFET amplifier by a factor of about 10. However, a MOSFET amplifier has a practically infinite input resistance.

1.8.2 BiCMOS Amplifiers

A BiCMOS amplifier combines the best features of BJT and MOSFET amplifiers. It consists of cascode-like connections of BJTs and MOSFETs. The basic half circuit for a BiCMOS configuration is shown in Fig. 1.49(a). The MOSFET M_1 acts as the driving device, and the BJT Q_1 acts as the load. The cascode configuration is shown in Fig. 1.49(b). The BJT Q_1 acts as the driving device, and the MOSFET M_1 and

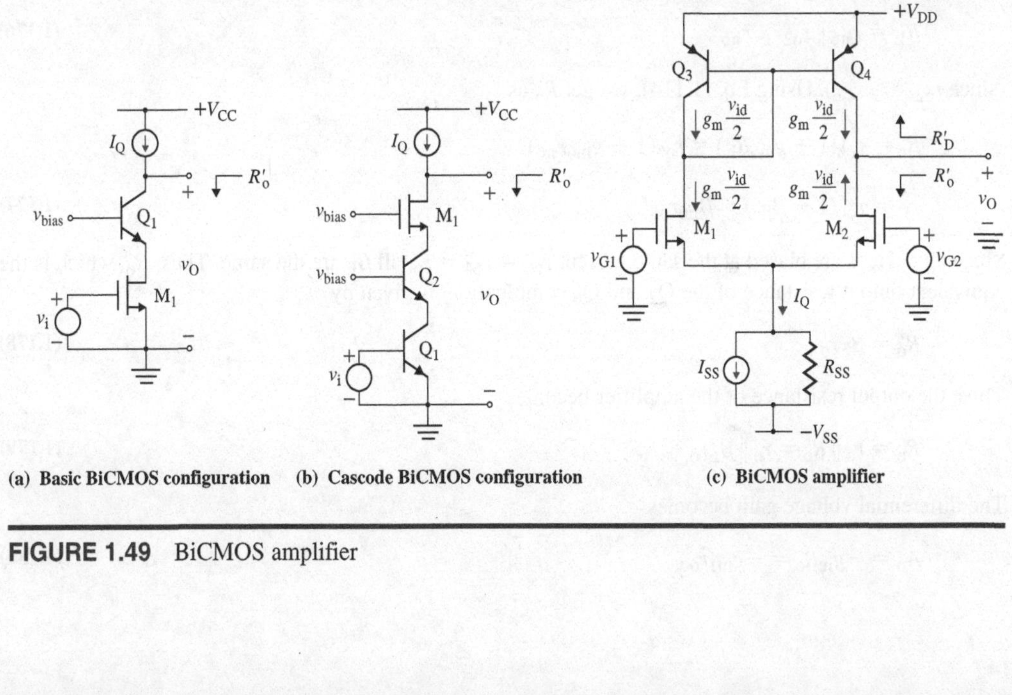

FIGURE 1.49 BiCMOS amplifier

the BJT Q_2 act as the load. A BiCMOS amplifier is shown in Fig. 1.49(c); this amplifier is identical to the CMOS amplifier in Fig. 1.14. As its abbreviation indicates, the current mirror in a BiCMOS is bipolar rather than unipolar, as in MOS devices. Transistors M_1 and M_2 are the amplifying devices. Transistor Q_4 acts as the load of transistor M_2. The output resistance R_o is the parallel combination of the output resistance $r_{o2} = 2V_M/I_Q$ for transistor M_2 and the output resistance $r_{o4} = 2V_A/I_Q$ for transistor Q_4; that is,

$$R_o = R'_o \| R'_D = r_{o2} \| r_{o4} \tag{1.172}$$

$$= \frac{2V_A V_M}{I_Q(V_A + V_M)} \tag{1.173}$$

The differential voltage gain A_d is given by

$$A_d = \frac{v_o}{v_{id}} = -g_m(r_{o2} \| r_{o4}) \tag{1.174}$$

where g_m, which is the transconductance of the driving MOSFET M_2, is given by Eq. (1.54). Substituting $(r_{o4} \| r_{o2}) = 2V_A V_M/I_Q(V_A + V_M)$ and $g_m = \sqrt{2K_n I_Q}$ into Eq. (1.174), we get

$$A_d = -\sqrt{2K_n I_Q}\,\frac{2V_M V_A}{I_Q(V_A + V_M)} \tag{1.175}$$

Since $r_{o4} > r_{o2}$, R_o and A_d will be greater for a BiCMOS amplifier than for a CMOS amplifier.

1.8.3 Cascode BiCMOS Amplifiers

Like the cascode BJT amplifier in Fig. 1.45, BiCMOS amplifiers can use cascodelike transistors to increase the voltage gain. This arrangement is shown in Fig. 1.50. Transistors Q_5 and Q_6 are connected in a CB configuration and form a CB differential stage. As shown in Fig. 1.46(c), the emitter resistance R'_E of Q_6 is the parallel combination of $r_{\pi 6}$ and r_{o2}; that is,

$$R'_E = r_{\pi 6} \| r_{o2} \approx r_{\pi 6} \tag{1.176}$$

(since $r_{o2} >> r_{\pi 6}$). Using Eq. (1.114), we get R'_o as

$$R'_o = r_{o6}(1 + g_{m6}R'_E) = r_{o6}(1 + g_{m6}r_{\pi 6})$$

$$= r_{o6}(1 + \beta_{F6}) \approx \beta_{F6} r_{o6} \tag{1.177}$$

Since all devices are biased at the same current $r_{o6} = r_{o4} = r_o$, all β_F are the same. Thus, R'_o, which is the equivalent output resistance of the Q_2 and Q_6 combination, is given by

$$R'_o = \beta_F r_o \tag{1.178}$$

Thus, the output resistance of the amplifier becomes

$$R_o = r_o \| R'_o = r_o \| \beta_F r_o \approx r_o \tag{1.179}$$

The differential voltage gain becomes

$$A_d = -g_m R_o = -g_{m1} r_o \tag{1.180}$$

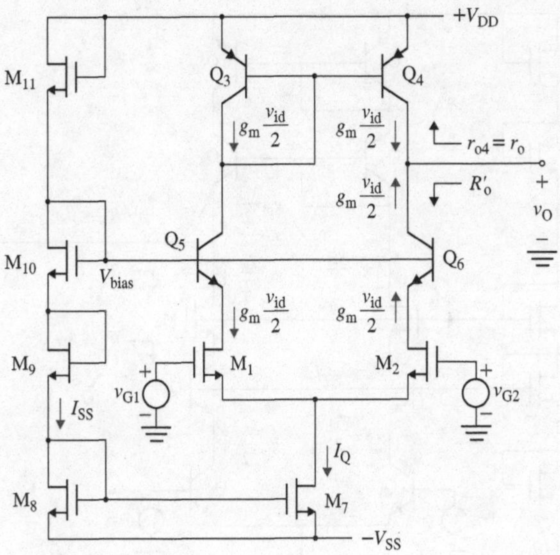

FIGURE 1.50 Cascode BiCMOS amplifier

Substituting $g_m = \sqrt{2K_nI_Q}$ and $r_o = 2V_A/I_Q$ into the preceding equation, we get

$$A_d = -\sqrt{2K_nI_Q}\,\frac{2V_A}{I_Q} = -2V_A\sqrt{\frac{2K_n}{I_Q}} \tag{1.181}$$

which gives a greater voltage gain than Eq. (1.175).

1.8.4 Double-Cascode BiCMOS Amplifiers

A very high gain can be obtained by double cascoding, as shown in Fig. 1.51. There are two cascode connections: the first connects BJTs Q_1, Q_2, Q_3, and Q_4, and the second connects MOSFETs M_5 and M_6. The resistance R'_S looking from the collector of Q_4 will be

$$R'_S = r_{o4}(1 + \beta_{F4}) \simeq \beta_{F4}r_{o4} \tag{1.182}$$

which will be the source resistance of M_6. Using Eq. (1.33), we can find the output resistance R'_o of M_6 as

$$R'_o \simeq r_{o6}(1 + g_{m6}R'_S) = r_{o6}(1 + g_{m6}\beta_{F4}r_{o4}) \tag{1.183}$$

$$\simeq r_{o6}g_{m6}\beta_{F4}r_{o4} \tag{1.184}$$

Like Q_2 and Q_4, transistors Q_8 and Q_{10} are also connected in cascode, and their equations will be similar to those for Q_2 and Q_4. From Eq. (1.177), the resistance looking from the collector of Q_8 is

$$R'_D = r_{o8}(1 + \beta_{F8}) \simeq \beta_{F8}r_{o8} \tag{1.185}$$

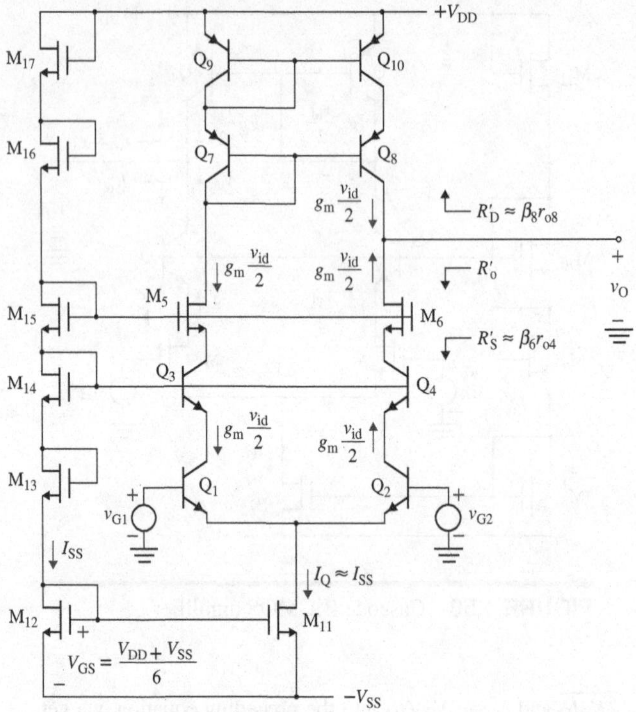

FIGURE 1.51 Double-cascode BiCMOS amplifier

Thus, the output resistance of the amplifier becomes

$$R_o = R'_D \| R'_S = (r_{o6} g_{m6} \beta_{F4} r_{o4}) \| (\beta_{F8} r_{o8}) \tag{1.186}$$

$$= \beta_F r_o \quad (\text{for } r_{o4} = r_{o8} = r_o, \beta_{F8} = \beta_F) \tag{1.187}$$

The differential voltage gain becomes

$$A_d = -g_m R_o = -g_m (r_{o6} g_{m6} \beta_{F4} r_{o4}) \| (\beta_{F8} r_{o8}) \tag{1.188}$$

$$= -g_m \beta_F r_o \tag{1.189}$$

where $g_m = I_Q/2V_T$. Equation (1.189) shows a considerable increase in the voltage gain.

EXAMPLE 1.15

Analyzing BiCMOS amplifiers The DC biasing current of a BiCMOS amplifier is kept constant at $I_Q = 10\ \mu A$. All bipolar transistors are identical, with $V_A = 50$ V and $\beta_F = 40$. Also, the MOS transistors are identical, with $V_M = 20$ V, $K_n = 25\ \mu A/V^2$, $W = 30\ \mu m$, and $L = 10\ \mu m$. Assume $V_T = 25.8$ mV.

Determine the differential voltage gain A_d for single-ended output **(a)** for the BiCMOS amplifier in Fig. 1.49(c), **(b)** for the cascode BiCMOS amplifier in Fig. 1.50, and **(c)** for the double-cascode BiCMOS amplifier in Fig. 1.51.

SOLUTION

$V_A = 50$ V, $\beta_F = 40$, $V_M = 20$ V, $K_n = 25\ \mu A/V^2$, $W = 30\ \mu m$, $L = 10\ \mu m$, and $I_D = I_C = I_Q/2 = 10\ \mu A/2 = 5\ \mu A$.

(a) We have

$$r_{o2} = \frac{2V_M}{I_Q} = \frac{2 \times 20\text{ V}}{10\ \mu\text{A}} = 4\text{ M}\Omega$$

$$r_{o4} = \frac{2V_A}{I_Q} = \frac{2 \times 50\text{ V}}{10\ \mu\text{A}} = 10\text{ M}\Omega$$

$$R_o = r_{o2} \| r_{o4} \equiv 4\text{ M}\Omega \| 10\text{ M}\Omega = 2.86\text{ M}\Omega$$

$$g_{m2} = \sqrt{2K_n I_Q} = \sqrt{2 \times 25\ \mu \times 10\ \mu} = 22.36\ \mu\text{A/V}$$

Thus, the differential voltage becomes

$$A_d = -g_{m2}R_o = -22.36\ \mu \times 2.86\text{ M} = -63.9$$

(b) We have

$$r_{o4} = r_{o6} = \frac{2V_A}{I_Q} = \frac{2 \times 50\text{ V}}{10\ \mu\text{A}} = 10\text{ M}\Omega$$

$$R_o' = \beta_F r_{o6} = 40 \times 10\text{ M}\Omega = 400\text{ M}\Omega$$

$$R_o = r_{o4} \| R_o' = 10\text{ M} \| 400\text{ M} = 10\text{ M}\Omega$$

$$g_{m2} = \sqrt{2K_n I_Q} = \sqrt{2 \times 25\ \mu \times 10\ \mu} = 22.36\ \mu\text{A/V}$$

Thus, the differential voltage becomes

$$A_d = -g_{m2}R_o = -22.36\ \mu \times 10\text{ M} = -223.6\text{ V/V}$$

(c) From Eq. (1.187),

$$R_o = \beta_F r_{o4} = 40 \times 10\text{ M} = 400\text{ M}\Omega$$

$$g_{m2} = \frac{I_Q}{2V_T} = \frac{10\ \mu\text{A}}{2 \times 25.8\text{ mV}} = 193.8\ \mu\text{A/V}$$

Thus, the differential voltage gain becomes

$$A_d = -g_{m2}R_o = -193.8\ \mu\text{A/V} \times 400\text{ M}\Omega = -77{,}520\text{ V/V}$$

which is considerably larger than for the other two configurations.

KEY POINTS OF SECTION 1.8

- A BiCMOS amplifier combines the advantages of BJT and MOS technologies to achieve the desirable circuit functions of infinite input resistance and a very large voltage gain and CMRR.
- BJT and MOS transistors can be connected in cascode to give an extremely large output resistance and voltage gain.

1.9 Frequency Response of Differential Amplifiers

The transistors (either MOSFETs or BJTs) in differential amplifiers have capacitance—hence, the gain of such amplifiers will be frequency dependent [5–7]. The techniques for analyzing frequency response can be applied to find the frequency response of differential amplifiers. In this section, we consider only the differential gain with passive and active loads. R_C represents the output resistance of the active source.

1.9.1 Frequency Response with Resistive Load

Let us first consider the differential half circuit shown in Fig. 1.52(a). Replacing the transistor by its frequency model gives the circuit in Fig. 1.52(b). The transistor can be either a BJT or a MOSFET. We can apply the same model and analysis for MOSFETs by substituting $r_\pi = \infty$, $C_\pi \approx C_{gs}$ and $C_\mu \approx C_{gd}$ in the derivations. Since the input signal is $v_{id}/2$, the r_π, C_π, and C_μ parameters of the transistor model are sealed by a factor of 2. Since $g_m(R_C \| r_{o1}) >> 1$, C_μ will dominate the high cutoff frequency ω_H. If we replace $C_\mu/2$ by its effective Miller capacitance C_M, we get the equivalent circuit shown in Fig. 1.52(c) and Eq. (G.3) in Appendix G; that is,

$$C_M = \left(\frac{C_\mu}{2}\right)[1 + g_m(R_C \| r_{o1})] \tag{1.190}$$

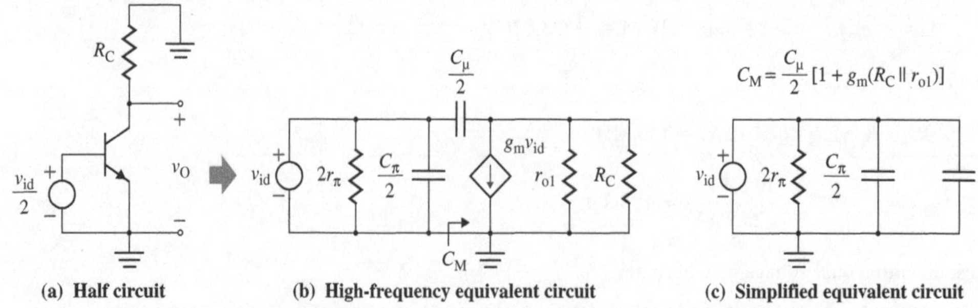

FIGURE 1.52 Differential half circuit and high-frequency equivalent

Thus, the high cutoff frequency ω_H is given by

$$\omega_H = \frac{1}{2r_\pi(C_\pi/2 + C_M)} = \frac{1}{2r_\pi\{C_\pi/2 + (C_\mu/2)[1 + g_m(R_C \| r_{o1})]\}} \tag{1.191}$$

which, for $R_C = r_{o1} = r_o$ and $g_m(R_C \| r_{o1}) >> 1$, can be approximated by

$$\omega_H = \frac{1}{2r_\pi(C_\mu/2)g_m r_o/2} = \frac{2}{r_\pi C_\mu g_m r_o} = \frac{2}{r_\pi C_\mu g_m R_C} \tag{1.192}$$

The signal source resistance $R_s = 0$. The effective resistance for $C_\pi/2$ will be $R_{C\pi} = (2r_\pi \| R_s) = 0$. Thus, the break frequency due to $C_\pi/2$ will be at infinity and will not influence ω_H in Eq. (1.192). Therefore, the frequency-dependent differential gain is given by

$$A_d(j\omega) = \frac{A_{do}}{1 + j\omega/\omega_H} \tag{1.193}$$

where $A_{do} = g_m(r_{o2} \| r_{o4})$ is the low-frequency differential gain with differential output voltage.

1.9.2 Frequency Response with Active Load

Now let us consider the output circuit for the amplifier with an active load, as shown in Fig. 1.42. The AC equivalent of that circuit is shown in Fig. 1.53(a). Replacing the transistors by their frequency model, we get Fig. 1.53(b). Since the effective transconductance is $2g_m$, the Miller capacitance C_M [from Eq. (G.3) in Appendix G] becomes

$$C_M = \left(\frac{C_\mu}{2}\right)[1 + 2g_m(r_{o2} \| r_{o4})] \tag{1.194}$$

Thus, the high cutoff frequency ω_H is given by

$$\omega_H = \frac{1}{2r_\pi\{C_\pi/2 + (C_\mu/2)[1 + 2g_m(r_{o2} \| r_{o4})]\}} \tag{1.195}$$

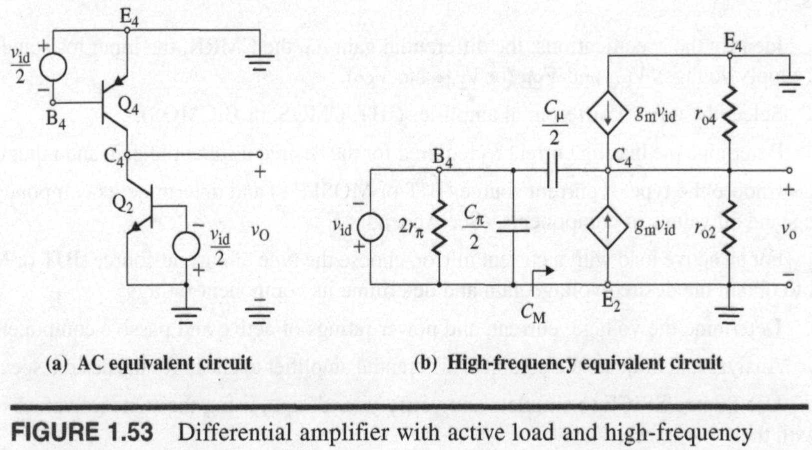

FIGURE 1.53 Differential amplifier with active load and high-frequency equivalent

which, for $r_{o2} = r_{o4} = r_o$ and $2g_m(r_{o2} \parallel r_{o4}) >> 1$, can be approximated by

$$\omega_H = \frac{1}{2r_\pi/(C_\mu/2)g_m r_o} = \frac{1}{r_\pi C_\mu g_m r_o} \tag{1.196}$$

Therefore, the frequency-dependent differential gain is given by

$$A_d(j\omega) = \frac{A_{do}}{1 + j\omega/\omega_H} \tag{1.197}$$

where $A_{do} = -2g_m(r_{o2} \parallel r_{o4})$ is the low-frequency differential gain with an active load and a single-sided output.

Notice from Eq. (1.192) that the high cutoff frequency is double that for an amplifier with an active load and single-sided output. However, this high cutoff frequency for an amplifier with a passive load is obtained at the expense of the output voltage gain. For this reason, the first stage of a wide-band amplifier often uses a balanced circuit with differential voltage when the CMRR is not the prime consideration.

KEY POINTS OF SECTION 1.9

- The transistors in differential amplifiers have capacitance; hence, the gain of such amplifiers will be frequency dependent and exhibit a high-pass characteristic.
- The high cutoff frequency for an amplifier with a passive load is obtained at the expense of the output voltage gain. For this reason, the first stage of a wide-band amplifier often uses a balanced circuit with differential voltage.

1.10 Design of Differential Amplifiers

Differential amplifiers are used as the input stage and are designed for a high differential gain, a high CMRR, and high differential and common-mode input resistances. The design of a differential stage involves the following steps:

Step 1. Identify the specifications: the differential gain A_d, the CMRR, the input resistance R_{id}, and the DC supply voltages V_{CC} and V_{EE} (or V_{DD} and V_{SS}).

Step 2. Select the type of differential amplifier (BJT, CMOS, or BiCMOS).

Step 3. Determine the biasing current I_Q required for the desired differential gain and input resistance.

Step 4. Choose the type of current source (BJT or MOSFET) and determine its component ratings. Use the standard values of components. (See Appendix E.)

Step 5. For an active load with a current mirror, choose the type of current source (BJT or MOSFET) needed to obtain the desired voltage gain and determine its component ratings.

Step 6. Determine the voltage, current, and power ratings of active and passive components.

Step 7. Analyze and evaluate the complete differential amplifier to check for the desired specifications.

Step 8. Use PSpice/SPICE to simulate and verify your design, using the standard values of components with their tolerances.

1.11 Electronic Noise

Noise is generated in electronic circuits, that is, an electronic circuit will have an output even without any input signal. Noise is generated by resistances and semiconductor devices. The capacitors and inductors do not generate noise.

1.11.1 Classification of Noise

Noise can be classified as one of five types:

Thermal Noise

Thermal noise is generated in resistors due to random motion produced by the thermal agitation of electrons. This noise is dependent on temperature. The equivalent circuit for thermal noise in resistors is shown in Fig. 1.54 [10]. The mean-square value of the noise generator is expressed as

$$V_t^2 = 4kTR\Delta f \quad \text{V}^2 \tag{1.198}$$

$$I_t^2 = 4kT\frac{\Delta f}{R} \quad \text{A}^2 \tag{1.199}$$

where k is the Boltzmann's constant (1.3806×10^{-23} J/K); T is the temperature, K (273 + T°C); R is the resistance, ohms; Δf is the noise bandwidth, Hz.

Shot Noise

Shot noise is generated by random fluctuations in the number of charged carriers that are emitted from a surface or diffused from a junction. This noise is always associated with a direct current flow and is present in bipolar transistors. The mean-square value of the noise current is expressed by

$$I_s^2 = 2qI_D\Delta f \quad \text{A}^2 \tag{1.200}$$

where Δf is the noise bandwidth, Hz; q is the electron charge (1.6022×10^{-16} C); I_D is the dc current, A.

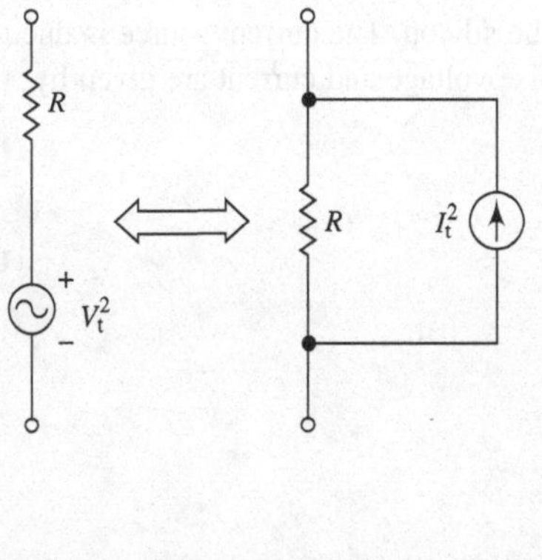

FIGURE 1.54 Equivalent circuit for noise in resistors

Flicker Noise

Flicker noise is generated due to surface imperfections resulting from the emission. This noise is associated with all active devices and some discrete passive elements such as carbon resistors. The mean square value of the noise current is expressed by

$$I_{\mathrm{f}}^2 = K_{\mathrm{f}} \frac{I_{\mathrm{D}}^{a}}{f} \Delta f \quad \mathrm{A}^2 \tag{1.201}$$

where Δf is the noise bandwidth, Hz; f is the operating frequency, Hz; I_D is the dc current, A; K_f is the flicker constant for a particular device in the range 0 to 1; a is the flicker-exponent constant in the range of 0.5 to 2.

Burst Noise

Burst noise is generated due to the presence of heavy metal ion contamination and is found in some integrated circuits and discrete transistors. The repetition rate of noise pulses is in the audio frequency range (a few kilohertz or less) and produces a "popping" sound when played through a speaker. This noise is also known as *popcorn noise*. The mean-square value of the noise current is expressed as

$$I_{\mathrm{b}}^2 = K_{\mathrm{b}} \frac{I_{\mathrm{D}}^{c}}{1+(f/f_{\mathrm{c}})} \Delta f \quad \mathrm{A}^2 \tag{1.202}$$

where Δf is the noise bandwidth, Hz; I_D is the direct current; K_b is the burst constant for a particular device; c is the burst-exponent constant in the range of 0.5 to 2; f_c is a particular frequency for a given noise in the range of 20 Hz to 15 kHz.

Avalanche Noise

Avalanche noise is produced by Zener or avalanche breakdown in *pn* junctions. The holes and electrons in the depletion region of a reverse-biased *pn* junction acquire sufficient energy to create hole–electron pairs by collision. This process is cumulative, resulting in the production of a random series of large noise spikes. This noise is associated with direct current and is much greater than shot noise for the same current. Zener diodes are normally avoided in circuits requiring low noise.

1.11.2 Noise in Diodes

The equivalent circuit for noise in diodes is shown in Fig. 1.55 [10]. There are two noise generators. The voltage generator is due to thermal noise in the resistance of the silicon. The current source is due to the shot noise and flicker noise. The mean-square values of the noise voltage and current are given by

$$V_{\mathrm{s}}^2 = 4kTr_{\mathrm{s}} \Delta f \quad \mathrm{V}^2 \tag{1.203}$$

$$I_{\mathrm{d}}^2 = 2qI_{\mathrm{D}} \Delta f + K_{\mathrm{f}} \frac{I_{\mathrm{D}}^{a}}{f} \Delta f \quad \mathrm{A}^2 \tag{1.204}$$

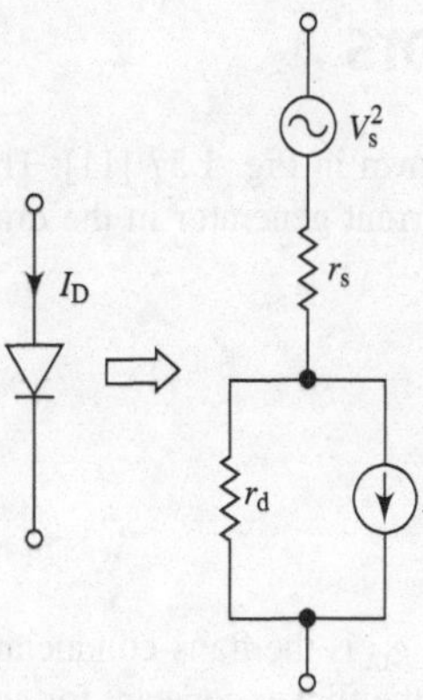

FIGURE 1.55 Equivalent circuit for noise in diodes

where I_D is the forward direct current, A; r_s is the resistance of the silicon, ohms; K_f is the flicker constant for a particular device in the range 0 to 1; a is the flicker-exponent constant in the range of 0.5 to 2.

1.11.3 Noise in Bipolar Transistors

The equivalent circuit for noise in bipolar transistors is shown in Fig. 1.56 [10]. The current generator in the collector is due to shot noise. The noise voltage generator in the base circuit is due to thermal noise in the base resistance. The current generator in the base circuit consists of shot noise, flicker noise, and burst noise. The mean-square values of the noise voltage and currents are expressed by

$$V_b^2 = 4kTr_b\Delta f \quad \text{V}^2 \tag{1.205}$$

$$I_c^2 = 2qI_C\Delta f \quad \text{A}^2 \tag{1.206}$$

$$I_b^2 = 2qI_B\Delta f + K_f\frac{I_B^a}{f}\Delta f + K_b\frac{I_B^c}{1+(f/f_c)}\Delta f \quad \text{A}^2 \tag{1.207}$$

where I_B is the base bias current, A; I_C is the collector bias current, A; r_b is the resistance at the transistor base, ohms; K_b is the burst constant for a particular device in the range 0 to 1; c is the burst-exponent constant in the range of 0.5 to 2; f_c is the a particular frequency for a given noise in the range of 20 Hz to 15 kHz; k_f is the flicker constant for a particular device in the range 0 to 1; a is the flicker-exponent constant in the range of 0.5 to 2.

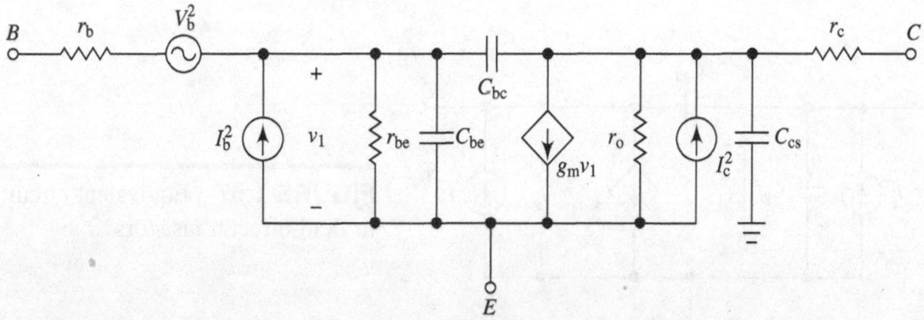

FIGURE 1.56 Equivalent circuit for noise in bipolar transistors

1.11.4 Noise in Field-Effect Transistors

The equivalent circuit for noise in field-effect transistors (FETs) is shown in Fig. 1.57 [11]. The current generator in the gate is due to shot noise, which is very small. The current generator in the drain circuit consists of thermal and flicker noise. The noise current is expressed by

$$I_g^2 = 2qI_G\Delta f \ \ \mathrm{A}^2 \tag{1.208}$$

$$I_d^2 = 4kT\left(\frac{2}{3}g_m\right)\Delta f + K_f\frac{I_D^a}{f}\Delta f \ \ \mathrm{A}^2 \tag{1.209}$$

where I_D is the drain bias current, A; I_G is the gate leakage current, A; g_m is the trans-conductance at the bias point, A/V; r_o is the resistance at the transistor output, ohms; K_f is the flicker constant for a particular device in the range 0 to 1; a is the flicker-exponent constant in the range of 0.5 to 2.

1.11.5 Equivalent Input Noise

Each noise generator contributes to the output of a circuit (Fig. 1.58). The effect of all the noise generators can be found by summing the mean-square value (but not the root-mean square value) of individual noise contributions. Once the total mean square noise output voltage is found, all the noises can be represented by an equivalent effective output voltage by the gain. The gain is the output noise voltage with respect to a defined input. It is time-consuming for hand calculations to determine the total noise-generated voltages and currents in electronic circuits. SPICE software can calculate the output noise and the equivalent input noise by the NOISE command [8–10]. It may be noted that PSpice calculates the noise in $\mathrm{V}/\sqrt{\mathrm{Hz}}$ or $\mathrm{A}/\sqrt{\mathrm{Hz}}$. Dividing the mean-square value of the noise output voltage $V_{out(noise)}$ by the noise bandwidth gives the output noise spectrum V_{out}, i.e.,

$$V_{out} = \frac{V_{out(noise)}}{\Delta f} \ \ \mathrm{V}/\sqrt{\mathrm{Hz}} \tag{1.210}$$

Dividing the output noise spectrum by the gain yields the equivalent input noise spectrum:

$$V_{in} = \frac{V_{out}}{G_v} \ \ \mathrm{V}/\sqrt{\mathrm{Hz}} \tag{1.211}$$

where G_v is the voltage gain with respect to the equivalent input source.

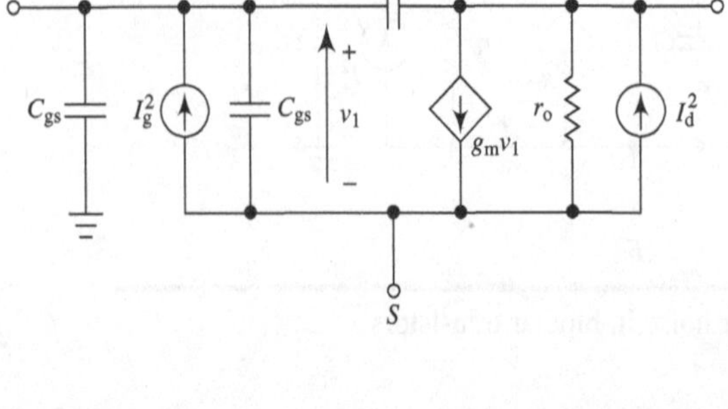

FIGURE 1.57 Equivalent circuit for noise in field-effect transistors

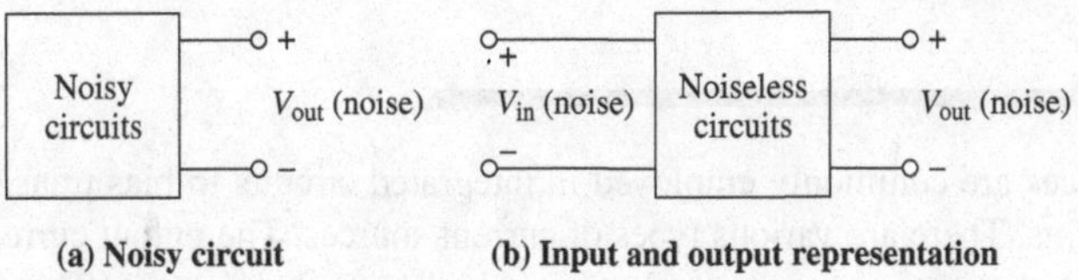

FIGURE 1.58 Equivalent noise input

If the equivalent input is a current source, the equivalent input current noise spectrum becomes

$$I_{in} = \frac{V_{out}}{R_i} \quad \text{V}/\sqrt{\text{Hz}} \tag{1.212}$$

where R_i is the trans-resistance with respect to the equivalent input source.

EXAMPLE 1.16

Determining the noise-generated voltages and currents The noise parameters are $q = 1.6022 \times 10^{-19}$, $k = 1.3806 \times 10^{-23}$, $T = 25°C$, $K_f = 0.5$, $K_b = 0.5$, $f = 1$ Hz, $f_c = 2.5$ kHz, $a = 1$, and $c = 1$. Determine the noise-generated voltages and currents for (a) the resistance in Fig. 1.54 if $R = 10$ kΩ, (b) the diode in Fig. 1.55 if $I_D = 1$ mA, and $r_s = 40$ Ω, and (c) the transistor in Fig. 1.56 if $I_C = 1$ mA, $I_B = 10$ μA, and $r_b = 40$ Ω.

SOLUTION

(a) $R = 10$ kΩ. From Eqs. (1.198) and (1.199), we get

$$V_f = \sqrt{4kTR\Delta f} = \sqrt{4\times1.3608\times10^{-23}\times(273+25)\times10\times10^3\times1} = 1.283\times10^{-8}\text{ V}$$

$$I_t = \sqrt{4kT\frac{\Delta f}{R}} = \sqrt{4\times1.3608\times10^{-23}\times(273+25)\times\frac{1}{10\times10^3}} = 1.283\times10^{-12}\text{ A}$$

(b) $I_D = 1$ mA and $r_s = 40$ Ω. From Eqs. (1.203) and (1.204), we get

$$V_s = \sqrt{4kTr_s\Delta f} = \sqrt{4\times1.3608\times10^{-23}\times(273+25)\times40\times1} = 8.113\times10^{-10}\text{ V}$$

$$I_d = \sqrt{2qI_D\Delta f + K_f\frac{I_D^a}{f}\Delta f} = \sqrt{2\times1.6022\times10^{-19}\times1\times10^{-3}\times1+0.5\times\frac{1\times10^{-3}}{10}}\times1 = 7.071\times10^{-3}\text{ A}$$

(c) $I_C = 1$ mA, $I_B = 10$ μA, and $r_b = 40$ Ω. From Eqs. (1.205), (1.206), and (1.207), we get

$$V_b = \sqrt{4kTr_b\Delta f} = \sqrt{4\times1.3608\times10^{-23}\times(273+25)\times40\times1} = 8.113\times10^{-10}\text{ V}$$

$$I_c = \sqrt{2qI_C\Delta f} = \sqrt{2\times1.6022\times10^{-19}\times1\times10^{-3}\times1} = 1.79\times10^{-11}\text{ A}$$

$$I_b = \sqrt{2qI_C\Delta f + K_f\frac{I_B^a}{f}\Delta f + K_b\frac{I_B^c}{1+(f/f_c)}\Delta f}$$

$$= \sqrt{2\times1.6022\times10^{-19}\times1\times10^{-3}\times1+0.5\times\frac{1\times10^{-3}}{10}\times1+0.5\times\frac{10\times10^{-6}}{1+(10/2.5\times10^3)^2}\times1} = 2.345\times10^{-3}\text{ A}$$

Summary

Active current sources are commonly employed in integrated circuits to bias transistor circuits at appropriate operating points. There are various types of current sources. The output current of a good source is independent of the transistor parameters and offers a high output resistance. The design of BJT current sources can be simplified by ignoring the base currents of transistors and by assuming a constant voltage drop between the base and the emitter. Diode-connected BJTs and MOSFETs rather than diodes are normally used in current sources to give matching mirror characteristics.

A current source can be either a source or a sink, depending on how it is connected. Current mirrors are often used as an active load to increase the output resistance and the voltage gain of a differential amplifier. Also, cascodelike connections increase the voltage gain.

Emitter-coupled (or source-coupled) amplifiers offer the advantage of direct coupling of cascoded stages. They are usually used as the input stage of differential amplifiers to give a high input resistance and a high CMRR. However, the range of the differential input voltage is very small (typically two to three times V_T for the BJT amplifier). The CMRR depends on the output resistance of the biasing current source. Thus, a current source with a high value of output resistance is highly desirable. Source-coupled differential pairs offer a higher input resistance than emitter-coupled pairs, but they have a low CMRR.

Noise is generated in electronic circuits, that is, an electronic circuit will have an output even without any input signal. Noise is generated by resistances and semiconductor devices. The capacitors and inductors do not generate noise.

References

1. P. E. Allen and D. R. Holberg, *CMOS Analog Circuit Design*. New York: Oxford University Press, 1996.
2. P. M. Chirlian, *Analysis and Design of Integrated Electronic Circuits*. New York: Wiley, 1986.
3. L. J. Giacoletto, *Differential Amplifiers*. New York: Wiley, 1970.
4. V. H. Grinich and H. G. Jackson, *Introduction to Integrated Circuits*. New York: McGraw-Hill, 1975.
5. P. R. Gray, P. J. Hurst, S. H. Lewis, and R. G. Meyer, *Analysis and Design of Analog Integrated Circuits*. New York: Wiley, 2001.
6. R. C. Jaeger and T. Blalock, *Microelectronic Circuit Design*. New York: McGraw-Hill, 2008.
7. B. Razavi, *Design of Analog CMOS Integrated Circuits*. New York: McGraw-Hill, 2001.
8. P. Antognetti and G. Massobrio, *Semiconductor Device Modeling with SPICE*. New York, NY: McGraw-Hill, 1988.
9. I. Getreu, *Modeling the Bipolar Transistor*, Tektronix, Inc. part# 062-2841-00.
10. M. H. Rashid, *Introduction to PSpice. Using OrCAD for Circuits and Electronics*. Upper Saddle River, NJ: Prentice Hall Inc., 3rd ed., 2003.

Review Questions

1. What are the advantages and disadvantages of a basic current source?
2. What are the advantages and disadvantages of a modified current source?
3. What are the advantages and disadvantages of a Widlar current source?

4. What are the advantages and disadvantages of a cascode current source?
5. What are the advantages and disadvantages of a Wilson current source?
6. What are the differences between Widlar and Wilson current sources?
7. What is a current mirror load?
8. What is the common-mode rejection ratio (CMRR)?
9. What are the advantages of an emitter-coupled pair?
10. What is the DC characteristic of an emitter-coupled pair?
11. What design criteria will yield a large value of CMRR in an emitter-coupled pair?
12. What are the advantages of a source-coupled pair?
13. What is the DC characteristic of a source-coupled pair?
14. What design criteria will yield a large value of CMRR in a source-coupled pair?
15. What is the purpose of cascodelike connections of transistors?
16. What is a thermal noise?
17. What is a shot noise?
18. What is a flicker noise?
19. What is a burst noise?
20. What is an avalanche noise?

Problems

The symbol **D** indicates that a problem is a design problem. The symbol **P** indicates that you can check the solution to a problem using PSpice/SPICE.

Assume that the device parameters are as follows: diodes, $I_S = 10^{-13}$ A and $I_{D(min)} = 1$ mA to ensure conduction; transistors, $\beta_F = h_{fe} = 50$, $V_{BE} = 0.7$ V, $I_S = 10^{-14}$ A, and $V_{CE(sat)} = 0.2$ V.

1.3 *MOSFET Current Sources*

1.1 The parameters of the MOSFET current source in Fig. 1.4 are $V_t = 1$ V, $I_O = 20$ μA, $I_R = 20$ μA, $V_{DD} =$ 15 V, and $V_M = 40$ V. The channel lengths are $L_1 = L_2 = 10$ μm and $L_3 = 100$ μm, and $K_x = 20\ \mu A/V^2$. Calculate the required values of **(a)** K_{n1}, W_1, **(b)** K_{n2}, W_2, **(c)** K_{n3}, W_3, and **(d)** the output resistance R_o of the current source. Assume $V_{GS1} = 1.5$ V and $V_{DS1} = 5$ V.

1.2 **D** **a.** Design the cascode current source in Fig. 1.6(a) to give $I_O = 10$ μA. Assume $V_{DD} = 10$ V. All MOSFETs are identical and have $L = 20$ μm, $W = 60$ μm, $V_t = 1$ V, $K_x = 20\ \mu A/V^2$, and $V_M = 40$ V.

b. Calculate the output resistance R_o and Thevenin's equivalent voltage V_{Th}.

1.3 **D** **a.** Design the modified Wilson current source in Fig. 1.7(c) to give $I_O = 10$ μA. Assume $V_{DD} = 10$ V and $V_M = 40$ V. All MOSFETs are identical and have $L = 10$ μm, $W = 40$ μm, $K_x = 20\ \mu A/V^2$, and $V_M = 40$ V.

b. Calculate the output resistance R_o and Thevenin's equivalent voltage V_{Th}.

1.4 **D** Design a MOSFET current source so that $R_o \geq 50$ kΩ at an output current of $I_O = 1$ mA.

1.5 **D** Design a MOSFET current source so that $R_o \geq 500$ kΩ at an output current of $I_O = 0.1$ mA.

1.6 The Widlar current source shown in Fig. P1.6 has I_{ref} = 50 μA, R = 2 kΩ, and V_{DD} = 12 V. The MOS parameters are K_n = 100 μA/V^2, V_t = 1 V, $|V_M|$ = 100 V, and $(W/L)_1 = (W/L)_2$ = 20. Determine **(a)** the output current I_o, **(b)** the output resistance r_{o2}, and **(c)** the value of R_{ref}.

D

FIGURE P1.6

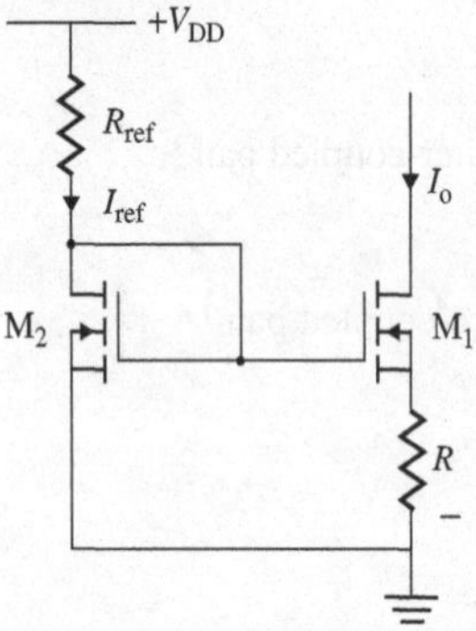

1.7 Design the modified Wilson current source in Fig. P1.7 to give I_O = 10 μA by determining the resistance R_{ref}. Assume V_{DD} = 12 V. All MOSFETs are identical and have L = 10 μm, W = 40 μm, K_x = 50 μA/V^2, and V_M = 100 V. Calculate the output resistance R_o and Thevenin's equivalent voltage V_{Th}.

D

FIGURE P1.7

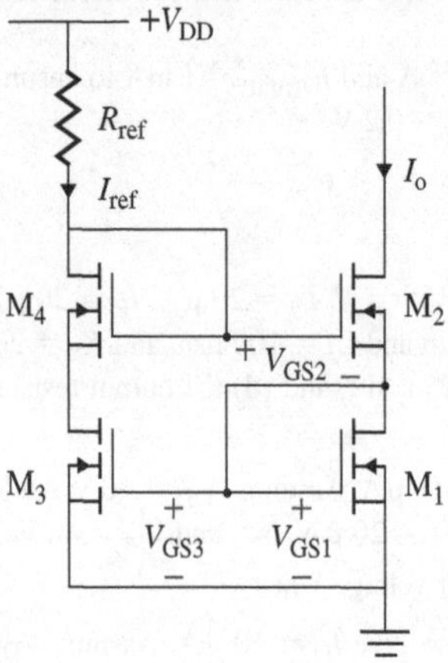

1.8 The current source shown in Fig. P1.8 has I_{ref} = 150 μA and V_{DD} = 12 V. The NMOS parameters are K_n = 200 μA/V^2, V_t = 1 V, V_M = −100 V, and $(W/L)_1 = (W/L)_2 = (W/L)_3$ = 20. Determine **(a)** the gate–source voltages (V_{GS1}, V_{GS2}, V_{GS3}), **(b)** the drain–source voltages (V_{DS1}, V_{DS2}, V_{DS3}) if the devices are to operate in the active region, and **(c)** the value of R_{ref}.

FIGURE P1.8

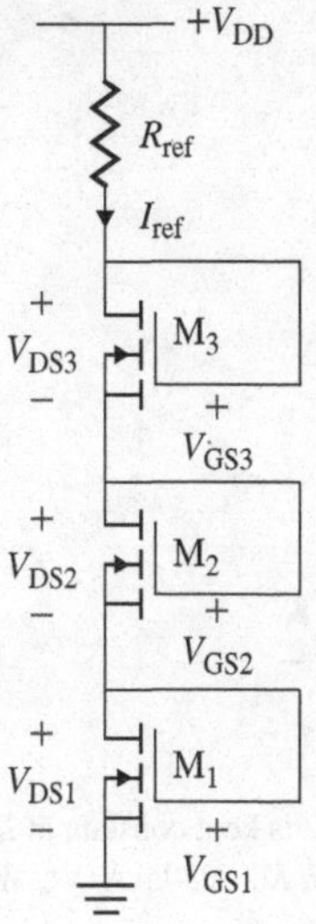

1.9 Repeat Prob. 1.8 if the current source in Fig. P1.8 has only two MOSFETs.

1.4 *MOS Differential Amplifiers*

1.10 The parameters of the MOS differential pair in Fig. 1.8 are R_{SS} = 150 kΩ, I_Q = 1 mA, V_{DD} = 15 V, and R_D = 5 kΩ. The MOSFETs are identical and have K_n = 1.5 mA/V^2, W/L = 1, and V_t = 1.2 V.

a. Calculate the DC drain currents through the MOSFETs if v_{id} = 10 mV.

b. Assuming $I_{D1} = I_{D2}$, calculate A_d, A_c, and CMRR; R_{id} and R_{ic}; the small-signal output voltage if v_{g1} = 5 mV and v_{g2} = 10 mV; and the drain voltage V_D.

1.11 Repeat Prob. 1.10 if the active biasing current source is replaced by resistance R_{SS} = 150 kΩ; that is, I_Q = 0.

1.12 **D** Design a MOS differential pair as shown in Fig. 1.13 in which one input terminal is grounded. The output is taken from the drain of transistor M_1. The DC biasing current is I_Q = 1 mA, and $V_{DD} = V_{SS}$ = 15 V. The MOSFETs are identical and have K_n = 1.5 mA/V^2 and V_t = 1.2 V.

a. For a small-signal voltage gain of A_1 = −20 V/V.

b. Calculate the design values of A_d, A_c, and CMRR.

1.13 **D** Design the CMOS amplifier shown in Fig. 1.16 by determining the W/L ratios of the MOSFETs and the threshold voltage V_t. The differential voltage gain should be A_d = 50 V/V at biasing current I_Q = 1 mA. Assume identical transistors whose channel modulation voltage is V_M = 40 V, channel constant is K_x = 10 μA/V^2, and channel length is L = 10 μm. The W/L ratio of the current source is unity, and V_{DD} = V_{SS} = 10 V.

1.14 A CMOS amplifier is shown in Fig. P1.14. The parameters for the NMOS are V_t = +2 V, V_M = −40 V, and V_{GS} = +4 V at I_D = 1 mA; the parameters for the PMOS are V_t = −3 V, V_M = 40 V, and V_{GS} = −6 V at I_D = 1 mA. Calculate **(a)** A_d, A_c, and CMRR and **(b)** R_{id} and R_{ic}.

FIGURE P1.14

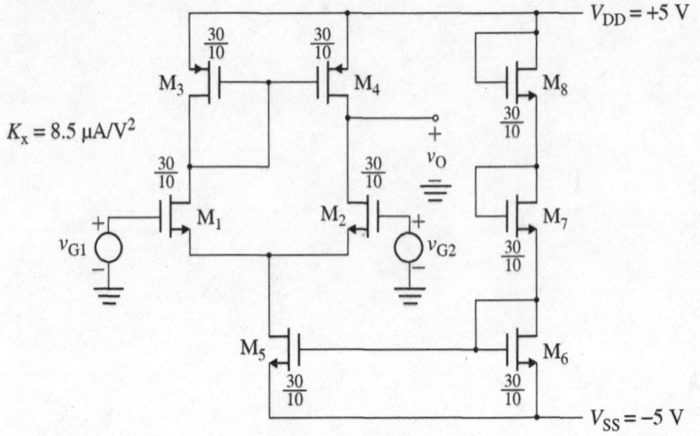

1.15 The DC biasing current of the cascoded MOS amplifier shown in Fig. 1.19(a) is kept constant at $I_Q = 50\ \mu A$ and $V_{DD} = V_{SS} = 5$ V. All MOS transistors are identical and have $V_M = 50$ V, $K_n = 50\ \mu A/V^2$, $W = 50\ \mu m$, and $L = 10\ \mu m$.

D

a. Determine the differential voltage gain A_d for single-ended output at the drain terminal of M_4.

b. Use PSpice to verify the result.

1.16 The MOS parameters of the CMOS amplifier shown in Figure P1.16 are $K_n = 100\ \mu A/V^2$, $V_{tN} = 1$ V, $V_{tP} = 1.5$ V, $|V_M| = 100$ V, $L = 10\ \mu m$, $(W/L)_1 = 60$, $(W/L)_7 = (W/L)_8 = 120$, and $(W/L)_2 = (W/L)_3 = (W/L)_4 = (W/L)_5 = (W/L)_6 = 30$. The DC supply voltages are $V_{DD} = V_{SS} = 5$ V, the reference current is to be set at $I_{ref} = 100\ \mu A$, and $C_L \approx \infty$.

a. Determine the value of R_{ref} and V_{GS6}.

b. Calculate A_d, A_c, and CMRR.

c. Determine the value of capacitor C_x to limit the upper frequency at $f_H = 500$ kHz.

FIGURE P1.16

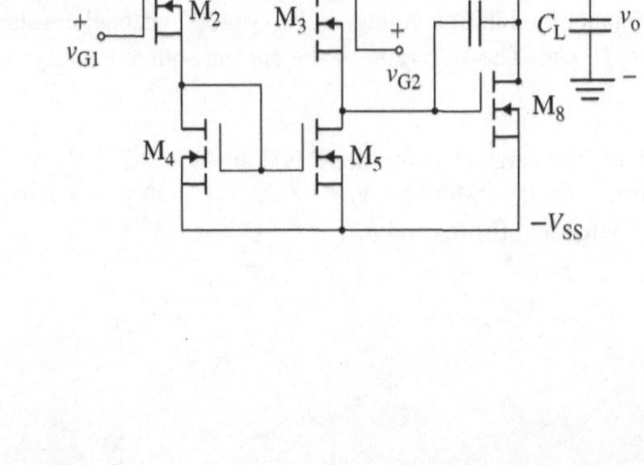

1.17 Design a CMOS amplifier as shown in Fig. P1.16 by determining the W/L ratios of the MOSFETs and the threshold voltage V_t. The small-signal voltage gain should be $A_d = 250$ V/V at biasing current $I_{ref} = 100$ μA. Assume identical transistors whose channel modulation voltage is $V_M = 70$ V, channel constant is $K_x = 10$ μA/V^2, and channel length $L = 10$ μm. Also, the width $(W/L)_6 = 1$ and $V_{DD} = V_{SS} = 5$ V.

D

1.5 *Depletion MOS Differential Amplifiers*

1.18 The parameters of the depletion MOS differential pair in Fig. 1.22 are $R_{SS} = 50$ kΩ, $I_Q = 1$ mA, $V_{DD} = V_{SS} = 30$ V, and $R_D = 2$ kΩ. The MOSFETs are identical and have $V_p = -4$ V and $I_{DSS} = 20$ mA.

a. Calculate the DC drain currents through the MOSFETs if $v_{id} = 30$ mV.

b. Assuming $I_{D1} = I_{D2}$, calculate A_d, A_c, and CMRR; R_{id} and R_{ic}; and the small-signal output voltage if $v_{g1} = 50$ mV and $v_{g2} = 20$ mV.

1.19 Repeat Prob. 1.18 if the transistor current source is replaced by resistance $R_{SS} = 100$ kΩ; that is, $I_{SS} = 0$.

1.20 **a.** Design a depletion MOS differential pair as shown in Fig. 1.13 in which one input terminal is grounded. The output is taken from the drain of transistor M_2. The DC biasing current is $I_Q = 5$ mA, and $V_{DD} = V_{SS} = 15$ V. The MOSFETs are identical and have $V_p = -4$ V and $I_{DSS} = 20$ mA. A small-signal voltage gain of $|A_2| = 20$ V/V is required.

D P

b. Calculate the design values of A_d, A_c, and CMRR.

1.21 A depletion MOS amplifier is shown in Fig. P1.21. The MOS parameters are $V_p = 4$ V, $I_{DSS} = -400$ μA, $V_M = 40$ V, $I_Q = 200$ μA, and $V_{DD} = V_{SS} = 15$ V.

a. Calculate the DC drain currents through the MOSFETs if $v_{id} = 10$ mV.

b. Assuming $I_{D1} = I_{D2}$, calculate A_d, A_c, and CMRR; R_{id} and R_{ic}; and the small-signal output voltage if $v_{g1} = 20$ mV and $v_{g2} = 10$ mV.

FIGURE P1.21

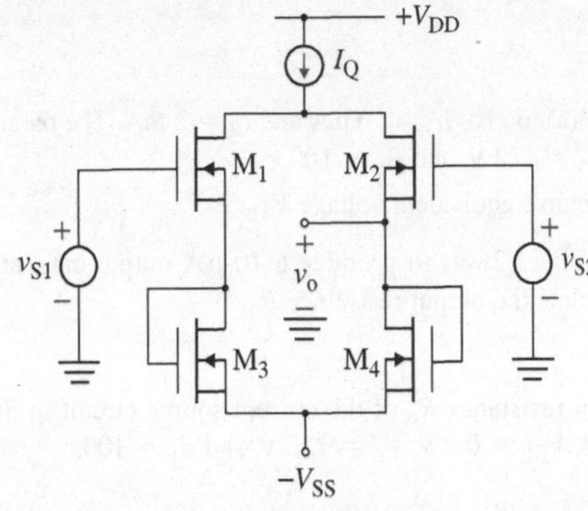

1.6 *BJT Current Sources*

1.22 The parameters of the basic current source in Fig. 1.25(a) are $\beta_F = 150$, $R_1 = 20$ kΩ, $V_{CC} = 15$ V, $V_{BE1} = V_{BE2} = 0.7$ V, and $V_A = 100$ V. Calculate **(a)** the output current $I_O = I_{C2}$, **(b)** the output resistance R_o, **(c)** Thevenin's equivalent voltage V_{Th}, and **(d)** the collector current ratio I_{C2}/I_{C1} if $V_{CE2} = 30$ V.

1.23 **a.** Design the basic current source in Fig. 1.25(a) to give an output current of $I_O = 200\ \mu A$. The transistor parameters are $\beta_F = 100$, $V_{CC} = 30$ V, $V_{BE1} = V_{BE2} = V_{CE1} = 0.7$ V, and $V_A = 150$ V.

b. Calculate the output resistance R_o, Thevenin's equivalent voltage V_{Th}, and the collector current ratio if $V_{CE2} = 30$ V.

1.24 The parameters of the modified current source in Fig. 1.27(a) are $\beta_F = 150$, $R_1 = 10\ k\Omega$, $V_{CC} = 15$ V, $V_{BE1} = V_{BE2} = V_{BE3} = 0.7$ V, and $V_A = 100$ V. Calculate **(a)** the output current $I_O = I_{C2}$, **(b)** the output resistance R_o, **(c)** Thevenin's equivalent voltage V_{Th}, and **(d)** the collector current ratio I_{C2}/I_{C1} if $V_{CE2} = 30$ V.

1.25 **a.** Design the modified basic current source in Fig. 1.27(a) to give an output current of $I_O = 50\ \mu A$. The transistor parameters are $\beta_F = 150$, $V_{CC} = 30$ V, $V_{BE1} = V_{BE2} = V_{BE3} = 0.7$ V, and $V_A = 100$ V.

b. Calculate the output resistance R_o, Thevenin's equivalent voltage V_{Th}, and the collector current ratio if $V_{CE2} = 20$ V.

1.26 The multiple transistors of the current source in Fig. P1.26 have $\beta_F = 150$, $R_1 = 10\ k\Omega$, $V_{CC} = 15$ V, and $V_A = 100$ V. The B-E voltages are equal, $V_{BE} = 0.7$ V. Calculate **(a)** the output current I_O, **(b)** the output resistance R_o, **(c)** Thevenin's equivalent voltage V_{Th}, and **(d)** the collector current ratio if $V_{CE2} = 15$ V.

FIGURE P1.26

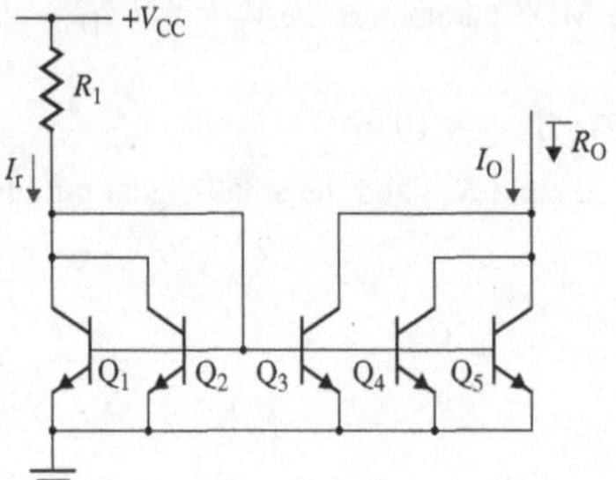

1.27 **a.** Design the Widlar current source in Fig. 1.28(a) to give $I_O = 10\ \mu A$ and $I_R = 2$ mA. The parameters are $V_{CC} = 30$ V, $V_{BE1} = 0.7$ V, $V_T = 26$ mV, $V_A = 150$ V, and $\beta_F = 100$.

b. Calculate the output resistance R_o and Thevenin's equivalent voltage V_{Th}.

1.28 Design a Widlar current source as shown in Fig. 1.28(a) to produce a 10-μA output current. Assume $\beta_F = 100$, $V_{CC} = 30$ V, and $R_1 = 30\ k\Omega$. Calculate the output resistance R_o.

1.29 Determine the output current I_O and the output resistance R_o of the current source circuit in Fig. P1.29. Assume $V_{CC} = 30$ V, $R_1 = 20\ k\Omega$, $R_2 = 10\ k\Omega$, $V_{BE} = 0.7$ V, $V_A = 150$ V, and $\beta_F = 100$.

FIGURE P1.29

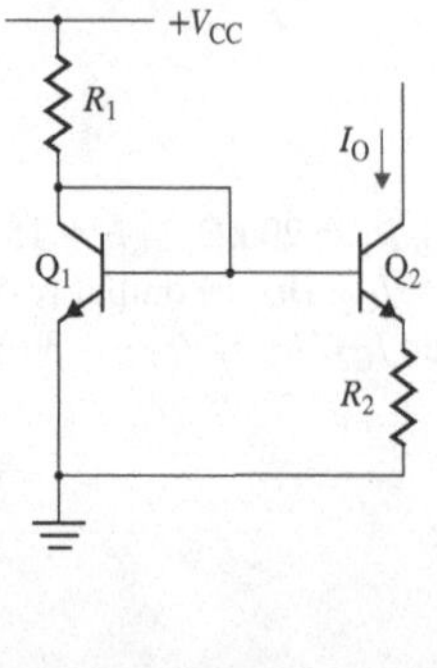

1.30 **a.** Design the Wilson current source in Fig. 1.30(a) to give $I_O = 10\ \mu A$. The parameters are $V_{CC} = 30$ V, $V_{BE} = 0.7$ V, $V_T = 26$ mV, $V_A = 100$ V, and $\beta_F = 150$.

D P **b.** Calculate the output resistance R_o and Thevenin's equivalent voltage V_{Th}.

1.31 For the Wilson current source in Fig. P1.31, determine the output current I_O and the output resistance R_o. Assume $V_{CC} = 20$ V, $V_{BE} = 0.7$ V, $V_T = 26$ mV, $V_A = 150$ V, and $\beta_F = 150$.

FIGURE P1.31

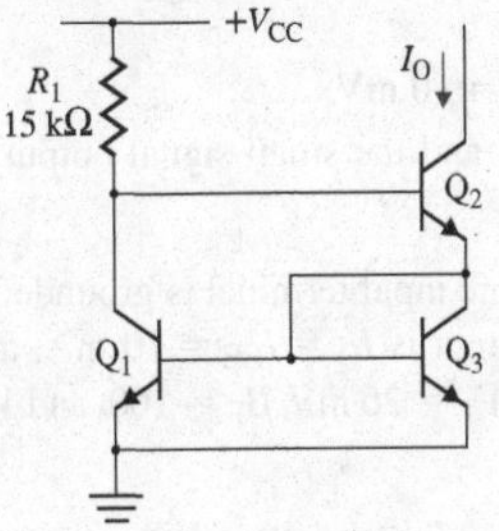

1.32 Repeat Prob. 1.31 for $V_{CC} = 30$ V.

1.33 For the current source in Fig. P1.33, determine the output resistance R_o and Thevenin's equivalent voltage V_{Th}. Assume $V_{CC} = 30$ V, $V_{BE} = 0.7$ V, $V_T = 26$ mV, $V_A = 150$ V, and $\beta_F = 150$.

FIGURE P1.33

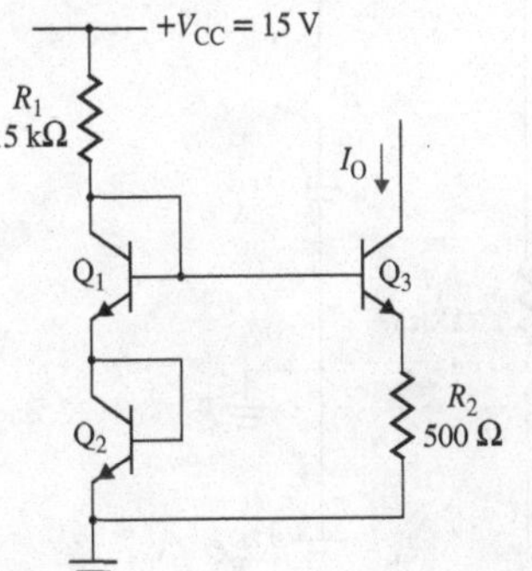

1.34 Determine the sensitivity S of output current I_O to supply voltage V_{CC} for the circuit in Fig. P1.34. S is defined as

$$S = \frac{V_{CC}/I_O}{\delta I_O/\delta V_{CC}}$$

FIGURE P1.34

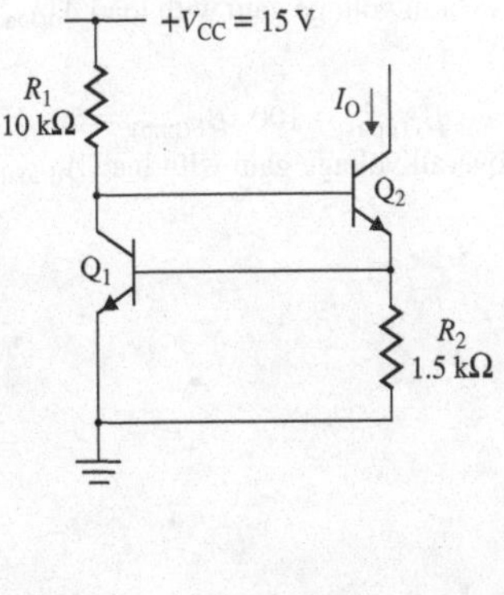

1.35 Design a BJT current source so that $R_o \geq 50\ \text{k}\Omega$ at an output current of $I_O = 1$ mA.
D

1.36 Design a BJT current source so that $R_o \geq 500\ \text{k}\Omega$ at an output current of $I_O = 1$ mA.
D

1.7 BJT Differential Amplifiers

1.37 The parameters of the emitter-coupled pair in Fig. 1.33 are $\beta_F = 150$, $R_{EE} = 20\ \text{k}\Omega$, $I_Q = 0.25$ mA, $V_{CC} =$ 12 V, and $R_C = 10\ \text{k}\Omega$.
P

a. Calculate the DC collector currents through the transistors if $v_{id} = 10$ mV.

b. Assuming $I_{C1} = I_{C2}$, calculate A_d, A_c, and CMRR; R_{id} and R_{ic}; and the small-signal output voltage if $v_{B1} = 30$ mV and $v_{B2} = 20$ mV. Assume $V_T = 26$ mV.

1.38 **a.** Design an emitter-coupled pair as shown in Fig. 1.41, in which one input terminal is grounded. The output is taken from the collector of transistor Q_2. The biasing current is $I_Q = I_{EE} = 10$ mA, and $V_{CC} = -V_{EE} = 12$ V. The transistors are identical. Assume $V_{BE} = 0.7$ V, $V_T = 26$ mV, $\beta_F = 100$, and $V_A = 40$ V. A small-signal voltage gain of $A_1 = -150$ V/V is required.
D

b. Calculate the design values of A_d, A_c, and CMRR.

1.39 A differential amplifier is shown in Fig. P1.39. The transistors are identical. Assume $V_{BE} = 0.7$ V, $V_T = 26$ mV, $\beta_F = 50$, and $V_A = 40$ V. Calculate the values of A_d, R_{id}, A_c, R_{ic}, and CMRR.
P

FIGURE P1.39

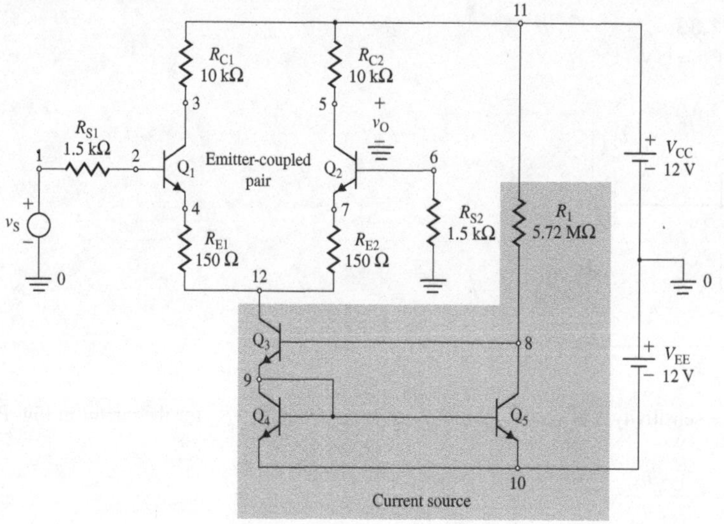

1.40 The parameters of the differential amplifier in Fig. 1.44 are $\beta_{F(npn)} = 100$, $\beta_{F(pnp)} = 50$, $V_A = 40$ V, $I_Q = 10\ \mu$A, and $V_{CC} = 10$ V. Calculate A_d, R_{id}, R_o, and the overall voltage gain with load $A_{d(load)}$. Assume $V_T = 26$ mV.

1.41 The parameters of the differential amplifier in Fig. 1.45 are $\beta_{F(npn)} = 100$, $\beta_{F(pnp)} = 50$, $V_A = 40$ V, $I_Q = 5\ \mu$A, and $V_{CC} = 10$ V. Calculate A_d, R_{id}, R_o, and the overall voltage gain with load $A_{d(load)}$. Assume $V_T = 26$ mV.

1.42 A differential amplifier is shown in Fig. P1.42. The transistors are identical. Assume $V_{BE} = 0.7$ V, $V_T = 26$ mV, $\beta_{F(npn)} = 100$, $\beta_{F(pnp)} = 50$, $V_A = 40$ V, and $V_{CC} = 10$ V. Calculate the values of R_1, R_2, A_d, R_{id}, A_c, R_{ic}, and CMRR.

FIGURE P1.42

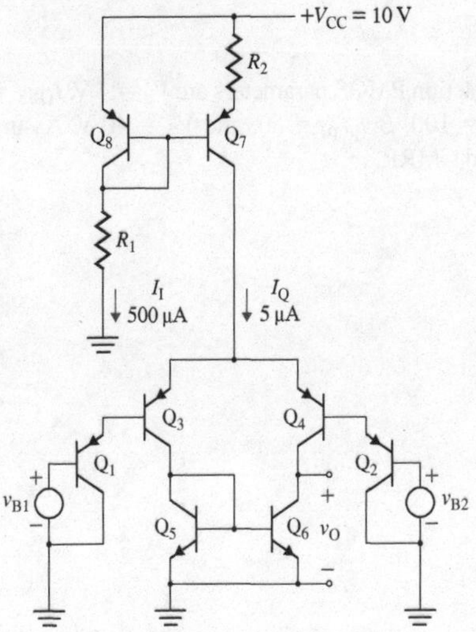

1.43 The parameters of the BJT amplifier shown in Fig. P1.43 are $\beta_{FN} = 100$, $\beta_{FP} = 120$, $|V_A| = 100$ V, and $V_{BEN} = -V_{BEP} = 0.65$ V. The DC supply voltages are $V_{CC} = V_{EE} = 5$ V, the reference current is to be set at $I_{ref} = 100$ μA, and $R_L = 50$ kΩ.

a. Determine the value of R_{ref}.

b. Calculate A_d, A_c, and CMRR.

c. Determine the value of capacitor C_x to limit the upper frequency at $f_H = 500$ kHz.

FIGURE P1.43

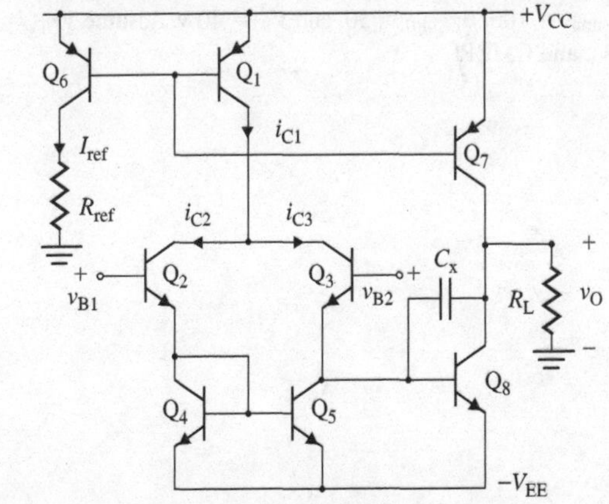

1.44 (D) Design a differential amplifier as shown in Fig. 1.44 to satisfy the following conditions: a differential voltage gain of $A_o \geq 1500$ V/V, CMRR of ≥ 3500, and load of $R_L = 50\ \text{k}\Omega$, with a coupling capacitor of $C_L = \infty$. Use all BJTs and the cascoded current source, similar to the one in Fig. 1.29. Use PSpice to verify your design with a $v_{id} = 5\ \mu\text{V}$ and $v_{ic} = 10$ mV. Supply voltages are $V_{CC} = -V_{EE} = 15$ V.

1.8 BiCMOS Differential Amplifiers

1.45 A BiCMOS amplifier is shown in Fig. P1.45. The depletion PMOS parameters are $V_p = 4$ V, $I_{DSS} = 400\ \mu\text{A}$, and $V_M = 100$ V. The BJT parameters are $\beta_{F(npn)} = 100$, $\beta_{F(pnp)} = 50$, and $V_A = 40$ V. Assume $V_{DD} = V_{EE} = 15$ V and $I_Q = 200\ \mu\text{A}$. Calculate A_d, A_c, and CMRR.

FIGURE P1.45

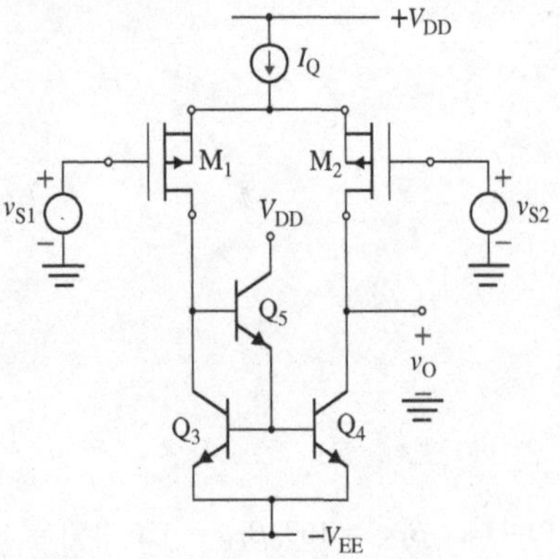

1.46 The DC biasing current of a BiCMOS amplifier is kept constant at $I_Q = 5\ \mu\text{A}$. All bipolar transistors are identical, with $\beta_{F(npn)} = 100$, $\beta_{F(pnp)} = 50$, and $V_A = 40$ V. Also, the MOS transistors are identical, with $|V_M| = 20$ V. For the NMOS, $V_t = +2$ V and $V_{GS} = +4$ V at $I_D = 1$ mA; for the PMOS, $V_t = -3$ V and $V_{GS} = -6$ V at $I_D = 1$ mA. Determine the differential voltage gain A_d for single-ended output **(a)** for the basic BiCMOS amplifier in Fig. 1.49(c), **(b)** for the cascode BiCMOS amplifier in Fig. 1.50, and **(c)** for the double-cascode BiCMOS amplifier in Fig. 1.51.

1.47 A BiCMOS amplifier is shown in Fig. P1.47. The PMOS parameters are $V_t = -3$ V and $V_{GS} = -6$ V at $I_D = 1$ mA. The BJT parameters are $\beta_{F(npn)} = 100$, $\beta_{F(pnp)} = 50$, and $V_A = 40$ V. Assume $V_{DD} = -V_{EE} = 15$ V and $I_Q = 200\ \mu\text{A}$. Calculate A_d, A_c, and CMRR.

FIGURE P1.47

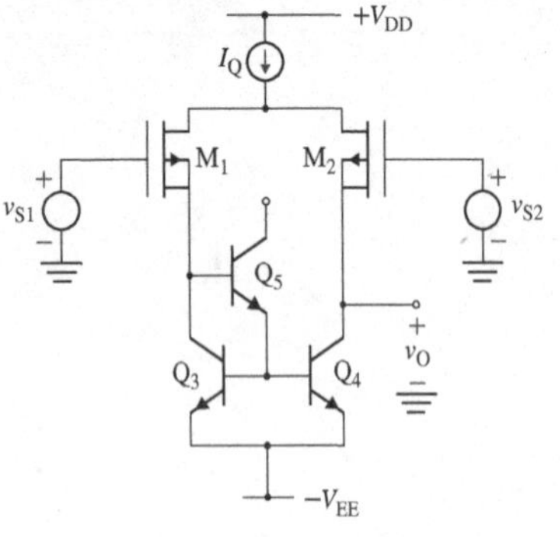

1.48 (D) Design a differential amplifier as shown in Fig. 1.51 to satisfy the following conditions: a differential voltage gain of $A_o \geq 1500$ V/V, CMRR of ≥ 3500, and load of $R_L = 50$ kΩ, with a coupling capacitor of $C_L = \infty$. Use all BJTs and NMOSs. Use PSpice to verify your design with $v_d = 5$ µV and $v_c = 10$ mV. Supply voltages are $V_{CC} = V_{EE} = 12$ V.

1.9 *Frequency Response of Differential Amplifiers*

1.49 The emitter-coupled pair in Fig. 1.33 has $R_{EE} = 20$ kΩ, $I_Q = 5$ mA, $V_{CC} = 12$ V, and $R_C = 10$ kΩ. The small-signal transistor parameters are $C_\pi = 5$ pF, $C_\mu = 2$ pF, $\beta_{F(npn)} = 100$, and $\beta_{F(pnp)} = 50$. Find the frequency-dependent gains $A_d(j\omega)$ and $A_c(j\omega)$.

1.50 The emitter-coupled pair in Fig. P1.50 has $I_Q = 5$ mA, $V_{CC} = -V_{EE} = 15$ V, and $R_C = 10$ kΩ. The small-signal transistor parameters are $C_\pi = 5$ pF, $C_\mu = 2$ pF, $\beta_{F(npn)} = 100$, and $\beta_{F(pnp)} = 50$. Find the frequency-dependent gains $A_d(j\omega)$ and $A_c(j\omega)$.

FIGURE P1.50

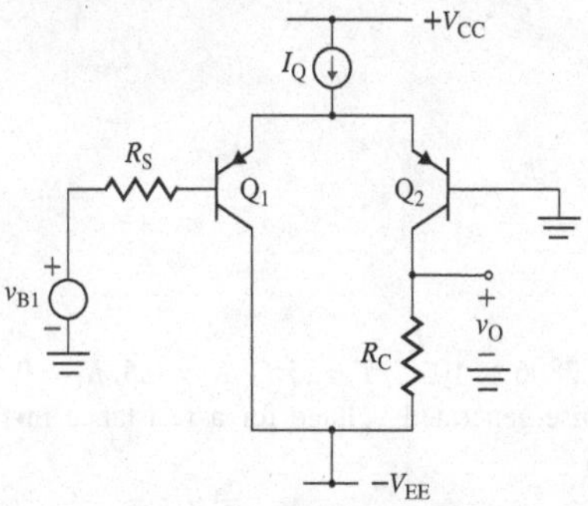

1.51 The differential amplifier in Fig. 1.42 has $V_A = 40$ V, $I_Q = 10$ µA at $R_{EE} = 50$ kΩ, and $V_{CC} = -V_{EE} = 15$ V. The small-signal transistor parameters are $C_\pi = 5$ pF, $C_\mu = 2$ pF, $\beta_{F(npn)} = 100$, and $\beta_{F(pnp)} = 50$. Find the frequency-dependent gains $A_d(j\omega)$ and $A_c(j\omega)$.

1.52 The source-coupled pair in Fig. 1.8 has $R_{SS} = 50$ kΩ, $I_Q = 10$ mA, $V_{DD} = 30$ V, and $R_D = 5$ kΩ. The MOSFETs are identical and have $K_n = 1.5$ mA/V^2, $W/L = 1$, $|V_M| = 100$ V, and $V_t = 1.2$ V. The small-signal MOS parameters are $C_{gs} = 5$ pF and $C_{gd} = 2$ pF. Find the frequency-dependent gains $A_d(j\omega)$ and $A_c(j\omega)$.

1.53 The MOS amplifier shown in Fig. 1.14 has $R_{SS} = 50$ kΩ, $I_Q = 1$ mA, and $V_{DD} = 15$ V. The MOSFETs are identical and have $K_n = 1.5$ mA/V^2, $W/L = 1$, $|V_M| = 100$ V, and $V_t = 1.2$ V. The small-signal MOS parameters are $C_{gs} = 5$ pF and $C_{gd} = 2$ pF. Find the frequency-dependent gains $A_d(j\omega)$ and $A_c(j\omega)$.

1.54 (D) (P) Design a differential amplifier configuration as shown in Fig. P1.54 to obtain a small-signal differential gain, with $A_d = 250$ V/V, CMRR $\geq 10^3$, and $f_b \geq 250$ kHz. Use any suitable transistors, BJTs, or MOSFETs. Assume $V_{CC} = V_{EE} = 12$ V. The configuration should be the one that is least expensive. Use PSpice to verify your design and give the component estimates.

FIGURE P1.54

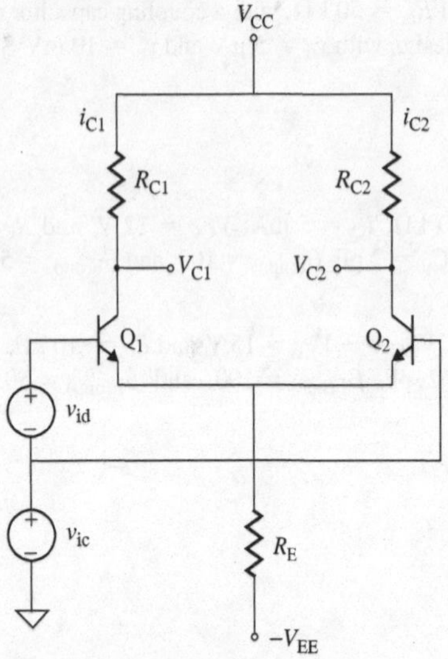

1.11 *Electronic Noise*

1.55 The noise parameters are $q = 1.6022 \times 10^{-19}$, $k = 1.3806 \times 10^{-23}$, $T = 25°C$, $K_f = 0.5$, $K_b = 0.5$, $f = 1$ Hz, $f_c = 2.5$ kHz, $a = 1$, and $c = 1$. Determine the noise-generated voltage for a resistance in Fig. 1.54 if $R = 10$ MΩ.

1.56 The noise parameters are $q = 1.6022 \times 10^{-19}$, $k = 1.3806 \times 10^{-23}$, $T = 25°C$, $K_f = 0.5$, $K_b = 0.5$, $f = 1$ Hz, $f_c = 2.5$ kHz, $a = 1$, and $c = 1$. Determine the noise-generated voltages and currents for a diode in Fig. 1.55 if $I_D = 10$ mA and $r_s = 50$ Ω.

1.57 The noise parameters are $q = 1.6022 \times 10^{-19}$, $k = 1.3806 \times 10^{-23}$, $T = 25°C$, $K_f = 0.5$, $K_b = 0.5$, $f = 1$ Hz, $f_c = 2.5$ kHz, $a = 1$, and $c = 1$. Determine the noise-generated voltages and currents for a BJT transistor in Fig. 1.56 if $I_C = 10$ mA, $I_B = 100$ μA, and $r_b = 100$ Ω.

1.58 The noise parameters are $q = 1.6022 \times 10^{-19}$, $k = 1.3806 \times 10^{-23}$, $T = 25°C$, $K_f = 0.5$, $K_b = 0.5$, $f = 1$ Hz, $f_c = 2.5$ kHz, $a = 1$, and $c = 1$. Determine the noise-generated voltages and currents for a FET in Fig. 1.57 if $I_D = 1$ mA, $I_G = 100$ nA, and $g_m = 1$ mA/V.

1.59 The noise parameters are $q = 1.6022 \times 10^{-19}$, $k = 1.3806 \times 10^{-23}$, $T = 25°C$, $K_f = 0.5$, $K_b = 0.5$, $f = 1$ Hz, $f_c = 2.5$ kHz, $a = 1$, and $c = 1$. Determine the noise-generated voltages and currents for a FET in Fig. 1.57 if $I_D = 10$ mA, $I_G = 100$ nA, and $g_m = 1.5$ mA/V.

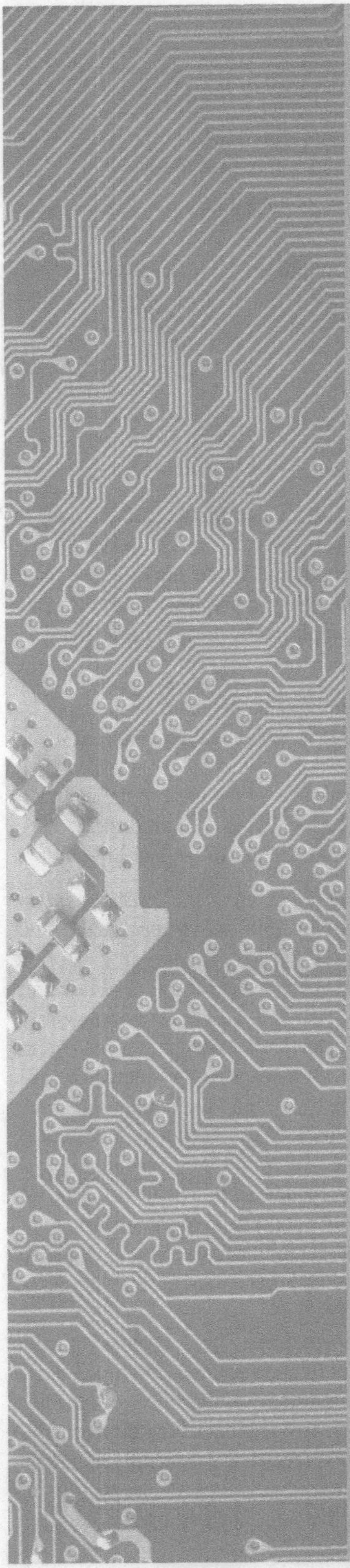

CHAPTER 2

FEEDBACK AMPLIFIERS

Learning Outcomes

After completing this chapter, students should be able to do the following:

- Describe the types and the properties of feedback amplifiers.
- Describe the different feedback configurations, their properties, and circuit implementations.
- Select the feedback configuration and analyze feedback amplifiers to meet specific requirements.
- Design a feedback network to meet desired closed-loop gain, the input impedance, and the output impedance.
- Determine the stability conditions of feedback amplifiers.
- Apply the compensation techniques to stabilize an unstable amplifier.

Symbols and Their Meanings

Symbol	Meaning
A, A_f	Open-loop and closed-loop gain of an amplifier
A_o, A_{of}	Low-frequency open-loop and closed-loop gain of an amplifier
i_i, i_e, i_f	Input, error, and feedback current signals
R_i, R_{if}	Input resistances of an amplifier without feedback and with feedback

Symbol	**Meaning**
R_o, R_{of}	Output resistances of an amplifier with feedback and without feedback
R_{ie}, R_{oe}	Equivalent input resistance and output resistance of an amplifier
v_i, v_e, v_f	Input, error, and feedback voltage signals
ω_o, ω_p, ω_g	Oscillation, pole, and crossover frequencies of an amplifier
f, f_D	Pole and dominant pole frequency of an amplifier
ϕ, T_L	Phase angle and transmission loop
β, β_F	Feedback factor and current gain of a BJT

2.1 Introduction

Feedback is commonly used in amplifier circuits. A signal that is proportional to the output is compared with an input or a reference signal so that a desired output is obtained from the amplifier. The difference between the input and the feedback signals, called the *error signal*, is amplified by the amplifier. There are two types of feedback:

- In *negative feedback*, the output signal (or a fraction of it) is continuously fed back to the input side and is subtracted from the input signal to create an error signal, which is then corrected by the amplifier to produce the desired output signal.
- In *positive feedback*, the output signal (or a fraction of it) is continuously fed back to the input side and added to the input signal to create a larger error signal, which is then amplified to produce a larger output until the output reaches the saturation voltage limit of the amplifier.

In negative feedback, the signal that is fed back to the input side is known as the *feedback signal*, and its polarity is opposite that of the input signal (i.e., it is out of phase by 180° with respect to the input signal). Negative feedback in an amplifier has four major benefits: (1) It stabilizes the overall gain of the amplifier with respect to parameter variations due to temperature, supply voltage, and so on; (2) it increases or decreases the input and output impedances; (3) it reduces the distortion and the effect of nonlinearity; and (4) it increases the bandwidth. There are two disadvantages of negative feedback: (1) The overall gain is reduced almost in direct proportion to the benefits, and it is often necessary to compensate for the decrease in gain by adding an extra amplifier stage; (2) the circuit may tend to oscillate, in which case careful design is required to overcome this problem. Negative feedback is also known as *degenerative feedback* because it degenerates (or reduces) the output signal.

The op-amp circuits use negative feedback. The amplifier gain A_f is almost independent of the op-amp gain A; it depends on the external circuit elements only. For example, the gain of the inverting amplifier is $-R_F/R_1$, which is independent of the op-amp gain A, and the input impedance is approximately R_1. The gain of the noninverting op-amp amplifier is $(1 + R_F/R_1)$, and its input impedance is very large. The output impedance of both amplifiers is very small.

In positive feedback, the feedback signal is in phase with the input signal. Thus, the error signal is the algebraic sum of the input and feedback signals, and it is amplified by the amplifier. Thus, the output may continue to increase, resulting in an unstable situation, and the circuit may oscillate between the limits of the power supplies at the resonant frequency of the amplifier. Positive feedback is often referred to as *regenerative feedback* because it increases the output signal. Positive feedback is generally applied in oscillator circuits, which we will study in Chapter 5. Note that positive feedback does not necessarily imply oscillations. In fact, positive feedback is quite useful in some applications, such as active filters.

2.2 Feedback

Consider the noninverting op-amp amplifier shown in Fig. 2.1(a). The voltages v_s, v_f, and v_e are related as follows:

$$v_e = v_s - v_f$$

$$v_o = Av_e$$

$$v_f = \frac{R_1}{R_1 + R_F} v_o$$

The input and output relationships described by these equations can be represented by a block diagram as shown in Fig. 2.1(b). The voltage v_e, which is the difference between v_s and v_f, is amplified by the voltage gain A. The *feedback signal* v_f is proportional to the output voltage and is fed back to the input side. Thus, the amplifier feeds the output voltage back to the input side and compares the voltages. There are two circuits: the amplifier circuit and the feedback circuit (or network) consisting of R_1 and R_F. The voltages v_s, v_f, and v_e form a series circuit at the input side as shown in Fig. 2.1(a), whereas v_o is applied directly to the feedback network. That is, the noninverting amplifier uses series-shunt, or *voltage-sensing/voltage-comparing*, feedback.

Now consider the inverting op-amp amplifier shown in Fig. 2.2(a). There is a common node at the input side where the input current i_i and the feedback current i_f meet. The voltages and currents are related as follows:

$$i_e = i_i - i_f$$

$$v_e = -R_i i_e$$

$$v_o = Av_e$$

$$i_f = \frac{-v_e - v_o}{R_F} \approx \frac{-v_o}{R_F}$$

since $v_e \approx 0$ and $v_o >> v_e$. The input and output relationships described by these equations are shown by the block diagram in Fig. 2.2(b). The current i_e, which is the difference between i_i and i_f, is amplified by the transimpedance gain $-R_iA$. The *feedback current signal* i_f is proportional to the output voltage. Thus,

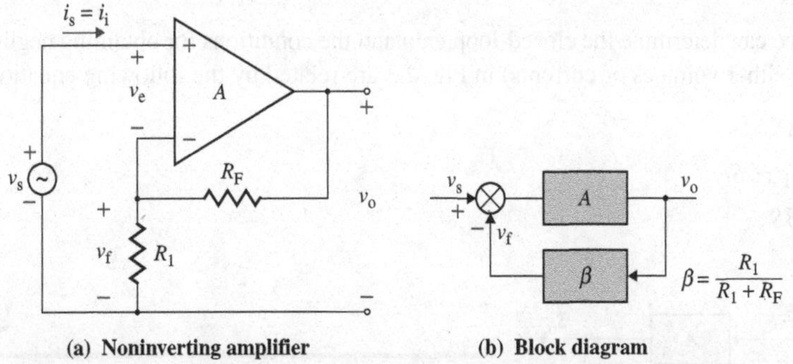

FIGURE 2.1 Feedback representation of noninverting op-amp amplifier

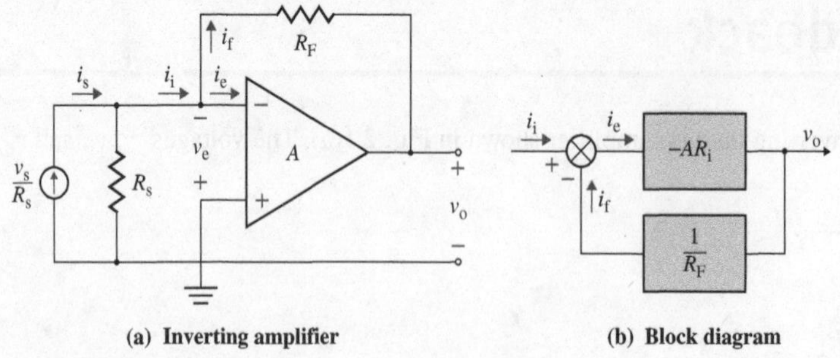

FIGURE 2.2 Feedback representation of inverting op-amp amplifier

the amplifier feeds the output voltage back to the input side and compares the currents. Here, the feedback network consists of R_F. The currents i_i, i_f, and i_e form a shunt connection at the input side, whereas v_o is applied directly to the feedback network. Thus, the inverting amplifier uses shunt-shunt, or *voltage-sensing/current-comparing*, feedback.

2.3 Characteristics of Feedback

In the inverting and noninverting amplifiers in Figs. 2.1 and 2.2, the output voltage is fed back directly to the input side and is compared to either the voltage or the current at the input side. Feedback can be represented by the general configuration shown in Fig. 2.3, where β is called the *feedback ratio* (or *factor*) and A is the amplifier gain. The units of A could be V/V, A/V, A/A, or V/A, and the units of β will be the reciprocal of those of A. For the noninverting amplifier, A is in V/V and β is in V/V; for the inverting amplifier, A is in V/A and β is in A/V.

▶ **NOTE** β is the feedback ratio, whereas β_f ($=\beta_F$) is the forward current gain of a bipolar transistor. To distinguish the feedback factor β from the BJT transistor current gain β_F, in this chapter we use the hybrid parameter h_{fe} for the BJT current gain.

2.3.1 Closed-Loop Gain

From Fig. 2.3, we can determine the closed-loop gain and the conditions for obtaining negligible error. The various signals (either voltages or currents) in Fig. 2.3 are related by the following equations:

$$S_o = AS_e \tag{2.1}$$

$$S_e = S_i - S_f \tag{2.2}$$

$$S_f = \beta S_o \tag{2.3}$$

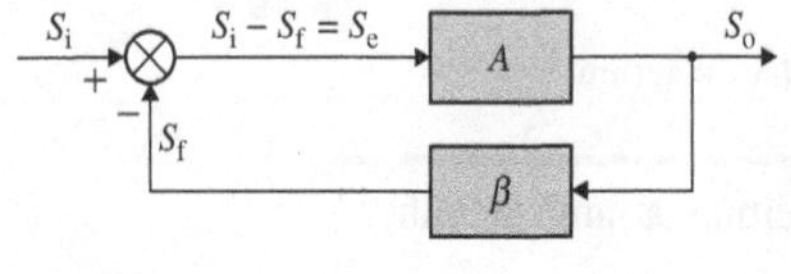

FIGURE 2.3 General feedback configuration

99

where A = open-loop gain of the amplifier
S_o = output signal
S_e = error signal
S_f = feedback signal
S_i = input signal
β = feedback factor

Substituting S_e from Eq. (2.2) into Eq. (2.1) gives

$$S_o = AS_e = AS_i - AS_f \tag{2.4}$$

Substituting S_f from Eq. (2.3) into Eq. (2.4) yields

$$S_o = AS_i - \beta AS_o$$

which gives the overall gain A_f with negative feedback as

$$A_f = \frac{S_o}{S_i} = \frac{A}{1 + \beta A} \tag{2.5}$$

A_f is often known as the *closed-loop gain*. Equation (2.5) is derived for negative feedback. βA is the gain around the feedback loop, known as the *loop gain* or the *loop transmission*. Let us define $T_L = \beta A$. If $T_L >> 1$, Eq. (2.5) becomes

$$A_f \approx \frac{1}{\beta} \tag{2.6}$$

That is, for large values of loop gain T_L, the closed-loop gain A_f is independent of the open-loop gain A and depends on the feedback factor β only.

Substituting S_f from Eq. (2.3) and S_o from Eq. (2.5) into Eq. (2.2) gives the error signal S_e:

$$S_e = S_i - S_f = S_i - \beta S_o = S_i - \frac{\beta AS_i}{1 + \beta A} = \frac{S_i}{1 + \beta A} = \frac{S_i}{1 + T_L} \tag{2.7}$$

As T_L becomes much greater than 1, S_e becomes much smaller than S_i, and $S_i \approx S_f$. Substituting S_o from Eq. (2.5) into Eq. (2.3) gives

$$S_f = \frac{\beta A}{1 + \beta A} S_i = \frac{T_L}{1 + T_L} S_i \tag{2.8}$$

If $T_L >> 1$, $S_f \approx S_i = \beta S_o$. That is, the output signal S_o is the amplified version of the input signal S_i, provided $\beta < 1$.

With positive feedback, the sign of βA changes and the closed-loop gain becomes

$$A_f = \frac{S_o}{S_i} = \frac{A}{1 - \beta A} \tag{2.9}$$

This confirms that the output will continue to increase until the saturation limit. Therefore, the amplifier can be made to oscillate by making $\beta A = 1$.

2.3.2 Gain Sensitivity

In most practical amplifiers, the open-loop gain A is dependent on temperature and the operating conditions of active devices. The effect of variations in the open-loop gain A can be determined from the sensitivity of the closed-loop gain A_f. Differentiating A_f in Eq. (2.5) with respect to A gives

$$\frac{dA_f}{dA} = \frac{(1 + \beta A) - \beta A}{(1 + \beta A)^2} = \frac{1}{(1 + \beta A)^2} \tag{2.10}$$

If A changes by δA, then A_f will change by δA_f. Thus, Eq. (2.10) yields

$$\delta A_f = \frac{\delta A}{(1 + \beta A)^2} \quad \text{(assuming } \delta A_f \approx dA_f) \tag{2.11}$$

which gives the approximate value of δA_f for finite increments in δA. The fractional change in A_f is given by

$$\frac{\delta A_f}{A_f} = \frac{1 + \beta A}{A} \frac{\delta A}{(1 + \beta A)^2} = \frac{\delta A/A}{1 + \beta A} \quad \text{(assuming } \delta A_f \approx dA_f) \tag{2.12}$$

which shows that a fractional change in A of $(\delta A/A)$ causes a fractional change in A_f of $(\delta A_f/A_f)$ such that

$$\frac{\delta A_f}{A_f} = \frac{1}{1 + \beta A}\left(\frac{\delta A}{A}\right)$$

If $\delta A/A$ is, say, 10%, the change in A_f is

$$\frac{\delta A_f}{A_f} = \frac{10}{1 + \beta A}\%$$

only. Thus, the sensitivity of the closed-loop gain A_f to the open-loop gain A is defined as

$$S_A^{A_f} = \frac{\delta A_f/A_f}{\delta A/A} = \frac{1}{1 + \beta A} \tag{2.13}$$

For $\beta A >> 1$, which is generally the case, the sensitivity of A_f to A becomes very small. Thus, a significant change in A will cause only a small change in A_f.

2.3.3 Feedback Factor Sensitivity

We can see from Eq. (2.6) that the closed-loop gain A_f depends on the feedback factor β only. The effect of variations of the feedback factor β on the gain A_f can be determined. Differentiating A_f in Eq. (2.5) with respect to β gives

$$\frac{dA_f}{d\beta} = -\frac{A^2}{(1 + \beta A)^2} = -A_f^2 \tag{2.14}$$

If β changes by $\delta\beta$, then A_f will change by δA_f. From Eq. (2.14), we get

$$\delta A_f = -A_f^2 \delta\beta \quad \text{(assuming } \delta A_f \approx dA_f) \tag{2.15}$$

Thus, the fractional change in A_f is given by

$$\frac{\delta A_f}{A_f} = -A_f \delta\beta \tag{2.16}$$

101

The sensitivity of closed-loop gain A_f to the feedback factor β is defined as

$$S_{\beta}^{A_f} = \frac{\delta A_f/A_f}{\delta\beta/\beta} = -A_f\beta = -\frac{\beta A}{1+\beta A} \tag{2.17}$$

For $\beta A >> 1$, Eq. (2.17) can be reduced to

$$S_{\beta}^{A_f} = -1 \tag{2.18}$$

Therefore, the gain A_f is directly sensitive to any change in the feedback factor β. The negative sign in Eq. (2.18) signifies that an increase in β will cause a decrease in A_f.

▶ **NOTE** Although the value of A can be positive or negative depending on the circuit configuration, we will use only the absolute value of A in equations for negative feedback such as Eq. (2.5).

EXAMPLE 2.1

Finding the effect of changes in the open-loop gain on the closed-loop gain The open-loop gain of an amplifier is $A = 250$, and the feedback factor is $\beta = 0.8$.

(a) Determine the closed-loop gain $A_f = S_o/S_i$.

(b) If the open-loop gain A changes by $+20\%$, determine the percentage change in the closed-loop gain A_f and its value.

(c) If the feedback factor β changes by -20%, determine the percentage change in the closed-loop gain A_f and its value.

SOLUTION

$A = 250$, $\beta = 0.8$, and $T_L = \beta A = 250 \times 0.8 = 200$.

(a) From Eq. (2.5),

$$A_f = \frac{250}{1+200} = 1.2438$$

(b) $\delta A/A = 20\%$. From Eq. (2.13),

$$\frac{\delta A_f}{A_f} = \frac{20\%}{1+200} = 0.1\%$$

$$\delta A_f = 0.1\% \times 1.2438 = 0.00124$$

$$A_f = 1.2438 + 0.00124 = 1.245$$

(c) $\delta\beta/\beta = 20\%$. From Eq. (2.18),

$$\frac{\delta A_f}{A_f} = -20\%$$

$$\delta A_f = -20\% \times 1.2438 = -0.249$$

$$A_f = 1.2439 - 0.249 = 0.995$$

NOTE: The overall gain A_f does not change much with a wide variation in the open-loop gain A. But the gain A_f changes directly with the feedback factor β. In designing a feedback amplifier, special care should be taken to ensure that the variation in the feedback factor is kept to a minimum.

2.3.4 Frequency Response

Negative feedback increases the bandwidth of an amplifier. To prove this, let us consider a simple amplifier whose open-loop gain A is dependent on the frequency, which can be expressed in Laplace's domain as

$$A(s) = \frac{A_o}{1 + s/(2\pi f_H)} \tag{2.19}$$

where A_o is open-loop low-frequency gain and f_H is open-loop 3-dB break frequency, in Hertz. From Eq. (2.5), the overall gain is given by

$$A_f(s) = \frac{A(s)}{1 + \beta A(s)} \tag{2.20}$$

where the feedback factor β is independent of frequency. Substituting $A(s)$ from Eq. (2.19) into Eq. (2.20) yields

$$A_f(s)\frac{A_o/[1 + s/(2\pi f_H)]}{1 + \beta A_o/[1 + s/(2\pi f_H)]} = \frac{A_o}{1 + \beta A_o} \frac{1}{1 + s/[2\pi f_H(1 + \beta A_o)]} \tag{2.21}$$

which gives the low-frequency closed-loop gain A_{of} as

$$A_{of} = \frac{A_o}{1 + \beta A_o} = \frac{A_o}{1 + T_{Lo}} \tag{2.22}$$

where $T_{Lo} = \beta A_o$ is called the *low-frequency loop gain*. From Eq. (2.21), the 3-dB break frequency f_{Hf} with feedback becomes

$$f_{Hf} = f_H(1 + \beta A_o) \tag{2.23}$$

Thus, *without feedback*, the following equations apply:

$$\begin{aligned} &\text{Low-frequency gain} = A_o \\ &\text{Bandwidth BW} = f_H \\ &\text{Gain–bandwidth product GBW} = A_o f_H \end{aligned} \tag{2.24}$$

With feedback, these equations apply:

$$\begin{aligned} &\text{Low-frequency gain } A_{of} = \frac{A_o}{1 + \beta A_o} \\ &\text{Bandwidth BW}_f = f_H(1 + \beta A_o) \\ &\text{Gain–bandwidth product GBW}_f = A_{of}\text{BW}_f = A_o f_H \end{aligned} \tag{2.25}$$

We can conclude from Eqs. (2.22) and (2.23) that feedback reduces the low-frequency gain by a factor of $(1 + \beta A_o)$ but increases the 3-dB frequency by the same amount $(1 + \beta A_o)$. However, the gain–bandwidth product remains constant at $A_o f_H$. Negative feedback allows the designer to trade gain for bandwidth, and it is widely used as a method for designing broadband amplifiers. The gain reduction is generally compensated for by adding more amplifying stages, which may also be feedback amplifiers. The plots of magnitude against frequency for A_f and A are shown in Fig. 2.4.

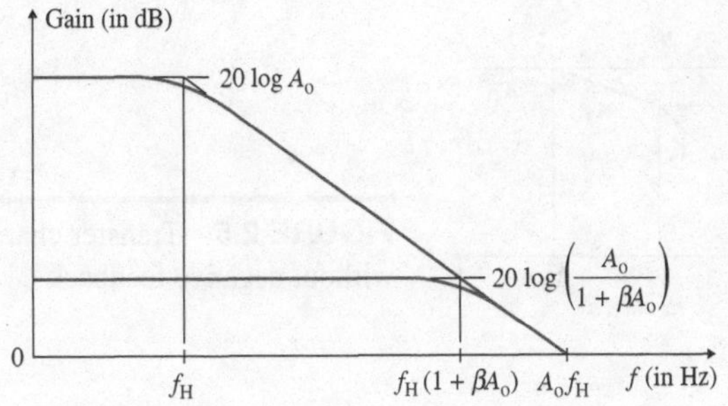

FIGURE 2.4 Plots of magnitude against frequency

EXAMPLE 2.2

Finding the effect of feedback on the frequency of an amplifier The feedback factor of a closed-loop amplifier is $\beta = 0.8$. The open-loop gain is expressed as

$$A(s) = \frac{250}{1 + s/(2\pi \times 100)}$$

Determine **(a)** the closed-loop low-frequency gain A_{of}, **(b)** the closed-loop bandwidth BW_f, and **(c)** the gain–bandwidth product GBW.

SOLUTION

$A_o = 250$, $\beta = 0.8$, and $f_H = 100$.

(a) From Eq. (2.22),

$$A_{of} = \frac{250}{(1 + 250 \times 0.8)} = 1.24378$$

(b) From Eq. (2.23),

$$B_{wf} = 100 \times (1 + 250 \times 0.8) = 20.1 \text{ kHz}$$

(c) From Eq. (2.24),

$$\text{Gain–bandwidth product GBW} = A_o f_H = 250 \times 100 = 25 \times 10^3$$

2.3.5 Distortion

An amplifier contains nonlinear devices such as transistors. As a result, the plot of the output signal S_o against the input signal S_i will not be linear. Thus, if the input signal is a sinusoidal waveform, the output signal will not be sinusoidal; that is, the output signal will be distorted. The effect of distortion in an amplifier is to reduce the gain of the open-loop transfer function. The closed-loop gain A_f, however, remains almost independent of the open-loop gain, as shown by Eq. (2.6). Therefore, negative feedback can reduce the effect of slope changes on the open-loop transfer function.

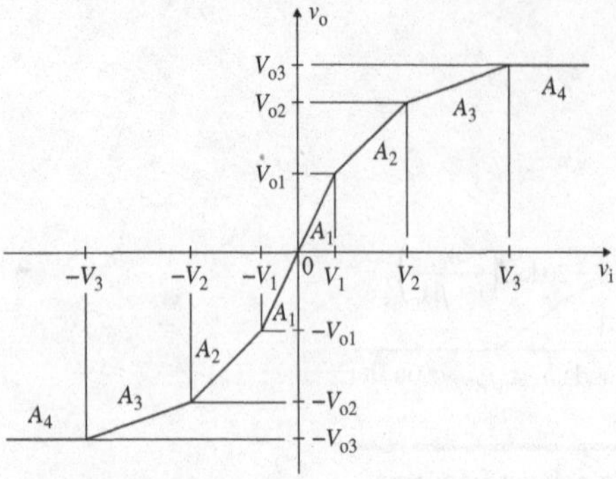

FIGURE 2.5 Transfer characteristic without negative feedback

Let us consider an amplifier whose transfer characteristic (v_o versus v_i) is nonlinear, as shown in Fig. 2.5. There are four regions of constant gain: A_1, A_2, A_3, and A_4. If negative feedback is applied with a feedback factor of β, Eq. (2.6) can be used as follows to calculate the closed-loop gains corresponding to the four regions:

$$A_{f1} = \frac{A_1}{1 + \beta A_1} \approx \frac{1}{\beta} \quad \text{for } \beta A_1 >> 1$$

$$A_{f2} = \frac{A_2}{1 + \beta A_2} \approx \frac{1}{\beta} \quad \text{for } \beta A_2 >> 1$$

$$A_{f3} = \frac{A_3}{1 + \beta A_3} \approx \frac{1}{\beta} \quad \text{for } \beta A_3 >> 1$$

$$A_{f4} = \frac{A_4}{1 + \beta A_4} \approx \frac{1}{\beta} \quad \text{for } \beta A_4 >> 1$$

Therefore, the slopes of the transfer characteristic in the four regions will be almost equal, and the transfer characteristic will be as shown in Fig. 2.6. The transfer characteristic with negative feedback is much less nonlinear than that of the original amplifier without negative feedback.

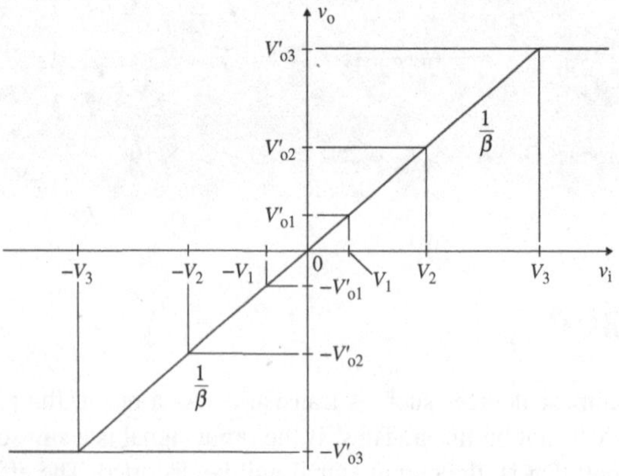

FIGURE 2.6 Transfer characteristic with negative feedback

EXAMPLE 2.3

Finding the effect of amplifier nonlinearity on the closed-loop gain The transfer characteristic of an amplifier without feedback is approximated by the following values of open-loop gain for given ranges of input voltage v_i:

$$A = \begin{cases} 1000 & \text{for } 0 < v_i \leq 0.5 \text{ mV} \\ 500 & \text{for } 0.5 \text{ mV} < v_i \leq 1 \text{ mV} \\ 250 & \text{for } 1 \text{ mV} < v_i \leq 2 \text{ mV} \\ 0 & \text{for } v_i > 2 \text{ mV} \end{cases}$$

If the feedback factor is $\beta = 0.5$, determine the closed-loop gains of the transfer characteristic.

SOLUTION

From Eq. (2.5), the closed-loop gains become

$$A_f = \begin{cases} \dfrac{1000}{1 + 0.5 \times 1000} = 1.996 & \text{for } 0 < v_i \leq 0.5 \text{ mV} \\ \dfrac{500}{1 + 0.5 \times 500} = 1.992 & \text{for } 0.5 \text{ mV} < v_i \leq 1 \text{ mV} \\ \dfrac{250}{1 + 0.5 \times 250} = 1.984 & \text{for } 1 \text{ mV} < v_i \leq 2 \text{ mV} \\ 0 & \text{for } v_i > 2 \text{ mV} \end{cases}$$

NOTE: As the input voltage v_i varies from 0 to 2 mV, the slope of output voltage versus input voltage (i.e., gain) varies from 1000 to 250 in a nonlinear fashion. But the closed-loop gain A_f remains almost constant.

KEY POINTS OF SECTION 2.3

- Feedback reduces the gain of a feedback amplifier. However, the bandwidth is widened proportionally. Also, feedback reduces the effect of amplifier distortion, nonlinearity, and variations in amplifier parameters.
- For a large value of loop-gain T_L, the closed-loop gain A_f is inversely proportional to the feedback factor β. That is, A_f is sensitive to changes in the feedback network parameters.

2.4 Feedback Topologies

The feedback configuration in Fig. 2.3 represents a general form; it does not indicate whether the input and output signals are voltages or currents. In practical amplifiers, the input and output signals can be either voltages or currents. If the output voltage is the feedback signal, it can be either compared with the input voltage to generate the error voltage signal (as shown in Fig. 2.1) or compared with the input current to generate the error current signal (as shown in Fig. 2.2). Similarly, the output current can be fed back and either compared with the input voltage to generate the error voltage signal or compared with the input current to generate the error current signal. Therefore, there are four feedback configurations, depending on whether the input and output signals are voltages or currents [1].

2.4.1 Feedback Configurations

There are four feedback configurations: series-shunt, series-series, shunt-shunt, and shunt-series. A potential divider, that is, a resistor (or capacitor) in series with another resistor (or capacitor) can sense a voltage, and placing a small resistor in the wire and sensing the voltage across it can sense a current. For a voltage subtraction, apply the input and the feedback signals to two distinct nodes, whereas for current subtraction, apply them to a single node. If the feedback signal is taken from the output terminal—for example, as in Fig. 2.7(c)—then the output side is a shunt feedback; if the feedback signal is taken from the voltage across a resistor—for example, as in Fig. 2.8(c)—then it is a series connection on the output sides.

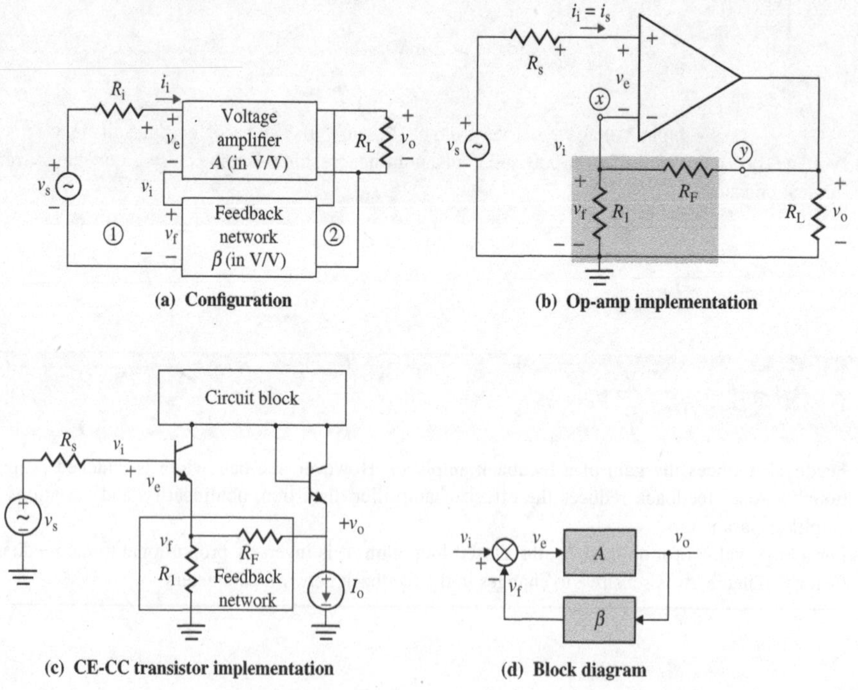

FIGURE 2.7 Series-shunt feedback

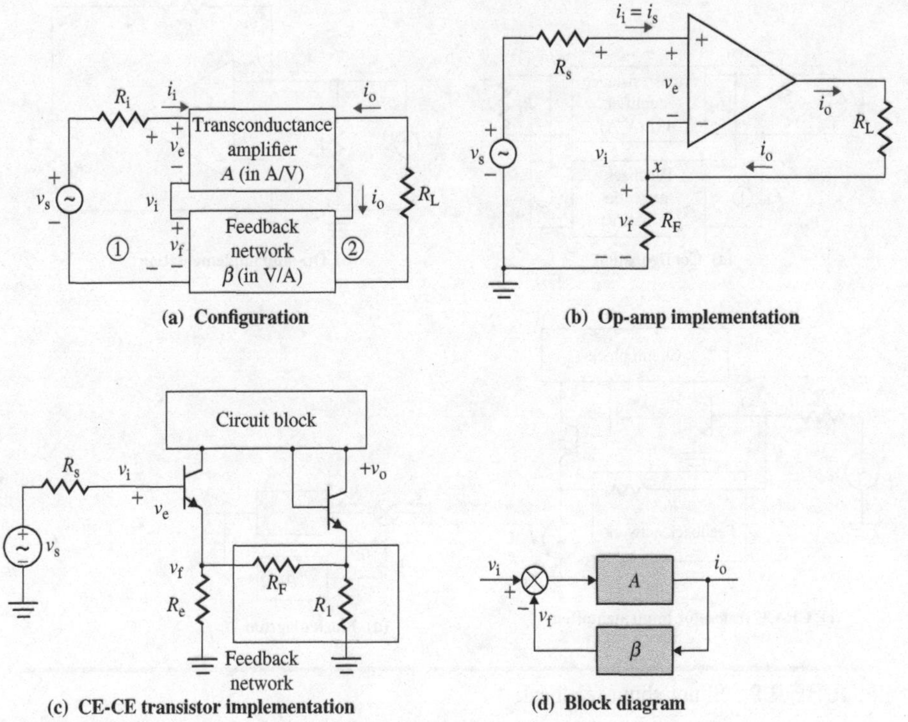

FIGURE 2.8 Series-series feedback

In series-shunt (*voltage-sensing/voltage-comparing*) feedback, shown in Fig. 2.7(a), the output voltage v_o is the input to the feedback network, and the feedback voltage v_f is proportional to the output voltage v_o. The feedback network forms a series circuit with the input voltage v_i but a parallel circuit with the output voltage v_o. The input current i_i flows through the loop formed by the input voltage, the amplifier, and the feedback network; that is, $v_i - v_f = v_e$. A series-shunt implementation using an op-amp is shown in Fig. 2.7(a). The discrete transistor implementation is shown in Fig. 2.7(c) and the feedback block as shown in Fig. 2.7(d).

In series-series (*current-sensing/voltage-comparing*) feedback, shown in Fig. 2.8(a), the output current i_o is the input to the feedback network, and the feedback voltage v_f is proportional to the output current i_o. The feedback network forms a series circuit with the input voltage and the output current. The input current flows through the loop formed by the input voltage, the amplifier, and the feedback network; that is, $v_i - v_f = v_e$. A series-series implementation using an op-amp is shown in Fig. 2.8(b). The discrete transistor implementation is shown in Fig. 2.8(c) and the feedback block as shown in Fig. 2.8(d).

In shunt-shunt (*voltage-sensing/current-comparing*) feedback, shown in Fig. 2.9(a), the output voltage v_o is also the input to the feedback network, and the feedback current i_f is proportional to the output voltage v_o. The feedback network is in parallel with both the input and the output voltages. The input current i_i is shared by the amplifier and the feedback network; that is, $i_i - i_f = i_e$. A shunt-shunt implementation using an op-amp is shown in Fig. 2.9(b). The discrete transistor implementation is shown in Fig. 2.9(c) and the feedback block in Fig. 2.9(d).

In shunt-series (*current-sensing/current-comparing*) feedback, shown in Fig. 2.10(a), the output current i_o is the input to the feedback network, and the feedback current i_f is proportional to the output

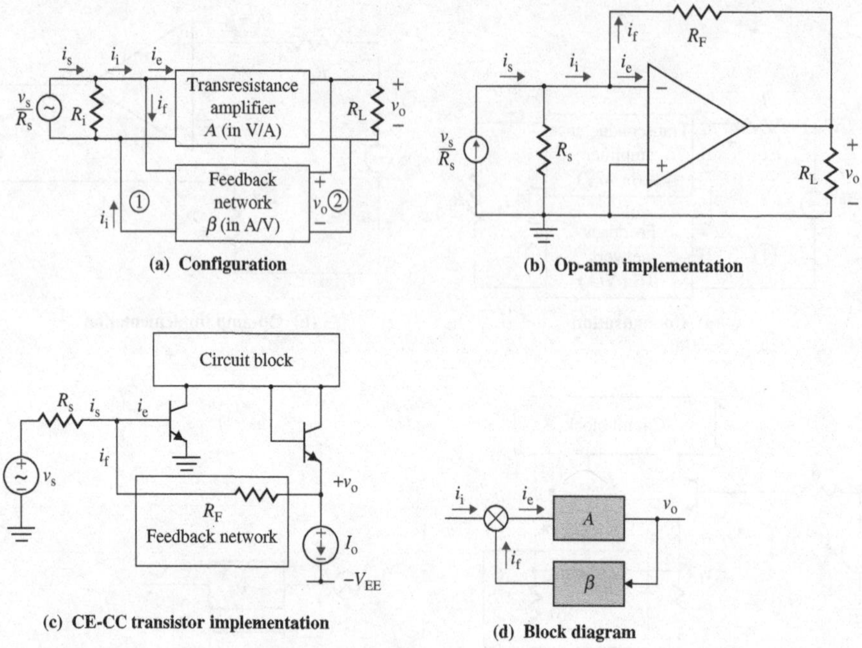

FIGURE 2.9 Shunt-shunt feedback

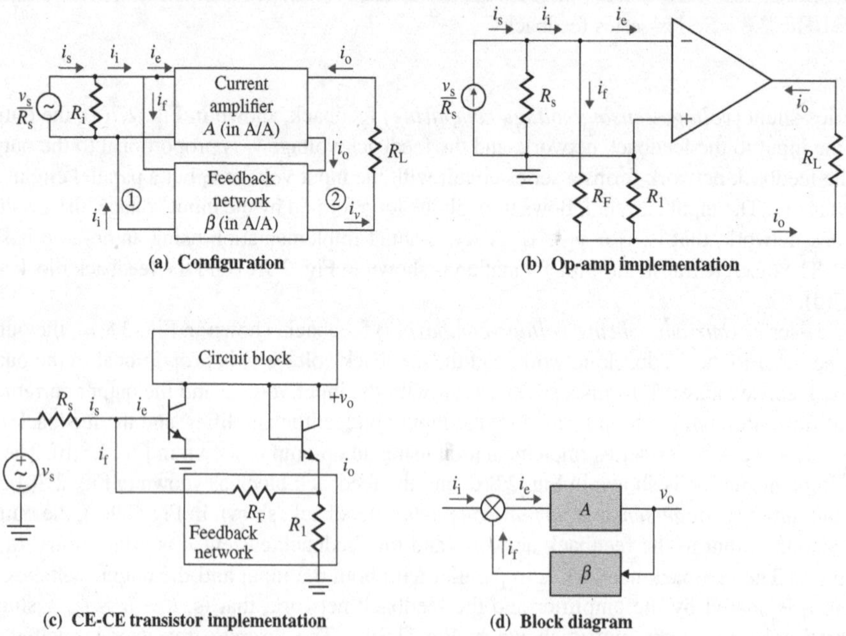

FIGURE 2.10 Shunt-series feedback

TABLE 2.1 Feedback relationships

	Gain	Input Resistance	Output Resistance
Without feedback	A	R_i	R_o
Series-shunt A (V/V) β (V/V)	$A_f = \dfrac{A}{1 + \beta A}$	$R_{if} = R_i(1 + \beta A)$	$R_{of} = \dfrac{R_o}{1 + \beta A}$
Series-series A (A/V or ℧) β (V/A or Ω)	$A_f = \dfrac{A}{1 + \beta A}$	$R_{if} = R_i(1 + \beta A)$	$R_{of} = R_o(1 + \beta A)$
Shunt-shunt A (V/A or Ω) β (A/V or ℧)	$A_f = \dfrac{A}{1 + \beta A}$	$R_{if} = \dfrac{R_i}{1 + \beta A}$	$R_{of} = \dfrac{R_o}{1 + \beta A}$
Shunt-series A (A/A) β (A/A)	$A_f = \dfrac{A}{1 + \beta A}$	$R_{if} = \dfrac{R_i}{1 + \beta A}$	$R_{of} = R_o(1 + \beta A)$

current i_o. The feedback network is in parallel with the input voltage but in series with the output current. The input current is shared by the amplifier and the feedback network; that is, $i_i - i_f = i_e$. A shunt-series implementation using an op-amp is shown in Fig. 2.10(b). The discrete transistor implementation is shown in Fig. 2.10(c) and the feedback block in Fig. 2.10(d).

2.4.2 Feedback Relationships

There are two circuits in a feedback amplifier: the amplifier circuit (or A circuit) and the feedback circuit (or β circuit). The effective gain is always decreased by a factor of $(1 + \beta A)$. In series-type arrangements, both A and β circuits are connected in series, and the effective resistance is increased by a factor of $(1 + \beta A)$. In shunt-type arrangements, A and β circuits are connected in parallel, and the effective resistance is decreased by a factor of $(1 + \beta A)$. The effects of different types of feedback are summarized in Table 2.1. Depending on the type of feedback, an amplifier is normally represented by one of four amplifier topologies: voltage, current, transconductance, or transresistance. A in Eq. (2.5) simply represents gain, which could be a voltage gain, a current gain, transconductance, or transresistance of the amplifier under the open-loop condition. Thus, A could be in units of V/V, A/A, A/V, or V/A.

KEY POINTS OF SECTION 2.4

- A feedback amplifier can be connected in one of four possible configurations: series-shunt, series-series, shunt-shunt, or shunt-series.
- With series feedback, the effective resistance is increased by a factor of $(1 + \beta A)$, whereas with shunt feedback the effective resistance is reduced by a factor of $(1 + \beta A)$.

2.5 Analysis of Feedback Amplifiers

In op-amp circuits, the open-loop gain A is independent of the feedback factor β; that is, the feedback network does not influence A. The first step in analyzing a feedback amplifier is to identify the main amplifier and its feedback network. However, in BJT and MOSFET amplifiers, the feedback is realized internally, and the feedback network cannot be separated from the main amplifier without affecting the open-loop gain A. Thus, the feedback network does influence A. The analysis of a feedback amplifier can be simplified by following these steps:

Step 1. Identify the feedback network.

Step 2. Identify the type of feedback on the input and output sides.

Step 3. Take into account the effects of the feedback network on the open-loop gain A by modifying the amplifier as follows:

a. Short-circuit the shunt feedback side to the ground so that there is no voltage signal to the feedback network. For example, terminal y of R_F in Fig. 2.7(b) would be connected to the ground so that R_1 became parallel to R_F.

b. Sever the series feedback side so that there is no current signal to the feedback network. For example, the inverting terminal x of the op-amp in Fig. 2.8(b) would be disconnected so that there was no current flowing into the feedback circuit and R_1 became in series with R_F.

Step 4. Represent the modified amplifier (from step 3) using one of the following equivalent amplifier topologies:

a. Voltage amplifier for series-shunt feedback
b. Transconductance amplifier for series-series feedback
c. Transresistance amplifier for shunt-shunt feedback
d. Current amplifier for shunt-series feedback

Calculate the values of the input resistance R_i, the output resistance R_o, and the open-loop gain A (representing transconductance, voltage, transresistance, or current) of the amplifier.

Step 5. The output of the amplifier is the input to the feedback network. Find the feedback factor β from one of the following two-port representations of the feedback network:

a. Voltage gain (V/V) representation for series-shunt feedback
b. Transresistance (V/A) representation for series-series feedback
c. Transconductance (A/V) representation for shunt-shunt feedback
d. Current gain (A/A) representation for shunt-series feedback

Step 6. Calculate the input resistance with feedback from one of the following equations:

$$R_{if} = R_i(1 + \beta A) \quad \text{for series-series and series-shunt feedback} \tag{2.26}$$

$$R_{if} = \frac{R_i}{1 + \beta A} \quad \text{for shunt-series and shunt-shunt feedback} \tag{2.27}$$

Step 7. Calculate the output resistance with feedback from one of the following equations:

$$R_{of} = R_o(1 + \beta A) \quad \text{for shunt-series and series-series feedback} \tag{2.28}$$

$$R_{of} = \frac{R_o}{1 + \beta A} \quad \text{for series-shunt and shunt-shunt feedback} \tag{2.29}$$

TABLE 2.2 V-I representations of amplifiers and feedback networks

Feedback Type	Amplifier	Units of A	Feedback Network	Units of β
Series-shunt	V-V	V/V	V-V	V/V
Shunt-series	I-I	A/A	I-I	A/A
Series-series	V-I	℧	I-V	Ω
Shunt-shunt	I-V	Ω	V-I	℧

Step 8. Use the following feedback equation to find the closed-loop gain A_f:

$$A_f = \frac{A}{1 + \beta A} \tag{2.30}$$

where β is the feedback factor representing the voltage gain, current gain, transconductance, or transresistance of the feedback network.

The V-I representations of the amplifier and its feedback network for different types of feedback are shown in Table 2.2. It is important to note that the units of A and β are different in each type of representation. To use the generalized equations for R_{if}, R_{of}, and A_f, the specific representation is essential.

▶ **NOTE** Since the units of A can be different depending on the type of feedback configuration, we will use the symbol μ_g for the open-loop voltage gain of the amplifier.

2.6 Series-Shunt Feedback

Series-shunt feedback is normally applied to a voltage amplifier. The amplifier in Fig. 2.7(a) is replaced by a voltage amplifier with an input resistance of R_i, an output resistance of R_o, and an open-loop voltage gain of A (in V/V). This arrangement is shown in Fig. 2.11. The feedback voltage v_f is in series with the input voltage v_i and is proportional to the output voltage v_o. The feedback network can be considered as a two-port network, and it can be modeled as a voltage gain circuit, as in Fig. 2.11, with an input resistance of R_y, an output resistance of R_x, and an open-loop voltage gain of β.

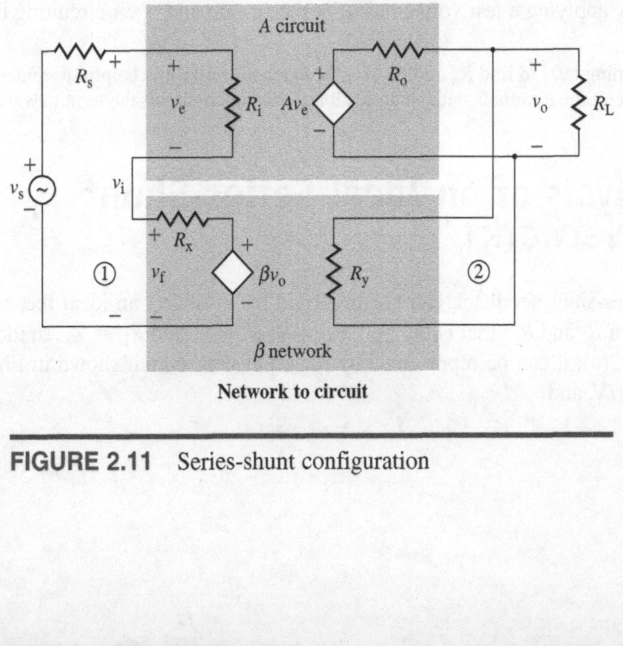

FIGURE 2.11 Series-shunt configuration

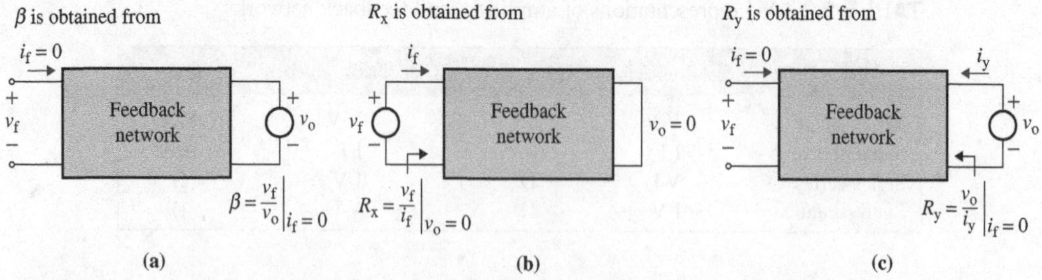

FIGURE 2.12 Test conditions for determining the parameters of a series-shunt feedback network

Determination of the model parameters requires separating the feedback network and representing it as a two-port network of four terminals. The test conditions for determining the parameters of the feedback network are shown in Fig. 2.12. The model parameters are obtained from three equations. The first equation is

$$R_x = \frac{v_f}{i_f}\Big|_{v_o=0} \quad \text{(short-circuited output side)} \tag{2.31}$$

which is obtained by applying a test voltage of v_f at the v_f side and shorting the v_o side. Note that the voltage across a short circuit is zero, and no current flows through an open circuit. The second equation is

$$R_y = \frac{v_o}{i_y}\Big|_{i_f=0} \quad \text{(open-circuited input side)} \tag{2.32}$$

which is obtained by applying a test voltage of v_o at the v_o side and open-circuiting the v_f side. The third equation is

$$\beta = \frac{v_f}{v_o}\Big|_{i_f=0} \quad \text{(open-circuited input side)} \tag{2.33}$$

which is obtained by applying a test voltage of v_o at the v_o side and open-circuiting the v_f side.

▶ **NOTE** In performing tests to find R_x and R_y of a feedback network, it is helpful to remember the following general rule: Short-circuit the terminals with shunt feedback, and open-circuit the terminals with series feedback.

2.6.1 Analysis of an Ideal Series-Shunt Feedback Network

The analysis of series-shunt feedback can be simplified by assuming an ideal feedback network and neglecting the effects of R_s and R_L; that is, $R_x = 0$, $R_y = \infty$, $R_s = 0$, and $R_L = \infty$. An ideal feedback network is shown in Fig. 2.13(a); it can be represented by the equivalent circuit shown in Fig. 2.13(b). The feedback factor β is in V/V, and

$$v_o = Av_e \tag{2.34}$$

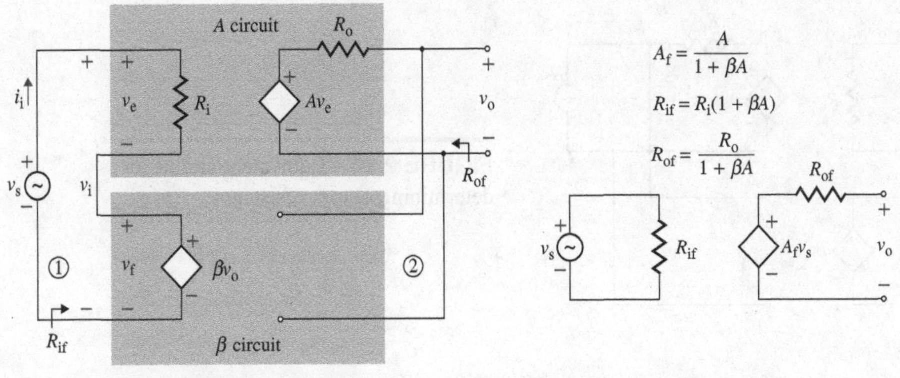

(a) Ideal feedback circuit (b) Equivalent circuit

FIGURE 2.13 Series-shunt configuration with an ideal feedback network

The feedback signal v_f, which is proportional to the output voltage v_o, is

$$v_f = \beta v_o \tag{2.35}$$

and $$v_e = v_s - v_f \tag{2.36}$$

From Eqs. (2.34), (2.35), and (2.36), the equation for the closed-loop voltage gain A_f becomes

$$A_f = \frac{A}{1 + \beta A} \tag{2.37}$$

which is similar to Eq. (2.5). From Eq. (2.36),

$$v_s = v_e + v_f$$

Substituting v_f from Eq. (2.35) and v_o from Eq. (2.34) into the above equation gives

$$v_s = v_e + \beta v_o = v_e + \beta A v_e = v_e(1 + \beta A) \tag{2.38}$$

The input current i_i is

$$i_i = \frac{v_e}{R_i} \tag{2.39}$$

Substituting v_s from Eq. (2.38) gives the input resistance R_{if} with feedback:

$$R_{if} = \frac{v_s}{i_i} = \frac{v_e(1 + \beta A)}{v_e/R_i} = (1 + \beta A)R_i \tag{2.40}$$

The input resistance R_{if} is always increased by a factor of $(1 + \beta A)$ with series feedback at the input side.

The output resistance with feedback, which is Thevenin's equivalent resistance, can be obtained by applying a test voltage v_x to the output side and shorting the input source. The equivalent circuit for determining Thevenin's equivalent resistance is shown in Fig. 2.14. We have

$$v_e + v_f = v_e + \beta v_x = 0 \quad \text{or} \quad v_e = -\beta v_x \tag{2.41}$$

and $$i_x = \frac{v_x - Av_e}{R_o} \tag{2.42}$$

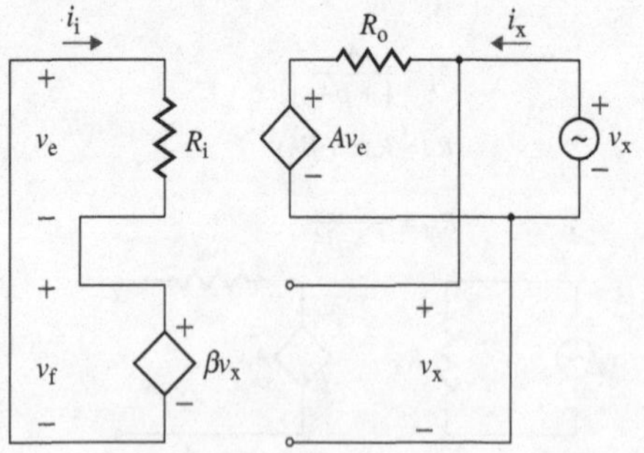

FIGURE 2.14 Equivalent circuit for determining output resistance

Substituting v_e from Eq. (2.41) into Eq. (2.42) yields

$$i_x = \frac{v_x - A(-\beta v_x)}{R_o} = \frac{(1 + \beta A)v_x}{R_o} \tag{2.43}$$

which gives the output resistance R_{of} with feedback as

$$R_{of} = \frac{R_o}{1 + \beta A} \tag{2.44}$$

Thus, the output resistance R_{of} at the output side is reduced by a factor of $(1 + \beta A)$. Shunt feedback at the output side always lowers the output resistance by a factor of $(1 + \beta A)$.

Series-shunt feedback increases the input resistance by $(1 + \beta A)$ and reduces the output resistance by $(1 + \beta A)$. This type of feedback is normally applied to a voltage amplifier. The input impedance, output impedance, and overall voltage gain can be written in generalized form in Laplace's domain of s as follows:

$$Z_{if}(s) = [1 + \beta A(s)]Z_i(s) \tag{2.45}$$

$$Z_{of}(s) = \frac{Z_o(s)}{1 + \beta A(s)} \tag{2.46}$$

$$A_f(s) = \frac{A(s)}{1 + \beta A(s)} \tag{2.47}$$

▶ **NOTES**

1. A is the open-loop voltage gain of the amplifier, in V/V.
2. β is the voltage gain of the feedback network and is less than or equal to 1 V/V.
3. $T_L = \beta A$ is the loop gain, which is dimensionless.
4. If the source has an impedance, then the overall voltage gain is reduced. The effective input voltage v_i to the amplifier can be found from

$$v_i = \frac{Z_i}{Z_s + Z_i} v_s \tag{2.48}$$

where Z_s is source impedance, Z_i is input impedance of the amplifier, and v_s is signal source voltage.

2.6.2 Analysis of a Practical Series-Shunt Feedback Network

The feedback network in Fig. 2.11 has a finite input resistance R_y and an output resistance R_x, which load the original amplifier, thereby affecting the performance of the feedback amplifier. The analysis in the previous section did not take into account the loading effect of the feedback network. The loading effect can be taken into account by including R_s, R_x, R_y, and R_L in the A circuit, as shown in Fig. 2.15(a). The open-loop parameters are modified accordingly, as shown in Fig. 2.15(b). These modifications allow us to apply the equations for ideal feedback.

The equivalent input resistance R_{ie} is given by

$$R_{ie} = R_i + R_x + R_s \tag{2.49}$$

where R_s is the source resistance. The equivalent output resistance R_{oe} is given by

$$R_{oe} = R_o \parallel R_y \parallel R_L \tag{2.50}$$

Using the voltage divider rule in Fig. 2.15(a), we see that the output voltage v_o is given by

$$v_o = \frac{R_y \parallel R_L}{(R_y \parallel R_L) + R_o} A v_e \tag{2.51}$$

where v_e is the voltage across R_i but not across R_{ie}. We need to find the voltage across R_i. Using the voltage divider rule gives v_e as

$$v_e = \frac{R_i}{R_i + R_x + R_s} v_{e1}$$

Substituting v_e into Eq. (2.51) gives the modified open-loop gain A_e:

$$A_e = \frac{v_o}{v_{e1}} = \frac{R_y \parallel R_L}{(R_y \parallel R_L) + R_o} \times \frac{R_i}{R_i + R_x + R_s} A \tag{2.52}$$

If R_{ie}, R_{oe}, and A_e are substituted for R_i, R_o, and A, respectively, Eqs. (2.34) through (2.48) can be applied to calculate the closed-loop parameters R_{if}, R_{of}, and A_f.

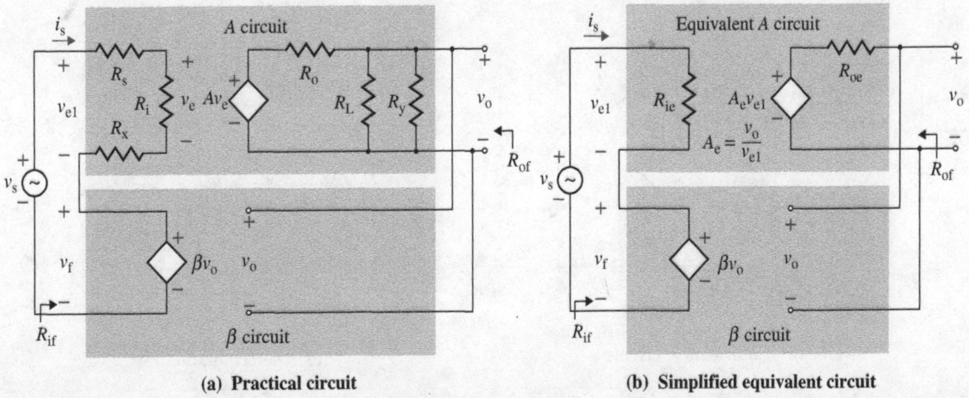

FIGURE 2.15 Practical series-shunt feedback

EXAMPLE 2.4

Finding the performance of a noninverting amplifier with series-shunt feedback The noninverting amplifier shown in Fig. 2.7(b) has $R_L = 10\ k\Omega$ and $R_s = 5\ k\Omega$. The feedback resistors are $R_1 = 10\ k\Omega$ and $R_F = 90\ k\Omega$. The op-amp parameters are $R_i = 2\ M\Omega$ and $R_o = 75\ \Omega$, and the open-loop voltage gain is $\mu_g = 2 \times 10^5$.

(a) Determine the input resistance seen by the source $R_{if} = v_s/i_s$, the output resistance R_{of}, and the closed-loop voltage gain $A_f = v_o/v_s$.

(b) Use PSpice/SPICE to verify your results.

SOLUTION

$R_L = 10\ k\Omega$, $R_s = 5\ k\Omega$, $R_1 = 10\ k\Omega$, $R_F = 90\ k\Omega$, $R_i = 2\ M\Omega$, $R_o = 75\ \Omega$, and $\mu_g = 2 \times 10^5$. Replacing the op-amp by its equivalent circuit gives the amplifier shown in Fig. 2.16(a).

(a) The steps in analyzing the feedback network are as follows:

Step 1. R_1 and R_F, which constitute the feedback network as shown in Fig. 2.16(a), produce a feedback voltage v_f proportional to the output voltage v_o.

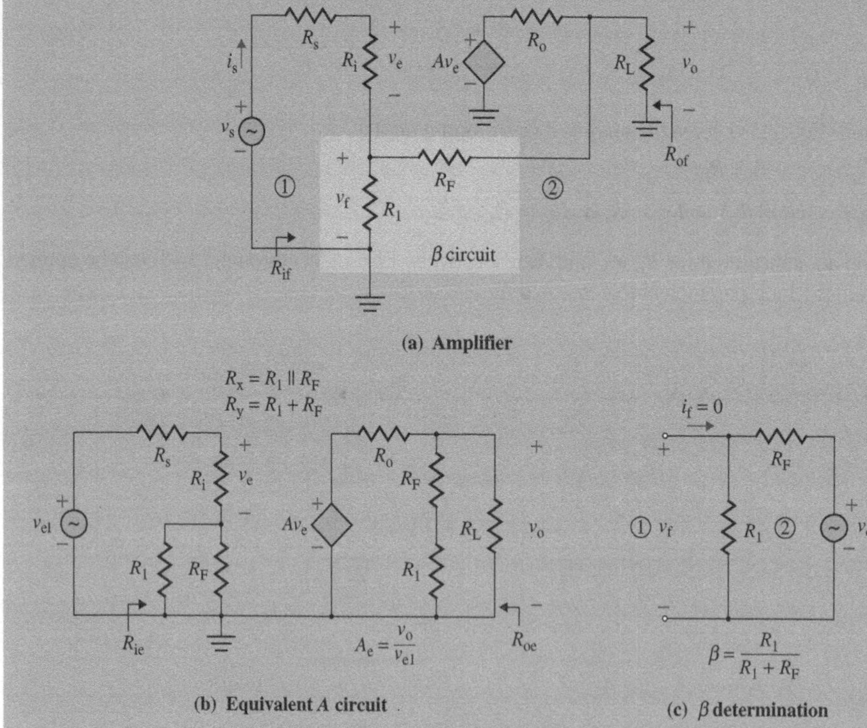

FIGURE 2.16 Noninverting amplifier with series-shunt feedback

Step 2. The amplifier uses series-shunt feedback. Thus, A must be in V/V; $A = \mu_g = 2 \times 10^5$.

Step 3. The effect of the feedback network at the input side is taken into account by short-circuiting the shunt feedback at the output side. Similarly, the effect at the output side is taken into account by severing the series feedback at the input side. We ground R_F on side 2 and sever the feedback circuit on side 1 from R_i. This modification is shown in Fig. 2.16(b). Then

$$R_x = R_1 \| R_F = 10\text{ k} \| 90\text{ k} = 9\text{ k}\Omega$$

$$R_y = R_1 + R_F = 10\text{ k} + 90\text{ k} = 100\text{ k}\Omega$$

Step 4. If we represent the amplifier of Fig. 2.16(b) by an equivalent voltage amplifier, the input resistance is

$$R_{ie} = R_s + R_i + (R_1 \| R_F) = R_s + R_i + R_x = 5\text{ k} + 2\text{ M} + 9\text{ k} = 2014\text{ k}\Omega$$

and the output resistance is

$$R_{oe} = R_o \| (R_1 + R_F) \| R_L = 75 \| (10\text{ k} + 90\text{ k}) \| 10\text{ k} = 74.4\ \Omega$$

From Eq. (2.52), the modified open-loop gain A_e is

$$A_e = \frac{(R_1 + R_F) \| R_L}{(R_1 + R_F) \| R_L + R_o} \times \frac{R_i}{R_s + R_i + (R_1 \| R_F)} A$$

$$= \frac{[(10\text{ k} + 90\text{ k}) \| 10\text{ k}] \times 2000\text{ k} \times 2 \times 10^5}{[(10\text{ k} + 90\text{ k}) \| 10\text{ k} + 75]\,(5\text{ k} + 2000\text{ k} + 9\text{ k})} = 1.9698 \times 10^5$$

Step 5. From Fig. 2.16(c), the feedback factor β is given by

$$\beta = \left.\frac{v_f}{v_o}\right|_{i_f=0} = \frac{R_1}{R_1 + R_F} = \frac{10\text{ k}}{10\text{ k} + 90\text{ k}} = 0.1\text{ V/V}$$

Step 6. The input resistance (seen by the source) with feedback is

$$R_{if} = \frac{v_s}{i_s} = R_{ie}(1 + \beta A_e) = 2014\text{ k} \times (1 + 0.1 \times 1.9698 \times 10^5) = 39.67\text{ G}\Omega$$

Step 7. The output resistance with feedback is

$$R_{of} = \frac{R_{oe}}{1 + \beta A_e} = \frac{74.4}{1 + 0.1 \times 1.9698 \times 10^5} = 3.778\text{ m}\Omega$$

Step 8. The closed-loop voltage gain A_f is

$$A_f = \frac{v_o}{v_s} = \frac{A_e}{1 + \beta A_e} = \frac{1.9698 \times 10^5}{1 + 0.1 \times 1.9698 \times 10^5} = 9.999\text{ V/V}$$

which is very close to the closed-loop gain we would get:

$$A_f = 1 + \frac{R_F}{R_1} = 1 + \frac{90\text{ k}}{10\text{ k}} = 10$$

(b) The series-shunt feedback circuit for PSpice simulation is shown in Fig. 2.17. The results of PSpice simulation are shown below, with hand calculations to the right:

`V(5)/VS=9.999E+00=9.999`	$A_f = 9.99$
`INPUT RESISTANCE AT VS=3.967E+10=39.67 GΩ`	$R_{if} = 39.67\text{ G}\Omega$
`OUTPUT RESISTANCE AT V(5)=3.776E-02=37.7 mΩ`	$R_{of} = 37.8\text{ m}\Omega$

The PSpice results are very close to the hand-calculated values.

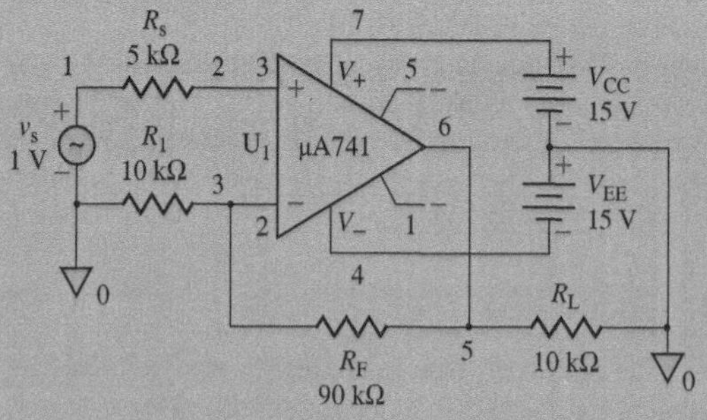

FIGURE 2.17 Series-shunt feedback network for PSpice simulation

NOTE: The main objective of PSpice/SPICE simulation in this chapter is to verify the techniques for analyzing feedback amplifiers. To verify hand-calculated results, we will use the simple DC op-amp model and the simple π model for BJTs. If we were to run simulations using the model provided in PSpice/SPICE, the results would differ slightly.

EXAMPLE 2.5

Finding the performance of a BJT amplifier with series-shunt feedback The AC equivalent circuit of a BJT amplifier is shown in Fig. 2.18(a). The transistor can be modeled as shown in Fig. 2.18(b). The DC bias currents of the transistors are $I_{C1} = 0.5$ mA, $I_{C2} = 1$ mA, and $I_{C3} = 5$ mA. The transistor parameters are $h_{fe} = h_{fe1} = h_{fe2} = h_{fe3} = 100$ and $r_\mu = r_o = \infty$.

(a) Use the techniques of feedback analysis to calculate the input resistance R_{if}, the output resistance R_{of}, and the closed-loop voltage gain A_f.

(b) Use PSpice/SPICE to check your results.

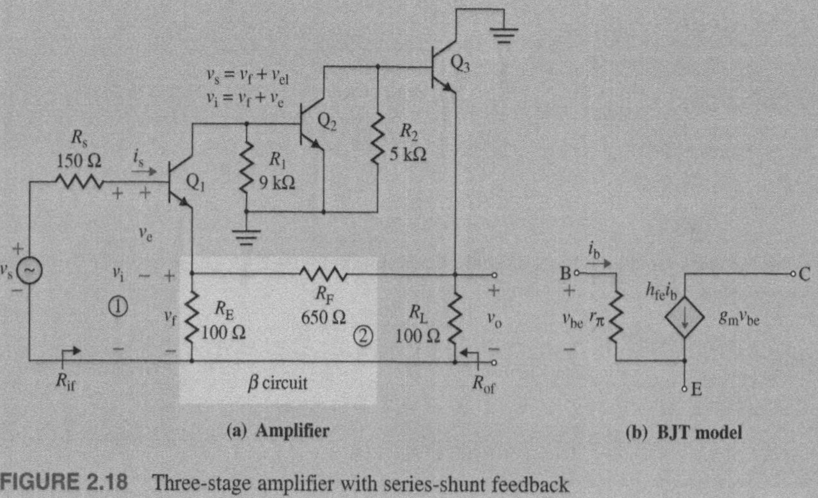

FIGURE 2.18 Three-stage amplifier with series-shunt feedback

NOTE: The following analysis can also be applied to MOSFET amplifiers by substituting a very large value of the B-E resistance r_π tending to infinity, $1/r_\pi \simeq 0$.

SOLUTION

For $I_{C1} = 0.5$ mA,

$$r_{\pi 1} = \frac{h_{fe} \times 25.8 \text{ mV}}{I_{C1}} = \frac{100 \times 25.8 \text{ mV}}{0.5 \text{ mA}} = 5.16 \text{ k}\Omega$$

$$g_{m1} = \frac{I_{C1}}{V_T} = \frac{0.5 \text{ mA}}{25.8 \text{ mV}} = 19.38 \text{ mA/V}$$

For $I_{C2} = 1$ mA,

$$r_{\pi 2} = \frac{h_{fe} \times 25.8 \text{ mV}}{I_{C2}} = \frac{100 \times 25.8 \text{ mV}}{1 \text{ mA}} = 2.58 \text{ k}\Omega$$

$$g_{m2} = \frac{I_{C2}}{V_T} = \frac{1 \text{ mA}}{25.8 \text{ mV}} = 38.76 \text{ mA/V}$$

For $I_{C3} = 5$ mA,

$$r_{\pi 3} = \frac{h_{fe} \times 25.8 \text{ mV}}{I_{C3}} = \frac{100 \times 25.8 \text{ mV}}{5 \text{ mA}} = 516 \ \Omega$$

$$g_{m3} = \frac{I_{C3}}{V_T} = \frac{5 \text{ mA}}{25.8 \text{ mV}} = 193.8 \text{ mA/V}$$

(a) The steps in analyzing the feedback network are as follows:

Step 1. R_E and R_F, which constitute the feedback network, produce a feedback voltage proportional to the output voltage. The input voltage v_i is compared with the feedback signal v_f. The error voltage $v_e = v_i - v_f$ is the B-E voltage of transistor Q_1. The block diagram representing the feedback mechanism is shown in Fig. 2.19(a). The feedback network is shown in Fig. 2.19(b).

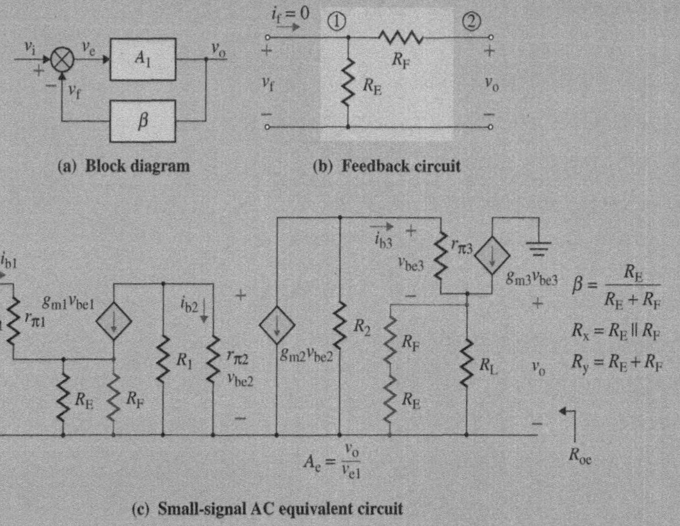

FIGURE 2.19 Equivalent circuits for Example 2.5

Step 2. The amplifier uses series-shunt feedback. Thus, the voltage gain A must be expressed in V/V.

Step 3. The effect of the feedback network is taken into account at the input side by shorting R_F at side 2 to the ground, as shown in Fig. 2.19(b), and at the output side by removing the feedback network from the emitter of Q_1. These modifications are shown in the small-signal AC equivalent circuit in Fig. 2.19(c).

Step 4. If we represent the amplifier of Fig. 2.19(c) by an equivalent voltage amplifier, the input resistance at the base of Q_1 is

$$R_i = \frac{v_i}{i_{b1}} = r_{\pi 1} + (1 + g_{m1}r_{\pi 1})(R_E \| R_F) = 5.16\text{ k} + 101 \times (100 \| 650) = 13.91\text{ k}\Omega$$

The input resistance is

$$R_{ie} = \frac{v_{el}}{i_{b1}} = R_s + R_i = R_s + r_{\pi 1} + (1 + g_{m1}r_{\pi 1})(R_E \| R_F) = 150 + 13.91\text{ k} = 14.06\text{ k}\Omega$$

The output resistance is

$$R_{oe} = R_L \| (R_F + R_E) \| \left(\frac{R_2 + r_{\pi 3}}{1 + g_{m1}r_{\pi 1}}\right) = 100 \| 750 \| \left(\frac{5\text{ k} + 516}{101}\right) = 33.74\ \Omega$$

Thus,

$$V_{be1} = \left(\frac{r_{\pi 1}}{R_i}\right)v_i = \left(\frac{5.16\text{ k}}{13.91\text{ k}}\right)v_i = 0.371v_i$$

$$v_{b2} = v_{be2} = -(R_1 \| r_{\pi 2})g_{m1}v_{be1} = -(9\text{ k} \| 2.58\text{ k}) \times 19.38\text{ m} \times 0.371v_i = -14.412v_i$$

$$v_{b3} = -\{R_2 \| [r_{\pi 3} + (1 + g_{m3}r_{\pi 3})(R_L \| (R_F + R_E))]\}g_{m2}v_{be2}$$

$$= -\{5\text{ k} \| [516 + (1 + 193.8\text{ m} \times 516)(100 \| 750)]\} \times 38.76\text{ m} \times (-14.412v_i) = 1.825 \times 10^3 v_i$$

$$v_o = \frac{v_{b3} \times (1 + g_{m3}r_{\pi 3})[R_L \| (R_F + R_E)]}{\{r_{\pi 3} + (1 + g_{m3}r_{\pi 3})[R_L \| (R_F + R_E)]\}}$$

$$= \frac{(1.825 \times 10^3 \times v_i) \times (1 + 193.8\text{ m} \times 516)(100 \| 750)}{[516 + (1 + 193.8\text{ m} \times 516)(100 \| 750)]} = 1725.7v_i$$

Therefore, the open-loop voltage gain A is given by $A = v_o/v_i = 1725.7$ V/V

and $$A_e = \frac{v_o}{v_{el}} = \frac{AR_i}{R_s + R_i} = \frac{1725.7 \times 13.91\text{ k}}{150 + 13.91\text{ k}} = 1707.3\text{ V/V}$$

Step 5. From Fig. 2.19(b) and Eq. (2.33), the feedback factor β is given by

$$\beta = \left.\frac{v_f}{v_o}\right|_{i_f=0} = \frac{R_E}{R_E + R_F} = \frac{100}{100 + 650} = 0.1333\text{ V/V}$$

Step 6. The input resistance (seen by the source) with feedback is

$$R_{if} = \frac{v_s}{i_s} = R_{ie}(1 + \beta A_e) = 14.06\text{ k}\Omega \times (1 + 0.1333 \times 1707.3) = 3.21\text{ M}\Omega$$

Step 7. The output resistance with feedback is

$$R_{of} = \frac{R_{oe}}{1 + \beta A_e} = \frac{33.74}{1 + 0.1333 \times 1707.3} = 0.147\ \Omega$$

Step 8. The closed-loop voltage gain A_f is

$$A_f = \frac{v_o}{v_s} = \frac{A_e}{1 + \beta A_e} = \frac{1707.3}{1 + 0.1333 \times 1707.3} = 7.47\text{ V/V}$$

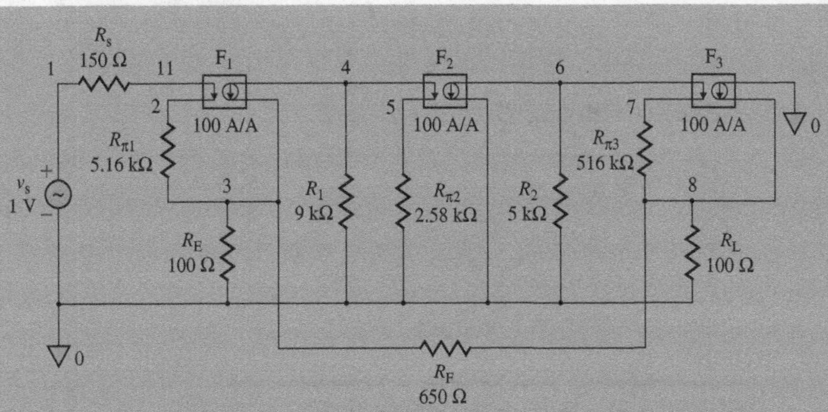

FIGURE 2.20 Series-shunt feedback network for PSpice simulation

(b) The series-shunt feedback network for PSpice simulation is shown in Fig. 2.20. The results of PSpice simulation (.TF analysis) are shown below, with hand calculations to the right:

`V(8)/VS=7.468E+00=7.468 V/V`	$A_f = 7.47$ V/V
`INPUT RESISTANCE AT VS=3.214E+06=3.214 MΩ`	$R_{if} = 3.21$ MΩ
`OUTPUT RESISTANCE AT V(8)=1.460E-01=0.146 Ω`	$R_{of} = 0.147$ Ω

Note the close agreement between the results obtained by PSpice and the hand calculations.

KEY POINTS OF SECTION 2.6

- Series-shunt feedback is applied to voltage amplifiers. This type of feedback increases the input resistance and reduces the output resistance by a factor of $(1 + \beta A)$.
- The amplifier is represented by a voltage amplifier and the feedback network as a voltage gain. Both A and β are in units of V/V.
- The loading effect of a feedback network can be taken into account by remembering the following general rule: Short-circuit the terminals with shunt feedback, and open-circuit the terminals with series feedback.

2.7 Series-Series Feedback

Series-series feedback is normally applied to a transconductance amplifier. For the series-series feedback amplifier in Fig. 2.8(a), we obtain the equivalent circuit shown in Fig. 2.21 if we represent the op-amp by its transconductance model. A is the open-loop transconductance gain of the op-amp in A/V. That is, $Av_e = \mu_g v_e / R_o$ and $A = \mu_g / R_o$. The feedback voltage v_f is proportional to the load current i_o. The feedback network can also be modeled in transimpedance form with an input resistance of R_y, an output resistance of R_x, and a transimpedance of β in V/A.

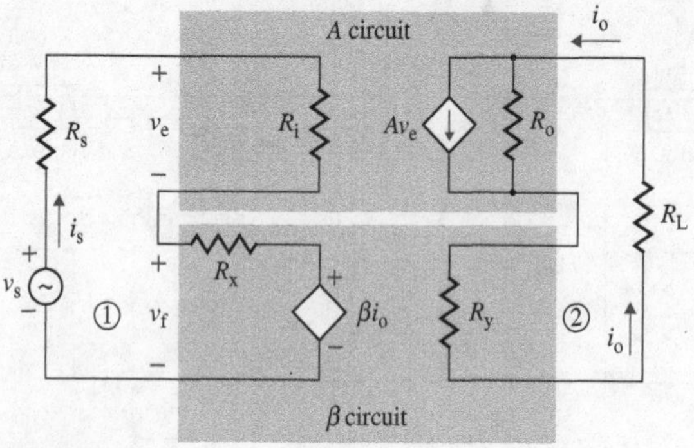

FIGURE 2.21 Series-series configuration

The test conditions for determining the parameters of the feedback network are shown in Fig. 2.22. The parameters of the model can be obtained from three equations. The first equation is

$$R_x = \left.\frac{v_f}{i_f}\right|_{i_o=0} \quad \text{(open-circuited output side)} \tag{2.53}$$

which is obtained by applying a test voltage of v_f at side 1 and open-circuiting side 2. The second equation is

$$R_y = \left.\frac{v_y}{i_o}\right|_{i_f=0} \quad \text{(open-circuited input side)} \tag{2.54}$$

which is obtained by applying a test voltage of v_y at side 2 and open-circuiting side 1. The third equation is

$$\beta = \left.\frac{v_f}{i_o}\right|_{i_f=0} \quad \text{(open-circuited input side)} \tag{2.55}$$

which is obtained by applying a test voltage of v_y at side 2 and open-circuiting side 1.

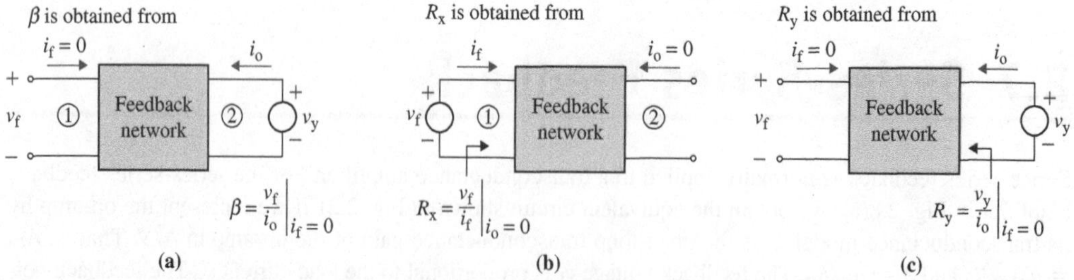

FIGURE 2.22 Test conditions for determining the parameters of a series-series feedback network

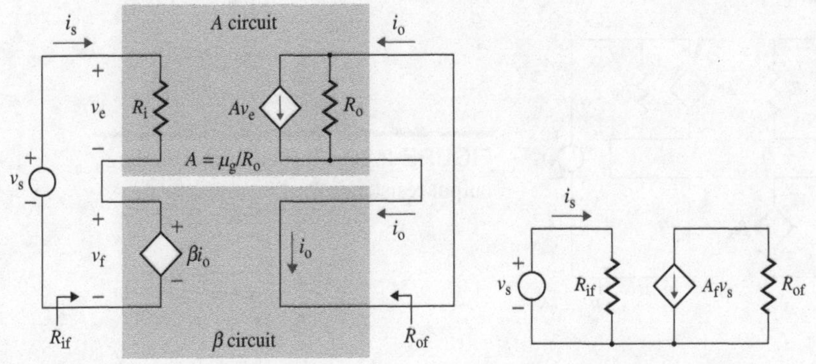

(a) Ideal feedback circuit (b) Equivalent circuit

FIGURE 2.23 Ideal series-series feedback network

2.7.1 Analysis of an Ideal Series-Series Feedback Network

Let us assume an ideal series-series feedback network—that is, $R_x = 0$, $R_y = 0$, $R_s = 0$, and $R_L = 0$. The feedback amplifier in Fig. 2.21 can be simplified to the one in Fig. 2.23(a), which can be represented by the equivalent circuit shown in Fig. 2.23(b). For this circuit,

$$i_o = Av_e$$

$$v_f = \beta i_o$$

$$v_s = v_e + v_f = \frac{i_o}{A} + \beta i_o = \frac{1 + \beta A}{A} i_o$$

which gives the closed-loop transconductance gain A_f as

$$A_f = \frac{i_o}{v_s} = \frac{A}{1 + \beta A} \tag{2.56}$$

Using KVL around the input side, we get

$$v_s = R_i i_s + v_f = R_i i_s + \beta i_o = R_i i_s + \beta A v_e = R_i i_s + \beta A R_i i_s$$

which gives the closed-loop input resistance R_{if} as

$$R_{if} = \frac{v_s}{i_s} = R_i(1 + \beta A) \tag{2.57}$$

To determine the output resistance with feedback, let us apply a test voltage v_x, as shown in Fig. 2.24:

$$v_x = R_o(i_o - Av_e) = R_o(i_o + \beta A i_o) = R_o(1 + \beta A)i_o$$

which gives the closed-loop output resistance R_{of} as

$$R_{of} = \frac{v_x}{i_x} = R_o(1 + \beta A) \tag{2.58}$$

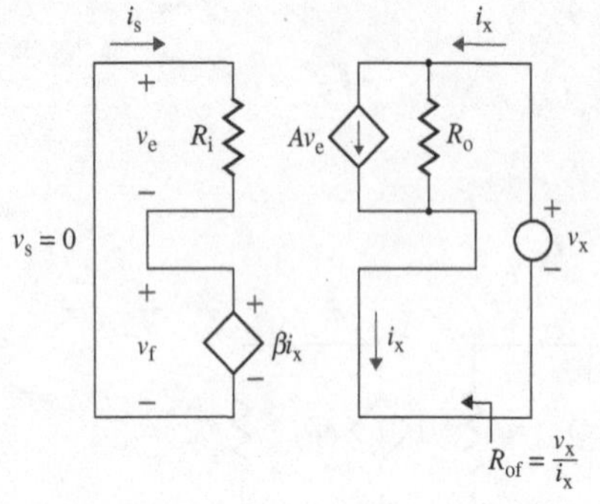

FIGURE 2.24 Test circuit for finding output resistance R_{of}

▶ **NOTES**

1. A is the open-loop transconductance of the amplifier, in A/V.
2. β is the transresistance of the feedback network, in V/A.
3. $T_L = \beta A$ is the loop gain, which is dimensionless.

2.7.2 Analysis of a Practical Series-Series Feedback Network

The analysis in the previous section did not take into account the loading effect of the feedback network. The open-loop parameters of the amplifier in Fig. 2.25(a) can be modified to include the loading effect due to R_s, R_x, R_y, and R_L, providing the equivalent ideal feedback network in Fig. 2.25(b). The modified parameters are given by

$$R_{ie} = R_s + R_i + R_x \tag{2.59}$$

$$R_{oe} = R_o + R_y + R_L \tag{2.60}$$

▶ **NOTE** It may appear from Fig. 2.25(b) that R_o would have to be in parallel with $(R_y + R_L)$ for us to find R_{oe} in Eq. (2.60), but if we convert the current source Av_e to a voltage source, we see that the effective resistance used to find the current i_o becomes $(R_o + R_y + R_L)$.

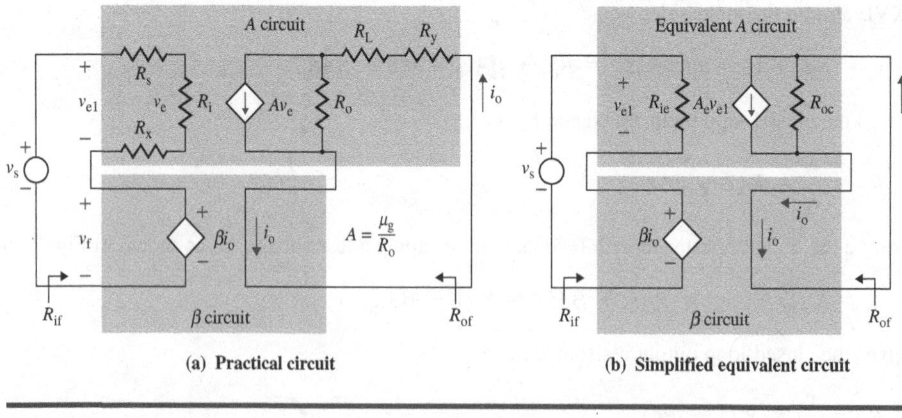

(a) Practical circuit **(b) Simplified equivalent circuit**

FIGURE 2.25 Practical series-series feedback amplifier

$$i_o = \frac{R_o}{R_o + R_L + R_y} A v_e \tag{2.61}$$

$$v_e = \frac{R_i}{R_i + R_x + R_s} v_{e1} \tag{2.62}$$

Substituting v_e from Eq. (2.62) into Eq. (2.61) gives the modified open-loop transconductance A_e:

$$A_e = \frac{i_o}{v_{e1}} = \frac{R_o R_i}{(R_o + R_L + R_y)(R_s + R_i + R_x)} A \tag{2.63}$$

If the values of R_i, R_o, and A are replaced by R_{ie}, R_{oe}, and A_e, respectively, then Eqs. (2.56) through (2.58) can be applied to calculate the closed-loop parameters R_{if}, R_{of}, and A_f.

EXAMPLE 2.6

Finding the performance of a noninverting amplifier with series-series feedback The noninverting amplifier shown in Fig. 2.8(b) has $R_L = 4\ \Omega$ and $R_s = 5\ \text{k}\Omega$. The feedback resistance is $R_F = 5\ \Omega$. The op-amp parameters are $R_i = 2\ \text{M}\Omega$ and $R_o = 75\ \Omega$, and the open-loop voltage gain is $\mu_g = 2 \times 10^5$.

(a) Determine the input resistance seen by the source $R_{if} = v_s/i_s$, the output resistance R_{of}, and the closed-loop transconductance gain $A_f = i_o/v_s$.

(b) Use PSpice/SPICE to verify your results.

SOLUTION

$R_L = 4\ \Omega$, $R_s = 5\ \text{k}\Omega$, $R_F = 5\ \Omega$, $R_i = 2\ \text{M}\Omega$, $R_o = 75\ \Omega$, and $\mu_g = 2 \times 10^5$. Replacing the op-amp in Fig. 2.8(b) by its equivalent circuit gives the amplifier shown in Fig. 2.26(a).

(a) The steps in analyzing the feedback network are as follows:

Step 1. R_F constitutes the feedback network, and it produces a feedback voltage v_f proportional to the output current i_o.

Step 2. The amplifier uses series-series feedback. Thus, A must be in A/V:

$$A = \frac{\mu_g}{R_o} = \frac{2 \times 10^5}{75} = 2.67\ \text{kA/V}$$

Step 3. The effect of the feedback network is taken into account by severing R_F from the op-amp at side 1 and from R_L at side 2. This modification is shown in Fig. 2.26(b). Then

$$R_x = R_F = 5\ \Omega$$

$$R_y = R_F = 5\ \Omega$$

Step 4. If we represent the amplifier of Fig. 2.26(b) by an equivalent voltage amplifier, the input resistance is

$$R_{ie} = R_s + R_i + R_x = R_s + R_i + R_F = 5\ \text{k} + 2\ \text{M} + 5 \approx 2005\ \text{k}\Omega$$

and the output resistance is

$$R_{oe} = R_o + R_F + R_L = 75 + 5 + 4 = 84\ \Omega$$

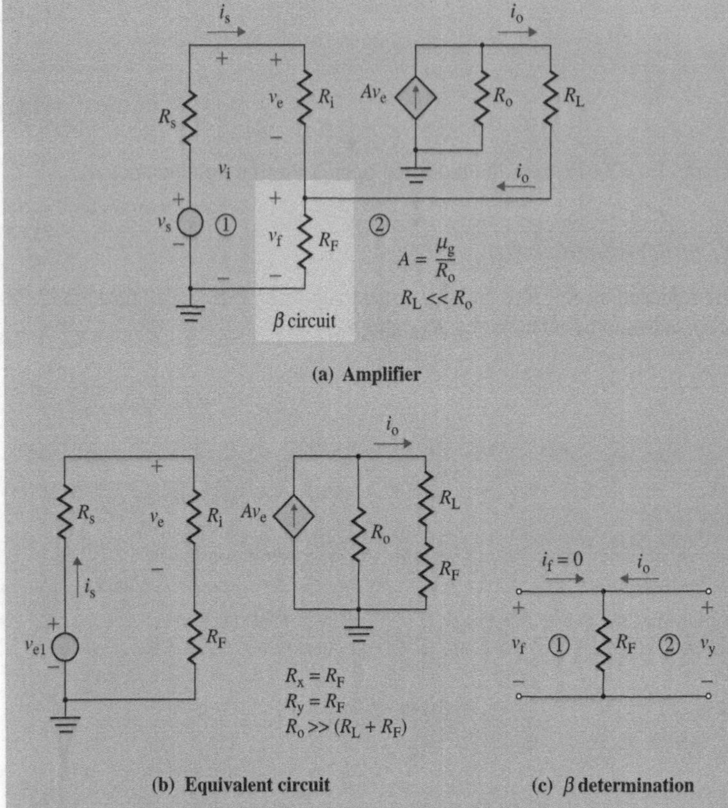

FIGURE 2.26 Noninverting amplifier with series-series feedback

We find the modified open-loop transconductance gain A_e from Eq. (2.63):

$$A_e = \frac{R_o R_i}{(R_o + R_L + R_F)(R_s + R_i + R_F)} A$$

$$= \frac{75 \times 2000\text{ k} \times 2.67\text{ kA/V}}{(75 + 4 + 5)(5\text{ k} + 2000\text{ k} + 5)} = 2.375 \times 10^3\text{ A/V}$$

Step 5. From Fig. 2.26(c), the feedback factor β is given by

$$\beta = \left.\frac{v_f}{i_y}\right|_{i_o=0} = R_F = 5\ \Omega$$

Step 6. The input resistance (seen by the source) with feedback is

$$R_{if} = \frac{v_s}{i_s} = R_{ie}(1 + \beta A_e) = 2005\text{ k} \times (1 + 5 \times 2.375 \times 10^3) = 23.81\text{ G}\Omega$$

Step 7. The output resistance with feedback is

$$R_{of} = R_{oe}(1 + \beta A_e) = 84 \times (1 + 5 \times 2.375 \times 10^3) = 997.6\text{ k}\Omega$$

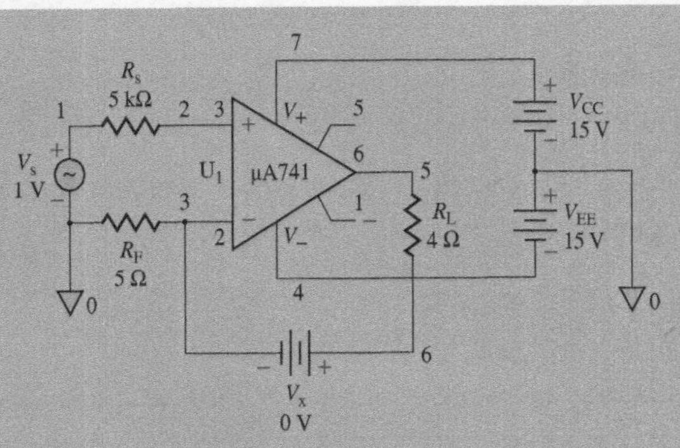

FIGURE 2.27 Series-series feedback network for PSpice simulation

Step 8. The closed-loop transconductance gain A_f is

$$A_f = \frac{i_o}{v_s} = \frac{A_e}{1 + \beta A_e} = \frac{2.375 \times 10^3}{1 + 5 \times 2.375 \times 10^3} = 200 \text{ mA/V}$$

(b) The series-series feedback circuit for PSpice simulation is shown in Fig. 2.27. The results of PSpice/SPICE simulation (.TF analysis) are shown below, with hand calculations to the right:

`I(VX)/VS=2.000E-01=0.2`	$A_f = 200$ mA/V
`INPUT RESISTANCE AT VS=2.381E+10=23.81 GΩ`	$R_{if} = 23.81$ GΩ
`OUTPUT RESISTANCE AT I(VX)=9.976E+05=997.6 kΩ`	$R_{of} = 997.6$ kΩ

The PSpice results are very close to the hand-calculated values.

EXAMPLE 2.7

Finding the performance of a BJT amplifier with series-series feedback The AC equivalent circuit of a feedback amplifier is shown in Fig. 2.28. The DC bias currents of the transistors are $I_{C1} = 0.5$ mA, $I_{C2} = 1$ mA, and $I_{C3} = 5$ mA. The transistor parameters are $h_{fe} = h_{fe1} = h_{fe2} = h_{fe3} = 100$ and $r_\mu = r_o = \infty$.

(a) Use the techniques of feedback analysis to calculate the input resistance R_{if}, the output resistance R_{of}, and the closed-loop transconductance gain $A_f = i_o/v_s$.

(b) Use PSpice/SPICE to check your results.

NOTE: The following analysis can also be applied to MOSFET amplifiers by substituting a very large value of the B-E resistance r_π tending to infinity, $1/r_\pi \simeq 0$.

SOLUTION

For $I_{C1} = 0.5$ mA,

$$r_{\pi 1} = \frac{h_{fe} \times 25.8 \text{ mV}}{I_{C1}} = \frac{100 \times 25.8 \text{ mV}}{0.5 \text{ mA}} = 5.16 \text{ k}\Omega$$

$$g_{m1} = \frac{I_{C1}}{V_T} = \frac{0.5 \text{ mA}}{25.8 \text{ mV}} = 19.38 \text{ mA/V}$$

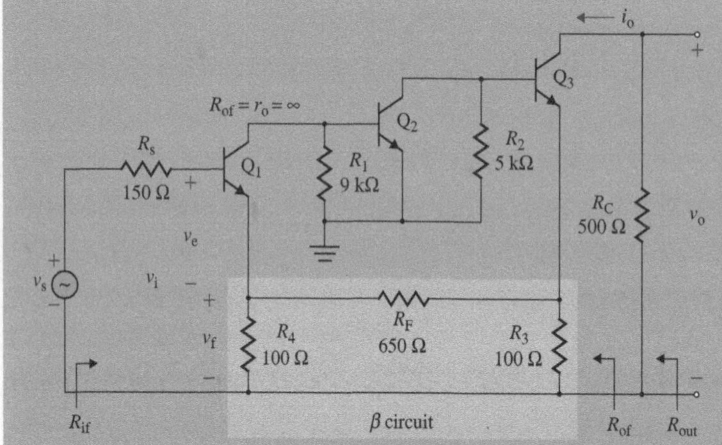

FIGURE 2.28 Three-stage amplifier with series-series feedback

For $I_{C2} = 1$ mA,

$$r_{\pi 2} = \frac{h_{fe} \times 25.8 \text{ mV}}{I_{C2}} = \frac{100 \times 25.8 \text{ mV}}{1 \text{ mA}} = 2.58 \text{ k}\Omega$$

$$g_{m2} = \frac{I_{C2}}{V_T} = \frac{1 \text{ mA}}{25.8 \text{ mV}} = 38.76 \text{ mA/V}$$

For $I_{C3} = 5$ mA,

$$r_{\pi 3} = \frac{h_{fe} \times 25.8 \text{ mV}}{I_{C3}} = \frac{100 \times 25.8 \text{ mV}}{5 \text{ mA}} = 516 \ \Omega$$

$$g_{m3} = \frac{I_{C3}}{V_T} = \frac{5 \text{ mA}}{25.8 \text{ mV}} = 193.8 \text{ mA/V}$$

(a) The steps in analyzing the feedback network are as follows:

Step 1. R_3, R_F, and R_4 constitute the feedback network. The input voltage v_i is compared with the feedback signal v_f. An error voltage $v_e = v_i - v_f$ is the input to the base of the transistor Q_3. The block diagram representing the feedback mechanism is shown in Fig. 2.29(a). The feedback network is shown in Fig. 2.29(b).

Step 2. The amplifier uses series-series feedback. Thus, the transconductance gain A must be expressed in A/V.

Step 3. The effect of the feedback network is taken into account at the input side by removing the network from the emitter of Q_1 at side 2, as shown in Fig. 2.29(b), and at the output side by removing the network from the emitter of Q_1 at side 1. These modifications are shown in the small-signal equivalent circuit in Fig. 2.29(c).

Step 4. If we represent the amplifier in Fig. 2.29(c) by an equivalent transconductance amplifier, the input resistance is

$$R_i = r_{\pi 1} + (1 + g_{m1} r_{\pi 1})[R_4 \| (R_F + R_3)] = 5.16 \text{ k} + 101 \times (100 \| 750) = 14.07 \text{ k}\Omega$$
$$R_{ie} = R_i + R_s = 14.07 \text{ k} + 150 = 14.22 \text{ k}\Omega$$

The output resistance seen at the collector of Q_3 is

$$R_o = r_o = \infty$$

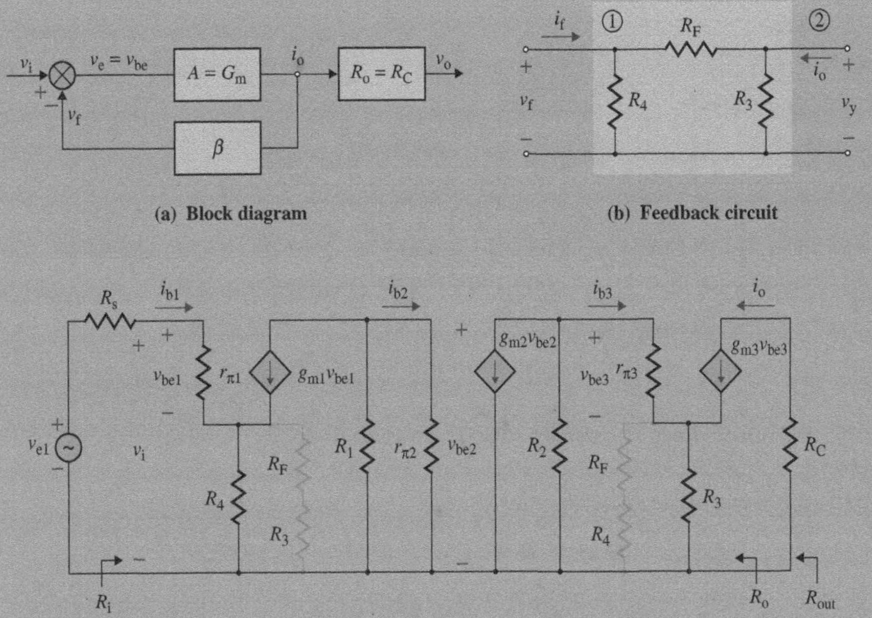

FIGURE 2.29 Equivalent circuits for Example 2.7

From Fig. 2.29(c), we get

$$V_{be1} = \left(\frac{r_{\pi1}}{R_i}\right)v_i = \left(\frac{5.16\text{ k}}{14.07\text{ k}}\right)v_i = 0.367v_i$$

$$v_{b2} = v_{be2} = -(R_1 \| r_{\pi2})g_{m1}v_{be1} = -(9\text{ k}\|2.58\text{ k}) \times 19.38\text{ m} \times 0.367v_i = -14.25v_i$$

$$v_{b3} = -\{R_2\|[r_{\pi3} + (1 + g_{m3}r_{\pi3})(R_3\|(R_F + R_4))]\}g_{m2}v_{be2}$$

$$= -\{5\text{ k}\|[516 + (1 + 193.8\text{ m} \times 516)(100\|750)]\} \times 38.76\text{ m} \times (-14.25v_i)$$

$$= 1.805 \times 10^3 v_i$$

$$v_o = \frac{v_{b3} \times r_{\pi3}}{\{r_{\pi3} + (1 + g_{m3}r_{\pi3})[R_3\|(R_F + R_4)]\}}$$

$$= \frac{(1.825 \times 10^3 \times v_i) \times 516}{[516 + (1 + 193.8\text{ m} \times 516)(100\|750)]} = 98.766v_i$$

$$i_o = g_{m3}v_{be3} = 193.8\text{ m} \times 98.766v_i = 19.14v_i$$

Therefore, the open-loop transconductance A is given by $A = i_o/v_i = 19.14$ A/V.

Step 5. From Eq. (2.55), the feedback factor β is given by

$$\beta = \left.\frac{v_f}{i_o}\right|_{i_f=0} = \frac{R_3R_4}{R_3 + R_4 + R_F} = \frac{100 \times 100}{100 + 100 + 650} = 11.76\text{ V/A}$$

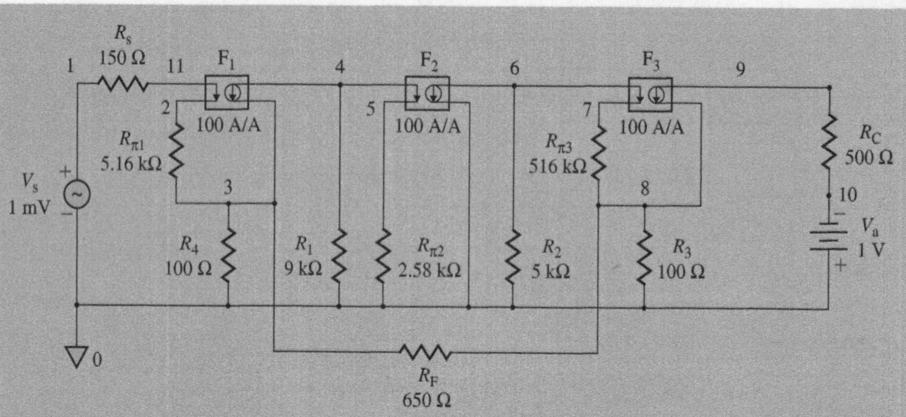

FIGURE 2.30 Series-series feedback network for PSpice simulation

Step 6. The input resistance with feedback is

$$R_{if} = \frac{v_s}{i_s} = R_{ie}(1 + \beta A_e) = 14.22 \text{ k}\Omega \times (1 + 11.76 \times 18.94) = 3.18 \text{ M}\Omega$$

Step 7. The output resistance with feedback is

$$R_{of} = R_o(1 + \beta A) = \infty$$

Since the output is taken across R_C, the output resistance of the amplifier is

$$R_{out} = R_{of} \| R_C = R_C = 500 \ \Omega$$

Step 8. The closed-loop transconductance gain A_f is

$$A_f = \frac{i_o}{v_s} = \frac{18.94}{1 + 11.76 \times 18.94} = 84.66 \text{ mA/V}$$

(b) The series-series feedback amplifier for PSpice simulation is shown in Fig. 2.30. The results of PSpice simulation (.TF analysis) are shown below, with hand calculations to the right.

`I(VU)/VS=8.379E-02=83.79 mA/V`	$A_f = 84.66$ mA/V
`INPUT RESISTANCE AT VS=3.215E+06=3.215 MΩ`	$R_{if} \approx 3.18$ MΩ
`OUTPUT RESISTANCE AT I(VU)=8.639E+13 Ω`	$R_{of} = \infty$

The PSpice results are very close to the hand-calculated values.

KEY POINTS OF SECTION 2.7

- Series-series feedback is applied to transconductance amplifiers. This type of feedback increases both the input resistance and the output resistance by a factor of $(1 + \beta A)$.
- The amplifier is represented by a transconductance amplifier and the feedback network as a transresistance gain. A is in units of A/V, and β is in units of V/A.

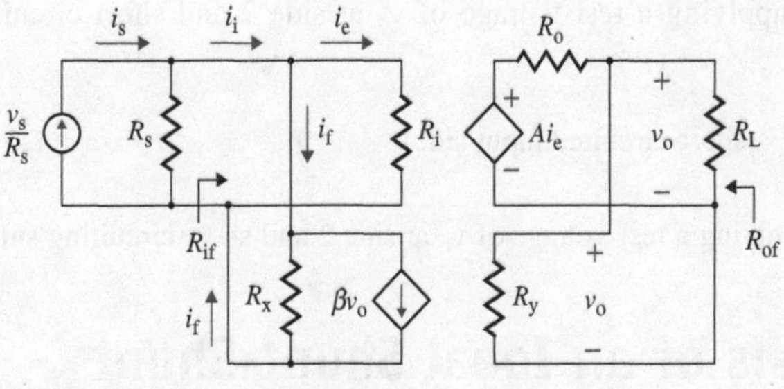

FIGURE 2.31 Shunt-shunt feedback

2.8 Shunt-Shunt Feedback

In shunt-shunt feedback, as shown in Fig. 2.9(a), the feedback circuit is in parallel with the amplifier. Although any amplifier can be represented by any of the four types, the analysis of shunt-shunt feedback can be simplified by representing the amplifier in transresistance form. This is shown in Fig. 2.31, in which the amplifier has an input resistance of R_i, an output resistance of R_o, and an open-loop transresistance gain of A (in V/A). That is, $A = \mu_g R_i$. The feedback current i_f is proportional to the output voltage v_o. The feedback network is modeled in transconductance form with an input resistance of R_y, an output resistance of R_x, and an open-loop transconductance gain of β.

The test conditions for determining the parameters of the feedback network are shown in Fig. 2.32. Given that the voltage across a short circuit is zero and no current flows through an open circuit, the model parameters can be defined by three equations. The first equation is

$$R_x = \left.\frac{v_f}{i_f}\right|_{v_o=0} \quad \text{(short-circuited output side)} \tag{2.64}$$

which is obtained by applying a test voltage of v_f at side 1 and short-circuiting side 2. The second equation is

$$R_y = \left.\frac{v_o}{i_y}\right|_{v_f=0} \quad \text{(short-circuited input side)} \tag{2.65}$$

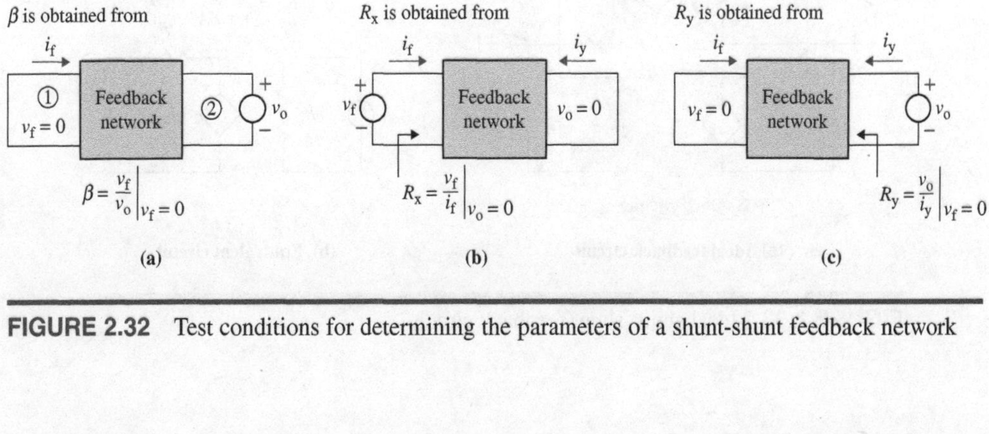

FIGURE 2.32 Test conditions for determining the parameters of a shunt-shunt feedback network

which is obtained by applying a test voltage of v_o at side 2 and short-circuiting side 1. The third equation is

$$\beta = \left.\frac{i_f}{v_o}\right|_{v_f=0} \quad \text{(short-circuited input side)} \tag{2.66}$$

which is obtained by applying a test voltage of v_o at side 2 and short-circuiting side 1.

2.8.1 Analysis of an Ideal Shunt-Shunt Feedback Network

Let us assume an ideal shunt-shunt feedback network—that is, $R_s = \infty$, $R_x = \infty$, $R_y = \infty$, and $R_L = 0$. The feedback amplifier in Fig. 2.31 is simplified to that shown in Fig. 2.33(a), which can be represented by the equivalent circuit shown in Fig. 2.33(b). The output voltage v_o becomes

$$v_o = Ai_e \tag{2.67}$$

The feedback current i_f is proportional to the output voltage v_o. That is,

$$i_f = \beta v_o \tag{2.68}$$

and

$$i_e = i_s - i_f \tag{2.69}$$

Substituting i_e from Eq. (2.69) and i_f from Eq. (2.68) into Eq. (2.67), we get the closed-loop transresistance A_f as

$$A_f = \frac{v_o}{i_s} = \frac{A}{1 + \beta A} \tag{2.70}$$

From Eq. (2.69),

$$i_s = i_e + i_f$$

Substituting i_f from Eq. (2.68) and v_o from Eq. (2.67) into the above equation gives

$$i_s = i_e + \beta v_o = i_e + \beta A i_e = i_e(1 + \beta A) \tag{2.71}$$

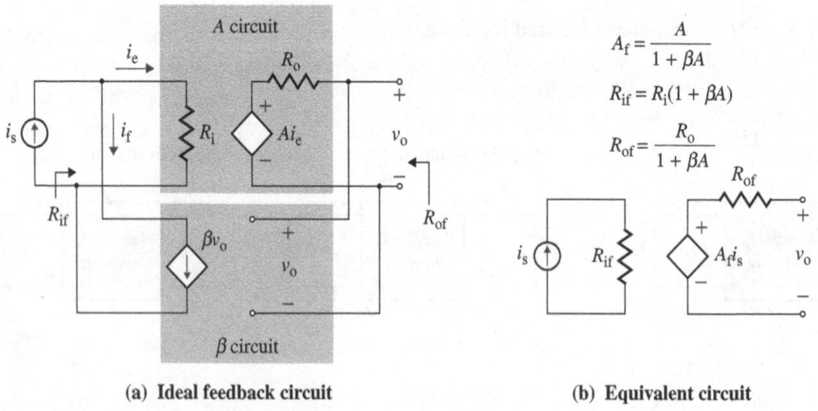

FIGURE 2.33 Ideal shunt-shunt feedback circuit

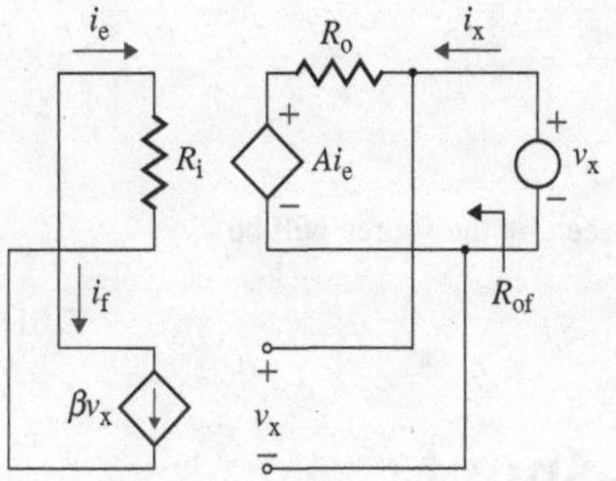

FIGURE 2.34 Equivalent circuit for determining output resistance

The error current i_e is related to v_i by

$$v_i = R_i i_e \tag{2.72}$$

Using i_s from Eq. (2.71) and v_i from Eq. (2.72), we can find the input resistance with feedback R_{if}:

$$R_{if} = \frac{v_i}{i_s} = \frac{i_e R_i}{i_e(1 + \beta A)} = \frac{R_i}{1 + \beta A} \tag{2.73}$$

The input resistance of an amplifier with shunt feedback at the input side is always decreased by a factor of $(1 + \beta A)$.

The output resistance with feedback R_{of}, which is Thevenin's equivalent resistance, can be obtained by applying a test voltage v_x to the output side and open-circuiting the input current source. The equivalent circuit for determining Thevenin's equivalent output resistance is shown in Fig. 2.34. We have

$$i_e = -i_f = -\beta v_x \tag{2.74}$$

and $$i_x = \frac{v_x - 4i_e}{R_o} \tag{2.75}$$

Substituting i_e from Eq. (2.74) into Eq. (2.75) yields

$$i_x = \frac{v_x + \beta A v_x}{R_o} = \frac{(1 + \beta A)v_x}{R_o} \tag{2.76}$$

which gives the output resistance with feedback R_{of} as

$$R_{of} = \frac{R_o}{1 + \beta A} \tag{2.77}$$

Thus, shunt feedback at the output always lowers the output resistance by a factor of $(1 + \beta A)$.

The input impedance, the output impedance, and the overall gain can be written in generalized form in Laplace's domain as follows:

$$Z_{if}(s) = \frac{Z_i(s)}{1 + \beta A(s)} \tag{2.78}$$

$$Z_{of}(s) = \frac{Z_o(s)}{1 + \beta A(s)} \tag{2.79}$$

$$A_f(s) = \frac{A(s)}{1 + \beta A(s)} \tag{2.80}$$

▶ **NOTES**

1. A is the open-loop transresistance of the amplifier, in V/A.
2. β is the transconductance of the feedback network, in A/V.
3. $T_L = \beta A$ is the loop gain, which is dimensionless.
4. If the source has an input impedance of Z_s, then the total impedance Z_{in} seen by the source will be

$$Z_{in}(s) = \frac{v_s}{i_s} = Z_{if}(s) + Z_s(s) \tag{2.81}$$

2.8.2 Analysis of a Practical Shunt-Shunt Feedback Network

The open-loop parameters of the amplifier in Fig. 2.35(a) can be modified to include the loading effect due to R_s, R_x, R_y, and R_L, producing the equivalent ideal feedback circuit in Fig. 2.35(b). The modified parameters are given by

$$R_{ie} = R_s \,\|\, R_i \,\|\, R_x \tag{2.82}$$

$$R_{oe} = R_o \,\|\, R_y \,\|\, R_L \tag{2.83}$$

and
$$v_o = \frac{R_y \,\|\, R_L}{(R_y \,\|\, R_L) + R_o} A i_e \tag{2.84}$$

where i_e is the current through R_i only, not through R_i and $R_s \,\|\, R_x$. Thus, by the current divider rule, i_e is given by

$$i_e = \frac{R_s \,\|\, R_x}{R_s \,\|\, R_x + R_i} i_{e1} \tag{2.85}$$

Substituting i_e from Eq. (2.85) into Eq. (2.84) gives the modified open-loop transresistance gain A_e:

$$A_e = \frac{v_o}{i_{e1}} = \frac{R_y \,\|\, R_L}{R_y \,\|\, R_L + R_o} \times \frac{R_s \,\|\, R_x}{R_s \,\|\, R_x + R_i} A \tag{2.86}$$

With these values of R_{ie}, R_{oe}, and A_e, we can use the equations in the preceding section to calculate the closed-loop parameters R_{if}, R_{of}, and A_f.

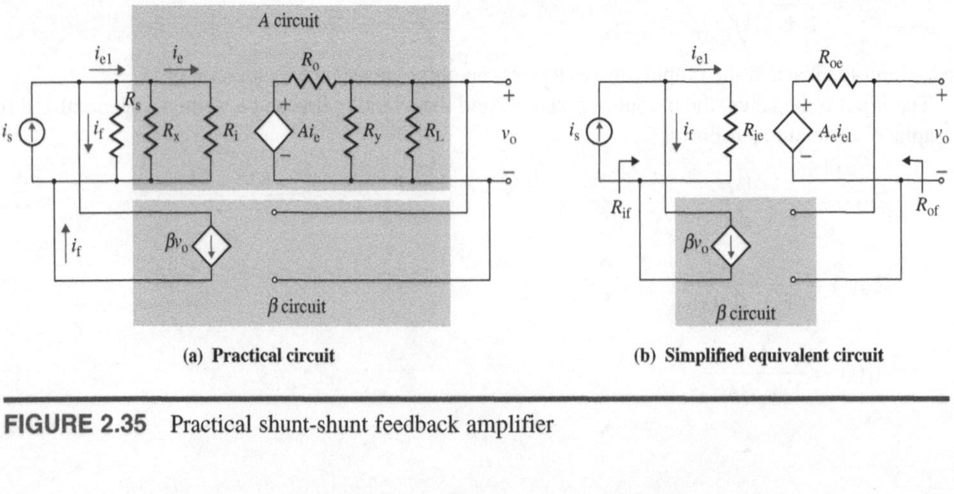

FIGURE 2.35 Practical shunt-shunt feedback amplifier

EXAMPLE 2.8

Finding the performance of an inverting amplifier with shunt-shunt feedback The inverting amplifier shown in Fig. 2.9(b) has $R_s = 2\ \text{k}\Omega$, $R_L = 5\ \text{k}\Omega$, and $R_F = 8\ \text{k}\Omega$. The op-amp parameters are $R_i = 2\ \text{M}\Omega$ and $R_o = 75\ \Omega$, and the open-loop voltage gain is $\mu_g = 2 \times 10^5$.

(a) Determine the input resistance seen by the source $R_{if} = v_i/i_s$, the output resistance R_{of}, the closed-loop transresistance $A_f = v_o/i_s$, and the voltage gain $A_{vf} = v_o/v_s$.

(b) Use PSpice/SPICE to verify your results.

SOLUTION

$R_L = 5\ \text{k}\Omega$, $R_s = 2\ \text{k}\Omega$, $R_F = 8\ \text{k}\Omega$, $R_i = 2\ \text{M}\Omega$, $R_o = 75\ \Omega$, and $\mu_g = 2 \times 10^5$. Replacing the op-amp in Fig. 2.9(b) by its equivalent circuit gives the amplifier shown in Fig. 2.36(a).

(a) The steps in analyzing the feedback network are as follows:

Step 1. R_F constitutes the feedback network, and it produces a feedback current i_f proportional to the output voltage v_o.

Step 2. The amplifier uses shunt-shunt feedback. Thus, A must be in V/A. Converting the voltage-controlled voltage source to a current-controlled voltage source, we get

$$v_o = -\mu_g v_e = -\mu_g R_i i_e = A i_e$$

which gives the open-loop transresistance A:

$$A = -\mu_g R_i = -2 \times 10^5 \times 2 \times 10^6 = -4 \times 10^{11}\ \text{V/A}$$

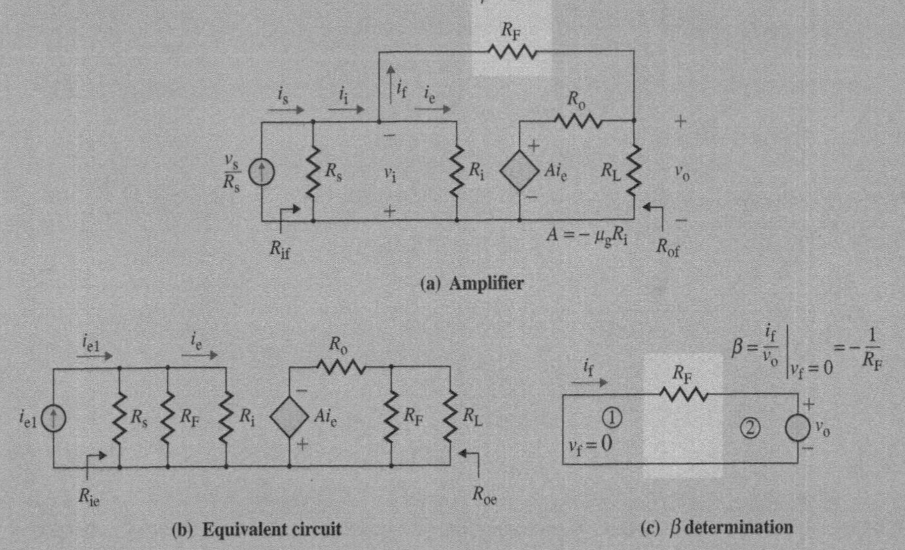

FIGURE 2.36 Inverting op-amp amplifier

Step 3. The effect of the feedback network is taken into account by shorting R_F to the ground at sides 1 and 2. This arrangement is shown in Fig. 2.36(b). Then

$$R_x = R_F = 8\ k\Omega$$

and $$R_y = R_F = 8\ k\Omega$$

Step 4. If we represent the amplifier of Fig. 2.36(a) by an equivalent transresistance amplifier, as shown in Fig. 2.36(b), we have

$$R_{ie} = R_s \| R_i \| R_F = 2\ k \| 2\ M \| 8\ k = 1.6\ k\Omega$$

and $$R_{oe} = R_o \| R_F \| R_L = 75 \| 8\ k \| 5\ k = 73.2\ \Omega$$

From Eq. (2.86), we find that the modified open-loop transresistance gain A_e is

$$A_e = \frac{v_o}{i_{e1}} = \frac{-R_F \| R_L}{(R_F \| R_L) + R_o} \times \frac{R_s \| R_F}{R_s \| R_F + R_i} A$$

$$= \frac{-(8\ k \| 5\ k) \times (2\ k \| 8\ k) \times 4 \times 10^{11}}{(8\ k \| 5\ k + 75)(2\ k \| 8\ k + 2\ M)} = -312.1\ M\Omega$$

Step 5. From Fig. 2.36(c), the feedback factor β is given by

$$\beta = \left.\frac{i_f}{v_o}\right|_{v_i=0} = -\frac{1}{R_F} = -125\ \mu\mho$$

The loop gain is

$$T_L = \beta A_e = 125\ \mu\mho \times 312.1 \times 10^6 = 39.02\ k\Omega$$

Step 6. The input resistance at the input side of the op-amp is

$$R_{if} = \frac{R_{ie}}{1 + \beta A_e} = \frac{R_{ie}}{1 + T_L}$$

$$= \frac{1.6\ k\Omega}{1 + 39.02 \times 10^3} = 41.0\ m\Omega$$

Step 7. The output resistance with feedback is

$$R_{of} = \frac{R_{oe}}{1 + \beta A_e} = \frac{R_{oe}}{1 + T_L} = \frac{73.2\ \Omega}{1 + 39.02 \times 10^3} = 1.876\ m\Omega$$

Step 8. The closed-loop transresistance gain A_f is

$$A_f = \frac{v_o}{i_s} = \frac{A_e}{1 + \beta A_e} = \frac{A_e}{1 + T_L}$$

$$= \frac{-312.1 \times 10^6}{1 + 39.02 \times 10^3} = -8\ k\Omega$$

$$v_o = \left(\frac{v_o}{v_s}\right) v_s = \left(\frac{v_o}{i_s}\right)\left(\frac{i_s}{v_s}\right) v_s = \left(\frac{-8\ k}{2\ k}\right) v_s = -4 v_s$$

Therefore, the overall voltage gain is $v_o/v_s = -4$.

NOTE: Substituting $R_1 = R_s$ gives the voltage gain of the inverting amplifier as $R_F/R_s = -8\ k/2\ k = -4$.

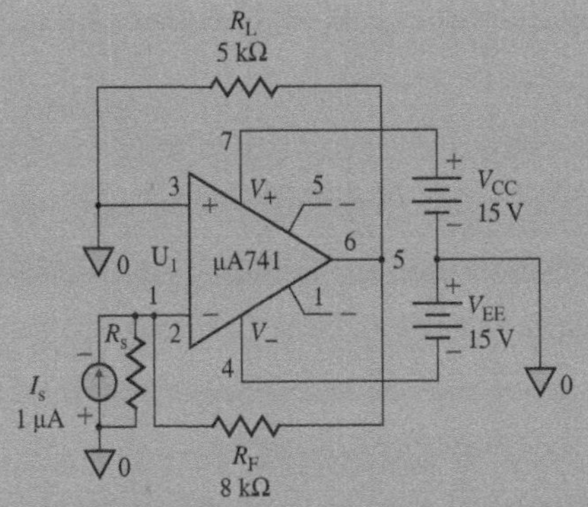

FIGURE 2.37 Shunt-shunt feedback circuit for PSpice simulation

(b) The shunt-shunt feedback circuit for PSpice simulation is shown in Fig. 2.37. The results of the simulation (.TF analysis) are shown below, with hand calculations to the right:

`V(5)/Is=-8.000E+03=-8 kΩ`	$A_f = -8\ \text{k}\Omega$
`INPUT RESISTANCE AT IS=4.097E-02=40.97 mΩ`	$R_{if} = 41.0\ \text{m}\Omega$
`OUTPUT RESISTANCE AT V(5)=1.876E-03=1.876 mΩ`	$R_{of} = 1.876\ \text{m}\Omega$

The PSpice results are very close to the hand-calculated values.

EXAMPLE 2.9

Finding the performance of a BJT amplifier with shunt-shunt feedback The parameters of the amplifier in Fig. 2.38 are $R_{C1} = 5\ \text{k}\Omega$, $R_E = 2.5\ \text{k}\Omega$, $R_{C2} = 5\ \text{k}\Omega$, $R_F = 4\ \text{k}\Omega$, and $R_s = 200\ \Omega$. The DC bias currents of the transistors are $I_{C1} = 0.5$ mA and $I_{C2} = 1$ mA. The transistor parameters are $h_{fe} = h_{fe1} = h_{fe2} = 150$ and $r_\mu = r_o = \infty$.

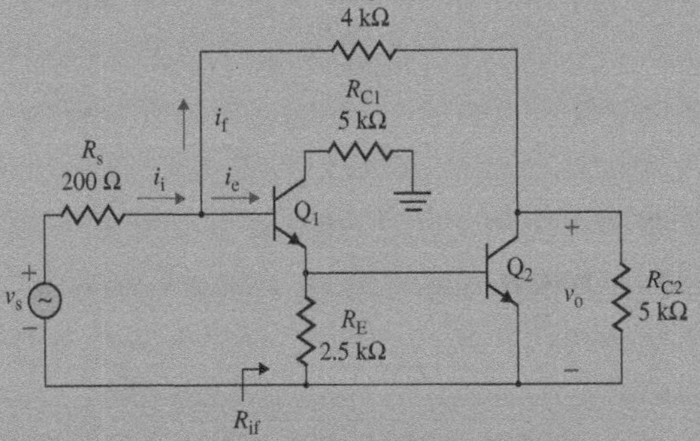

FIGURE 2.38 AC equivalent of two-stage amplifier with shunt-shunt feedback

(a) Use the techniques of feedback analysis to calculate the input resistance R_{if}, the output resistance R_{of}, and the closed-loop transresistance gain A_f.

(b) Use PSpice/SPICE to check your results.

NOTE: The following analysis can also be applied to MOSFET amplifiers by substituting a very large value of the B-E resistance r_π tending to infinity, $1/r_\pi \simeq 0$.

SOLUTION

For $I_{C1} = 0.5$ mA,

$$r_{\pi 1} = \frac{h_{fe} \times 25.8 \text{ mV}}{I_{C1}} = \frac{150 \times 25.8 \text{ mV}}{0.5 \text{ mA}} = 7.74 \text{ k}\Omega$$

$$g_{m1} = \frac{I_{C1}}{V_T} = \frac{0.5 \text{ mA}}{25.8 \text{ mV}} = 19.38 \text{ mA/V}$$

For $I_{C2} = 1$ mA,

$$r_{\pi 2} = \frac{h_{fe} \times 25.8 \text{ mV}}{I_{C2}} = \frac{150 \times 25.8 \text{ mV}}{1 \text{ mA}} = 3.87 \text{ k}\Omega$$

$$g_{m2} = \frac{I_{C2}}{V_T} = \frac{1 \text{ mA}}{25.8 \text{ mV}} = 38.76 \text{ mA/V}$$

The feedback current i_f is proportional to the output voltage v_o. The effective input current to the amplifier is $i_e = i_i - i_f$.

(a) The steps in analyzing the feedback network are as follows:

Step 1. R_F acts as the feedback network. The feedback current i_f is proportional to the output voltage v_o. The effective input current to the amplifier is $i_e = i_i - i_f$. The functional block diagram is shown in Fig. 2.39(a).

Step 2. The amplifier uses shunt-shunt feedback, as shown in Fig. 2.39(b). Thus, the units of the gain A must be V/A.

Step 3. The effect of the feedback network can be taken into account at the input side by short-circuiting R_F to the ground at side 2 in Fig. 2.39(b), and at the output side by short-circuiting R_F to the ground at side 1. These modifications are shown in the small-signal AC equivalent circuit in Fig. 2.39(c).

Step 4. If we represent the amplifier in Fig. 2.39(c) by an equivalent transresistance amplifier, the resistance at the base of the amplifier is

$$R_b = \frac{v_i}{i_{b1}} = r_{\pi 1} + (1 + g_{m1} r_{\pi 1})(R_E \| r_{\pi 2}) = 7.74 \text{ k} + 151(2.5 \text{ k} \| 3.87 \text{ k}) = 237 \text{ k}\Omega$$

The input resistance of the amplifier is

$$R_{ie} = \frac{v_i}{i_s} = R_s \| R_F \| [r_{\pi 1} + (1 + g_{m1} r_{\pi 1})(R_E \| r_{\pi 2})]$$
$$= 200 \| 4 \text{ k} \| [7.74 \text{ k} + 151(2.5 \text{ k} \| 3.87 \text{ k})] = 190.3 \ \Omega$$

The output resistance of the amplifier is

$$R_{oe} = R_F \| R_{C2} = 4 \text{ k} \| 5 \text{ k} = 2.22 \text{ k}\Omega$$

Thus,

$$v_{b1} = R_{ie} i_s = 190.3 i_s$$

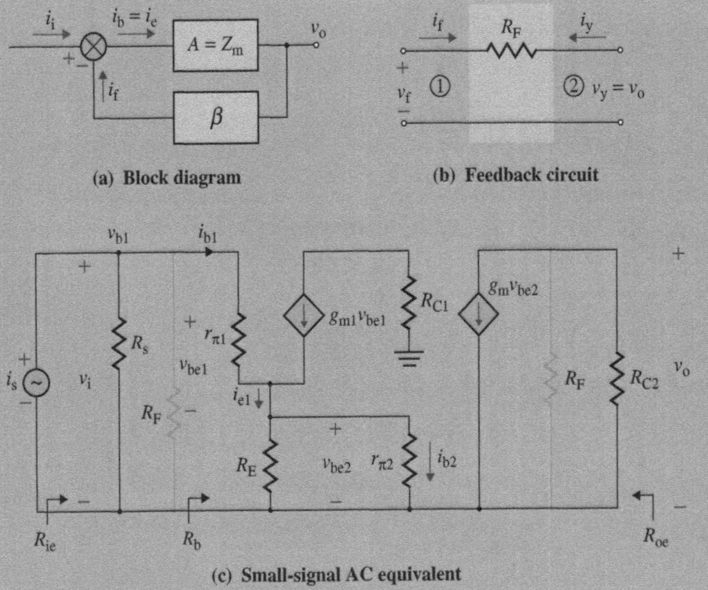

FIGURE 2.39 Equivalent circuits for Example 2.9

$$v_{be2} = \frac{v_{b1} \times (1 + g_{m1}r_{\pi1})(R_E \| r_{\pi2})}{[r_{\pi1} + (1 + g_{m1}r_{\pi1})(R_E \| r_{\pi2})]}$$

$$= \frac{(190.3 \times i_s) \times (1 + 193.8 \text{ m} \times 7.74 \text{ k})(2.5 \text{ k} \| 3.87 \text{ k})}{[7.74 \text{ k} + (1 + 193.8 \text{ m} \times 7.74 \text{ k})(2.5 \text{ k} \| 3.87 \text{ k})]} = 184.11 i_s$$

$$i_o = g_{m2}v_{be2} = 38.76 \text{ m} \times 184.11 i_s = 7.136 i_s$$

$$v_o = -(R_F \| R_{C2}) i_o = -(4 \text{ k} \| 5 \text{ k}) \times 7.136 i_s = -15.865 \times 10^3 \times i_s$$

Therefore, the open-loop transresistance A is given by $A = v_o/i_s = -15.865$ kA/V.

Step 5. Using Eq. (2.66), we find that the feedback factor β is

$$\beta = \left.\frac{i_f}{v_o}\right|_{v_f=0} = -\frac{1}{R_F} = -\frac{1}{4 \text{ k}} = -0.25 \text{ mA/V}$$

and

$$\beta A_e = -15.865 \text{ kV/A} \times \left(-\frac{1}{4 \text{ kV/A}}\right) = 3.966$$

Step 6. The input resistance seen by the current source is

$$R_{if} = \frac{v_i}{i_s} = \frac{R_{ie}}{1 + \beta A_e} = \frac{190.3\ \Omega}{1 + 0.25 \text{ mA/V} \times 15.865 \text{ kV/A}} = 38.3\ \Omega$$

Step 7. The output resistance with feedback is

$$R_{of} = \frac{R_{oe}}{1 + \beta A_e} = \frac{2.22 \text{ k}\Omega}{1 + 0.25 \text{ mA/V} \times 15.865 \text{ kV/A}} = 447\ \Omega$$

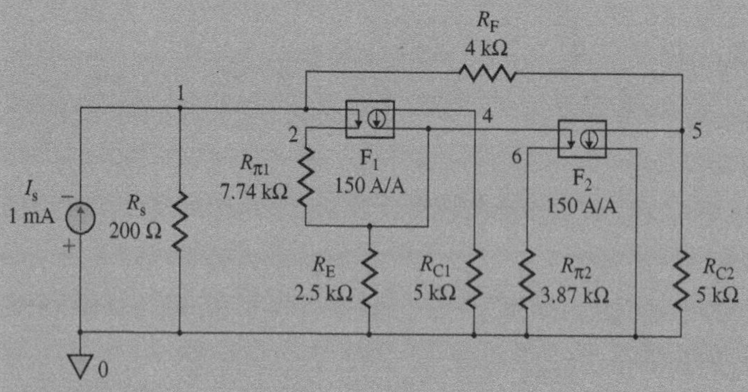

FIGURE 2.40 Two-stage shunt-shunt feedback network for PSpice simulation

Step 8. The closed-loop transresistance gain A_f is

$$A_f = \frac{v_o}{i_s} = \frac{-15.865 \text{ kV/A}}{1 + 0.25 \text{ mA/V} \times 15.865 \text{ kV/A}} = -3.195 \text{ kV/A}$$

(b) The shunt-shunt feedback amplifier for PSpice simulation is shown in Fig. 2.40. The results of PSpice/SPICE simulation (.TF analysis) are shown below, with hand calculations to the right:

`V(5)/Is=-3.19E+03=-3.19 kV/A`	$A_f = -3.195$ kV/A
`INPUT RESISTANCE AT Is=3.854E+01=38.54 Ω`	$R_{in} = 38.3$ Ω
`OUTPUT RESISTANCE AT V(5)=4.5E+02=450 Ω`	$R_{of} = 447$ Ω

The PSpice results are very close to the hand-calculated values.

KEY POINTS OF SECTION 2.8

- Shunt-shunt feedback is applied to transresistance amplifiers. This type of feedback reduces both the input and the output resistances by a factor of $(1 + \beta A)$.
- The amplifier is represented by a transresistance amplifier and the feedback network as a transconductance gain. A is in units of V/A, and β is in units of A/V.

2.9 Shunt-Series Feedback

In shunt-series feedback, as shown in Fig. 2.10(a), the feedback network is in parallel with the amplifier at the input side and in series with the amplifier at the output side. The amplifier is represented as a current amplifier. This is shown in Fig. 2.41, in which the amplifier has an input resistance of R_i, an output resistance of R_o, and an open-loop current gain of A (in A/A); that is, $A = \mu_g R_i / R_o$.

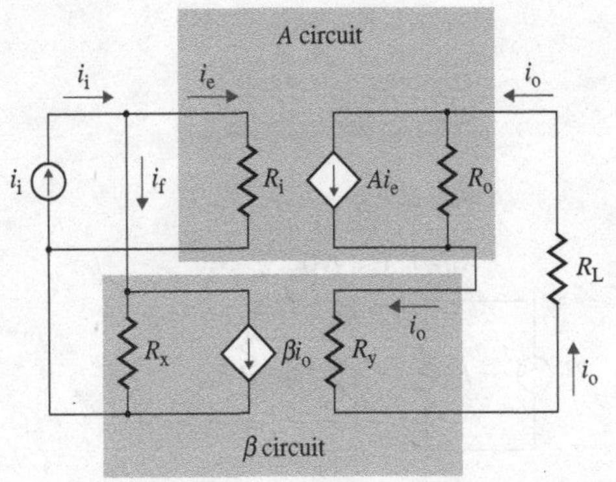

FIGURE 2.41 Shunt-series feedback amplifier

The feedback current i_f is proportional to the output current i_o. The feedback network is modeled in gain form, with an input resistance of R_y, an output resistance of R_x, and a current gain of β. The test conditions for determining the parameters of the feedback network are shown in Fig. 2.42. The parameters of the model are defined by three equations. The first equation is

$$R_x = \left.\frac{v_f}{i_f}\right|_{i_y=0} \quad \text{(open-circuited output side)} \tag{2.87}$$

which is obtained by applying a test voltage of v_f at side 1 and open-circuiting side 2. The second equation is

$$R_y = \left.\frac{v_o}{i_y}\right|_{v_f=0} \quad \text{(short-circuited input side)} \tag{2.88}$$

which is obtained by applying a test voltage of v_y at side 2 and short-circuiting side 1. The third equation is

$$\beta = \left.\frac{i_f}{i_y}\right|_{v_f=0} \quad \text{(short-circuited input side)} \tag{2.89}$$

which is obtained by applying a test voltage of v_y at side 2 and short-circuiting side 1.

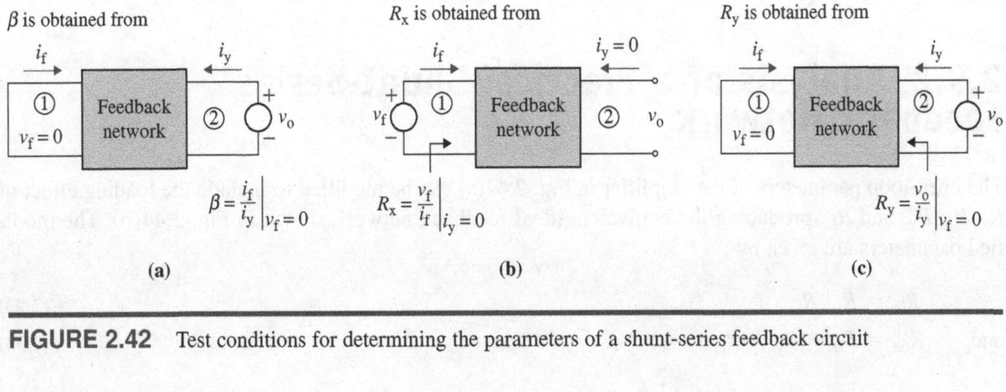

FIGURE 2.42 Test conditions for determining the parameters of a shunt-series feedback circuit

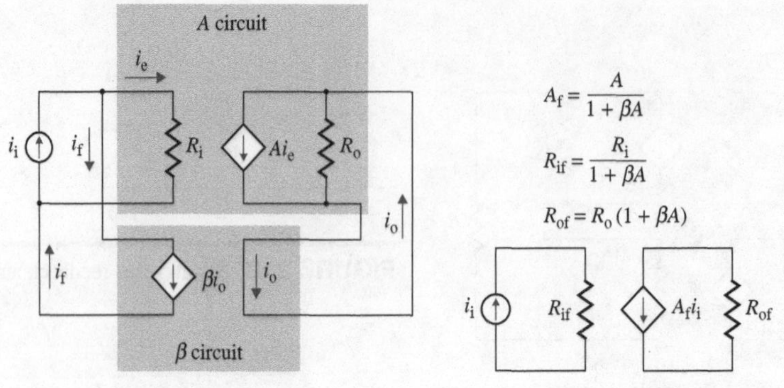

(a) Ideal feedback circuit **(b) Equivalent circuit**

FIGURE 2.43 Ideal shunt-series feedback amplifier

2.9.1 Analysis of an Ideal Shunt-Series Feedback Network

Let us assume an ideal feedback network—that is, $R_x = \infty$, $R_y = 0$, and $R_L = 0$. The feedback amplifier in Fig. 2.41 can be simplified to the one in Fig. 2.43. The feedback factor β is in A/A.

If $R_L << R_o$, it can be shown that

$$R_{if} = \frac{v_i}{i_i} = \frac{R_i}{1 + \beta A} \tag{2.90}$$

$$R_{of} = R_o(1 + \beta A) \tag{2.91}$$

$$A_f = \frac{i_o}{i_i} = \frac{A}{1 + \beta A} \tag{2.92}$$

▶ **NOTES**

1. A is the open-loop current gain of the amplifier, in A/A.
2. β is the current gain of the feedback network, in A/A.
3. $T_L = \beta A$ is the loop gain, which is dimensionless.

2.9.2 Analysis of a Practical Shunt-Series Feedback Network

The open-loop parameters of the amplifier in Fig. 2.44(a) can be modified to include the loading effect of R_s, R_x, R_y, and R_L, producing the equivalent ideal feedback network shown in Fig. 2.44(b). The modified parameters are given by

$$R_{ie} = R_i \,\|\, R_x \tag{2.93}$$

and

$$R_{oe} = R_o + R_y + R_L \tag{2.94}$$

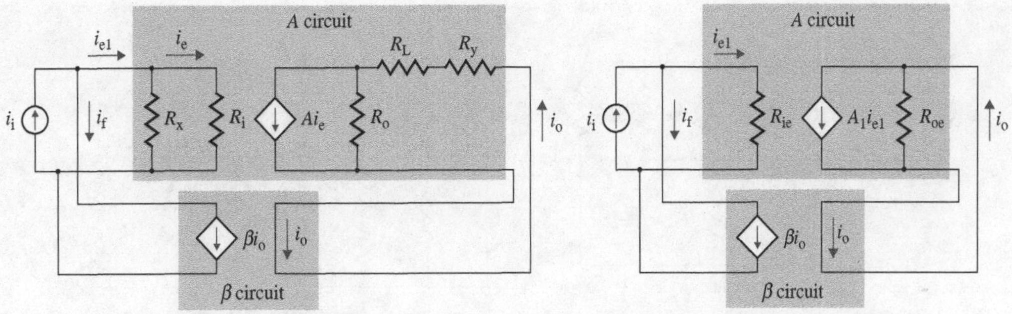

(a) Practical circuit **(b) Simplified equivalent circuit**

FIGURE 2.44 Practical shunt-series feedback amplifier

It may appear from Fig. 2.44(b) that R_o would have to be in parallel with $(R_y + R_L)$ for us to find R_{oe} in Eq. (2.94), but if we convert the current source Ai_e to a voltage source, we see that the effective resistance used to find the current i_o becomes $(R_o + R_y + R_L)$:

$$i_o = \frac{R_o}{R_y + R_o + R_L} Ai_e \tag{2.95}$$

$$i_e = \frac{R_x}{R_i + R_x} i_{e1} \tag{2.96}$$

From Eqs. (2.95) and (2.96), we can derive the modified open-loop current gain A_e:

$$A_e = \frac{i_o}{i_{e1}} = \frac{R_o R_x}{(R_y + R_L + R_o)(R_i + R_x)} A \tag{2.97}$$

If the values of R_i, R_o, and A are replaced by R_{ie}, R_{oe}, and A_e, respectively, Eqs. (2.90) through (2.92) can be applied to calculate the closed-loop parameters R_{if}, R_{of}, and A_f.

EXAMPLE 2.10

Finding the performance of an inverting amplifier with shunt-series feedback The parameters of the shunt-series feedback amplifier shown in Fig. 2.10(b) are $R_L = 5\ \Omega$, $R_s = 2.5\ \text{k}\Omega$, $R_F = 200\ \Omega$, and $R_1 = 5\ \Omega$. The op-amp parameters are $R_i = 2\ \text{M}\Omega$, $R_o = 75\ \Omega$, and $\mu_g = 2 \times 10^5$.

(a) Determine the input resistance at the op-amp input $R_{if} = v_i/i_i$, the output resistance R_{of}, and the closed-loop current gain $A_f = i_o/i_i$.

(b) Use PSpice/SPICE to check your results.

SOLUTION

$R_s = 2.5\ \text{k}\Omega$, $R_F = 200\ \Omega$, $R_1 = 5\ \Omega$, $R_L = 5\ \Omega$, $R_i = 2\ \text{M}\Omega$, $R_o = 75\ \Omega$, and $\mu_g = 2 \times 10^5$. Replacing the op-amp by its equivalent circuit gives the amplifier shown in Fig. 2.45(a).

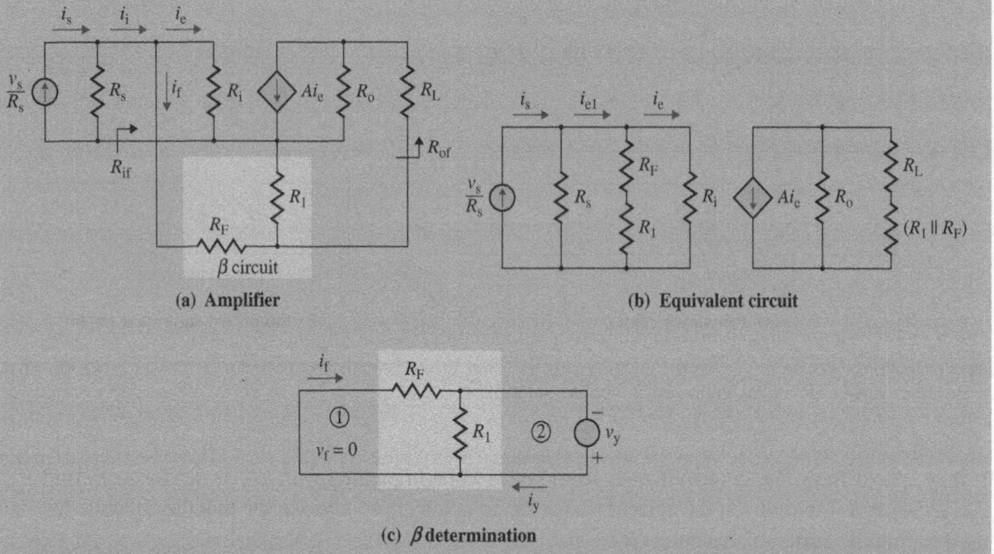

FIGURE 2.45 Op-amp with shunt-series feedback

(a) Converting the voltage-controlled voltage source to a current-controlled current source, we get

$$i_o = \frac{\mu_g v_i}{R_o} = \frac{\mu_g R_i i_e}{R_o} = Ai_e$$

which gives

$$A = \frac{\mu_g R_i}{R_o} = \frac{2 \times 10^5 \times 2 \times 10^6}{75} = 53.33 \times 10^8 \text{ A/A}$$

The effect of the feedback network is taken into account by severing R_1 from the load at side 2 and shorting R_F to the ground at side 1, as shown in Fig. 2.45(b). Then

$$R_x = R_F + R_1 = 200 + 5 = 205 \ \Omega$$

$$R_y = R_F \| R_1 = 200 \| 5 = 4.88 \ \Omega$$

$$R_{ie} = R_i \| (R_F + R_1) = R_i \| R_x = 2 \text{ M} \| 205 = 204.98 \ \Omega$$

$$R_{oe} = R_o + R_y + R_L = \text{R}_o + (R_F \| R_1) + R_L = 75 + 4.88 + 5 = 84.88 \ \Omega$$

Equation (2.89) gives the feedback factor β:

$$\beta = \left.\frac{i_f}{i_y}\right|_{v_f=0} = \frac{R_1}{R_F + R_1} = \frac{5}{200 + 5} = 24.39 \times 10^{-3} \text{ A/A}$$

Equation (2.97) gives the modified open-loop current gain A_e:

$$A_e = \frac{75 \times 205 \times 53.33 \times 10^8}{(4.88 + 5 + 75)(2 \text{ M} + 205)} = 483 \times 10^3 \text{ A/A}$$

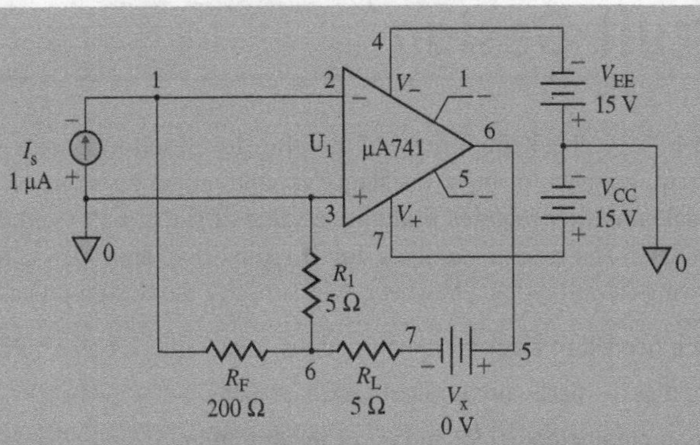

FIGURE 2.46 Shunt-series feedback circuit for PSpice simulation

The loop gain is

$$T_L = \beta A_e = 24.39 \times 10^{-3} \times 483 \times 10^3 = 11.76 \times 10^3$$

The resistances are

$$R_{if} = \frac{R_{ie}}{1 + T_L} = \frac{204.98}{1 + 11.76 \times 10^3} = 17.4 \text{ m}\Omega$$

$$R_{in} = \frac{v_s}{i_s} = R_{if} + R_s = 17.4 \text{ m} + 2.5 \text{ k} \approx 2.5 \text{ k}\Omega$$

$$R_{of} = R_{oe} \times (1 + T_L) = 84.88 \times (1 + 11.76 \times 10^3) = 999.6 \text{ k}\Omega$$

Then

$$A_f = \frac{i_o}{i_i} = \frac{483 \times 10^3}{1 + 11.763 \times 10^3} \approx 41$$

(b) The shunt-series feedback circuit for PSpice simulation is shown in Fig. 2.46. The results of PSpice simulation (.TF analysis) are shown below, with hand calculations to the right:

`I(VX)/Is=4.100E+01=41`	$A_f = 41$
`INPUT RESISTANCE AT Is=1.740E-02=17.4 mΩ`	$R_{if} = 17.4$ mΩ
`OUTPUT RESISTANCE AT I(VX)=1.000E+06=1 MΩ`	$R_{of} = 999.6$ kΩ

The PSpice results are very close to the hand-calculated values.

KEY POINTS OF SECTION 2.9

- Shunt-series feedback is applied to current amplifiers. This type of feedback reduces the input resistance and increases the output resistance by a factor of $(1 + \beta A)$.
- The amplifier is represented by a current amplifier and the feedback network as a current gain. Both A and β are in units of A/A.

2.10 Feedback Circuit Design

The feedback factor β is the key parameter of a feedback amplifier and modifies its closed-loop gain A_f. The type of feedback used depends on the requirements for input resistance R_{if} and output resistance R_{of}. If A is independent of β, designing a feedback amplifier requires finding the value of β that will yield the desired value of A_f, R_{if}, R_{of}, or bandwidth BW. The design becomes cumbersome if A depends on the feedback network—that is, on β. The following iterative steps, however, will simplify the design process:

Step 1. Decide on the type of feedback needed to meet the specifications. Use Table 2.1 as a guide.

Step 2. Disconnect the feedback link; that is, make sure no feedback is present.

Step 3. Find the approximate open-loop parameters A, R_i, and R_o of the amplifier.

Step 4. Find the values of feedback factor β and feedback resistance(s) that will satisfy the closed-loop requirement. Use the relations in Table 2.1.

Step 5. Using the feedback resistance(s), recalculate the open-loop parameters A, R_i, and R_o.

Step 6. Find the closed-loop parameters A_f, R_{if}, and R_{of}.

Step 7. Repeat steps 3 through 5 until the desired closed-loop condition is satisfied. Normally, a number of iterations will be required.

EXAMPLE 2.11

D

Designing a series-shunt feedback circuit Feedback is applied to a voltage amplifier whose open-loop parameters are $R_i = 4.5\ \text{k}\Omega$, $R_o = 500\ \Omega$, low-frequency voltage gain $A = -450$ V/V, and bandwidth BW $= f_H = 10$ kHz. The load resistance is $R_L = 10\ \text{k}\Omega$. Determine the values of the feedback network so that the following specifications are satisfied:

(a) Bandwidth with feedback $f_{Hf} = 1$ MHz, $R_{if} > R_i$, and $R_{of} < R_o$

(b) $R_{if} = 50R_i$ and $R_{of} < R_o$

(c) $R_{if} > R_i$ and $R_{of} = R_o/250$

SOLUTION

$R_i = 4.5\ \text{k}\Omega$, $R_o = 500\ \Omega$, $A = -450$ V/V, and $f_H = 10$ kHz.

(a) Since the input resistance R_{if} should increase and the output resistance R_{of} should decrease, we can see from Table 2.1 that the feedback must be of the series-shunt type. Thus, we can use the equations derived in Sec. 2.6. The feedback network consists of R_1 and R_F, as shown in Fig. 2.47. Since A is negative, the value of v_f will be negative. Thus, v_f is added to v_s rather than subtracted from it (i.e., $v_s + v_f = v_e$), and this is accomplished by connecting the feedback signal to the negative terminal of v_s. To minimize the loading effects, $(R_1 + R_F)$ must be much larger than R_L. The condition

$$(R_1 + R_F) >> R_L \tag{2.98}$$

is generally satisfied by choosing $R_1 + R_F = 10R_L$. Since the gain–bandwidth product remains constant, Eq. (2.24) gives $Af_H = A_f f_{Hf}$. That is,

$$450 \times 10\ \text{kHz} = A_f \times 1\ \text{MHz}$$

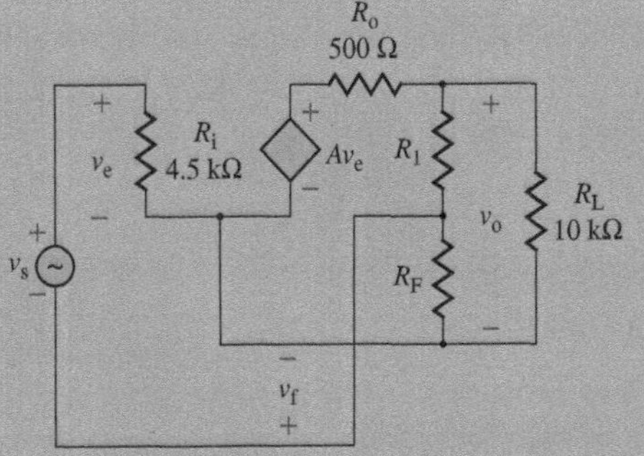

FIGURE 2.47 Amplifier with series-shunt feedback

which gives $A_f = 450/100 = 4.5$ V/V. From Eq. (2.37),

$$4.5 = \frac{450}{1 + 450\beta}$$

which gives $\beta = 0.22$. The feedback factor β is related to R_1 and R_F by

$$\beta = \frac{R_F}{R_1 + R_F} = \frac{R_F}{10R_L} \tag{2.99}$$

which, for $\beta = 0.22$ and $R_L = 10$ kΩ, gives

$$R_F = 10\beta R_L = 10 \times 0.22 \times 10 \text{ k} = 22 \text{ k}\Omega$$

and $$R_1 = 10R_L - R_F = 100 \text{ k} - 22 \text{ k} = 78 \text{ k}\Omega$$

(b) Since $R_{if} = 50R_i$ and $R_{if} = R_i(1 + \beta A)$ from Eq. (2.40), we get

$$50 = 1 + \beta A \qquad \text{(for negative feedback)}$$

which, for $|A| = 450$, gives $\beta = 0.109$. For $\beta = 0.109$ and $R_L = 10$ kΩ, Eq. (2.99) gives

$$R_F = 10\beta R_L = 10 \times 0.109 \times 10 \text{ k} \approx 11 \text{ k}\Omega$$

and $$R_1 = 10R_L - R_F = 100 \text{ k} - 11 \text{ k} = 89 \text{ k}\Omega$$

(c) Since $R_{of} = R_o/250$ and $R_{of} = R_o/(1 + \beta A)$ from Eq. (2.44), we get

$$250 = 1 + \beta A$$

which, for $|A| = 450$, gives $\beta = 0.5533$. For $\beta = 0.5533$ and $R_L = 10$ kΩ, Eq. (2.99) gives

$$R_F = 10\beta R_L = 10 \times 0.5533 \times 10 \text{ k} \approx 55 \text{ k}\Omega$$

and $$R_1 = 10R_L - R_F = 100 \text{ k} - 55 \text{ k} = 45 \text{ k}\Omega$$

NOTE: The above solution gives approximate values for the designer to start with. More accurate values of A_f, R_{if}, and R_{of} could be found by considering the loading effects of the feedback network and the load resistance (as in Example 2.4). The design steps should be repeated until the desired specifications are met.

EXAMPLE 2.12

D

Designing a shunt-series feedback circuit An amplifier with shunt-series feedback is shown in Fig. 2.48. The amplifier has $R_1 = 6.6\ \text{k}\Omega$, $R_2 = 2\ \text{k}\Omega$, $R_{C1} = 5\ \text{k}\Omega$, $R_3 = 5\ \text{k}\Omega$, $R_4 = 10\ \text{k}\Omega$, $R_E = 500\ \Omega$, and $R_{C2} = 5\ \text{k}\Omega$. The transistor parameters are $h_{fe} = 150$, $r_\pi = r_{\pi 1} = r_{\pi 2} = 2.58\ \text{k}\Omega$, and $r_o = \infty$.

(a) Determine the value of feedback resistor R_F so that the closed-loop current gain A_f is 10% of the open-loop current gain A.

(b) Use PSpice/SPICE to check your results.

NOTE: The following analysis can also be applied to MOSFET amplifiers by substituting a very large value of the B-E resistance r_π, tending to infinity, $1/r_\pi \simeq 0$.

SOLUTION

The feedback current i_f is proportional to the emitter voltage $v_e = v_f$, which in turn is proportional to the output current $i_o \approx -i_c$. The feedback mechanism is shown in Fig. 2.49(a) and the feedback network in Fig. 2.49(b). Replacing the transistors by their small-signal model gives the small-signal AC equivalent circuit of the amplifier shown in Fig. 2.49(c),

(a) We have

$$g_m = \frac{h_{fe}}{r_\pi} = \frac{150}{2.58\ \text{k}} = 5.814\ \text{mA/V}$$

$$R_{B1} = R_1 \| R_2 = 6.6\ \text{k} \| 2\ \text{k} = 1.53\ \text{k}\Omega$$

$$R_{B2} = R_3 \| R_4 \| R_{C1} = 5\ \text{k} \| 10\ \text{k} \| 5\ \text{k} = 2\ \text{k}\Omega$$

Step 1. Assume that there is no feedback—that is, $R_F = \infty$.

Step 2. Find the open-loop parameters A, R_i, and R_o. The input resistance of the amplifier is

$$R_i = \frac{v_s}{i_s} = R_{B1} \| r_{\pi 1} = 1.53\ \text{k} \| 2.58\ \text{k} = 962.4\ \Omega$$

The output resistance of the amplifier is

$$R_o = r_o = \infty \quad \text{and} \quad R_{oe} = R_o$$

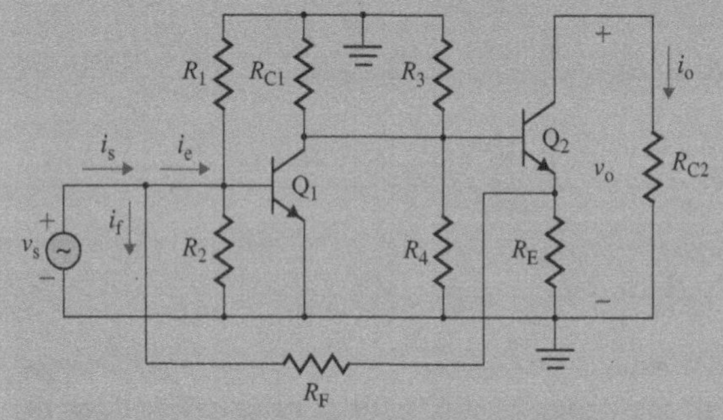

FIGURE 2.48 Two-stage amplifier with shunt-series feedback

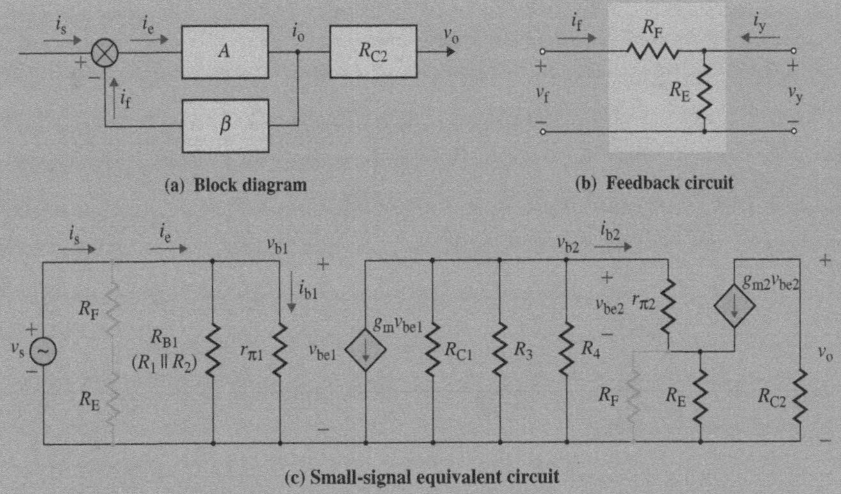

FIGURE 2.49 Equivalent circuits for Example 2.12

From Fig. 2.49(c), we get

$$v_{b1} = v_{be1} = R_i i_s = 962.4 i_s$$

$$v_{b2} = -\{R_{B2} \| [r_{\pi 2} + (1 + g_{m2} r_{\pi 2}) R_E]\} g_{m1} v_{be1}$$

$$= -\{2\,\text{k} \| [2.58\,\text{k} + (1 + 58.14\,\text{m} \times 2.58\,\text{k}) \times 500]\} \times 58.14\,\text{m} \times (962.4 i_s) = -109.1 \times 10^3 i_s$$

$$v_{be2} = \frac{v_{b2} \times r_{\pi 2}}{[r_{\pi 2} + (1 + g_{m2} r_{\pi 2}) R_E]}$$

$$= \frac{(-109.1 \times 10^3 i_s) \times 2.58\,\text{k}}{[2.58\,\text{k} + (1 + 58.14\,\text{m} \times 2.58\,\text{k}) \times 500]} = -3.605 i_s$$

$$i_o = g_{m2} v_{be2} = 58.14\,\text{m} \times (-3.605 i_s) = -209.2 i_s$$

Therefore, the open-loop current gain A is given by $A = i_o / i_s = -209.2$ A/A.

Step 3. Find the values of feedback factor β and resistance R_F. Since the closed-loop current gain A_f is 10% of the open-loop current gain A, we can write

$$A_f = \frac{|A|}{1 + \beta |A|} = 0.1\,A$$

That is,

$$1 + \beta |A| = 10 \quad \text{or} \quad \beta\,|A| = 9 \quad \text{and} \quad \beta = \frac{9}{|A|} = \frac{9}{209.2} = 0.043$$

Since the feedback network in Fig. 2.49(b) must be represented by a current amplifier, the feedback factor β is given by

$$\beta = \left.\frac{i_f}{i_y}\right|_{v_f=0} = \frac{R_E}{R_E + R_F} = \frac{500}{500 + R_F}$$

which, for $\beta = 0.043$, gives $R_F = 11.13\,\text{k}\Omega$.

Step 4. Find the new open-loop parameters A, R_i, and R_o. R_F is included by short-circuiting the shunt feedback side and severing the series feedback side. This arrangement is shown in Fig. 2.49(c) by the gray lines. The input resistance of the amplifier is

$$R_i = \frac{v_s}{i_s} = (R_F + R_E) \| R_{B1} \| r_{\pi 1} = (11.13\text{ k} + 500) \| 1.53\text{ k} \| 2.58\text{ k} = 887.2\ \Omega$$

The output resistance of the amplifier is

$$R_o = r_o = \infty \quad \text{and} \quad R_{oe} = R_o = \infty$$

From Fig. 2.49(c), we get

$$v_{b1} = v_{be1} = R_i i_s = 887.2 i_s$$

$$v_{b2} = -\{R_{B2} \| [r_{\pi 2} + (1 + g_{m2} r_{\pi 2})(R_E \| R_F)]\} g_{m1} v_{be1}$$

$$= -\{2\text{ k} \| [2.58\text{ k} + (1 + 58.14\text{ m} \times 2.58\text{ k}) \times (11.13\text{ k} \| 500)]\} \times 58.14\text{ m} \times (887.2 i_s)$$

$$= -100.7 \times 10^3 i_s$$

$$v_{be2} = \frac{v_{b2} \times r_{\pi 2}}{[r_{\pi 2} + (1 + g_{m2} r_{\pi 2})(R_E \| R_F)]}$$

$$= \frac{(-100.7 \times 10^3 i_s) \times 2.58\text{ k}}{[2.58\text{ k} + (1 + 58.14\text{ m} \times 2.58\text{ k}) \times (11.13\text{ k} \| 500)]} = -3.471 \times 10^3 i_s$$

$$i_o = g_{m2} v_{be2} = 58.14\text{ m} \times (-3.471 \times 10^3 i_s) = -201.4 i_s$$

Therefore, the open-loop current gain A is given by $A = i_o/i_s = -201.4$ A/A.

The closed-loop parameters are as follows:

$$R_{if} = \frac{R_i}{1 + \beta A} = \frac{887.2}{1 + 0.043\text{ A/A} \times 201.4\text{ A/A}} = 91.8\ \Omega$$

The feedback will not affect the output resistance; that is, $R_{of} = \infty$ and

$$A_f = \frac{201.4}{1 + 0.043\text{ A/A} \times 201.4\text{ A/A}} = 20.85\text{ A/A}$$

Step 5. Repeating steps 3 to 5 for the second iteration with $A = 201.4$ A/A, we get the following values:

$$\beta = \frac{9}{|A|} = \frac{9}{201.4} = 0.0447$$

$$R_F = 10.69\text{ k}\Omega$$

$$R_i = \frac{v_s}{i_s} = (R_F + R_E) \| R_{B1} \| r_{\pi 1} = (10.69\text{ k} + 500) \| 1.53\text{ k} \| 2.58\text{ k} = 886.2\ \Omega$$

$$R_o = r_o = \infty$$

$$R_{oe} = R_o = \infty$$

$$v_{b1} = v_{be1} = R_i i_s = 886.2 i_s$$

$$v_{b2} = -\{R_{B2} \| [r_{\pi 2} + (1 + g_{m2} r_{\pi 2})(R_E \| R_F)]\} g_{m1} v_{be1}$$

$$= -\{2\text{ k} \| [2.58\text{ k} + (1 + 58.14\text{ m} \times 2.58\text{ k}) \times (11.13\text{ k} \| 500)]\} \times 58.14\text{ m} \times (886.2 i_s)$$

$$= -100.4 \times 10^3 i_s$$

$$v_{be2} = \frac{v_{b2} \times r_{\pi 2}}{[r_{\pi 2} + (1 + g_{m2}r_{\pi 2})(R_E \| R_F)]}$$

$$= \frac{(-100.4 \times 10^3 i_s) \times 2.58\text{ k}}{[2.58\text{ k} + (1 + 58.14\text{ m} \times 2.58\text{ k}) \times (11.13\text{ k} \| 500)]} = -3.466 \times 10^3 i_s$$

$$i_o = g_{m2}v_{be2} = 58.14\text{ m} \times (-3.466 \times 10^3 i_s) = -201.1 i_s$$

Therefore, the open-loop current gain A is given by $A = i_o/i_s = -201.1$ A/A.

$$R_{if} = \frac{R_i}{1 + \beta A} = \frac{886.2}{1 + 0.0447\text{ A/A} \times 201.1\text{ A/A}} = 88.5\ \Omega$$

The output resistance with feedback is

$$R_{of} = R_o(1 + \beta A) = \infty$$

Since the output is taken across R_C, the output resistance of the amplifier is

$$R_{out} = R_{of} \| R_C = R_C = 5\text{ k}\Omega$$

and $$A_f = \frac{201.1}{1 + 0.0447\text{ A/A} \times 201.1\text{ A/A}} = 20.13\text{ A/A}$$

which is 10% of $A = 201.1$. Thus, there is no need for further iterations.

(b) The shunt-series feedback amplifier for PSpice simulation is shown in Fig. 2.50. The results of PSpice simulation (.TF analysis) are shown below, with hand calculations to the right:

`I(VZ)/IS=2.002E+01=20.02`	$A_f = 20.13$ A/A
`INPUT RESISTANCE AT IS=8.803E+01=88.03 Ω`	$R_{if} = 88.5\ \Omega$
`OUTPUT RESISTANCE AT I(VZ)=7.991E+13=79.91 TΩ`	$R_{of} = \infty$

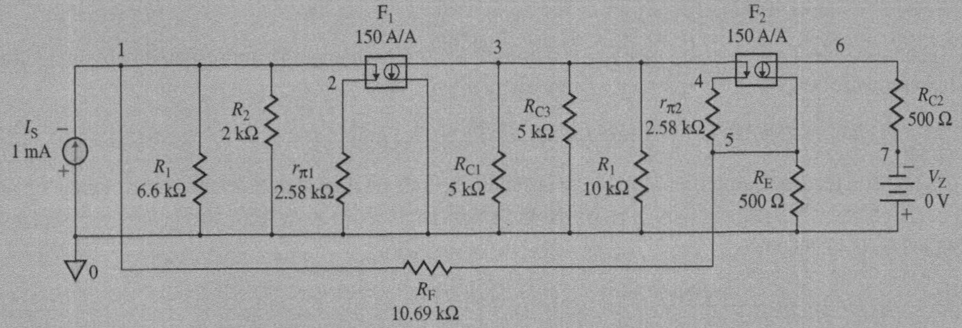

FIGURE 2.50 Two-stage shunt-series feedback amplifier for PSpice simulation

NOTE: The output current is measured through a dummy voltage source V_Z connected across R_{C2}. If V_Z were connected in series with R_{C2}, PSpice would give a very large output resistance as a result of the ideal current source of transistor Q_2.

KEY POINTS OF SECTION 2.10

- Designing feedback amplifiers requires determining the type of feedback and the feedback network.
- It is also necessary to find the component values of the network. A designer normally begins by finding the value of feedback factor β that gives the desired closed-loop gain, with the assumption of an ideal feedback network. Once initial estimates of the component values have been made, the normal analysis is performed to verify the closed-loop gain. Normally, several iterations are required to come to the final solution.

2.11 Stability Analysis

Negative feedback modifies the gain, the input resistance, and the output resistance of an amplifier. It also improves the performance parameters; for example, it reduces both the sensitivity of the gain to amplifier parameter changes and the distortion due to nonlinearities. However, negative feedback may become positive, thereby causing oscillation and instability. So far in this chapter we have assumed that the feedback network is resistive and the feedback factor β remains constant. But β can depend on frequency. In such cases, the closed-loop transfer function of a negative feedback circuit in Laplace's domain of s is given by

$$A_f(s) = \frac{S_o(s)}{S_i(s)} = \frac{A(s)}{1 + A(s)\beta(s)} \tag{2.100}$$

2.11.1 Closed-Loop Frequency and Stability

The open-loop gain $A(s)$ and the feedback factor $\beta(s)$ are dependent on the frequency. For physical systems, $s = j\omega$, and Eq. (2.100) can be written in the frequency domain as follows:

$$A_f(j\omega) = \frac{A(j\omega)}{1 + A(j\omega)\beta(j\omega)} \tag{2.101}$$

The loop gain $T_L(j\omega) = A(j\omega)\beta(j\omega)$ is a complex number that can be represented by its magnitude and phase, as follows:

$$T_L(j\omega) = A(j\omega)\beta(j\omega) = |A(j\omega)\beta(j\omega)|e^{j\phi(\omega)} = |T_L(j\omega)| \angle\phi \tag{2.102}$$

Whether a feedback amplifier is stable or unstable depends on the magnitude and phase of the loop gain $T_L(j\omega)$. Consider the frequency ω_{180} at which the phase angle $\phi(\omega) = \pm 180°$, so $T_L(j\omega) = -|T_L(j\omega)|$. Equation (2.101) becomes

$$A_f(j\omega_{180}) = \frac{A(j\omega_{180})}{1 - |T_L(j\omega_{180})|} \tag{2.103}$$

and the feedback becomes positive. However, the value of $A_f(j\omega_{180})$ will depend on the following conditions:

1. If $|T_L(j\omega_{180})| < 1$, the denominator $(1 - |T_L(j\omega_{180})|)$ will be less than unity; that is, $|A_f(j\omega_{180})| > |A(j\omega_{180})|$. In this case, the feedback system will be *stable*.
2. If $|T_L(j\omega_{180})| = 1$, the denominator $(1 - |T_L(j\omega_{180})|)$ will be zero, and $A_f(j\omega_{180})$ will be infinite; that is, the amplifier will have an output with zero input voltage, and the loop will oscillate without

any external input signal. In this case, the feedback system will be an *oscillator*. Let us assume that there is no input signal in Fig. 2.3; that is, $S_i = 0$. We get

$$S_f = A(j\omega_{180})\beta(j\omega_{180})S_e = |T_L(j\omega_{180})|S_e = -S_e \tag{2.104}$$

Since S_f is multiplied by -1 in the summer block at the input side, the feedback causes the signal S_e at the amplifier input to be sustained. Thus, there will be sinusoidal signals of frequency ω_{180} at the input and output of the amplifier. Under these conditions, the amplifier is said to oscillate at the frequency ω_{180}.

3. If $|T_L(j\omega_{180})| > 1$, the denominator $(1 - |T_L(j\omega_{180})|)$ will be greater than unity; that is, $|A_f(j\omega_{180})| < |A(j\omega_{180})|$. In this case, the feedback system will be *unstable*. We have

$$S_f = A(j\omega_{180})\beta(j\omega_{180})S_e = |T_L(j\omega_{180})|S_e \tag{2.105}$$

Since S_f is multiplied by -1 in the summer block at the input side, the amplifier oscillates, and the oscillations grow in amplitude until some nonlinearity reduces the magnitude of the loop gain $|T_L(j\omega_{180})|$ to exactly unity. In practical amplifiers, nonlinearity is always present in some form, and sustained oscillations will be obtained. This type of feedback condition is employed in oscillator circuits. Oscillators use positive feedback with a loop gain greater than unity; nonlinearity reduces the loop gain to unity.

We have seen that oscillations can occur in a negative feedback amplifier, depending on the frequency. The stability of an amplifier is determined by its response to an input or a disturbance. An amplifier is said to be *absolutely stable* if the output eventually settles down after a small disturbance. An amplifier is *absolutely unstable* if a small disturbance causes the output to build up (i.e., to increase continuously) until it reaches the saturation limits of the amplifier. The stability of an amplifier depends on its poles.

2.11.2 Poles and Instability

In the above analysis, the denominator of Eq. (2.100) was the key parameter in characterizing the response of an amplifier. The closed-loop gain $A_f(s)$ of an amplifier will be infinite if

$$1 + A(s)\beta(s) = 0 \tag{2.106}$$

Equation (2.106) is called the *characteristic equation*, and its roots determine the response of the amplifier. For example, if

$$1 + A(s)\beta(s) = s^2 + 2s + 5 = 0$$

the roots are $s = -1 \pm j2$. The roots of the characteristic equation are the poles of a system.

2.11.3 Transient Response and Stability

Consider an amplifier with a pole pair at $s = \sigma_o \pm j\omega_n$. If there is any disturbance (such as the closing of the DC power-supply switch), the equation of the transient response (after conversion from Laplace's s domain to the time domain) will contain terms of the following form:

$$v_o(t) = e^{\sigma_o t}[e^{+j\omega_n t} + e^{-j\omega_n t}] = 2e^{\sigma_o t}\cos(\omega_n t)$$

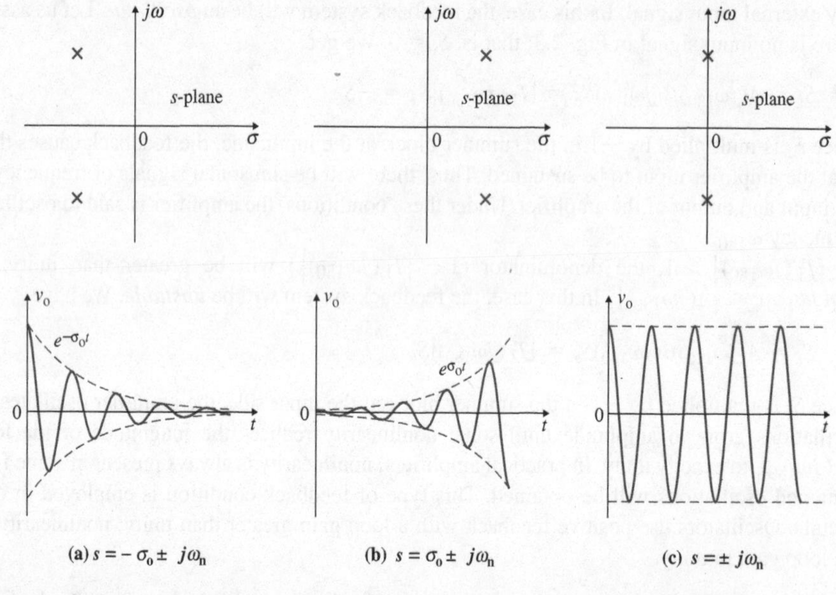

FIGURE 2.51 Relationship between pole location and transient response

This is a sinusoidal output with an envelope of exp $(\sigma_o t)$. Thus, the relationship between stability and the roots of this characteristic equation may be stated as follows:

1. If the poles are in the left half of the *s*-plane, as shown in Fig. 2.51(a), such that the denominator is of the form $(s - \sigma_o + j\omega_n)(s - \sigma_o - j\omega_n)$, all roots of the characteristic equation will have negative real parts, and σ_o will be negative. Thus, the response due to initial conditions or disturbances will decay exponentially to zero as time approaches infinity. The plot of such a response is also shown in Fig. 2.51(a). Such a system will be *stable*.
2. If the poles are in the right half of the *s*-plane, as shown in Fig. 2.51(b), such that the denominator is of the form $(s + \sigma_o + j\omega_n)(s + \sigma_o - j\omega_n)$, all roots will have positive real parts, and σ_o will be positive. The response will increase exponentially in magnitude as time increases until some nonlinearity limits its growth. The plot of such a response is also shown in Fig. 2.51(b). Such an amplifier will be *unstable*.
3. If the poles are on the $j\omega$-axis, as shown in Fig. 2.51(c), such that the denominator is of the form $(s + j\omega_n)(s - j\omega_n)$, the characteristic equation will not have roots with real parts, and σ_o will be zero. The response will be sustained oscillations, as shown in the plot in Fig. 2.51(c). The output will be sinusoidal in response to an initial condition or a disturbance. For a feedback amplifier, this will mean instability. However, for an oscillator circuit, this will mean the normal output.

2.11.4 Closed-Loop Poles and Stability

The location of the poles of a closed-loop system on a complex plane determines the transient response of a system. If the poles lie in the left half plane, all time-domain exponential terms decay to zero. If the poles lie in the right half plane, the system is likely to oscillate because its time-domain response will be growing exponentially. The plot of the poles as the loop gain varies on a complex plane, is called the *root locus* and it indicates how close the systems are to oscillate the system. Let us take an amplifier with one pole of the open-loop gain relation as given by

$$A(s) = \frac{A_o}{1 + s/\omega_o} \tag{2.107}$$

Therefore, the closed-loop gain becomes

$$A_f(s) = \frac{A(s)}{1 + \beta A(s)} = \frac{A_o/1 + \beta A_o}{1 + s/\omega_o(1 + \beta A_o)} \tag{2.108}$$

This gives the closed-loop pole $\omega_p = -\omega_o(1 + \beta A_o)$, which has a real-valued pole in the left half plane. The pole as shown in Fig. 2.52(a) moves away from the origin as the loop gain increases.

Let us consider an open-loop gain with two poles as given by

$$A(s) = \frac{A_o}{\left(1 + s/\omega_{p1}\right)\left(1 + s/\omega_{p1}\right)} \tag{2.109}$$

This gives the closed-loop gain as

$$\begin{aligned} A_f(s) &= \frac{A(s)}{1 + \beta A(s)} = \frac{A_o}{\left(1 + s/\omega_{p1}\right)\left(1 + s/\omega_{p1}\right) + \beta A_o} \\ &= \frac{A_o\omega_{p1}\omega_{p2}}{s^2 + (\omega_{p1} + \omega_{p2})s + (1 + \beta A_o)\omega_{p1}\omega_{p2}} \end{aligned} \tag{2.110}$$

This can be solved for the closed-loop two poles as given by

$$s_1, s_2 = \frac{-(\omega_{p1} + \omega_{p2}) \pm \sqrt{(\omega_{p1} + \omega_{p2})^2 - 4(1 + \beta A_o)\omega_{p1}\omega_{p2}}}{2} \tag{2.111}$$

for $\beta = 0$, $s_1 = -\omega_{p1}$, and $s_2 = -\omega_{p2}$. As β increases, the poles move toward each other and become equal at $\beta = \beta_1$. For $\beta > \beta_1$, they become complex. The pole plot is shown in Fig. 2.52(b).

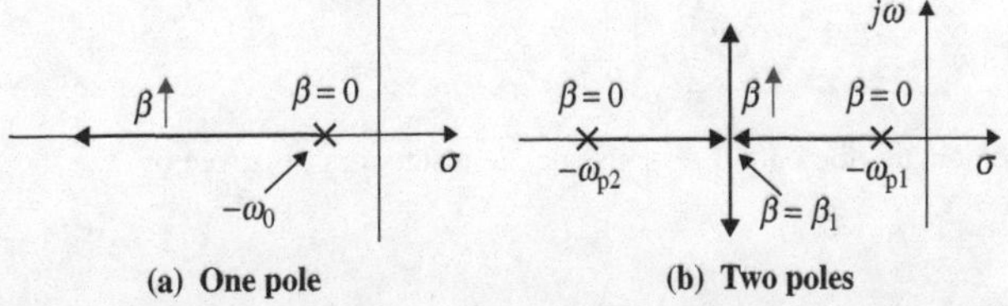

FIGURE 2.52 Closed-loop pole plots

EXAMPLE 2.13

Finding the stability due to a step input The open-loop gain of an amplifier is given by

$$A(s) = \frac{s}{s^2 - 2}$$

Determine the stability of the closed-loop response due to a step input signal. Assume feedback factor $\beta(s) = 1$.

SOLUTION

For a step input in Laplace's domain of s, $S_i(s) = 1/s$. From Eq. (2.5), the closed-loop gain is

$$A_f(s) = \frac{S_o(s)}{S_i(s)} = \frac{s/(s^2 - 2)}{1 + s/(s^2 - 2)} = \frac{s}{s^2 + s - 2} = \frac{s}{(s - 1)(s + 2)}$$

For a step input, $S_i(s) = 1/s$ and the output response is given by

$$S_o(s) = A_f(s)S_i(s) = \frac{A_f(s)}{s} = \frac{1}{(s - 1)(s + 2)} = \frac{1}{3}\left(\frac{1}{s - 1} - \frac{1}{s + 2}\right)$$

Thus, in the time domain, the output response due to a step input becomes

$$s_o(t) = \frac{1}{3}(e^t - e^{-2t})$$

Therefore, at $t = \infty$, $s_o(t) = \infty$. The amplifier will be unstable because it has a positive pole, $s = 1$.

EXAMPLE 2.14

Finding the step response of a feedback amplifier The open-loop gain of an amplifier is given by

$$A(s) = \frac{1}{s^2 + 3}$$

Determine the closed-loop step response of the amplifier. Assume feedback factor $\beta(s) = 1$.

SOLUTION

For a step input in Laplace's domain of s, $S_i(s) = 1/s$. From Eq. (2.5), the closed-loop gain is

$$A_f(s) = \frac{S_o(s)}{S_i(s)} = \frac{1/(s^2 + 3)}{1 + 1/(s^2 + 3)} = \frac{1}{s^2 + 4}$$

For a step input, the closed-loop response is

$$S_o(s) = A_f(s)S_f(s) = \frac{1}{(s^2 + 4)s} = \frac{1}{4}\left(\frac{1}{s} - \frac{s}{s^2 + 4}\right)$$

Therefore, in the time domain, the output response due to a step input is given by

$$s_o(t) = \frac{1}{4}(1 - \cos 2t)$$

Therefore, the response due to a step input is *oscillatory*. The amplifier will be unstable because it has poles on the $j\omega$-axis, $s = \pm j2$.

2.11.5 Nyquist Stability Criterion

If the loop gain $T_L(j\omega) = A(j\omega)\beta(j\omega)$ in Eq. (2.102) becomes -1, the closed-loop gain will tend to be infinite, and the amplifier will be unstable. Therefore, a feedback amplifier will be unstable if the loop gain $T_L(j\omega)$ becomes

$$T_L(j\omega) = A(j\omega)\beta(j\omega) = -1 \tag{2.112}$$

To satisfy the condition of Eq. (2.112), the magnitude and phase angle of the loop gain must be unity and $\pm 180°$, respectively; that is,

$$|T_L(j\omega)| = 1 \tag{2.113}$$

$$\phi = \angle T_L(j\omega) = \pm 180° \tag{2.114}$$

Therefore, investigating the stability of an amplifier requires determining the frequency response and characterizing the magnitude and phase angle against the frequency. The frequency response can be represented by a series of phasors, each at a different frequency. Joining the extremities (i.e., the magnitude points) of these phasors gives the frequency locus, as shown in Fig. 2.53, which is a plot of magnitude versus phase in polar coordinates as the frequency ω is varied from 0 to $\pm\infty$. This plot is known as a *Nyquist plot*. The $T_L(j\omega)$ plot for negative frequencies is the mirror image through the Re-axis (real axis) of the plot for positive frequencies. The Nyquist plot intersects the negative Re-axis at the frequency ω_{180}. If the intersection occurs to the left of the point $(-1, 0)$, the magnitude of the loop gain at this frequency is greater than unity and the amplifier will be unstable. On the other hand, if the intersection occurs to the right of the point $(-1, 0)$, the amplifier will be stable. The point on the locus where the phase is $-180°$ is called the *phase crossover frequency* ω_p $(=\omega_{180})$; the point on the locus where the gain is unity is called the *gain crossover frequency* ω_g.

The stability of a system can be determined from the Nyquist criterion for stability, which can be stated as follows: If the Nyquist plot encircles the point $(-1, 0)$, the amplifier is unstable. This criterion tests for the poles of loop gain $T_L(s)$ in the right half plane [2]. If the Nyquist plot encircles the point $(-1, 0)$, the amplifier has poles in the right half plane, and the circuit will oscillate. The number of times the plot encircles the point $(-1, 0)$ gives the number of poles in the right half plane. There are two in Fig. 2.53.

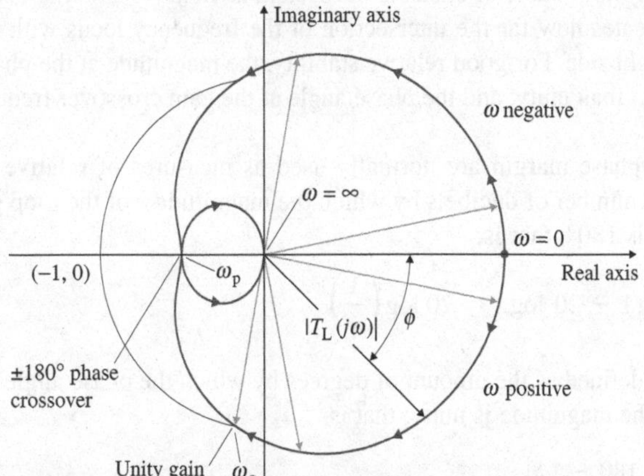

FIGURE 2.53 Typical Nyquist plot, or polar plot

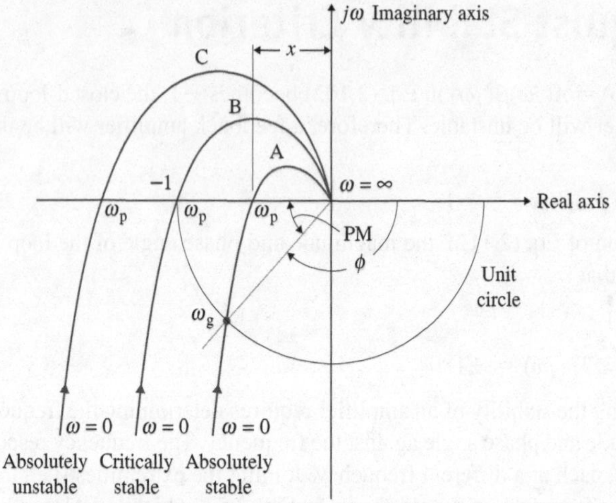

FIGURE 2.54 Nyquist plots

The Nyquist plots of amplifiers with three different characteristics are shown in Fig. 2.54. For plot A, the loop gain is less than unity. Therefore, oscillations will not build up on the closed loop, and the amplifier will be absolutely stable. For plot B through $(-1, 0)$, the loop gain is unity, and the poles are on the imaginary axis. The amplifier will be critically stable on the closed loop. For plot C, the loop gain is greater than unity, and the amplifier will be absolutely unstable on the closed loop.

2.11.6 Relative Stability

The term "absolutely stable" or "absolutely unstable" tells us what will eventually happen to a system. However, it does not say *how* stable or unstable the system is. *Relative stability* is a measure of the degree of stability, and it indicates how far the intersection of the frequency locus with the Re-axis is from the point $(-1, 0)$ on the right side. For good relative stability, the magnitude at the phase crossover frequency should have a value less than unity, and the phase angle at the gain crossover frequency should not have a value near $\pm 180°$.

Gain margin and phase margin are normally used as measures of relative stability. *Gain margin* (GM) is defined as the number of decibels by which the magnitude x of the loop gain falls short of unity when the phase angle is 180°; that is,

$$\text{GM} = 20\log 1 - 20\log x = 20\log\left(\frac{1}{x}\right) \tag{2.115}$$

Phase margin (PM) is defined as the amount of degrees by which the phase angle ϕ of the loop gain falls short of $\pm 180°$ when the magnitude is unity; that is,

$$\text{PM} = \phi_m = 180 - |\phi| \tag{2.116}$$

For adequate relative stability, the gain margin and phase margin should be greater than 10 dB and 45°, respectively.

EXAMPLE 2.15

Finding the phase margin and gain margin of a feedback amplifier Determine the phase margin and gain margin of a feedback amplifier whose loop gain is given by

$$T_L(j\omega) = \frac{2}{(1 + j\omega)^3}$$

SOLUTION

The locus will cross the negative Re-axis when the phase is $\phi = -180°$ at $\omega = \omega_{180} = \omega_p$. Thus,

$$3 \tan^{-1}(\omega_p) = 180° \quad \text{or} \quad \tan^{-1}\omega_p = 60° \quad \text{or} \quad \omega_p = \sqrt{3} \text{ rad/s}$$

The magnitude at $\omega = \omega_p$ becomes

$$x = |T_L(j\omega)| = \frac{2}{(\sqrt{1 + 3})^3} = 0.25$$

That is, $x < 1$. Thus, the frequency locus will not encircle the point $(-1, 0)$, and the amplifier will be absolutely stable on the closed loop. Using Eq. (2.115), we have

$$\text{GM} = 20 \log\left(\frac{1}{0.25}\right) = 12.04 \text{ dB}$$

The gain crossover frequency ω_g is found when the gain is unity:

$$x = |T_L(j\omega_g)| = \frac{2}{(\sqrt{1 + \omega_g^2})^3} = 1$$

so $\quad (1 + \omega_g^2)^3 = 2^2 \quad$ or $\quad 1 + \omega_g^2 = 1.5874 \quad$ or $\quad \omega_g = 0.7664$ rad/s

The phase angle ϕ (at $\omega = \omega_g$) is

$$\phi_g = 3 \tan^{-1}\omega_g = 3 \tan^{-1}(0.7664) = 3 \times 37.47° = 112.4°$$

Using Eq. (2.116), we have

$$\text{PM} = \phi_m = 180° - 112.4° = 67.6°$$

2.11.7 Effects of Phase Margin

At the gain crossover frequency ω_g, the magnitude of the loop gain is unity. That is,

$$|T_L(j\omega_g)| = |A(j\omega_g)|\beta = 1$$

or

$$|A(j\omega_g)| = \frac{1}{\beta} \tag{2.117}$$

where β is assumed to remain constant and is independent of frequency. The phase margin influences the transient and frequency responses of a feedback amplifier, and its effects can be determined from Eq. (2.101). The effects of the phase margin will be illustrated through the analysis of four cases.

Case 1. The phase margin is PM = $\phi_m = 30°$, and $|\phi| = 180° - 30° = 150°$. Substituting Eq. (2.117) into Eq. (2.101) yields

$$A_f(j\omega_g) = \frac{A(j\omega_g)}{1 + 1\angle\phi} = \frac{A(j\omega_g)}{1 + e^{j\phi}} = \frac{A(j\omega_g)}{1 + e^{-j150°}} = \frac{A(j\omega_g)}{1 - 0.866 - j0.5} = \frac{A(j\omega_g)}{0.134 - j0.5}$$

which gives the magnitude of the closed-loop gain as

$$|A_f(j\omega_g)| = \frac{|A(j\omega_g)|}{0.517} = 1.93|A(j\omega_g)| = \frac{1.93}{\beta} \tag{2.118}$$

Therefore, there will be a peak of 1.93 times the low-frequency gain of $1/\beta$.

Case 2. The phase margin is PM = $\phi_m = 45°$, and $\phi = 180° - 45° = 135°$.

$$A_f(j\omega_g) = \frac{A(j\omega_g)}{1 + e^{-j135°}} = \frac{A(j\omega_g)}{1 - 0.71 - j0.71} = \frac{A(j\omega_g)}{0.29 - j0.707}$$

and $$|A_f(j\omega_g)| = \frac{|A(j\omega_g)|}{0.765} = 1.306|A(j\omega_g)| = \frac{1.306}{\beta} \tag{2.119}$$

In this case, there will be a peak of 1.306 times the low-frequency gain of $1/\beta$.

Case 3. The phase margin is PM = $\phi_m = 60°$, and $\phi = 180° - 60° = 120°$.

$$A_f(j\omega_g) = \frac{A(j\omega_g)}{1 + e^{-j120°}} = \frac{A(j\omega_g)}{1 - 0.5 - j0.866} = \frac{A(j\omega_g)}{0.5 - j0.866}$$

and $$|A_f(j\omega_g)| = |A(j\omega_g)| = \frac{1}{\beta} \tag{2.120}$$

In this case, there will be no peak above the low-frequency gain of $1/\beta$.

Case 4. The phase margin is PM = $\phi_m = 90°$, and $\phi = 180° - 90° = 90°$.

$$A_f(j\omega_g) = \frac{A(j\omega_g)}{1 + e^{-j90°}} = \frac{A(j\omega_g)}{1 - j1.0}$$

and $$|A_f(j\omega_g)| = \frac{|A(j\omega_g)|}{\sqrt{2}} = 0.707|A(j\omega_g)| = \frac{0.707}{\beta} \tag{2.121}$$

In this case, there will be a gain reduction below the low-frequency gain of $1/\beta$.

The frequency responses of a transfer function with one pole and a transfer function with two poles are shown in Fig. 2.55 for various values of the phase margin. For a 90° phase margin, the transfer function has only one pole. In a two-pole system, as the phase margin is reduced, the gain peak increases until the gain approaches infinity, and oscillation occurs at a phase margin of $\phi_m = 0$. After the gain peak (at a normalized frequency of $f/f_g = 1$), the gain decays with a slope of -40 dB/decade because there are two poles in the transfer function.

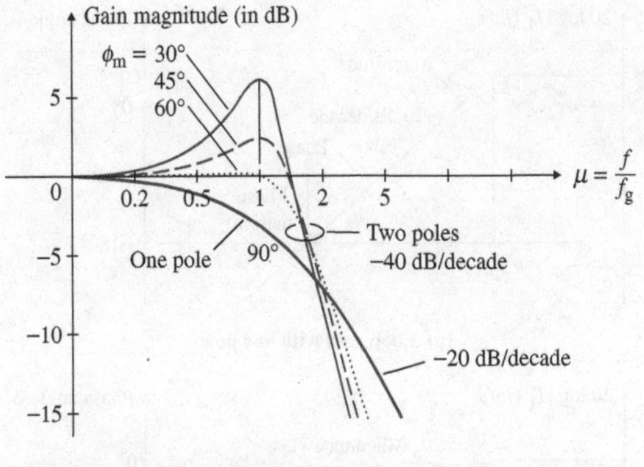

FIGURE 2.55 Effect of phase margin on frequency response

2.11.8 Stability Using Bode Plots

A *Bode plot*, which is a plot of magnitude and phase against frequency, is a very convenient method of determining the stability of an amplifier. The loop gain is plotted in decibels, and the frequency is plotted on a logarithmic scale. Consider the loop gain with a single pole frequency given by

$$T_L(j\omega) = A(j\omega)\beta(j\omega) = \frac{A_o}{1 + j\omega/\omega_{p1}} = \frac{A_o}{1 + jf/f_{p1}} \tag{2.122}$$

whose phase angle is $\phi = \angle T_L(j\omega) = -\tan^{-1}(f/f_{p1})$. The typical Bode plot is shown in Fig. 2.56(a). The magnitude of the loop gain is 20 log $|T_L(j\omega)|$ until $f = f_{p1}$, at which point the loop gain falls at a rate of 20 dB/decade and the phase angle is $\phi = -45°$. *Phase margin* is the difference between 180° and the phase angle when the gain is 0 dB. *Gain margin*, which is the value of the magnitude in dB when the phase angle is −180°, can be read from the plot. Thus, for a single-pole amplifier, $\phi = -90°$ at $T_L(j\omega_g) = 1$, and the phase margin is $\phi_m = 180 - |\phi| = 90°$. There is no phase crossover, and the gain margin is infinite. An amplifier with negative feedback will always be stable.

Consider the loop gain with three-pole frequencies given by

$$T_L(j\omega) = A(j\omega)\beta(j\omega) = \frac{A_o}{(1 + jf/f_{p1})(1 + jf/f_{p2})(1 + jf/f_{p3})} \tag{2.123}$$

whose phase angle can be found from

$$\phi = \angle T_L(j\omega) = -\tan^{-1}\left(\frac{f}{f_{p1}}\right) - \tan^{-1}\left(\frac{f}{f_{p2}}\right) - \tan^{-1}\left(\frac{f}{f_{p3}}\right) \tag{2.124}$$

The typical Bode plot is shown in Fig. 2.56(b). Assuming that the poles are widely separated (i.e., by a decade: $f_{p3} = 10f_{p2} = 100f_{p1}$), the phase angle ϕ is approximately −45° at the first pole frequency f_{p1}, (−90° − 45° =) −135° at the second pole frequency f_{p2}, and (−180° − 45° =) −225° at the third pole frequency f_{p3}. The phase angle at $T_L(j\omega_g) = 1$ is $\phi = -270°$, which gives a phase margin of $\phi_m = 180° - 270° = -90°$; that is, there is no phase margin, and the amplifier will be absolutely unstable. However, at $\omega = \omega_p = \omega_{180}$, the loop gain is positive; that is, there is a gain margin.

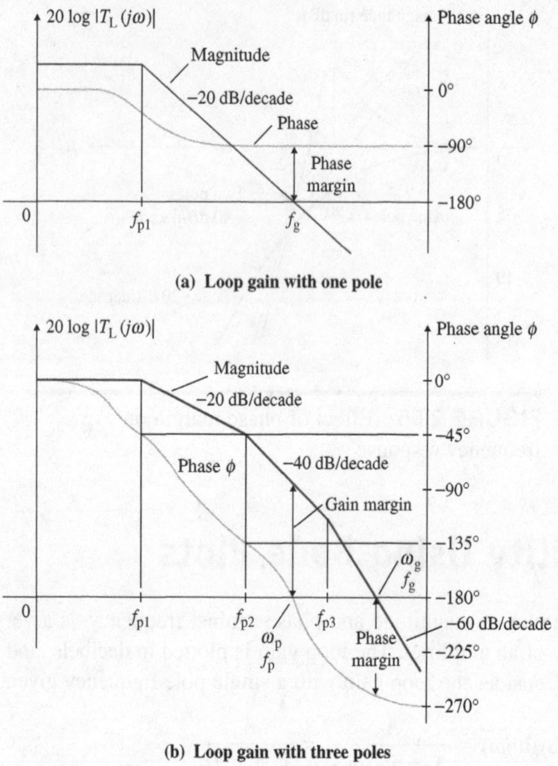

FIGURE 2.56 Typical Bode plots

Rather than plotting the loop gain 20 log $|T_L(j\omega)|$ directly, we can choose to plot 20 log $|A(j\omega)|$ and 20 log $|1/\beta(j\omega)|$ separately. This approach is shown in Fig. 2.57 for two resistive values of β. The difference between the two curves gives

$$20 \log |A(j\omega)| - 20 \log \left|\frac{1}{\beta}(j\omega)\right| = 20 \log |A(j\omega)| + 20 \log |\beta(j\omega)| = 20 \log |T_L(j\omega)|$$

The sum of the two phase angles gives

$$\angle A(j\omega) - \angle \frac{1}{\beta}(j\omega) = \phi$$

This approach has the advantage of allowing us to investigate the stability of an amplifier for a variety of feedback networks simply by drawing the lines for 20 log $|1/\beta(j\omega)|$. For zero gain margin,

$$20 \log |A(j\omega)| - 20 \log \left|\frac{1}{\beta}(j\omega)\right| = 0$$

That is, the intersection of the 20 log $|1/\beta(j\omega)|$ plot with the 20 log $|A(j\omega)|$ plot gives the critical value of β.

We can see from Fig. 2.57 that the $-180°$ phase occurs on the -40 dB/decade segment of the Bode plot. Therefore, for stability, the 20 log $|1/\beta(j\omega)|$ plot should intersect the 20 log $|A(j\omega)|$ plot at a point on the -20 dB/decade segment. As a general rule, the difference of slopes at the intersection—which is called the *rate of closure*—should not exceed 20 dB/decade.

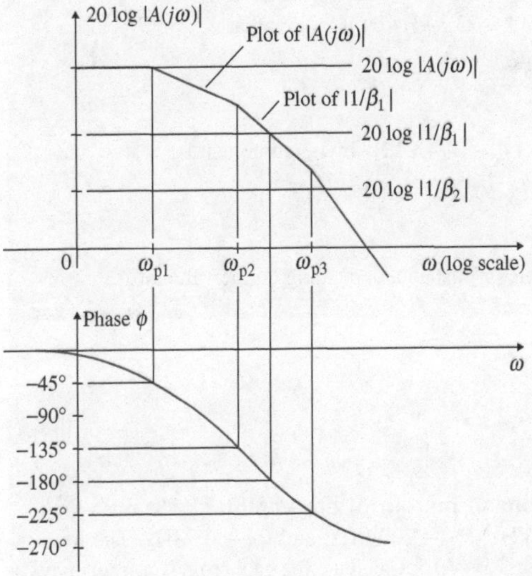

FIGURE 2.57 Separate Bode plots for $A(j\omega)$ and $1/\beta(j\omega)$

EXAMPLE 2.16

Finding the phase crossover frequency and the phase margin of an amplifier The open-loop gain of an amplifier has break frequencies at $f_{p1} = 10$ kHz, $f_{p2} = 100$ kHz, and $f_{p3} = 1$ MHz. The low-frequency (or DC) gain is $A_o = 2 \times 10^5$. Calculate **(a)** the phase crossover frequency f_p and **(b)** the phase margin ϕ_m for $\beta = 0.01$, 0.001, and 0.0001.

SOLUTION

The magnitude of the open-loop gain is given by

$$A(j\omega) = \frac{2 \times 10^5}{(1 + jf/10^4)(1 + jf/10^5)(1 + jf/10^6)}$$

$$|A(j\omega)| = \frac{2 \times 10^5}{[1 + (f/10^4)^2]^{1/2}[1 + (f/10^5)^2]^{1/2}[1 + (f/10^6)^2]^{1/2}}$$

and the phase angle is

$$\phi = -\tan^{-1}\left(\frac{f}{10^4}\right) - \tan^{-1}\left(\frac{f}{10^5}\right) - \tan^{-1}\left(\frac{f}{10^6}\right)$$

(a) The frequency at which $\phi = -180°$ is found by iteration as $f_p = f_{180} = 333$ kHz.

(b) For $\beta = 0.01$ and $|A(j\omega)| = 1/\beta = 100$, we get $f_g = 1144.8$ kHz by iteration; that is,

$$\phi = -223.4° \quad \text{and} \quad \phi_m = 180 - |\phi| = -43.4°$$

Thus, the amplifier will be unstable.

For $\beta = 0.001$ and $|A(j\omega)| = 1/\beta = 1000$, we get $f_g = 423.3$ kHz by iteration; that is,

$$\phi = -188.3^\circ \quad \text{and} \quad \phi_m = 180 - |\phi| = -8.3^\circ$$

Thus, the amplifier will be unstable.

For $\beta = 0.0001$ and $|A(j\omega)| = 1/\beta = 10{,}000$, we get $f_g = 124.1$ kHz by iteration; that is,

$$\phi = -143.6^\circ \quad \text{and} \quad \phi_m = 180 - |\phi| = 36.4^\circ$$

Thus, the amplifier will be stable.

We can conclude that a feedback amplifier can be stable or unstable, depending on the value of β.

EXAMPLE 2.17

Finding the phase crossover frequency and the phase margin of an amplifier The open-loop gain of an amplifier has break frequencies at $f_{p1} = 100$ kHz, $f_{p2} = 200$ kHz, and $f_{p3} = 1$ MHz. The low-frequency (or DC) gain is $A_o = 800$, and the feedback factor is $\beta = 0.5$. Calculate the gain crossover frequency f_g and the phase margin ϕ_m.

SOLUTION

The low-frequency loop gain is $A_o\beta = 800 \times 0.5 = 400$. The magnitude of the loop gain is given by

$$|T_L(j\omega)| = x = \frac{400}{[1 + (f/f_{P1})^2]^{1/2}[1 + (f/f_{p2})^2]^{1/2}[1 + (f/f_{p3})^2]^{1/2}}$$

and the phase angle is

$$\phi = -\tan^{-1}\left(\frac{f}{f_{p1}}\right) - \tan^{-1}\left(\frac{f}{f_{p2}}\right) - \tan^{-1}\left(\frac{f}{f_{p3}}\right)$$

At the gain crossover $|T_L(j\omega_g)| = 1$, the gain crossover frequency f_g can be determined from the Bode plot or by iteration. By iteration, the value of frequency that gives a loop gain of unity is found to be $f_g = 1917.03$ kHz, and the corresponding value of phase angle is $\phi = -233.5^\circ$. Thus, the phase margin is

$$\phi_m = 180^\circ - |\phi| = 180^\circ - 233.5^\circ = -53.5^\circ$$

which is negative, so the amplifier will be unstable.

KEY POINTS OF SECTION 2.11

- An amplifier with negative feedback can be stable or unstable, depending on the frequency and the feedback factor β. If the loop gain is $|T_L(j\omega)| \geq 1$ and its phase angle is $\phi = \pm 180^\circ$, the amplifier will be unstable.
- Nyquist and Bode plots can be used to determine the stability of a feedback amplifier. The degree of stability is normally measured by the gain margin and the phase margin. A phase margin of 45° is usually adequate to limit the peak to 30% of the low-frequency gain.

2.12 Compensation Techniques

We know that if an amplifier has more than two poles, the phase angle of the loop gain could exceed $-180°$ beyond a certain frequency. An amplifier with negative feedback can be unstable, depending on the frequency ω and the amount of feedback β. The process of stabilizing an unstable feedback amplifier is called *compensation*. The basic amplifier of a feedback circuit should be designed with as few stages as possible because each stage of gain adds more poles to the transfer function, making the compensation problem more difficult. An amplifier can be stabilized by adding a dominant pole, by changing the dominant pole, by Miller compensation, or by modifying the feedback path [3–5].

2.12.1 Addition of a Dominant Pole

A dominant pole can be introduced into the amplifier so that the phase shift is less than $-180°$ when the loop gain is unity. Consider a feedback amplifier with loop gain of the form

$$T_L(j\omega) = A(j\omega)\beta = \frac{A_o}{(1 + j\omega/\omega_{p1})(1 + j\omega/\omega_{p2})(1 + j\omega/\omega_{p3})} \tag{2.125}$$

where $\omega_{p1} = 2\pi f_{p1}$ rad/s, $\omega_{p2} = 2\pi f_{p2}$ rad/s, and $\omega_{p3} = 2\pi f_{p3}$ rad/s. The Bode plot is shown in Fig. 2.58(a). To compensate the amplifier, a new dominant pole $\omega_D = 2\pi f_D$ is introduced so that $\omega_D < \omega_{p1} < \omega_{p2} < \omega_{p3}$, and the resultant loop gain becomes

$$T_L(j\omega) = A(j\omega)\beta = \frac{A_o}{(1 + j\omega/\omega_D)(1 + j\omega/\omega_{p1})(1 + j\omega/\omega_{p2})(1 + j\omega/\omega_{p3})} \tag{2.126}$$

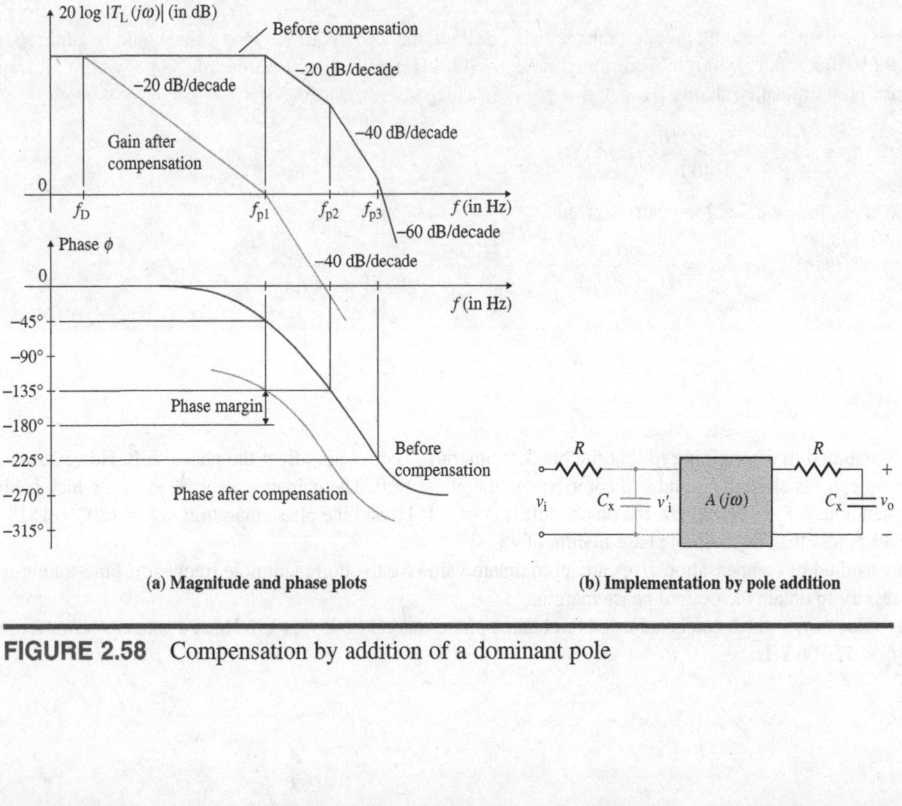

FIGURE 2.58 Compensation by addition of a dominant pole

The Bode plot of this modified loop gain is indicated in Fig. 2.58(a) by lighter lines. The introduction of the dominant pole causes the loop gain to fall at a rate of 20 dB/decade until frequency f_{p1} is reached. If the frequency f_D of the dominant pole is chosen so that the loop gain is unity at frequency f_{p1}, then the phase shift at frequency f_{p1} due to the dominant pole f_D is $-90°$, and the phase shift due to the first pole f_{p1} is $-45°$. At $f = f_{p1}$, the total phase shift is $(-90° - 45° =) -135°$, and the phase margin is $\phi_m = 180° - |\phi| = 180° - 135° = 45°$, which means that the amplifier will be stable. The original amplifier would have been unstable in the feedback connection. This compensation is achieved at the expense of bandwidth reduction. The uncompensated unity-gain bandwidth f_{p1} is much higher than the compensated unity-gain bandwidth f_D. This method of compensation involves a sacrifice of the frequency capability of the amplifier and is often known as *narrowbanding*.

A dominant pole can be implemented by adding a capacitor C_x in such a manner that it adds a break frequency to the basic amplifier. With this approach, illustrated in Fig. 2.58(b) showing two possible locations, f_D is given by

$$f_D = \frac{1}{2\pi R C_x} \tag{2.127}$$

where R is the resistance seen by capacitor C_x.

EXAMPLE 2.18

Stabilizing an amplifier by adding a dominant pole Stabilize the amplifier in Example 2.17 by adding a dominant pole so that the phase margin is 45°.

SOLUTION

Since the gain–bandwidth product must remain constant, the low-frequency loop gain should be reduced from $A_o\beta$ (or $800 \times 0.5 = 400$) at f_D to unity at f_{p1} ($=100$ kHz) with a -20 dB/decade slope, which indicates a direct proportionality; that is, $f_D \times A_o\beta = f_{p1} \times 1$, which gives

$$f_D = \frac{f_{p1}}{A_o\beta} = \frac{100 \times 10^3}{400} = 250 \text{ Hz}$$

Therefore, the modified loop gain is given by

$$T_L(j\omega) = A(j\omega)\beta = \frac{400}{(1 + jf/f_D)(1 + jf/f_{p1})(1 + jf/f_{p2})(1 + jf/f_{p3})}$$

▶ NOTES

1. It is assumed in determining f_D that the break frequency f_{p2} does not affect the phase shift. However, in this example, f_{p2} is close to f_{p1} and *will* contribute to the phase shift. The gain crossover frequency, which is obtained by iteration, is $f_g = 74{,}795$ Hz, the phase shift is $\phi = -151°$, and the phase margin is $\phi_m = 180° - 151° = 29°$, which is less than the desired phase margin of 45°.
2. This method of compensation gives an approximate value for the dominant pole frequency. Fine-tuning is necessary to obtain the desired phase margin.
3. The value of f_D, which can be adjusted to obtain a phase margin of 45°, is 152 Hz at a gain crossover frequency of $f_g = 52.106$ kHz.

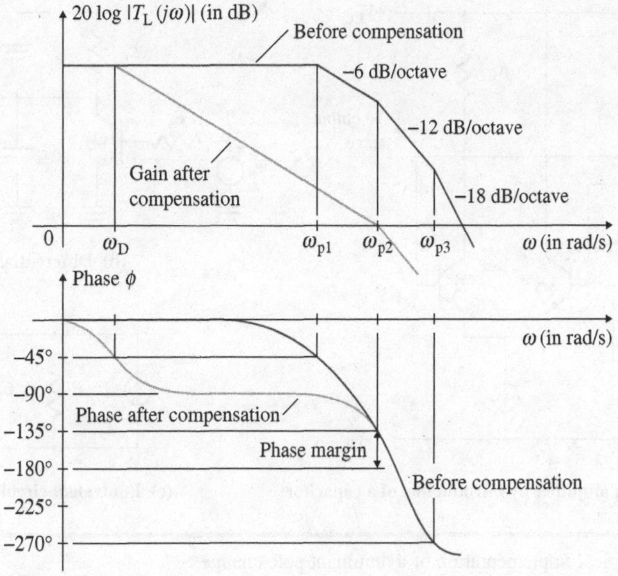

FIGURE 2.59 Compensation by changing the dominant pole

2.12.2 Changing the Dominant Pole

In the compensation method just discussed in Sec. 2.12.1, a dominant pole was added to the amplifier, and the original amplifier poles were assumed to be unaffected. This approach reduces the bandwidth considerably. A second method of compensation is to change the dominant pole, adding a capacitor to the amplifier in such a way that the original dominant frequency f_{p1} is reduced so that it performs the compensation function. That is, the original pole f_{p1} is moved to the left so that $f_D = f_{p1}$. This modification is shown in Fig. 2.59. For a 45° phase margin in a unity feedback amplifier, f_D must cause the gain to fall to unity at f_{p2}. Thus, f_{p2} will become the unity-gain bandwidth. Since f_{p2} is five or ten times the frequency f_{p1}, this method gives a substantial improvement in bandwidth.

The dominant pole is changed by adding a capacitor internally to the amplifier. One way of doing so is shown in Fig. 2.60(a) for the differential stage of an op-amp. The differential half circuit is shown in Fig. 2.60(b), and the small-signal equivalent circuit is shown in Fig. 2.60(c). The current i_s is the collector current of Q_2, C_i is the internal capacitance, and R_i is the effective resistance; that is, $C_i = C_{\pi 4}$, and $R_i = R_{C1} \parallel r_{o2} \parallel r_{\pi 4}$. Thus, the dominant pole frequency can be found from

$$f_D = f_{p1} = \frac{1}{2\pi R_i(C_i + C_x)} \tag{2.128}$$

where C_x is the additional capacitance that is added to give the desired dominant frequency. The disadvantage of this method is that the value of C_x is usually quite large (typically >1000 pF). Thus, it will be difficult—if not impossible—to realize a compensating capacitor in an IC chip. The maximum practical size of a monolithic capacitor is about 100 pF.

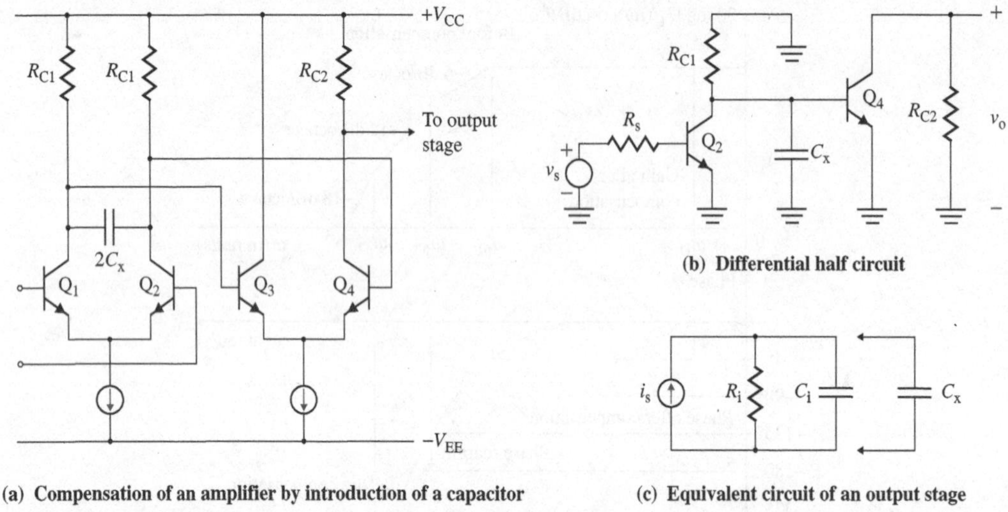

FIGURE 2.60 Typical implementation of a dominant pole change

We have assumed that the addition of C_X will reduce the original dominant pole frequency f_{p1} so that it performs the compensation function. However, the higher-frequency poles of the amplifier will also be changed by the addition of C_X. In practice, the effect of C_X on the pole positions is usually evaluated by computer simulation. Another estimate of C_X is made on the basis of the new data, and this process of trial and error continues until the desired stability condition is reached, usually after several iterations.

EXAMPLE 2.19

Finding the compensating capacitance to modify the dominant pole The equivalent circuit of an output stage is shown in Fig. 2.60(c), where $R_i = 24\ \text{k}\Omega$ and $C_i = 20$ pF.

(a) Find the break frequency f_P.

(b) Find the additional capacitance C_X that will move the break frequency to $f_D = 40$ kHz.

SOLUTION

(a) The break frequency f_P is given by

$$f_P = \frac{1}{2\pi C_i R_i} = \frac{1}{2\pi \times 20 \times 10^{-12} \times 24 \times 10^3} = 331.6\ \text{kHz}$$

(b) An additional capacitance C_X will move the break frequency to f_D; that is,

$$f_D = \frac{1}{2\pi (C_i + C_X) R_i}$$

which, for $f_D = 40$ kHz, gives $C_i + C_X = 165.8$ pF; that is, $C_X = 165.8 - 20 = 145.8$ pF.

2.12.3 Miller Compensation and Pole Splitting

In a third compensation method, a small capacitance (say, C_x) is connected between the input and output of a gain stage in a multistage amplifier. Compensation is achieved using Miller multiplication of the capacitance. Figure 2.61(a) illustrates this form of compensation for the μA741 op-amp. C_x constitutes a shunt-shunt feedback, and the feedback factor β depends on the frequency. The Darlington pair consisting of Q_{16} and Q_{17} can be replaced by a single equivalent transistor Q, as shown in Fig. 2.61(b); the simplified equivalent circuit is shown in Fig. 2.61(c). R_i and C_i represent the total resistance and total capacitance, respectively, between node B and the ground. R_o and C_o represent the total resistance and total capacitance, respectively, between node C and the ground. C_i includes the Miller capacitance due to C_μ of the transistor and the output capacitance of the preceding stage. Similarly, C_o includes the Miller capacitance due to C_μ and the input capacitance of the succeeding stage. In the absence of the compensating capacitance (i.e., when $C_x = 0$), the two poles will be

$$f_{p1} = \frac{1}{2\pi C_i R_i} \tag{2.129}$$

and
$$f_{p2} = \frac{1}{2\pi C_o R_o} \tag{2.130}$$

The analysis of Fig. 2.61(c) will be similar to that of Fig. 2.23(b). Using Eqs. (H.8) and (H.9) in Appendix H, we get as the new poles

$$\omega'_{p1} \approx \frac{1}{g_m C_x R_i R_o} \tag{2.131}$$

$$\omega'_{p2} \approx \frac{g_m C_x}{C_i C_o + C_x(C_i + C_o)} \tag{2.132}$$

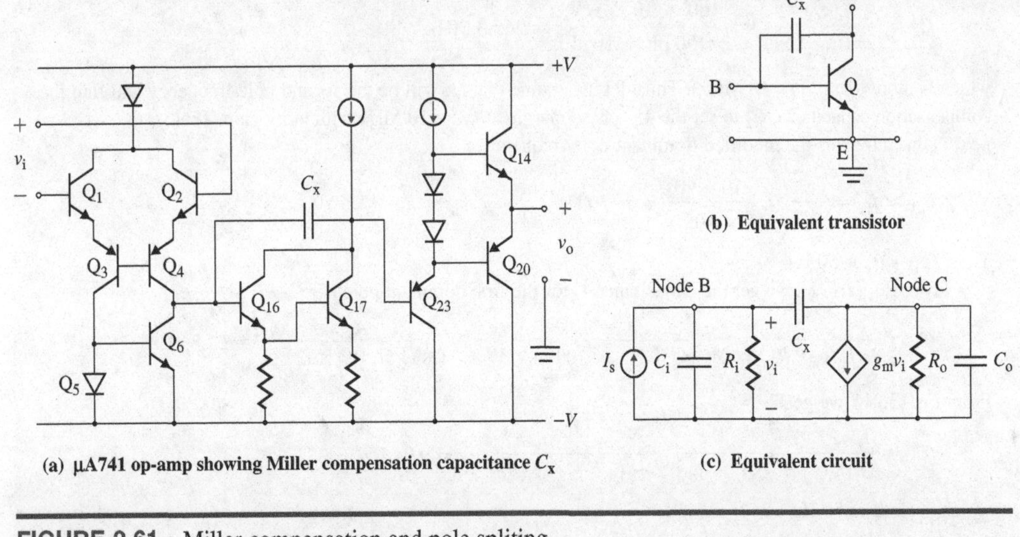

(b) Equivalent transistor

(a) μA741 op-amp showing Miller compensation capacitance C_x

(c) Equivalent circuit

FIGURE 2.61 Miller compensation and pole spliting

If $C_x >> C_o$, Eq. (2.132) can be approximated by

$$\omega'_{p2} = \frac{g_m}{C_i(C_o/C_x) + (C_i + C_o)} \approx \frac{g_m}{C_i + C_o} \tag{2.133}$$

From the preceding equations, we can see that as C_x is increased, ω'_{p1} is reduced, and ω'_{p2} is increased. This separation of poles is called *pole splitting*. The increase in ω'_{p2} will move the pole frequency for a 45° phase margin to the right and will hence widen the bandwidth. In Eq. (2.131), C_x is multiplied by a factor $g_m R_o$, and the effective capacitance is much larger: $g_m R_o C_x$. Thus, the value of C_x will be much smaller than it is when compensation is achieved by adding or changing a pole.

EXAMPLE 2.20

D

Finding the compensating capacitance by pole splitting The open-loop gain of an amplifier is 2×10^5 and has break frequencies at $f_{p1} = 100$ kHz, $f_{p2} = 1$ MHz, and $f_{p3} = 10$ MHz. The gain stage of the amplifier has the equivalent circuit shown in Fig. 2.61(c), whose parameters are $g_m = 100$ mA/V, $C_i = 50$ pF, and $C_o = 10$ pF. Determine the value of compensating capacitance C_x that will give a closed-loop phase margin of 45° with a resistive feedback of up to $\beta = 1$.

SOLUTION

$g_m = 100 \times 10^{-3}$, $C_i = 50$ pF, $C_o = 10$ pF, $f_{p1} = 100$ kHz, $f_{p2} = 1$ MHz, and $f_{p3} = 10$ MHz. We can find the values of R_i and R_o as follows:

$$R_i = \frac{1}{2\pi f_{p1} C_i} = \frac{1}{2\pi \times 100\text{ kHz} \times 50\text{ pF}} = 31.83\text{ k}\Omega$$

$$R_o = \frac{1}{2\pi f_{p2} C_o} = \frac{1}{2\pi \times 1\text{MHz} \times 10\text{ pF}} = 15.92\text{ k}\Omega$$

From Eq. (2.133), we can find the modified value f'_{p2}; that is,

$$f'_{p2} \approx \frac{g_m}{2\pi(C_i + C_o)} = \frac{100 \times 10^{-3}\text{ A/V}}{2\pi(50\text{ pF} + 10\text{ pF})} = 265.3\text{ MHz}$$

which is more than f_{p3} (=10 MHz). Thus, let us assume that f_{p3} will be the second pole frequency and find the compensation capacitance C_x to set the 45° phase margin at $f_{p3} = 10$ MHz with unity gain. That is, $f_D \times A_o\beta = f_{p3} \times 1$, which gives the modified dominant pole frequency as

$$f'_{p1} \approx f_D = \frac{f_{p3}}{A_o\beta} = \frac{10 \times 10^6}{2 \times 10^5 \times 1} = 50\text{ Hz}$$

Thus, $f'_{p1} \approx f_D = 50$ Hz.

From Eq. (2.131), we get the capacitance C for the first dominant pole f'_{p1}:

$$C_x \approx \frac{1}{2\pi f'_{p1} g_m R_i R_o} = \frac{1}{2\pi \times 50\text{ Hz} \times 100\text{ mA/V} \times 31.83\text{ k}\Omega \times 15.92\text{ k}\Omega} = 62.8\text{ pF}$$

From Eq. (2.91), we get

$$f'_{p1} = \frac{1}{2\pi[R_i(C_i + C_x) + R_o(C_x + C_o) + g_m C_x R_i R_o]} = 49.94\text{ Hz}$$

which is close to 50 Hz.

2.12.4 Modification of the Feedback Path

Op-amps are normally compensated by adding a capacitor internally. This type of compensation, however, wastes bandwidth because the bandwidth is reduced considerably. For example, the dominant pole frequency of the μA741 op-amp is only 10 Hz.

Compensation can also be accomplished by modifying the feedback network so that the feedback factor β becomes frequency dependent and has a zero that cancels the pole of the original amplifier. This method is generally used in the compensation of fixed-gain amplifiers, where achieving a wide bandwidth is of prime concern.

Let us consider a feedback network with one pole and one zero. Then the loop gain of an amplifier with three poles will be of the form

$$A(j\omega)\beta(j\omega) = \frac{A_o(1 + j\omega/\omega_z)}{(1 + j\omega/\omega_p)(1 + j\omega/\omega_{p1})(1 + j\omega/\omega_{p2})(1 + j\omega/\omega_{p3})} \quad (2.134)$$

where ω_p and ω_z are the pole and the zero of the feedback network, respectively.

A typical implementation is shown in Fig. 2.62(a) for a shunt-series feedback amplifier. The feedback network includes a capacitor C_F. The loading effects of C_F at the input and output sides of the amplifier are shown in Fig. 2.62(b). The capacitor C_F will have only a minor effect on the amplifier transfer func-

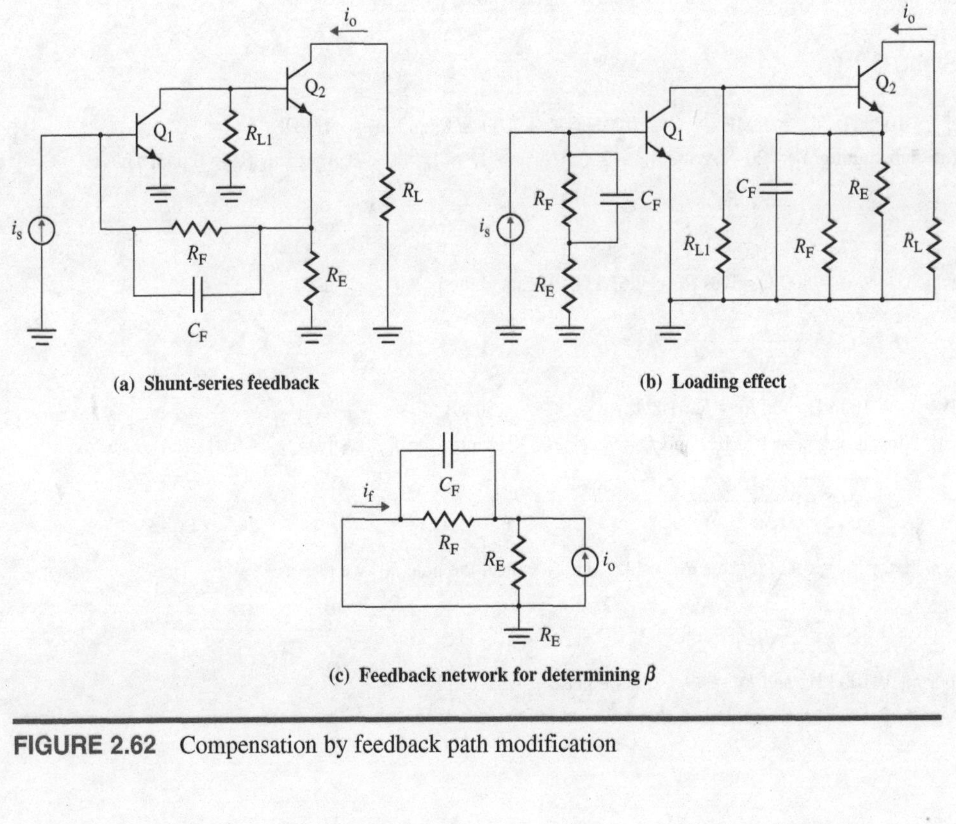

FIGURE 2.62 Compensation by feedback path modification

tion $A(s)$. The feedback network is shown in Fig. 2.62(c), from which we can find the feedback transfer function as

$$\beta(s) = \frac{i_f}{i_o} = \frac{R_E}{R_E + R_F} \times \frac{1 + sR_F C_F}{1 + sC_F R_F R_E/(R_E + R_F)} \tag{2.135}$$

$$= \frac{\beta_o(1 + s/\omega_z)}{1 + s/\omega_p} \tag{2.136}$$

where $\beta_o = R_E/(R_E + R_F)$ = low-frequency feedback factor
$\omega_z = 1/(R_F C_F)$ = zero of the feedback network
$\omega_p = (R_E + R_F)/(R_E R_F C_F)$ = pole of the feedback network

EXAMPLE 2.21

D

Compensating by feedback path modification The open-loop gain of the amplifier in Example 2.12 has break frequencies at f_{p1} = 100 kHz, f_{p2} = 1 MHz, and f_{p3} = 10 MHz. The low-frequency gain is A_o = 200 A/A, and the emitter resistance is R_E = 500 Ω. Determine the values of compensating capacitance C_F and resistance R_F **(a)** to give a low-frequency closed-loop gain of A_f = 20 A/A and cancel the pole f_{p1} = 100 kHz and **(b)** to add a pole of f_p = 10 MHz and cancel the pole f_{p1} = 100 kHz.

SOLUTION

f_{p1} = 100 kHz, f_{p2} = 1 MHz, f_{p3} = 10 MHz, A_o = 200 A/A, and R_E = 500 Ω.

(a) Substituting A_f = 20 A/A and A_o = 200 A/A into $A_f = A_o/(1 + \beta_o A_o)$ gives $\beta_o = 0.045$. Thus,

$$\beta_o = 0.045 = \frac{R_E}{R_E + R_F}$$

which, for R_E = 500 Ω, gives R_F = 2.61 kΩ. To cancel the pole f_{p1}, we use

$$f_z = f_{p1} = \frac{1}{2\pi R_F C_F}$$

For f_{p1} = 100 kHz and R_F = 2.61 kΩ, we get C_F = 150 pF.

(b) Substituting f_p = 10 MHz and $f_z = f_{p1}$ = 100 kHz into $f_p = f_z/\beta_o$ gives $\beta_o = 0.01$. Thus,

$$\beta_o = 0.01 = \frac{R_E}{R_E + R_F}$$

which, for R_E = 500 Ω, gives R_F = 49.5 kΩ. To cancel the pole f_{p1}, we use

$$f_z = f_{p1} = \frac{1}{2\pi R_F C_F}$$

For f_{p1} = 100 kHz and R_F = 49.5 kΩ, we get C_F = 32.15 pF.

EXAMPLE 2.22

D

Compensating by feedback path modification

(a) A common-emitter amplifier with shunt-shunt feedback is shown in Fig. 2.63. The capacitance C_F and resistance R_F form the feedback network. The voltage gain without feedback is $A = |A_v| = v_o/v_s \approx 100$. Design the feedback network to meet the following specifications:

The bandwidth with feedback must be increased by a factor of 10; that is, $BW_f = 10BW$.
The voltage gain with feedback must be $|A_f| = v_o/v_s = 10$.

(b) Use PSpice/SPICE to check your results.

SOLUTION

(a) Shunt-shunt feedback will reduce both the input resistance and the output resistance, but it should widen the bandwidth. The value of C_F should be such that it is virtually short-circuited over the frequency range of the amplifier. The feedback is taken from the output side because the output is out of phase with the input voltage v_S. Since the gain of the amplifier is $A = 10^2$, the closed-loop gain will be almost independent of A. We get

$$\frac{v_o}{i_s} = -\frac{1}{\beta} = -R_F \tag{2.137}$$

and $$i_s = \frac{v_s}{R_s}$$

which gives the closed-loop gain as

$$A_f = \frac{v_o}{v_s} = -\frac{R_F}{R_s} \approx -\frac{1}{\beta R_s} \quad \text{(V/V)} \tag{2.138}$$

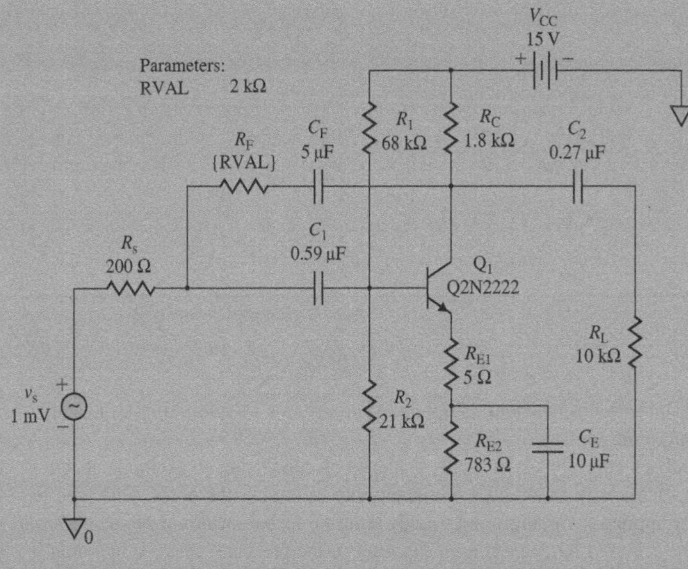

FIGURE 2.63 Common-emitter amplifier with shunt-shunt feedback

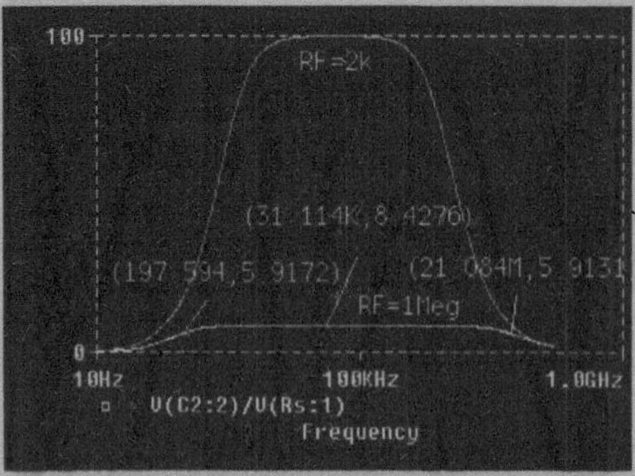

FIGURE 2.64 Frequency response for Example 2.22

For shunt-shunt feedback, $BW_f = BW(1 + \beta A)$. The gain must decrease by a factor of 10. Thus, $\beta R_s = 1/10 = 0.1$.

$$\frac{1}{\beta R_s} = \frac{R_F}{R_s} = 10$$

which, for $R_s = 200\ \Omega$, gives $R_F = 2\ k\Omega$. We want to choose a capacitor that will ensure that C_F is virtually short-circuited at low frequency, $f_L = 1$ kHz: let's choose $C_F = 5\ \mu F$. Thus,

$$X_{CF} \le \frac{1}{2\pi \times 1\ kHz \times 5\ \mu F} = 31.8\ \Omega$$

(b) The PSpice plots of the frequency response are shown in Fig. 2.64 for $R_F = 2\ k\Omega$ and 1 MΩ. For $R_F = 1\ M\Omega$ (for almost no feedback, $\beta \approx 0$), we get $|A| = 100$, $f_L = 1089$ Hz, $f_H = 1838$ kHz, and

$$BW = f_H - f_L = 1838\ k - 1.09\ k = 1836.91\ kHz$$

With feedback of $R_F = 2\ k\Omega$ (for $\beta = 0.1$), we get $|A_f| = 8.43$ (expected value is 10), $f_{Lf} = 198$ Hz, $f_{Hf} = 21{,}084$ kHz, and

$$BW_f = f_{Hf} - f_{Lf} = 21{,}084\ k - 0.2\ k = 21{,}083.8\ kHz$$

(expected value is $10BW = 10 \times 1836.91$ kHz $= 18{,}369$ kHz). The difference between the PSpice values and the expected values is caused by the fact that we did not include the effects of resistances such as R_F and R_C in hand calculations.

KEY POINTS OF SECTION 2.12

- An amplifier may be unstable when feedback is applied. The process of stabilizing an unstable amplifier is called compensation.
- The compensation of feedback amplifiers is normally achieved by connecting an external capacitor to the basic amplifier in such a way as to add a dominant pole or to split the poles. This type of compensation is normally applied to op-amps and usually reduces the bandwidth of the amplifier.
- Compensation can also be achieved by connecting a capacitor to the feedback network such that it adds a pole and a zero to the loop gain. This type of compensation is normally applied to fixed-gain amplifiers so as to yield a wide bandwidth.

TABLE 2.3 Effects of feedback topologies

	Series-shunt	Shunt-shunt	Shunt-series	Series-series
Z_{if}	$R_i(1+\beta A)$	$R_i/(1+\beta A)$	$R_i/(1+\beta A)$	$R_i(1+\beta A)$
Z_{of}	$R_o/(1+\beta A)$	$R_o/(1+\beta A)$	$R_o(1+\beta A)$	$R_o(1+\beta A)$
A_f	$A/(1+\beta A)$	$A/(1+\beta A)$	$A/(1+\beta A)$	$A/(1+\beta A)$

Summary

There are two types of feedback: negative feedback and positive feedback. Negative feedback is normally used in amplifier circuits, and positive feedback is applied exclusively in oscillators. Negative feedback has certain advantages, such as stabilization of overall gain with respect to parameter variations, reduction of distortion, reduction of the effects of nonlinearity, and increase in bandwidth. However, these advantages are obtained at the expense of gain reduction, and additional amplifier stages may be required to make up the gain reduction. If the loop gain $\beta A >> 1$, the overall (or closed-loop) gain depends inversely on the feedback factor β and is directly sensitive to changes in the feedback factor.

The gain–bandwidth product of feedback amplifiers remains constant. If the gain is reduced by negative feedback, then the bandwidth is increased by the same amount. Depending on its implementation in electronic circuits, feedback can be classified as one of four types: series-shunt, shunt-shunt, series-series, or shunt-series. A shunt connection reduces the input (or output) impedance by a factor of $(1 + \beta A)$, and a series connection increases the input (or output) impedance by a factor of $(1 + \beta A)$. The closed-loop gain is always decreased by a factor of $(1 + \beta A)$. Table 2.3 summarizes the effects of various feedback topologies.

The stability of an amplifier depends on the poles of the transfer function. For a stable amplifier, the characteristic equation should not have any roots with positive real parts. The Nyquist criterion is one of the methods for determining the stability of feedback systems. The stability of an amplifier is normally measured in terms of phase margin and gain margin. A Bode plot can also be used to determine the stability of an amplifier. An inherently unstable amplifier can be made stable by introducing a dominant pole in the transfer function; this is generally achieved in electronic circuits by connecting a feedback capacitor.

References

1. B. Razavi, *Design of Analog CMOS Integrated Circuits.* New York: McGraw-Hill, 2001.
2. B. C. Kuo, *Automatic Control Systems*. Englewood Cliffs, NJ: Prentice Hall, 1982.
3. P. R. Gray and R. G. Meyer, *Analysis and Design of Integrated Circuits.* New York: Wiley, 1992, Chapters 8 and 9.
4. P. M. Chirlian, *Analysis and Design of Analog Integrated Electronic Circuits*. New York: Harper & Row, 1981, Chapter 16.
5. P. R. Gray, P. J. Hurst, S. H. Lewis, and R. G. Meyer, *Analysis and Design of Analog Integrated Circuits.* New York: Wiley, 2001.

Review Questions

1. What are the two types of feedback?
2. What are the advantages of feedback?
3. What are the disadvantages of feedback?
4. What is loop transmission?
5. What is gain sensitivity?
6. What is feedback factor sensitivity?
7. What is the difference between open-loop gain and closed-loop gain?
8. What are the four types of feedback topologies?
9. What are the features of series-shunt feedback?
10. What are the features of shunt-shunt feedback?
11. What are the features of shunt-series feedback?
12. What are the features of series-series feedback?
13. What are the effects of the feedback network on amplifier performance?
14. What are the effects of the source impedance on amplifier performance?
15. What are the effects of the load impedance on amplifier performance?
16. How do you take into account the β dependency of A?
17. How do you find the modified gain A of a feedback amplifier?
18. What are the steps in designing a feedback network?
19. What is a characteristic equation?
20. What are the effects of the poles on the stability of an amplifier?
21. What is the Nyquist criterion?
22. What are the conditions for instability?
23. What is gain crossover frequency?
24. What is phase crossover frequency?
25. What is relative stability?
26. What is a phase margin?
27. What is a gain margin?
28. What is the effect of phase margin on system response?
29. What is a Bode plot?
30. What is compensation?
31. What is the common method of compensation in amplifiers?
32. How is a dominant pole introduced in electronic circuits?

Problems

The symbol **D** indicates that a problem is a design problem. The symbol **P** indicates that you can check the solution to a problem using PSpice/SPICE or Electronics Workbench.

2.3 *Characteristics of Feedback*

2.1 Three voltage amplifiers are cascaded, as shown in Fig. P2.1.

a. Determine the value of β to give a closed-loop gain of $A_f = -100$.

b. If the gain of each stage increases by 10%, determine A_f. Use the β of part (a).

FIGURE P2.1

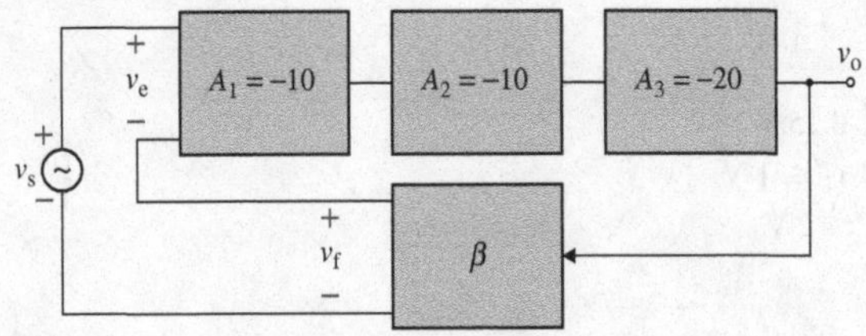

2.2 The open-loop gain of an amplifier with negative feedback is $A = 50$, and the feedback factor is $\beta = 0.5$.

a. Determine the closed-loop gain A_f.

b. If the open-loop gain A changes by +15%, determine the percentage change in the closed-loop gain A_f and its value.

c. If the feedback factor β changes by +15%, determine the percentage change in the closed-loop gain A_f and its value.

2.3 The open-loop gain of an amplifier is $A = 50$, and the feedback factor is $\beta = 0.8$. If the open-loop gain A changes by −20% and the feedback factor β changes by +15%, determine the closed-loop gain A_f.

2.4 Two feedback amplifiers are connected in series. Each amplifier has a distortion of 20%. Determine the overall gain and the overall distortion if **(a)** each amplifier has its own feedback, as shown in Fig. P2.4(a), and **(b)** only one feedback is applied to the cascaded amplifiers, as shown in Fig. P2.4(b).

FIGURE P2.4

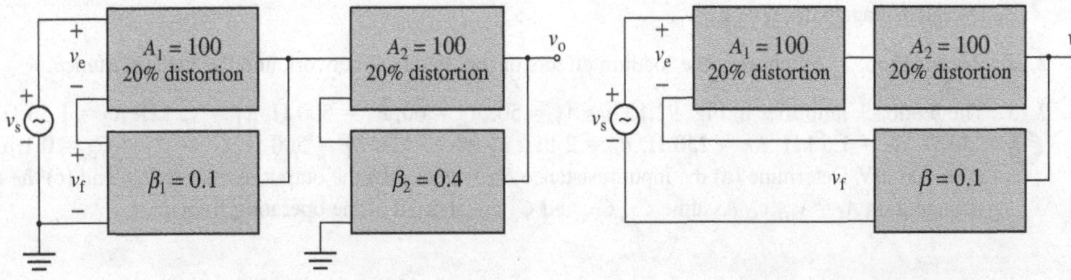

(a) Cascaded feedback **(b) Feedback on cascaded amplifiers**

2.5 A feedback amplifier is to have a closed-loop gain of $A_f = 60$ dB and a sensitivity of 10% to the open-loop gain A. Determine the open-loop gain with a unity feedback $\beta = 1$.

2.6 The feedback factor of an amplifier is $\beta = 0.5$. The open-loop gain A, which is frequency dependent, can be expressed as

$$A(j\omega) = \frac{2 \times 10^5}{1 + jf/10}$$

Determine **(a)** the closed-loop low-frequency gain A_{of}, **(b)** the closed-loop bandwidth BW, and **(c)** the gain–bandwidth product GBW.

2.7 The feedback factor of an amplifier is $\beta = 0.8$. The open-loop gain A can be expressed in Laplace's domain of s as

$$A(s) = \frac{250s}{(1 + 0.1s)(1 + 0.001s)}$$

Determine **(a)** the closed-loop low-frequency gain A_{of}, **(b)** the closed-loop bandwidth BW, and **(c)** the gain–bandwidth product GBW.

2.8 The transfer characteristic of an amplifier is described by the following values of open-loop gain A for given ranges of input voltage v_2:

$$A = \begin{cases} 50 & \text{for } 0 < v_i \leq 0.25 \text{ V} \\ 40 & \text{for } 0.25 \text{ V} < v_i \leq 1 \text{ V} \\ 10 & \text{for } 1 \text{ V} < v_i \leq 2 \text{ V} \\ 0 & \text{for } v_i > 2 \text{ V} \end{cases}$$

If the feedback factor is $\beta = 0.8$, determine the closed-loop gains of the transfer characteristic.

2.6 *Series-Shunt Feedback*

2.9 (P) The noninverting amplifier shown in Fig. 2.7(a) has $R_1 = 40$ kΩ, $R_f = 10$ kΩ, $R_L = 15$ kΩ, and $R_s = 0$. The op-amp parameters are $R_i = 2$ MΩ and $R_o = 50$ Ω, and the open-loop voltage gain is $\mu_g = 2 \times 10^5$. Determine **(a)** the input resistance seen by the source $R_{if} = v_s/i_s$, **(b)** the output resistance R_{of}, and **(c)** the closed-loop voltage gain $A_f = v_o/v_s$.

2.10 Repeat Prob. 2.9, ignoring the loading effects of the feedback network and the load resistance.

2.11 (P) A voltage amplifier with negative feedback is shown in Fig. 2.47. The amplifier has an open-loop voltage gain of $A = 250$, an input resistance of $R_i = 4.5$ kΩ, and an output resistance of $R_o = 500$ Ω. The load resistance is $R_L = 10$ kΩ. The feedback circuit has $R_1 = 24$ kΩ and $R_F = 8$ kΩ. The source has a resistance of $R_s = 2$ kΩ. Determine **(a)** the input resistance $R_{if} = v_s/i_s$, **(b)** the output resistance R_{of}, and **(c)** the overall voltage gain $A_f = v_o/v_s$.

2.12 Repeat Prob. 2.11, ignoring the loading effects of the feedback network and the load resistance.

2.13 (P) The feedback amplifier in Fig. P2.13 has $A_1 = 50$, $A_2 = 60$, $R_s = 500$ Ω, $R_1 = 15$ kΩ, $R_2 = 1.5$ kΩ, $R_3 = 250$ Ω, $R_4 = 1.5$ kΩ, $R_5 = 250$ Ω, $R_6 = 2$ kΩ, $R_L = 4.7$ kΩ, $R_F = 500$ Ω, $C_1 = C_2 = C_3 = 0.1$ μF, and $v_s = 100$ mV. Determine **(a)** the input resistance $R_{if} = v_s/i_s$, **(b)** the output resistance R_{of}, and **(c)** the overall voltage gain $A_f = v_o/v_s$. Assume C_1, C_2, and C_3 are shorted at the operating frequency.

FIGURE P2.13

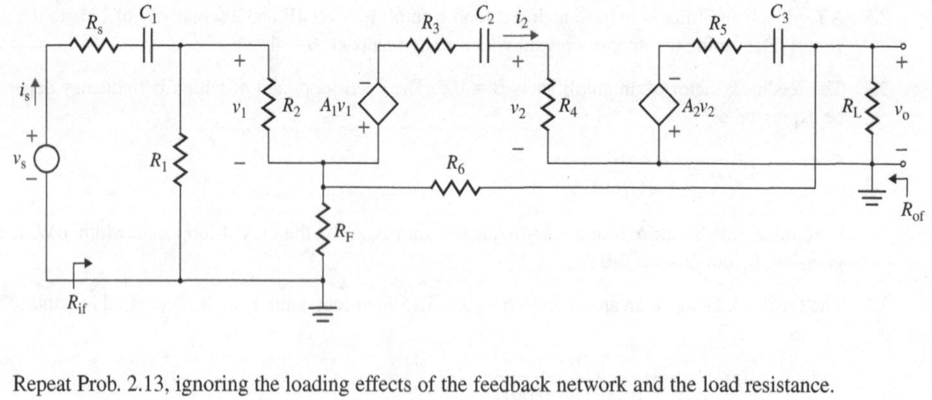

2.14 Repeat Prob. 2.13, ignoring the loading effects of the feedback network and the load resistance.

2.15 Use PSpice/SPICE to plot output impedance against frequency for the feedback amplifier in Prob. 2.13. The frequency should be varied from 10 Hz to 1 MHz in increments of 1 decade and 10 points per decade.

2.16 P The AC equivalent circuit of the CE amplifier in Fig. P2.16 has $R_1 = 10$ kΩ, $R_2 = 1.5$ kΩ, $R_C = 1.5$ kΩ, $R_E = 250$ Ω, $R_s = 200$ Ω, $R_3 = 24$ kΩ, $R_4 = 8$ kΩ, and $R_L = 1$ kΩ. The transistor parameters are $h_{fe} = 150$, $r_\pi = 2.5$ kΩ, and $r_o = 25$ kΩ. Use the techniques of feedback analysis to calculate the **(a)** input resistance R_{if}, **(b)** the output resistance R_{of}, and **(c)** the closed-loop voltage gain A_f.

FIGURE P2.16

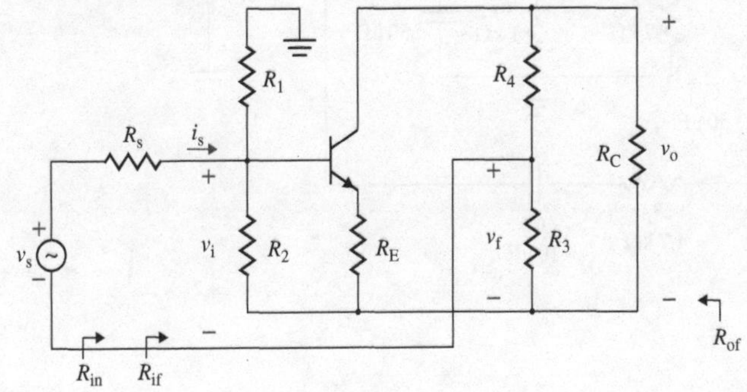

2.17 P Use the techniques of feedback analysis to calculate the input resistance R_{if}, the output resistance R_{of}, and the closed-loop voltage gain A_f of the amplifier in Fig. 2.18(a) with series-shunt feedback. The DC bias currents of the transistors are $I_{C1} = 0.1$ mA, $I_{C2} = 0.5$ mA, and $I_{C3} = 2$ mA. The transistor parameters are $h_{fe} = h_{fe1} = h_{fe2} = h_{fe3} = 150$, $r_o = 25$ kΩ, and $r_\mu = \infty$.

2.18 D The emitter follower in Fig. P2.18 has $R_B = 75$ kΩ, $R_E = 750$ Ω, $R_L = 10$ kΩ, and $R_s = 250$ Ω. The transistor parameters are $h_{fe} = 150$, $r_\pi = 250$ Ω, and $r_o = \infty$. Draw a block diagram of the feedback mechanism. Use the techniques of feedback analysis to calculate **(a)** the input resistance R_{if}, **(b)** the output resistance R_{of}, and **(c)** the closed-loop voltage gain A_f.

FIGURE P2.18

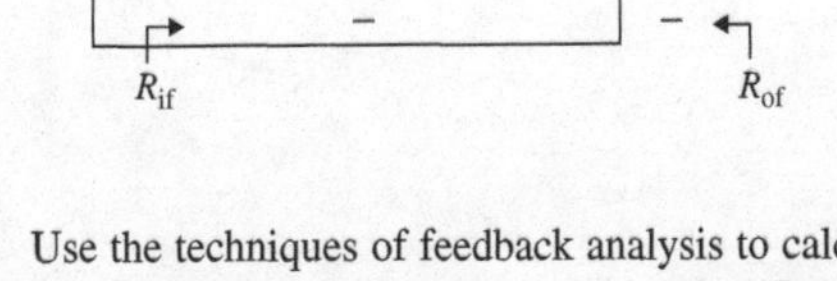

2.19 P Use the techniques of feedback analysis to calculate the input resistance R_{if}, the output resistance R_{of}, and the closed-loop voltage gain A_f of the amplifier in Fig. P2.19. The transistor parameters are $h_{fe} = h_{fe1} = h_{fe2} = 10$, $r_{\pi 1} = r_{\pi 2} = 250$ Ω, $r_o = 1.5$ kΩ, and $r_\mu = \infty$.

FIGURE P2.19

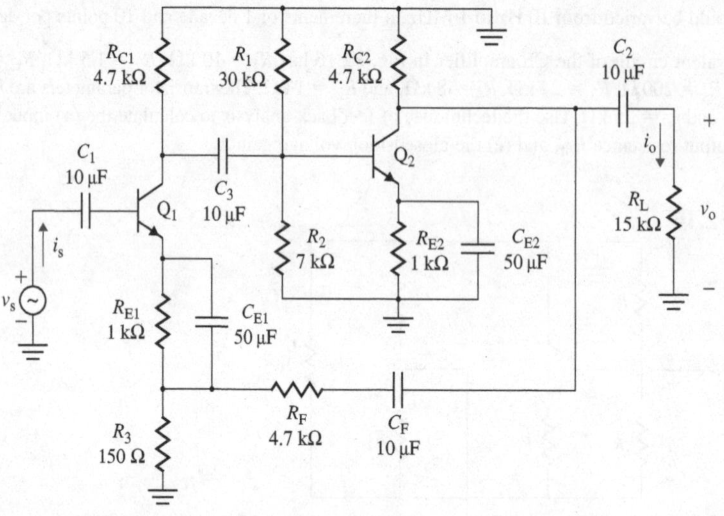

2.20 The AC equivalent circuit of a feedback amplifier is shown in Fig. 2.18(a). The DC bias currents of the transistors are $I_{C1} = 0.5$ mA, $I_{C2} = 1$ mA, and $I_{C3} = 5$ mA. The transistor parameters are $h_{fe} = 100$, $r_o = 25$ kΩ, and $r_\mu = \infty$. Use the techniques of feedback analysis to calculate **(a)** the input resistance R_{if}, **(b)** the output resistance R_{of}, and **(c)** the closed-loop voltage gain A_f.

2.21 For the amplifier in Fig. 2.18(a), determine the value of the feedback resistor R_F so that the closed-loop voltage gain A_f is 25% of the open-loop voltage gain A. The DC bias currents of the transistors are $I_{C1} = 0.5$ mA, $I_{C2} = 1$ mA, and $I_{C3} = 5$ mA. The transistor parameters are $h_{fe} = h_{fe1} = h_{fe2} = h_{fe3} = 100$, $r_o = 25$ kΩ, and $r_\mu = \infty$.

2.22 The MOS amplifier shown in Fig. P2.22 is biased to have the following small-signal MOS parameters: $g_{m1} = 1.2$ mA/V, $r_{o1} = 25$ kΩ, $g_{m2} = 1.6$ mA/V, and $r_{o1} = 25$ kΩ. If $R_{D1} = 1.5$ kΩ, then $R_{D2} = 1$ kΩ, $R_{SR} = 500$ Ω, $R_F = 5$ kΩ, and $C_F = 20$ pF. Determine **(a)** the voltage gain without feedback $A = v_o/v_s$, **(b)** the voltage gain with feedback A_f, and **(c)** the feedback capacitor C_F to limit the high frequency $f_H = 50$ kHz.

FIGURE P2.22

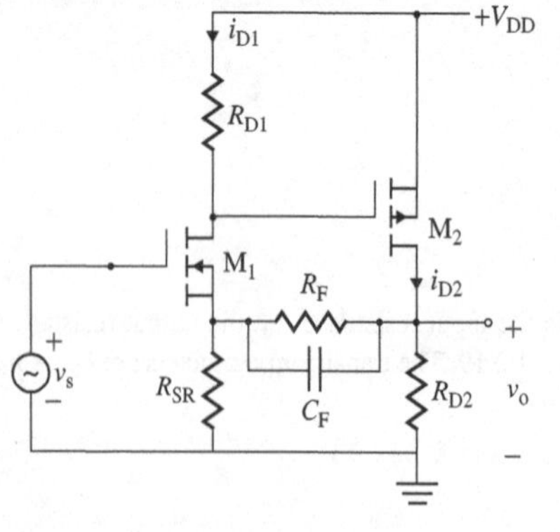

2.23 The MOS amplifier shown in Fig. P2.23 is biased to have the following small-signal MOS parameters: $g_{m1} = g_{m2} = 1\ \text{mA/V}$, $r_{o1} = r_{o2} = 20\ \text{k}\Omega$, $g_{m3} = g_{m4} = 1.2\ \text{mA/V}$, and $r_{o3} = r_{o4} = 22\ \text{k}\Omega$. If $C_F = 6$ nF and $C_1 = 4$ nF, determine **(a)** the voltage gain without feedback $A = v_o/v_s$, **(b)** the voltage gain with feedback A_f, and **(c)** the high cutoff frequency f_H.

FIGURE P2.23

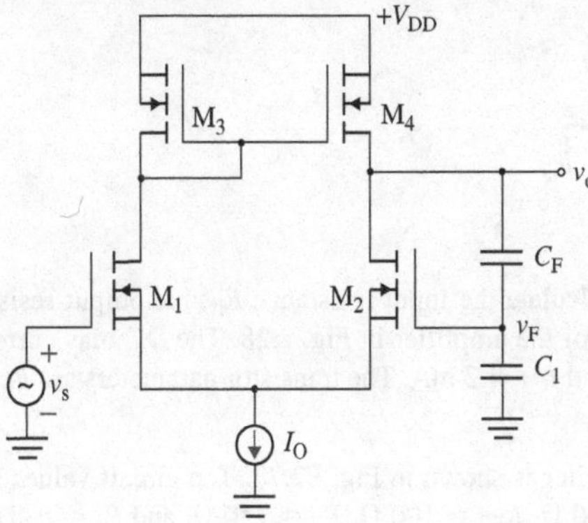

2.7 *Series-Series Feedback*

2.24 (P) A transconductance amplifier with negative feedback is shown in Fig. P2.24. The amplifier has an open-loop transconductance of $A = 50 \times 10^{-3}$ A/V, an input resistance of $R_i = 25\ \text{k}\Omega$, and an output resistance of $R_o = 50\ \text{k}\Omega$. The feedback circuit has $R_F = 2.5\ \text{k}\Omega$. The source resistance is $R_s = 1\ \text{k}\Omega$, and the load resistance is $R_L = 500\ \Omega$. Determine **(a)** the input resistance $R_{if} = v_s/i_s$, **(b)** the output resistance R_{of}, and **(c)** the overall transconductance gain $A_f = i_o/v_s$.

FIGURE P2.24

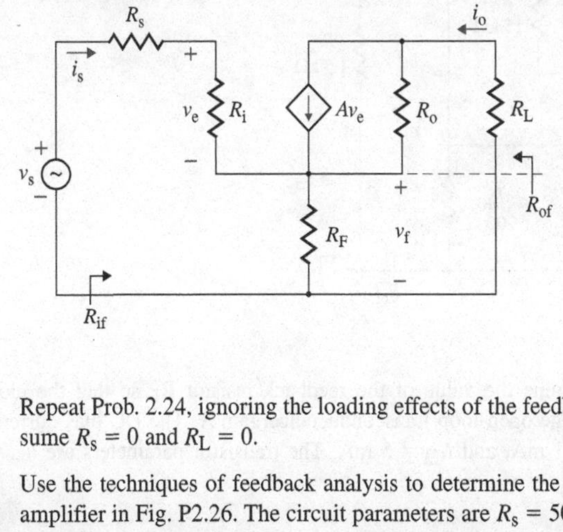

2.25 Repeat Prob. 2.24, ignoring the loading effects of the feedback network and the load resistance; that is, assume $R_s = 0$ and $R_L = 0$.

2.26 (P) Use the techniques of feedback analysis to determine the input and output resistance of the CE transistor amplifier in Fig. P2.26. The circuit parameters are $R_s = 500\ \Omega$, $R_E = 250\ \Omega$, $R_2 = 15\ \text{k}\Omega$, $R_1 = 5\ \text{k}\Omega$, $R_C = 5\ \text{k}\Omega$, and $R_L = 10\ \text{k}\Omega$. The π-model parameters are $r_o = 25\ \text{k}\Omega$, $h_{fe} = 150$, $r_\pi = 250\ \Omega$, $g_m = 0.3876$ A/V, and $r_\mu = \infty$.

FIGURE P2.26

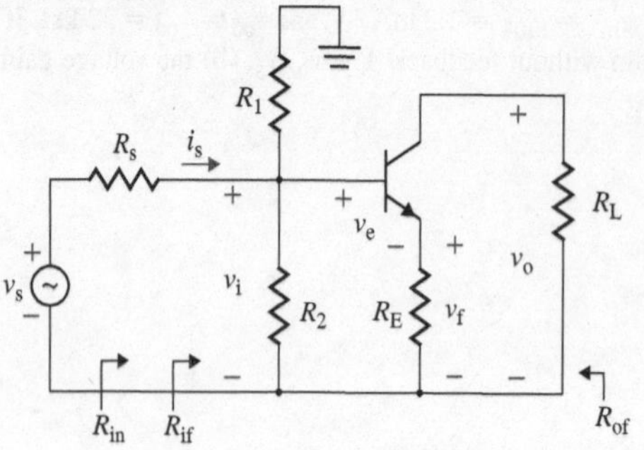

2.27 Use the techniques of feedback analysis to calculate the input resistance R_{if}, the output resistance R_{of}, and the closed-loop transconductance gain A_f of the amplifier in Fig. 2.28. The DC bias currents of the transistors are $I_{C1} = 0.1$ mA, $I_{C2} = 0.5$ mA, and $I_{C3} = 2$ mA. The transistor parameters are $h_{fe} = h_{fe1} = h_{fe2} = h_{fe3} = 150$, $r_o = 25$ kΩ, and $r_\mu = \infty$.
P

2.28 The AC equivalent circuit of a feedback amplifier is shown in Fig. P2.28. The circuit values are $R_{C1} = 2.5$ kΩ, $R_{C2} = 5$ kΩ, $R_{C3} = 1.5$ kΩ, $R_{E1} = 100$ Ω, $R_{E2} = 100$ Ω, $R_F = 750$ Ω, and $R_s = 0$. The transistor parameters are $h_{fe} = 100$, $r_\pi = 2.5$ kΩ, $r_o = 25$ kΩ, and $r_\mu = \infty$. Use the techniques of feedback analysis to calculate **(a)** the input resistance R_{if}, **(b)** the output resistance R_{of}, and **(c)** the closed-loop voltage gain A_f.
P

FIGURE P2.28

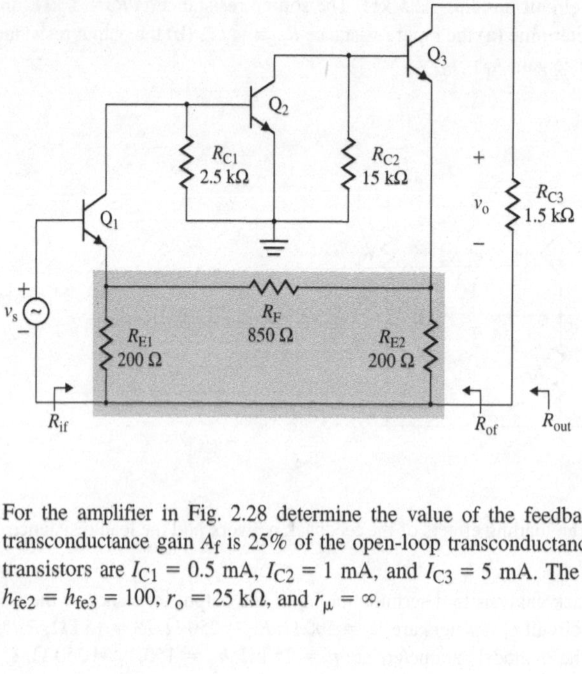

2.29 For the amplifier in Fig. 2.28 determine the value of the feedback resistor R_F so that the closed-loop transconductance gain A_f is 25% of the open-loop transconductance gain A. The DC bias currents of the transistors are $I_{C1} = 0.5$ mA, $I_{C2} = 1$ mA, and $I_{C3} = 5$ mA. The transistor parameters are $h_{fe} = h_{fe1} = h_{fe2} = h_{fe3} = 100$, $r_o = 25$ kΩ, and $r_\mu = \infty$.
D P

2.30 D P For the amplifier in Fig. P2.30, determine the value of the feedback resistor R_F so that the closed-loop transconductance gain A_f is 5 mA/V. The transistor parameters are $h_{fe} = 100$, $r_\pi = 250\ \Omega$, $r_o = 50\ k\Omega$, and $r_\mu = \infty$. Assume $R_i = 2\ M\Omega$, $R_o = 50\ \Omega$, and $A = 10^5$ V/V.

FIGURE P2.30

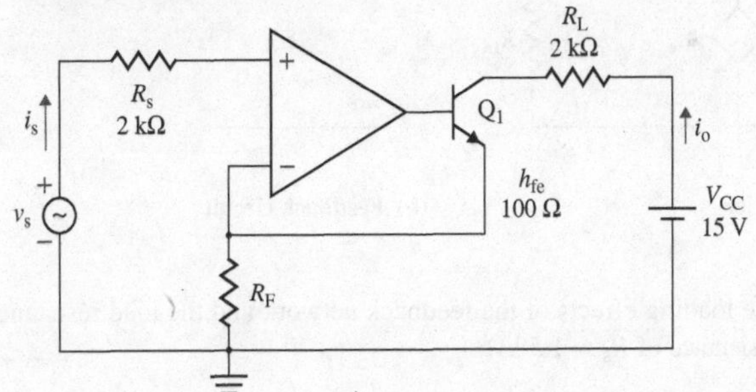

2.31 The MOS amplifier shown in Fig. P2.31 is biased to have the following small-signal MOS parameters: $g_{m1} = 1.2$ mA/V, $r_{o1} = 25\ k\Omega$, $g_{m2} = 1.6$ mA/V, and $r_{o2} = 25\ k\Omega$. If $R_D = 1.5\ k\Omega$, then $R_{SR1} = 500\ \Omega$, $R_{SR2} = 2\ k\Omega$, and $R_F = 8\ k\Omega$. Determine **(a)** the voltage gain without feedback $A = i_o/v_s$, **(b)** the voltage gain with feedback A_f, and **(c)** the feedback capacitor C_F to limit the high frequency $f_H = 50$ kHz.

FIGURE P2.31

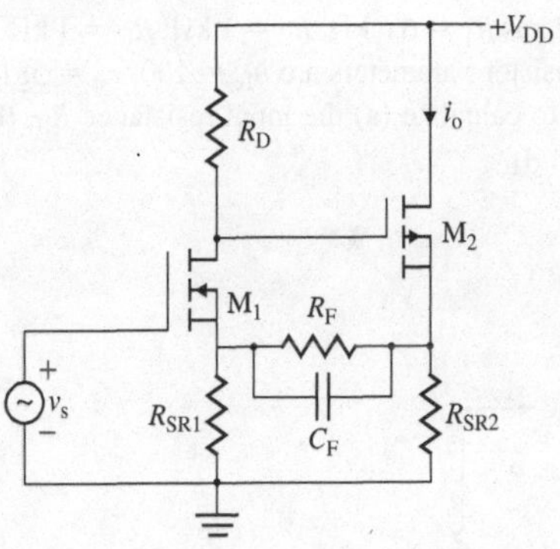

2.8 *Shunt-Shunt Feedback*

2.32 P The inverting op-amp amplifier shown in Fig. 2.9(c) has $R_F = 40\ k\Omega$ and $R_1 = 10\ k\Omega$. The op-amp has an input resistance of $R_i = 5\ M\Omega$, an output resistance of $R_o = 50\ \Omega$, and an open-loop voltage gain of $A = 2 \times 10^5$ V/V. The load resistance is $R_L = 15\ k\Omega$. Determine **(a)** the input resistance seen by the source $R_{in} = v_s/i_s$, **(b)** the output resistance R_{of}, **(c)** the closed-loop transresistance $A_f = v_o/i_i$, and **(d)** the overall voltage gain $A_{vf} = v_o/v_s$. Assume source resistance $R_s = 1\ k\Omega$.

2.33 P A transresistance amplifier with negative feedback is shown in Fig. P2.33. The open-loop transresistance is $A = 750\ k\Omega$, the input resistance is $R_i = 5.5\ k\Omega$, and the output resistance is $R_o = 500\ \Omega$. The feedback circuit has $R_F = 47\ k\Omega$. The source has a resistance of $R_s = 0$. Determine **(a)** the input resistance seen by the source $R_{in} = v_s/i_s$, **(b)** the output resistance R_{of}, and **(c)** the overall voltage gain $A_{vf} = v_o/v_s$. Assume load resistance $R_L = 1\ k\Omega$.

FIGURE P2.33

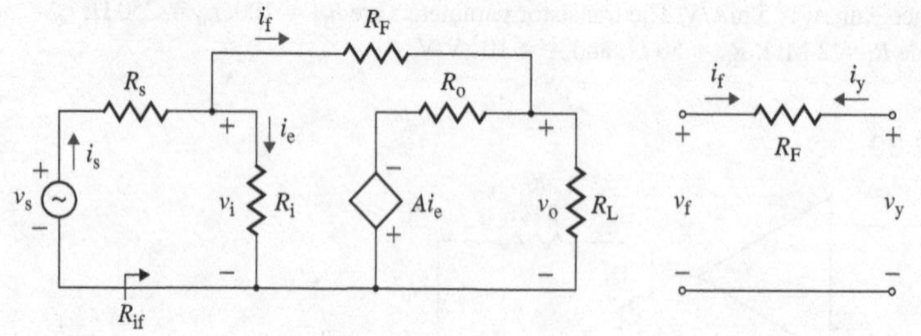

(a) **Amplifier** (b) **Feedback circuit**

2.34 Repeat Prob. 2.33, ignoring the loading effects of the feedback network and the load resistance and assuming that the source has a resistance of R_s = 1.5 kΩ.

2.35 P Use the techniques of feedback analysis to calculate the input resistance R_{if}, the output resistance R_{of}, and the closed-loop transresistance gain A_f of the amplifier in Fig. 2.38. The circuit values are R_{C1} = 5 kΩ, R_E = 2.5 kΩ, R_{C2} = 5 kΩ, R_F = 4 kΩ, and R_s = 200 Ω. The transistor parameters are h_{fe} = 150, r_π = 2 kΩ, r_o = 25 kΩ, and $r_\mu = \infty$.

2.36 P The AC equivalent circuit of a feedback amplifier is shown in Fig. 2.38. The circuit values are R_{C1} = 5 kΩ, R_E = 2.5 kΩ, R_{C2} = 5 kΩ, R_F = 4 kΩ, and R_s = 200 Ω. The transistor parameters are h_{fe} = 100, r_π = 2 kΩ, r_o = 25 kΩ, and $r_\mu = \infty$. Use the techniques of feedback analysis to calculate **(a)** the input resistance R_{if}, **(b)** the output resistance R_{of}, and **(c)** the closed-loop voltage gain A_f.

2.37 P The AC equivalent circuit of the amplifier in Fig. P2.37 has R_1 = 6.6 kΩ, R_2 = 1 kΩ, R_C = 1 kΩ, R_E = 100 Ω, R_s = 500 Ω, R_F = 8 kΩ, and R_L = 5 kΩ. The transistor parameters are h_{fe} = 100, r_π = 581 Ω, and r_o = 22.5 kΩ. Use the techniques of feedback analysis to calculate **(a)** the input resistance R_{if}, **(b)** the output resistance R_{of}, and **(c)** the closed-loop voltage gain A_f.

FIGURE P2.37

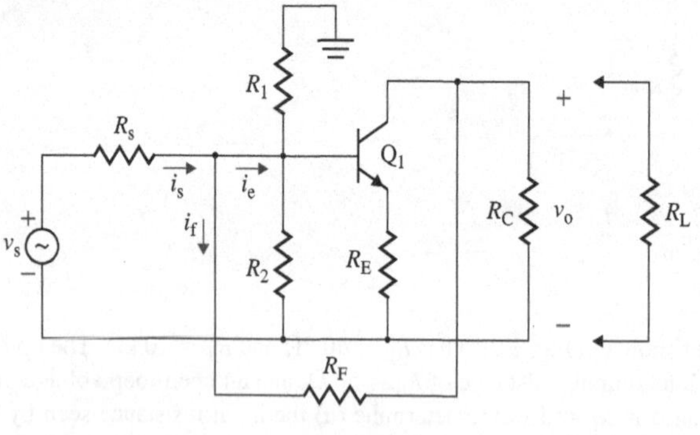

2.38 P The AC equivalent circuit of a feedback amplifier is shown in Fig. P2.38. The circuit values are R_s = 1 kΩ, R_C = 10 kΩ, and R_F = 24 kΩ. The transistor parameters are h_{fe} = 150, r_π = 500 Ω, r_o = 25 kΩ, and $r_\mu = \infty$. Use the techniques of feedback analysis to calculate **(a)** the input resistance R_{if}, **(b)** the output resistance R_{of}, and **(c)** the closed-loop voltage gain A_f.

FIGURE P2.38

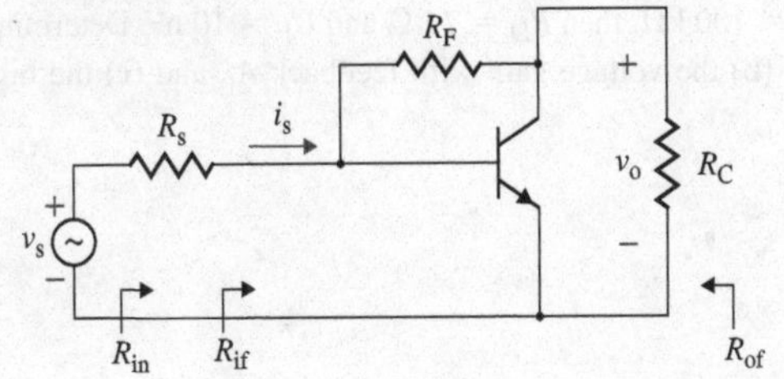

2.39 The AC equivalent circuit of a feedback amplifier is shown in Fig. P2.39. The circuit values are $R_D = 10\ k\Omega$, $R_F = 50\ k\Omega$, and $R_s = 1\ k\Omega$. The transistor parameters are $g_m = 1$ mA/V, $r_d = 50\ k\Omega$, and $r_\mu = 25\ k\Omega$. Use the techniques of feedback analysis to calculate **(a)** the input resistance R_{if}, **(b)** the output resistance R_{of}, and **(c)** the closed-loop voltage gain A_f.

FIGURE P2.39

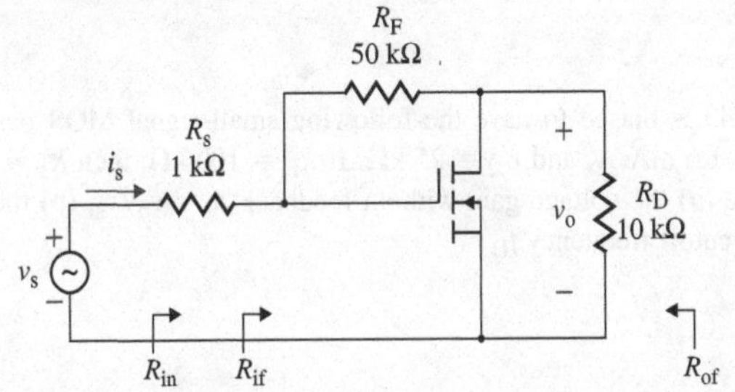

2.40 Determine the value of the feedback resistor R_F so that the closed-loop voltage gain A_f of the amplifier in Fig. 2.38 is 25% of the open-loop voltage gain A. The circuit values are $R_{C1} = 5\ k\Omega$, $R_E = 2.5\ k\Omega$, $R_{C2} = 5\ k\Omega$, and $R_s = 200\ \Omega$. The transistor parameters are $h_{fe} = 150$, $r_\pi = 2\ k\Omega$, $r_o = 25\ k\Omega$, and $r_\mu = \infty$.

2.41 Determine the value of the feedback resistor R_F so that the closed-loop current gain A_f of the amplifier in Fig. P2.41 is 10% of the open-loop voltage gain A. The transistor parameters are $h_{fe} = 150$, $r_\pi = 2\ k\Omega$, $r_o = 25\ k\Omega$, and $r_\mu = \infty$. Assume that the coupling capacitors can be considered as short-circuited at the operating frequency range.

FIGURE P2.41

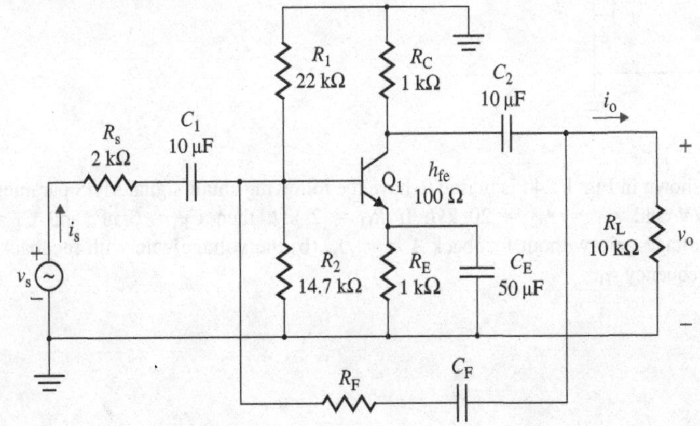

2.42 The MOS amplifier shown in Fig. P2.42 is biased to have the following small-signal MOS parameters: $g_{m1} = 1.2$ mA/V and $r_{o1} = 25$ kΩ. If $R_F = 100$ kΩ, then $R_D = 2$ kΩ and $C_F = 10$ nF. Determine **(a)** the voltage gain without feedback $A = v_o/i_s$, **(b)** the voltage gain with feedback A_f, and **(c)** the high cutoff frequency f_H.

FIGURE P2.42

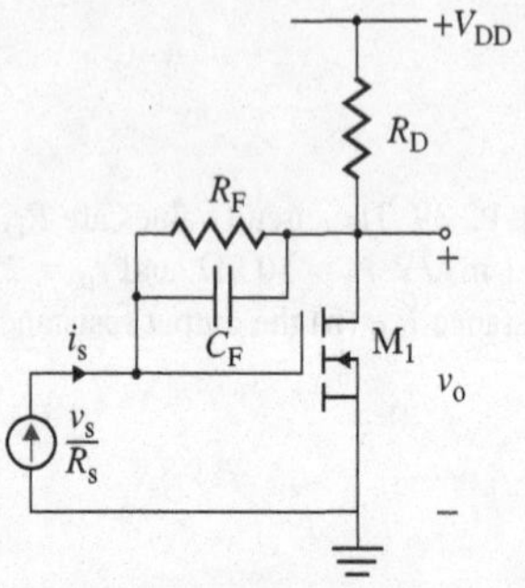

2.43 The MOS amplifier shown in Fig. P2.43 is biased to have the following small-signal MOS parameters: $g_{m1} = 1.2$ mA/V, $r_{o1} = 25$ kΩ, $g_{m2} = 1.6$ mA/V, and $r_{o2} = 25$ kΩ. If $R_F = 100$ kΩ, then $R_1 = 400$ kΩ, $R_s = 1$ kΩ, and $C_F = 10$ nF. Determine **(a)** the voltage gain without feedback $A = v_o/v_s$, **(b)** the voltage gain with feedback A_f, and **(c)** the high cutoff frequency f_H.

FIGURE P2.43

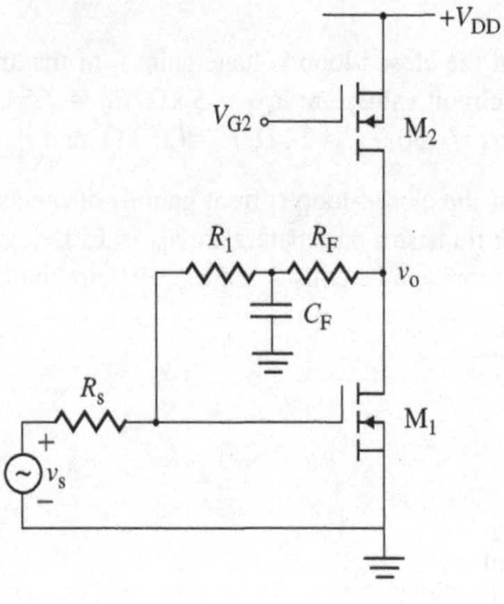

2.44 The MOS amplifier shown in Fig. P2.44 is biased to have the following small-signal MOS parameters: $g_{m1} = g_{m2} = 1.2$ mA/V and $r_{o1} = r_{o2} = 20$ kΩ. If $R_D = 2$ kΩ, then $C_F = 6$ nF, and $C_1 = 2$ nF. Determine **(a)** the voltage gain without feedback $A = v_o/i_s$, **(b)** the voltage gain with feedback A_f, and **(c)** the high cutoff frequency f_H.

FIGURE P2.44

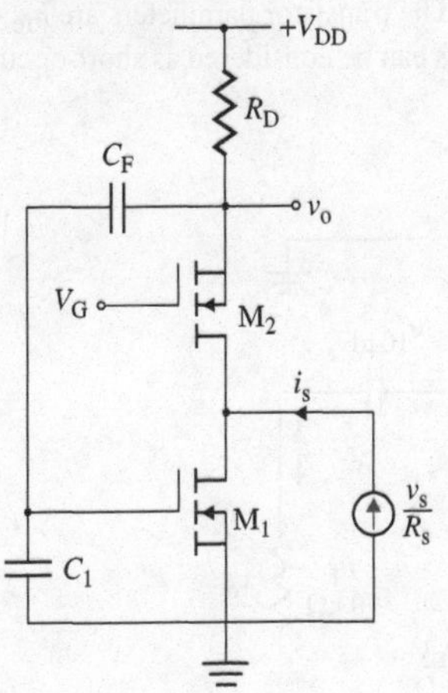

2.9 *Shunt-Series Feedback*

2.45 The current amplifier in Fig. P2.45 has negative feedback. The open-loop current gain is $A = 60$, the input resistance is $R_i = 500$ kΩ, and the output resistance is $R_o = 27$ kΩ. The feedback circuit has $R_F = 20$ kΩ and $R_1 = 2.5$ kΩ. The source resistance is $R_s = 500$ Ω, and the load resistance is $R_L = 100$ Ω. Determine **(a)** the input resistance $R_{if} = v_s/i_s$, **(b)** the output resistance R_{of}, and **(c)** the overall current gain $A_f = i_o/i_s$.

FIGURE P2.45

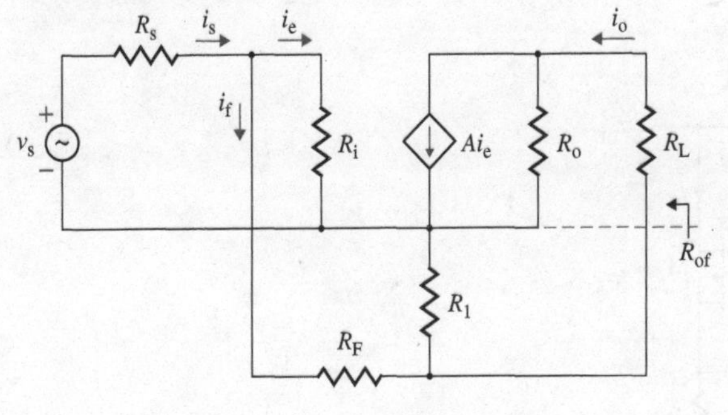

2.46 Repeat Prob. 2.45, ignoring the loading effects of the feedback network and the load resistance and assuming that the source has a resistance of $R_s = 5$ kΩ.

2.47 The CE amplifier in Fig. 2.48 has $R_F = 10$ kΩ, $R_1 = 6.6$ kΩ, $R_2 = 1$ kΩ, $R_{C1} = 5$ kΩ, $R_E = 500$ Ω, $R_3 = 5$ kΩ, $R_4 = 10$ kΩ, $R_{C2} = 5$ kΩ, and $R_s = 200$ Ω. The transistor parameters are $h_{fe} = 150$, $r_\pi = 258$ Ω, and $r_o = 25$ kΩ. Use the techniques of feedback analysis to calculate **(a)** the input resistance R_{if}, **(b)** the output resistance R_{of}, and **(c)** the closed-loop voltage gain A_f.

2.48 Determine the value of the feedback resistor R_F so that the closed-loop current gain A_f of the amplifier in Fig. 2.48 with shunt-series feedback is 25% of the open-loop current gain A. The amplifier has $R_1 = 6.6$ kΩ, $R_2 = 2$ kΩ, $R_{C1} = 5$ kΩ, $R_E = 500$ Ω, $R_3 = 5$ kΩ, $R_4 = 10$ kΩ, $R_{C2} = 5$ kΩ, and $R_s = 0$. The transistor parameters are $h_{fe} = 100$, $r_\pi = 2.58$ kΩ, and $r_o = 25$ kΩ.

2.49 Determine the value of the feedback resistor R_F so that the midfrequency closed-loop current gain A_f of the amplifier in Fig. P2.49 is 10% of the open-loop current gain A. The transistor parameters are $h_{fe} = 100$, $r_\pi = 2.58\ k\Omega$, and $r_o = 25\ k\Omega$. Assume that the coupling capacitors can be considered as short-circuited at the operating frequency range.

D P

FIGURE P2.49

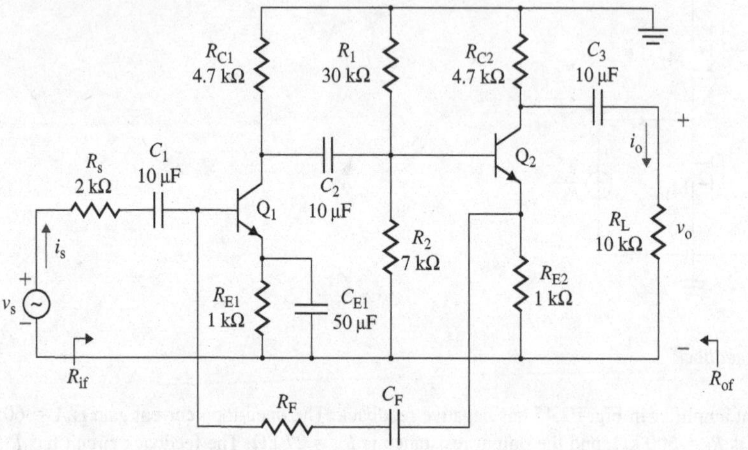

2.50 The MOS amplifier shown in Fig. P2.50 is biased to have the following small-signal MOS parameters: $g_{m1} = 1.2$ mA/V, $r_{o1} = 25\ k\Omega$, $g_{m2} = 1.6$ mA/V, and $r_{o2} = 25\ k\Omega$. If $R_D = 1.5\ k\Omega$, then $R_{SR1} = 500\ k\Omega$, $R_{SR2} = 2\ k\Omega$, and $R_F = 8\ k\Omega$. Determine **(a)** the voltage gain without feedback $A = i_o/v_s$, **(b)** the voltage gain with feedback A_f, and **(c)** the feedback capacitor C_F to limit the high frequency $f_H = 50$ kHz.

FIGURE P2.50

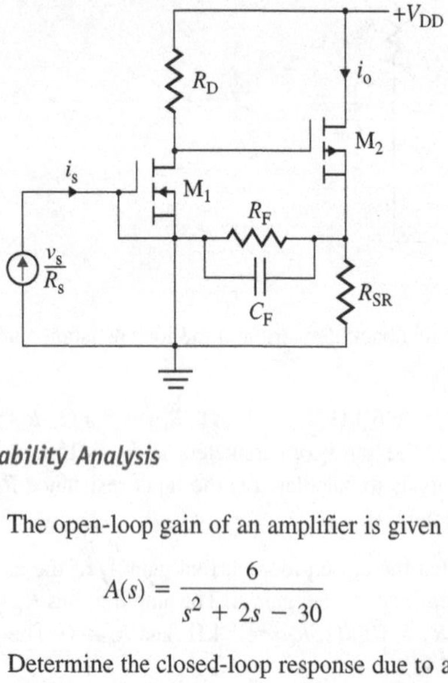

2.11 *Stability Analysis*

2.51 The open-loop gain of an amplifier is given by

$$A(s) = \frac{6}{s^2 + 2s - 30}$$

Determine the closed-loop response due to a step input. Assume feedback factor $\beta(s) = 1$.

2.52 The open-loop gain of an amplifier is given by

$$A(s) = \frac{s}{s^2 + 100}$$

Determine the closed-loop response due to a step input. Assume feedback factor $\beta(s) = 1$.

2.53 The loop gain of a feedback amplifier is given by

$$T_L(j\omega) = \frac{A_o}{j\omega(1 + j\omega T_1)(1 + j\omega T_2)}$$

where A_o is a gain constant and T_1 and T_2 are time constants. Determine the phase margin and gain margin of the amplifier.

2.54 The loop gain of a feedback amplifier is given by

$$T_L(j\omega) = \frac{A_o}{j\omega[(j\omega)^2 K_2 + j\omega K_1 + 1]}$$

where A_o is a gain constant and K_1 and K_2 are constants. Determine the phase margin and gain margin of the amplifier.

2.55 The loop gain of a feedback amplifier is given by

$$T_L(s) = \frac{10}{s(s + 1)(s + 2)}$$

Determine the phase margin and gain margin of the amplifier.

2.56 If the phase margin of an amplifier is PM = 40° and the magnitude of the open-loop gain is $|A(j\omega)| = 50$, find the magnitude of the closed-loop gain $|A_f(j\omega)|$.

2.57 The open-loop gain of an amplifier has break frequencies at f_{p1} = 10 kHz, f_{p2} = 100 kHz, and f_{p3} = 1 MHz. The low-frequency gain is $A_o = 250$, and the feedback factor is $\beta = 0.9$. Calculate the gain margin GM and the phase margin PM.

2.12 Compensation Techniques

2.58 For the amplifier in Prob. 2.57, determine the frequency at the dominant pole so that the phase margin is PM = 45°.

2.59 P The feedback amplifier in Fig. 2.60(a) has $g_m = 40 \times 10^{-3}$, R_i = 3.5 kΩ, C_i = 10 pF, R_o = 24 kΩ, and C_o = 5 pF.

a. Calculate the two pole frequencies for $C_x = 0$ and the value of feedback capacitance C_x so that the frequency of the dominant pole is f_D = 1.5 kHz.

b. Use PSpice/SPICE to plot the closed-loop transimpedance A_f and the input impedance Z_i against frequency.

2.60 D P The equivalent circuit of an output stage is shown in Fig. 2.60(c), where R_i = 22 kΩ and C_i = 18 pF.

a. Find the break frequency f_p.

b. Find the additional capacitance C_x that will move the break frequency to f_D = 50 kHz.

2.61 D P The open-loop gain of the multistage CE amplifier has break frequencies at f_{p1} = 805 kHz, f_{p2} = 9.6 MHz, f_{p3} = 13 MHz, and f_{p4} = 61 MHz. The midfrequency gain is A_{mid} = 9594. Determine the value of compensating capacitance C_x that will give a closed-loop phase margin of 45° with a resistive feedback of up to $\beta = 1$.

2.62 The open-loop gain of the multistage MOSFET amplifier has break frequencies at $f_{p1} = 13.75$ kHz, $f_{p2} = 248$ MHz, and $f_{p3} = 400$ MHz. The midfrequency gain is $A_{mid} = 311$. Determine the value of compensating capacitance C_x that will give a closed-loop phase margin of 45° with a resistive feedback of $\beta = 1$.

2.63 The open-loop gain of the CE amplifier has a midfrequency gain of $A_{mid} = -300$ V/V. Determine the values of compensating capacitance C_F and resistance R_F in a shunt-shunt feedback network to give a midfrequency closed-loop gain of $A_f = -30$ V/V. The low break frequency should be less than 1 kHz. Assume source resistance $R_s = 250\ \Omega$.

2.64 The open-loop gain of the amplifier in Example 2.12 has break frequencies at $f_{p1} = 10$ kHz, $f_{p2} = 100$ kHz, and $f_{p3} = 1$ MHz. The low-frequency gain is $A_o = 100$ A/A, and the emitter resistance is $R_E = 200\ \Omega$. Determine the values of compensating capacitance C_F and resistance R_F **(a)** to give a low-frequency closed-loop gain of $A_f = 20$ A/A and cancel the pole $f_{p1} = 10$ kHz and **(b)** to add a pole of $f_p = 10$ MHz and cancel the pole $f_{p1} = 10$ kHz.

2.65 The transistor Q_4 of the amplifier output stage as shown in Fig. 2.60(a) has an input resistance $r_{be4} = 20$ kΩ and an input capacitance $C_{j4} = 40$ pF.

a. Find the break frequency f_p.

b. Find the value of additional compensating capacitance C_x which will modify the dominant pole and will move the break frequency to $f_D = 25$ kHz.

2.66 The differential amplifier shown in Fig. 2.60(a) has an open-loop gain with break frequencies at $f_{p1} = 50$ kHz, $f_{p2} = 1$ MHz, and $f_{p3} = 10$ MHz. The gain stage of the amplifier has the equivalent circuit shown in Fig. 2.60(c) whose parameters are $g_m = 10$ mA/V, $C_i = 20$ pF, and $C_o = 5$ pF. Determine the value of compensating capacitance C_x by pole splitting that will give a closed-loop phase margin of 45° with a resistive feedback of up to $\beta = 1$.

2.67 The amplifier shown in Fig. P2.67 has an open-loop gain with break frequencies at $f_{p1} = 150$ kHz and $f_{p2} = 1$ MHz. The parameters of Q_4 are $g_{m3} = 1$ mA/V, and the output capacitance of $C_{o3} = 10$ pF. The input capacitance of Q_2 is $C_{i2} = 10$ pF. Determine the value of compensating capacitance C_x by pole splitting that will give a closed-loop phase margin of 50° with a resistive feedback of up to $\beta = 1$.

FIGURE P2.67

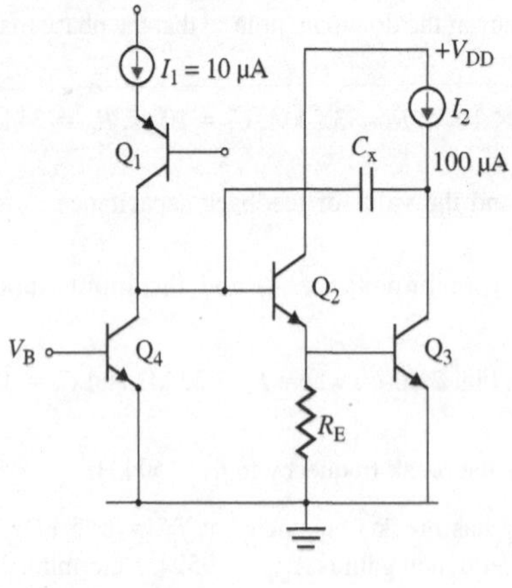

2.68 The open-loop gain of the amplifier shown in Fig. 2.61(a) has break frequencies at $f_{p1} = 50$ kHz, $f_{p2} = 1$ MHz, and $f_{p3} = 10$ MHz. The low-frequency gain is $A_o = 100$ A/A, and the emitter resistance $R_E = 400\ \Omega$. Determine the values of compensating capacitance C_F and resistance R_F by feedback path modification **(a)** to give a low-frequency closed-loop gain of $A_f = 20$ A/A and to cancel the pole $f_{p1} = 100$ kHz and **(b)** to add a pole of $f_p = 50$ MHz and to cancel the pole $f_{p1} = 50$ kHz.

2.69 A common-source amplifier with shunt-shunt feedback is shown in Fig. P2.69. The capacitor C_F and resistor R_F form the feedback network. The voltage gain without feedback is $A = |A_v| = v_o/v_s \approx 20$.

a. Design the feedback network to meet the following specifications: The bandwidth with feedback must be increased by 10 times; that is, $BW_f = 10$ BW. The voltage gain with feedback is $|A_f| = v_o/v_s = 5$.

b. Use PSpice to verify your results.

FIGURE P2.69

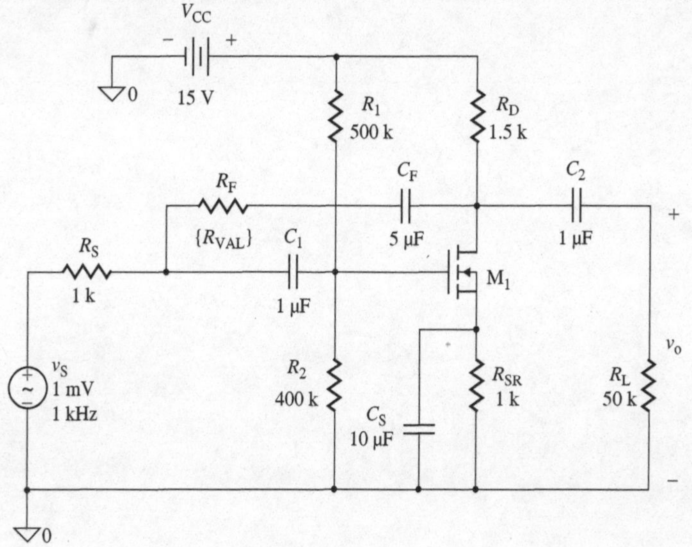

CHAPTER 3

POWER AMPLIFIERS

Learning Outcomes

After completing this chapter, students should be able to do the following:

- Describe the types of power amplifiers and their transfer characteristics.
- Select a power amplifier to meet certain requirements.
- Analyze and design power amplifiers.
- Design power amplifiers to meet certain specifications.
- Apply the methods for eliminating crossover distortion and reducing offsets and nonlinearities on the output voltage.
- Apply an active current source to bias the output stage.
- Identify and describe the internal structure of IC power op-amps.

Symbols and Their Meanings

Symbol	**Meaning**
v_i, v_o	Input and output voltages of an amplifier
I_p, V_p, V_{pp}	Peak current, peak voltage, and peak-to-peak voltage of a transistor
I_C, I_D	Collector and drain currents of a transistor
I_Q, I_R	Quiescent and reference currents of an amplifier
α, k	Conduction angle and duty cycle of a transistor

Symbol	Meaning
P_C, P_S, P_L	Collector power, DC supply power, and load power of an amplifier
i_S, i_O	Input signal and output currents of an amplifier
V_{BEN}, V_{BEP}	Base–emitter (B-E) voltages of an *npn* and a *pnp* transistor
h_{fe}	Hybrid current gain parameter of a transistor
η	Efficiency of an amplifier
ω_n	Natural frequency of an *LC* filter, radians per second

3.1 Introduction

The amplifiers were operated as input and/or intermediate stages to obtain a large voltage gain or current gain. The transistors within the amplifiers were operated in the active region so that their small-signal models were valid. These stages were not required to provide appreciable amounts of power, and the distortion of the output signal was negligible because the transistors operated in the active region.

The requirements for the output stages of audio-frequency power amplifiers are significantly different from those of small-signal low-power amplifiers. An output stage must deliver an appreciable amount of power and be capable of driving low-impedance loads such as loudspeakers. The distortion of the output signal must also be low. Distortion is measured by a quality factor known as *total harmonic distortion* (THD), which is the rms value of the harmonic components of the output signal, excluding the fundamental, expressed as a percentage of the rms of the fundamental component. The THD of high-fidelity audio amplifiers is usually less than 0.1%.

The DC power requirement of an audio amplifier must be as small as possible so that the efficiency of the amplifier is as high as possible. Increasing the efficiency of the amplifier reduces the amount of power dissipated by the transistors and the amount of power drawn from DC supplies, thereby reducing the cost of the power supply and prolonging the life of batteries in battery-powered amplifiers. Also, a low DC power requirement helps to keep the internal junction temperature of the transistors well below the maximum allowable temperature (in the range of 150°C to 200°C for silicon devices). As a result, a low DC power requirement minimizes the size of heat sinks and can eliminate the need for cooling fans. Therefore, an output stage should deliver the required amount of power to the load *efficiently*.

3.2 Classification of Power Amplifiers

Power amplifiers [1, 2] are generally classified into six types: A, B, AB, and C for analog designs and classes D and E for switching designs. This classification is based on the shape of the drain current i_D for a MOSFET or collector current i_C waveform for a BJT in response to a sinusoidal input signal. In an analog amplifier, the input signal to the amplifying devices causes a proportional output current to flow out of the output terminal, whereas in switching amplifiers the output current of the amplifying device is pulsating type. The output current and the load power come from the DC power supply. The load of power amplifiers is generally a speaker. The waveforms of the collector (or the drain) currents for various types of amplifiers are shown in Fig. 3.1. Table 3.1 shows the conduction angle and duty cycle of the input signal for analog circuits. The duty cycle for the class D and E amplifiers is generally variable up to 50% of the switching period.

In a class A amplifier, the DC biasing collector current I_C of a transistor is higher than the peak amplitude of the AC output current I_p. Thus, the transistor in a class A amplifier conducts during the entire cycle of the input signal, and the conduction angle is $\theta = \omega t = 360°$. That is, the collector current of a

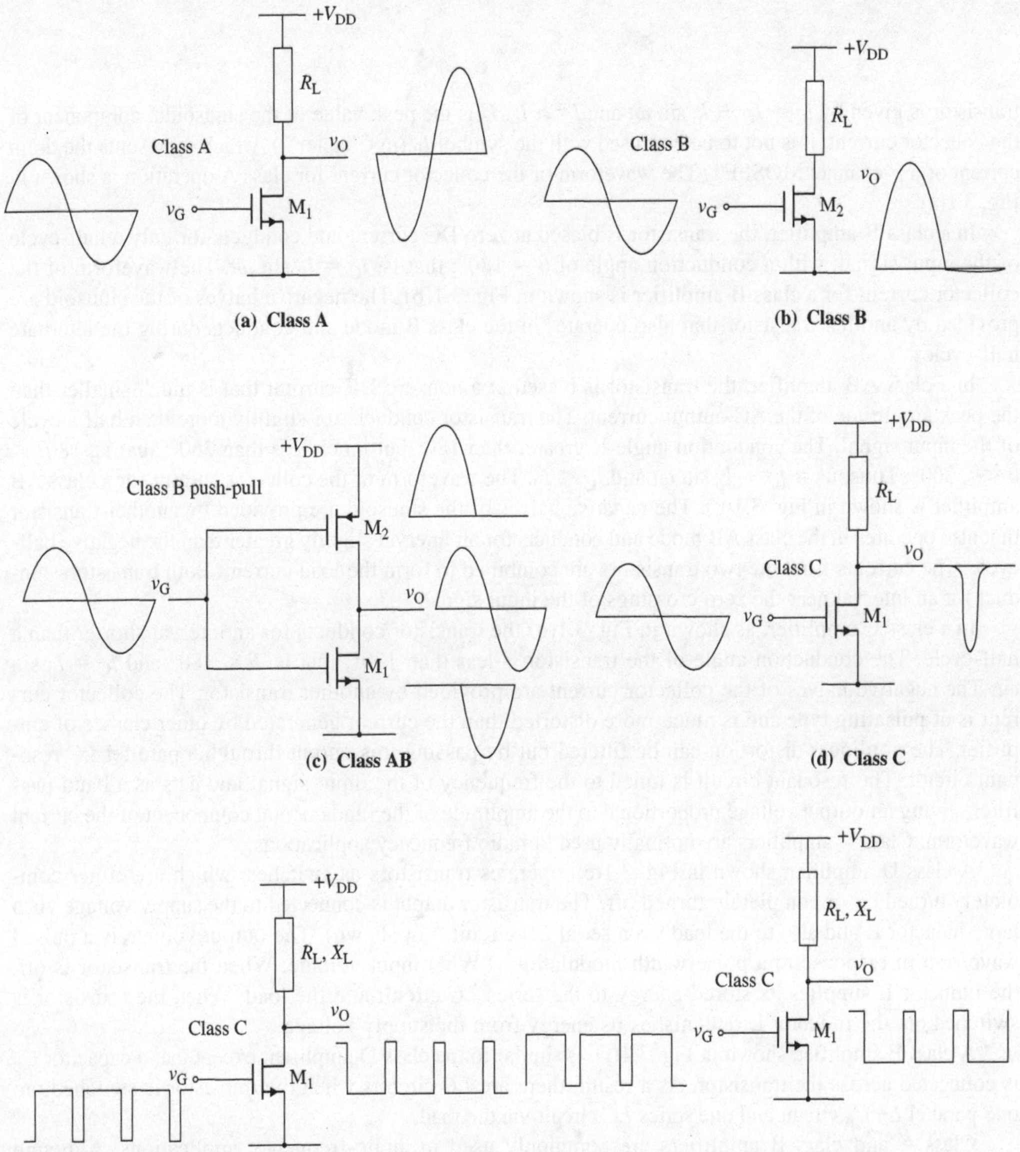

FIGURE 3.1 Output voltages for various classes of amplifiers

TABLE 3.1 Conduction angle and duty cycle of power transistors

	Class A	Class AB	Class B	Class C	Class D	Class E
Conduction angle α	$\alpha = 2\pi$ or 360°	$\pi < \alpha < 2\pi$ or 181° to 359°	$\alpha = \pi$ or 180°	$\alpha <$ 180° or 0° to 179°	$\alpha <$ 180° or 0° to 179°	$\alpha <$ 180° or 0° to 179°
Duty cycle k	k = 100%	50% $< k <$ 100%	k = 50%	$k <$ 50%	$k <$ 50%	$k <$ 50%

transistor is given by $i_C = I_C + I_p \sin \omega t$ and $I_C > I_p$. I_p is the peak value of the sinusoidal component of the collector current; it is not to be confused with the symbol I_P (in Chapter 2), which represents the drain current of a *p*-channel MOSFET. The waveform of the collector current for class A operation is shown in Fig. 3.1(a).

In a class B amplifier, the transistor is biased at zero DC current and conducts for only a half-cycle of the input signal, with a conduction angle of $\theta = 180°$; that is, $i_C = I_p \sin \omega t$. The waveform of the collector current for a class B amplifier is shown in Fig. 3.1(b). The negative halves of the sinusoid are provided by another transistor that also operates in the class B mode and conducts during the alternate half-cycles.

In a class AB amplifier, the transistor is biased at a nonzero DC current that is much smaller than the peak amplitude of the AC output current. The transistor conducts for slightly more than half a cycle of the input signal. The conduction angle is greater than 180° but much less than 360°; that is, $180° < \theta << 360°$. Thus, $i_C = I_C + I_p \sin \omega t$ and $I_C < I_P$. The waveform of the collector current for a class AB amplifier is shown in Fig. 3.1(c). The negative halves of the sinusoid are provided by another transistor that also operates in the class AB mode and conducts for an interval slightly greater than the negative half-cycle. The currents from the two transistors are combined to form the load current. Both transistors conduct for an interval near the zero crossings of the input signal.

In a class C amplifier, as shown in Fig. 3.1(d) the transistor conducts for an interval shorter than a half-cycle. The conduction angle of the transistor is less than 180°; that is, $\theta < 180°$ and $i_C = I_p \sin \omega t$. The negative halves of the collector current are provided by another transistor. The collector current is of pulsating type and is much more distorted than the current generated by other classes of amplifier. The nonlinear distortion can be filtered out by passing this output through a parallel *LC*-resonant circuit. The resonant circuit is tuned to the frequency of the input signal and acts as a band-pass filter, giving an output voltage proportional to the amplitude of the fundamental component of the current waveform. Class C amplifiers are normally used in radio frequency applications.

A class D amplifier, shown in Fig. 3.1(e), operates transistors as switches, which are either completely turned on or completely turned off. The transistor output is connected to the supply voltage via a large inductor L and also to the load via a serial *LC* circuit (not shown). The output voltage is a pulsed waveform in response to a pulse-width modulation (PWM) input voltage. When the transistor is off, the inductor L supplies its stored energy to the series *LC* circuit and the load. When the transistor is switched on, the inductor L replenishes its energy from the supply voltage.

A class E amplifier, shown in Fig. 3.1(f), is similar to the class D amplifier, except that a capacitor C_p is connected across the transistor. As a result, there are *LC* circuits when the transistor is switched on: one parallel $L-C_p$ circuit and one series *LC* circuit via the load.

Class A and class B amplifiers are commonly used in audio-frequency applications. Although there are many other types of amplifiers, we will consider the following kinds: emitter followers, class A amplifiers, class B push-pull amplifiers, complementary class AB push-pull amplifiers, quasi-complementary class AB push-pull amplifiers, transformer-coupled class AB push-pull amplifiers, and power op-amps.

KEY POINTS OF SECTION 3.2

- A power amplifier can be classified into one of six groups—A, B, AB, C, D, or E—depending on the conduction interval of the transistors used in the amplifier.
- To avoid distortion due to clipping, the maximum peak collector current of a transistor is limited to a specified value.

3.3 Power Transistors

The amplifying devices can be either MOSFETs or BJTs. The MOSFETs offer superior performances over the BJTs [3–5]. They require virtually zero input current, and they have faster switching times, no secondary breakdown, and stable gain and response time over a wide temperature range. The BJT current gain β_F can vary widely with temperature, and the variation of the MOSFET transconductance g_m with temperature is less than the variation of the BJT current gain β_F. As a result, power MOSFETs are replacing power BJTs in most applications.

There are no significant differences in the internal construction of small-signal and power BJTs. Although the mechanism of operation of power MOSFETs is same as that of small-signal MOSFETs, the power MOSFETs differ in their internal construction from the small-signal transistors in order to have more channel width for obtaining more current-carrying capability. To achieve a large channel width with good characteristics, power MOSFETs are fabricated with a repetitive pattern of small cells operating in parallel. The voltage ratings of power MOSFETs range from 50 V to 100 V, with current ratings from 10 A to 30 A.

There are two basic power MOSFET structures. The first is called a *DMOS* device, which uses a double-diffusion process; its cross section is shown in Fig. 3.2(a). It has two parallel current paths from the drain to the source. The *p*-substrate region is diffused deeper than the *n*+-source. The *n*-drift region must be moderately doped so that the drain breakdown voltage is sufficiently large and the thickness of the *n*-drift region is made as thin as possible to minimize drain resistance. The second structure, as shown in Fig. 3.2(b), uses a vertical channel known as a VMOS structure. In this case, the *p*-substrate diffusion is performed over the entire surface, and a V-shaped groove is then formed, extending through the *n*-drift region. For a high efficiency, on-resistance R_{on}, which should be low, can be found from

$$R_{DS(on)} = R_{SC} + R_{CH} + R_{DC} \tag{3.1}$$

where R_{SC} = source terminal contact resistance

R_{CH} = channel resistance

R_{DC} = drain terminal contact resistance

The values of the contact resistances R_{SC} and R_{DC} are proportional to the semiconductor resistivity. We can find the channel resistance R_{CH} in the linear region of operation as given by

$$R_{CH} = \frac{v_{DS}}{i_D} = \frac{L}{W\mu_n C_{ox}(v_{GS} - V_t)} \tag{3.2}$$

This explains why the channel length should be low and the width should be high to reduce on-resistance. An increase in the drain current increases the power loss and the junction temperature, which in turn increases the threshold voltage V_t and R_{CH}, thereby limiting the drain current.

For the sake of illustration of power amplifiers, we will show and analyze circuits using BJTs. The BJT analysis can be applied to MOSFET circuits by substituting $\beta_F = 0$ for the BJT current gain. Table 3.2 lists the circuit parameters for replacing a BJT by a MOSFET in a circuit.

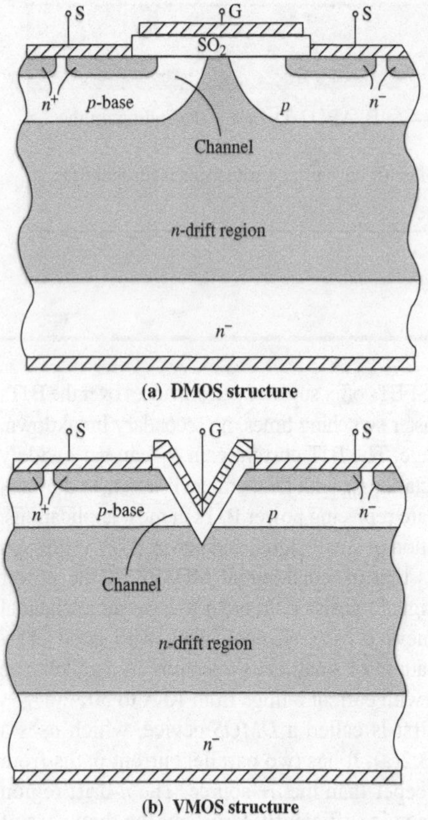

FIGURE 3.2 Cross section of a power MOSFET

TABLE 3.2 Circuit parameters of MOSFETs and BJTs

Circuit Parameters	MOSFETs	BJTs
	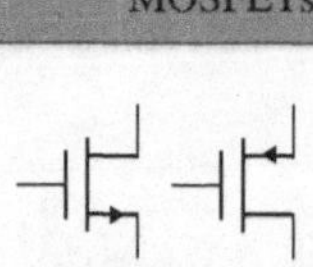	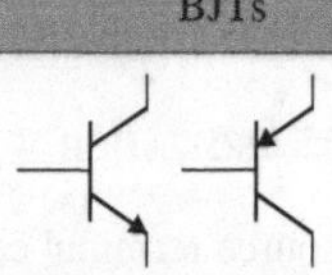
Supply voltage	Power supply $\pm V_{DD}$	$\pm V_{CC}$
Output current	Drain current i_D	Collector current i_C
Driving voltage	Gate–source voltage v_{GS}	Base–emitter voltage v_{BE}
Input voltage	Gate voltage v_G	Base voltage v_B
Transistor input current	$i_G \simeq 0$	Base current i_B
Current ratio	$\frac{i_D}{i_G} \simeq \infty$	$\frac{i_C}{i_B} = \beta_F$
Diode-connected transistor	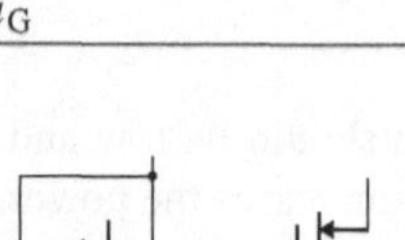	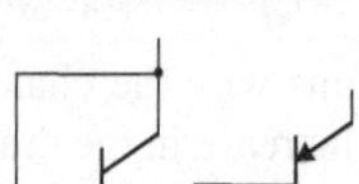

KEY POINTS OF SECTION 3.3

- The power MOSFETs offer superior performances over the power BJTs. MOSFETs require virtually zero input current and have faster switching times, no secondary breakdown, and stable gain and response time over a wide temperature range.
- There are two basic power MOSFET structures. The channel length of MOSFETs is made low, and the width is made high to reduce on-resistance.

3.4 Class A Amplifiers

The common-source (common-emitter) amplifiers and the source and emitter followers fall into the class A category. The transistors of class A amplifiers are always turned on. Therefore, the transistor can become hot, with most of the power provided by the power supply being dissipated as heat. Although efficiency is poor (around 20%), the accuracy of the signal amplifications is quite high. We have analyzed the amplifiers to derive the small-signal parameters without any consideration to the circuit efficiency. For power amplifiers, efficiency is a major performance parameter.

3.4.1 Emitter Followers

An emitter follower is a class A amplifier; its circuit diagram is shown in Fig. 3.3(a)The voltage gain and the DC current of transistor Q_1 are affected by the values of load resistance R_L. The peak-to-peak voltage swing is less than V_{CC}. If the emitter resistance R_E in Fig. 3.3(a) can be replaced by a current source, as shown in Fig. 3.3(b), the peak-to-peak voltage swing can be increased to a value larger than V_{CC}. The voltage gain can be maintained at almost unity even with a small load resistance, on the order of 100 Ω.

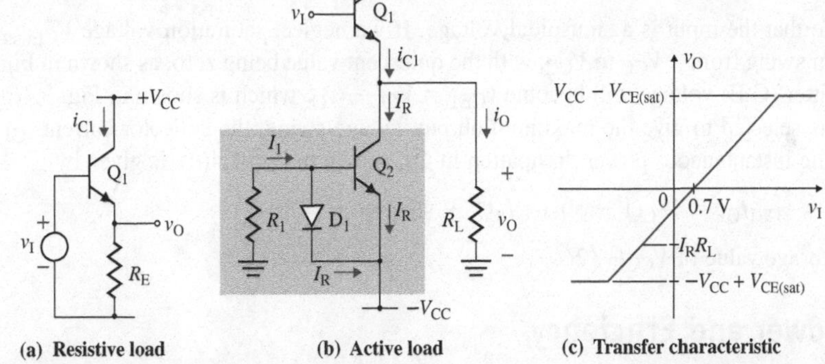

FIGURE 3.3 Emitter follower

Transfer Characteristic

Assuming that the diode drop is $V_{D1} = 0.7$ V, the B-E drop of a transistor is $V_{BE} = 0.7$ V, and the transistor current gain $\beta_F >> 1$, the reference current I_R can be approximated by I_1. Using KVL for the current source, we get

$$0 = I_R R_1 + V_{BE} - V_{CC}$$

which gives the reference current I_R as

$$I_R = \frac{V_{CC} - V_{BE}}{R_1} \tag{3.3}$$

The output voltage is given by

$$v_O = v_1 - V_{BE} \tag{3.4}$$

which yields the transfer characteristic shown in Fig. 3.3(c). At $v_I = 0$ (i.e., $v_O = -V_{BE}$), the characteristic has an offset voltage. Thus, for $i_{C1} > 0$, Q_1 will be on, and the positive peak value of output voltage is

$$+V_{O(max)} = V_{CC} - V_{CE(sat)} \tag{3.5}$$

Q_1 will be turned off when $i_{C1} = 0$ and $i_O = -I_R$. The output voltage can also be written as

$$v_O = i_O R_L = -I_R R_L$$

$$= -\frac{V_{CC} - V_{BE}}{R_1} R_L \tag{3.6}$$

If the value of R_L is less than R_1, the peak negative value of the output voltage will be less than the peak negative value of $V_{CC} - V_{CE(sat)}$ and the output will not be symmetrical. This will cause distortion (or clipping). Thus, the condition that avoids distortion and obtains the maximum output voltage swing is given by

$$R_L \geq R_1 \tag{3.7}$$

and the maximum voltage swing (peak to peak) without clipping is

$$V_{pp} = 2(V_{CC} - V_{CE(sat)}) \tag{3.8}$$

Signal Waveforms

Let us assume that the input is a sinusoidal voltage. If we neglect saturation voltage $V_{CE(sat)}$, the output voltage v_O can swing from $-V_{CC}$ to V_{CC}, with the quiescent value being zero, as shown in Fig. 3.4(a). The collector–emitter (C-E) voltage will become $v_{CE1} = V_{CC} - v_O$, which is shown in Fig. 3.4(b). Assuming that $I_R = I_Q$ is selected to give the maximum output voltage swing, the collector current i_{C1} is shown in Fig. 3.4(c). The instantaneous power dissipation in Q_1, shown in Fig. 3.4(d), is given by

$$P_{D1} \approx v_{CE1} i_{C1} = V_{CC}(1 - \sin \omega t) I_R (1 + \sin \omega t) \tag{3.9}$$

and has an average value of $V_{CC} I_R / 2$.

Output Power and Efficiency

The power efficiency of the output stage of an emitter follower is defined by

$$\eta = \frac{\text{Load power } P_L}{\text{Supply power } P_S} \tag{3.10}$$

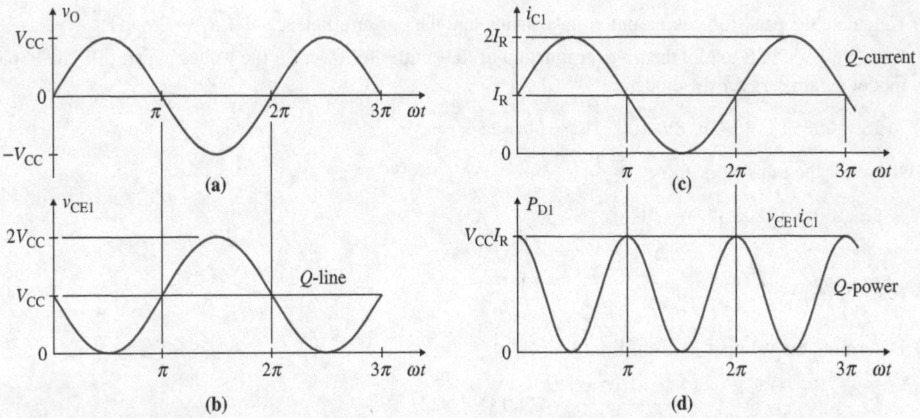

FIGURE 3.4 Signal waveforms of an emitter follower

Assuming that the output voltage is sinusoidal with a peak value of V_p, the average load power will be

$$P_L = \frac{V_p^2}{2R_L} \tag{3.11}$$

The average current drawn by transistor Q_1 will be I_R, and thus the average power drawn from the positive supply will be $V_{CC}I_R$. Since the current in transistor Q_2 remains constant at I_R, the power drawn from the negative supply will also be $V_{CC}I_R$, if we neglect the power drawn by the current source consisting of the diode D_1 and resistor R_1. Thus, the total average supply power will be

$$P_S = 2V_{CC}I_R \tag{3.12}$$

From Eqs. (3.11) and (3.12), we get the power efficiency as

$$\eta = \frac{V_p^2}{4R_L V_{CC} I_R} = \frac{1}{4}\left(\frac{V_p}{R_L I_R}\right)\left(\frac{V_p}{V_{CC}}\right) \tag{3.13}$$

which will give the maximum efficiency when

$$V_p = V_{CC} \leq R_L I_R$$

That is, $\eta_{max} = 25\%$, which is rather low. In practice, the peak output voltage is limited to less than V_{CC} to avoid transistor saturation and associated nonlinear distortion. Thus, the efficiency actually ranges from 10% to 20%. Emitter followers are generally used as output stages for high-frequency (≈10 MHz), low-power (≤1 W) amplifiers.

EXAMPLE 3.1

D

Designing an emitter follower

(a) Design the emitter follower of the circuit in Fig. 3.3(b). Assume $V_{CC} = 12$ V, $V_{BE} = 0.7$ V, $V_{CE(sat)} = 0.5$ V, $I_R = 5$ mA, and $R_L = 650\ \Omega$. Assume identical transistors of current gain $h_{fe} = \beta_F = 100$. Note that h_{fe} is the hybrid parameter and both h_{fe} and β_F represent the current gain of a BJT.

(b) Determine the critical value of load resistance to avoid clipping (or distortion).

(c) Calculate the peak-to-peak output voltage swing if $R_L = 650\ \Omega$.

(d) Calculate the peak-to-peak output voltage swing and the power efficiency η if $R_L = 2.5\ k\Omega$.

(e) Use PSpice/SPICE to plot the transfer function of the emitter follower for the values in part (a). The PSpice model parameters for the diode are

```
IS=100E-15 RS=16 BV=100 IBV=100E-15
```

and those for the transistors are

```
BF=100 VA=100
```

SOLUTION

(a) Determine the value of R_1 from Eq. (3.3):

$$R_1 = \frac{V_{CC} - V_{BE}}{I_R} = \frac{12\ V - 0.7\ V}{5\ mA} = 2260\ \Omega$$

(b) From Eq. (3.7), we find the critical value of load resistance to avoid clipping:

$$R_{L(crit)} = R_1 = 2260\ \Omega$$

(c) For $R_L = 650\ \Omega$, the negative peak output voltage is

$$-V_{O(max)} = -I_R R_L = -5\ mA \times 650\ \Omega = -3.25\ V$$

The positive peak output voltage is

$$+V_{O(max)} = V_{CC} - V_{CE(sat)} = 12\ V - 0.5\ V = 11.5\ V$$

Therefore, the peak-to-peak output voltage swing will be from 11.5 V to −3.25 V.

(d) For $R_L = 2.5\ k\Omega$ (which is greater than $R_{L(crit)} = 2260\ \Omega$), the negative peak output voltage will be limited to

$$-V_{O(max)} = -V_{CC} + V_{CE(sat)} = -12\ V + 0.5\ V = -11.5\ V$$

Therefore, the peak-to-peak output voltage swing will be from −11.5 V to +11.5 V. Thus, Eq. (3.13) gives

$$\eta = \frac{(11.5\ V)^2}{4 \times 2.5\ k\Omega \times 12\ V \times 5\ mA} = 22\%$$

(e) The emitter-follower circuit for PSpice simulation is shown in Fig. 3.5.

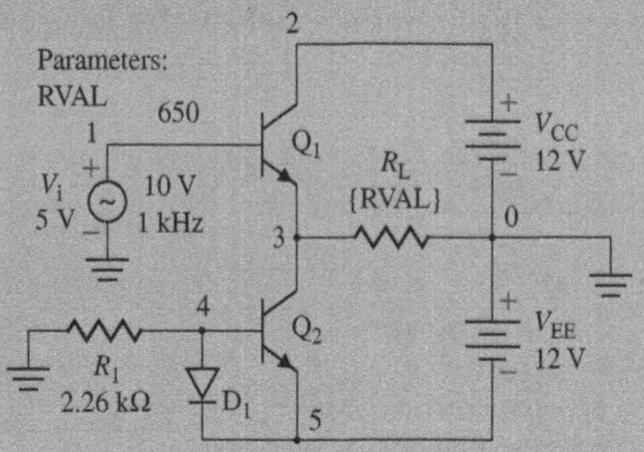

FIGURE 3.5 Emitter-follower circuit for PSpice simulation

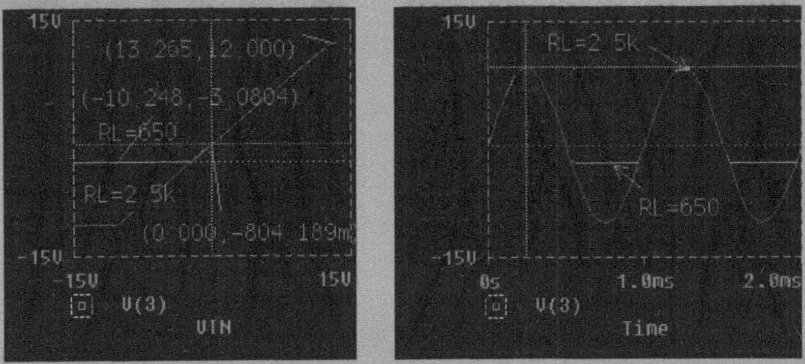

(a) Transfer characteristic **(b) Output voltage**

FIGURE 3.6 Transfer characteristic and output voltage for Example 3.1

The transfer characteristic is shown in Fig. 3.6(a). For $R_L = 650\ \Omega$, it gives $+V_{O(max)} = 12$ V (expected value is 11.5 V) and $-V_{O(max)} = -3.08$ V (expected value is -3.25 V). For $R_L = 2.5\ \text{k}\Omega$, it gives $+V_{O(max)} =$ 13.26 V (expected value is 11.5 V) and $-V_{O(max)} = -10.25$ V (expected value is -11.5 V). The output voltage, shown in Fig. 3.6(b), has an offset of 0.8 V (expected value is 0.7 V) and is clamped to a certain value, thereby introducing distortion.

The Fourier analysis (.FOUR) gives the following results (from the PSpice output file).

For $R_L = 650\ \Omega$,

```
DC COMPONENT=1.300972E+00
TOTAL HARMONIC DISTORTION=3.131888E+01 PERCENT
```

For $R_L = 2.5\ \text{k}\Omega$,

```
DC COMPONENT=-7.972313E-01
TOTAL HARMONIC DISTORTION=1.895552E-01 PERCENT
```

3.4.2 Basic Common-Emitter Amplifier

The simplest kind of class A amplifier is shown in Fig. 3.7(a). This configuration lacks biasing stability (i.e., emitter resistance R_E) and is not suitable for power amplifiers. However, we will use this circuit to derive the power efficiency of class A amplifiers. Let us assume that the nonlinearity introduced by the transistor is negligible so that the output signals will be sinusoidal for sinusoidal input signals. These assumptions will simplify the various power calculations. The waveforms of the collector current and voltages are shown in Fig. 3.7(b).

Transfer Characteristic

The input voltage v_I is related to the collector current I_C by

$$v_I = V_{BE} = V_T \ln\left(\frac{i_C}{I_S}\right) \tag{3.14}$$

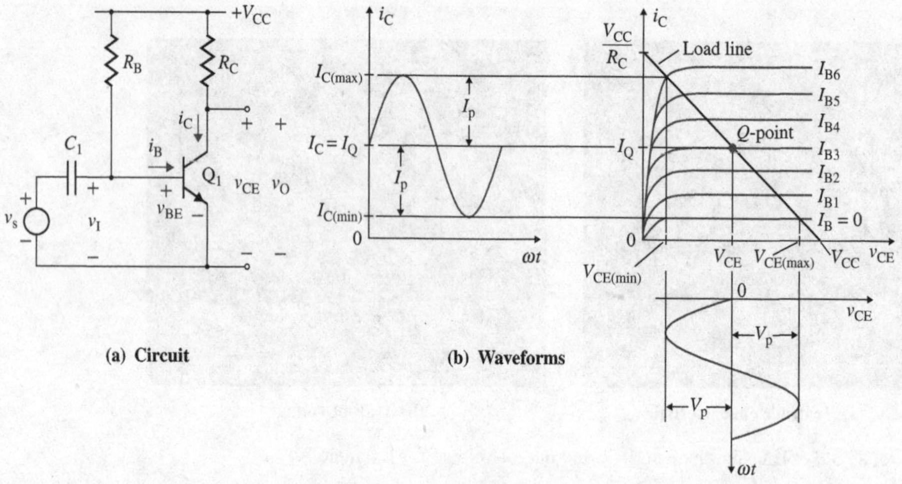

FIGURE 3.7 Basic common-emitter class A amplifier

Using i_C from Eq. (3.14), we can find the output voltage as

$$v_O = V_{CC} - R_C i_C = V_{CC} - R_C I_S \ln\left(\frac{v_I}{V_T}\right) \tag{3.15}$$

Therefore, the transfer characteristic (v_O versus v_I), which is shown in Fig. 3.8, is nonlinear.

Output Power and Efficiency

The average DC power required from the power supply is given by

$$P_S = V_{CC} I_C \tag{3.16}$$

The average load or output power is given by

$$P_L = \left(\frac{V_p}{\sqrt{2}}\right)\left(\frac{I_p}{\sqrt{2}}\right) = \frac{V_p I_p}{2} \tag{3.17}$$

$$= \frac{I_p R_L I_P}{2} = \frac{I_p^2 R_L}{2} \tag{3.18}$$

▶ **NOTE** $\sqrt{2}$ converts the peak value to an rms value.

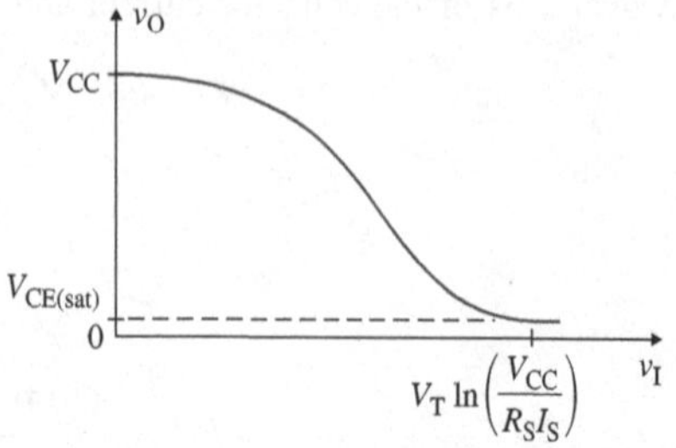

FIGURE 3.8 Transfer characteristic of a common-emitter class A amplifier

where V_p and I_p are the peak values of the AC output voltage and current, respectively. Using the minimum and maximum values of the output signals, we can express V_p and I_p as

$$V_p = \frac{V_{CE(max)} - V_{CE(min)}}{2} \tag{3.19}$$

$$I_p = \frac{I_{C(max)} - I_{C(min)}}{2} \tag{3.20}$$

where V_{CE} ideally extends over its full range. Using Eqs. (3.19) and (3.20), we can express Eq. (3.17) as

$$P_L = \frac{V_p I_p}{2} = \frac{(V_{CE(max)} - V_{CE(min)})(I_{C(max)} - I_{C(min)})}{8} \tag{3.21}$$

which will give the maximum load power $P_{L(max)}$ when $V_{CE(min)} = 0$, $I_{C(min)} = 0$, $V_{CE(max)} = V_{CC}$, and $I_{C(max)} = 2I_C$. Thus, Eq. (3.21) gives $P_{L(max)}$ as

$$P_{L(max)} = \frac{V_{CC}(2I_C)}{8} = \frac{V_{CC}I_C}{4} \tag{3.22}$$

The conversion efficiency, which is defined as the ratio of load power to the DC source power, is expressed as

$$\eta = \frac{P_L}{P_S} \times 100\% \tag{3.23}$$

Substituting Eqs. (3.16) and (3.21) into Eq. (3.23) gives the maximum efficiency as

$$\eta_{max} = \frac{(V_{CE(max)} - V_{CE(min)})(I_{C(max)} - I_{C(min)})}{8V_{CC}I_C} \tag{3.24}$$

which, for $V_{CE(min)} = 0$, $I_{C(min)} = 0$, $V_{CE(max)} = V_{CC}$, and $I_{C(max)} = 2I_C$, becomes

$$\eta_{max} = \frac{V_{CE(max)}I_{CE(max)}}{8V_{CC}I_C} = \frac{V_{CC}(2I_C)}{8V_{CC}I_C} = \frac{1}{4} = 25\% \tag{3.25}$$

Hence, the maximum efficiency for a class A amplifier under ideal conditions is 25%. Although in practice the actual efficiency will be less than 25%, this percentage is often used as a guideline for determining the biasing requirement I_C. For example, if $V_{CC} = 30$ V and $P_{L(max)} = 50$ W, then

$$P_S = \frac{P_{L(max)}}{\eta_{(max)}} = \frac{50}{0.25} = 200 \text{ W}$$

and

$$I_C = \frac{P_S}{V_{CC}} = \frac{200}{30} = 6.67 \text{ A}$$

The quality of an amplifier is often measured by the *figure of merit* F_m, which is defined by

$$F_m = \frac{\text{Maximum collector dissipation}}{\text{Maximum output power}} = \frac{P_{C(max)}}{P_{L(max)}} \tag{3.26}$$

The maximum collector dissipation is given by

$$P_{C(max)} = \frac{V_{CC}I_C}{2} \tag{3.27}$$

Substituting Eqs. (3.22) and (3.27) into Eq. (3.26) yields

$$F_m = \frac{V_{CC}I_C/2}{V_{CC}I_C/4} = 2 \tag{3.28}$$

Thus, the collector power dissipation is twice the maximum output power. That is, for a maximum output of 50 W, the collector must be able to dissipate at least 100 W. This requirement is the major disadvantage of class A amplifiers because it necessitates the use of a large and expensive heat sink to cool the transistors.

3.4.3 Common-Emitter Amplifiers with Active Load

Because of their high voltage gain, common-emitter stages are often used as output-stage drivers in integrated circuit design. A common-emitter stage is shown in Fig. 3.9(a). A current source consisting of Q_2 and Q_3 establishes the reference current I_R, which is given by

$$I_R = \frac{V_{CC} - V_{EB2}(=V_{EB3})}{R_1} \tag{3.29}$$

With no load, $R_L = \infty$, $i_O = 0$, and $i_{C1} = I_Q = I_R$. Thus, the load current i_O is given by

$$i_O = I_R - i_{C1} \tag{3.30}$$

where the collector current i_{C1} is related to the input voltage v_I by

$$i_{C1} = I_S \exp\left(\frac{v_I}{V_T}\right) \tag{3.31}$$

Transfer Characteristic

The output voltage v_O is given by

$$\begin{aligned} v_O &= R_L i_O = R_L(I_R - i_{C1}) \\ &= R_L\left[I_R - I_S \exp\left(\frac{v_I}{V_T}\right)\right] \end{aligned} \tag{3.32}$$

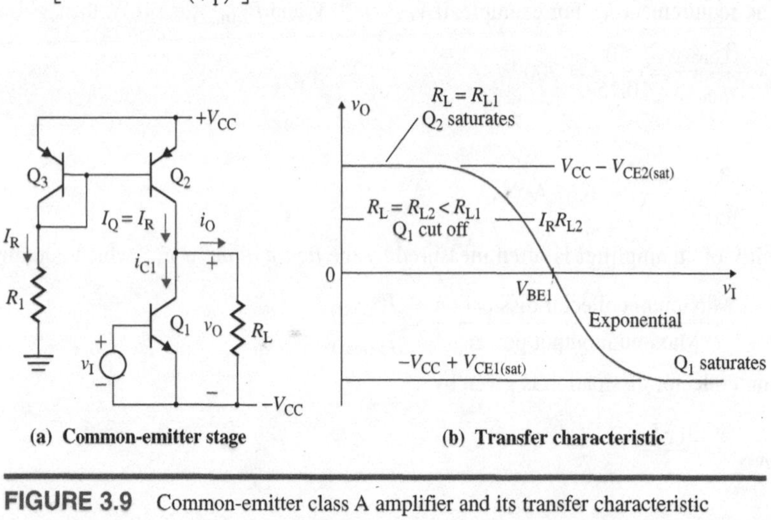

(a) Common-emitter stage (b) Transfer characteristic

FIGURE 3.9 Common-emitter class A amplifier and its transfer characteristic

which is the transfer characteristic (v_O versus v_I), shown in Fig. 3.9(b). When $i_{C1} = 0$, transistor Q_1 is cut off and transistor Q_2 supplies I_Q to the load. The output voltage becomes

$$v_O = R_L I_R = R_L \frac{V_{CC} - V_{BE2}}{R_1} \tag{3.33}$$

If this value of v_O is less than the maximum possible positive output voltage of $V_{CC} - V_{CE2(sat)}$, distortion will occur, as shown in Fig. 3.9(b). Thus, the condition for the maximum positive output voltage swing is $R_L \geq R_1$. As the input voltage v_I is increased, the current in transistor Q_1 increases and v_O becomes negative according to Eq. (3.32), until Q_1 saturates and the output voltage becomes

$$v_O = -V_{CC} + V_{CE1(sat)}$$

The transfer characteristic is basically a simple exponential and shows distortion on the output voltage. The input voltage required to produce the maximum output voltage swing is typically a few tens of millivolts or less.

Output Power and Efficiency

All the equations derived for the emitter follower and the basic common-emitter circuits apply to this common-emitter stage. Thus, the maximum efficiency is $\eta_{max} = 25\%$, and the figure of merit is $F_m = 2$. The breakdown voltage rating of Q_1 and Q_2 must be $2V_{CC}$.

3.4.4 Transformer-Coupled Load Common-Emitter Amplifier

The efficiency of an amplifier can be improved with a transformer-coupled load. A class A amplifier with a transformer-coupled load is shown in Fig. 3.10(a). The elimination of the collector resistance R_C, used for DC biasing in Fig. 3.7(a), accounts for the increase in efficiency. The transformer at the output stage provides an impedance match in order to transfer maximum power to the load. A load such as the impedance of a loudspeaker is usually very small, typically 4 Ω to 16 Ω.

The voltage and current relations of the output transformer are

$$V_{pL} = \left(\frac{n_p}{n_s}\right) V_{sL} \quad \text{and} \quad I_{pL} = \left(\frac{n_s}{n_p}\right) I_{sL}$$

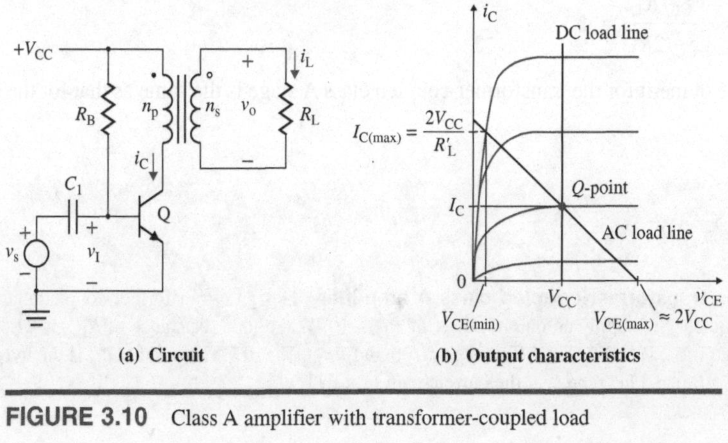

FIGURE 3.10 Class A amplifier with transformer-coupled load

where n_s and n_p = secondary and primary windings, respectively
V_{sL} and V_{pL} = secondary and primary voltages, respectively
I_{sL} and I_{pL} = the secondary and primary currents, respectively

The effective load resistance referred to the primary side can be found from

$$R'_L = \left(\frac{V_{pL}}{I_{pL}}\right) = \left(\frac{n_p}{n_s}\right)^2 \left(\frac{V_{sL}}{I_{sL}}\right) = \left(\frac{n_p}{n_s}\right)^2 R_L \tag{3.34}$$

The AC (dynamic) load line is determined by R'_L. The DC (static) load line is almost vertical because of the very small primary resistance of the transformer. Assuming $V_{CE(min)} = 0$ in Eq. (3.19) and $I_{C(min)} = 0$ in Eq. (3.20), the peak values of the output voltage and current at the primary side of the transformer are

$$V_p = \frac{V_{CE(max)}}{2} = V_{CC} \tag{3.35}$$

$$I_p = \frac{I_{C(max)}}{2} = I_C \tag{3.36}$$

Equation (3.21) gives the maximum efficiency as

$$\eta_{max} = \frac{V_{CE(max)} I_{C(max)}}{8 V_{CC} I_C} = \frac{2V_{CC}(2I_C)}{8 V_{CC} I_C} = \frac{1}{2} = 50\% \tag{3.37}$$

Hence, the maximum efficiency of a class A stage is doubled by using a transformer coupled to the load. The value of V_p for a transformer-coupled stage is V_{CC}; for the basic common-emitter amplifier, it is only $V_{CC}/2$.

Equation (3.18) gives the maximum load power as

$$P_{L(max)} = \frac{I_p^2 R'_L}{2} = \frac{V_{CC}^2}{2R'_L} \tag{3.38}$$

The maximum collector dissipation is given by

$$P_{C(max)} = V_{CC} I_C = V_{CC} I_p = \frac{V_{CC}^2}{R'_L} \tag{3.39}$$

Substituting Eqs. (3.38) and (3.39) into Eq. (3.26) yields

$$F_m = \frac{V_{CC}^2/R'_L}{V_{CC}^2/2R'_L} = 2 \tag{3.40}$$

Thus, the figure of merit for the transformer-coupled class A stage is the same as that for the basic common-emitter stage.

EXAMPLE 3.2

D **Designing a transformer-coupled class A amplifier** Design a transformer-coupled class A amplifier with high efficiency to supply an output power of $P_L = 10$ W at a load resistance of $R_L = 4\ \Omega$. Assume a DC supply voltage of 12 V and BJTs of $\beta_F = h_{fe} = 100$ and $V_{CE(sat)} = 0.7$ V. Note that h_{fe} is the hybrid parameter and both β_f ($\approx\beta_F$) and h_{fe} represent the current gain of a BJT.

SOLUTION

The design steps are as follows:

Step 1. Determine the maximum collector-to-emitter voltage of the transistors:

$$V_{CE(max)} \geq 2V_{CC} = 2 \times 12 = 24\text{ V}$$

Step 2. From Eq. (3.40), calculate the collector power dissipation, which must be at least twice the AC power:

$$P_C \geq F_m P_L = 2P_L = 2 \times 10 = 20\text{ W}$$

Step 3. Calculate the value of the quiescent collector current I_C:

$$I_C = \frac{P_C}{V_{CC}} = \frac{20}{12} = 1.67\text{ A}$$

Step 4. Calculate the slope of the load line to find the AC load resistance R'_L:

$$R'_L = \frac{V_{CC}}{I_C} = \frac{12}{1.67} = 7.19\ \Omega$$

Step 5. From Eq. (3.34), calculate the required turns ratio of the transformer:

$$\frac{n_p}{n_s} = \left(\frac{R'_L}{R_L}\right)^{1/2} = \left(\frac{7.19}{4}\right)^{1/2} = 1.34$$

Step 6. Calculate the peak collector current:

$$I_{C(max)} = 2I_C = 2 \times 1.67 = 3.34\text{ A}$$

Step 7. Calculate the quiescent base current I_B:

$$I_B = \frac{I_C}{h_{fe}} = \frac{1.67}{100} = 16.7\text{ mA}$$

Step 8. Calculate the base resistance R_B:

$$R_B = \frac{V_{CC} - V_{CE(sat)}}{I_B} = \frac{12\text{ V} - 0.7\text{ V}}{16.7\text{ mA}} = 677\ \Omega$$

KEY POINTS OF SECTION 3.4

- An emitter follower biased with an active current source is the most commonly used output stage.
- An emitter follower has a high input impedance and a low gain of almost unity. However, it exhibits an offset voltage of approximately $-V_{BE} \approx -0.7$ V at $V_i = 0$.
- In a class A amplifier, the transistor conducts a continuous DC biasing current. As a result, the maximum power efficiency is only 25%. The figure of merit, which is the ratio of maximum collector power dissipation to maximum output power, is 2.
- The power efficiency of a class A stage can be increased to 50% with a transformer-coupled load.

3.5 Class B Push-Pull Amplifiers

In a class B push-pull amplifier, two complementary transistors (one *npn* transistor and one *pnp* transistor) are employed to perform the push-pull operation [1, 6]. The efficiency is much higher than that of a class A amplifier, and the accuracy of the output is improved. It is the most commonly used audio-power amplifier. This section discusses two types of class B amplifiers.

3.5.1 Complementary Push-Pull Amplifiers

A complementary push-pull amplifier is shown in Fig. 3.11(a). For $v_I > 0$, transistor Q_P remains off and transistor Q_N operates as an emitter follower. For a sufficiently large value of v_I, Q_N saturates and the maximum positive output voltage becomes

$$V_{CE(max)} = V_{CC} - V_{CE1(sat)}$$

For $v_I < 0$, transistor Q_N remains off and transistor Q_P operates as an emitter follower. For a sufficiently large negative value of v_I, Q_P saturates and the maximum negative output voltage becomes

$$-V_{CE(max)} = -(V_{CC} - V_{CE2(sat)}) = -V_{CC} + V_{CE2(sat)}$$

Assuming identical transistors of $V_{BE1} = V_{BE2} = V_{BE}$, the output voltage is given by

$$v_O = v_I - V_{BE} \quad \text{for} \quad -0.7\text{ V} \geq v_I \geq 0.7\text{ V} \tag{3.41}$$

which gives the transfer characteristic of v_O versus v_I shown in Fig. 3.11(b). However, during the interval $-0.7\text{ V} \leq v_I \leq 0.7\text{ V}$, both Q_P and Q_N remain off, and $v_O = 0$. This causes a dead zone and crossover distortion on the output voltage, as illustrated in Fig. 3.12.

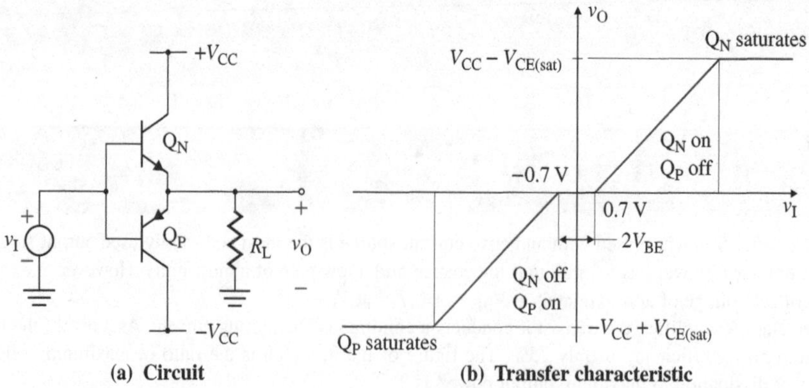

FIGURE 3.11 Complementary class B push-pull amplifier

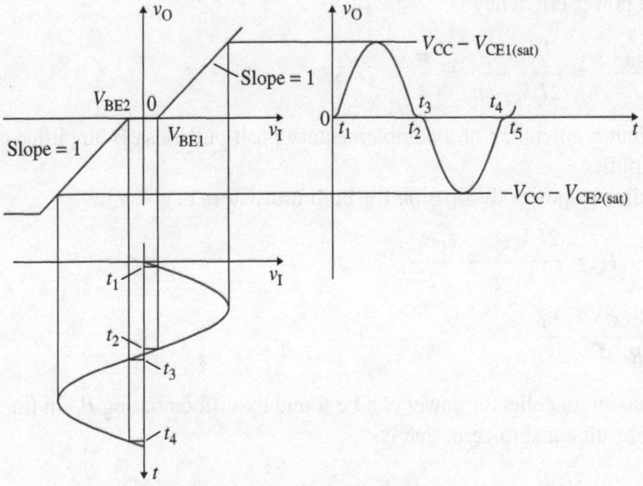

FIGURE 3.12 Crossover distortion on input and output waveforms

Output Power and Efficiency

Let us assume that $V_{CE1(sat)} = V_{CE2(sat)} = V_{CE(sat)} = 0$ and $I_{C(min)} = 0$. Assuming a sinusoidal variation of the collector current $i_{C1} = I_p \sin \omega t$, the average collector current of a transistor can be found from

$$I_{C1} = \frac{1}{2\pi}\int_0^{\pi} i_{c1}\, dt = \frac{1}{2\pi}\int_0^{\pi} I_p \sin(\omega t)\, d(\omega t) = \frac{I_p}{\pi} \tag{3.42}$$

The average current drawn from the DC supply source by transistors Q_N and Q_P is

$$I_{dc} = 2I_{C1} = \frac{2I_p}{\pi} \tag{3.43}$$

Thus, the average input power supplied from the DC source is

$$P_S = I_{dc}V_{CC} = \frac{2I_pV_{CC}}{\pi} \tag{3.44}$$

From Eq. (3.17), the output power is given by

$$P_L = \frac{I_p^2R_L}{2} = \frac{I_pV_p}{2}$$

Thus, the power efficiency becomes

$$\eta = \frac{P_L}{P_S} = \frac{I_pV_p/2}{2I_pV_{CC}/\pi} = \frac{\pi}{4}\left(\frac{V_p}{V_{CC}}\right) \tag{3.45}$$

which gives $\eta = 50\%$ at $V_p = 2V_{CC}/\pi$ and $\eta = 78.5\%$ at $V_p = V_{CC}$. At $V_p = V_{CC}$, the maximum output power is given by

$$P_{L(max)} = \frac{I_p^2R_L}{2} = \frac{I_pV_p}{2} = \frac{I_pV_{CC}}{2} = \frac{V_{CC}^2}{2R_L} \tag{3.46}$$

Thus, the maximum power efficiency is

$$\eta_{max} = \frac{P_{L(max)}}{P_S} = \frac{I_p V_{CC}/2}{2I_p V_{CC}/\pi} = \frac{\pi}{4} = 78.5\%$$

Therefore, the maximum efficiency of a complementary push-pull class B amplifier is much higher than that of a class A amplifier.

The average collector power dissipation for both transistors is given by

$$2P_C = P_S - P_L = \frac{2I_p V_{CC}}{\pi} - \frac{I_p^2 R_L}{2} \tag{3.47}$$

$$= \frac{2V_p V_{CC}}{\pi R_L} - \frac{V_p^2}{2R_L} \tag{3.48}$$

The condition for maximum collector power can be found by differentiating P_C in Eq. (3.47) with respect to I_p and setting the result equal to zero; that is,

$$\frac{dP_C}{dI_p} = \frac{2V_{CC}}{\pi} - \frac{2I_p R_L}{2} = 0$$

which gives the peak current for maximum collector power dissipation as

$$I_{p(max)} = \frac{2V_{CC}}{\pi R_L} \tag{3.49}$$

and the corresponding maximum peak voltage as

$$V_{p(max)} = I_p R_L = \frac{2V_{CC}}{\pi} \tag{3.50}$$

Substituting $I_{p(max)}$ from Eq. (3.49) and $V_{p(max)}$ from Eq. (3.50) into Eq. (3.47) gives the maximum collector dissipation as

$$2P_{C(max)} = \frac{4V_{CC}^2}{\pi^2 R_L} - \frac{2V_{CC}^2}{\pi^2 R_L} = \frac{2V_{CC}^2}{\pi^2 R_L} \tag{3.51}$$

Normalizing P_C in Eq. (3.48) with respect to $P_{C(max)}$ in Eq. (3.51), we get

$$\frac{P_C}{P_{C(max)}} = \pi\left(\frac{V_p}{V_{CC}}\right) - \frac{\pi^2}{4}\left(\frac{V_p}{V_{CC}}\right)^2 \tag{3.52}$$

which becomes 100% at $V_p = 2V_{CC}/\pi$ and 67.4% at $V_p = V_{CC}$. The normalized plot of P_C (with respect to $P_{C(max)}$) versus peak voltage V_p is shown in Fig. 3.13. The power dissipation is highest at 50% efficiency.

We can obtain the figure of merit from Eqs. (3.46) and (3.51) as follows:

$$F_m = \frac{P_{C(max)}}{P_{L(max)}} = \frac{V_{CC}^2/\pi^2 R_L}{V_{CC}^2/2R_L} = \frac{2}{\pi^2} \approx \frac{1}{5} = \frac{2}{10} = 20\% \tag{3.53}$$

Thus, the figure of merit for class B amplifiers exceeds that of class A amplifiers by a factor of 10. The power dissipation rating of the individual transistors is only approximately one-fifth of the output power, and this results in much smaller heat sinks, which are normally needed to keep the junction temperature of power transistors within the maximum permissible limit.

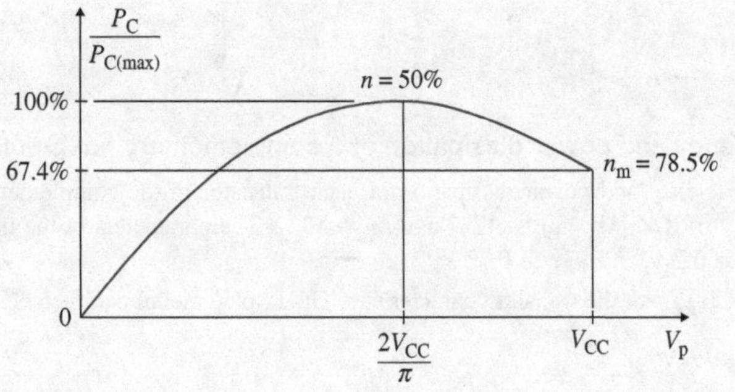

FIGURE 3.13 Power dissipation versus peak output voltage

Dead-Zone Minimization

The dead zone can be reduced to practically zero by using feedback with an op-amp, as shown in Fig. 3.14(a). The op-amp is connected in unity-gain mode with series-shunt feedback. With this arrangement, either Q_P or Q_N will be on if v_I and v_O differ by $\pm V_{BE}/A$, where A is the open-loop gain of the op-amp. Thus, for $A = 10^5$ and $V_{BE} = 0.7$, the dead zone will be reduced to less than $\pm(0.7/10^5) = \pm 7\ \mu V$. The transfer characteristic is shown in Fig. 3.14(b). Resistance R_1 limits the current drawn by the transistors from the op-amp output. It also provides the base current necessary for the load current I_L. Since the emitter current of a transistor is related to the base current by a factor of $(1 + h_{fe})$, the maximum value of R_1 can be found approximately from

$$I_B(1 + h_{fe}) = \frac{V_{BE}(1 + h_{fe})}{R_1} \geq \frac{V_{O(max)}}{R_L} = \frac{V_{CC} - V_{CE(sat)}}{R_L}$$

which gives the maximum value of R_1 as

$$R_1 \leq \frac{V_{BE}(1 + h_{fe})R_L}{V_{CC} - V_{CE(sat)}} \tag{3.54}$$

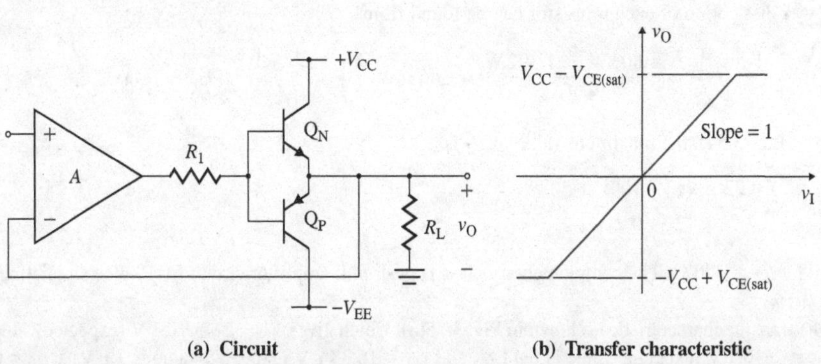

FIGURE 3.14 Minimization or elimination of the dead zone with feedback

EXAMPLE 3.3

Finding the efficiency and power dissipation of a complementary push-pull amplifier

(a) Calculate the efficiency and power dissipation of each transistor in the complementary push-pull output stage in Fig. 3.14(a) if $V_{CC} = V_{EE} = 12$ V and $R_L = 50\ \Omega$. The parameters of the transistors are $\beta_F = h_{fe} = 100$, $V_{CE(sat)} = 0.2$ V, and $V_{BE} = 0.72$ V.

(b) Use PSpice/SPICE to plot the transfer characteristic. The PSpice model parameters of the transistors are

```
BF=100 VJE=0.7V
```

SOLUTION

(a) The peak load voltage is

$$V_p = V_{CC} - V_{CE(sat)} = 12 - 0.2 = 11.8 \text{ V}$$

The peak load current is

$$I_p = \frac{V_p}{R_L} = \frac{11.8 \text{ V}}{50\ \Omega} = 0.236 \text{ A}$$

From Eq. (3.44), the DC power from the supply source is

$$P_S = \frac{2I_p V_{CC}}{\pi} = \frac{2 \times 0.236 \text{ A} \times 12 \text{ V}}{\pi} = 1.803 \text{ W}$$

From Eq. (3.17), the output power is

$$P_L = \frac{I_p V_p}{2} = \frac{0.236 \text{ A} \times 11.8 \text{ V}}{2} = 1.392 \text{ W}$$

Thus, the power efficiency is

$$\eta = \frac{P_L}{P_S} = \frac{1.392 \text{ W}}{1.803 \text{ W}} = 77.2\%$$

The power dissipation of each transistor can be found from

$$P_C = \frac{P_S - P_L}{2} = \frac{1.803 \text{ W} - 1.392 \text{ W}}{2} = 206 \text{ mW}$$

(b) From Eq. (3.54), the maximum value of R_1 is

$$R_1 \leq \frac{0.72 \times (1 + 100) \times 50}{12 - 0.2} = 308\ \Omega$$

We will let $R_1 = 300\ \Omega$. The complementary class B push-pull amplifier circuit for PSpice simulation is shown in Fig. 3.15.

The transfer characteristic is shown in Fig. 3.16(a), which gives $V_{O(max)} = 10.1$ V (expected value is 11.8), $V_{O(min)} = -10.1$ V (expected value is -11.8), and $v_O = 16.38\ \mu$V (expected value is 7 μV) at $v_I = 0$. The input and output voltages are shown in Fig. 3.16(b) for $v_I = 10 \sin(2000\pi t)$.

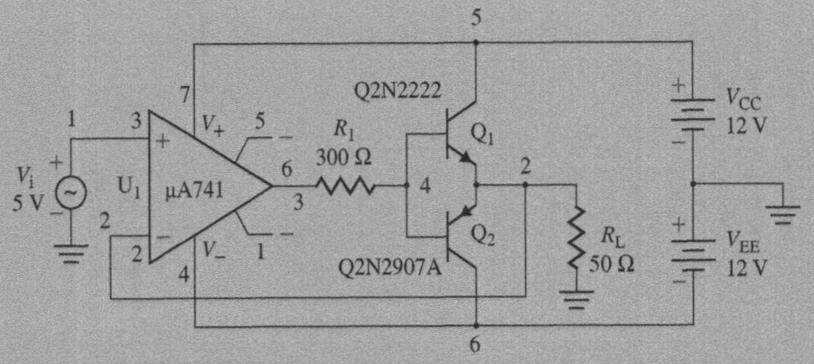

FIGURE 3.15 Complementary class B push-pull amplifier circuit for PSpice simulation for Example 3.3

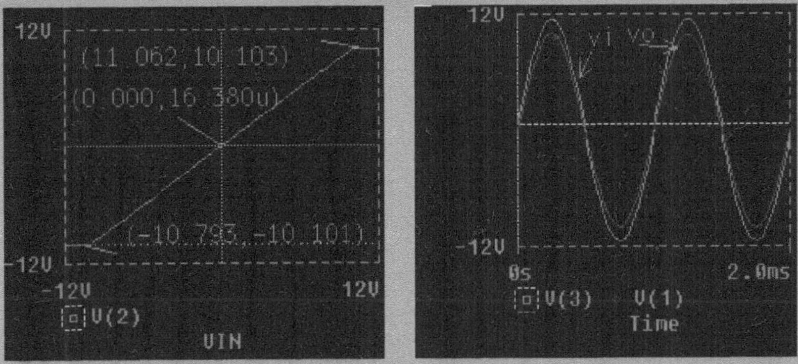

(a) Transfer characteristic **(b) Output voltage**

FIGURE 3.16 Transfer characteristic and output voltage for Example 3.3

3.5.2 Transformer-Coupled Load Push-Pull Amplifier

The power efficiency of an amplifier can be improved considerably by using a transformer-coupled class B push-pull configuration, as shown in Fig. 3.17(a). The amplifier has three stages: input transformer TX_1, the gain stage of transistors Q_1 and Q_2, and output transformer TX_2. Resistance R_B and a battery V_{BB} are used to provide DC-biasing voltage V_{BE} for the transistors. Two transistors are employed to perform the push-pull operation. For $v_I > 0$, the base of transistor Q_1 is positive and that of Q_2 becomes negative because of the transformer action. The *npn* transistor Q_2 remains off, and the *npn* transistor Q_1 operates as an amplifier. For $v_I < 0$, the base of transistor Q_2 is positive and that of Q_1 becomes negative because of the transformer action. Transistor Q_1 remains off, and transistor Q_2 operates as an amplifier. The input transformer TX_1 of the input stage supplies a virtually distortion-free input signal and matches the output impedance of the driver stage to the input impedance of the output stage.

Signal Waveforms

Assuming the input current is sinusoidal, the collector currents of the transistors are as shown in Fig. 3.17(b). The load current shown in Fig. 3.17(c) is composed of the two collector currents i_{C1} and i_{C2}. The load cur-

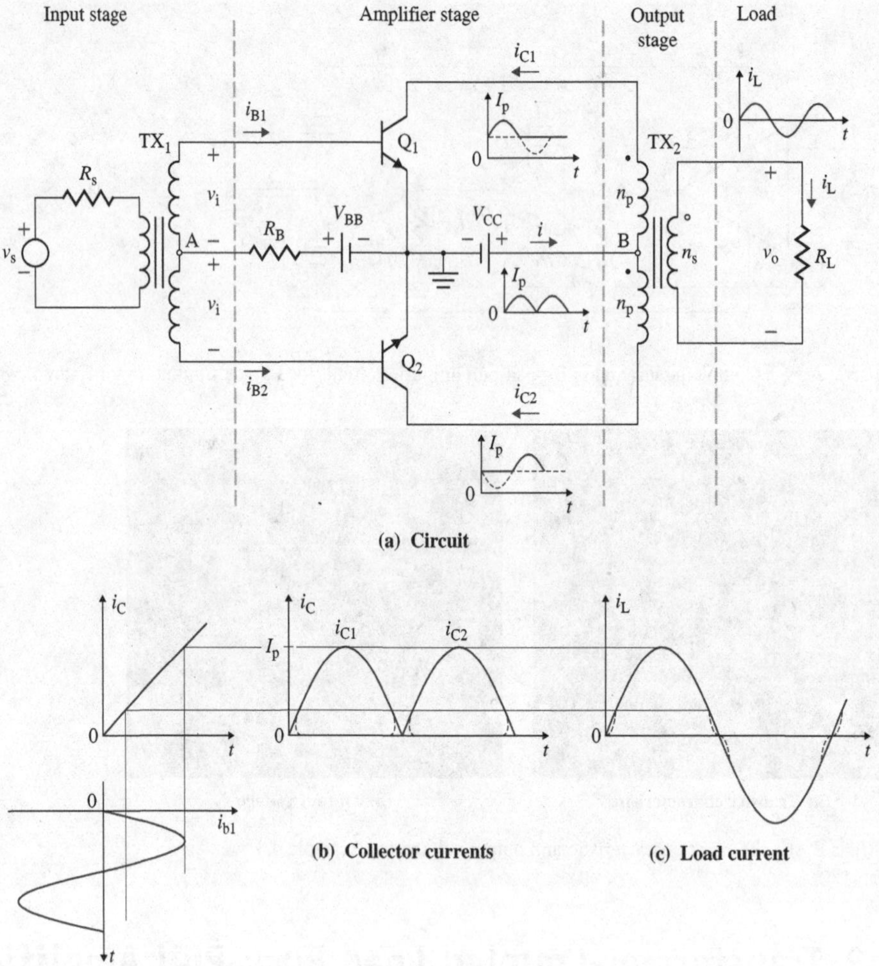

FIGURE 3.17 Transformer-coupled class B push-pull amplifier

rent is distorted near zero crossings because the transistors are nonlinear devices and because the base current $i_B = 0$ (and hence $i_C = 0$) for $V_{BE} \le 0.7$ V. This distortion, shown in Fig. 3.17(c) by dashed lines, is usually referred to as *crossover distortion.*

Output Power and Efficiency

The AC load line for a single transistor (Q_N) is shown in Fig. 3.18. The maximum peak current of a transistor is V_{CC}/R_L'. The average current drawn from the DC supply source by transistors Q_N and Q_P is

$$I_{dc} = 2I_{C1} = \frac{2I_p}{\pi} \tag{3.55}$$

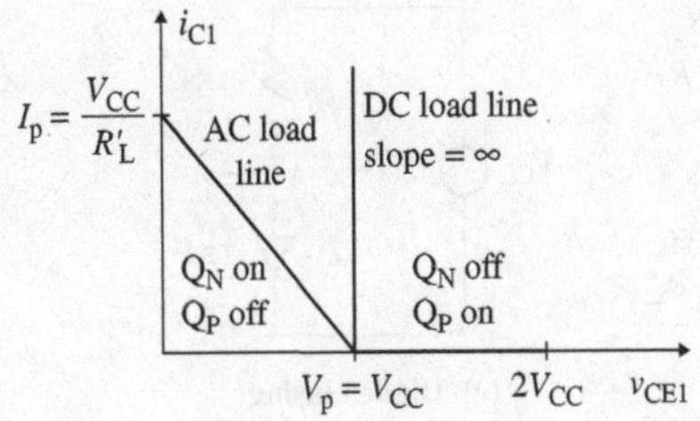

FIGURE 3.18 Load line of a single transistor

Thus, the average input power supplied from the DC source is

$$P_{dc} = I_{dc}V_{CC} = \frac{2I_pV_{CC}}{\pi} \tag{3.56}$$

From Eq. (3.18), the maximum output power is

$$P_{L(max)} = \frac{I_p^2R'_L}{2} = \frac{I_pV_p}{2} = \frac{I_pV_{CC}}{2} = \frac{V_{CC}^2}{2R'_L} \tag{3.57}$$

where the effective load resistance R'_L (referred to as the primary side of TX_2) is given by

$$R'_L = \left(\frac{n_p}{n_s}\right)^2 R_L \tag{3.58}$$

Thus, the maximum power efficiency is

$$\eta_{max} = \frac{P_{L(max)}}{P_{dc}} = \frac{I_pV_{CC}/2}{2I_pV_{CC}/\pi} = \frac{\pi}{4} = 78.5\% \tag{3.59}$$

Therefore, the maximum efficiency of a transformer-coupled push-pull class B amplifier is much higher than that of a class A amplifier.

The average collector power dissipation for both transistors is given by

$$2P_C = P_{dc} - P_L = \frac{2I_pV_{CC}}{\pi} - \frac{I_p^2R'_L}{2} \tag{3.60}$$

$$= \frac{2V_pV_{CC}}{\pi R'_L} - \frac{V_p^2}{2R'_L} \tag{3.61}$$

From Eq. (3.49), the peak current for maximum collector power dissipation is

$$I_{p(max)} = \frac{2V_{CC}}{\pi R'_L} \tag{3.62}$$

From Eq. (3.50), the peak voltage for maximum collector power dissipation is

$$V_{p(max)} = I_pR'_L = \frac{2V_{CC}}{\pi} \tag{3.63}$$

Substituting Eqs. (3.62) and (3.63) into Eq. (3.51) gives the maximum collector dissipation as

$$2P_{C(max)} = \frac{4V_{CC}^2}{\pi^2R'_L} - \frac{2V_{CC}^2}{\pi^2R'_L} = \frac{2V_{CC}^2}{\pi^2R'_L} \tag{3.64}$$

The power efficiency becomes 50% at $V_p = V_{p(max)}$. At $V_p = V_{CC}$, the maximum efficiency of 78.5% occurs.

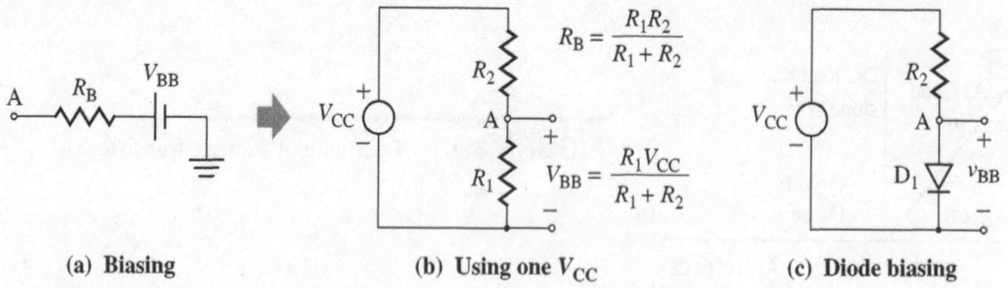

FIGURE 3.19 DC biasing of transformer-coupled class B push-pull amplifier

We can obtain the figure of merit from Eqs. (3.57) and (3.64) as follows:

$$F_m = \frac{P_{C(max)}}{P_{L(max)}} = \frac{V_{CC}^2/\pi^2 R_L'}{V_{CC}^2/2R_L'} = \frac{2}{\pi^2} \approx \frac{1}{5} = 20\% \tag{3.65}$$

which is the same as that for a complementary push-pull amplifier. The figure of merit for transformer-coupled amplifiers exceeds that of class A amplifiers by a factor of 10.

DC Biasing

Resistor R_B and battery V_{BB}, shown in Fig. 3.19(a), provide the B-E DC voltage V_{BE}, which is approximately 0.7 V for a silicon transistor. In practice, a V_{CC} supply with a suitable voltage divider is used rather than a separate supply, as shown in Fig. 3.19(b). R_1 and R_2 are chosen so that $V_{BE} \approx 0.7$ V (for silicon transistors). The parallel combination of R_1 and R_2 is kept as small as possible so that voltage drop $2R_B I_{B1} << V_{BE}$, which is generally satisfied by choosing $2R_B I_{B1} = 0.1V_{BE}$. Since the diode drop is similar to the B-E voltage of a transistor, a silicon diode is often used instead of resistance R_1, as shown in Fig. 3.19(c).

EXAMPLE 3.4

D

Designing a transformer-coupled class B amplifier Design a transformer-coupled class B push-pull amplifier, as shown in Fig. 3.17(a), to supply a maximum output power of $P_{L(max)} = 10$ W at a load resistance of $R_L = 4\ \Omega$. Assume a DC supply voltage of 15 V and transistors of $\beta_F = h_{fe} = 100$ and $V_{BE} = 0.7$ V.

SOLUTION

The design steps are as follows:

Step 1. Determine the maximum collector-to-emitter voltage of the transistors:

$$V_{CE(max)} \geq 2V_{CC} = 2 \times 15\text{ V} = 30\text{ V}$$

Step 2. From Eq. (3.57), calculate the effective load resistance:

$$R_L' = \frac{V_{CC}^2}{2P_{L(max)}} = \frac{(15\text{ V})^2}{2 \times 10\text{ W}} = 11.25\ \Omega$$

Step 3. Calculate the peak current of each transistor:

$$I_p = \frac{V_{CC}}{R'_L} = \frac{15\ \text{V}}{11.25\ \Omega} = 1.33\ \text{A}$$

Step 4. From Eq. (3.55), calculate the average current of each transistor:

$$I_{C1} = \frac{I_p}{\pi} = \frac{1.33\ \text{A}}{\pi} = 0.424\ \text{A}$$

Step 5. From Eq. (3.64), calculate the maximum collector power dissipation:

$$P_C = \left(\frac{2}{\pi^2}\right)P_{L(\max)} = \left(\frac{2}{\pi^2}\right) \times 10\ \text{W} = 2\ \text{W}$$

Step 6. Calculate the DC power from the source:

$$P_S = 2I_{C1}V_{CC} = 2 \times 0.424\ \text{A} \times 15\ \text{V} = 12.72\ \text{W}$$

Step 7. From Eq. (3.58), calculate the required turns ratio of the transformer:

$$\frac{n_p}{n_s} = \left(\frac{R'_L}{R_L}\right)^{1/2} = \left(\frac{11.25\ \Omega}{4\ \Omega}\right)^{1/2} = 1.68$$

Step 8. Calculate the required quiescent base current I_{B1}:

$$I_{B1} = \frac{I_{C1}}{h_{fe}} = \frac{0.424\ \text{A}}{100} = 4.24\ \text{mA}$$

Step 9. Calculate the biasing resistances R_1 and R_2. We have

$$V_{BB} - V_{BE} = R_B(I_{B1} + I_{B2}) = R_B(2I_{B1})$$

If we let $R_B(2I_{B1}) = 10\%$ of $V_{BE} = 0.1V_{BE} = 0.1 \times 0.7\ \text{V} = 0.07\ \text{V}$, then

$$R_B = \frac{0.07\ \text{V}}{2I_{B1}} = 8.25\ \Omega$$

and $$V_{BB} = V_{BE} + R_B(2I_{B1}) = 0.7\ \text{V} + 0.07\ \text{V} = 0.77\ \text{V}$$

Since $$V_{BB} = \frac{R_1 V_{CC}}{R_1 + R_2} = 0.77\ \text{V}$$

and $$R_B = \frac{R_1 R_2}{R_1 + R_2} = 8.25\ \Omega$$

Solving for R_1 and R_2, we get $$R_1 = \frac{R_B V_{CC}}{V_{CC} - V_{BB}} = \frac{8.25\ \Omega \times 15\ \text{V}}{15\ \text{V} - 0.77\ \text{V}} = 8.7\ \Omega$$

$$R_2 = \frac{R_B V_{CC}}{V_{BB}} = \frac{8.25\ \Omega \times 15\ \text{V}}{0.77\ \text{V}} = 161\ \Omega$$

KEY POINTS OF SECTION 3.5

- In a class B push-pull amplifier, a *pnp* and an *npn* transistor form a pair, and each transistor conducts for only 180°. The quiescent DC-biasing current is zero. As a result, the maximum power efficiency is 78.5%, and the maximum figure of merit is only 20%.
- Because of B-E voltage drops, a push-pull amplifier exhibits a dead zone in the transfer characteristic, which increases distortion of the output voltage. The crossover distortion and nonlinearities can be reduced practically to zero by applying feedback.

3.6 Complementary Class AB Push-Pull Amplifiers

The crossover distortion of a complementary class B push-pull amplifier is minimized or eliminated in a class AB amplifier, in which the transistors operate in the active region when the input voltage v_I is small ($v_I \approx 0$ V). The transistors are biased in such a way that each transistor conducts for a small quiescent current I_Q at $v_I = 0$ V. A biasing circuit is shown in Fig. 3.20(a). A biasing voltage V_{BB} is applied between the bases of Q_N and Q_P. For $v_I = 0$, a voltage $V_{BB}/2$ appears across the B-E junction of each Q_N and Q_P. Choosing $V_{BB}/2 = V_{BEN} = V_{EBP}$ will ensure that both transistors will be on the verge of conducting. That is, $v_O = 0$ for $v_I = 0$. A small positive input voltage v_I will then cause Q_N to conduct; similarly, a small negative input voltage will cause Q_P to conduct.

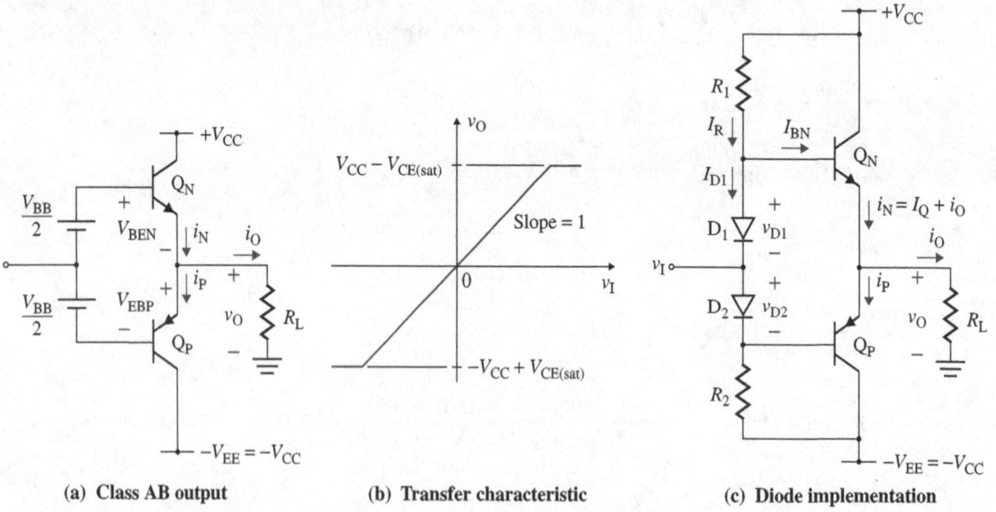

FIGURE 3.20 Elimination of the dead zone in a class AB amplifier

3.6.1 Transfer Characteristic

The output voltage v_O is given by

$$v_O = v_I + \frac{V_{BB}}{2} - V_{BEN}\,(=V_{EBP}) \tag{3.66}$$

which, for identical transistors of $V_{BEN} = V_{EBP}$ and $V_{BB}/2 = V_{BEN}$, gives $v_O = v_I$. Therefore, most of the crossover distortion is eliminated. The transfer characteristic is shown in Fig. 3.20(b). For positive v_O, a current i_O flows through R_L; that is,

$$i_N = i_P + i_O \tag{3.67}$$

Any increase in i_N will cause a corresponding increase in V_{BEN} above the quiescent value of $V_{BB}/2$. Since V_{BB} must remain constant, the increase in V_{BEN} will cause an equal decrease in V_{EBP} and hence in i_P. Thus,

$$V_{BB} = V_{BEN} + V_{EBP} \tag{3.68}$$

which, expressed in terms of saturation current I_S, becomes

$$2V_T \ln\left(\frac{I_Q}{I_S}\right) = V_T \ln\left(\frac{i_N}{I_S}\right) + V_T \ln\left(\frac{i_P}{I_S}\right)$$

After simplification, we get

$$I_Q^2 = i_N i_P \tag{3.69}$$

$$= i_N(i_N - i_O) = i_N^2 - i_N i_O \tag{3.70}$$

which can be solved for the current i_N for a given quiescent current I_Q. Thus, as i_N increases, i_P decreases by the same ratio. However, their product remains constant. As v_I becomes positive, Q_N acts as an emitter follower delivering output power, and Q_P conducts only a very small current. When v_I becomes negative, the opposite occurs: Q_P acts as an emitter follower, and v_O follows the input signal v_I. The circuit operates in class AB mode because both transistors remain on and operate in the active region.

3.6.2 Output Power and Efficiency

The power relationships in class AB amplifiers are identical to those in class B amplifiers, except that the class AB circuit dissipates a quiescent power of $I_Q V_{CC}$ per transistor. Thus, from Eq. (3.44), we can find the average power supplied from the DC source as

$$P_S = \frac{2I_p V_{CC}}{\pi} + I_Q V_{CC} = V_{CC}\left(I_Q + \frac{2I_p}{\pi}\right) \tag{3.71}$$

3.6.3 Biasing with Diodes

The biasing circuit in Fig. 3.20(a) has a serious problem when the temperatures of Q_N and Q_P increase as a result of their power dissipation. Recall that the value of V_{BE} for a given current falls with temperature at approximately 2.5 mV/°C. Thus, if the biasing voltage $V_{BB}/2$ remains constant with temperature, V_{BE}

(=$V_{BB}/2$) is also held constant, and the collector current will increase as temperature increases. The increase in the collector current increases the power dissipation, in turn increasing the collector current and causing the temperature to rise further. This phenomenon, in which a positive feedback mechanism leads to excessive temperature rise, is called *thermal runaway*. Thermal runaway can ultimately lead to the destruction of the transistors unless they are protected.

To avoid thermal runaway, the biasing voltages must decrease as the temperature increases. One solution is to use diodes that have a compensating effect, as shown in Fig. 3.20(c). The diodes must be in close contact with the output transistors so that their temperature will increase by the same amount as that of Q_N and Q_P. Therefore, in discrete circuits, the diodes should be mounted on the metal of Q_N or Q_P. Since resistances R_1 and R_2 provide the quiescent current I_Q for the transistors and also ensure that the diodes conduct, to guarantee the base biasing current for Q_N when the load current becomes maximum we must have

$$I_R = I_{D1} + \frac{i_N}{1 + h_{fe}} \approx I_{D1} + \frac{I_Q + i_O}{1 + h_{fe}}$$

Thus, the values of R_1 and R_2 can be found from

$$R_1 = R_2 = \frac{V_{CC} - V_{D1}(=V_{D2} = V_{BB}/2)}{I_{D1(\min)} + (I_Q + i_{O(\max)})/(1 + h_{fe})} \tag{3.72}$$

where $I_Q = I_S \exp(V_{BB}/2V_T)$ and $I_{D1(min)}$ is the minimum current needed to ensure diode conduction. Because I_Q is usually smaller than $i_{O(max)}$, I_Q can often be neglected in finding the values of R_1 and R_2.

EXAMPLE 3.5

D

Designing a biasing circuit for a class AB amplifier

(a) Design a biasing circuit for the class AB amplifier of Fig. 3.20(c) to supply the maximum output voltage at a load resistance of $R_L = 50\ \Omega$. The quiescent biasing current I_Q is 2 mA. Assume a DC supply voltage of 12 V. The diode parameters are $I_S = 10^{-13}$ A, $V_{D1} = V_{D2} = 0.7$ V, and $I_{D(min)} = 1$ mA to ensure conduction. The transistor parameters are $\beta_F = h_{fe} = 50$, $V_{BE} = 0.7$ V, $I_S = 10^{-14}$ A, and $V_{CE(sat)} = 0.2$ V.

(b) Find the biasing voltage V_{BB} for $v_O = 0$ and 11.8 V.

SOLUTION

(a) The maximum peak load voltage is

$$V_{p(max)} = V_{CC} - V_{CE(sat)} = 12 - 0.2 = 11.8\text{ V}$$

The maximum peak load current is

$$I_{p(max)} = \frac{V_{p(max)}}{R_L} = \frac{11.8}{50} = 236\text{ mA}$$

From Eq. (3.71), the maximum DC power from the supply source is

$$P_S \approx \frac{2I_{p(max)}V_{CC}}{\pi} + I_QV_{CC} = \frac{2 \times 236\text{ mA} \times 12\text{ V}}{\pi} + 2\text{ mA} \times 12\text{ V} = 1.83\text{ W}$$

From Eq. (3.17), the output power is

$$P_{L(max)} = \frac{I_{p(max)}V_{p(max)}}{2} = 236 \text{ mA} \times \frac{11.8 \text{ V}}{2} = 1.39 \text{ W}$$

Thus, the maximum power efficiency is

$$\eta_{max} = \frac{P_{L(max)}}{P_S} = \frac{1.39}{1.83} = 76.1\%$$

The power dissipation of each transistor can be found from

$$P_C = \frac{P_S - P_{L(max)}}{2} = \frac{1.83 \text{ W} - 1.39 \text{ W}}{2} = 220 \text{ mW}$$

From Eq. (3.72), we get

$$R_1 = R_2 = \frac{12 \text{ V} - 0.7 \text{ V}}{1 \text{ mA} + (2 \text{ mA} + 236 \text{ mA})/(1 + 50)} = 1.99 \text{ k}\Omega$$

(b) We have

$$I_R = I_{D1(min)} + \frac{I_Q + I_{p(max)}}{1 + h_{fe}} = 1 \text{ mA} + \frac{2 \text{ mA} + 236 \text{ mA}}{1 + 50} = 5.667 \text{ mA}$$

For $v_O = 0$, $i_N = I_Q = 2$ mA. Thus, the base current of the *npn* transistor is

$$I_{BN} = \frac{I_N}{1 + h_{fe}} = \frac{2 \text{ mA}}{1 + 50} = 0.039 \text{ mA}$$

and $\quad I_{D1} = I_R - I_{BN} = 5.667 \text{ mA} - 0.039 \text{ mA} = 5.628 \text{ mA}$

Therefore, the biasing voltage V_{BB} becomes

$$V_{BB} = 2V_T \ln\left(\frac{I_{D1}}{I_S}\right) = 2 \times 25.8 \text{ mV} \times \ln\left(\frac{5.628 \text{ mA}}{10^{-13} \text{ A}}\right) = 1.277 \text{ V}$$

For $v_O = 11.8$, $i_N = I_Q + I_{p(max)} = 2 \text{ mA} + 236 \text{ mA} = 238$ mA. Thus,

$$I_{BN} = \frac{i_N}{1 + h_{fe}} = \frac{238 \text{ mA}}{1 + 50} = 4.67 \text{ mA}$$

and $\quad I_{D1} = I_R - I_{BN} = 5.667 \text{ mA} - 4.67 \text{ mA} = 1 \text{ mA}$

Therefore, for $I_{D1} = 1$ mA, the biasing voltage V_{BB} becomes $V_{BB} = 1.19$ V.

3.6.4 Biasing with Diodes and an Active Current Source

The biasing technique in Fig. 3.20(a) is generally used in integrated circuits; however, an active current source is normally used rather than discrete resistance. This arrangement is shown in Fig. 3.21(a). In integrated circuits, collector-shorted transistors are usually used instead of diodes. If Q_N and Q_P are to handle large amounts of power, their geometry must also be large. However, the diodes can be smaller devices such that $I_R = I_Q/n$, where n is the ratio of the emitter junction area of Q_N and Q_P to the junction area of D_1 and D_2. That is, the saturation current I_S of Q_N and Q_P can be n times that of the biasing diodes.

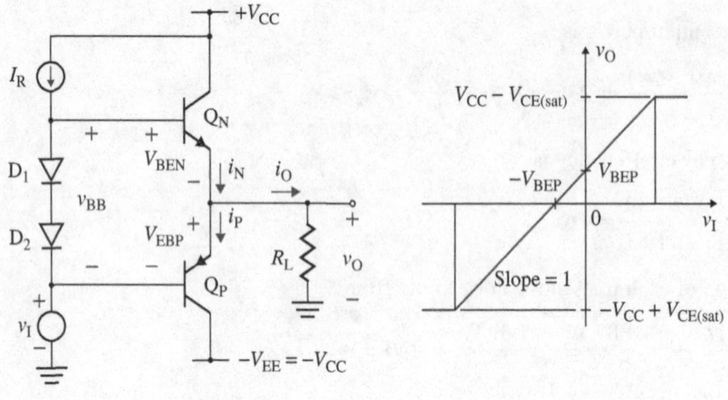

FIGURE 3.21 Biasing of a class AB amplifier with diodes and an active current source

Transfer Characteristic

The voltage between the bases of Q_P and Q_N is the same as the voltage drop across the two diodes. That is,

$$V_{BB} = V_{D1} + V_{D2} \approx 0.7 + 0.7 = 1.4 \text{ V}$$

The base-to-emitter voltage of Q_N is given by

$$V_{BEN} = V_{BB} - V_{EBP} = 1.4 - V_{EBP} \tag{3.73}$$

Thus, the B-E junctions of both Q_N and Q_P are always forward biased. Because of diodes D_1 and D_2, Q_N and Q_P remain in the active region when $v_I = 0$ V. The output voltage v_O is given by

$$\begin{aligned} v_O &= v_I + V_{BB} - V_{BEN} \qquad (3.74) \\ &= v_I + V_{EBP} = v_I - V_{BEP} \end{aligned}$$

which yields the transfer characteristic shown in Fig. 3.21(b). The dead zone is eliminated. However, there is an offset voltage of V_{EBP}, which can be reduced practically to zero by applying feedback similar to that applied to the class B amplifier in Fig. 3.14.

EXAMPLE 3.6

D

Designing an active current–source biasing circuit for a class AB amplifier

(a) Design an active current source for the class AB amplifier in Fig. 3.21(a) in order to provide the biasing current of $I_R = 5.67$ mA needed for Example 3.5. Assume $V_{CC} = V_{EE} = 12$ V, $V_{BE} = 0.7$ V, and $R_L = 50\ \Omega$.

(b) Use PSpice/SPICE to plot the transfer characteristic and the instantaneous i_N, i_P, and i_O for $v_I = 5 \sin (2000\pi t)$. The PSpice model parameters for the transistors are

```
IS=1E-14 BF=50 VJE=0.7
```

and for the diodes are

```
IS=1E-13 BV=100
```

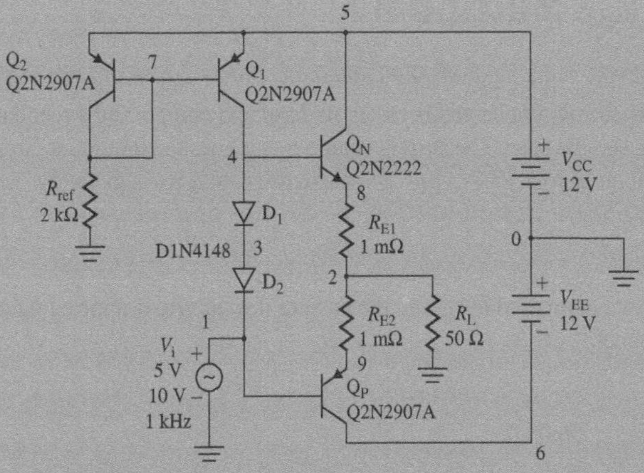

FIGURE 3.22 Complementary class AB push-pull amplifier circuit for PSpice simulation

SOLUTION

(a) The biasing current source I_R, shown in Fig. 3.22, can be produced by two *pnp* transistors Q_1 and Q_2 and a resistance R_1. Thus,

$$I_R \approx I_{ref} = \frac{V_{CC} - V_{EBP}}{R_{ref}} \tag{3.75}$$

which, for $I_R = 5.67$ mA, gives

$$R_{ref} = \frac{V_{CC} - V_{EBP}}{I_R} = \frac{12\text{ V} - 0.7\text{ V}}{5.67\text{ mA}} \approx 2\text{ k}\Omega$$

(b) In practice, emitter resistances R_{E1} and R_{E2}, shown in Fig. 3.22, are connected to ensure the stability of the biasing point.

The transfer characteristic for the circuit in Fig. 3.22 is shown in Fig. 3.23(a). It gives $V_{O(max)} \equiv V(2) = 11.139$ V and offset voltages of $v_O = 560.4$ mV at $v_I = 0$ and $v_O = 0$ at $v_I = -640.9$ mV. The plots of i_N, i_P, and i_O are shown in Fig. 3.23(b), which give $i_{N(peak)} \equiv I(RE1) = 109.66$ mA, $i_{P(peak)} \equiv I(RE2) = -83.92$ mA, and $i_{O(peak)} \equiv I(RL) = 109.59$ mA to -83.83 mA. Thus, because of the offset voltage, the load current is not symmetrical.

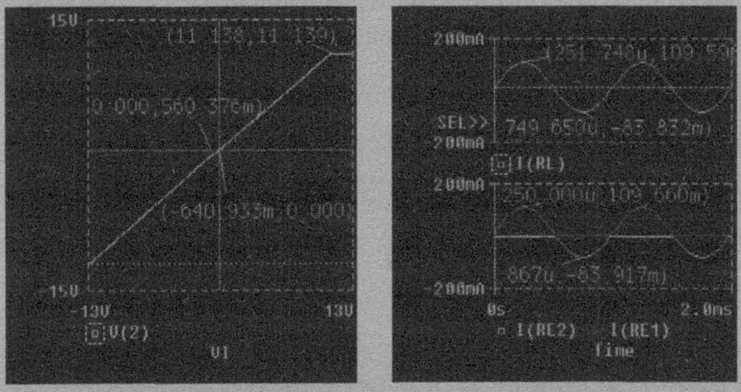

(a) Transfer characteristic (b) Output voltage

FIGURE 3.23 Transfer characteristic and output voltage for Example 3.6

3.6.5 Biasing with a V_{BE} Multiplier

A V_{BE} *multiplier circuit* that can automatically adjust the biasing voltage V_{BB} is shown in Fig. 3.24. The circuit consists of a transistor Q_1 with a resistor R_1 connected between its base and emitter and a feedback resistor R_F connected between the collector and the base. The current source I_R supplies the multiplier circuit and the base current for Q_N. Since the voltage across R_1 is V_{BE1}, the current through R_1 is given by

$$I_1 = \frac{V_{BE1}}{R_1} \tag{3.76}$$

The base current of Q_1 is generally negligible compared to I_1, and the current through R_F is approximately equal to I_1. Thus, the biasing voltage becomes

$$\begin{aligned} V_{BB} &= I_1(R_1 + R_F) \\ &= \frac{V_{BE1}}{R_1}(R_1 + R_F) = V_{BE1}\left(1 + \frac{R_F}{R_1}\right) \end{aligned} \tag{3.77}$$

Therefore, the circuit multiplies V_{BE1} by the factor $(1 + R_F/R_1)$—hence the name V_{BE} multiplier. By selecting the ratio R_F/R_1, one can set the value of V_{BB} required to give a desired quiescent current I_Q. For $R_F/R_1 = 1$, $V_{BB} = 2V_{BE1}$. The value of V_{BE1} is related to i_{C1} by

$$V_{BE1} = V_T \ln\left(\frac{i_{C1}}{I_{S1}}\right) \tag{3.78}$$

where I_{S1} is the saturation current of Q_1 and

$$i_{C1} = I_R - I_1 - \frac{I_Q + i_O}{1 + h_{fe}} \tag{3.79}$$

Under quiescent conditions, $i_O = 0$ and the base current of Q_N is normally small enough that it can be neglected. That is, $I_Q/(1 + h_{fe}) \approx 0$, and Q_1 carries the maximum current: $I_{C1(max)} \approx I_R - I_1$. However, at the peak value of v_O, the base current of Q_N will be maximum and the current available to the multiplier will be minimum; that is,

$$I_{C1(min)} = I_R - I_1 - \frac{I_Q + i_{O(max)}}{1 + h_{fe}} \tag{3.80}$$

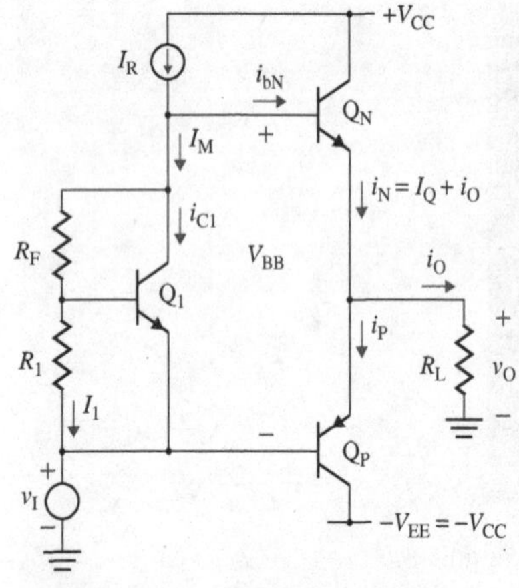

FIGURE 3.24 Biasing of a class AB amplifier with a V_{BE} multiplier

Thus, i_{C1} can vary widely from $I_{C1(max)}$ to $I_{C1(min)}$. However, according to Eq. (3.78), a large change in i_{C1} will cause only a small change in V_{BE1}. Thus, I_1 and V_{BB} will remain almost constant. As with diode biasing, the transistor Q_1 must be in close contact with Q_N and Q_P in order to provide a thermal compensating effect.

EXAMPLE 3.7

D

Designing a V_{BE} multiplier for a class AB amplifier

(a) Design a V_{BE} multiplier for the class AB amplifier in Fig. 3.24 in order to provide the biasing current of $I_R = 5.67$ mA needed for Example 3.5. Assume $h_{fe} = 50$, $I_S = 10^{-14}$ A, $V_{CC} = V_{EE} = 12$ V, $V_{BE} = 0.7$ V, and $R_L = 50\ \Omega$. Assume a minimum current of $I_{M(min)} = 1$ mA to the multiplier, $I_Q = 2$ mA, and peak load current $I_{p(max)} = 236$ mA.

(b) Use PSpice/SPICE to plot the transfer characteristic and the instantaneous i_N, i_P, and i_O for $v_I = 5 \sin (2000\pi t)$. The PSpice model parameters for the transistors are

```
IS=1E-14 BF=50 VJE=0.7
```

and for the diodes are

```
IS=1E-13 BV=100
```

SOLUTION

(a) Since the current source must provide the base current when the load current is maximum,

$$I_R = I_{M(min)} + \frac{I_Q + I_{p(max)}}{1 + h_{fe}} = 1\text{ mA} + \frac{2\text{ mA} + 236\text{ mA}}{1 + 50} = 5.67\text{ mA}$$

The biasing voltage V_{BB} required to yield a quiescent current of $I_Q = 2$ mA is

$$V_{BB} = 2V_T \ln\left(\frac{I_Q}{I_S}\right) = 2 \times 25.8\text{ mV} \times \ln\left(\frac{2\text{ mA}}{10^{-14}\text{ mA}}\right) = 1.343\text{ V}$$

The minimum current through the multiplier must be $I_{M(min)} = 1$ mA. Let $I_{1(min)} = I_{M(min)}/2 = 0.5$ mA and $I_{C(min)} = I_{M(min)}/2 = 0.5$ mA. If $I_{C(min)}$ is too small, transistor Q_1 will be off, which is not desirable. Equation (3.77) gives

$$R_1 + R_F = \frac{V_{BB}}{I_{1(min)}} = \frac{1.343\text{ V}}{0.5\text{ mA}} \approx 2.7\text{ k}\Omega$$

The current source must be designed to supply $I_R = 5.67$ mA. However, when the output voltage is zero, then $i_O = 0$, and $I_R = 5.67$ mA must flow through the multiplier. That is, transistor Q_1 must carry $i_{C1} = I_R - I_{1(min)} = 5.67\text{ mA} - 0.5\text{ mA} = 5.17$ mA, and the corresponding B-E voltage will be

$$V_{BE1} = V_T \ln\left(\frac{i_{C1}}{I_S}\right) = 25.8\text{ mV} \times \ln\left(\frac{5.17\text{ mA}}{10^{-14}\text{ mA}}\right) = 0.696\text{ V}$$

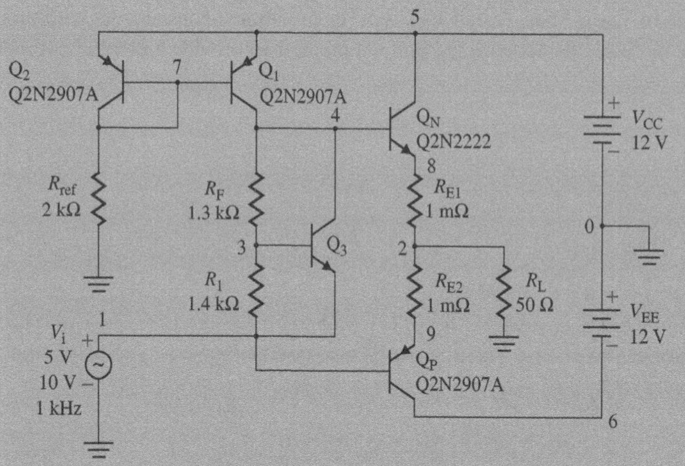

FIGURE 3.25 Complementary class AB push-pull amplifier circuit, with V_{BE} multiplier, for PSpice simulation

Thus, the value of R_1 can be found from Eq. (3.76) as

$$R_1 = \frac{V_{BE1}}{I_1} = \frac{0.696\ \text{V}}{0.5\ \text{mA}} = 1.39\ \text{k}\Omega \approx 1.4\ \text{k}\Omega$$

Therefore, the value of R_F becomes

$$R_F = 2.7\ \text{k}\Omega - R_1 = 2.7\ \text{k}\Omega - 1.4\ \text{k}\Omega = 1.3\ \text{k}\Omega$$

(b) The circuit file for the PSpice simulation is similar to that for Example 3.6, except that the diodes are replaced by the V_{BE} multiplier. This configuration is shown in Fig. 3.25.

The transfer characteristic is shown in Fig. 3.26(a). It gives $v_{O(max)} = 11.13$ V and offset voltages of $v_O = 794.3$ mV at $v_I = 0$ and $v_O = 0$ at $v_I = -796$ mV. The plots of the input and output voltages are shown in Fig. 3.26(b). Their waveforms are almost identical, except that the magnitudes are shifted by 763 mV.

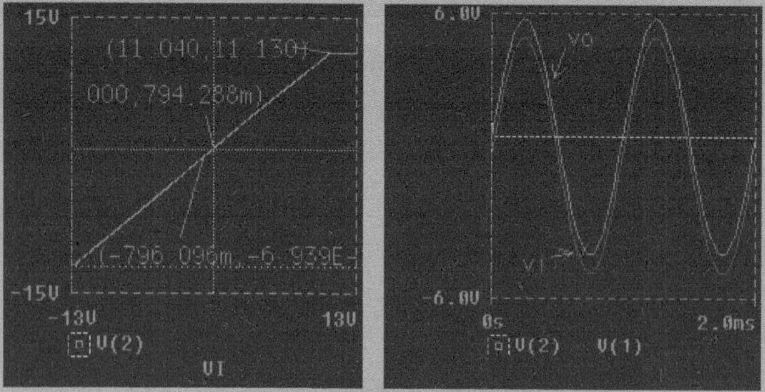

(a) Transfer characteristic **(b) Output voltage**

FIGURE 3.26 Transfer characteristic and output voltage for Example 3.7

3.6.6 Quasi-Complementary Class AB Amplifiers

Because *pnp* transistors have limited current-carrying capability, the complementary output stage is suitable only for delivering load power on the order of a few hundred milliwatts or less. If output power of several watts or more is required, *npn* transistors should be used. A composite *pnp* transistor can be made from a *pnp* transistor Q_P and a high-power *npn* transistor Q_{N1}. This arrangement, called a quasi-complementary output stage, is shown in Fig. 3.27(a).

The pair Q_P-Q_{N1} is equivalent to a *pnp* transistor, as shown in Fig. 3.27(b). The collector current of Q_P is given by

$$I_{CP} = I_S \exp\left(\frac{V_{EBP}}{V_T}\right) \tag{3.81}$$

The composite collector current I_C is the emitter current of Q_{N1}; that is,

$$I_C = (1 + h_{fe})I_{CP} = (1 + h_{fe})I_S \exp\left(\frac{V_{EBP}}{V_T}\right) \tag{3.82}$$

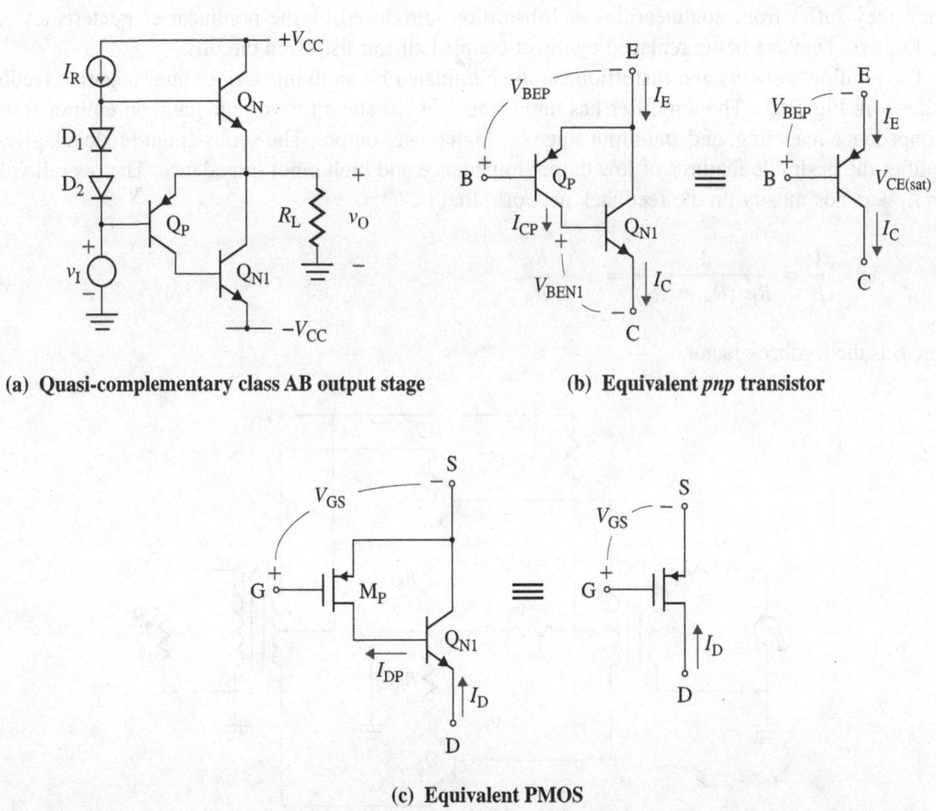

(a) Quasi-complementary class AB output stage
(b) Equivalent *pnp* transistor
(c) Equivalent PMOS

FIGURE 3.27 Quasi-complementary class AB output stage

which is the same as the relationship for a normal *pnp* transistor. However, the *npn* transistor carries most of the current and the *pnp* transistor carries only a small amount of the current. The saturation voltage of the composite *pnp* will be $V_{CEP(sat)} + V_{BEN1}$, which is higher than that of a normal *pnp* transistor.

A composite *pnp* transistor can be replaced by a MOS-bipolar combination [4], known as a composite PMOS, as shown in Fig. 3.27(c). The overall transfer characteristic of the composite PMOS is given by

$$I_D = -(1 + h_{fe})I_{DP} = -(1 + h_{fe})\frac{\mu_n C_o}{2}\left(\frac{W}{L}\right)(V_{GS} - V_t)^2 \qquad (3.83)$$

Thus, the composite PMOS has a W/L ratio that is $(1 + h_{fe})$ times larger than that of a normal PMOS device.

3.6.7 Transformer-Coupled Class AB Amplifiers

The class AB circuit shown in Fig. 3.28 is identical to the class B circuit in Fig. 3.17(a), except that it is biased slightly into conduction so that a quiescent current I_Q flows through Q_1 and Q_2. This current is achieved by making V_{BB} slightly greater than $V_{BE} = V_{BE1}$ $(= V_{BE2} \approx 0.7\text{ V})$. Resistors R_1 and R_2 can be selected to give the desired value of V_{BB}. Although transformer-coupled amplifiers offer high power efficiency, they suffer from nonlinearities and distortion introduced by the nonlinear characteristics of the transformers. They are being replaced by direct-coupled all-transistorized circuits.

The nonlinear effects and distortion can be eliminated by applying series-shunt negative feedback, as shown in Fig. 3.29. The amplifier has three stages: a CE stage for voltage gain, an emitter follower for impedance matching, and an output stage for high power output. The series-shunt feedback gives the amplifier the desirable features of low output impedance and high input impedance. The overall voltage gain A_f depends mostly on the feedback network; that is,

$$A_f \approx \frac{1}{\beta} = \frac{1}{R_E/(R_E + R_F)} = 1 + \frac{R_F}{R_E} \qquad (3.84)$$

where β is the feedback factor.

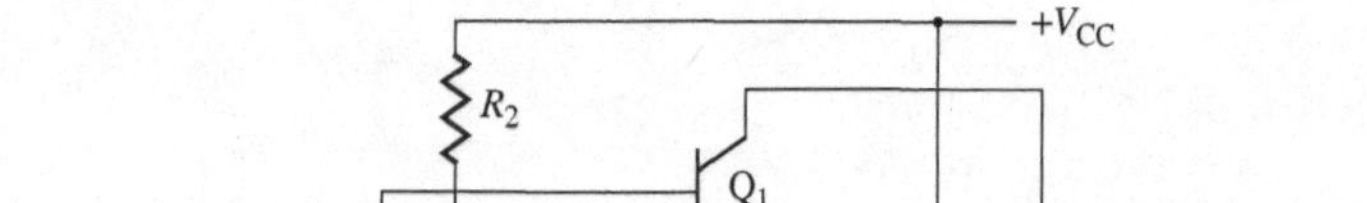

FIGURE 3.28 Transformer-coupled class AB push-pull amplifier

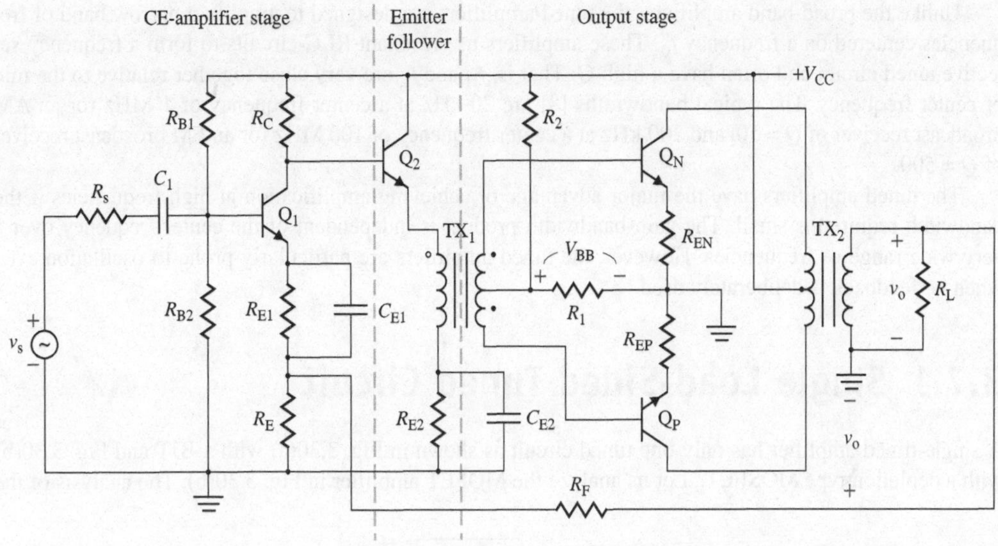

FIGURE 3.29 Transformer-coupled class A amplifier with series-shunt feedback

KEY POINTS OF SECTION 3.6

- The complementary class AB amplifier is the most commonly used output stage. Its circuit operation is similar to that of a class B amplifier, except that the transistors have a slightly positive bias so that a quiescent DC current flows even when the input voltage is zero.
- The B-E DC voltage of each transistor is usually set to approximately V_{BE}, which is the voltage required to yield the desired quiescent current. The amplifier is commonly biased with diodes and an active current source or with a V_{BE} multiplier.
- Class AB amplifiers exhibit an offset output voltage at zero input voltage. However, feedback can be applied to reduce the offset voltage.
- A quasi-complementary amplifier uses a composite *pnp* transistor, which can deliver higher output power than a normal *pnp* device.

3.7 Tuned Amplifiers

The communication systems generally require frequency-selective networks where the information is transmitted by modulating either the amplitude or the frequency of a high-frequency carrier wave. If the total bandwidth of the system is less than one-tenth of the carrier frequency, it is called a *narrowband* system. In applications, such as radio and television, amplifiers with narrow bandwidths are needed to select one signal from the large number that may be present such as from an antenna. The frequencies of interest are typically well above the unity gain frequency of operational amplifiers.

Unlike the broad-band amplifiers, the tuned amplifiers are designed to amplify a narrow band of frequencies centered on a frequency f_O. These amplifiers use resonant RLC-circuits to form a frequency selective tuned circuit and often have a high-Q. That is, f_H and f_L are very close together relative to the mid or center frequency. The typical bandwidths [8] are 20 kHz at a center frequency of 1 MHz for an AM broadcast receiver of $Q = 50$, and 200 kHz at a center frequency of 100 MHz for an FM broadcast receiver of $Q = 500$.

The tuned amplifiers have the major advantage of achieving amplification at high frequencies if the bandwidth required is small. The gain-bandwidth product is independent of the center frequency over a very wide range of frequencies. However, the tuned amplifiers are particularly prone to oscillation even when no feedback is deliberately used.

3.7.1 Single Load-Sided Tuned Circuit

A single-timed amplifier has only one tuned circuit as shown in Fig. 3.30(a) with a BJT and Fig. 3.30(b) with a depletion type MOSFET. Let us analyze the MOFET amplifier in Fig. 3.30(b). The analysis of the

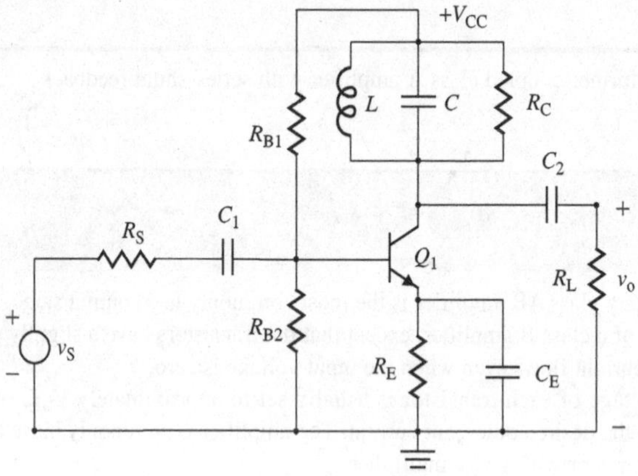

(a) BJT Amplifier

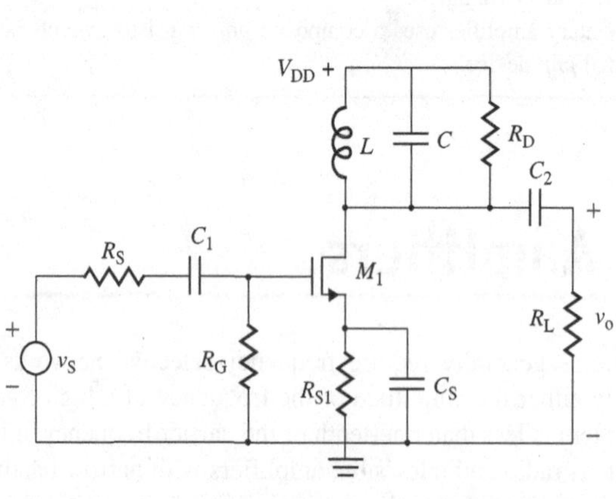

(a) Depletion MOSFET Amplifier

FIGURE 3.30 Single-tuned amplifiers

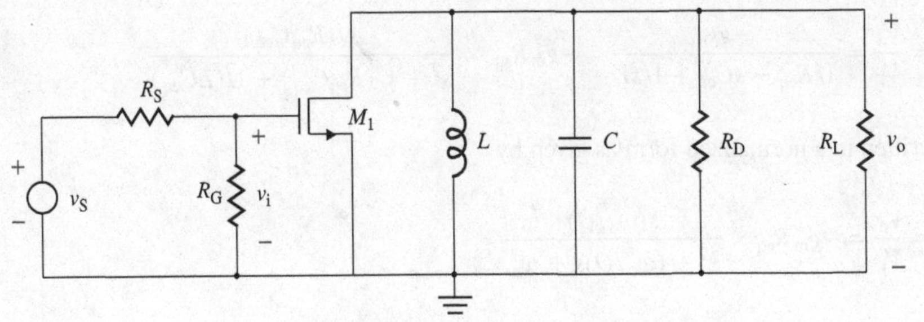

(a) High-frequency ac equivalent circuit

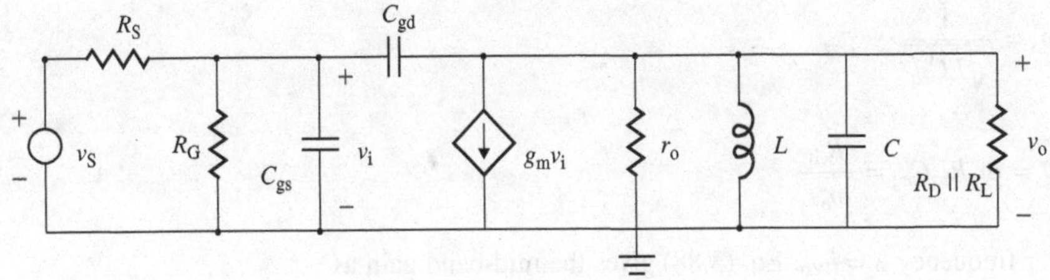

(b) Small-signal ac equivalent circuit

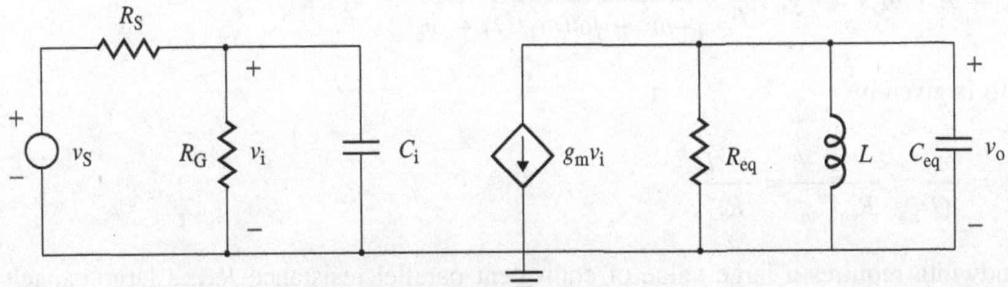

(c) Simplified ac equivalent circuit

FIGURE 3.31 AC equivalent circuits of single-tuned amplifiers

BJT amplifier will be similar. The ac equivalent circuit of Fig. 3.30(b) is shown in Fig. 3.31(a). Replacing the MOSFET by its ac-model, the small-signal ac equivalent circuit is shown in Fig. 3.31(b) which can be simplified to Fig. 3.31(c). The equivalent circuit parameters are:

$$C_i = C_{gs} \text{ for MOSFETs}; \quad C_i = C_{be} \text{ for BJTs}; \quad R_{eq} = r_o \parallel R_D \parallel R_L \tag{3.85}$$

$$R_B = R_G \text{ for MOSFET}; \quad R_B = R_{B1} \parallel R_{B2} \parallel r_\pi \text{ for BJT}$$

Using Eq. (2.99), we can include the effect of C_{gd} in C so that

$$C_{eq} = C + C_{gd}\left(1 + \frac{1}{g_m R_{eq}}\right) \cong C + C_{gd} \tag{3.86}$$

Figure 3.31(c) represents a parallel resonant circuit and its characteristics are derived in Section B.10.4 of Appendix B. The voltage gain is given by

$$A_v = \frac{V_o}{V_i} = \frac{-g_m}{1/R_{eq} + sC_{eq} + 1/sL} = -g_m R_{eq} \times \frac{s/(R_{eq}C_{eq})}{s^2 + s/(R_{eq}C_{eq}) + 1/(LC_{eq})} \tag{3.87}$$

This can be written in a normalized form as given by

$$A_v = \frac{V_o}{V_i} = -g_m R_{eq} \times \frac{s(\omega_o/Q)}{s^2 + (\omega_o/Q)s + \omega_o^2} \tag{3.88}$$

The center frequency ω_o and Q are given by

$$\omega_o = \frac{1}{\sqrt{LC_{eq}}} \tag{3.89a}$$

$$Q = \omega_o R_{eq} C_{eq} = \frac{R_{eq}}{\omega_o L} \tag{3.89b}$$

At the center frequency, $s = j\omega_o$, Eq. (3.88) gives the mid-band gain as

$$A_{mid} = A_v(j\omega_o) = -g_m R_{eq} \times \frac{j\omega(\omega_o/Q)}{-\omega_o^2 + j\omega(\omega_o/Q) + \omega_o} = -g_m R_{eq} \tag{3.90}$$

The bandwidth is given by

$$BW = \frac{\omega_o}{Q} = \frac{1}{R_{eq}C_{eq}} = \frac{\omega_o^2 L}{R_{eq}} \tag{3.91}$$

A narrow-bandwidth requires a large value of equivalent parallel resistance R_{eq}, a large capacitor C_{eq}, and/or a small inductance L. The maximum value of $R_{eq} = r_o$. Thus, Q is limited by the output resistance of the transistor which depends on the choice of operating point of the transistor.

3.7.2 Single Input-Sided Tuned Amplifier

If the amplifier is driven from relatively large impedance, the tuned circuit can also be placed at the input side of the amplifier as shown in Fig. 3.32. The Norton model of the source has been used and the source resistance R_s is included into the tuned circuit.

$$C_i = C_{gs} \text{ for MOSFETs;} \quad C_i = C_{be} \text{ for BJTs;} \quad R_B = R_G \text{ for MOSFET;}$$
$$R_B = R_{B1} \| R_{B2} \| r_\pi \text{ for BJT;} \quad R_{eq} = R_s \| R_B; \quad R_o' = r_o \| R_D \| R_L \tag{3.92}$$

Using Miller's theorem in Eq. (G.8) in Appendix G, we can include the effect of C_{gd} in C_{eq} so that

$$C_{eq} = C + C_i + C_{gd}(1 + g_m R_{eq}) \tag{3.93}$$

The trans-impedance gain is given by

$$Z_i = \frac{V_o}{I_s} = -g_m R_o' R_{eq} \times \frac{s/(R_{eq}C_{eq})}{s^2 + s/(R_{eq}C_{eq}) + 1/(LC_{eq})} \tag{3.94}$$

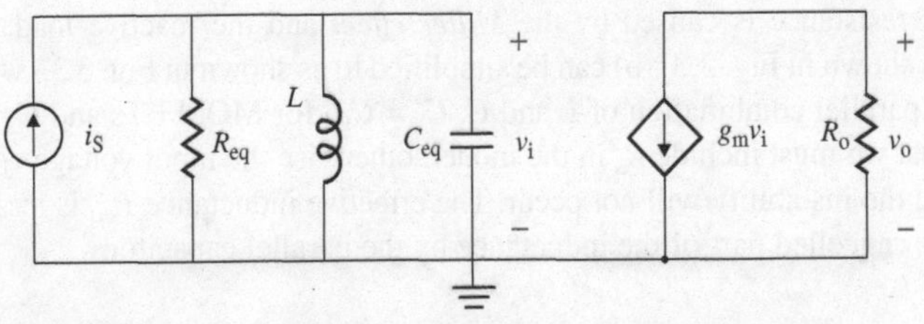

FIGURE 3.32 Simplified ac-equivalent circuit for single input-sided tuned amplifiers

This equation has the same form as Eq. (3.87), except the mid-gain gain is different, $A_{mid} = -g_m R_{eq} R_o$. Therefore, Eq. (3.89) for the center frequency and Eq. (3.91) for the bandwidth are applicable to this circuit.

EXAMPLE 3.8

Designing a single-tuned amplifier Design a tuned amplifier as shown in Fig. 3.31(a) to give a mid-gain of $A_{mid} = -50$, a center frequency of $f_o = 1$ MHz, and a bandwidth of $BW = 10$ kHz. The circuit parameters are: $g_m = 5$ mA/V, $r_o = 20$ kΩ, $R_L = 100$ kΩ, $C_i = 40$ pF and $C_x = 10$ pF.

SOLUTION

$\omega_o = 2\pi \times 10^6 = 6.283 \times 10^6$ rad/s

From Eq. (3.90), $R_{eq} = -A_{mid}/g_m = -(-50/5 \times 10^{-3}) = 10$ kΩ

From Eq. (3.85), $R_{eq} = 10\text{ k}\Omega = r_o \| R_D \| R_L = 20\text{ k} \| R_D \| 100\text{ k}$, which gives $R_D = 25$ kΩ

From Eq. (3.91), we get

$$C_{eq} = \frac{1}{2\pi \times BWR_{eq}} = \frac{1}{2\pi \times 10 \times 10^3 \times 10 \times 10^3} = 1.592 \text{ nF}$$

$$C = C_{eq} - C_x = 1.592 \times 10^{-9} - 10 \times 10^{-12} = 1.582 \text{ nF}$$

From Eq. (3.89a), we get

$$L = \frac{1}{\omega_o^2 C_{eq}} = \frac{1}{(6.283 \times 10^6)^2 \times 1.592 \times 10^{-12}} = 15.92 \text{ µH}$$

3.7.3 Stability of Single-Tuned Amplifiers

The load-side tuned circuit is prone to being unstable. For frequencies below ω_o, $|j\omega L|$ becomes less than $|1/j\omega C|$ and more current flows through the inductor than the capacitor. As a result, the tuned circuit appears inductive. Thus, the input impedance of the transistor can become negative for frequencies below ω_o.

This negative input resistance is caused by the *Miller effect* and the reactive load. The small-signal ac equivalent circuit as shown in Fig. 3.31(b) can be simplified to as shown in Fig. 3.33 where L_{eq} is the equivalent inductance of parallel combination of L and C. $C_x = C_{gd}$ for MOSFETs and $C_x = C_{cb}$ for BJTs. It is important to note that we must include R_s in the model, otherwise the input voltage v_i to the transistor will not be changing and the instability will not occur. The effective inductance L_{eq} is less than the inductance L. This is due to the cancelled part of the inductance by the parallel capacitor.

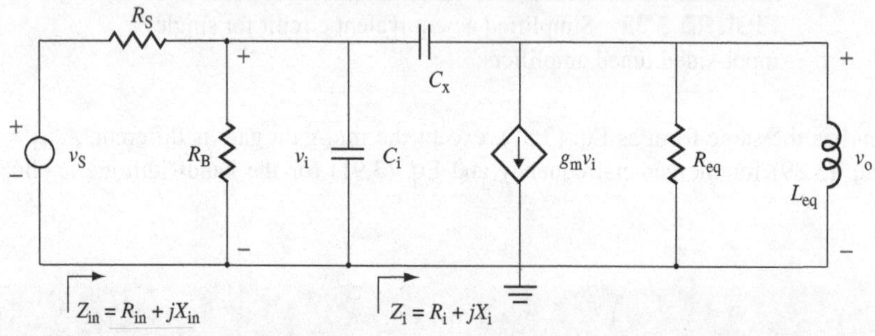

FIGURE 3.33 Simplified ac-equivalent circuit showing the transistor input impedance

The admittance of the equivalent inductance L_{eq} must be equal to the admittances of the parallel combination of L and C.

$$Y_L(j\omega) = \frac{1}{j\omega L_{eq}} = \frac{1}{j\omega L} + j\omega C \tag{3.95}$$

This can be solved for L_{eq} as given by

$$L_{eq} = \frac{L}{1 - \omega^2 LC} \tag{3.96}$$

which is dependent on the frequency and can be either inductive or capacitive. Using Eq. (G.2) in Appendix G for the Miller's theorem, the equivalent input impedance of the transistor is given by

$$Z_i(j\omega) = \frac{Z_{cx}(j\omega)}{1 - A_v(j\omega)} = \frac{1/j\omega C_x}{1 - A_v(j\omega)} \tag{3.97}$$

Since the current through C_x is negligible in comparison to the current source $g_m v_i$, the voltage gain can be approximated as

$$A_v(j\omega) = \frac{V_o(j\omega)}{V_i(j\omega)} \cong -g_m Z_{eq}(j\omega) = -g_m(R_{eq} \| j\omega L_{eq}) = \frac{-j\omega g_m L_{eq}}{1 + j\omega L_{eq}/R_{eq}} \tag{3.98}$$

Substituting Eq. (3.98) into Eq. (3.97) gives

$$Z_i(j\omega) = -\frac{g_m R_{eq}^2}{\omega^2 L_{eq} C_x (1 + g_m R_{eq})^2 + R_{eq}^2 C_x / L_{eq}} - j\frac{R_{eq}^2 + \omega^2 L_{eq}^2 (1 + g_m R_{eq})}{\omega^3 L_{eq}^2 C_x (1 + g_m R_{eq})^2 + \omega R_{eq}^2 C_x} \tag{3.99}$$

which has a real part and an imaginary part such that $Z_i(j\omega) = R_i(j\omega) + jX_i(j\omega)$. The negative real component $R_i(j\omega)$ can cause instability. The values of the effective input resistance and input reactance seen by the input signal v_s are given by

$$R_{in} = R_s + R_B \| R_i(j\omega); \quad X_{in} = \frac{1}{j\omega C_i} \| jX_i(j\omega) \tag{3.100}$$

Therefore, the input reactance X_{in} can be capacitive or inductive, depending on the component values and the frequency. If the total input reactance is capacitive, the input circuit has a pole at $s = -1/C_{in}R_{in}$ in the complex plane. If the input reactance is inductive, then there is a pole at $s = -R_{in}/L_{in}$. In either case, if R_{in} is negative, the pole is in the right half plane and the circuit is unstable having a natural response of exponentially increasing.

3.7.4 Autotransformer Coupled-Tuned Amplifiers

A practical inductor has a fine series resistance R_c as shown in Fig. 3.34(a). The Q of an inductor coil is given by $Q_c \cong \omega L/R_c$. The Q_c may not be high enough to achieve the desired selectivity of the center frequency. The series resistance can be transformed into an equivalent parallel resistance as shown in Figure 3.34(b). The admittance of the series circuit is given by

$$Y(j\omega) = \frac{1}{R_c + j\omega L} = \frac{R_c - j\omega L}{R_c^2 + \omega^2 L^2} \tag{3.101}$$

And the admittance of the parallel circuit is

$$Y(j\omega) = \frac{1}{R_p} + \frac{1}{j\omega L} \tag{3.102}$$

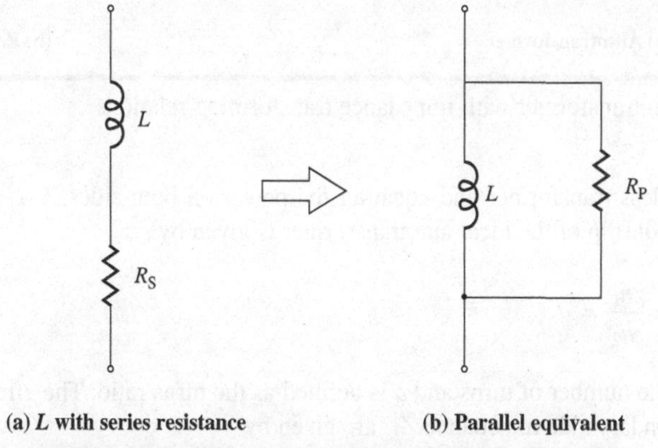

FIGURE 3.34 An inductor with series resistance converted to an equivalent parallel circuit

Assuming a high Q-coil, $\omega_L >> R_c$, and Eq. (3.101) can be approximated to

$$Y(j\omega) = \frac{R_c}{\omega^2 L^2} - j\frac{1}{\omega L} = \frac{R_c}{\omega^2 L^2} + \frac{1}{j\omega L} \tag{3.103}$$

Equating Eq. (3.103) with Eq. (3.102) gives R_p as

$$R_p = \frac{\omega^2 L^2}{R_c} = \omega L Q_c \tag{3.104}$$

Therefore, R_p should be included in Figs. 3.31(c), 3.32 and 3.33. If the Q_c of the inductor is not high enough, this parallel resistor may not be high enough to make the overall tuned circuit Q high enough and the selectivity of the tuned amplifier will be decreased. An autotransformer as shown in Fig. 3.35(a) can be used as an impedance matching network. The autotransformer has a single winding with tapping in one or more places. The coils are typically wound on a ferrite core, and the coefficient of coupling is very close to unity.

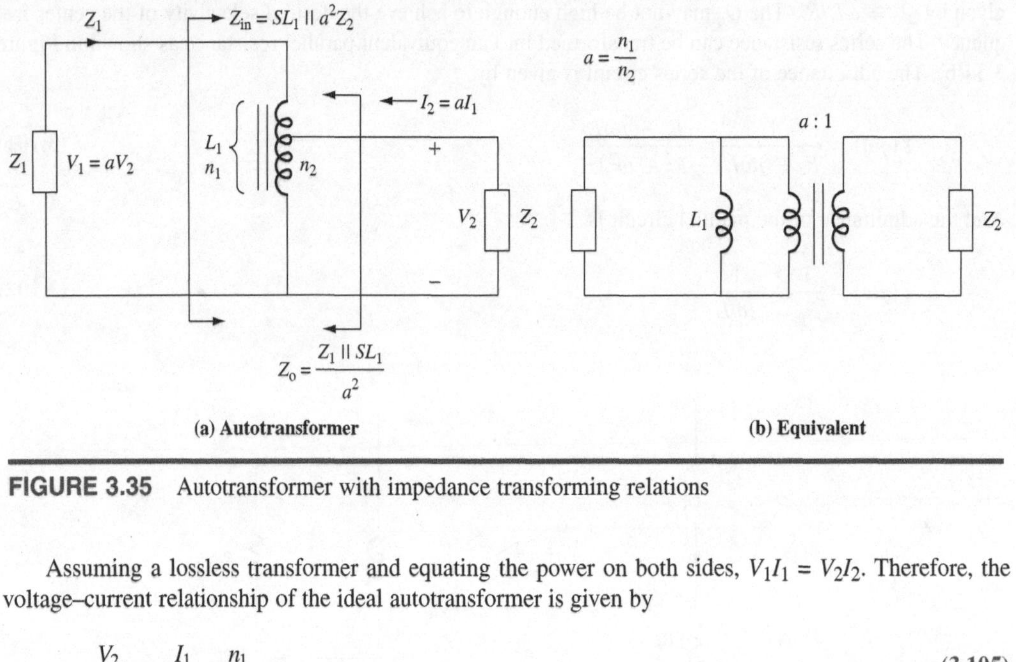

FIGURE 3.35 Autotransformer with impedance transforming relations

Assuming a lossless transformer and equating the power on both sides, $V_1 I_1 = V_2 I_2$. Therefore, the voltage–current relationship of the ideal autotransformer is given by

$$\frac{V_2}{V_1} = -\frac{I_1}{I_2} = \frac{n_1}{n_2} = a \tag{3.105}$$

where n_1 and n_2 are the number of turns and a is defined as the turns ratio. The effective primary side impedance Z_1' and secondary side impedance Z_2' are given by

$$V_1 = aV_2 = -aI_1 Z_2 = -a(-aI_2)Z_2 = (a^2 Z_2)I_1$$

or $$Z_1' = \frac{V_1}{I_1} = a^2 Z_2 \tag{3.106a}$$

$$V_2 = \frac{V_1}{a} = \frac{-I_1 Z_1}{a} = \frac{-(-I_2/a)Z_1}{a} = \left(\frac{Z_1}{a^2}\right) I_2$$

or $$Z_2' = \frac{V_2}{I_2} = \frac{Z_1}{a^2} \tag{3.106b}$$

Therefore, the autotransformer can be considered as a two-port network as shown in Fig. 3.35(b) where L_1 is the total inductance and a is the turns ratio. The impedance seen looking into port 1 is equal to L_1 in parallel with a^2 times the impedance connected to port 2. Similarly, the impedance looking into port 2 is equal to Z_2 in parallel with the impedance connected to port 1, all divided by a^2.

For a MOSFET amplifier, the impedance of the gate-drain capacitance C_{gd}, the output capacitance C_o, and the output resistance r_o of the transistor can be small enough in magnitude to degrade the characteristics of the tuned amplifier. The problem can be solved by connecting the transistor to the tuned circuit through an autotransformer as shown in Fig. 3.36(a) and the equivalent tuned circuit in Fig. 3.36(b). L_1 represents the total inductance of the transformer primary side. The equivalent output capacitance of the transistor is reduced by the factor of a^2, and the output resistance r_o is increased by the same factor. Thus, a much higher Q can be obtained, and the center frequency is not expected to shift significantly by changes in the value of C_{gd} for MOSFETs or C_{cb} for BJTs.

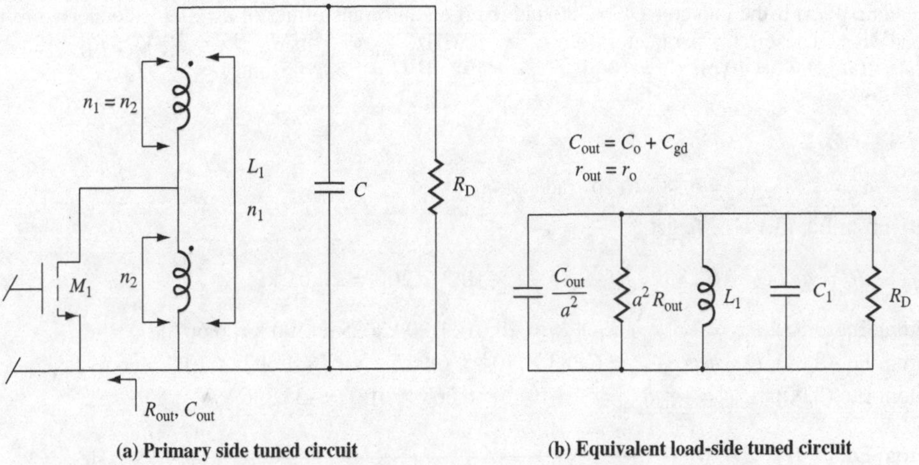

FIGURE 3.36 Output-side tuned MOSFET-amplifier with an autotransformer.

If the tuned circuit is placed at the input of the amplifier, the input impedance of the transistor may be small enough in magnitude to degrade the characteristics of the tuned amplifier. The input capacitance resulting from the Miller effect can be quite large and the input resistance of a BJT can be quite low. A BJT amplifier is connected to the input-side tuned circuit through an autotransformer as shown in Fig. 3.37(a) and the equivalent tuned circuit in Fig. 3.37(b). L_2 represents the total inductance of the transformer primary side.

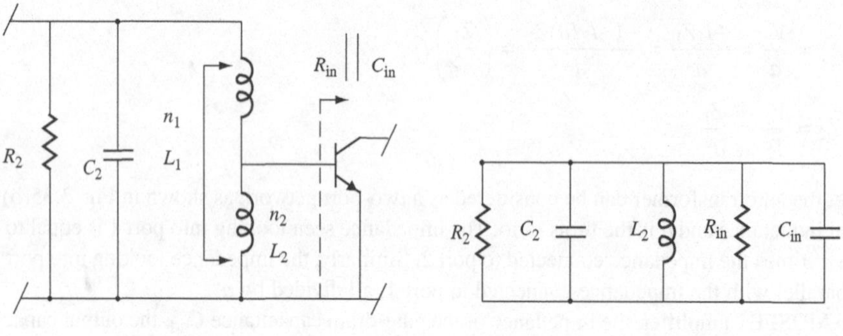

(a) Primary side tuned circuit

(b) Equivalent input-side tuned circuit

FIGURE 3.37 Input-side tuned BJT-amplifier with an autotransformer

EXAMPLE 3.9

Finding the effects of inductor *Q* and autotransformer For Example 3.8, find the mid-gain A_{mid}, Q and BW (a) if the inductor $Q_C = 200$ and (b) if an autotransformer of $a = 0.9$ is connected to the load side. The circuit parameters are: f_o = 1 MHz, g_m = 5 mA/V, r_o = 20 kΩ, R_D = 25 kΩ, R_L = 100 kΩ, C_i = 40 pF, C_x = 10 pF, C_{eq} = 1.592 nF, C = 1.582 nF, and L = 15.92 μH.

SOLUTION

$$\omega_o = 2\pi \times 10^6 = 6.283 \times 10^6 \text{ rad/s}$$

(a) From Eq. (3.104), we get

$$R_p = \omega_o L Q_c = 6.283 \times 10^6 \times 15.92 \times 10^{-6} \times 200 = 20.01 \text{ k}\Omega$$

From Eq. (3.85), $R_{eq} = R_p \parallel r_o \parallel R_D \parallel R_L = 20.01\text{k} \parallel 20 \text{ k} \parallel 25 \text{ k} \parallel 100 \text{ k} = 6.667 \text{ k}\Omega$
From Eq. (3.89), $Q = \omega_o R_{eq} C_{eq} = 6.283 \times 10^6 \times 6.667 \times 10^3 \times 1.592 \times 10^{-9} = 66.692$
From Eq. (3.90), $A_{mid} = -g_m R_{eq} = 5 \times 10^{-3} \times 6.667 \times 10^{-3} = -33.336$ V/V

From Eq. (3.91), $BW = \dfrac{1}{2\pi R_{eq} C_{eq}} = \dfrac{1}{2\pi \times 6.667 \times 10^3 \times 1.592 \times 10^{-9}} = 14.99$ KHz

Note: The gain is reduced while the bandwidth is increased due to the inductor Q.

(b) From Eqs. (3.85) and (106a),

$$R_{eq} = R_p \parallel a^2 r_o \parallel R_D \parallel R_L = 20.01 \text{ k} \parallel 0.9^2 \times 20 \text{ k} \parallel 25 \text{ k} \parallel 100 \text{ k} = 9.952 \text{ k}\Omega$$

From Eq. (3.90), $A_{mid} = -g_m R_{eq} = -5 \times 10^{-3} \times 9.952 \times 10^3 = -49.758$ V/V

$$C_{eq} = C + C_x/a^2 = 1.582 \times 10^{-9} \times 10 \times 10^{-12}/0.9^2 = 1.582 \times 10^{-9}$$

From Eq. (3.89), $Q = \omega_o R_{eq} C_{eq} = 6.283 \times 10^6 \times 9.952 \times 10^3 \times 1.582 \times 10^{-9} = 99.488$

From Eq. (3.91), $\dfrac{1}{2\pi R_{eq} G_{eq}} = \dfrac{1}{2\pi \times 9.952 \times 10^3 \times 1.582 \times 10^{-9}} = 10.11$ kHz

Note: The autotransformer has the effect of increasing the gain while reducing the bandwidth and the effect of inductor Q_C is compensated.

3.7.5 Synchronous-Tuned Amplifiers

Tuned amplifiers may have more than one tuned circuits [8,9,10]. If each tuned circuit has the same center frequency and bandwidth, then this amplifier is called as *synchronous tuned.* On the other hand, if the tuned circuits are different and arranged so that the overall effect is to produce a desired narrowband response, then the amplifier is known as *stagger tuned.* The overall bandwidth of a synchronously tuned amplifier will be lower than that of one of the tuned circuits alone. Assuming that there is no interaction with each other, the cutoff frequencies for cascode N-identical tuned circuits can be determined by setting the square of the magnitude of the transfer function equal to one-half. That is, for $s = j\omega$, Eq. (3.88) gives

$$\left(\frac{1}{1 + Q^2\left(\dfrac{\omega}{\omega_o} - \dfrac{\omega_o}{\omega}\right)^2}\right)^N = \frac{1}{2} \tag{3.107}$$

This equation leads to a quadratic equation for ω, and the two solutions are given by

$$\omega_{1,2} = \frac{\omega_o}{2Q}\sqrt{2^{1/N} - 1}\left(1 \pm \sqrt{1 + \frac{4Q^2}{2^{1/N} - 1}}\right) \tag{3.108}$$

Taking the positive sign inside the parentheses in Eq. (3.108) leads to the positive value of high cut-off ω_{CH} and taking the negative sign leads to the low cut-off $-\omega_{CL}$. Therefore, the bandwidth is given by

$$\begin{aligned} BW_N = \omega_{CH} - \omega_{CL} &= \frac{\omega_o}{2Q}\sqrt{2^{1/N} - 1}\left[\left(1 + \sqrt{1 + \frac{4Q^2}{2^{1/N} - 1}}\right) + \left(1 - \sqrt{1 + \frac{4Q^2}{2^{1/N} - 1}}\right)\right] \\ &= \frac{\omega_o}{Q}\sqrt{2^{1/N} - 1} = BW\sqrt{2^{1/N} - 1} \end{aligned} \tag{3.109}$$

where BW is the bandwidth of a single stage as given by Eq. (3.91). The amount by which the bandwidth is reduced for a cascade of N-identical tuned circuits is sometimes called the *shrinkage factor*, $S_N = \sqrt{2^{1/N} - 1}$.

A synchronous tuned amplifier with tuned circuits at both the amplifier input and output is shown in Fig. 3.38(a). The high-frequency equivalent circuit is shown in Fig. 3.38(b). The source resistor R_{sr} is bypassed by capacitor C_S, and C_C is a coupling capacitor shorted at a high frequency. There are two significant problems with the synchronous tuning as shown in Fig. 3.38. First, an alignment of the two tuned circuits is difficult because of the interaction between the two tuned circuits due to the Miller multiplication of C_{gd}. Second, the amplifier can easily become an oscillator due to the feedback from the output of the amplifier back to the input through C_{gd}.

A cascode-type BJT amplifier as shown in Fig. 3.39(a) can resolve these problems. The common-base transistor Q_2 can isolate the load circuit from the common-emitter transistor Q_1. This can eliminate effectively the Miller multiplication of capacitor C_x. The circuit also isolates the tuned circuits at both the input and the output so that these circuits operate independently. The small-signal ac equivalent is shown in Fig. 3.39(b).

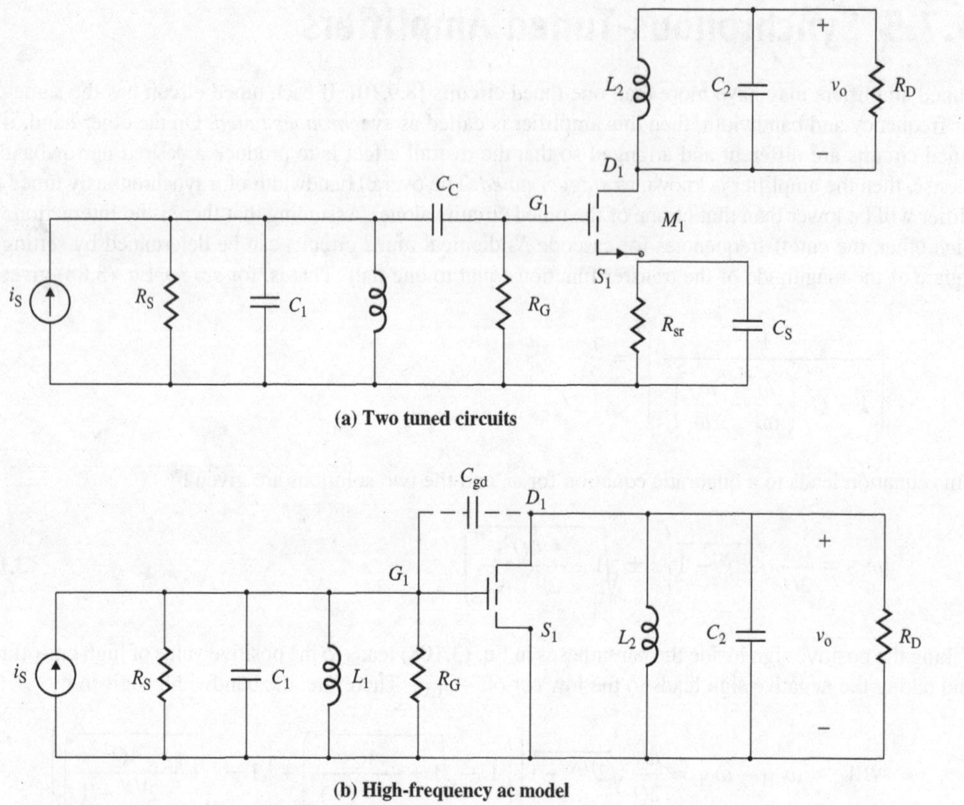

FIGURE 3.38 Synchronous-tuned amplifiers

The circuit parameters of Fig. 3.39(b) are:

$$R_1 = R_s \,\|\, R_1' \,\|\, r_{\pi 1}; \quad C_1 = C_1' + C_i + C_{M\,in} \tag{3.110a}$$

$$R_2 \simeq r_{o2} \,\|\, R_2' \,\|\, R_L; \quad C_2 = C_2' + C_{cb2} \tag{3.110b}$$

Since the capacitance C_{be2}, which is very small, is in parallel with a very small resistance, $r_{\pi 2}$, its effect can be neglected at the frequency of interest. Using Eq. (3.87), we can find the voltage gain of the amplifier as given by

$$A_v = \frac{V_o}{V_s} = \frac{V_o}{I_{e2}} \times \frac{I_{e2}}{V_i} \times \frac{V_i}{V_s} = \left(\frac{-\alpha_{i2} s/C_2}{s^2 + s/R_2C_2 + 1/L_2C_2} \right) g_m \left(\frac{s/C_1}{s^2 + s/R_1C_1 + 1/L_1C_1} \right) \frac{1}{R_s}$$

$$= \frac{-g_m \alpha_{i2} R_1 R_2}{R_s} \left(\frac{s/R_2C_2}{s^2 + s/R_2C_2 + 1/L_2C_2} \right) g_m \left(\frac{s/R_1C_1}{s^2 + s/R_1C_1 + 1/L_1C_1} \right) \tag{3.111}$$

For two stages, $N = 2$, the shrinkage factor is $S_2 = \sqrt{2^{1/2} - 1} = 0.64$. To achieve an overall bandwidth, the individual bandwidths should be equal to $BW = BW_2/0.64 = 1.56\, BW_2$. Therefore, we get

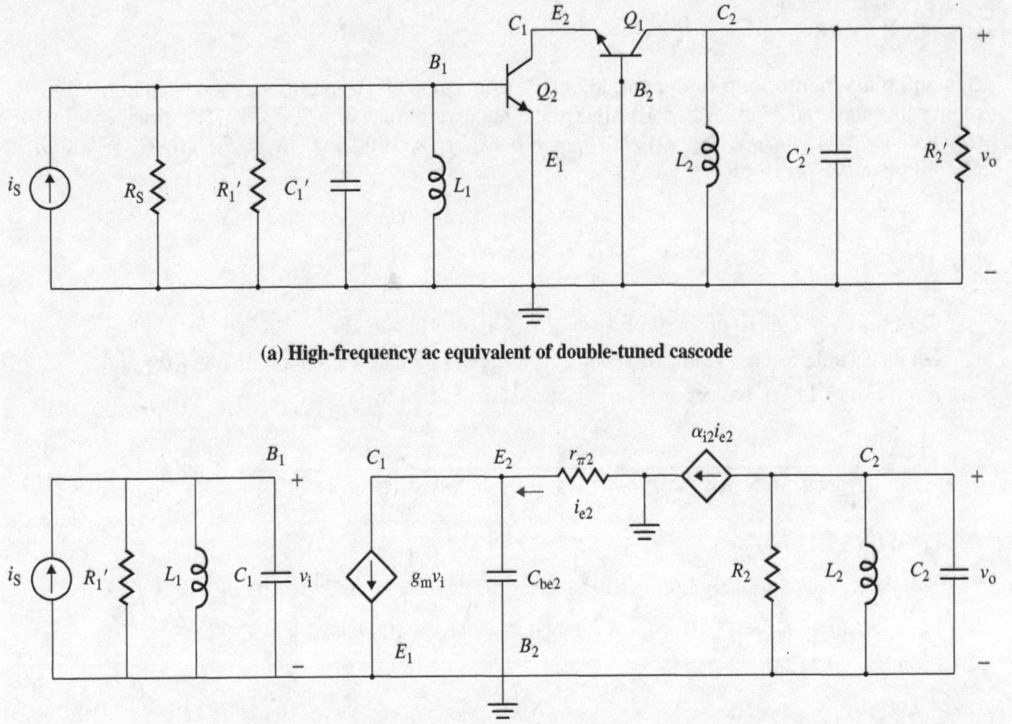

FIGURE 3.39 Double-tuned cascaded-tuned amplifiers

$$\frac{1}{R_1C_1} = \frac{1}{R_2C_2} = 1.56BW_2 \tag{3.112a}$$

$$\omega_o = \frac{1}{\sqrt{L_1C_1}} = \frac{1}{\sqrt{L_2C_2}} \tag{3.112b}$$

The voltage gain at the center frequency is given by

$$A_{\text{mid}} = \frac{-g_m\alpha_{i2}R_1R_2}{R_s} \tag{3.113}$$

Depending on the source and the load impedances, autotransformers may be needed at both the input and the output sides of cascade to satisfy the design constraints.

EXAMPLE 3.10

Designing a synchronous-tuned amplifier Design a tuned BJT amplifier as shown in Fig. 3.39(a) to give a center frequency of $f_o = 10$ MHz and a bandwidth of $BW = 200$ kHz. The circuit parameters are: $g_m = 15$ mA/V, $a_{i2} = 0.9$, $r_{\pi 1} = 8.6$ kΩ, $r_o = 1000$ kΩ, $R_L = 20$ kΩ, $R_s = 500$ Ω, $C_i = 40$ pF and $C_x = 10$ pF.

SOLUTION

$C_{Min} = 2C_x = 2 \times 10$ pF = 20 pF. $\omega_o = 2\pi \times 10 \times 10^6 = 6.283 \times 10^7$ rad/s

Let us assume $R_{1p} = 47$ kΩ . $R_1 = R_{1p} \parallel R_s \parallel r_{\pi 1} = 47\text{ k} \parallel 500 \parallel 8.6\text{ k} = 467.824$ Ω

From Eq. (3.112a), we get

$$C_1 = \frac{1}{2\pi \times 1.56 \times BW_2 \times R_1} = \frac{1}{2\pi \times 1.56 \times 200 \times 10^3 \times 467.824} = 1.09 \text{ nF}$$

From Eq. (3.110a), we get

$$C_{1p} = C_1 - C_i - C_{Min} = 1.09 \times 10^{-9} - 40 \times 10^{-12} - 20 \times 10^{-12} = 1.03 \text{ nF}$$

Let us assume $R_{2p} = 47$ kΩ, $R_2 = R_{2p} \parallel r_o \parallel R_L = 47\text{ k} \parallel 100\text{ k} \parallel 20\text{ k} = 12.3$ kΩ

From Eq. (3.112a), we get

$$C_2 = \frac{1}{2\pi \times 1.56 \times BW_2 \times R_2} = \frac{1}{2\pi \times 1.56 \times 200 \times 10^3 \times 12.3 \times 10^3} = 41.46 \text{ pF}$$

From Eq. (3.110b), we get

$$C_{2p} = C_2 - C_x = 41.46\ 10^{-12} - 10 \times 10^{-12} = 31.46 \text{ pF}$$

From Eq. (3.112b), we get

$$L_1 = \frac{1}{\omega_o^2 C_1} = \frac{1}{(6.283 \times 10^7)^2 \times 1.09 \times 10^{-9}} = 0.2323\ \mu\text{H}$$

$$L_2 = \frac{1}{\omega_o^2 C_2} = \frac{1}{(6.283 \times 10^7)^2 \times 41.46 \times 10^{-12}} = 6.11\ \mu\text{H}$$

From Eq. (3.113), we get

$$A_{mid} = \frac{-g_m \alpha_{i2} R_1 R_2}{R_s} = \frac{-15 \times 10^{-3} \times 0.9 \times 467.824 \times 12.3 \times 10^3}{500} = -155.411 \text{ V/V}$$

3.7.6 Stagger-Tuned Amplifiers

The shrinkage factor can be avoided by the use of stagger tuning. In stagger tuning, each circuit resonates at a different frequency. Eq. (3.111) can be expressed in factorized form as given by

$$A_v(s) = \frac{-g_m \alpha_{i22}}{R_s C_1 C_2}\left(\frac{s}{(s-p_1)(s-p_1^*)}\right)\left(\frac{s}{(s-p_2)(s-p_2^*)}\right) \quad (3.114)$$

The locations of the poles and zeros are shown in Fig. 3.40(a). The magnitude response for each pole pair is shown in Fig. 3.40(b). In order to achieve a Butterworth response with a bandwidth equal to *BW*, the poles are placed on a circle of radius *BW* and at 45° angles with respect to the real and imaginary axes as shown in Fig. 3.40(a) [9]. Therefore, the two low-pass poles can be expressed as

$$p_{L1,2} = \frac{BW}{\sqrt{2}}(-1 \pm j) \quad (3.115)$$

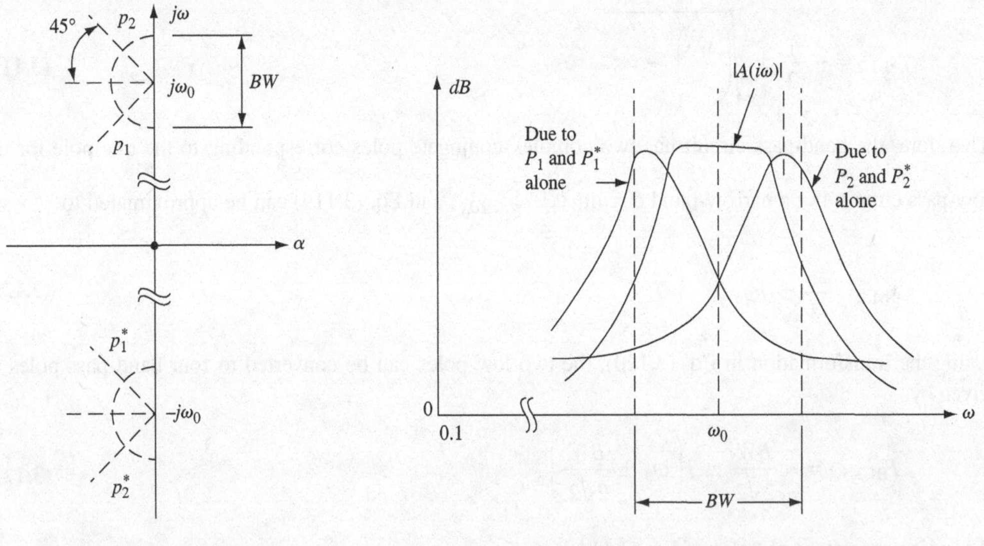

(a) Plot of pole–zero locations

(b) Responses for the individual tuned circuits

FIGURE 3.40 Pole–zero locations and magnitude responses of stagger-tuned amplifier

A low-pass pole can be transformed into two equivalent band-pass poles by equating the admittances of LC and RLC circuits. The low-pass admittance of the LC ciruit in Fig. 3.41(a) is

$$Y(s_L) = G + s_L C \quad (3.116)$$

The band-pass admittance of the RLC circuit in Fig. 3.41(b) is

$$Y(s_H) = G + s_B C + \frac{1}{s_B C} = G + C\left(s_B + \frac{\omega_o^2}{s_B}\right) \quad (3.117)$$

Equating (3.116) and (3.117) gives the relation of the low-pass pole to the band-pass pole as

$$s_L = s_B + \frac{\omega_o^2}{s_B} \quad (3.118)$$

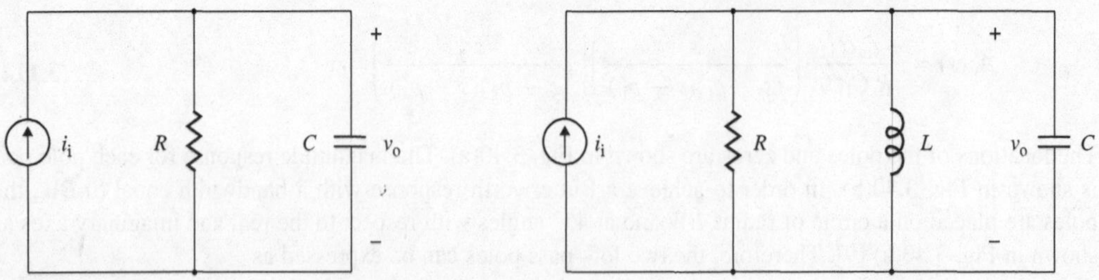

FIGURE 3.41 RLC and RC circuits

which gives two band-pass poles as

$$s_{B1,2} = -\frac{s_L}{2} \pm \sqrt{\left(\frac{s_L}{2}\right)^2 - \omega_o^2} \tag{3.119}$$

Therefore, the band-pass circuit has two complex-conjugate poles corresponding to the one pole for the low-pass circuit. For a narrow-band circuit, $\frac{|S_L|}{2} << \omega_o$, and Eq. (3.119) can be approximated to

$$s_{B1,2} \approx \frac{s_L}{2} \pm j\omega_o \tag{3.120}$$

Using the transformation in Eq. (3.120), the two low poles can be converted to four band-pass poles as given by

$$P_{B1,2,3,4} = -\frac{BW}{2\sqrt{2}} \pm j\left(\omega_o \pm \frac{BW}{2\sqrt{2}}\right) \tag{3.121}$$

Assigning the values of poles of Eq. (3.114), we get

$$p_1, p_1^* = -\frac{BW}{2\sqrt{2}} \pm j\left(\omega_o - \frac{BW}{2\sqrt{2}}\right) \tag{3.122a}$$

$$p_2, p_2^* = -\frac{BW}{2\sqrt{2}} \pm j\left(\omega_o + \frac{BW}{2\sqrt{2}}\right) \tag{3.122b}$$

These pole pairs can be converted to two-quadratic equations as given by

$$s^2 + s\frac{BW}{\sqrt{2}} + \omega_o^2\left(1 - \frac{BW}{\omega_o\sqrt{2}}\right) + \left(\frac{BW}{\sqrt{2}}\right)^2 \tag{3.123a}$$

$$s^2 + s\frac{BW}{\sqrt{2}} + \omega_o^2\left(1 + \frac{BW}{\omega_o\sqrt{2}}\right) + \left(\frac{BW}{\sqrt{2}}\right)^2 \tag{3.123b}$$

Equating the coefficients of polynomials in Eq. (3.123a) and (3.123b) to the denominators of Eq. (3.111), we get

$$R_1C_1 = R_2C_2 = \frac{\sqrt{2}}{BW} \tag{3.124a}$$

$$\frac{1}{L_1C_1} = \omega_o^2\left(1 - \frac{BW}{\omega_o\sqrt{2}}\right) + \frac{BW^2}{4} \tag{3.124b}$$

$$\frac{1}{L_2C_2} = \omega_o^2\left(1 + \frac{BW}{\omega_o\sqrt{2}}\right) + \frac{BW^2}{4} \tag{3.124c}$$

For frequencies greater than the lower cut-off, $\omega >> \omega_{CL}$, the following approximation can be made:

$$\frac{\omega}{\left|j\omega - p^*\right|} \approx \frac{1}{2}$$

which gives the magnitude of the voltage gain in Eq. (3.111) as

$$|A_v(j\omega)| \approx \frac{g_m\alpha_{i2}}{4R_sC_1C_2}\frac{1}{|j\omega_o - p_1|\,|j\omega_o - p_2|} \tag{3.125}$$

From Fig. 3.40(a), the denominator of Eq. (3.125) can be approximated as [9]

$$|j\omega_o - p_1| = |j\omega_o - p_2| = \frac{BW}{2} \tag{3.126}$$

Thus, Eq. (3.125) becomes

$$|A_v(j\omega)| \approx \frac{g_m\alpha_{i2}}{R_sC_1C_2}\frac{1}{BW^2} \tag{3.127}$$

which after substituting for BW in Eq. (3.124a) gives

$$|A_v(j\omega)| \approx \frac{g_m\alpha_{i2}R_1R_2}{2R_s} \tag{3.128}$$

This is one-half of the gain in Eq. (3.113) of the synchronously-tuned amplifier.

EXAMPLE 3.11

Designing a stagger-tuned amplifier Design a stagger-tuned BJT amplifier as shown in Fig. 3.39(a) to give a center frequency of $f_o = 10$ MHz and a bandwidth of $BW_2 = 1$ MHz. The circuit parameters are: $g_m = 15$ mA/V, $\alpha_{i2} = 0.9$, $r_{\pi 1} = 8.6$ kΩ, $r_o = 100$ kΩ, $R_L = 20$ kΩ, $R_s = 500$ Ω, $C_i = 40$ pF, and $C_x = 10$ pF.

SOLUTION

$C_{mn} = 2C_x = 2 \times 10$ pF $= 20$ pF. $\omega_o = 2\pi \times 10 \times 10^6 = 6.283 \times 10^7$ rad/s

Let us assume $R_{1p} = 2.7\text{ k}\Omega$. $R_1 = R_{1p} \parallel Rs \parallel r_{\pi 1} = 2.7\text{ k} \parallel 500 \parallel 8.6\text{ k} = 402.148\ \Omega$
From Eq. (3.124a), we get

$$C_1 = \frac{\sqrt{2}}{2\pi \times 1.56 \times BW_2 \times R_1} = \frac{\sqrt{2}}{2\pi \times 1.56 \times 10 \times 10^6 \times 402.148} = 559.7\text{ pF}$$

From Eq. (3.110a), we get

$$C_{1p} = C_1 - C_i - C_{Min} = 559.7 \times 10^{-12} - 40 \times 10^{-12} - 20 \times 10^{-12} = 499.7\text{ pF}$$

Let us assume $R_{2p} = 2.7\text{ k}\Omega$, $R_2 = R_{2p} \parallel r_o \parallel R_L = 2.7\text{ k} \mid 100\text{ k} \parallel 20\text{ k} = 2.324\text{ k}\Omega$
From Eq. (3.124a), we get

$$C_2 = \frac{\sqrt{2}}{2\pi \times 1.56 \times BW_2 \times R_2} = \frac{1}{2\pi \times 1.56 \times 1 \times 10^6 \times 2.324 \times 10^3} = 96.87\text{ pF}$$

From Eq. (11.110b), we get

$$C_{2p} = C_2 - C_x = 96.87 \times 10^{-12} - 10 \times 10^{-12} = 86.87\text{ pF}$$

From Eqs. (3.124b) and (3.124c), we get

$$L_1 = \frac{1}{\left[\omega_o^2\left(1 - \frac{BW_2}{\omega_o\sqrt{2}}\right) + \frac{BW_2^2}{4}\right]C_1}$$

$$= \frac{1}{\left[(6.283 \times 10^7)^2 \times \left(1 - \frac{2 \times \pi \times 1 \times 10^6}{6.283 \times 10^7 \times \sqrt{2}}\right) + \frac{(2 \times \pi \times 1 \times 10^6)^2}{4}\right] \times 5.597 \times 10^{-10}}$$

$$= 4.875\ \mu\text{H}$$

$$L_2 = \frac{1}{\left[\omega_o^2\left(1 + \frac{BW_2}{\omega_o\sqrt{2}}\right) + \frac{BW_2^2}{4}\right]C_2}$$

$$= \frac{1}{\left[(6.283 \times 10^7)^2 \times \left(1 + \frac{2 \times \pi \times 1 \times 10^6}{6.283 \times 10^7 \times \sqrt{2}}\right) + \frac{(2 \times \pi \times 1 \times 10^6)^2}{4}\right] \times 96.87 \times 10^{-12}}$$

$$= 2.437\ \mu\text{H}$$

From Eq. (3.113), we get

$$A_{mid} = \frac{-g_m \alpha_{i2} R_1 R_2}{R_s} = \frac{-15 \times 10^{-3} \times 0.9 \times 402.148 \times 2.324 \times 10^3}{500} = -12.615$$

3.8 Class C Amplifiers

Class C amplifiers conduct less than 50% of the input signal, and the efficiencies can be as high as 90%, but the output has a high amount of distortion. These amplifiers can find applications in radio frequency transmitters where the distortion can be considerably reduced by using tuned loads on the amplifying devices. Figure 3.42(a) shows the simplified arrangement of a class C amplifier with a parallel-connected load resistance R_L. To isolate the load from the DC power supply and for impedance matching with the external load, R_L is generally connected through a transformer. The class C amplifier can also be operated as a push-pull configuration consisting of an NMOS and a PMOS with a common gate connection for the input signal.

When the gate signal $v_G = V_G + V_m \sin \omega t$ exceeds the MOS threshold voltage V_t, the MOS device conducts, and the corresponding drain current flows through the parallel load. The small-signal equivalent circuit is shown in Fig. 3.42(b). In the linear region, the transistor will behave as a switch with a drain–source resistance, which can be found from Eq. (3.2) as given by

$$R_{DS} = \frac{v_{DS}}{i_D} \simeq \frac{L}{W\mu_n C_{ox}(V_G + V_m \sin \omega t - V_t)} \tag{3.129}$$

The AC gate–source voltage $v_{gs} = V_m \sin \omega t$ will vary the drain current ($i_d = g_m v_{gs}$), which is the source for the *RLC* load. From Fig. 3.42(b), we can find the small-signal AC output voltage as given by

$$v_o = g_m v_{gs}(R_L \parallel X_L \parallel X_C) \tag{3.130}$$

where $X_L(=j\omega L)$ and $X_C(=1/j\omega C)$ are the impedances of L and C, respectively. If the gate–source voltage is a pulsed DC voltage of PWM, the drain current will also be of the same form.

The values of L and C can be selected either to oscillate at a specified damping factor or to form a tuned circuit by choosing the values of L and C such that $\omega L = 1/\omega C$ at the tuned frequency ω_n. The tuned circuit will resonate only at particular frequencies, so the unwanted frequencies are dramatically suppressed, and the desirable full signal (sine wave) will be abstracted by the tuned load. Provided that the transmitter is not required to operate over a very wide band of frequencies, this arrangement works extremely well. Other residual harmonics can be removed using a filter.

The load R_L also can be connected to form a series resonant circuit as shown in Fig. 3.42(c). When the transistor is switched on, the drain and the inductor currents will rise through the path formed by V_{DD}, L, and M_1. When the transistor is switched off, the drain current will be zero, and the inductor current

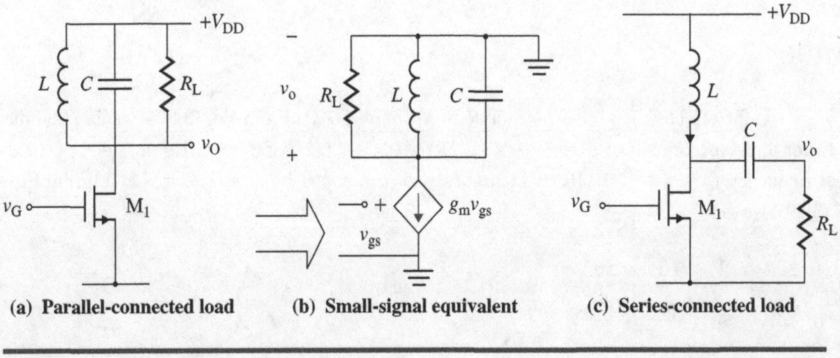

FIGURE 3.42 Class C amplifier

will fall through the path formed by V_{DD}, L, C, and R_L. The energy stored in the inductor during the transistor on-time will be transferred to the load R_L via the capacitor C. The damping factor of series and parallel *RLC* circuits (see Appendix B) can be found from

$$\delta = \frac{1}{2R_L}\sqrt{\frac{L}{C}} \quad \text{(for series resonant)} \tag{3.131}$$

$$= \frac{1}{2R_L}\sqrt{\frac{L}{C}} \quad \text{(for parallel resonant)}$$

The values of L and C are related to the damping ratio δ and the resonant frequency f_n as given by

$$L = \frac{R_L}{2\delta\omega_n} \quad \text{(for series resonant)} \tag{3.132}$$

$$= \frac{2\delta R_L}{\omega_n} \quad \text{(for parallel resonant)}$$

$$C = \frac{2\delta}{R_L\omega_n} \quad \text{(for series resonant)} \tag{3.133}$$

$$= \frac{1}{2\delta R_L\omega_n} \quad \text{(for parallel resonant)}$$

EXAMPLE 3.12

Finding the values of *L* and *C* for a class C amplifier The gate voltage of the class C amplifier in Fig. 3.42(a) is $v_G = 2 + 1.5\sin(2\pi f_s t)$ with $f_s = 1$ MHz. The DC supply voltages are $V_{DD} = \pm 15$ V. The load resistance $R_L = 50\ \Omega$, which is connected through a transformer of turns ratio of $n = 1{:}1$.

(a) Find the values of L and C to form a tuned resonant circuit.

(b) Use PSpice to plot the voltage $v_o(t)$, the transistor drain current $i_D(t)$, and the transistor drain voltage $v_D(t)$ for varying the gate voltage component $V_m = 3$ V to 4 V with an increment of 0.5 V. Use the PSpice parameters of the NMOS IRF150 whose $V_t = 2.84$ V.

SOLUTION

(a) For $n = 1{:}1$, the effective load resistance is $R = R_L/n^2 = 50\ \Omega$. From Eq. (3.87), we can find the damping factor for a tuned circuit of $L = C$, $\delta = (1/2R)\sqrt{L/C} = 1/2 \times 50 = 0.01$. For $f_s = 1 \times 10^6$, the resonant frequency $f_o = f_s = 1 \times 10^6$, which is related to L and C by $f_o = 1/(2\pi\sqrt{LC})$. From Eqs. (3.88) and (3.89), we get

$$L = \frac{2\delta R}{2\pi f_s} = \frac{2 \times 0.01 \times 50}{2 \times \pi \times 10^6} = 159.2\text{ nH} \simeq 160\text{ nH}$$

$$C = \frac{1}{2\delta R 2\pi f_s} = \frac{1}{2 \times 0.01 \times 50 \times 2 \times \pi \times 10^6} = 159.2\text{ nF} \simeq 160\text{ nF}$$

(b) The PSpice schematic is shown in Fig. 3.43. A small resistance $R_x = 1\ \mu\Omega$ is connected to avoid a convergence problem of the zero-resistance DC loop. The PSpice plot of the transistor drain current $i_D(t)$ is shown in Fig. 3.44(a), the transistor drain voltage $v_D(t)$ in Fig. 3.44(b), and the output load voltage $v_o(t)$ in Fig. 3.44(c) for V_m = 3 V, 3.5 V, and 4 V. For $v_G(t) < V_t = 2.84$ V, the transistor is off. The output voltage increases with the gate voltage as expected as shown in Fig. 3.44(c). The drain current becomes negative at a higher value because the inductive load L induces a negative voltage (Ldi/dt) effect due to the abrupt turning off of the transistor; this negative voltage forward biases the built-in source–channel pn junction. The peak drain voltage reaches $2V_{DD}$ due to the Ldi/dt effect.

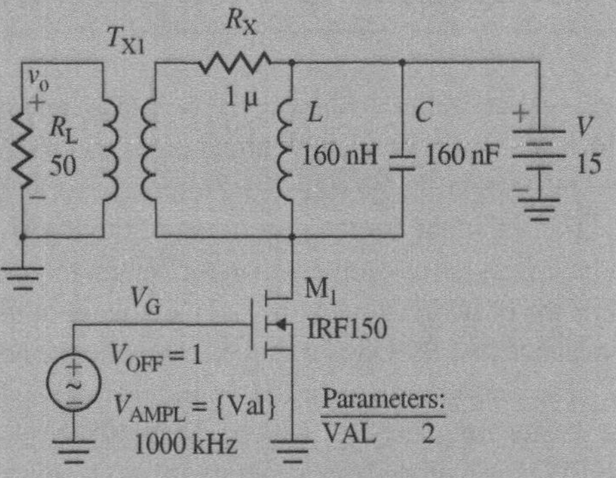

FIGURE 3.43 PSpice schematic for Example 3.12

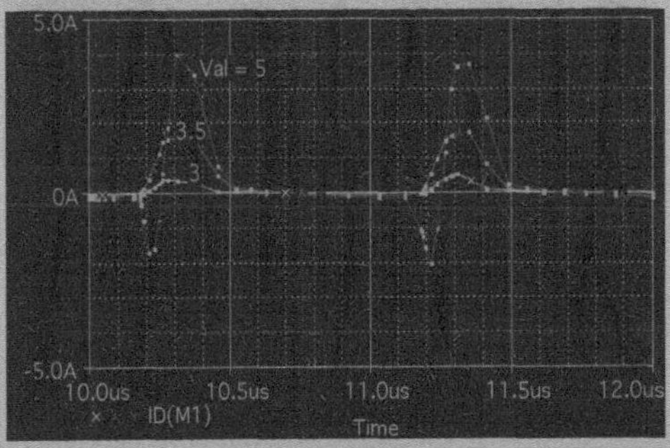

(a) Transistor drain current

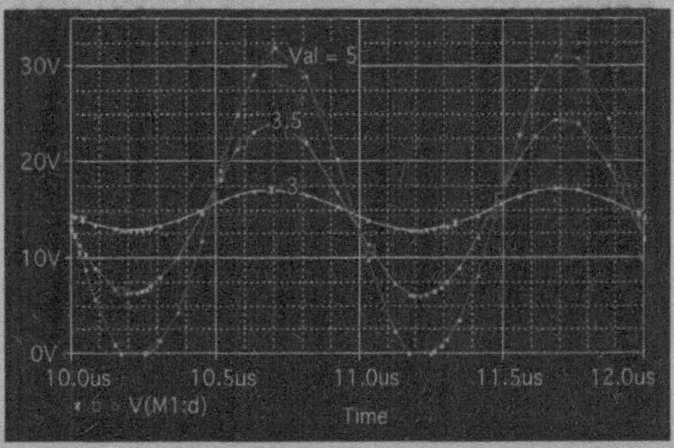

(b) Transistor drain voltage

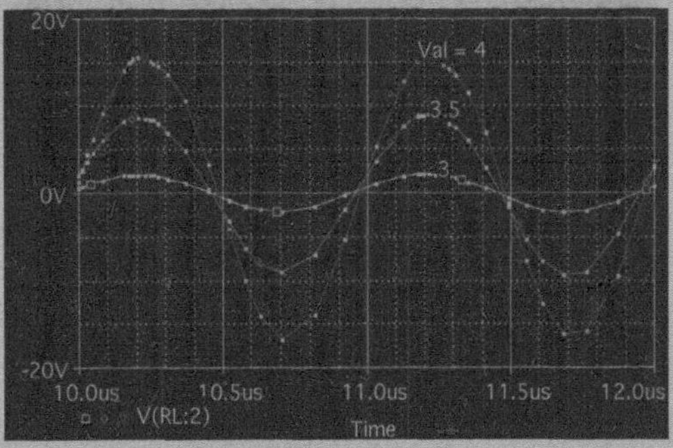

(c) Output load voltage

FIGURE 3.44 PSpice voltages and currents of class C amplifier for Example 3.12

3.9 Class D Amplifiers

A class D amplifier as shown in Fig. 3.45(a) operates transistors M_1 and M_2 as complementary switches—which are either completely turned on or completely turned off with a PWM gate signal—which are generated by comparing the sinusoidal reference (or input) signal voltage v_r with a triangular wave carrier signal v_{cr}. The technique for the generation of the PWM waveform is shown in Fig. 3.45(b), which can be implemented with an op-amp comparator. When the gate–source voltage is positive, the NMOS is switched on while the PMOS is turned off, and the voltage at the transistor drain is $+V_{DD}$. When the gate–source voltage is negative, the PMOS is switched on while the NMOS is turned off, and the transistor output is $-V_{SS}$.

The output voltage at the drain terminal is a pulsed waveform switching between $+V_{DD}$ and V_{SS} at the carrier frequency f_{cr}, which is a multiple (in the range of 10 to 30) of the signal frequency f_s. Since the PWM waveform was generated from a sinusoidal reference signal, a low-pass *LC* filter as shown in Fig. 3.45(c) is used to restore the original sinusoidal component from the high-frequency modulated voltage. One additional *LC* filter may be used to reduce the amount of distortion. The class D amplifier is highly efficient (around 95%), which can be even closer to 100%, and finds applications in such portable devices as laptop computers and MP3 players. Class D amplifiers also find applications as high-quality audio amplifiers. The pulse width of the modulated output depends on the ratio of the peak reference signal to the peak carrier signal, commonly known as the *modulation index* [7], as given by

$$m = \frac{V_r}{V_{cr}} \tag{3.134}$$

The characteristic of the *LC* filter can be determined from the input–output relationship in the Laplace's domain of *s* as given by

$$G_f(s) = \frac{V_o(s)}{V_i(s)} = \frac{\omega_n^2}{s^2 + 2\delta\omega_n s + \omega_n^2} \tag{3.135}$$

where the natural resonant frequency is $\omega_n = 1/\sqrt{LC}$, and the damping factor $\delta = (1/2R)\sqrt{L/C}$. By substituting $s \equiv j\omega$ in Eq. (3.135), we can find the filter transfer function in the frequency domain as given by

$$G_f(j\omega) = \frac{\omega_n^2}{(j\omega)^2 + j\omega 2\delta\omega_n + \omega_n^2} = \frac{\omega_n^2}{\omega_n^2 - \omega^2 + j\omega 2\delta\omega_n} \tag{3.136}$$

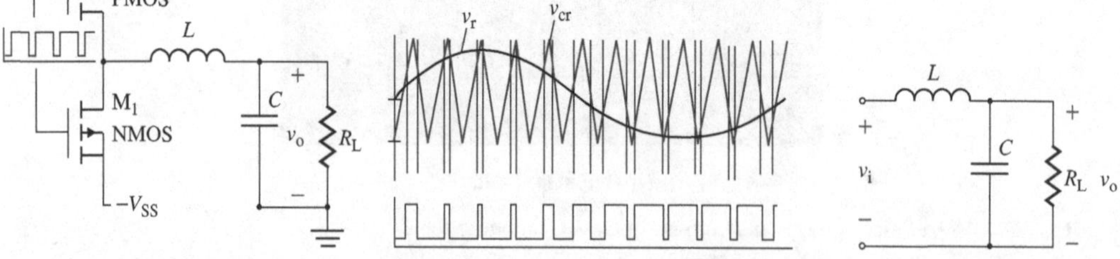

(a) Circuit arrangement (b) Generation of PWM gate signal (c) Output filter

FIGURE 3.45 Class D amplifier

which can be expressed in polar form with normalized frequency ratio $u = \omega/\omega_n$ as given by

$$G_f(j\omega) = \frac{1}{1 - u^2 + j2\delta u} = \frac{1}{\sqrt{(1-u^2)^2 + (2\delta u)^2}} \angle -\tan^{-1}\left(\frac{1-u^2}{2\delta u}\right) \quad (3.137)$$

which shows that $u = \omega/\omega_n$ affects the amount of phase shift present at the output. A high ω_n will decrease the phase shift, whereas a low ω_n will increase it. To minimize the amount of ripple present at the output, ω_n has to be smaller than the carrier frequency $\omega_{cr}(=1/2\pi f_{cr})$, and it should ideally be equal to the reference frequency $\omega_r(=2\pi f_r)$.

EXAMPLE 3.13

Finding the values of *L* and *C* for a class D amplifier The gate voltage of the class D amplifier in Fig. 3.45(a) is generated by modulating a sinusoidal reference $V_r = 1$ V at $f_r = 20$ kHz with a carrier signal $V_{cr} = 1$ V at $f_{cr} = 30f_s$. The DC supply voltages are $V_{DD} = \pm 5$ V. The load resistance $R_L = 8\ \Omega$ speaker.

(a) Find the values of L and C to form a tuned resonant circuit.

(b) Use PSpice to plot the gate-modulating voltage $v_G(t)$, the transistor output voltage $v_D(t)$, and the transistor output voltage $v_D(t)$ by a modulation index of $m = 0.5$ and 0.9. Use the PSpice parameters of the NMOS IRF150 whose $V_t = 2.84$ V and the PMOS IRF9140 whose $V_t = 3.67$ V.

SOLUTION

(a) The lower-order harmonic component is $f_L = 2f_s = 2 \times 20 \times 10^3 = 40 \times 10^3$. For the impedance of the filter capacitor, C must be much smaller than R_L so that the higher-order harmonic components pass through C. Using a 10:1 ratio, we can write $R_L >> 1/(2\pi f_L C)$ or

$$C = \frac{10}{2\pi f_L R_L} = \frac{10}{2\pi \times 40 \times 10^3 \times 8} \simeq 2.5\ \mu\text{F}$$

Let us make the filter cutoff frequency $f_n = f_L = 1/(2\pi\sqrt{LC})$, which gives L for $C = 2.5\ \mu$F as

$$L = \frac{1}{C(2\pi f_n)^2} = \frac{1}{2.5 \times 10^{-6}(2\pi \times 40 \times 10^3)^2} \simeq 6.4\ \mu\text{H}$$

(b) The PSpice schematic is shown in Fig. 3.46. The PSpice plot of the gate modulating voltage $v_G(t)$ is shown in Fig. 3.47(a), the transistor drain voltage $v_D(t)$ in Fig. 3.47(b), and the output load voltage $v_o(t)$ in Fig. 3.47(c) for $m = 0.8$ and 1.0. For $v_G(t) < V_t = 2.84$ V (MMOS) and $v_G(t) > V_t = 3.67$ V (PMOS), the transistors are off. The output voltage increases with the modulation index as expected. It should be noted that the transistor output is $2V_{DD}$. From the output file, we can find the THD of the output voltage, which is a measure of the waveform quality, by performing the Fourier analysis: THD = 4.847% for $m = 0.8$, and THD = 3.0% for $m = 1$.

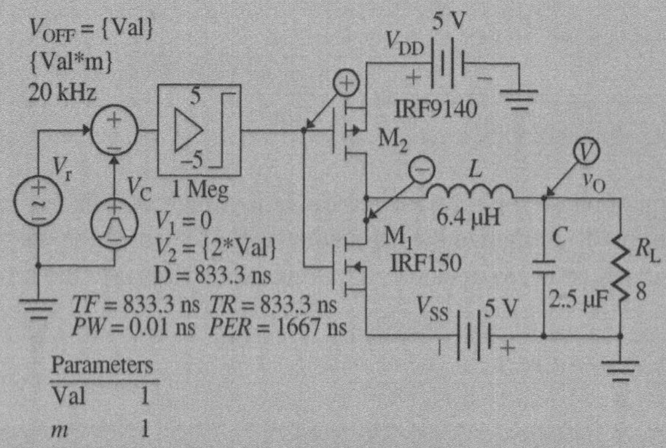

FIGURE 3.46 PSpice schematic for Example 3.13

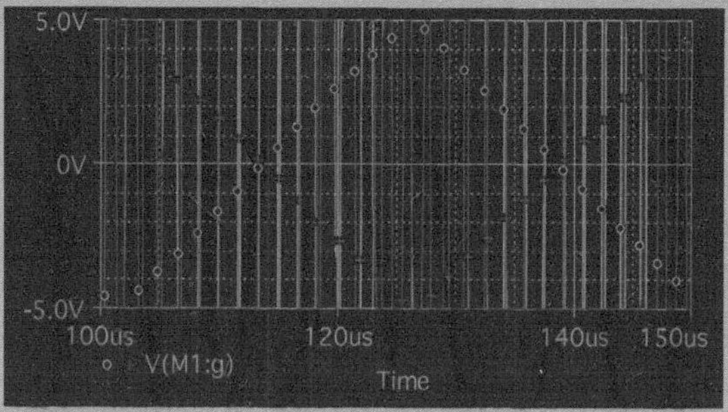

(a) Gate modulated voltage

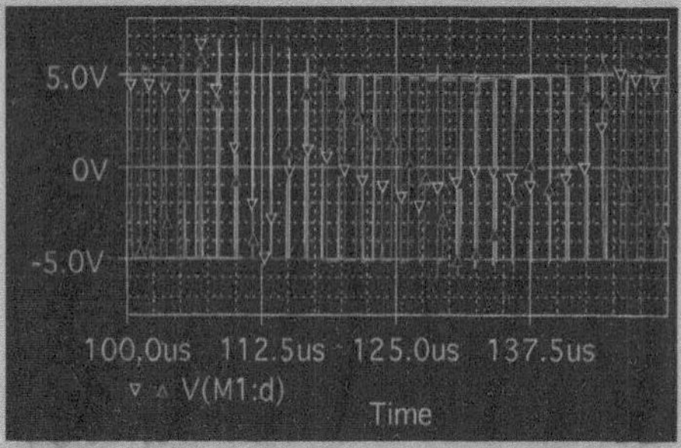

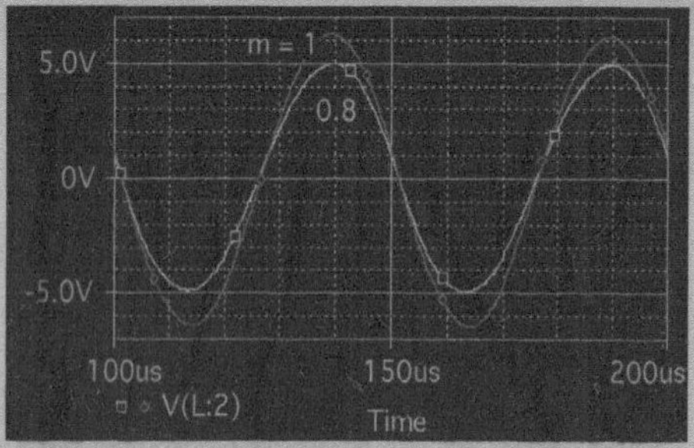

(b) Transistor drain voltage **(c) Output load voltage**

FIGURE 3.47 PSpice voltages and currents of a class D amplifier for Example 3.13

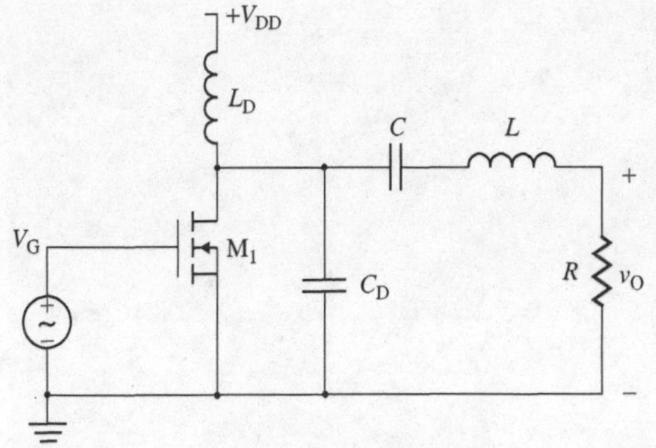

FIGURE 3.48 Class E amplifier

3.10 Class E Amplifiers

As shown in Fig. 3.48, a class E amplifier is similar to the class D amplifier, except that a capacitor C_D is connected across the transistor. It uses only one transistor as a switch and has low switching losses, yielding a high efficiency of more than 95%. It is used for applications requiring less than 100 W. When the transistor is switched on, the current through the inductor L_D rises; when the transistor is turned off, L_D forms a resonant circuit with C_D and oscillates on. As a result, the transistor output is almost resonant voltage. To minimize switching losses and increase efficiency, the transistor should be gated when the resonance current through the inductor L_D reaches its minimum low (ideally zero). That means the transistor will be turned on at zero current and will be subjected to minimum switching losses, increasing the amplifier efficiency.

To obtain an almost sinusoidal current through the load, the values of L and C are chosen to have a high quality factor, $Q \geq 7$, and a low damping ratio, usually $\delta \leq 0.072$. The optimum parameters to give the maximum efficiency can be found from [7]

$$L_D = \frac{0.4001R}{\omega_s} \tag{3.138}$$

$$C_D = \frac{2.165}{R\omega_s} \tag{3.139}$$

$$\omega_s L_D - \frac{1}{\omega_s C_D} = 0.3533R \tag{3.140}$$

EXAMPLE 3.14

Finding the values of L_D, C_D, L, and C for a class E amplifier The gate voltage of the class E amplifier in Fig. 3.48 is a sinusoidal voltage of 4 V (peak) at 100 kHz. The DC supply voltages are $V_{DD} = 12$ V. The load resistance $R = 10\ \Omega$.

(a) Find the optimum values of L_D, C_D, L, and C.

(b) Use PSpice to plot the gate voltage $v_G(t)$, the transistor drain current $i_D(t)$, the transistor output voltage $v_D(t)$, and the output voltage $v_o(t)$ for peak gate voltages of 4 V and 4.5 V. Use the PSpice parameters of the NMOS IRF150 whose $V_t = 2.84$ V.

SOLUTION

(a) For $f_s = 100$ kHz,

$$\omega_s = 2\pi f_s = 2\pi \times 100 \text{ k} = 6.283 \times 10^5 \text{ rad/s}$$

From Eq. (3.94),

$$L_D = \frac{0.4001R}{\omega_s} = \frac{0.4001 \times 10}{6.283 \times 10^5} \simeq 6.4\ \mu\text{H}$$

From Eq. (3.95),

$$C_D = \frac{2.165}{R\omega_s} = \frac{2.165}{10 \times 6.283 \times 10^5} \simeq 345 \text{ nF}$$

For choosing $Q = 7$,

$$L_D = \frac{QR}{\omega_s} = \frac{7 \times 10}{6.283 \times 10^5} = 111\ \mu\text{H}$$

From Eq. (3.96),

$$C_D = \frac{1}{(\omega_s L - 0.3533\ R)\omega_s} = \frac{1}{(6.283 \times 10^5 \times 111 \times 10^{-6} - 0.3533 \times 10) \times 6.283 \times 10^5} \simeq 24 \text{ nF}$$

(b) The PSpice schematic is shown in Fig. 3.49. The PSpice plots of the gate voltage $v_G(t)$ and the drain current i_D are shown in Fig. 3.50(a), the transistor output voltage $v_D(t)$ and the inductor current in Fig. 3.38(b), and the output load voltage $v_o(t)$ in Fig. 3.50(c) for peak gate voltages of 4 V and 4.5 V. For $v_G(t) < V_t = 2.84$ V (MMOS), the transistors are off. The output voltage increases with the gate voltage as expected, and the transistor output is $2V_{DD}$. From the output file, we can find the THD of the output voltage, which is a measure of the waveform quality, by performing the Fourier analysis: THD = 4.279% for Val = 4 V and THD = 3.078% for Val = 4.5 V.

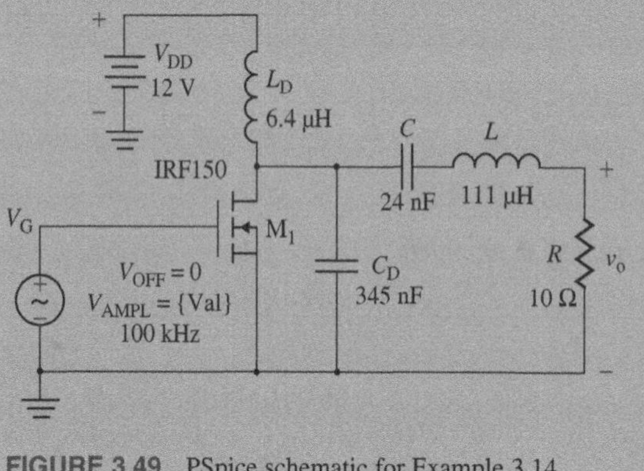

FIGURE 3.49 PSpice schematic for Example 3.14

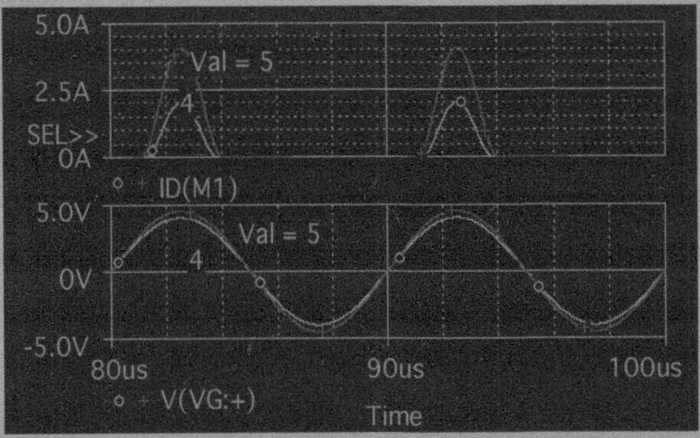

(a) Gate voltage and drain current

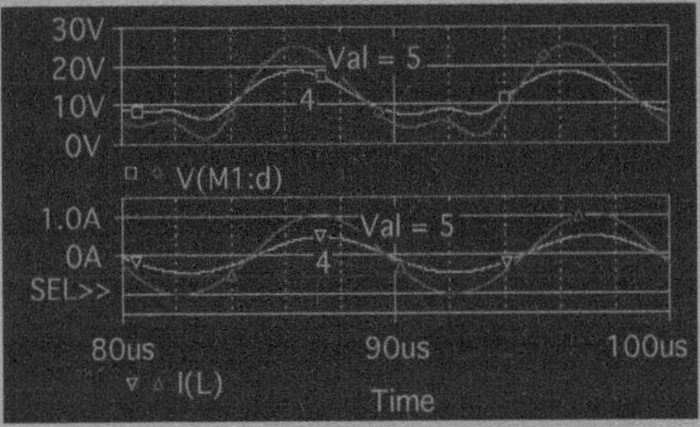

(b) Transistor output voltage and inductor current

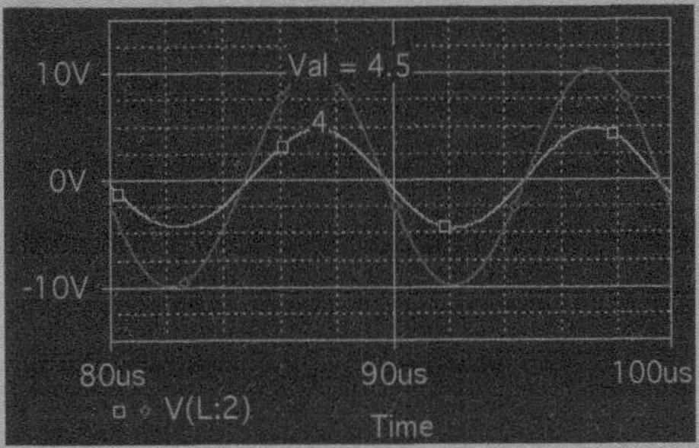

(c) Output load voltage

FIGURE 3.50 PSpice voltages and currents of a class E amplifier for Example 3.14

3.11 Short-Circuit and Thermal Protection

An output stage is normally protected against short-circuiting and excessive temperature rise. A class AB amplifier with such protection is shown in Fig. 3.51.

3.11.1 Short-Circuit Protection

The circuit used in Fig. 3.51 for short-circuit protection consists of transistor Q_1 and resistor R_{E1}. If a short circuit occurs at the load while Q_N is conducting, a large current will flow through R_{E1}, and a voltage V_{RE1} proportional to the short-circuit current will develop across R_{E1}. When voltage V_{RE1} becomes large enough, transistor Q_1 will turn on and carry most of the biasing current I_{R1}. Thus, the base current of Q_N will be reduced to a safe level. The voltage drops across the emitter resistors will reduce the output volt-

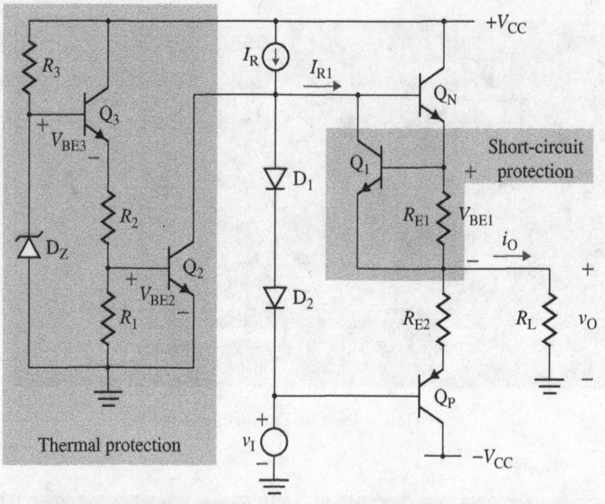

FIGURE 3.51 Short-circuit and thermal protection in a class AB amplifier

age by the same amount. Therefore, the values of R_{E1} and R_{E2} should be as low as possible (on the order of milliohms). Their values are determined from

$$R_{RE1} = R_{RE2} = \frac{V_{BE1}}{i_{O(short)}} \tag{3.141}$$

where $i_{O(short)}$ is the permissible short-circuiting current to turn on transistor Q_1. For $V_{BE1} = 0.7$ V and $i_{O(short)} = 200$ mA, $R_{RE1} = R_{RE2} = 0.7/200 \text{ mA} = 3.5\ \Omega$. Although R_{E1} and R_{E2} reduce the output voltage swing, they will give biasing stability to the quiescent current I_Q and protect Q_N and Q_P against thermal runaway (i.e., excessive junction temperature rise).

3.11.2 Thermal Protection

The circuit used in Fig. 3.51 for thermal protection consists of two transistors (Q_2 and Q_3), three resistors (R_1, R_2, and R_3), and a zener diode. Normally, transistor Q_2 is off. If the temperature rises, V_{BE3} will fall because of the negative temperature coefficient of Q_3, and the zener voltage V_Z will rise because of the positive temperature coefficient of zener diode D_Z. As a result, the voltage at the emitter of Q_3 will rise, and the voltage at the base of Q_2 will also rise. If the temperature rise is adequate, Q_2 will turn on, diverting the reference current I_R from the amplifier and hence shutting down the amplifier. The maximum permissible temperature rise ΔT can be determined from

$$[\Delta T(K_{DZ} + K_{Q3}) + V_{BE3}]\frac{R_1}{R_1 + R_2} = V_{BE2} = V_T \ln\left(\frac{I_R}{I_S}\right) \tag{3.142}$$

where K_{DZ} = temperature coefficient of the zener diode, in V/°C

K_{Q3} = temperature coefficient of transistor Q_3, in V/°C

V_{BE3} = quiescent B-E voltage of Q_3

KEY POINTS OF SECTION 3.11

- The transistors of an output stage are normally protected from the excessive current that results from short-circuiting. Protection can be provided by adding an extra transistor and a small collector resistor for each of the output transistors.
- Thermal protection is accomplished by taking advantage of the negative temperature coefficient of a transistor and the positive temperature coefficient of a zener diode.

3.12 Power Op-Amps

Op-amps have some desirable characteristics, such as a very high open-loop gain ($>10^5$), a very high input impedance (up to $10^9\ \Omega$), and a very low input biasing current. However, the AC output power of op-amps is generally low. High power can be obtained from a power amplifier consisting of an op-amp followed by a class AB buffer. The general structure of a power op-amp is shown in Fig. 3.52. The buffer stage consists of transistors Q_1, Q_2, Q_3, and Q_4. R_1 and R_2 bias transistors Q_1 and Q_2 such that $-V_{BE1} + V_{BE3} \approx 0$ and $V_{BE2} - V_{BE4} = 0$. Transistor Q_3 supplies the positive load current until the voltage across R_3 is sufficiently large to turn on Q_5. Then Q_5 supplies additional load current. Similarly, transistors Q_4 and Q_6 supply the negative load current. The stage formed by Q_5 and Q_6 supplies the additional load current and acts as a *current booster*. Emitter resistors R_{E1} and R_{E2} are used for biasing stability.

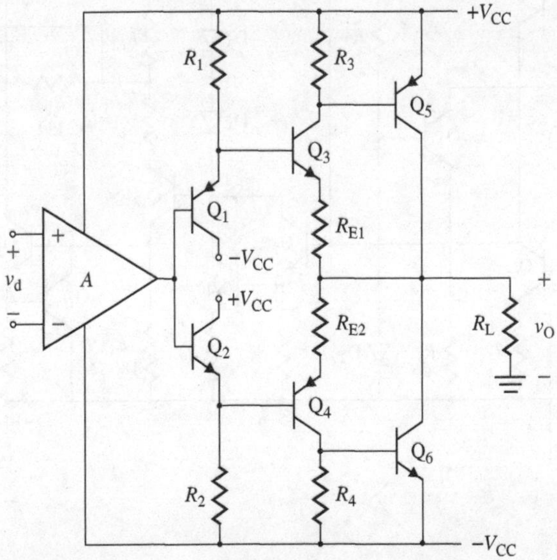

FIGURE 3.52 General structure of a power op-amp

3.12.1 IC Power Amplifiers

A variety of power amplifiers that combine a conventional op-amp chip with a current booster are available commercially. Some have internal negative feedback already applied to give a fixed closed-loop voltage gain; others do not have on-chip feedback. We will consider two such representative op-amps: the LH0021 op-amp and the LM380 op-amp, both manufactured by National Semiconductor.

Power Op-Amp LH0021

The schematic of power op-amp LH0021 is shown in Fig. 3.53. The LH0021 is designed to operate from a power supply of ±25 V. It is capable of a peak output voltage swing of about 12 V into a 10-Ω load over the entire frequency range of 15 kHz. The distortion of the output voltage is less than 1.6%. The circuit can be divided into three stages: a differential stage, a gain stage, and an output stage.

The differential input stage consists of transistors Q_1 through Q_4 biased by Q_7, which sinks the base currents of Q_3 and Q_4. Transistors Q_5 and Q_6 serve as the current mirror active load.

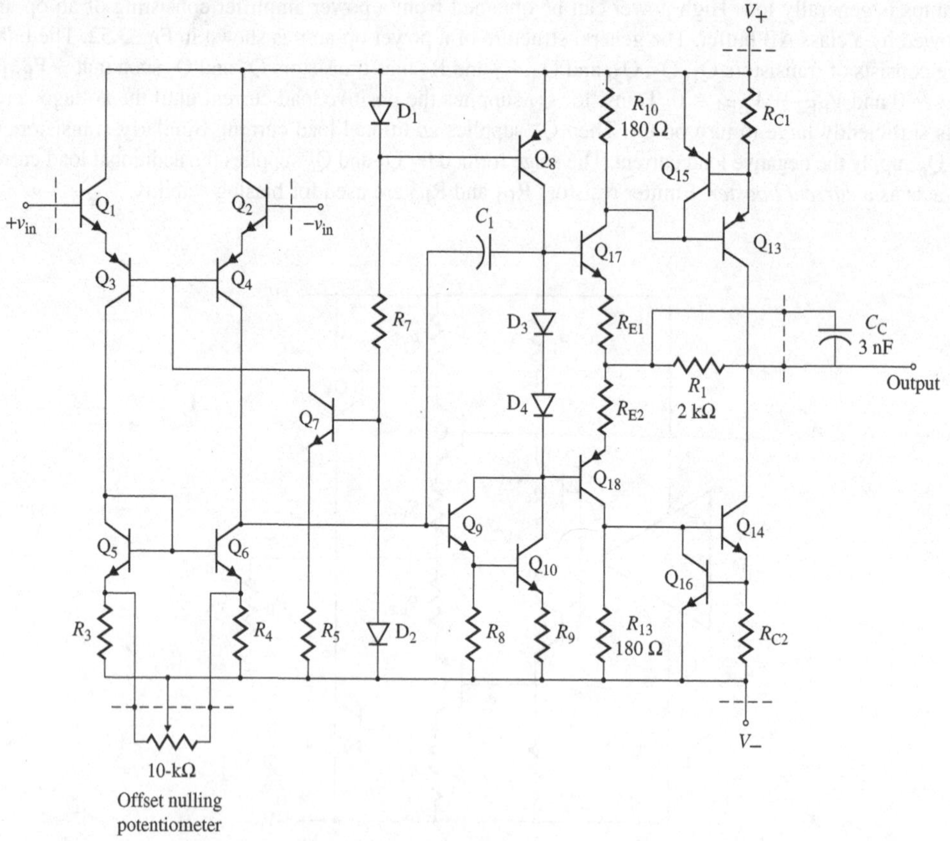

FIGURE 3.53 Power op-amp LH0021 (Courtesy of National Semiconductor, Inc.)

The gain stage is a common-emitter configuration and consists of Q_9 and Q_{10} connected as a Darlington pair. Transistor Q_8 serves as an active current source for this stage. Capacitor C_1 is the pole-splitting compensating capacitor connected in shunt-shunt feedback.

The output stage is a complementary class AB push-pull circuit. It consists of transistors Q_{13}, Q_{14}, Q_{15}, Q_{16}, Q_{17}, and Q_{18}. Diodes D_3 and D_4 provide the biasing voltage for class AB operation in order to minimize crossover distortion. Transistors Q_{13} and Q_{14} act as the current booster. Resistors R_{C1} and R_{C2} limit the currents through Q_{13} and Q_{14}, respectively, by turning on Q_{15} and Q_{16}. Resistor R_1 protects Q_{17} and Q_{18} by limiting the current flow through them.

A small external capacitor C_C is connected to offer a low impedance to a capacitive load. The combination of R_1 and the load capacitor does not form a low-pass *RC* network, and any phase delay of the output voltage can be avoided.

Power Op-Amp LM380

The schematic of power op-amp LM380 is shown in Fig. 3.54. The LM380 is designed to operate from a single power supply in the range of 12 V to 22 V. The output power can be as high as 5 W onto a 10-Ω load. The distortion of the output voltage is less than 3%. The circuit can be divided into three stages: a differential stage, a gain stage, and an output stage. An external capacitor C_x can be used to bypass the current source to improve the low-frequency response.

The differential input stage consists of *pnp* transistors Q_3 through Q_6. Transistor Q_3 is biased by Q_{10}, whereas transistor Q_4 is biased by a DC current from the output terminal through R_2. Under quiescent conditions (i.e., with an input voltage), the biasing currents for Q_3 and Q_4 will be equal. Thus, the current through and the voltage across R_3 will be zero. Transistors Q_5 and Q_6 serve as the current

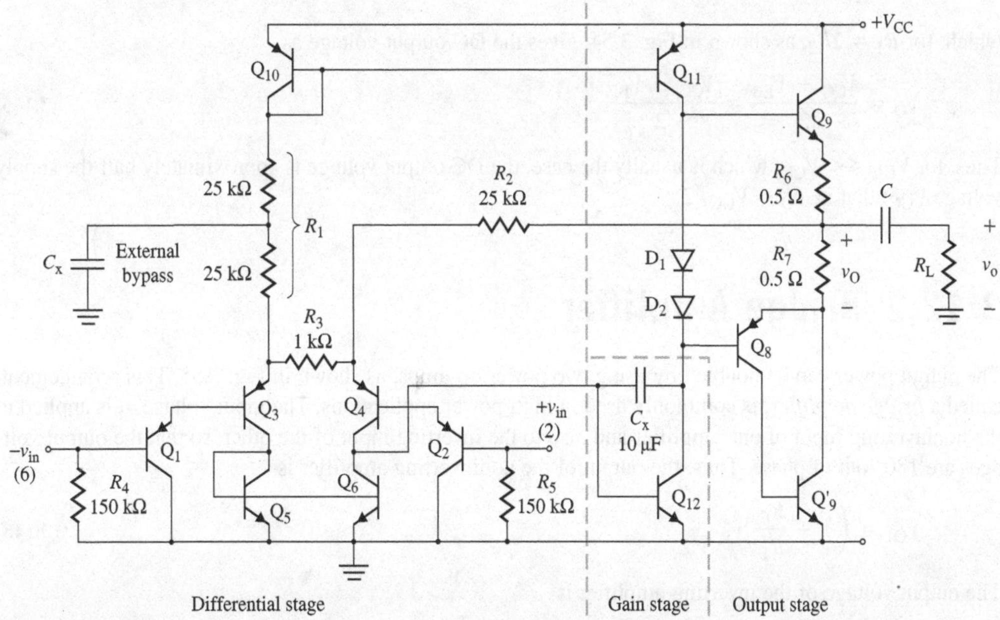

FIGURE 3.54 Power op-amp LM380 (Courtesy of National Semiconductor, Inc.)

mirror active load for this stage. The *pnp* transistors Q_1 and Q_2 act as emitter followers for input buffering. Resistors R_1 and R_2 provide DC paths to the ground for the base currents of Q_3 and Q_4.

The gain stage consisting of Q_{12} has the common-emitter configuration. Transistor Q_{11} serves as a current source active load for the gain stage. Capacitor C is the pole-splitting compensating capacitor intended to yield a wide bandwidth.

The output stage is a quasi-complementary class AB push-pull circuit. It consists of transistors Q_8, Q_9, and Q_9'. Diodes D_1 and D_2 provide the biasing voltage for class AB operation. Emitter resistors R_6 and R_7 give biasing stability.

Resistor R_2 provides DC feedback from the DC output voltage v_O to the emitter of Q_4. If for some reason v_O increases, then there will be a corresponding increase in the current through R_2 and the emitter current I_{E4} of Q_4, causing an increase in the collector current of Q_4. As a result, the voltage at the base of Q_{12} and its base current will increase. This, in turn, will increase the collector current of Q_{12} and reduce the base current of Q_9. Thus, v_O will be reduced.

To find v_O, let us assume that all transistors are identical and the base currents are negligible compared to the emitter currents. The emitter biasing current of Q_3 can be found approximately from

$$I_{E3} = \frac{V_{CC} - V_{EB10} - V_{EB3} - V_{EB1}}{R_1} = \frac{V_{CC} - 3V_{EB}}{R_1} \tag{3.143}$$

Also, the emitter current of Q_4 can be found from

$$I_{E4} = \frac{v_O - V_{EB4} - V_{EB2}}{R_2} = \frac{v_O - 2V_{EB}}{R_2} \tag{3.144}$$

where v_O is the DC output voltage. For $I_{E3} = I_{E4}$ (so that no current flows through R_3), we get

$$\frac{V_{CC} - 3V_{EB}}{R_1} = \frac{v_O - 2V_{EB}}{R_2}$$

which, for $R_1 = 2R_2$, as shown in Fig. 3.54, gives the DC output voltage as

$$v_O = \frac{V_{CC} + V_{EB}}{2} = \frac{V_{CC} - V_{BE}}{2} \tag{3.145}$$

Thus, for $V_{BE} << V_{CC}$, which is usually the case, the DC output voltage is approximately half the supply voltage V_{CC}; that is, $v_O \approx V_{CC}/2$.

3.12.2 Bridge Amplifier

The output power can be doubled by using two power op-amps, as shown in Fig. 3.55. This arrangement, called a *bridge amplifier*, is commonly used in high-power applications. The input voltage v_I is applied to the noninverting input of one amplifier and also to the inverting input of the other, so that the output voltages are 180° out of phase. Thus, the output of the noninverting amplifier is

$$v_{O1} = \left(1 + \frac{R_F}{R_1}\right) v_I \tag{3.146}$$

The output voltage of the inverting amplifier is

$$v_{O2} = -\frac{R_3}{R_2} v_I \tag{3.147}$$

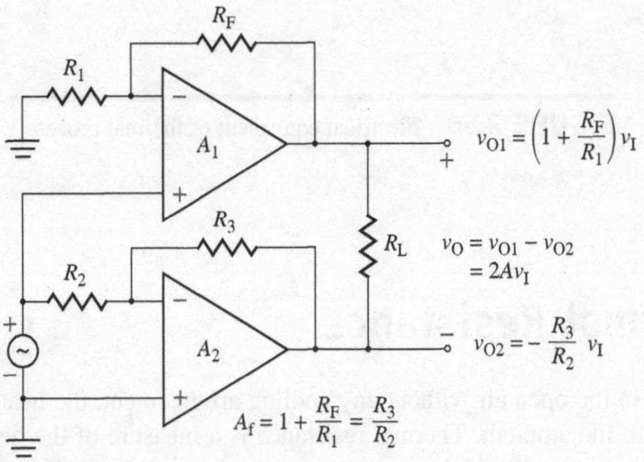

FIGURE 3.55 Bridge amplifier

The voltage across the load becomes

$$v_O = v_{O1} - v_{O2} = \left(1 + \frac{R_F}{R_1}\right)v_I + \frac{R_3}{R_2}v_I$$

which, for $R_3/R_2 = 1 + R_F/R_1 = A_f$, becomes

$$v_O = 2A_f v_I \tag{3.148}$$

where A_f is the closed-loop voltage gain of each amplifier.

KEY POINTS OF SECTION 3.12

- IC power amplifiers are known as power op-amps. They consist of a differential stage, a gain stage, and an output stage with a current booster.
- Power op-amps are normally used with feedback and are compensated for frequency response by internal capacitance in the gain stage.

3.13 Thermal Considerations

Power transistors dissipate a large amount of power. The power dissipation is converted to heat, which causes the temperature of the collection junction to rise. The physical structure, packaging, and specifications of transistors differ depending on their current-handling capability and power dissipation. The current rating of power transistors can go as high as 500 A, with power dissipation of up to 200 W, especially when the transistors are used as switching elements for power converters [6]. Power transistors must be protected from excessive temperature rise. The junction temperature T_J must be kept within a specified maximum $T_{J(max)}$ to avoid damage to the transistor. For silicon transistors, $T_{J(max)}$ is in the range of 150°C to 200°C.

T_J = junction temperature
T_A = ambient temperature

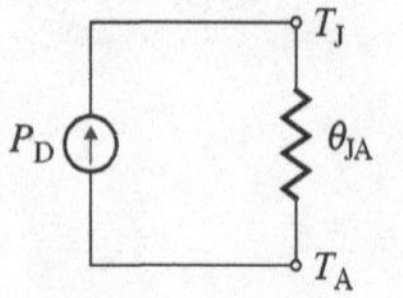

FIGURE 3.56 Electrical equivalent of thermal process

3.13.1 Thermal Resistance

If a transistor operates in the open air without any cooling arrangement, the heat will be transferred from the transistor junction to the ambient. Thermal resistance is a measure of the heat transfer. It is the temperature drop divided by power dissipation under steady-state conditions. Thus, the units are °C/W. The thermal resistance θ_{JA} for heat flow from the junction to the ambient is given by

$$\theta_{JA} = \frac{T_J - T_A}{P_D} \text{ (in °C/W)} \tag{3.149}$$

where T_J = junction temperature, in °C
T_A = ambient temperature, in °C

Equation (3.105) represents the heat-transfer process and is analogous to Ohm's law. Power dissipation corresponds to current, temperature difference to voltage difference, and thermal resistance to electrical resistance. Thus, the thermal process can be represented by an analogous electrical circuit, as shown in Fig. 3.56.

3.13.2 Heat Sink and Heat Flow

To keep the junction temperature below $T_{J(max)}$, the transistor is normally mounted on a heat sink, which facilitates the removal of heat from the device to the surrounding air. Typical heat sinks with devices attached are shown in Fig. 3.57. Heat is transferred from the device to the air by one of three methods:

1. Conduction from junction to case with thermal resistance θ_{JC} and from case to heat sink with thermal resistance θ_{CS}. The values of θ_{JC} and θ_{CS} depend on the cross section, length, and temperature difference across the conducting medium. The imperfect matching of adjacent surfaces will increase θ_{CS}.

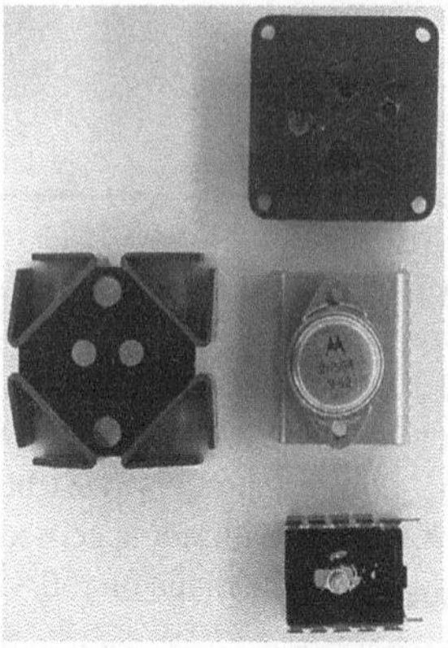

FIGURE 3.57 Transistors mounted on heat sinks

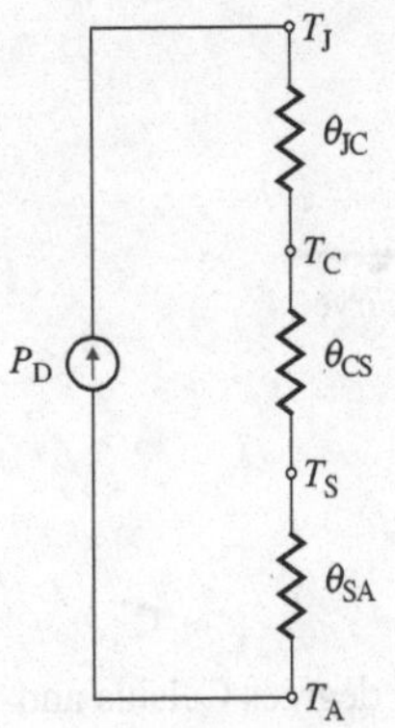

FIGURE 3.58 Thermal equivalent circuit

The value of θ_{CS} can be made small by coating the mated surfaces of the transistor and sink with a thermally conducting compound.

2. Convection from case to ambient with thermal resistance θ_{CA} and from heat sink to ambient with thermal resistance θ_{SA}. The values of θ_{CA} and θ_{SA} depend on surface condition, type of convecting fluid, velocity and characteristic of the fluid, and temperature difference between surface and fluid.
3. Radiation from cooling fins to the air. The heat transfer will depend on surface emissivity and area, as well as temperature difference between the radiating fins and the air.

The thermal equivalent circuit of a transistor on a heat sink is shown in Fig. 3.58. The power dissipation is related to the junction temperature T_J and the ambient temperature T_A by

$$T_J - T_A = P_D(\theta_{JC} + \theta_{CS} + \theta_{SA}) \tag{3.150}$$

where θ_{JC} = thermal resistance from junction to case, in °C/W

θ_{CS} = thermal resistance from case to sink, in °C/W

θ_{SA} = thermal resistance from sink to ambient, in °C/W

The values of θ_{JC} and θ_{CS} are specified in the manufacturer's data sheet. The circuit designer has to specify the required value of θ_{SA} for the heat sink. The value of θ_{SA} found from Eq. (3.150) may be too small to match any standard heat sink with natural air cooling. In this case, it might be necessary to cool the heat sink using liquid or forced air from a fan.

3.13.3 Power Dissipation versus Temperature

The ambient temperature T_A and the case temperature T_C are related to the power dissipation P_D by

$$T_C - T_A = P_D(\theta_{CS} + \theta_{SA}) \tag{3.151}$$

The transistor manufacturer normally specifies the maximum junction temperature $T_{J(max)}$, the maximum power dissipation $P_{D(max)}$ at a specified case temperature T_{C0} (usually 25°C), and the thermal resistance θ_{JC}. The manufacturer often provides a power-derating curve, as shown in Fig. 3.59, which gives the maximum allowable power dissipation if the case temperature is higher than 25°C. T_{C0} is the case temperature

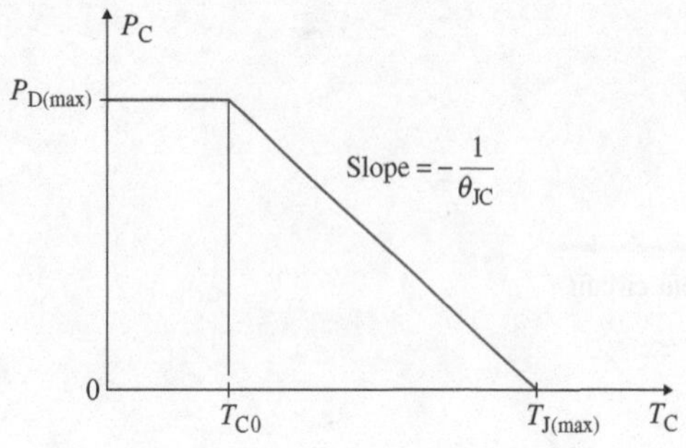

FIGURE 3.59 Power derating curve

at which the derating begins. $T_{C(max)}$ is the maximum value of the case temperature in degrees Celsius and is equal to $T_{J(max)}$. The power dissipation at a case temperature of T_C can be found from

$$P_D(T = T_C) = P_{D(max)} - \frac{P_{D(max)}}{T_{J(max)} - T_{C0}}(T_C - T_{C0}) \quad \text{for } T_C \geq T_{C0} \tag{3.152}$$

EXAMPLE 3.15

Finding the power dissipation of a transistor The power dissipation of a transistor is specified as $P_{D(max)} = 150$ W at $T_{C0} = 25°C$. The transistor is mounted on a heat sink. The thermal resistances are $\theta_{JC} = 0.5°C/W$, $\theta_{CS} = 0.2°C/W$, and $\theta_{SA} = 1.5°C/W$. If $T_{J(max)} = 200°C$ and $T_A = 45°C$, calculate the maximum permissible power dissipation of the transistor.

SOLUTION

We can find the maximum power dissipation P_D from Eq. (3.106):

$$200 - 45 = P_D(0.5 + 0.2 + 1.5) = 2.2P_{D(max)}$$

which gives

$$P_D = \frac{200 - 45}{2.2} = 70.5 \text{ W}$$

From $P_D = 70.5$ W, Eq. (3.107) gives the case temperature as

$$T_C = 45 + 70.5 \times (0.2 + 1.5) = 164.9°C$$

The corresponding power dissipation becomes

$$P_D(T_C = 164.9°C) = 150 - \frac{150}{200 - 25} \times (164.9 - 25) = 30.1 \text{ W}$$

In other words, for $P_D = 70.5$ W, $T_C = 164.9°C$, which in turn limits the power dissipation to 30.1 W. Therefore, P_D cannot be 70.5 W. We need to find the actual power. Substituting T_C from Eq. (3.107) into Eq. (3.108) yields

$$P_D = P_{D(max)} - \frac{P_{D(max)}}{T_{J(max)} - T_{C0}}[T_A + P_D(\theta_{CS} + \theta_{SA})]$$

which gives

$$P_D\left[1 + \frac{P_{D(max)}}{T_{J(max)} - T_{CO}}(\theta_{CS} + \theta_{SA})\right] = P_{D(max)}\left[1 - \frac{T_A}{T_{J(max)} - T_{CO}}\right]$$

Since

$$P_D\left[1 + \frac{150}{200 - 25}(0.2 + 1.5)\right] = 150 \times \left[1 - \frac{45}{200 - 25}\right]$$

the permissible power dissipation is $P_D = 45.35$ W.

▶ **NOTE** $\theta_{JC} = P_{D(max)}/(T_{J(max)} - T_{CO}) = 150/(200 - 25) = 0.857°C/W$. But in Example 3.15, $\theta_{JC} = 0.5°C/W$ was specified in order to illustrate the method for finding the actual power dissipation.

KEY POINTS OF SECTION 3.13

- Power dissipates from a transistor in the form of heat, which causes the temperature of the collection junction to rise. Power transistors or power op-amps are normally mounted on a heat sink, which provides a low thermal impedance for the heat flow.
- The maximum power dissipation is specified by the manufacturer at a specified case temperature T_{CO} (usually 25°C), which is not normally attainable. It is often necessary to determine the allowable power dissipation from a derating curve.

3.14 Design of Power Amplifiers

Since class B and class AB amplifiers eliminate the dead zone, they are generally used as the output stages of practical amplifiers. Thus, the design of a power amplifier is mainly the design of the output stage, and it involves the following steps:

Step 1. Identify the specifications of the output stage (e.g., output power P_L, load resistance R_L, and DC supply voltages V_{CC} and V_{EE} or V_{DD} and V_{SS}).

Step 2. Select the type of output operation, usually class B or class AB operation.

Step 3. Determine the voltage and current ratings of all transistors.

Step 4. Determine the values and power ratings of all resistors. Also, determine the turns ratio(s) of transformers in case of a transformer-coupled load.

Step 5. Select the type of DC-biasing circuit. Determine the specifications of active and passive components.

Step 6. Select the power transistors that will meet the voltage, current, and power requirements. Find their maximum junction temperature $T_{J(max)}$ and thermal resistances θ_{JC} and θ_{CS}.

Step 7. Determine the power dissipation of the transistors, and find the desired thermal resistance of the heat sink.

Step 8. Use PSpice/SPICE to simulate and verify your design, using the standard values of components with their tolerances.

Summary

A power amplifier generally forms the output stage of an audio-frequency amplifier. The requirements of power amplifiers are different from those of small-signal low-power amplifiers. Power amplifiers must deliver an appreciable amount of power to a low-impedance load, while at the same time creating very little distortion in the output signal. Power amplifiers are generally classified into six types: class A, class B, class AB, class C, class D, and class E. The efficiency of a class A amplifier is only 25%, that of a class B type is 50%, and that of a class AB push-pull type is 78.5%. Class D and E amplifiers can give 90% or higher efficiency. Complementary class AB push-pull amplifiers eliminate or reduce the distortion and dead zone in the output signal. The output stage is normally biased by an active current source in order to supply quiescent biasing current at zero input signal. A quasi-complementary amplifier, which uses a composite *pnp* transistor consisting of a *pnp* and an *npn* transistor, can increase the output power. Power IC op-amps generally consist of an op-amp followed by a class AB buffer. A power transistor must be mounted on a heat sink to limit the junction temperature to an acceptable value.

The communication systems generally require frequency-selective networks where the information is transmitted by modulating either the amplitude or the frequency of a high-frequency carrier wave. If the total bandwidth of the system is less than one-tenth of the carrier frequency, it is called a *narrowband* system. An auto-transformer is often used at the input and/or output side(s) for impedance matching.

References

1. P. R. Gray, P. J. Hurst, S. H. Lewis, and R. G. Meyer, *Analysis and Design of Analog Integrated Circuits*. New York: Wiley, 2001.
2. H. L. Krauss, C. W. Bostian, and F. H. Raab, *Solid State Radio Engineering*. New York: Wiley, 1981.
3. E. S. Oxner, *Power FETs and Their Applications*. Englewood Cliffs, NJ: Prentice Hall, 1982.
4. D. Foty, *MOSFET Modeling with SPICE.* Upper Saddle River, NJ: Prentice Hall, 1997.
5. M. H. Rashid, *Introduction to PSpice Using OrCAD for Circuits and Electronics*. Upper Saddle River, NJ: Prentice Hall, 2003.
6. S. Soclof, *Design and Applications of Analog Integrated Circuits*. Englewood Cliffs, NJ: Prentice Hall, 1991, Chapter 12.
7. M. H. Rashid, *Power Electronics—Circuits, Devices, and Applications*. Englewood Cliffs, NJ: Prentice Hall, 2003.
8. R. C. Jaeger and T. N. Blalock, *Microelectronic Circuit Design.* New York: McGraw Hill, 2008.
9. R. R. Spencer and Mohammed S. Ghausi, *Introduction to Electronic Circuit Designs.* Upper Saddle River, NJ: Prentice Hall, 2003.
10. R. A. Coiclaser, D. A. Neamen and C. F. Hawkins, *Electronic Circuit Analysis—Basic Principles*. New York, NY: John Wiley & Sons, 1984.

Review Questions

1. What are the six types of power amplifiers?
2. What are the major differences among the six types of power amplifiers?
3. What are the advantages of an emitter follower with active current–source biasing?
4. What is the limiting design condition for avoiding clipping on the output voltage?

5. What is a figure of merit for an amplifier?
6. What is a common-emitter class A amplifier?
7. What are the advantages and disadvantages of a common-emitter class A amplifier?
8. What is the maximum efficiency of a common-emitter class A amplifier?
9. What is a figure of merit for a common-emitter class A amplifier?
10. What is a transformer-coupled load class A amplifier?
11. What are the advantages and disadvantages of a transformer-coupled load class A amplifier?
12. What is the maximum efficiency of a transformer-coupled load class A amplifier?
13. What is a figure of merit for a transformer-coupled load class A amplifier?
14. What is a complementary class B push-pull amplifier?
15. What are the advantages and disadvantages of a complementary class B push-pull amplifier?
16. What is the maximum efficiency of a complementary class B push-pull amplifier?
17. What is a figure of merit for a complementary class B push-pull amplifier?
18. What is the cause of crossover distortion in a complementary class B push-pull amplifier?
19. What is a transformer-coupled load class B push-pull amplifier?
20. What are the advantages and disadvantages of a transformer-coupled load class B push-pull amplifier?
21. What is the maximum efficiency of a transformer-coupled load class B push-pull amplifier?
22. What is a figure of merit for a transformer-coupled load class B push-pull amplifier?
23. What are the methods of eliminating or minimizing crossover distortion in a complementary class B push-pull amplifier?
24. What is a complementary class AB push-pull amplifier?
25. What are the advantages and disadvantages of a complementary class AB push-pull amplifier?
26. What is the maximum efficiency of a complementary class AB push-pull amplifier?
27. What is a figure of merit for a complementary class AB push-pull amplifier?
28. What are the methods for DC biasing an output stage?
29. What are the methods for providing short-circuit and thermal protection for an output stage?
30. What is an IC power amplifier?
31. What is the thermal equivalent circuit of a transistor?
32. What is the power derating curve of a power transistor?
33. What is the purpose of a heat sink for a power transistor?
34. What is class C amplifier?
35. What is class D amplifier?
36. What is class E amplifier?
37. What is tuned amplifier?
38. What are the disadvantages of tuned amplifiers?
39. What are the conditions for the instability of tuned amplifiers?
40. What is the function of autotransformer in tuned amplifiers?
41. What are the differences between the synchronous and the stagger-tuned amplifiers?
42. What are the advantages and disadvantages of the synchronous-tuned amplifiers?
43. What are the advantages and disadvantages of the stagger-tuned amplifiers?

Problems

The symbol **D** indicates that a problem is a design problem. The symbol **P** indicates that you can check the solution to a problem using PSpice/SPICE or Electronics Workbench.

Assume that the PSpice model parameters for the diodes are

```
IS=100E-15 BV=100 IBV=100E-13
```

and for the transistors are

```
IS=100E-15 IBV=100E-14 BF=100 VJE=0.8 VA=100
```

(unless specified).

3.4 *Class A Amplifiers*

3.1 D P a. Design an emitter follower as shown in Fig. 3.3(b). Assume $V_{CC} = 15$ V, $V_{BE} = 0.7$ V, $V_{CE(sat)} = 0.5$ V, $I_R = 10$ mA, and $R_L = 1$ kΩ. Assume identical transistors of $\beta_F = h_{fe} = 100$. Find the voltage, current, and power rating of the transistors.

b. Use PSpice/SPICE to check your design by plotting the transfer function.

3.2 For the emitter follower in Prob. 3.1, determine the critical value of load resistance to avoid clipping (or distortion), and calculate the peak-to-peak output voltage swing V_{pp} if $R_L = 2$ kΩ.

3.3 The parameters of the emitter follower in Fig. 3.3(b) are $V_{CC} = 12$ V, $R_1 = 2.5$ kΩ, and $R_L = 750$ Ω. The transistors are identical, and their parameters are $V_{BE} = 0.7$ V, $V_{CE(sat)} = 0.5$ V, and $\beta_F = h_{fe} = 100$. Determine **(a)** the peak voltage and current of transistor Q_1 and **(b)** the average output power P_L.

3.4 For the emitter follower in Prob. 3.3, calculate **(a)** the power efficiency, **(b)** the value of R_L for maximum efficiency, and **(c)** the corresponding value of efficiency η_{max}.

3.5 D P a. Design a class A amplifier, as shown in Fig. 3.7(a), for high efficiency. The audio output power is 48 W. The available supply voltage is 12 V. Assume transistors of $\beta_F = h_{fe} = 100$. Find the voltage, current, and power rating of the transistors.

b. Use PSpice/SPICE to check your design by plotting the collector current and voltage.

3.6 The parameters of the class A amplifier in Fig. 3.9(a) are $V_{CC} = 15$ V, $R_1 = 15$ kΩ, and $R_L = 1$ kΩ. The transistors are identical, and their parameters are $V_{BE} = 0.7$ V, $V_{CE(sat)} = 0.5$ V, and $\beta_F = h_{fe} = 100$. Determine **(a)** the peak voltage and current of transistor Q_1, **(b)** the average output power P_L, and **(c)** the efficiency η.

3.7 D P a. Design a transformer-coupled load class A amplifier, as shown in Fig. 3.10(a), to have high efficiency and to supply an output power of $P_L = 32$ W at a load resistance of $R_L = 8$ Ω. Assume a DC supply voltage of $V_{CC} = 15$ V. Find the voltage, current, and power rating of the transistors and the transformer turns ratio. Assume $\beta_F = 100$ and $V_{CE(sat)} = 0.5$ V.

b. Use PSpice/SPICE to check your design by plotting the collector current and voltage.

3.8 The parameters of the transformer-coupled load class A amplifier in Fig. 3.10(a) are $V_{CC} = 15$ V and $R_L = 16$ Ω. The transistor parameters are $V_{BE} = 0.7$ V, $V_{CE(sat)} = 0.5$ V, and $h_{fe} = 100$. The transformer turns ratio is 2:1. Determine **(a)** the peak voltage and current of transistor Q_1, **(b)** the average output power P_L, and **(c)** the efficiency η.

3.5 *Class B Push-Pull Amplifiers*

3.9 D P a. Design a complementary class B push-pull amplifier, as shown in Fig. 3.11(a), to supply an output power of $P_L = 16$ W at a load resistance of $R_L = 4$ Ω. Assume a DC supply voltage of $V_{CC} = 15$ V and transistors of $\beta_F = h_{fe} = 100$ and $V_{BE} = 0.7$ V. Find the voltage, current, and power rating of the transistors.

b. Use PSpice/SPICE to check your design by plotting the transfer function, the output voltage, and the load current.

3.10 For the amplifier in Prob. 3.9, calculate the efficiency and power dissipation of each transistor.

3.11 Calculate the power efficiency η and power dissipation P_D of each transistor in the complementary class B push-pull output stage in Fig. 3.11(a) if $V_{CC} = 15$ V and $R_L = 10\ \Omega$. The parameters of the transistors are $\beta_F = h_{fe} = 150$, $V_{CE(sat)} = 0.2$ V, and $V_{BE} = 0.7$.

3.12 Calculate the power efficiency η and power dissipation P_D of each transistor in the complementary class B push-pull output stage in Fig. 3.11(a) if $V_{CC} = 12$ V and $R_L = 50\ \Omega$. The parameters of the transistors are $\beta_F = h_{fe} = 100$, $V_{CE(sat)} = 0.2$ V, and $V_{BE} = 0.7$.

3.13 Design a transformer-coupled class B push-pull amplifier, as shown in Fig. 3.17(a), to supply an output power of $P_L = 32$ W at a load resistance of $R_L = 8\ \Omega$. Assume a DC supply voltage of $V_{CC} = 15$ V and transistors of $\beta_F = h_{fe} = 100$ and $V_{BE} = 0.7$ V.

3.14 The parameters of the transformer-coupled class B push-pull amplifier in Fig. 3.17(a) are $V_{CC} = 15$ V and $R_L = 8\ \Omega$. The transistor parameters are $V_{BE} = 0.7$ V, $V_{CE(sat)} = 0.5$ V, and $\beta_F = h_{fe} = 100$. The transformer turns ratio is 2:1. Determine **(a)** the peak voltage and current of transistor Q_1, **(b)** the average output power P_L, and **(c)** the efficiency η.

3.6 *Complementary Class AB Push-Pull Amplifiers*

3.15 **a.** Design a biasing circuit for the class AB amplifier of Fig. 3.20(c) to supply a maximum output power of $P_{L(max)} = 20$ W. The quiescent biasing current I_Q is 2 mA. Assume a DC supply voltage of $V_{CC} = 15$ V. The diode parameters are $I_S = 10^{-13}$ A and $I_{D(min)} = 1$ mA to ensure conduction. The transistor parameters are $h_{fe} = 50$, $V_{BE} = 0.7$ V, and $V_{CE(sat)} = 0.2$ V.

b. Find the biasing voltage V_{BB} for $v_O = 0$ and 3.8 V.

3.16 **a.** Design an active current source for the class AB amplifier in Fig. 3.21(a) in order to provide a biasing current of $I_R = 10$ mA. Assume $V_{CC} = 15$ V, $V_{BE} = 0.7$ V, and $R_L = 50\ \Omega$. The diode parameters are $I_S = 10^{-13}$ A and $I_{D(min)} = 1$ mA to ensure conduction. The transistor parameters are $h_{fe} = 50$, $V_{BE} = 0.7$ V, and $V_{CE(sat)} = 0.2$ V.

b. Use PSpice/SPICE to plot the transfer characteristic and the instantaneous i_N, i_P, and i_O for $v_I = 5 \sin (2000\pi t)$.

3.17 **a.** Design a V_{BE} multiplier for the class AB amplifier in Fig. 3.24 in order to provide a biasing current of $I_R = 10$ mA. Assume $V_{CC} = 15$ V, $V_{BE} = 0.7$ V, and $R_L = 50\ \Omega$. Assume a minimum current of $I_{M(min)} = 1$ mA to the multiplier and $I_Q = 2$ mA. The transistor parameters are $\beta_F = h_{fe} = 50$, $V_{BE} = 0.7$ V, and $V_{CE(sat)} = 0.2$ V.

b. Use PSpice/SPICE to plot the transfer characteristic and the instantaneous i_N, i_P, and i_O for $v_I = 5 \sin (2000\pi t)$.

3.11 *Tuned Amplifiers*

3.18 Design a tuned amplifier as shown in Fig. 3.31(a) to give a mid-gain of $A_{mid} = -40$, a center frequency of $f_o = 1$ MHz, and a bandwidth of $BW = 20$ kHz. The circuit parameters are: $g_m = 4$ mA/V, $r_o = 40$ kΩ, $R_L = 100$ kΩ, $C_i = 40$ pF, and $C_x = 10$ pF.

3.19 Design a tuned amplifier as shown in Fig. 3.31(a) to give a mid-gain of $A_{mid} = -40$, a center frequency of $f_o = 1$ MHz, and a bandwidth of $BW = 10$ kHz. The circuit parameters are: $g_m = 4$ mA/V, $r_o = 40$ kΩ, $R_L = 50$ kΩ, $C_i = 40$ pF and $C_x = 10$ pF.

3.20 For Example 3.8, find the mid-gain A_{mid}, Q and BW (a) if the inductor has $Q = 100$ and (b) an autotransformer of $a = 10$ is connected to the load side. The circuit parameters are: $f_o = 1$ MHz, $g_m = 4$ mA/V, $r_o = 40$ kΩ, $R_D = 25$ kΩ, $R_L = 100$ kΩ, $C_i = 40$ pF, $C_x = 10$ pF, $R_{eq} = 10$ kΩ, $C_{eq} = 1.592$ nF, $C = 1.892$ nF, and $L = 15.92$ μH.

3.21 Design a tuned BJT amplifier as shown in Fig. 3.39(a) to give a center frequency of $f_o = 10$ MHz and a bandwidth of $BW = 100$ kHz. The circuit parameters are: $g_m = 15$ mA/V, $\alpha_{i2} = 0.9$, $r_{\pi 1} = 8.6$ kΩ, $r_o = 40$ kΩ, $R_L = 100$ kΩ, $R_s = 500$ Ω, $C_i = 40$ pF, and $C_x = 10$ pF.

3.22 Design a stagger-tuned BJT amplifier as shown in Fig. 3.39(a) to give a center frequency of $f_o = 10$ MHz and a bandwidth of $BW_2 = 500$ kHz. The circuit parameters are: $g_m = 4$ mA/V, $\alpha_{i2} = 0.9$, $r_{\pi 1} = 8.6$ kΩ, $r_o = 100$ kΩ, $R_L = 40$ kΩ, $R_s = 500$ Ω, $C_i = 40$ pF, and $C_x = 10$ pF.

3.8–3.10 Class C, D, and E Amplifiers

3.23 The gate voltage of the class C amplifier in Fig. 3.42(a) is $v_G = 2.5 + 1.0 \sin(2\pi f_s t)$ with $f_s = 1.5$ MHz. The DC supply voltages are $V_{DD} = \pm 15$ V. The load resistance $R_L = 50$ Ω, which is connected through a transformer of a turns ratio of $n = 1{:}1$.

a. Find the values of L and C to form a tuned resonant circuit.

b. Use PSpice to plot the voltage $v_o(t)$, the transistor drain current $i_D(t)$, and the transistor drain voltage $v_D(t)$for varying the gain voltage component $V_m = 3$ V to 4 V with an increment of 0.5 V.

c. Use the PSpice parameters of the NMOS IRF150 whose $V_t = 2.84$ V.

3.24 The gate voltage of the class C amplifier in Fig. 3.42(a) is $v_G = 2.5 + 1.0 \sin(2\pi f_s t)$ with $f_s = 1.5$ MHz. The DC supply voltages are $V_{DD} = \pm 15$ V. The load resistance $R_L = 150$ Ω, which is connected through a transformer of a turns ratio of $n = 2{:}1$.

a. Find the values of L and C to form a tuned resonant circuit.

b. Use PSpice to plot the voltage $v_o(t)$, the transistor drain current $i_D(t)$, and the transistor drain voltage $v_D(t)$ for varying the gain voltage component $V_m = 3$ V to 4 V with an increment of 0.5 V.

c. Use the PSpice parameters of the NMOS IRF150 whose $V_t = 2.84$ V.

3.25 The gate voltage of the class C amplifier in Fig. 3.42(a) is $v_G = 2.5 + 1.5 \sin(2\pi f_s t)$ with $f_s = 1.5$ MHz. The DC supply voltages are $V_{DD} = \pm 15$ V. The load resistance $R_L = 50$ Ω, which is connected through a transformer of a turns ratio of $n = 2{:}1$.

a. Find the values of L and C to form a tuned resonant circuit.

b. Use PSpice to plot the voltage $v_o(t)$, the transistor drain current $i_D(t)$, and the transistor drain voltage $v_D(t)$ for varying the gain voltage component $V_m = 3$ V to 4 V with an increment of 0.5 V.

c. Use the PSpice parameters of the NMOS IRF150 whose $V_t = 2.84$ V.

3.26 The gate voltage of the class D amplifier in Fig. 3.45(a) is generated by modulating a sinusoidal reference $V_r = 1$ V at 25 kHz with a carrier signal $V_{cr} = 1$ V at $f_{cr} = 25 f_s$. The DC supply voltages are $V_{DD} = \pm 5$ V. The load resistance $R_L = 8$ Ω speaker.

a. Find the values of L and C to form a tuned resonant circuit.

b. Use PSpice to plot the gate-modulating voltage $v_G(t)$, the transistor output voltage $v_D(t)$, and the transistor output voltage $v_D(t)$ by a modulation index of $m = 0.5$ and 0.9.

c. Use the PSpice parameters of the NMOS IRF150 whose $V_t = 2.84$ V and the PMOS IRF9140 whose $V_t = 3.67$ V.

3.27 The gate voltage of the class D amplifier in Fig. 3.45(a) is generated by modulating a sinusoidal reference $V_r = 1$ V at 40 kHz with a carrier signal $V_{cr} = 1$ V at $f_{cr} = 30 f_s$. The DC supply voltages are $V_{DD} = \pm 5$ V. The load resistance $R_L = 16$ Ω speaker.

a. Find the values of L and C to form a tuned resonant circuit.

b. Use PSpice to plot the gate-modulating voltage $v_G(t)$, the transistor output voltage $v_D(t)$, and the transistor output voltage $v_D(t)$ by a modulation index of $m = 0.5$ and 0.9.

c. Use the PSpice parameters of the NMOS IRF150 whose $V_t = 2.84$ V and the PMOS IRF9140 whose $V_t = 3.67$ V.

3.28 The gate voltage of the class D amplifier in Fig. 3.45(a) is generated by modulating a sinusoidal reference $V_r = 1$ V at 20 kHz with a carrier signal $V_{cr} = 1$ V at $f_{cr} = 20f_s$. The DC supply voltages are $V_{DD} = \pm 5$ V. The load resistance $R_L = 16\ \Omega$ speaker.

a. Find the values of L and C to form a tuned resonant circuit.

b. Use PSpice to plot the gate-modulating voltage $v_G(t)$, the transistor output voltage $v_D(t)$, and the transistor output voltage $v_D(t)$ by a modulation index of $m = 0.5$ and 0.9.

c. Use the PSpice parameters of the NMOS IRF150 whose $V_t = 2.84$ V and the PMOS IRF9140 whose $V_t = 3.67$ V.

3.29 The gate voltage of the class E amplifier in Fig. 3.48 is a sinusoidal voltage of 5 V (peak) at 100 kHz. The DC supply voltages are $V_{DD} = 15$ V. The load resistance $R = 16\ \Omega$.

a. Find the optimum values of L_D, C_D, L, and C.

b. Use PSpice to plot the gate voltage $v_G(t)$, the transistor drain current $i_D(t)$, the transistor output voltage $v_D(t)$, and the output voltage $v_o(t)$ for peak gate voltages of 4 V and 4.5 V.

c. Use the PSpice parameters of the NMOS IRF150 whose $V_t = 2.84$ V.

3.30 The gate voltage of the class E amplifier in Fig. 3.48 is a sinusoidal voltage of 4 V (peak) at 50 kHz. The DC supply voltages are $V_{DD} = 12$ V. The load resistance $R = 8\ \Omega$.

a. Find the optimum values of L_D, C_D, L, and C.

b. Use PSpice to plot the gate voltage $v_G(t)$, the transistor drain current $i_D(t)$, the transistor output voltage $v_D(t)$, and the output voltage $v_o(t)$ for peak gate voltages of 4 V and 4.5 V.

c. Use the PSpice parameters of the NMOS IRF150 whose $V_t = 2.84$ V.

3.31 The gate voltage of the class E amplifier in Fig. 3.48 is a sinusoidal voltage of 4 V (peak) at 150 kHz. The DC supply voltages are $V_{DD} = 12$ V. The load resistance $R = 8\ \Omega$.

a. Find the optimum values of L_D, C_D, L, and C.

b. Use PSpice to plot the gate voltage $v_G(t)$, the transistor drain current $i_D(t)$, the transistor output voltage $v_D(t)$, and the output voltage $v_o(t)$ for peak gate voltages of 4 V and 4.5 V.

c. Use the PSpice parameters of the NMOS IRF150 whose $V_t = 2.84$ V.

3.11 *Short-Circuit and Thermal Protection*

3.32 **D** Design a thermal protection circuit as shown in Fig. 3.51 to limit the maximum temperature rise to $\Delta T = 110$°C. The temperature coefficients of the diode and the transistors are $K_{DZ} = K_{Q3} = 2.5$ mV/°C. The normal biasing current is $I_R = 5$ mA, and the collector current of Q_3 is 1 mA. Assume $V_{BEQ2} = 0.7$ V and $V_{CC} = 15$ V.

3.12 *Power Op-Amps*

3.33 **D** Use LM380 power op-amps to design a bridge amplifier as shown in Fig. 3.55. The output power is $P_L = 20$ W at $R_L = 4\ \Omega$. Assume $V_{CC} = 15$ V and $v_I = 100$ mV (peak).

3.34 The gain and output stages of an LM380 power op-amp are shown in Fig. P3.34. Determine the small-signal differential voltage gain $A_{id} = v_O/v_I$, the differential input resistance R_{id}, and the output resistance R_o. The diode parameters are $I_S = 10^{-13}$ A and $I_{D(min)} = 1$ mA to ensure conduction. The transistor parameters are $\beta_F = h_{fe} = 50$, $V_{BE} = 0.7$ V, $V_A = 40$ V, and $V_{CE(sat)} = 0.2$ V. Assume a DC supply voltage of $V_{CC} = 15$ V and $R_L = 10$ kΩ.

FIGURE P3.34

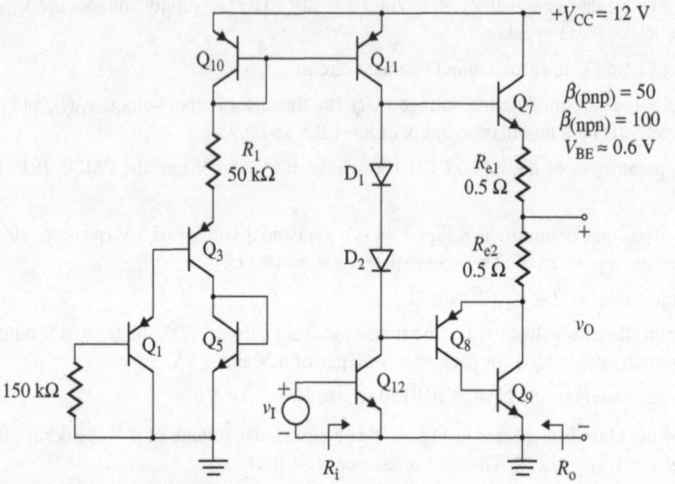

3.35 The power op-amp shown in Fig. 3.52 uses DC supply voltage $V_{CC} = -V_{EE} = 15$ V and transistors of type Q2N2222 having $I_s = 14.34 \times 10^{-15}$ and $\beta_F = 255.9$.

a. Determine the output voltage v_O and the input resistance R_i for an input voltage of $v_{in} = 5$ V.

b. Use PSpice to verify the results in part (a) and determine the high cutoff frequency f_H by plotting the frequency response.

3.36 Design the bridge amplifier shown in Fig. 3.55 to obtain a voltage gain $A_f = 250$ V/V. Assume (D) $V_{CC} = -V_{EE} = -12$ V. Use PSpice to verify the results.

3.13 *Thermal Considerations*

3.37 The power dissipation of a transistor is specified as $P_{D(max)} = 250$ W at $T_{CO} = 25$°C. The transistor is mounted on a heat sink. The thermal resistances are $\theta_{JC} = 0.7$°C/W, $\theta_{CS} = 0.2$°C/W, and $\theta_{SA} = 0.8$°C/W. If $T_{J(max)} = 200$°C and $T_A = 45$°C, calculate the maximum permissible power dissipation of the transistor.

3.38 The power dissipation of a transistor is specified as $P_{D(max)} = 290$ W at $T_{CO} = 25$°C. The transistor is mounted on a heat sink. The thermal resistances of the transistor are $\theta_{JC} = 0.6$°C/W and $\theta_{CS} = 0.2$°C/W. If $T_{J(max)} =$ 200°C, $T_C = 150$°C, and $T_A = 45$°C, calculate the required thermal resistance of the heat sink θ_{SA}.

CHAPTER 4

ACTIVE FILTERS

Learning Outcomes

After completing this chapter, students should be able to do the following:

- List the advantages of active filters over passive filters.
- Describe the characteristics and types of active filters.
- Determine the pole locations for the Butterworth function.
- Analyze active filters.
- Select an active filter to meet desired frequency requirements.
- Design an active filter to meet desired specifications.

Symbols and Their Meanings

Symbol	Meaning
V_i, V_o	Input and output voltages of a filter
BW	Bandwidth of a filter
f_{clk}	Clock frequency, in hertz
$p_1, \ldots, p_n$, $z_1, \ldots, z_n$	Poles and zeros of a transfer function
$H(s)$, $D(s)$, $N(s)$	Transfer function, and denominator and numerator of a transfer function
H_n	Transfer function of an nth-order filter
Q, K	Quality factor and constant parameter of a filter

Symbol	Meaning
f_o, f_c, f_L, f_H	Cutoff, center, low cutoff, and high cutoff frequencies of a filter
R_n, C_n	Normalized resistance and capacitance of a filter
θ_p, θ_z	Pole and zero angles

4.1 Introduction

In electrical engineering, a filter is a frequency-selective circuit that passes a specified band of frequencies and blocks or attenuates signals of frequencies outside this band. These signals are usually voltages. Filters that employ only passive elements such as capacitors, inductors, and resistors are called *passive filters*. Filters that make use of the properties of op-amps in addition to resistors and capacitors are called *active filters* or more often *analog filters*, in contrast to *digital filters*. Both analog and digital filters can be implemented in the same IC chip. This chapter introduces active filters and deals with the analysis and design of simple circuit topologies. Because of their practical importance, analog filters are often covered in a single course [1, 2].

4.2 Active versus Passive Filters

Both active and passive filters are used in electronic circuits. However, active filters offer the following advantages over passive filters:

- *Flexibility of gain and frequency adjustment*: Since op-amps can provide a voltage gain, the input signal in active filters is not attenuated as it is in passive filters. It is easy to adjust or tune active filters.
- *No loading effect*: Because of the high input resistance and low output resistance of op-amps, active filters do not cause loading of the input source or the load.
- *Cost and size*: Active filters are less expensive than passive filters because of the availability of low-cost op-amps and the absence of inductors.
- *Parasitics*: Parasitics are reduced in active filters because of their smaller size.
- *Digital integration*: Analog filters and digital circuitry can be implemented on the same IC chip.
- *Filtering functions*: Active filters can realize a wider range of filtering functions than passive filters.
- *Gain*: An active filter can provide gain, whereas a passive filter often exhibits a significant loss.

Active filters also have some disadvantages:

- *Bandwidth*: Active components have a finite bandwidth, which limits the applications of active filters to the audio-frequency range. Passive filters do not have such an upper-frequency limitation and can be used up to approximately 500 MHz.
- *Drifts*: Active filters are sensitive to component drifts due to manufacturing tolerances or environmental changes; in contrast, passive filters are less affected by such factors.
- *Power supplies*: Active filters require power supplies, whereas passive filters do not.
- *Distortion*: Active filters can handle only a limited range of signal magnitudes; beyond this range, they introduce unacceptable distortion.
- *Noise*: Active filters use resistors and active elements, which produce electrical noise.

In general, the advantages of active filters outweigh their disadvantages in voice and data communication applications. Active filters are used in almost all sophisticated electronic systems for communication and signal processing, such as television, telephone, radar, space satellite, and biomedical equipment. However, passive filters are still widely used.

4.3 Types of Active Filters

Let $V_i \angle 0$ be the input voltage to the filtering circuit shown in Fig. 4.1. The output voltage V_o and its phase shift θ will depend on the frequency ω. If two voltages are converted to Laplace's domain of s, the ratio of the output voltage $V_o(s)$ to the input voltage $V_i(s)$ is referred to as the *voltage transfer function H*(s):

$$H(s) = \frac{V_o(s)}{V_i(s)}$$

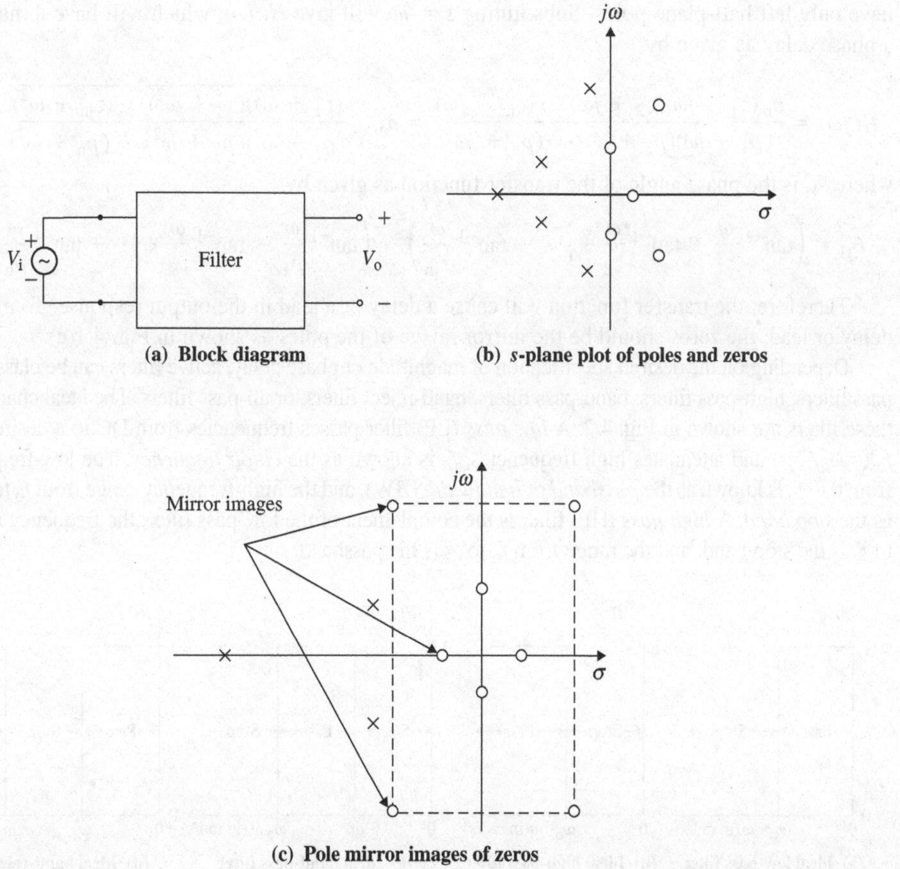

FIGURE 4.1 Filtering circuit

The transfer function $H(s)$ can be expressed in the general form

$$H(s) = \frac{a_m s^m + \cdots + a_2 s^2 + a_1 s + a_0}{s^n + \cdots + b_2 s^2 + b_1 s + b_0} \quad (\text{for } n \geq m) \tag{4.1}$$

whose coefficients are determined so as to meet the desired filter specifications. Substituting $s = j\omega$ will give $H(j\omega)$, which will have a magnitude and a phase delay. The denominator and the numerator of Eq. (4.1) are polynomials in s with real coefficients. If these polynomials are factored, $H(s)$ can be written as [3, 4]

$$H(s) = \frac{N(s)}{D(s)} = \frac{a_m(s + z_1)(s + z_2)\cdots(s + z_m)}{(s + p_1)(s + p_2)\cdots(s + p_n)} \tag{4.2}$$

The expression $z_1, z_2, \ldots, z_m$ is referred to as the *zeros* of $H(s)$ because $H(s) = 0$ when $s = -z_m$ and the expression $p_1, p_2, \ldots, p_n$ is referred to as the *poles* of $H(s)$ because $H(s) = \infty$ when $s = -p_n$. The poles and zeros can be plotted in the complex s-plane as shown in Fig. 4.1(b), where $s = \sigma + j\omega$ is considered as one set. Since the coefficients of $H(s)$ are all real, the poles and zeros occur in conjugate pairs.

For stability all poles must lie in the left half-plane. When the poles lie on the $j\omega$-axis and are simple, the network oscillates; when the poles lie in the right half-plane, the responses grow exponentially with time. Thus, for stability reasons, the characteristic polynomial $D(s)$ of a realizable system must have only left half-plane poles. Substituting $s = j\omega$ will give $H(j\omega)$, which will have a magnitude and a phase delay as given by

$$H(j\omega) = \frac{a_m(z_1 + j\omega)(z_2 + j\omega)\cdots(z_m + j\omega)}{(p_1 + j\omega)(p_2 + j\omega)\cdots(p_n + j\omega)} = a_m\sqrt{\frac{(z_1^2 + \omega^2)(z_2 + \omega^2)\cdots(z_m + \omega^2)}{(p_1 + \omega^2)(p_2 + \omega^2)\cdots(p_n + \omega^2)}}\angle\theta_H \tag{4.3}$$

where θ_H is the phase angle of the transfer function as given by

$$\theta_H = \left(\tan^{-1}\frac{\omega}{z_1} + \tan^{-1}\frac{\omega}{z_2} + \cdots + \tan^{-1}\frac{\omega}{z_m}\right) - \left(\tan^{-1}\frac{\omega}{p_1} + \tan^{-1}\frac{\omega}{p_2} + \cdots + \tan^{-1}\frac{\omega}{p_n}\right) \tag{4.4}$$

Therefore, the transfer function will cause a delay or a lead in the output response. To minimize this delay or lead, the zeros should be the mirror image of the poles as shown in Fig. 4.1(c).

Depending on the desired specification of magnitude or phase delay, active filters can be classified as low-pass filters, high-pass filters, band-pass filters, band-reject filters, or all-pass filters. The ideal characteristics of these filters are shown in Fig. 4.2. A *low-pass* (LP) filter passes frequencies from DC to a desired frequency f_o $(=\omega_o/2\pi)$ and attenuates high frequencies. f_o is known as the *cutoff frequency*. The low-frequency range from 0 to f_o is known as the *passband* or *bandwidth* (BW), and the high-frequency range from f_o to ∞ is known as the *stop band*. A *high-pass* (HP) filter is the complement of the low-pass filter; the frequency range from 0 to f_o is the stop band, and the range from f_o to ∞ is the passband.

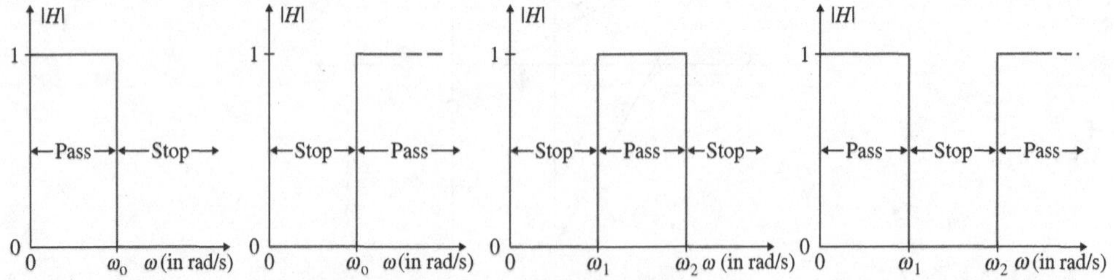

(a) Ideal low-pass filter **(b) Ideal high-pass filter** **(c) Ideal band-pass filter** **(d) Ideal band-reject filter**

FIGURE 4.2 Ideal filter characteristics

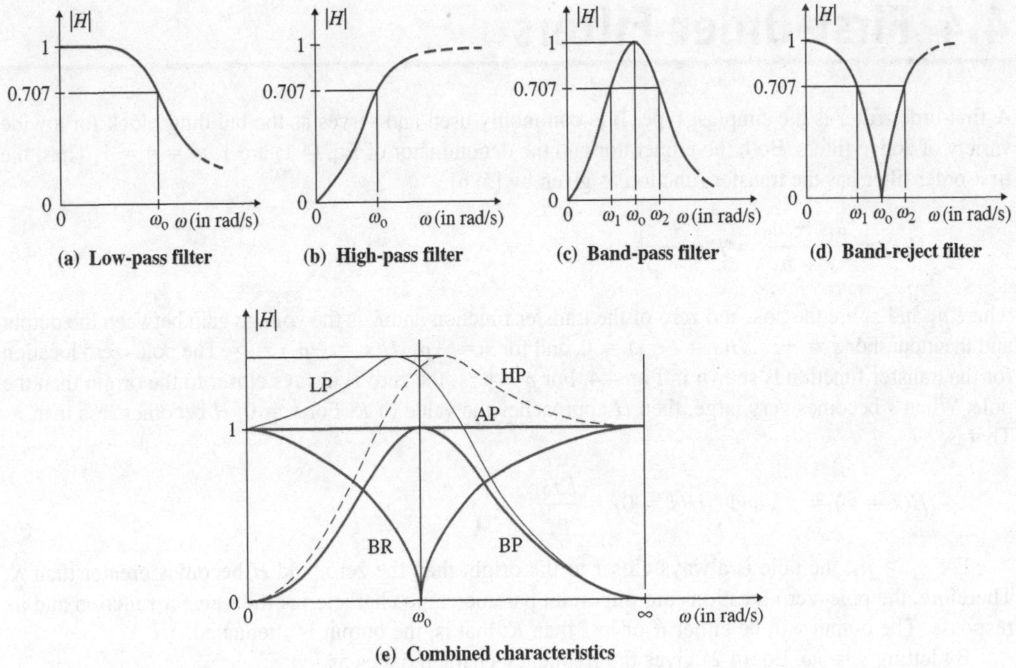

FIGURE 4.3 Realistic filter characteristics

A *band-pass* (BP) filter passes frequencies from f_1 to f_2 and stops all other frequencies. A band-reject (BR) filter is the complement of a BP filter; the frequencies from f_1 to f_2 are stopped, and all other frequencies are passed. BR filters are sometimes known as *band-stop filters*. An *all-pass* (AP) filter passes all frequencies from 0 to ∞, but it provides a phase delay.

It is impossible to create filters with the ideal characteristics shown in Fig. 4.2. Instead of abrupt changes from pass to stop behavior and from stop to pass behavior, practical filters exhibit a gradual transition from stop band to passband. Realistic filter characteristics are shown in Fig. 4.3 [(a)–(d)]. All characteristics are combined in Fig. 4.3(e). The cutoff frequency corresponds to the frequency at which the gain is 70.7% of its maximum value. The sharpness of the transition or the rate at which the characteristic changes is known as the *roll-off* or the *falloff rate*. If the frequency is plotted on a logarithmic scale, the plot is known as a *Bode plot*, and the roll-off, or asymptotic slope, is measured in multiples of ± 6 dB per octave or ± 20 dB per decade.

KEY POINTS OF SECTION 4.3

- Depending on the frequency characteristic, filters can be classified as low-pass, high-pass, band-pass, band-reject, or all-pass.
- It is not possible to create filters with the ideal characteristics of abrupt changes from pass to stop behavior and from stop to pass behavior. Practical filters exhibit a gradual transition from stop band to passband.

4.4 First-Order Filters

A first-order filter is the simplest type. It is commonly used and serves as the building block for a wide variety of active filters. Both the numerator and the denominator of Eq. (4.1) are 1, $m = n = 1$. Thus, the first-order filter has the transfer function as given by [5, 6]

$$H(s) = \frac{a_1 s + a_0}{s + b_0} = K\frac{s + z_1}{s + p_1} \tag{4.5}$$

where p_1 and z_1 are the pole and zero of the transfer function and K is the voltage gain between the output and the input. For $s = -z_1$, $H(s = -z_1) = 0$, and for $s = -p_1$, $H(s = -p_1) = \infty$. The pole–zero location for the transfer function is shown in Fig. 4.4. For $p_1 > z_1$, the zero is always closer to the origin than the pole. When s becomes very large, then H approaches the value of K. For $s = 0$, H becomes less than K. That is,

$$H(s = \infty) = 1 \quad \text{and} \quad H(s = 0) = \frac{Kz_1}{p_1}$$

For $z_1 > p_1$, the pole is always closer to the origin than the zero, and H becomes greater than K. Therefore, the pole–zero locations are important parameters to characterize the transfer function and its response. The output will be either K or less than K; that is, the output is attenuated.

By letting $s = j\omega$, Eq. (4.2) gives the frequency characteristics as

$$H(j\omega) = K\frac{z_1 + j\omega}{p_1 + j\omega} = \frac{Kz_1}{p_1} \times \frac{1 + (j\omega/z_1)}{1 + (j\omega/p_1)} \tag{4.6}$$

which gives the magnitude as

$$|H(j\omega)| = \frac{Kz_1}{p_1}\sqrt{\frac{1 - (\omega/z_1)^2}{1 - (\omega/p_1)^2}} \tag{4.7}$$

The phase angle θ is given by

$$\theta = \theta_z - \theta_p = \tan^{-1}\left(\frac{\omega}{z_1}\right) - \tan^{-1}\left(\frac{\omega}{p_1}\right) \tag{4.8}$$

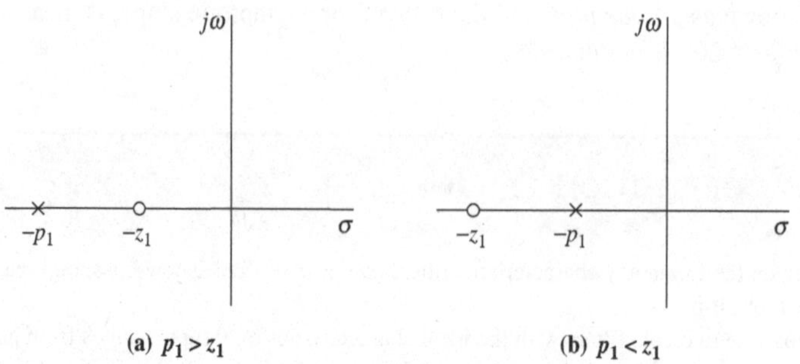

FIGURE 4.4 Pole and zero locations of first-order filters

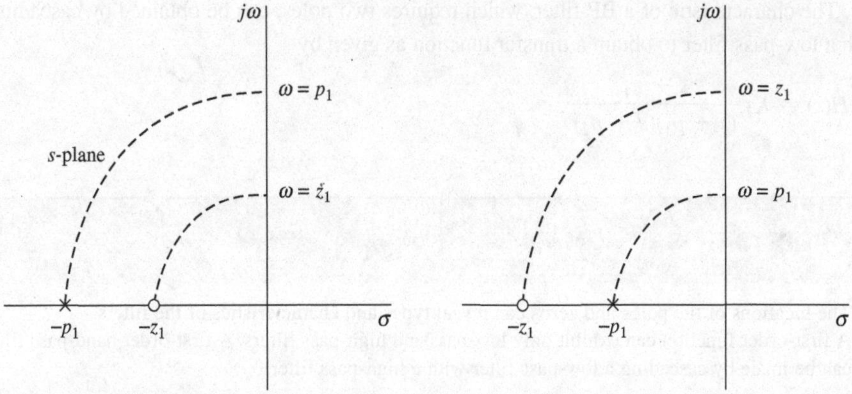

(a) $p_1 > z_1$ **(b)** $p_1 < z_1$

FIGURE 4.5 Poles and zeros for $s = z_1$ and $s = z_p$

The phase angle θ can be leading, having a positive value, or lagging, having a negative value. The phase angle for zero becomes $\theta_z = \tan^{-1}(\omega/z_1) = 45°$ at $\omega = z_1$, as shown in Fig. 4.5(a), and the phase angle for pole is $\theta_p = \tan^{-1}(\omega/z_p) = 45°$ at $\omega = z_p$, as shown in Fig. 4.5(b). Depending on the values of z_1 and p_1, Eq. (4.3) can exhibit the characteristics of a lead low-pass filter, a lag low-pass filter, a lead HP filter, and a lag HP filter. The conditions for z_1 and p_1 to obtain these filters are shown in Table 4.1. A pole 1 in the denominator gives the low-pass characteristic, whereas the addition of a zero at the origin gives the HP characteristic. If $p_1 > z_1$, it makes an HP characteristic, whereas $p_1 < z_1$ makes a low-pass char-

TABLE 4.1 Poles and zeros for high- and low-pass filters

Pole and Zero	$T(s)$	Classification
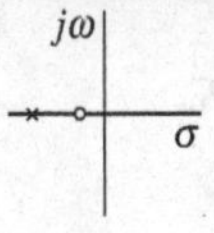	$\dfrac{s+z_1}{s+p_1}$ $(p_1 > z_1)$	Lag high-pass
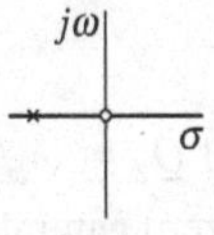	$\dfrac{s}{s+p_1}$ $\boxed{z=0}$	Lead high-pass
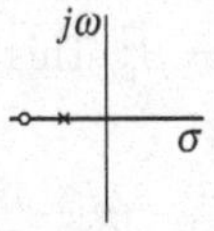	$\dfrac{s+z_1}{s+p_1}$ $(z_1 > p_1)$	Lead low-pass
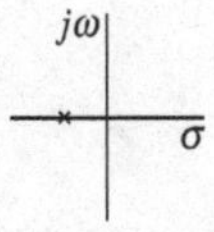	$\dfrac{p_1}{s+p_1}$ $\boxed{z=0}$	Lag low-pass

acteristic. The characteristic of a BP filter, which requires two poles, can be obtained by cascading an HP filter with a low-pass filter to obtain a transfer function as given by

$$H(s) = K_1 \frac{s + z_1}{(s + p_1)(s + p_2)} \tag{4.9}$$

KEY POINTS OF SECTION 4.4

- The locations of the poles and zeros can reveal types and characteristics of the filters.
- A first-order function can exhibit only low-pass and high-pass filters. A first-order band-pass filter can be made by cascading a low-pass filter with a high-pass filter.

4.5 The Biquadratic Function

For an active filter with $n > 2$, Eq. (4.1) will become complex. Thus, a second-order transfer function (i.e., a function with $n = 2$) is most commonly used. For $n = 2$, Eq. (4.1) becomes

$$H(s) = \frac{a_2 s^2 + a_1 s + a_0}{s^2 + b_1 s + b_0} = K \frac{(s + z_1)(s + z_2)}{(s + p_1)(s + p_2)} \tag{4.10}$$

The poles (or zeros) can be determined from the roots of a quadratic equation from

$$p_1, p_2 = \frac{-b_1 \pm \sqrt{b_1^2 - 4b_0}}{2} = -\frac{b_1}{2} \pm \sqrt{\left(\frac{b_1}{2}\right)^2 - b_0} \tag{4.11}$$

where b_0 is generally greater than $b_1^2/4$ and the poles are complex such that $p_1, p_2 = \alpha_p \pm j\beta_p$, which can be substituted in Eq. (4.7) to give the transfer function of the form as given by

$$H(s) = K \frac{(s + \alpha_z + j\beta_z)(s + \alpha_z - j\beta_z)}{(s + \alpha_p + j\beta_p)(s + \alpha_p - j\beta_p)} = K \frac{s^2 + 2\alpha_z s + \omega_z^2}{s^2 + 2\alpha_p s + \omega_p^2} \tag{4.12}$$

$$= K \frac{s^2 + (\omega_z/Q_z)s + \omega_z^2}{s^2 + (\omega_p/Q_p)s + \omega_p^2} \tag{4.13}$$

where $Q_p = \omega_p/2\alpha_p = \omega_p/2\mathrm{Re}(p_1)$ is called the *pole quality factor* and $Q_z = \omega_z/2\alpha_z = \omega_z/2\mathrm{Re}(z_1)$ is called the *zero quality factor*. $\omega_p(=\sqrt{b_0})$ and $\omega_z(=\sqrt{a_0})$ are the pole and zero natural resonant frequencies, respectively. The imaginary part β_p is related to α_p and ω_p by $\omega_p^2 = \alpha_p^2 + \beta_p^2$. This can be solved for β_p as given by

$$\beta_p = \sqrt{\omega_p^2 - \alpha_p^2} = \omega_p \sqrt{1 - \frac{1}{4Q_p^2}} \tag{4.14}$$

The pole angle ϕ_p as shown in Fig. 4.6(a) can be found from

$$\phi_p = \cos^{-1}\left(\frac{\alpha_p}{\omega_p}\right) = \cos^{-1}\left(\frac{1}{2Q_p}\right) \tag{4.15}$$

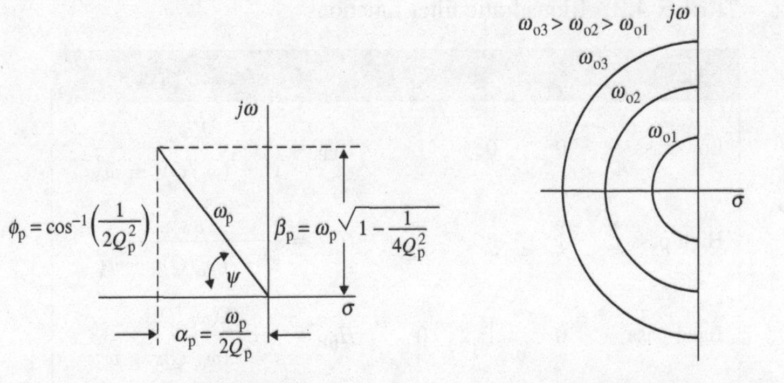

(a) Pole angle **(b) Loci of pole angles for $\omega = \omega_o$**

FIGURE 4.6 Pole angles and their loci

which shows that the phase delay depends on the quality factor Q_p, whose value is chosen to be greater than 1. If $Q_p > 5$, Eq. (4.14) is approximated to $\beta_p \simeq \omega_p$ with an error less than 1%. As a rule of thumb, $Q_p = 7$ is generally used for second filter designs.

Equation (4.10) is known as the *biquadratic function*, which serves as the building block for a wide variety of active filters and has the general form given by

$$H(s) = K\frac{k_2 s^2 + k_1(\omega_o/Q)s + k_o\omega_o^2}{s^2 + (\omega_o/Q)s + \omega_o^2} \tag{4.16}$$

where ω_o = *undamped natural* (or *resonant*) *frequency*

Q = *quality factor* or *figure of merit*

K = *DC gain*

Constants k_2, k_1, and k_o are ± 1 or 0. Table 4.2 shows their possible values for each type of filter. Substituting $s = j\omega$ into Eq. (4.16) will give the frequency domain $H(j\omega)$, which will have a magnitude and a phase delay:

$$H(j\omega) = \frac{-k_2\omega^2 + jk_1(\omega_o/Q)\omega + k_o\omega_o^2}{-\omega^2 + j(\omega_o/Q)\omega + \omega_o^2} = \frac{(k_o\omega_o^2 - k_2\omega^2) + jk_1(\omega_o/Q)\omega}{(\omega_o^2 - \omega^2) + j(\omega_o/Q)\omega} \tag{4.17}$$

where $\omega = 2\pi f$, in radians per second, and f = supply frequency, in hertz.

Normalizing the frequency ω with respect to the natural frequency ω_o so that $u = \omega/\omega_o$, Eq. (4.15) becomes

$$H(ju) = \frac{(k_o - k_2u^2) + (jk_1u/Q)}{(1 - u^2) + (ju/Q)} \tag{4.18}$$

where ω_o = natural frequency, in radians per second.

It can be shown (Appendix B) that Q is related to the bandwidth BW and to ω_o by

$$Q = \frac{\omega_o}{\text{BW}} = \frac{\omega_o}{\omega_H - \omega_L} \tag{4.19}$$

where ω_H = high cutoff frequency, in radians per second, and ω_L = low cutoff frequency, in radians per second.

TABLE 4.2 Biquadratic filter functions

Filter	k_2	k_1	k_0	Transfer Function
Low-pass	0	0	1	$H_{LP} = \dfrac{K\omega_o^2}{s^2 + (\omega_o/Q)s + \omega_o^2}$
High-pass	1	0	0	$H_{HP} = \dfrac{Ks^2}{s^2 + (\omega_o/Q)s + \omega_o^2}$
Band-pass	0	1	0	$H_{BP} = \dfrac{K(\omega_o/Q)s}{s^2 + (\omega_o/Q)s + \omega_o^2}$
Band-reject	1	0	1	$H_{BR} = \dfrac{K(s^2 + \omega_o^2)}{s^2 + (\omega_o/Q)s + \omega_o^2}$
All-pass	1	−1	1	$H_{AP} = K\dfrac{s^2 - (\omega_o/Q)s + \omega_o^2}{s^2 + (\omega_o/Q)s + \omega_o^2}$

Depending on the values of the constants k_0, k_1, and k_2, there are four possible cases whose transfer function and phase angles in frequency domain can be described as follows:

Case 1. For $k_0 = 1$ and $k_1 = k_2 = 0$, there is no zero. Equation (4.18) gives the low-pass characteristic as shown in Table 4.3(a), and its transfer function in frequency domain is given by

$$H_{LP}(s) = K\frac{1}{(1 - u^2) + (ju/Q)} = \left|\frac{K}{\sqrt{(1 - u^2)^2 + (u/Q)^2}}\right| \angle - \tan^{-1}\left(\frac{u/Q}{1 - u^2}\right) \tag{4.20}$$

Case 2. For $k_1 = 1$ and $k_0 = k_2 = 0$, there is one zero on the $j\omega$-axis. Equation (4.18) gives the BP characteristic as shown in Table 4.3(b), and its transfer function in frequency domain is given by

$$H_{BP}(s) = K\frac{ju/Q}{(1 - u^2) + (ju/Q)} = \left|\frac{(u/Q)K}{\sqrt{(1 - u^2)^2 + (u/Q)^2}}\right| \angle \left[\frac{\pi}{2} - \tan^{-1}\left(\frac{u/Q}{1 - u^2}\right)\right] \tag{4.21}$$

Case 3. For $k_0 = k_2 = 1$ and $k_1 = 0$, there are two zeros at the origin. Equation (4.18) gives the BR characteristic as shown in Table 4.3(c), and its transfer function in frequency domain is given by

$$H_{BR}(s) = K\frac{(1 - u^2)}{(1 - u^2) + (ju/Q)} = \left|\frac{(1 - u^2)K}{\sqrt{(1 - u^2)^2 + (u/Q)^2}}\right| \angle - \tan^{-1}\left(\frac{u/Q}{1 - u^2}\right) \tag{4.22}$$

Case 4. For $k_2 = 1$ and $k_0 = k_1 = 0$, there are two zeros at the origin. Equation (4.18) gives the HP characteristic as shown in Table 4.3(d), and its transfer function in frequency domain is given by

$$H_{HP}(s) = K\frac{-u^2}{(1 - u^2) + (ju/Q)} = \left|\frac{u^2K}{\sqrt{(1 - u^2)^2 + (u/Q)^2}}\right| \angle \left[\pi - \tan^{-1}\left(\frac{u/Q}{1 - u^2}\right)\right] \tag{4.23}$$

TABLE 4.3 Characteristics of second-order filters

Filter Type	Transfer Function	Frequency Response	Poles and Zeros
(a) Low-pass	$H_{LP}(s) = \dfrac{\omega_o^2}{s^2 + (\omega_o/Q)s + \omega_o^2}$	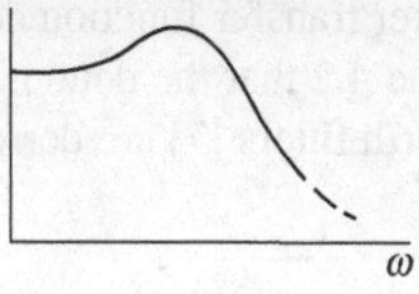	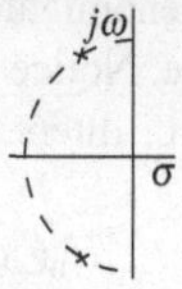
(b) Low-band-pass	$H_{BP}(s) = \dfrac{(\omega_o/Q)s}{s^2 + (\omega_o/Q)s + \omega_o^2}$	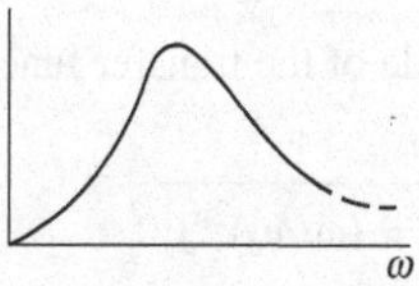	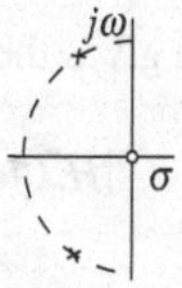
(c) Band-reject notch	$H_{BS}(s) = \dfrac{s^2 + \omega_o^2}{s^2 + (\omega_o/Q)s + \omega_o^2}$	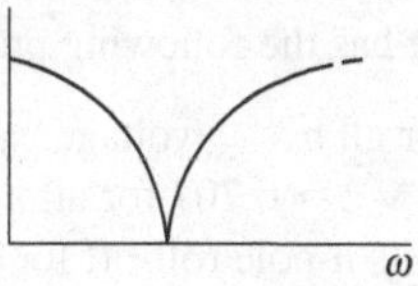	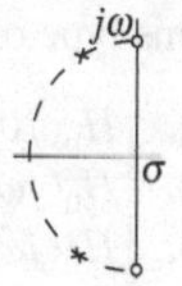
(d) High-pass	$H_{HP}(s) = \dfrac{s^2}{s^2 + (\omega_o/Q)s + \omega_o^2}$	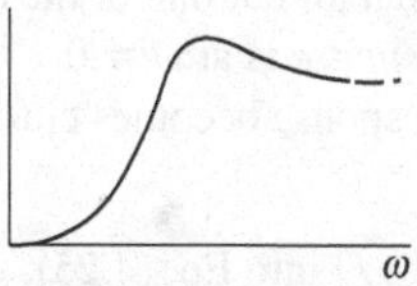	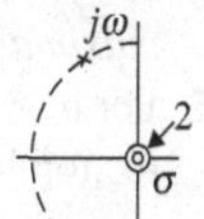
(e) All-pass	$H_{AP}(s) = \dfrac{s^2 - (\omega_o/Q)s + \omega_o^2}{s^2 + (\omega_o/Q)s + \omega_o^2}$	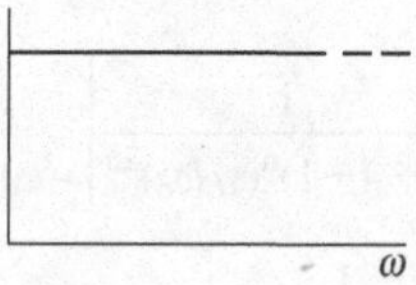	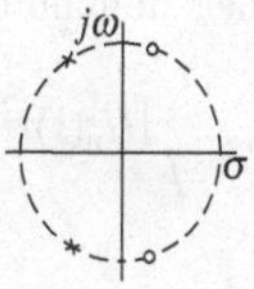

Case 5. For $k_0 = k_2 = 1$ and $k_1 = -1$, two zeros are identical to the two poles. Equation (4.18) gives the AP characteristic as shown in Table 4.3(e), and its transfer function in frequency domain is given by

$$H_{AP}(s) = K\frac{(1 - u^2) + (ju/Q)}{(1 - u^2) + (ju/Q)} = |K| \angle 0° \tag{4.24}$$

KEY POINTS OF SECTION 4.5

- The quality factor Q is a measure of the bandwidth of a filter. The lower the value of Q, the more selective the filter will be.
- The denominator of all quadratic filter functions is the same, but the numerator depends on the type of filter.

4.6 Butterworth Filters

The denominator of a filter transfer function determines the poles and the falloff rate of the frequency response. Notice from Table 4.2 that the denominator of the biquadratic function has the same form for all types of filters. Butterworth filters [7] are derived from the magnitude-squared function

$$|H_n(j\omega)|^2 = \frac{1}{(1 + \omega/\omega_o)^{2n}} \tag{4.25}$$

which gives the magnitude of the transfer function as

$$|H_n(j\omega)| = \frac{1}{[1 + (\omega/\omega_o)^{2n}]^{1/2}} \tag{4.26}$$

Plots of this response, known as the *Butterworth response*, are shown in Fig. 4.7 for $n = 1, 2, 4, 6, 8$, and 10. This type of response has the following properties:

1. $|H_n(j0)| = 1$ for all n (voltage gain at zero frequency—that is, DC voltage gain at $\omega = 0$).
2. $|H_n(j\omega_o)| = 1/\sqrt{2} \approx 0.707$ for all n (voltage gain at $\omega = \omega_o$).
3. $|H_n(j\omega_o)|$ exhibits n-pole roll-off for $\omega > \omega_o$.
4. It can be shown that all but one of the derivatives of $|H_n(j\omega)|$ equal zero near $\omega = 0$. That is, the *maximally flat response* is at $\omega = 0$.
5. For $n > 10$, the response becomes close to the ideal characteristic of abrupt change from passband to stop band.

By substituting $\omega = s/j$ into Eq. (4.25), we can express the transfer function for Butterworth filters in the s domain by

$$|H_n(s)|^2 = \left|\frac{1}{1 + (-1)^n(s/\omega_o)^{2n}}\right| \tag{4.27}$$

$$= \left|\frac{1}{D_n(s)D_n(-s)}\right| \tag{4.28}$$

where $D_n(s)$ is a polynomial in s, all of whose roots have negative real parts, and $|D_n(s)| = |D_n(-s)|$.

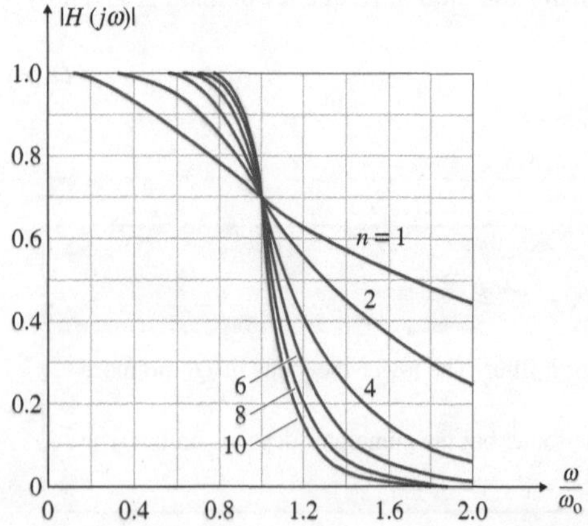

FIGURE 4.7 Butterworth response

4.6.1 Butterworth Function for $n = 1$

For $n = 1$ and letting $\omega_o = 1$, Eq. (4.27) becomes

$$|H_1(s)|^2 = \left|\frac{1}{1 - s^2}\right| = \left|\frac{1}{D_1(s)D_1(-s)}\right| \tag{4.29}$$

Factorizing $1 - s^2$, we get $D_1(s)$ and $D_1(-s)$ as

$$D_1(s)D_1(-s) = 1 - s^2 = (1 + s)(1 - s)$$

which gives

$$D_1(s) = (1 + s) \quad \text{and} \quad D_1(-s) = (1 - s)$$

Since $|D_1(s)| = |D_1(-s)|$, Eq. (4.24) gives the Butterworth function with negative real parts. That is, for $D_1(s)$ only, we get the general form

$$H_1(s) = \frac{1}{(s/\omega_o) + 1} = \frac{\omega_o}{s + \omega_o}$$

The pole angles are $\phi_p = 0°$ and $180°$. For $n = 1$, it becomes a first-order function that is not generally used. For the Butterworth filter, the minimum is a second-order function, $n \geq 2$.

4.6.2 Butterworth Function for $n = 2$

If we let $\omega_o = 1$, for $n = 2$ Eq. (4.27) becomes

$$|H_2(s)|^2 = \left|\frac{1}{1 + s^4}\right| = \left|\frac{1}{D_2(s)D_2(-s)}\right| \tag{4.30}$$

Factoring $1 + s^4$ gives $D_2(s)D_2(-s)$ as

$$D_2(s)D_2(-s) = s^4 + 1 = \left(s - \frac{-1 - j}{\sqrt{2}}\right)\left(s - \frac{1 + j}{\sqrt{2}}\right)\left(s - \frac{1 - j}{\sqrt{2}}\right)\left(s - \frac{1 + j}{\sqrt{2}}\right)$$

which gives

$$D_2(s) = \left(s - \frac{-1 - j}{\sqrt{2}}\right)\left(s - \frac{-1 + j}{\sqrt{2}}\right) = s^2 + \sqrt{2}s + 1$$

and $$D_2(-s) = \left(s - \frac{1 - j}{\sqrt{2}}\right)\left(s - \frac{1 + j}{\sqrt{2}}\right) = s^2 - \sqrt{2}s + 1$$

Since $|D_2(s)| = |D_2(-s)|$, Eq. (4.30) gives the Butterworth function with negative real parts. That is, for $D_2(s)$ only, we get the general form

$$H_2(s) = \frac{1}{(s/\omega_o)^2 + \sqrt{2}(s/\omega_o) + 1} = \frac{\omega_o^2}{s^2 + \sqrt{2}\omega_o s + \omega_o^2} \tag{4.31}$$

which has $Q = 1/\sqrt{2} = 0.707$. Thus, for $n = 2$, a Butterworth filter will exhibit the frequency characteristic of a second-order system (Appendix B), and the frequency response will fall at a rate of -40 dB/decade or -12 dB/octave.

The angles for all k (=$2n$ = 4) poles are

$$\phi_k = \frac{180^\circ + 360^\circ \times k}{2 \times n} = 45^\circ, 135^\circ, 225^\circ, 315^\circ$$

4.6.3 Butterworth Function for $n = 3$

If we let $\omega_o = 1$, for $n = 3$ Eq. (4.27) becomes

$$|H_3(s)|^2 = \left|\frac{1}{1 - s^6}\right| = \left|\frac{1}{D_3(s)D_3(-s)}\right| \tag{4.32}$$

Factoring $1 - s^6$, we get

$$D_3(s)D_3(-s) = 1 - s^6 = (s^2 + s + 1)(s^2 - s + 1)(s + 1)(-s + 1)$$

which gives $D_3(s)$, whose roots have negative real parts, as

$$D_3(s) = (s^2 + s + 1)(s + 1) = s^3 + 2s^2 + 2s + 1$$

The transfer function for $n = 3$ is given by

$$H_3(s) = \frac{1}{(s/\omega_o)^3 + 2(s/\omega_o)^2 + 2(s/\omega_o) + 1} \tag{4.33}$$

$$= \frac{\omega_o^3}{s^3 + 2\omega_o s^2 + 2\omega_o^2 s + \omega_o^3} \tag{4.34}$$

Thus, for $n = 3$, a Butterworth filter will exhibit the frequency characteristic of a third-order system, and the frequency response will fall at a rate of −60 dB/decade or −18 dB/octave.

The angles for all k (= $2n$ = 6) poles are

$$\phi_k = \frac{360^\circ \times k}{2 \times n} = 0^\circ, 60^\circ, 120^\circ, 180^\circ, 240^\circ, 300^\circ$$

4.6.4 Butterworth Function for Higher-Order Filters

A Butterworth filter will exhibit the frequency characteristic of an nth-order system and will fall at a rate of $20n$ dB/decade or $6n$ dB/octave. The denominator $D_n(s)$ depends on the LHS poles that will be phase shifted by 180° with respect to the poles of $D_n(s)$ on the RHS. Each pole p_k has it complex conjugate $\overline{p}_k$ as shown in Fig. 4.8 in the s-plane for $n = 1$ to 5. Taking the LHS real-axis as the reference, we can find the LHS pole angles for an nth-order filter from the following expression:

$$\phi_k = 180^\circ - 90^\circ\left(\frac{2k + n - 1}{n}\right) \quad (\text{for } k = 1, 2, \ldots, n) \tag{4.35}$$

Assuming $\omega_o = 1$, the corresponding real and imaginary parts of the poles are given by

$$\alpha_k = \cos\phi_k \quad \text{and} \quad \beta_k = \sin\phi_k$$

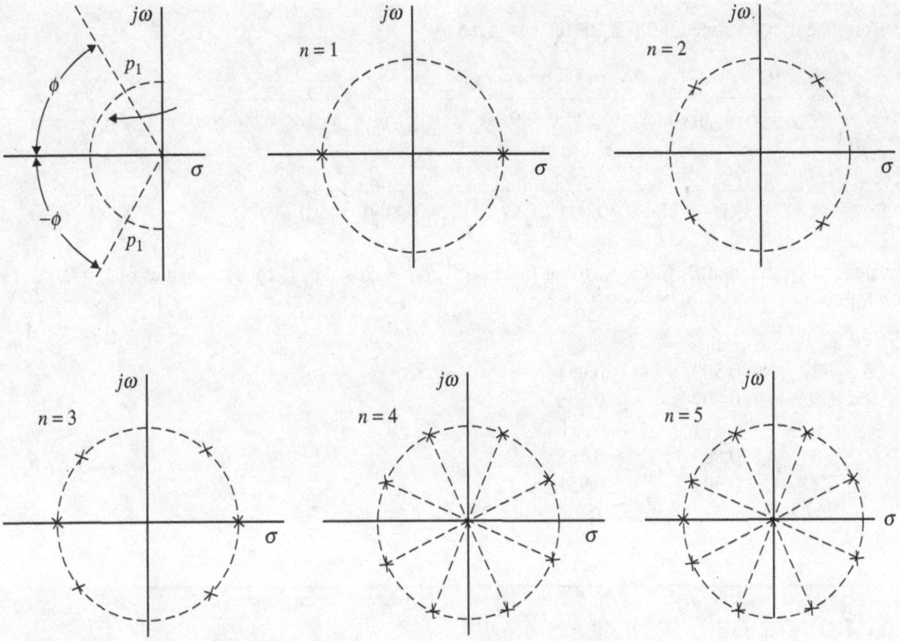

FIGURE 4.8 Pole locations for the Butterworth function

Substituting α_k in Eq. (4.12), we can find the function $D(s)$, which is the reciprocal of $H(s)$, from

$$D_n(s) = (1 + s)\prod_k(s^2 + 2\alpha_k s + 1) \quad \text{(for odd values of } k = 1, 3, \ldots, n) \tag{4.36}$$

$$= \prod_k(s^2 + 2\alpha_k s + 1) \quad \text{(for even values of } k = 2, 4, \ldots, n)$$

EXAMPLE 4.1

Finding the transfer function for the Butterworth response Find the transfer function for the fifth-order Butterworth response of a BP filter.

SOLUTION

$n = 5$. From Eq. (4.35) we get the pole angles $\phi_k = 0°, \pm 36°$, and $\pm 72°$. The real values are $\alpha_k = \cos\phi_k = 1, 0.809, 0.309, 0.809$, and 0.309. Therefore,

$$D_5(s) = (1 + s)(s^2 + 2\alpha_1 s + 1)(s^2 + 2\alpha_2 s + 1)$$

$$= (1 + s)(s^2 + 2 \times 0.309s + 1)(s^2 + 2 \times 0.809s + 1)$$

Therefore, the transfer function for a BP filter is given by

$$H(s) = \frac{Ks^5}{1 + (s/\omega_o)[(s/\omega_o)^2 + 2 \times 0.309s/\omega_o + 1][(s/\omega_o)^2 + 2 \times 0.809s/\omega_o + 1]}$$

$$= \frac{Ks^5\omega_o^5}{(1 + s)(s^2 + 2 \times 0.309s + \omega_o^2)(s^2 + 2 \times 0.809s + \omega_o^2)}$$

The values of Q_k, α_k, and β_k of the factorial function $\Pi_k(s^2 + 2\alpha_k s + 1)$ for even values of $k = 2, 4, \ldots, n$ are as follows:

k	ϕ_k	α_k	β_k
1	72	0.309	0.951
2	36	0.809	0.588
3	0	1	0
4	−36	0.809	−0.588
5	−72	0.309	−0.951

KEY POINTS OF SECTION 4.6

- Butterworth filters can give a maximally flat response.
- For $n > 10$, the response becomes close to the ideal characteristic of abrupt change from passband to stop band. However, a filter with $n = 2$ is quite satisfactory for most applications.

4.7 Transfer Function Realization

The poles and zeros of a transfer function for active filters are generally obtained from series and parallel connection impedances as shown in Fig. 4.9. The transfer function describing the relationship between the output and input voltages can be described by [8]

$$H(s) = \frac{V_o(s)}{V_i(s)} = \frac{Z_2(s)}{Z_1(s) + Z_2(s)} \tag{4.37}$$

where the impedances $Z_1(s)$ and $Z_2(s)$ consist of series and parallel connections of Rs and Cs. The basic elements for generating poles and zeros are listed in Table 4.4.

The transfer function is a dimensionless quantity. It is convenient to normalize all circuit elements so that they become unitless quantities. The relationships among the circuit elements remain the same, irrespective

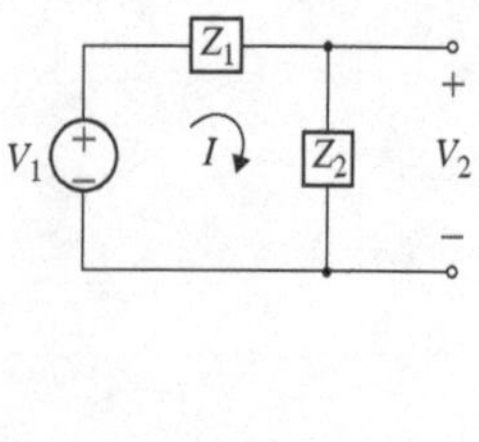

FIGURE 4.9 Impedances for obtaining transfer function

TABLE 4.4 Circuit elements for generating poles and zeros

Z_1 or Z_2 Elements	Impedance	Admittance
	R	$\frac{1}{R}$
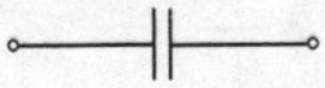	$\frac{1}{sC}$	sC
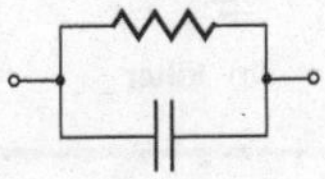	$\frac{1}{sC + 1/R}$	$sC + \frac{1}{R}$
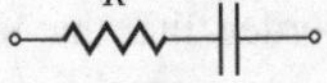	$R + \frac{1}{sC}$	$\frac{1}{R + (1/sC)}$

of their values. Normalization allows scaling of all values and has the advantages of (a) removing the dimensions from the circuit variables, (b) the relative relationship independent of impedance level, and (c) allowing the designer to select convenient and practical values of circuit elements. Frequency normalization is done by dividing the frequency variable by a normalizing frequency, usually ω_o, and the impedances are normalized by dividing all impedances in the circuit by a normalizing resistance R_o. The normalized s becomes $s_n = s/\omega_o$, and the normalized frequency becomes $\omega_n = \omega/\omega_o$. Thus, the normalized impedances for resistance R and capacitance C become

$$R_n = \frac{R}{R_o} \quad \text{and} \quad \frac{1}{s_n C_n} = \left(\frac{1}{s/\omega_o}\right)\left(\frac{1}{\omega_o C R_o}\right)$$

Therefore, the normalized element values are given by

$$R_n = \frac{R}{R_o} \quad \text{and} \quad C_n = C\omega_o R_o \tag{4.38}$$

4.8 Low-Pass Filters

Depending on the order of the biquadratic polynomial in Eq. 4.16, low-pass filters can be classified into two types: first-order and second-order.

4.8.1 First-Order Low-Pass Filters

The transfer function of a first-order low-pass filter has the general form

$$H(s) = \frac{K\omega_o}{s + \omega_o} \tag{4.39}$$

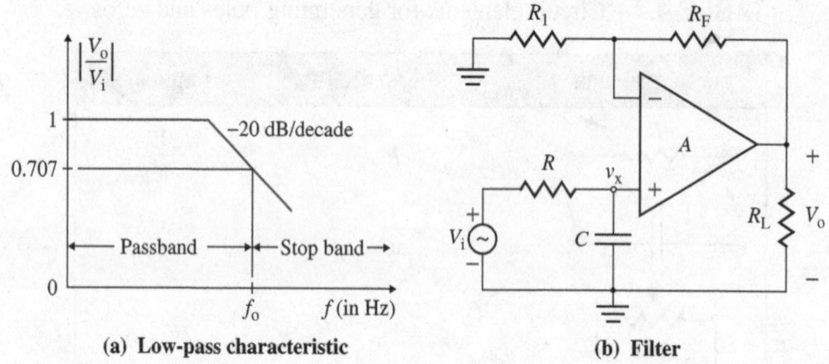

FIGURE 4.10 First-order low-pass filter with $K = 1$

A typical frequency characteristic is shown in Fig. 4.10(a). A first-order filter that uses an RC network for filtering is shown in Fig. 4.10(b). The op-amp operates as a noninverting amplifier, which has the characteristics of a very high input impedance and a very low output impedance.

The voltage (V_X in Laplace's domain of s) at the noninverting terminal of the op-amp can be found by the voltage divider rule:

$$V_X(s) = \frac{1/sC}{R + 1/sC} V_i(s) = \frac{1}{1 + sRC} V_i(s)$$

The output voltage of the noninverting amplifier is

$$V_o(s) = \left(1 + \frac{R_F}{R_1}\right) V_X(s) = \left(1 + \frac{R_F}{R_1}\right) \frac{1}{1 + sRC} V_i(s)$$

which gives the voltage transfer function $H(s)$ as

$$H(s) = \frac{V_o(s)}{V_i(s)} = \frac{K}{1 + sRC} \tag{4.40}$$

where the DC gain is

$$K = 1 + \frac{R_F}{R_1} \tag{4.41}$$

Substituting $s = j\omega$ into Eq. (4.40), we get

$$H(j\omega) = \frac{V_o(j\omega)}{V_i(j\omega)} = \frac{K}{1 + j\omega RC} \tag{4.42}$$

which gives the cutoff frequency f_o at 3-dB gain as

$$f_o = \frac{1}{2\pi RC} \tag{4.43}$$

The magnitude and phase angle of the filter gain can be found from

$$|H(j\omega)| = \frac{K}{[1 + (\omega/\omega_o)^2]^{1/2}} = \frac{K}{[1 + (f/f_o)^2]^{1/2}} \tag{4.44}$$

and $\phi = -\tan^{-1}(f/f_o)$ **(4.45)**

where f = frequency of the input signal, in hertz.

EXAMPLE 4.2

D

Designing a first-order low-pass filter

(a) Design a first-order low-pass filter to give a high cutoff frequency of $f_o = 1$ kHz with a pass-band gain of 4. If the desired frequency is changed to $f_n = 1.5$ kHz, calculate the new value of R_n.

(b) Use PSpice/SPICE to plot the frequency response of the filter designed in part (a) from 10 Hz to 10 kHz.

SOLUTION

(a) The high cutoff frequency is $f_o = 1$ kHz. Choose a value of C less than or equal to 1 μF: let $C = 0.01$ μF. Using Eq. (4.43), calculate the value of R:

$$R = \frac{1}{2\pi f_o C} = \frac{1}{2\pi \times 1\text{ kHz} \times 0.01\ \mu\text{F}} = 15{,}916\ \Omega \quad \text{(use a 20-k}\Omega\text{ potentiometer)}$$

Choose values of R_1 and R_F to meet the pass-band gain K. From Eq. (4.41), $K = 1 + R_F/R_1$. Since $K = 4$,

$$\frac{R_F}{R_1} = 4 - 1 = 3$$

If we let $R_1 = 10$ kΩ, $R_F = 30$ kΩ.

Calculate the frequency scaling factor, FSF $= f_o/f_n$:

$$\text{FSF} = \frac{f_o}{f_n} = \frac{1\text{ kHz}}{1.5\text{ kHz}} = 0.67$$

Calculate the new value of R_n = FSF × R:

$$R_n = \text{FSF} \times R = 0.67 \times 15{,}916 = 10{,}664\ \Omega \quad \text{(use a 15-k}\Omega\text{ potentiometer)}$$

NOTE: The exact numbers of the resistances are used to demonstrate the validity of the calculated values with the PSpice simulation results. In a practical design, the commercially available standard values will be used (i.e., 16 kΩ instead of 15,916 Ω; see Appendix E).

(b) A low-pass filter with the calculated values of the circuit parameters and the LF411 op-amp is shown in Fig. 4.11.

The plot of the voltage gain is shown in Fig. 4.12, which gives $K = 4.0$ (expected value is 4) and $f_o \approx 998$ Hz (expected value is 1 kHz) at $|H(j\omega)| = 0.707 \times 4 = 2.828$. Thus, the results are close to the expected values. If we use a linear op-amp model in Fig. 3.8, the results will differ slightly.

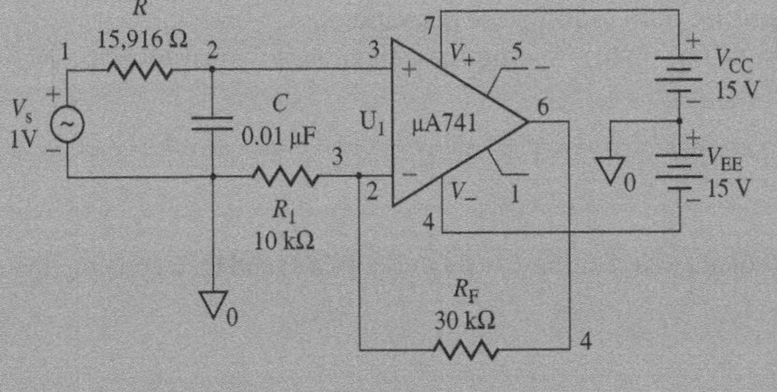

FIGURE 4.11 Low-pass filter for PSpice simulation

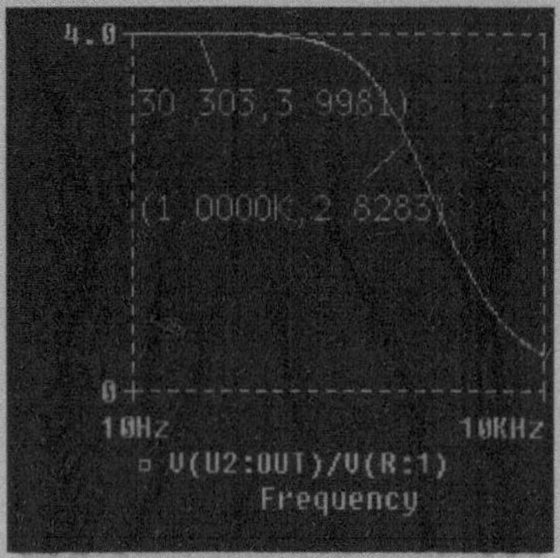

FIGURE 4.12 PSpice frequency plot for Example 4.2

4.8.2 Second-Order Low-Pass Filters

The roll-off of a first-order filter is only −20 dB/decade in the stop band. A second-order filter exhibits a stop-band roll-off of −40 dB/decade and thus is preferable to a first-order filter. In addition, a second-order filter can be the building block for higher-order filters ($n = 4, 6, \ldots$). Substituting $k_2 = k_1 = 0$ and $k_0 = 1$ into Eq. (4.16), we get the general form

$$H(s) = \frac{K\omega_o^2}{s^2 + (\omega_o/Q)s + \omega_o^2} \tag{4.46}$$

where K is the DC gain. A typical frequency characteristic is shown in Fig. 4.13(a); for high values of Q, overshoots will be exhibited at the resonant frequency f_o. For frequencies above f_o, the gain rolls off at the rate of −40 dB/decade. A first-order filter can be converted to a second-order filter by adding an additional RC network, known as the *Sallen–Key* circuit, as shown in Fig. 4.13(b). The input RC network is shown in Fig. 4.13(c); the equivalent circuit appears in Fig. 4.13(d). The transfer function of the filter network is

$$H(s) = \frac{V_o(s)}{V_i(s)} = \frac{K/R_2R_3C_2C_3}{s^2 + s(R_3C_3 + R_2C_3 + R_2C_2 - KR_2C_2)/R_2R_3C_2C_3 + 1/R_2R_3C_2C_3} \tag{4.47}$$

where $K = (1 + R_F/R_1)$ is the DC gain. (See Prob. 4.16 for the derivation.)

Equation (4.47) is similar in form to Eq. (4.46). Setting the denominator equal to zero gives the characteristic equation

$$s^2 + s\frac{R_3C_3 + R_2C_3 + R_2C_2 - KR_2C_2}{R_2R_3C_2C_3} + \frac{1}{R_2R_3C_2C_3} = 0 \tag{4.48}$$

which will have two real parts and two equal roots. Setting $s = j\omega$ in Eq. (4.48) and then equating the real parts to zero, we get

$$-\omega^2 + \frac{1}{R_2R_3C_2C_3} = 0$$

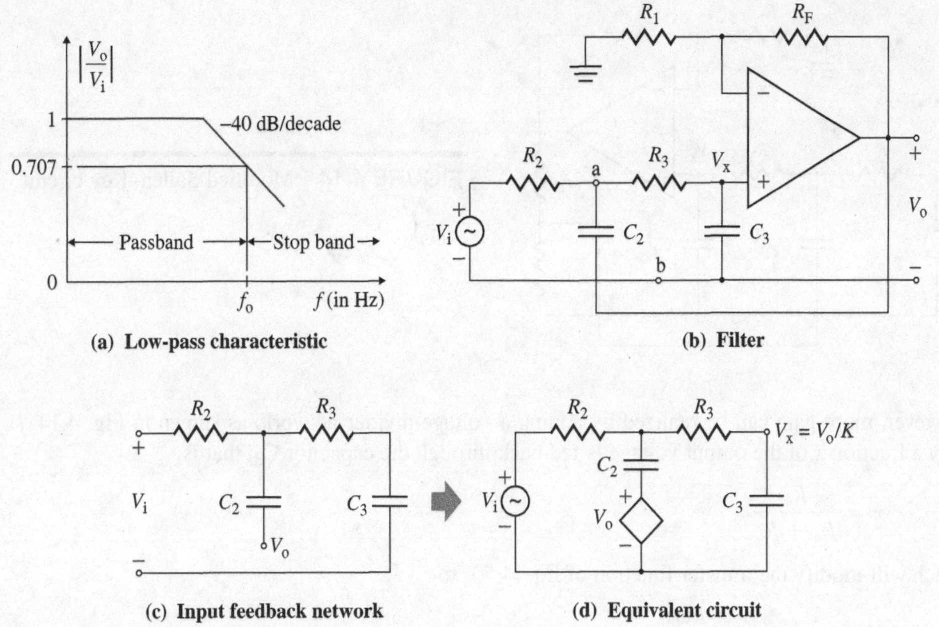

FIGURE 4.13 Second-order low-pass filter with $K = 1$

which gives the cutoff frequency as

$$f_o = \frac{\omega_o}{2\pi} = \frac{1}{2\pi\sqrt{R_2R_3C_2C_3}} \tag{4.49}$$

To simplify the design of second-order filters, equal resistances and capacitances are normally used—that is, $R_1 = R_2 = R_3 = R$, $C_2 = C_3 = C$. Then Eq. (4.47) can be simplified to

$$H(s) = \frac{K\omega_o^2}{s^2 + (3 - K)\omega_o s + \omega_o^2} \tag{4.50}$$

Comparing the denominator of Eq. (4.50) with that of Eq. (4.46) shows that Q can be related to K by

$$Q = \frac{1}{3 - K} \tag{4.51}$$

or $$K = 3 - \frac{1}{Q} \tag{4.52}$$

The frequency response of a second-order system at the 3-dB point will depend on the damping factor ζ such that $Q = 1/2\zeta$. A Q-value of $1/\sqrt{2}$ (= 0.707), which represents a compromise between the peak magnitude and the bandwidth, causes the filter to exhibit the characteristics of a flat passband as well as a stop band, and gives a fixed DC gain of $K = 1.586$:

$$K = 1 + \frac{R_F}{R_1} = 3 - \sqrt{2} = 1.586 \tag{4.53}$$

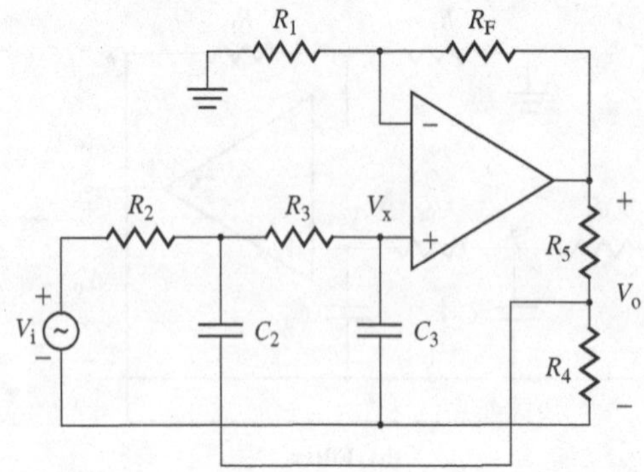

FIGURE 4.14 Modified Sallen–Key circuit

However, more gain can be realized by adding a voltage-divider network, as shown in Fig. 4.14, so that only a fraction x of the output voltage is fed back through the capacitor C_2; that is,

$$x = \frac{R_4}{R_4 + R_5} \tag{4.54}$$

which will modify the transfer function of Eq. (4.50) to

$$H(s) = \frac{K\omega_o^2}{s^2 + (3 - xK)\omega_o s + \omega_o^2} \tag{4.55}$$

and the quality factor Q of Eq. (4.51) to

$$Q = \frac{1}{3 - xK} \tag{4.56}$$

Thus, for $Q = 0.707$, $xK = 1.586$, allowing a designer to realize more DC gain K by choosing a lower value of x, where $x < 1$.

EXAMPLE 4.3

D

Designing a second-order low-pass filter

(a) Design a second-order low-pass filter as in Fig. 4.14, to give a high cutoff frequency of $f_H = f_o = 1$ kHz, a pass-band gain of $K = 4$, and $Q = 0.707$, 1, 2, and ∞.

(b) Use PSpice/SPICE to plot the frequency response of the output voltage of the filter designed in part (a) from 10 Hz to 10 kHz.

SOLUTION

(a) To simplify the design calculations, let $R_1 = R_2 = R_3 = R_4 = R$ and let $C_2 = C_3 = C$. Choose a value of C less than or equal to 1 μF: Let $C = 0.01$ μF. For $R_2 = R_3 = R$ and $C_2 = C_3 = C$, Eq. (4.49) is reduced to

$$f_o = \frac{1}{2\pi RC}$$

which gives the value of R as

$$R = \frac{1}{2\pi f_o C} = \frac{1}{2\pi \times 1\text{ kHz} \times 0.01\ \mu\text{F}} = 15{,}916\ \Omega \quad \text{(use a 20-k}\Omega\text{ potentiometer)}$$

Then $\quad R_F = (K-1)R_1 = (4-1) \times 15{,}916 = 47{,}748\ \Omega$

For $Q = 0.707$ and $K = 4$, Eq. (4.56) gives $x = 1.586/K = 1.586/4 = 0.396$. From Eq. (4.54), we get

$$\frac{R_5}{R_4} = \frac{1}{x} - 1 = \frac{1-x}{x} \tag{4.57}$$

which, for $x = 0.396$ and $R_4 = R = 15{,}916\ \Omega$, gives

$$R_5 = 1.525 \times 15{,}916 = 24{,}275\ \Omega \quad \text{(use a 30-k}\Omega\text{ potentiometer)}$$

For $Q = 1$ and $K = 4$, Eq. (4.56) gives $3 - xK = 1$ or $x = 2/K = 0.5$, and

$$R_5 = R = 15{,}916\ \Omega$$

For $Q = 2$ and $K = 4$, Eq. (4.56) gives $3 - xK = 1/2$ or $x = 2.5/K = 0.625$, and

$$R_5 = 0.6R = 9550\ \Omega$$

For $Q = \infty$ and $K = 4$, Eq. (4.56) gives $3 - xK = 1/Q = 0$ or $x = 3/K = 0.75$, and

$$R_5 = 0.333R = 5305\ \Omega$$

NOTE: The exact numbers of the resistances are used to demonstrate the validity of the calculated values with the PSpice simulation results. In a practical design, the commercially available standard values will be used (i.e., 48 kΩ instead of 47,748 Ω and 25 kΩ instead of 24,275 Ω; see Appendix E).

(b) The low-pass filter, with the designed values of the circuit parameters and a simple DC model of the op-amp, is shown in Fig. 4.15.

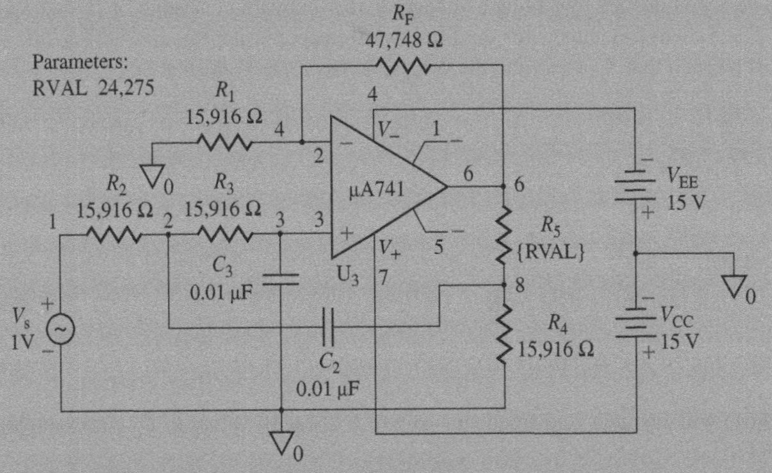

FIGURE 4.15 Second-order low-pass filter for PSpice simulation

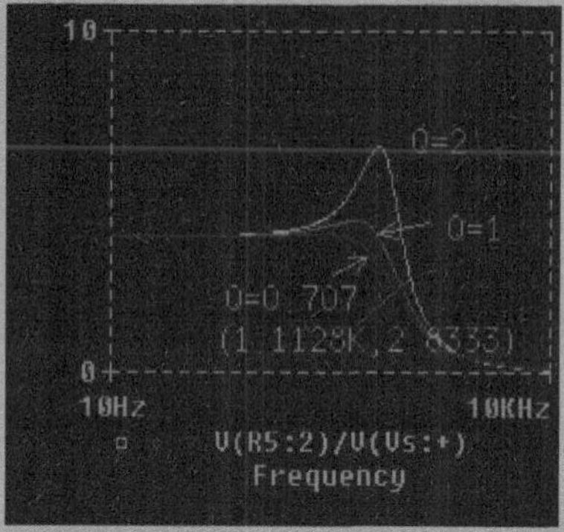

FIGURE 4.16 PSpice frequency response for Example 4.3

The PSpice plot of the voltage gain A_v [=V(R5:2)/V(Vs:+)] (using linear op-amp model) is shown in Fig. 4.16. For $Q = 0.707$, we get $f_o = 758$ Hz (expected value is 1 kHz) at a gain of 2.833 (estimated value is $4 \times 0.707 = 2.828$). The error in the frequency is caused by the finite frequency-dependent gain of the op-amp. If we use an ideal op-amp, the simulation will be very close to the expected value. The peaking of the gain increases with higher values of Q; however, the bandwidth also increases slightly ($f_o = 1113$ Hz for $Q = 2$). The plot for $Q = \infty$ is not shown.

4.8.3 Butterworth Low-Pass Filters

The Butterworth response requires that $|H(j0)| = 1$ (or 0 dB); the transfer function in Eq. (4.50) for the Sallen–Key circuit gives $|H(j0)| = K$ to achieve a Butterworth response with Sallen–Key topology. Therefore, we must reduce the gain by $1/K$. Consider the portion of the circuit to the left of the terminals a and b in Fig. 4.13(b). The resistance R_2 is in series with the input voltage V_i, as shown in Fig. 4.17(a). The gain reduction can be accomplished by adding a voltage-divider network consisting of R_a and R_b, as shown in Fig. 4.17(b). The Sallen–Key circuit for the Butterworth response appears in Fig. 4.17(c). The values of R_a and R_b must be such that $R_{in} = R_2$ and the voltage across R_b is V_i/K; that is,

$$\frac{R_a R_b}{R_a + R_b} = R_2 \tag{4.58}$$

$$\frac{R_b}{R_a + R_b} = \frac{1}{K} \tag{4.59}$$

Solving for R_a and R_b, we get

$$R_a = KR_2 \qquad \text{for } |H(j0)| = 1 \text{(or 0 dB)} \tag{4.60}$$

$$R_b = \frac{K}{K-1} R_2 \quad \text{for } |H(j0)| = 1 \text{ (or 0 dB)} \tag{4.61}$$

Equations (4.60) and (4.61) will ensure a zero-frequency gain of 0 dB at all values of Q. For example, if $K = 4$ and $R_2 = 15{,}916\ \Omega$,

$$R_a = 4 \times 15{,}916 = 63{,}664\ \Omega \quad \text{and} \quad R_b = \frac{4 \times 15{,}916}{(4-1)} = 21{,}221\ \Omega$$

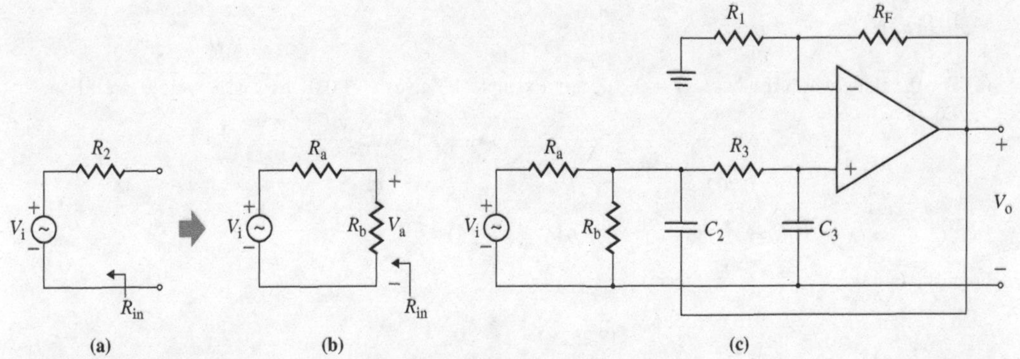

FIGURE 4.17 Sallen–Key circuit for the Butterworth response

It is more desirable, however, to have a gain of 0 dB at the resonant frequency ω_o—that is, $|H(j\omega_o)| = 1$ (or 0 dB). Substituting $s = j\omega_o$ into Eq. (4.50) gives a gain magnitude of $K/(3 - K)$, which gives the required gain reduction of $(3 - K)/K$; that is,

$$\frac{R_b}{R_a + R_b} = \frac{3 - K}{K} \tag{4.62}$$

Solving Eqs. (4.58) and (4.62) for R_a and R_b, we get

$$R_a = R_2 \frac{K}{3 - K} \quad \text{for } |H(j\omega_o)| = 1 \text{ (or 0 dB)} \tag{4.63}$$

$$R_b = R_2 \frac{K}{2K - 3} \quad \text{for } |H(j\omega_o)| = 1 \text{ (or 0 dB)} \tag{4.64}$$

Therefore, we can design an active filter to yield a gain of 0 dB at either $\omega = 0$ or $\omega = \omega_o$. In the case where $|H(j\omega_o)| = 1$ (or 0 dB) is specified, the zero-frequency gain will be reduced by a factor $(3 - K)/K$; that is,

$$|H(j0)| = 3 - K \quad \text{for } |H(j\omega_o)| = 1 \text{ (or 0 dB)} \tag{4.65}$$

For $Q = \sqrt{2}$ and $K = 3 - 1/Q = 1.586$, Eq. (4.65) gives $|H(j0)| = 3 - K = 1.414$, provided that we design the filter for $|H(j\omega_o)| = 1$ (or 0 dB).

EXAMPLE 4.4

D

Designing a second-order low-pass Butterworth filter for $|H(j\omega_o)| = 1$

(a) Design a second-order Butterworth low-pass filter as in Fig. 4.17(c) to yield $|H(j\omega_o)| = 1$ (or 0 dB), a cutoff frequency of $f_o = 1$ kHz, and $Q = 0.707$.

(b) Use PSpice/SPICE to plot the frequency response of the output voltage of the filter designed in part (a) from 10 Hz to 10 kHz.

SOLUTION

(a) For the Butterworth response, $Q = 0.707$, and Example 4.3 gives $C = 0.01\ \mu\text{F}$ and $R = 15{,}916\ \Omega$. From Eq. (4.52),

$$K = 3 - \frac{1}{Q} = 3 - \frac{1}{0.707} = 1.586$$

and $\quad R_F = (K - 1)R_1 = (1.586 - 1) \times 15{,}916 = 9327\ \Omega$

From Eq. (4.63),

$$R_a = \frac{RK}{3 - K} = \frac{15{,}916 \times 1.586}{3 - 1.586} = 17{,}852\ \Omega$$

From Eq. (4.64),

$$R_b = \frac{RK}{2K - 3} = \frac{15{,}916 \times 1.586}{2 \times 1.586 - 3} = 146{,}760\ \Omega$$

From Eq. (4.65),

$$|H(j0)| = 3 - K = 3 - 1.586 = 1.414$$

NOTE: The exact numbers of the resistances are used to demonstrate the validity of the calculated values with the PSpice simulation results. In a practical design, the commercially available standard values will be used (i.e., 147 kΩ instead of 146,760 Ω and 18 kΩ instead of 17,852 Ω).

(b) For PSpice simulation, the circuit in Fig. 4.15 can be modified by removing R_4 and R_5, replacing R_2 by R_a, and adding R_b. This modified circuit is shown in Fig. 4.18.

The PSpice plot of the voltage gain is shown in Fig. 4.19, which gives $|H(j\omega_o)| = 1.0$ at $f_o = 1$ kHz and $|H(j0)| = 1.414$, both of which correspond to the expected values.

NOTE: The simulation was run with the DC linear model.

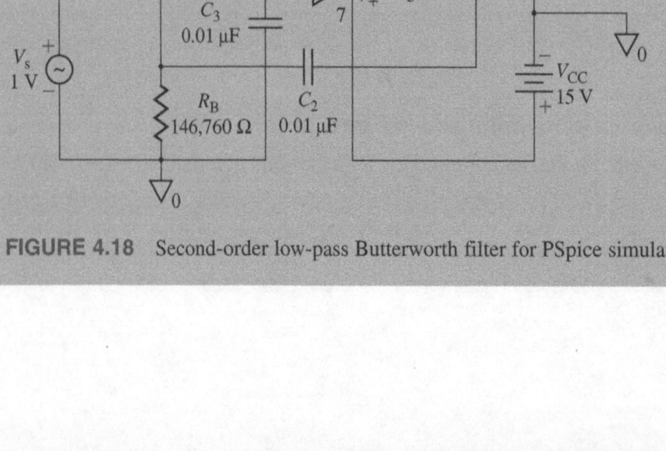

FIGURE 4.18 Second-order low-pass Butterworth filter for PSpice simulation

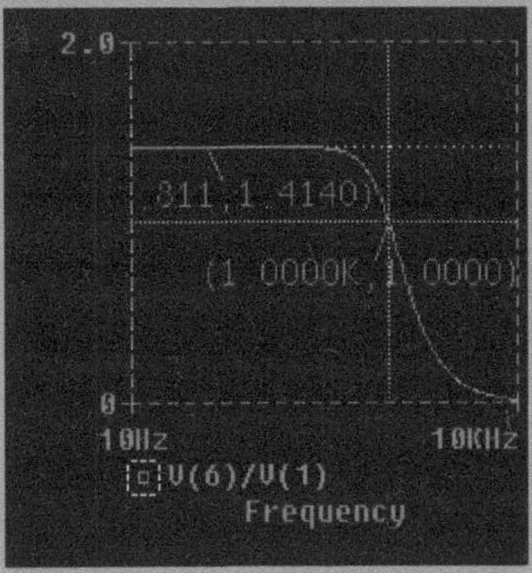

FIGURE 4.19 PSpice frequency response for Example 4.4

KEY POINTS OF SECTION 4.8

- A second-order filter with a fall rate of 40 dB/decade is preferred over a first-order filter with a fall rate of 20 dB/decade. First- and second-order filters can be used as building blocks for higher-order filters.
- The Sallen–Key circuit is a commonly used second-order filter. This circuit can be designed to exhibit the characteristics of a flat passband as well as a stop band and can be modified to give pass-band gain as well as the Butterworth response.

4.9 High-Pass Filters

High-pass filters can be classified broadly into two types: first-order and second-order. Higher-order filters can be synthesized from these two basic types. Since the frequency scale of a low-pass filter is 0 to f_o and that of a high-pass filter is f_o to ∞, their frequency scales have a reciprocal relationship. Therefore, if we can design a low-pass filter, we can convert it to a high-pass filter by applying an *RC-CR* transformation. This transformation can be accomplished by replacing R_n by C_n and C_n by R_n. The op-amp, which is modeled as a voltage-controlled voltage source, is not affected by this transformation. The resistors that are used to set the DC gain of the op-amp circuit are not affected either.

4.9.1 First-Order High-Pass Filters

The transfer function of a first-order high-pass filter has the general form

$$H(s) = \frac{sK}{s + \omega_o} \tag{4.66}$$

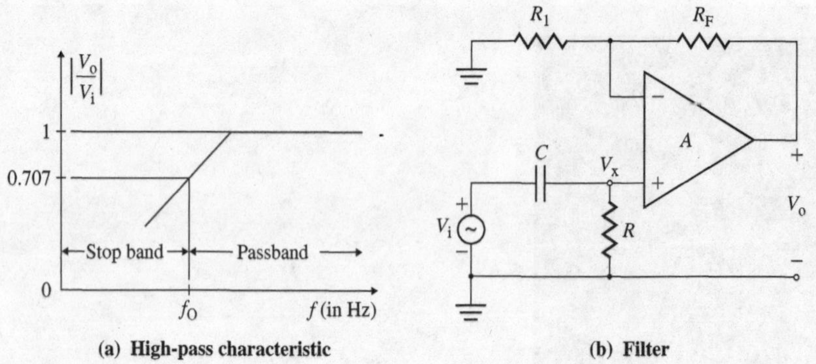

FIGURE 4.20 First-order high-pass filter

A typical high-pass frequency characteristic is shown in Fig. 4.20(a). A first-order high-pass filter can be formed by interchanging the frequency-dependent resistor and capacitor of the low-pass filter of Fig. 4.10(b). This arrangement is shown in Fig. 4.20(b). The voltage at the noninverting terminal of the op-amp can be found by the voltage divider rule. That is,

$$V_x(s) = \frac{R}{R + 1/sC} V_i(s) = \frac{S}{s + 1/RC} V_i(s)$$

The output voltage of the noninverting amplifier is

$$V_o(s) = \left(1 + \frac{R_F}{R_1}\right) V_x(s) = \left(1 + \frac{R_F}{R_1}\right) \frac{s}{s + 1/RC} v_i(s)$$

which gives the voltage gain as

$$H(s) = \frac{V_o(s)}{V_i(s)} = \frac{sK}{s + 1/RC} \tag{4.67}$$

where $K = 1 + R_F/R_1$ is the DC *voltage gain.*

Substituting $s = j\omega$ into Eq. (4.67), we get

$$H(j\omega) = \frac{V_o(j\omega)}{V_i(j\omega)} = \frac{j\omega K}{j\omega + 1/RC} = \frac{j\omega K}{j\omega + \omega_o} \tag{4.68}$$

which gives the cutoff frequency f_o at 3-dB gain as

$$f_o = \frac{\omega_o}{2\pi} = \frac{1}{2\pi RC} \tag{4.69}$$

as in Eq. (4.43). The magnitude and phase angle of the filter gain can be found from

$$|H(j\omega)| = \frac{(\omega/\omega_o)K}{[1 + (\omega/\omega_o)^2]^{1/2}} = \frac{(f/f_o)K}{[1 + (f/f_o)^2]^{1/2}} \tag{4.70}$$

and $$\phi = 90° - \tan^{-1}(f/f_o) \tag{4.71}$$

This filter passes all signals with frequencies higher than f_o. However, the high-frequency limit is determined by the bandwidth of the op-amp itself. The gain–bandwidth product of a practical μ741-type op-amp is 1 MHz.

EXAMPLE 4.5

D

Designing a first-order high-pass filter Design a first-order high-pass filter with a cutoff frequency of $f_o = 1$ kHz and a pass-band gain of 4.

SOLUTION

High-pass filters are formed simply by interchanging R and C of the input RC network, so the design and frequency scaling procedures for low-pass filters are also applicable. Since $f_o = 1$ kHz, we can use the values of R and C that were determined for the low-pass filter of Example 4.2—that is,

$C = 0.01\ \mu$F

$R = 15{,}916\ \Omega$ (use a 20-kΩ potentiometer)

Similarly, we use $R_1 = 10$ kΩ and $R_F = 30$ kΩ to yield $K = 4$.

A PSpice simulation that confirms the design values can be run by interchanging the locations of R and C in Fig. 4.11 so that the statements for R and C read as follows:

```
C  1  2  0.01UF   ; For C connected between nodes 1 and 2
R  2  0  15916    ; For R connected between nodes 2 and 0
```

4.9.2 Second-Order High-Pass Filters

A second-order high-pass filter has a stop-band characteristic of 40 dB/decade rise. The general form of a second-order high-pass filter is

$$H(s) = \frac{s^2K}{s^2 + (\omega_o/Q)s + \omega_o^2} \tag{4.72}$$

where K is the high-frequency gain. Figure 4.21(a) shows a typical frequency response. As in the case of the first-order filter, a second-order high-pass filter can be formed from a second-order low-pass filter by interchanging the frequency-dominant resistors and capacitors. Figure 4.21(b) shows a second-order high-pass filter derived from the Sallen–Key circuit of Fig. 4.13(b). The transfer function can be derived by applying the RC-to-CR transformation and substituting $1/s$ for s in Eq. (4.47). For $R_1 = R_2 = R_3 = R$

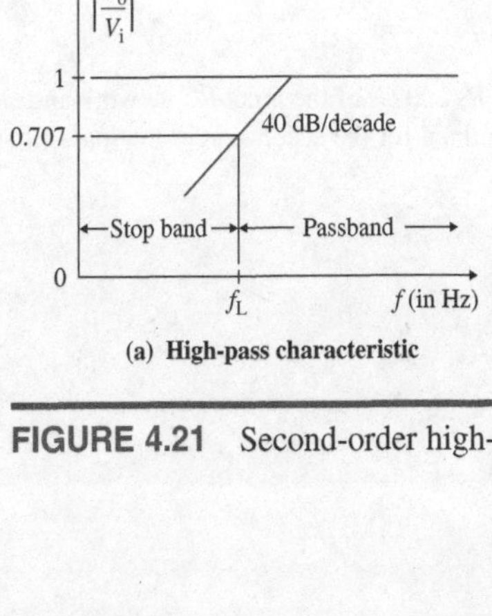

(a) High-pass characteristic

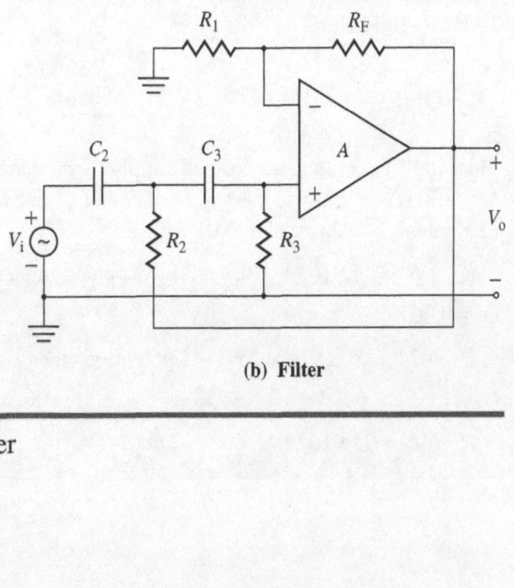

(b) Filter

FIGURE 4.21 Second-order high-pass filter

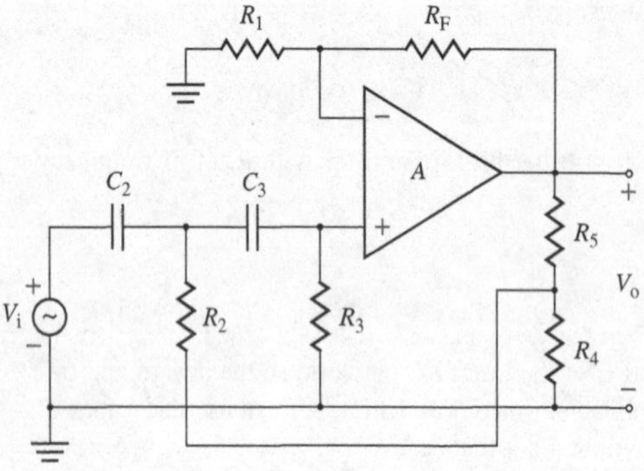

FIGURE 4.22 Modified second-order high-pass filter

and $C_2 = C_3 = C$, the transfer function becomes

$$H(s) = \frac{s^2K}{s^2 + (3 - K)\omega_o s + \omega_o^2} \tag{4.73}$$

and Eq. (4.49) gives the cutoff frequency as

$$f_o = \frac{\omega_o}{2\pi} = \frac{1}{2\pi\sqrt{R_2R_3C_2C_3}} = \frac{1}{2\pi RC} \tag{4.74}$$

Q and K of the circuit remain the same. A voltage-divider network can be added, as shown in Fig. 4.22, so that only a fraction x of the output voltage is fed back through resistor R_2. The transfer function of Eq. (4.73) then becomes

$$H(s) = \frac{s^2K}{s^2 + (3 - xK)\omega_o s + \omega_o^2} \tag{4.75}$$

EXAMPLE 4.6

D

Designing a second-order high-pass filter

(a) Design a second-order high-pass filter as in Fig. 4.22, with a cutoff frequency of $f_o = 1$ kHz, a pass-band gain of $K = 4$, and $Q = 0.707$, 1, 2, and ∞.

(b) Use PSpice/SPICE to plot the frequency response of the output voltage of the filter designed in part (a) from 10 Hz to 100 kHz.

SOLUTION

(a) Since high-pass filters are formed simply by interchanging the Rs and Cs of the input RC network and since $f_o = 1$ kHz, we can use the values of R and C that were determined for the second-order low-pass filter of Example 4.3—that is, $C = 0.01$ μF, and

$R_4 = R = 15{,}916\ \Omega$ (use a 20-kΩ potentiometer)

For $Q = 0.707$,

$R_5 = 24{,}275\ \Omega$ (use a 30-kΩ potentiometer)

For $Q = 1$,

$R_5 = R = 15{,}916\ \Omega$

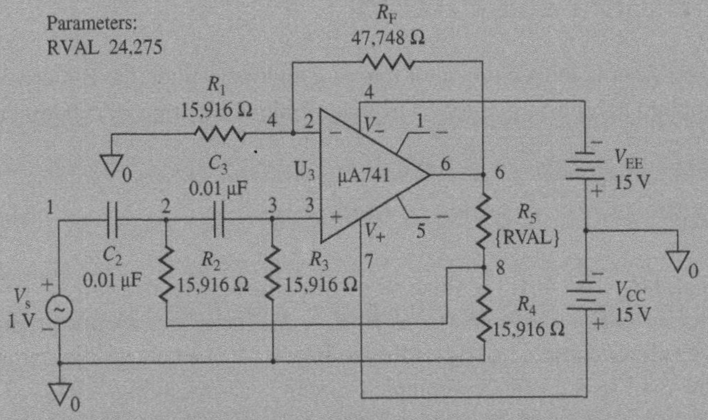

FIGURE 4.23 Second-order high-pass filter for PSpice simulation

For $Q = 2$,

$$R_5 = 0.6R = 9550\ \Omega$$

For $Q = \infty$,

$$R_5 = 0.3333R = 5305\ \Omega$$

NOTE: The exact numbers of the resistances are used to demonstrate the validity of the calculated values with the PSpice simulation results. In a practical design, the commercially available standard values will be used (i.e., 10 kΩ instead of 9550 Ω; see Appendix E).

(b) Figure 4.23 shows the circuit obtained by interchanging the locations of R and C in Fig. 4.15. The PSpice plots are shown in Fig. 4.24. As expected, the voltage gain shows increased peaking for a higher value of Q. The PSpice statements for R and C read as follows:

```
C2  1  2  0.01UF  ; For C2 connected between nodes 1 and 2
R2  2  8  15916   ; For R2 connected between nodes 2 and 8
C3  2  3  0.01UF  ; For C3 connected between nodes 2 and 3
R3  3  0  15916   ; For R3 connected between nodes 3 and 0
```

NOTE: We will notice a high-end roll-off due to the internal capacitances of the μA741 op-amp.

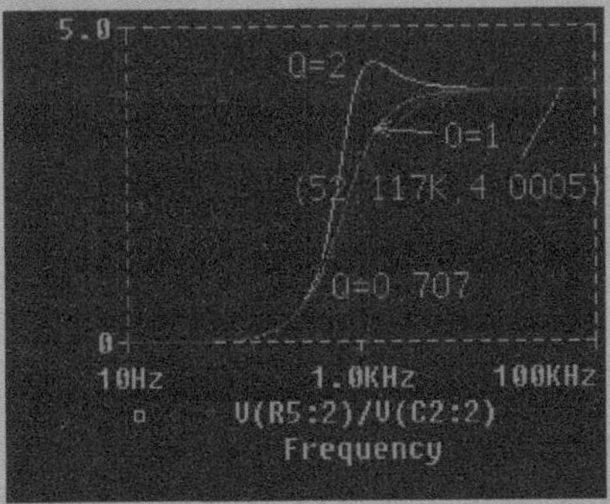

FIGURE 4.24 PSpice plots of frequency response for Example 4.6

4.9.3 Butterworth High-Pass Filters

Since the frequency scale of a low-pass filter is the reciprocal of that of a high-pass filter, the Butterworth response of Eq. (4.26) can also be applied to high-pass filters. The magnitude of the transfer function becomes

$$|H_n(j\omega)| = \frac{1}{[1 + (\omega_o/\omega)^{2n}]^{1/2}} \tag{4.76}$$

where $|H_n(j\infty| = 1$ for all n, rather than $|H_n(j0)| = 1$. The Butterworth response requires that $|H(j\infty)| = 1$ (or 0 dB); however, the transfer function in Eq. (4.73) gives $|H(j\infty)| = K$. Therefore, we must reduce the gain by $1/K$. The gain reduction can be accomplished by adding to Fig. 4.25(a) a voltage-divider network consisting of C_a and C_b, as shown in Fig. 4.25(b). The complete circuit is shown in Fig. 4.25(c). The values of C_a and C_b must be such that $C_{in} = C_2$ and the voltage across C_b is V_i/K; that is,

$$C_a + C_b = C_2 \tag{4.77}$$

$$\frac{C_a}{C_a + C_b} = \frac{1}{K} \tag{4.78}$$

Solving for C_a and C_b, we get

$$C_a = \frac{C_2}{K} \qquad [\text{for } |H(j\infty)| = 1 \text{ (or 0 dB)}] \tag{4.79}$$

$$C_b = C_2\frac{K-1}{K} \qquad [\text{for } |H(j\infty)| = 1 \text{ (or 0 dB)}] \tag{4.80}$$

Equations (4.79) and (4.80) will ensure a high-frequency gain of 0 dB at all values of Q. For $C_2 = 0.01\ \mu\text{F}$ and $K = 4$, we get

$$C_a = \frac{0.01\ \mu\text{F}}{4} = 2.5\text{ nF} \quad \text{and} \quad C_b = \frac{0.01\ \mu\text{F} \times (4-1)}{4} = 7.5\text{ nF}$$

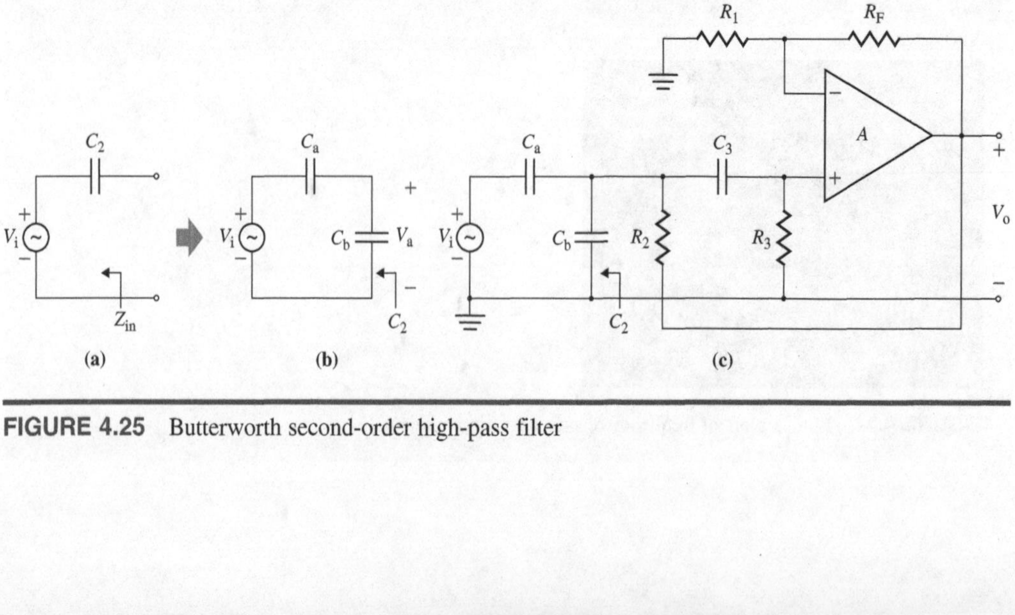

FIGURE 4.25 Butterworth second-order high-pass filter

As with low-pass filters, however, it is more desirable to have a gain of 0 dB at the resonant frequency ω_o—that is, for $|H(j\omega_o)| = 1$ (or 0 dB). Substituting $s = j\omega_o$ into Eq. (4.73) yields a gain magnitude of $K/(3 - K)$, which gives the required gain reduction of $(3 - K)/K$. Thus, Eq. (4.78) becomes

$$\frac{C_a}{C_a + C_b} = \frac{3 - K}{K} \tag{4.81}$$

Solving Eqs. (4.77) and (4.81) for C_a and C_b, we get

$$C_a = C_2\frac{3 - K}{K} \quad [\text{for } |H(j\omega_o)| = 1 \text{ (or 0 dB)}] \tag{4.82}$$

$$C_b = C_2\frac{2K - 3}{K} \quad [\text{for } |H(j\omega_o)| = 1 \text{ (or 0 dB)}] \tag{4.83}$$

Therefore, we can design a Butterworth high-pass filter to yield a gain of 0 dB at either $\omega = \infty$ or $\omega = \omega_o$. However, in the case where $|H(j\omega_o)| = 1$ (or 0 dB) is specified, the high-frequency gain will be reduced by a factor $(3 - K)/K$; that is,

$$|H(j\infty)| = 3 - K \quad [\text{for } |H(j\omega_o)| = 1 \text{ (or 0 dB)}] \tag{4.84}$$

For $Q = \sqrt{2}$ and $K = 3 - 1/Q = 1.586$, Eq. (4.84) gives $|H(j\omega_o)| = 3 - K = 1.414$, provided that we design the filter for $|H(j\omega_o)| = 1$ (or 0 dB).

EXAMPLE 4.7

D

Designing a second-order high-pass Butterworth filter for $|H(j\infty)| = 1$

(a) Design a second-order Butterworth high-pass filter as in Fig. 4.25(c) to yield $|H(j\infty)| = 1$ (or 0 dB), a cut-off frequency of $f_o = 1$ kHz, and $Q = 0.707$.

(b) Use PSpice/SPICE to plot the frequency response of the output voltage of the filter designed in part (a) from 10 Hz to 100 kHz.

SOLUTION

(a) For $Q = 0.707$, Example 4.6 gives $C = 0.01\ \mu\text{F}$ and $R = 15{,}916\ \Omega$. From Eq. (4.52),

$$K = \frac{3 - 1}{Q} = \frac{3 - 1}{0.707} = 1.586$$

and $\quad R_F = (K - 1)R_1 = (1.586 - 1) \times 15{,}916 = 9327\ \Omega$

From Eq. (4.79),

$$C_a = \frac{C}{K} = \frac{0.01\ \mu\text{F}}{1.586} = 6.305\text{ nF}$$

From Eq. (4.80),

$$C_b = \frac{C(K - 1)}{K} = \frac{0.01\ \mu\text{F} \times (1.586 - 1)}{1.586} = 3.695\text{ nF}$$

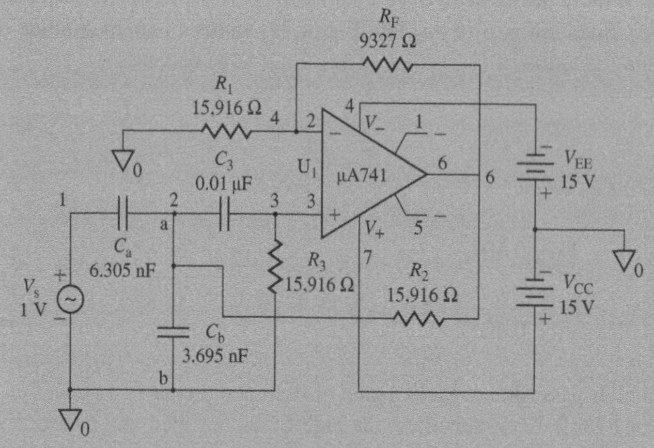

FIGURE 4.26 Second-order high-pass Butterworth filter for PSpice simulation

NOTE: The exact numbers of the resistances and capacitances are used to demonstrate the validity of the calculated values with the PSpice simulation results. In a practical design, the commercially available standard values will be used (i.e., 9.3 kΩ instead of 9327 Ω and C_b = 4 nF instead of 3.695 nF; see Appendix E).

(b) For PSpice simulation, the circuit of Fig. 4.18 can be transformed into the circuit of Fig. 4.26 by removing R_4 and R_5, interchanging the locations of *R* and *C*, replacing C_2 by C_a, and adding C_b between nodes a and b. The PSpice statements for *R* and *C* read as follows:

```
CA  1  2  6.305NF   ; For CA connected between nodes 1 and 2
CB  2  0  3.695NF   ; For CB connected between nodes 2 and 0
R2  2  6  15916     ; For R2 connected between nodes 2 and 6
C3  2  3  0.01UF    ; For C3 connected between nodes 2 and 3
R3  3  0  15916     ; For R3 connected between nodes 3 and 0
```

The PSpice plot of the gain is shown in Fig. 4.27, which gives $|H(j\infty)| = 1.0$ at $\omega = \infty$.

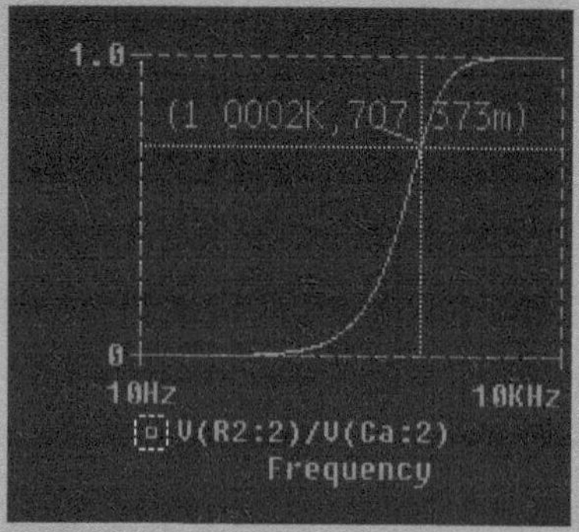

FIGURE 4.27 PSpice plot of frequency response for Example 4.7

KEY POINTS OF SECTION 4.9

- A low-pass filter can be converted to a high-pass filter by applying the *RC*-to-*CR* transformation: R_n is replaced by C_n, and C_n is replaced by R_n.
- The Sallen–Key low-pass circuit can be modified to exhibit second-order high-pass characteristics with a pass-band gain as well as a Butterworth response.

4.10 Band-Pass Filters

A band-pass filter has a passband between two cutoff frequencies f_L and f_H such that $f_H > f_L$. Any frequency outside this range is attenuated. The transfer function of a BP filter has the general form

$$H_{BP}(s) = \frac{K_{PB}(\omega_C/Q)s}{s^2 + (\omega_C/Q)s + \omega_C^2} \tag{4.85}$$

where K_{PB} is the pass-band gain and ω_C is the center frequency in radians per second. There are two types of band-pass filters: wide band pass and narrow band pass. Although there is no dividing line between the two, it is possible to identify them from the value of the *quality factor Q*. A filter may be classified as wide band pass if $Q \le 10$ and narrow band pass if $Q > 10$. The higher the value of Q, the more selective the filter or the narrower its bandwidth (BW) will be. Thus, Q is a measure of the selectivity of a filter. The relationship of Q to 3-dB bandwidth and center frequency f_C is given by

$$Q = \frac{\omega_C}{\text{BW}} = \frac{f_C}{f_H - f_L} \tag{4.86}$$

For a wide-band-pass filter, the center frequency f_C can be defined as

$$f_C = \sqrt{f_L f_H} \tag{4.87}$$

where f_L is the low cutoff frequency, in hertz, and f_H is the high cutoff frequency, in hertz. In a narrow-band-pass filter, the output peaks at the center frequency f_C.

4.10.1 Wide-Band-Pass Filters

The frequency characteristic of a wide-band-pass filter is shown in Fig. 4.28(a), where $f_H > f_L$. This characteristic can be obtained by implementing Eq. (4.85), which may not give a flat midband gain over a wide bandwidth. An alternative arrangement is to use two filters: one low-pass filter and one high-pass filter. The output is obtained by multiplying the low-frequency response by the high-frequency response, as shown in Fig. 4.28(b); this solution can be implemented simply by cascading the first-order (or second-order) high-

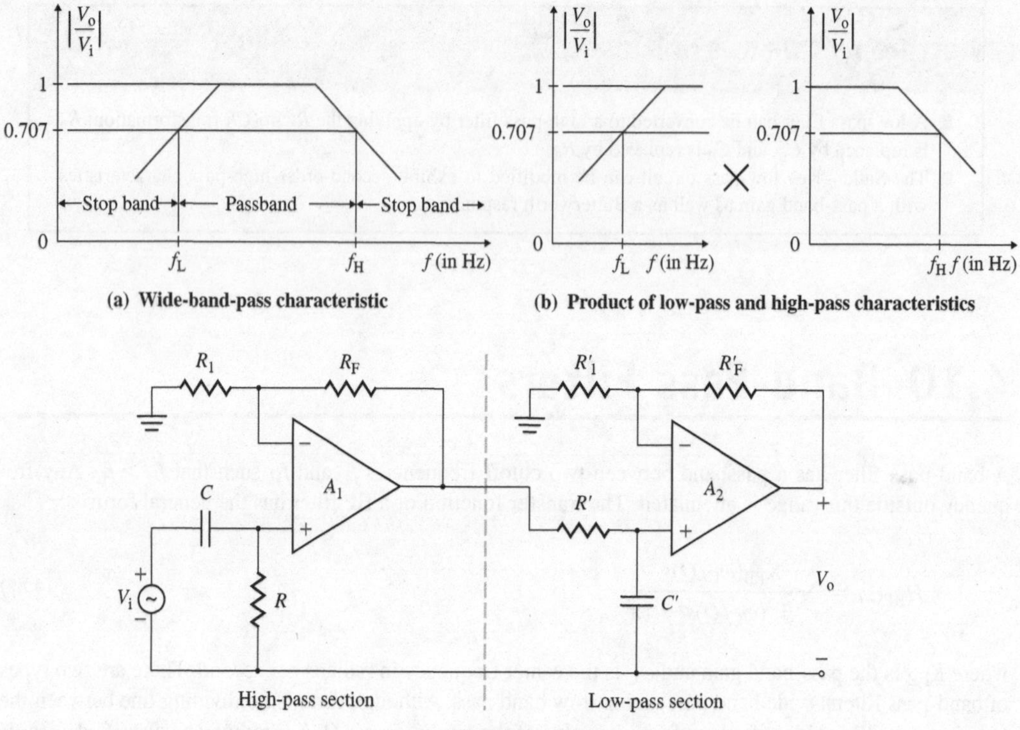

(c) Filter

FIGURE 4.28 Wide-band-pass filter

pass and low-pass sections. The order of the band-pass filter depends on the order of the high-pass and low-pass sections. This arrangement has the advantage that the falloff, rise, and midband gain can be set independently. However, it requires more op-amps and components.

Figure 4.28(c) shows a ±20 dB/decade wide-band-pass filter implemented with first-order high-pass and first-order low-pass filters. In this case, the magnitude of the voltage gain is equal to the product of the voltage gain magnitudes of the high-pass and low-pass filters. From Eqs. (4.40) and (4.67), the transfer function of the wide-midband filter for first-order implementation becomes

$$H(s) = \frac{K_{PB}\omega_H s}{(s + \omega_L)(s + \omega_H)} \tag{4.88}$$

Using Eqs. (4.46) and (4.72) gives the transfer function for second-order implementation:

$$H(s) = \frac{K_{PB}\omega_H^2 s^2}{[s^2 + (\omega_L/Q)s + \omega_L^2][s^2 + (\omega_H/Q)s + \omega_H^2]} \tag{4.89}$$

where K_{PB} = overall pass-band gain = high-pass gain K_H × low-pass gain K_L.

EXAMPLE 4.8

D

Designing a wide-band-pass filter

(a) Design a wide-band-pass filter with $f_L = 10$ kHz, $f_H = 1$ MHz, and a pass-band gain of $K_{PB} = 16$.

(b) Calculate the value of Q for the filter.

(c) Use PSpice/SPICE to plot the frequency response of the filter designed in part (a) from 100 Hz to 10 MHz.

SOLUTION

(a) Let the gain of the high-pass section be $K_H = 4$. For the first-order high-pass section, $f_L = 10$ kHz. Following the steps in Example 4.5, we let $C = 1$ nF. Then

$$R = \frac{1}{2\pi \times 10\,\text{kHz} \times 1\,\text{nF}} = 15.915\ \text{k}\Omega$$

and $$K_H = 1 + \frac{R_F}{R_1} = 4 \quad \text{or} \quad \frac{R_F}{R_1} = 4 - 1 = 3$$

If we let $R_1 = 10\ \text{k}\Omega$, $R_F = 3R_1 = 30\ \text{k}\Omega$.

For the first-order low-pass section, $f_H = 1$ MHz and the desired gain is $K_L = K_{PB}/K_H = 16/4 = 4$. Following the steps in Example 4.2, we let $C' = 10$ pF. Then

$$R' = \frac{1}{2\pi \times 1\,\text{MHz} \times 10\,\text{pF}} = 15.915\ \text{k}\Omega$$

and $$K_L = 1 + \frac{R'_F}{R'_1} = 4 \quad \text{or} \quad \frac{R'_F}{R'_1} = 4 - 1 = 3$$

If we let $R'_1 = 10\ \text{k}\Omega$, $R'_F = 3R'_1 = 30\ \text{k}\Omega$.

(b) From Eq. (4.87),

$$f_C = \sqrt{10\ \text{kHz} \times 1\ \text{MHz}} = 100\ \text{kHz}$$

and $$\text{BW} = 1\ \text{MHz} - 10\ \text{kHz} = 990\ \text{kHz}$$

From Eq. (4.86), we can find

$$Q = \frac{100\,\text{kHz}}{1\,\text{MHz} - 10\,\text{kHz}} = 0.101$$

NOTE: The exact numbers of the resistances are used to demonstrate the validity of the calculated values with the PSpice simulation results. In a practical design, the commercially available standard values will be used (i.e., 16 kΩ instead of 15.915 kΩ; see Appendix E).

(c) The wide-band-pass filter with the calculated values is shown in Fig. 4.29.

The frequency response (with a linear model) is shown in Fig. 4.30, which gives $K_{PB} = 15.842$ (expected value is 16), $f_L = 10.04$ kHz (expected value is 10 kHz), and $f_H = 997$ kHz (expected value is 1 MHz). For a low value of bandwidth, the response due to the high-pass filter may not reach the expected gain before the low-pass filter becomes effective. As a result, the pass-band gain may be much lower than 16.

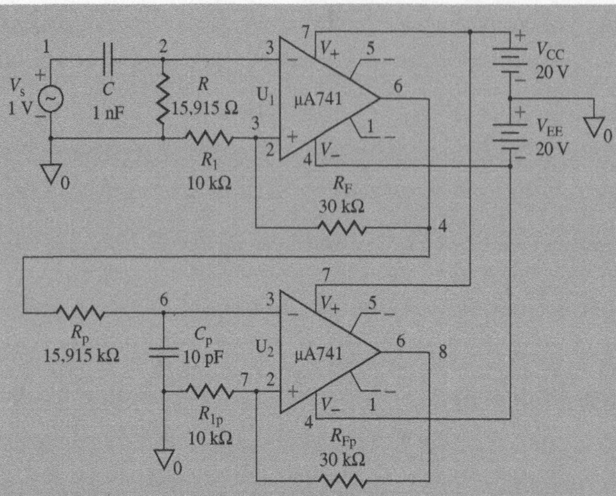

FIGURE 4.29 First-order band-pass filter for PSpice simulation

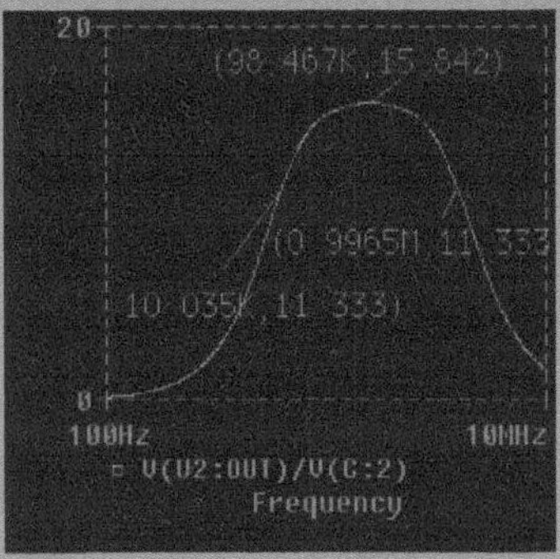

FIGURE 4.30 PSpice plot of frequency response for Example 4.8

NOTE: If we use the linear model, the high-end roll-off will be sooner due to the internal capacitances of the μA741 op-amp.

4.10.2 Narrow-Band-Pass Filters

A typical frequency response of a narrow-band-pass filter is shown in Fig. 4.31(a). This characteristic can be derived by setting a high Q-value for the band-pass filter shown in Fig. 4.31(b). This filter uses only one op-amp in the inverting mode. Because it has two feedback paths, it is also known as a *multiple feedback filter*. For a low Q-value, it can also exhibit the characteristic of a wide-band-pass filter.

A narrow-band-pass filter is generally designed for specific values of f_C and Q or f_C and BW. The op-amp, along with C_2 and R_2, can be regarded as an inverting differentiator such that $V_o(s) = (-sC_2R_2)V_x(s)$; the equivalent filter circuit is shown in Fig. 4.31(c). The transfer function of the filter network is

$$H_{BP}(s) = \frac{V_o(s)}{V_i(s)} = \frac{(-1/R_1C_1)s}{s^2 + (1/R_2)(1/C_1 + 1/C_2)s + 1/R_1R_2C_1C_2} \tag{4.90}$$

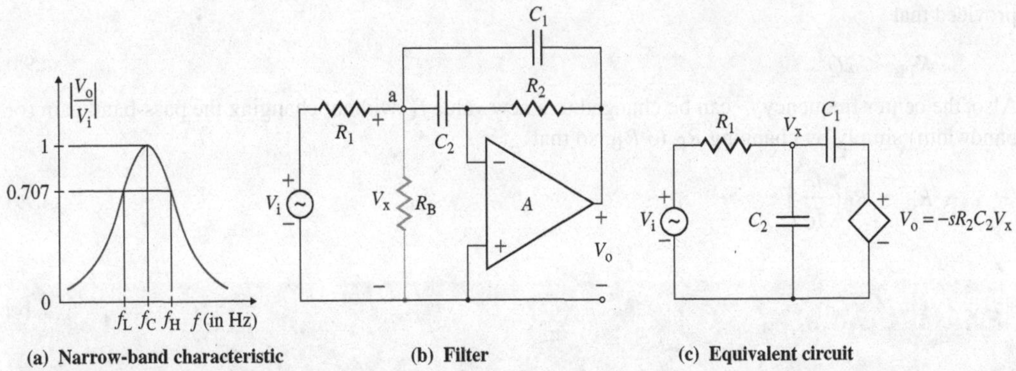

FIGURE 4.31 Narrow-band-pass filter

which is similar in form to Eq. (4.85). (See Prob. 4.29 for the derivation.) For $C_1 = C_2 = C$, Eq. (4.90) gives

$$\omega_C = \frac{1}{\sqrt{R_1 R_2 C_1 C_2}} = \frac{1}{C\sqrt{R_1 R_2}} \tag{4.91}$$

$$Q = \frac{1}{2}\sqrt{\frac{R_2}{R_1}} \tag{4.92}$$

$$K_{PB}\left(\frac{\omega_C}{Q}\right) = \frac{1}{R_1 C_1} \tag{4.93}$$

Solving these equations, we can find the component values:

$$R_1 = \frac{Q}{2\pi f_C C K_{PB}} \tag{4.94}$$

$$R_2 = \frac{Q}{\pi f_C C} \tag{4.95}$$

$$K_{PB} = \frac{R_2}{2R_1} = 2Q^2 \tag{4.96}$$

Resistance R_1 can be replaced by R_A, and resistance R_B can be connected between nodes a and 0 so that the design specification $|H_{BP}(j\omega_C)| = 1$ (or 0 dB) is met for the Butterworth response. The method of calculating the values of R_A and R_B for a gain reduction of $1/K_{PB}$ $(=1/2Q^2)$ is explained in Sec. 4.8.

Notice from Eq. (4.96) that, for a known value of Q, the value of K_{PB} is fixed. It is, however, possible to have different values of K_{PB} and Q by choosing only the value of R_B without changing the value of R_1. The new value of the gain K_{PB} is related to $2Q^2$ by

$$\frac{K_{PB}}{2Q^2} = \frac{R_B}{R_1 + R_B}$$

which gives the value of R_B as

$$R_B = \frac{R_1 K_{PB}}{2Q^2 - K_{PB}} = \frac{Q}{2\pi f_C C(2Q^2 - K_{PB})} \tag{4.97}$$

provided that

$$K_{PB} < 2Q^2 \tag{4.98}$$

Also, the center frequency f_C can be changed to a new value f'_C without changing the pass-band gain (or bandwidth) simply by changing R_B to R'_B, so that

$$R'_B = R_B\left(\frac{f_C}{f'_C}\right)^2 \tag{4.99}$$

EXAMPLE 4.9

D

Designing a narrow-band-pass filter

(a) Design a narrow-band-pass filter as in Fig. 4.31(b) such that $f_C = 1$ kHz, $Q = 4$, and $K_{PB} = 8$.

(b) Calculate the value of R_B required to change the center frequency from 1 kHz to 1.5 kHz.

(c) Use PSpice/SPICE to plot the frequency response of the narrow-band-pass filter designed in part (a) from 100 Hz to 1 MHz.

SOLUTION

(a) $f_C = 1$ kHz and $Q = 4$. Let $C_1 = C_2 = C = 0.0047$ μF. Check that the condition in Eq. (4.98) is satisfied; that is, $2Q^2 = 2 \times 4^2 = 32$, which is greater than $K_{PB} = 8$. Thus, we must use R_B in Fig. 4.32. Using Eqs. (4.94), (4.95), (4.96), and (4.97), we get

$$R_1 = \frac{Q}{2\pi f_C C K_{PB}} = \frac{4}{2\pi \times 1\text{ kHz} \times 0.0047\,\mu\text{F} \times 8} = 16.93\text{ k}\Omega$$

$$R_2 = \frac{Q}{\pi f_C C} = \frac{4}{\pi \times 1\text{ kHz} \times 0.0047\,\mu\text{F}} = 270.9\text{ k}\Omega$$

$$K_{PB} = \frac{R_2}{2R_1} = \frac{270.9\text{ k}\Omega}{2 \times 16.93\text{ k}\Omega} = 8$$

$$R_B = \frac{Q}{2\pi f_C C(2Q^2 - K_{PB})} = \frac{4}{2\pi \times 1\text{kHz} \times 0.0047\,\mu\text{F} \times (2 \times 4^2 - 8)} = 5.64\text{ k}\Omega$$

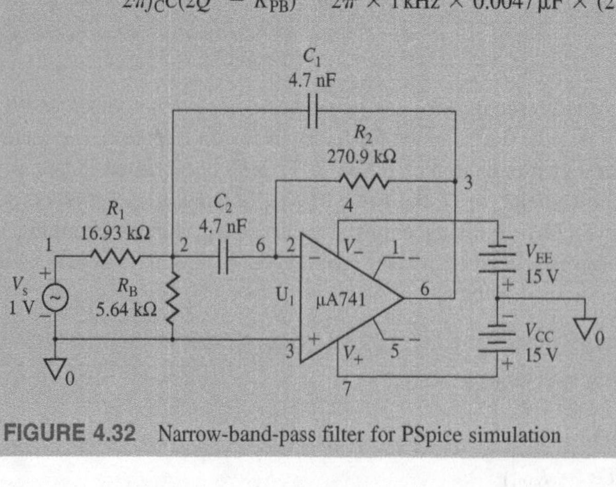

FIGURE 4.32 Narrow-band-pass filter for PSpice simulation

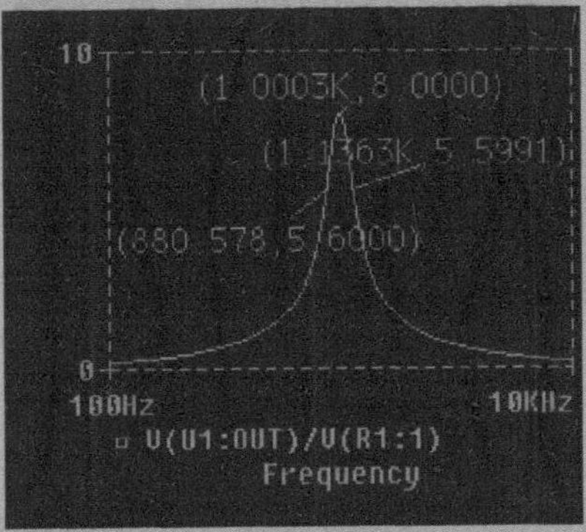

FIGURE 4.33 PSpice plot of frequency response for Example 4.9

(b) From Eq. (4.99), we find that the new value of R'_B is

$$R'_B = R_B\left(\frac{f_C}{f'_C}\right)^2 = 5.64\text{ k}\Omega\left(\frac{1\text{ kHz}}{1.5\text{ kHz}}\right)^2 = 2.51\text{ k}\Omega$$

NOTE: The exact numbers of the resistances are used to demonstrate the validity of the calculated values with the PSpice simulation results. In a practical design, the commercially available standard values will be used (i.e., 17 kΩ instead of 16.93 kΩ and 270 kΩ instead of 270.9 kΩ; see Appendix E).

(c) The narrow-band-pass filter with the designed values is shown in Fig. 4.32.

The frequency response is shown in Fig. 4.33, which gives $f_C \approx 1$ kHz (expected value is 1 kHz) and $K_{PB} = 8$.

KEY POINTS OF SECTION 4.10

- The wide-band-pass characteristic can be obtained by cascading a high-pass filter with a low-pass filter.
- A narrow-band-pass filter has a sharply tuned center frequency and can be implemented with only one op-amp in inverting mode of operation.

4.11 Band-Reject Filters

A band-reject filter attenuates signals in the stop band and passes those outside this band. It is also called a *band-stop* or *band-elimination filter*. The transfer function of a second-order band-reject filter has the general form

$$H_{BR}(s) = \frac{K_{PB}(s^2 + \omega_C^2)}{s^2 + (\omega_C/Q)s + \omega_C^2} \tag{4.100}$$

where K_{PB} is the pass-band gain. Band-reject filters can be classified as wide band reject or narrow band reject. A narrow-band-reject filter is commonly called a *notch filter*. Because of its higher Q (>10), the bandwidth of a narrow-band-reject filter is much smaller than that of a wide-band-reject filter.

4.11.1 Wide-Band-Reject Filters

The frequency characteristic of a wide-band-reject filter is shown in Fig. 4.34(a). This characteristic can be obtained by adding a low-pass response to a high-pass response, as shown in Fig. 4.34(b); the solution can be implemented by summing the responses of a first-order (or second-order) high-pass section and low-pass section through a summing amplifier. This arrangement is shown in Fig. 4.34(c). The order of the band-reject filter depends on the order of the high-pass and low-pass sections. For a band-reject response to be realized, the cutoff frequency f_L of the high-pass filter must be larger than the cutoff frequency f_H of the low-pass filter. In addition, the pass-band gains of the high-pass and low-pass sections must be equal. With an inverting summer (A_3), the output will be inverted. Note that R_{OM} in Fig. 4.34(c) has no function but to minimize the op-amp offsets.

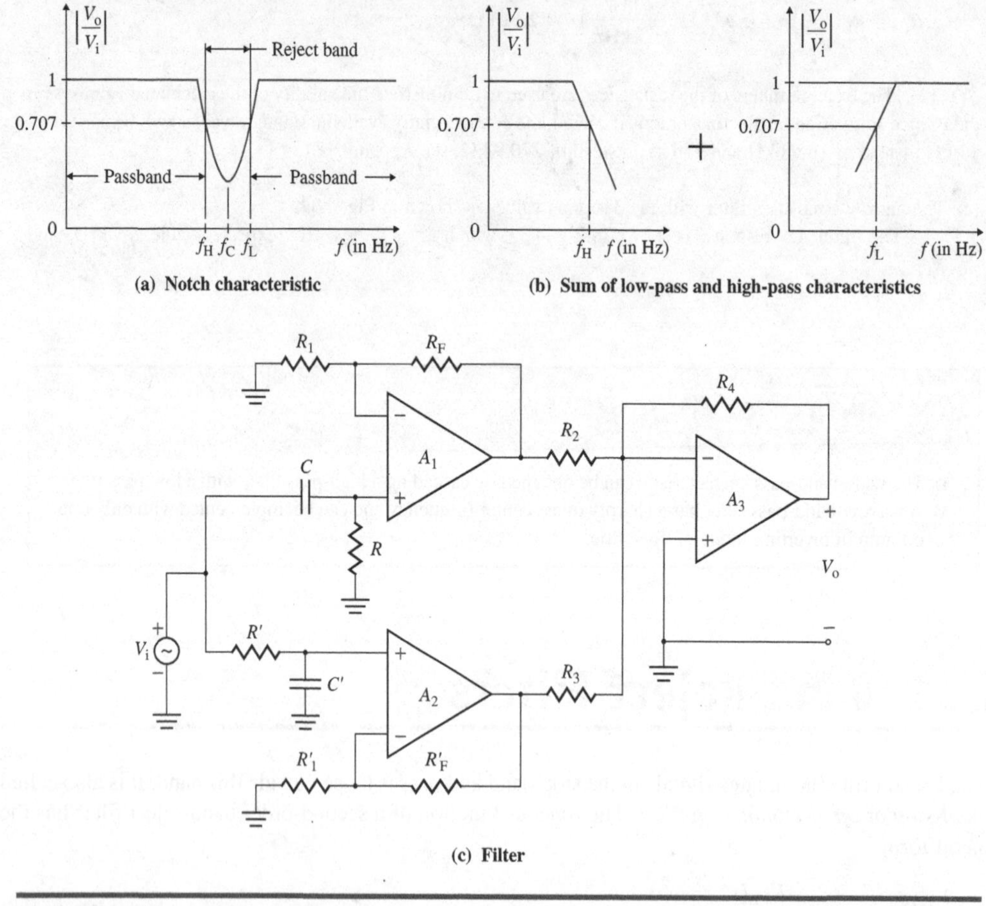

FIGURE 4.34 Wide-band-reject filter

EXAMPLE 4.10

Designing a wide-band-reject filter

(a) Design a wide-band-reject filter as shown in Fig. 4.34(c) with f_L = 100 kHz, f_H = 10 kHz, and a pass-band gain of K_{PB} = 4.

(b) Calculate the value of Q for the filter.

(c) Use PSpice/SPICE to plot the frequency response of the filter designed in part (a) from 10 Hz to 10 MHz.

SOLUTION

(a) In Example 4.8, we designed a wide-band-pass filter with f_L = 10 kHz and f_H = 1 MHz. In this example, we have f_L = 100 kHz and f_H = 10 kHz; that is, $f_L > f_H$. However, we can follow the design steps in Example 4.8 to find the component values, provided that we interchange the high-pass and low-pass sections. Thus, for the high-pass section of f_L = 100 kHz, C = 100 pF and R = 15.915 kΩ, and for the low-pass section of f_H = 10 kHz, C' = 1 nF and R' = 15.915 kΩ. For a pass-band gain of K_{PB} = 4, use $R_1 = R_1'$ = 10 kΩ and $R_F = R_F'$ = 30 kΩ. For the summing amplifier, set a gain of 1. Choose $R_2 = R_3 = R_4$ = 10 kΩ.

(b) From Eq. (4.87),

$$f_C = \sqrt{10 \text{ kHz} \times 100 \text{ kHz}} = 31.623 \text{ kHz}$$

and $\text{BW} = 100 \text{ kHz} - 10 \text{ kHz} = 90 \text{ kHz}$

From Eq. (4.86), we can find

$$Q = \frac{31.623 \text{ kHz}}{(100 \text{ kHz} - 10 \text{ kHz})} = 0.351$$

(c) The circuit for PSpice simulation of the wide-band-reject filter is shown in Fig. 4.35.

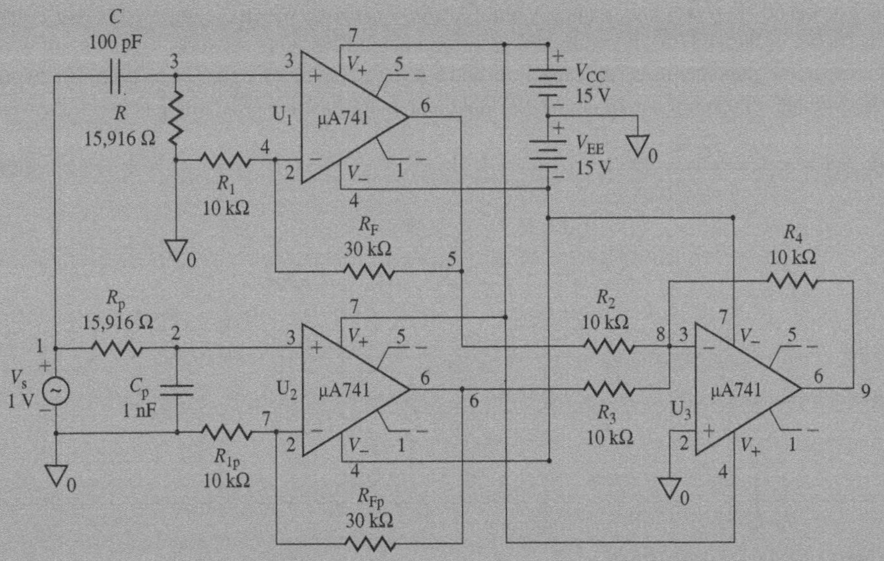

FIGURE 4.35 Wide-band-reject filter for PSpice simulation

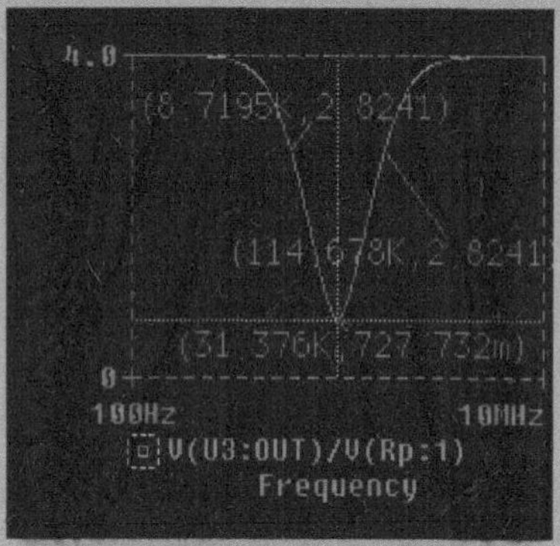

FIGURE 4.36 PSpice plot of frequency response for Example 4.10 (using linear op-amp model)

The frequency response is shown in Fig. 4.36, which gives $f_C \approx 31.376$ kHz (expected value is 31.623 kHz) and $K_{PB} = 4$ (expected value is 4).

NOTE: If we use the linear op-amp model, there will be a serious high-end roll-off with μA741 op-amp.

4.11.2 Narrow-Band-Reject Filters

A typical frequency response of a narrow-band-reject filter is shown in Fig. 4.37(a). This filter, often called a *notch filter*, is commonly used in communication and biomedical instruments to eliminate undesired frequencies such as the 60-Hz power line frequency hum. A *twin-T network*, which is composed of two T-shaped networks, as shown in Fig. 4.37(b), is commonly used for a notch filter. One network is made up of two resistors and a capacitor; the other uses two capacitors and a resistor. To increase the Q of a twin-T network, it is used with a voltage follower. It can be shown [9] that the transfer function of a twin-T network is given by

$$H_{NF}(s) = \frac{K_{PB}(s^2 + \omega_n^2)}{s^2 + (\omega_o/Q)s + \omega_o^2} \tag{4.101}$$

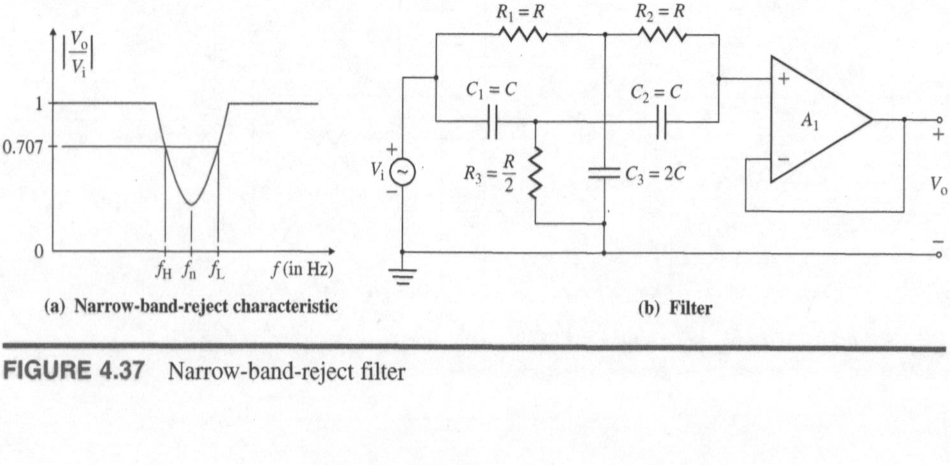

(a) Narrow-band-reject characteristic (b) Filter

FIGURE 4.37 Narrow-band-reject filter

where $\omega_n = 1/RC$

$\omega_o = 1/\sqrt{3}RC$

$Q = \sqrt{3/4}$

$K_{PB} = 1$

Therefore, the *notch-out frequency*, which is the frequency at which maximum attenuation occurs, is given by

$$f_n = \frac{1}{2\pi RC} \tag{4.102}$$

EXAMPLE 4.11

D

Designing a narrow-band-reject (notch) filter

(a) Design a notch filter as in Fig. 4.37(b) with $f_n = 60$ Hz.

(b) Use PSpice/SPICE to plot the frequency response of the filter designed in part (a) from 1 Hz to 1 kHz.

SOLUTION

(a) $f_n = 60$ Hz. Choose a value of C less than or equal to 1 μF: Let $C = 0.047$ μF. Then, from Eq. (4.102),

$$R = \frac{1}{2\pi f_n C} = \frac{1}{2\pi \times 60 \text{ Hz} \times 0.047 \ \mu\text{F}} = 56.44 \text{ k}\Omega$$ (use a 59-kΩ standard resistor of 10% tolerance)

$$R_3 = \frac{R}{2} = 28.22 \text{ k}\Omega$$ (use two 59-kΩ resistors in parallel)

$$C_3 = 2C = 0.094 \ \mu\text{F}$$ (use two 0.047-μF capacitors in parallel)

(b) The circuit of the notch filter for PSpice simulation is shown in Fig. 4.38.

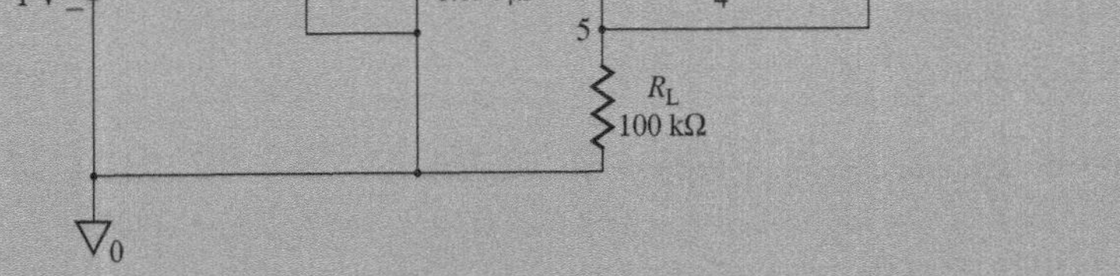

FIGURE 4.38 Narrow-band-reject filter for PSpice simulation

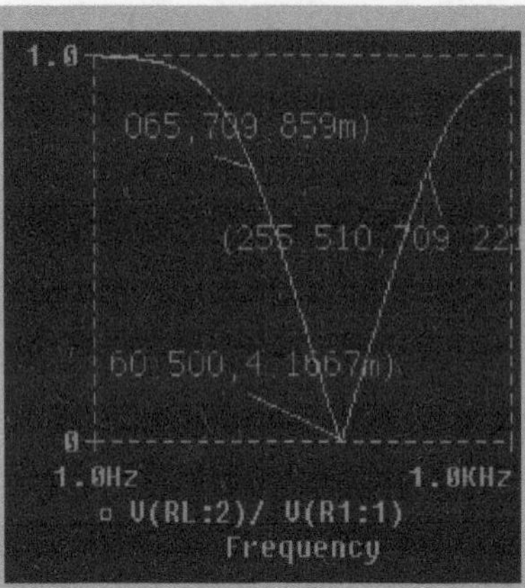

FIGURE 4.39 PSpice plot of frequency response for Example 4.11

The frequency response of the filter is shown in Fig. 4.39, which gives f_n = 60.9 Hz (expected value is 60 Hz) and K_{PB} = 1 (expected value is 1).

KEY POINTS OF SECTION 4.11

- The wide-band-reject characteristic can be obtained by adding the output of a low-pass filter to that of a high-pass filter through a summing amplifier.
- A narrow-band-reject filter has a sharply tuned reject frequency and can be implemented with only one op-amp in the noninverting mode of operation.

4.12 Delyiannis-Friend Circuit

The Delyiannis-Friend circuit as shown in Fig. 4.40 can exhibit the characteristic of a band-pass filter as listed in category (b) in Table 4.3. This circuit can be used as a module and cascaded to realize the Butterworth band-pass specifications. It consists of a differentiating circuit shown in Fig. 4.40(b) and its equivalent circuit shown in Fig. 4.40(c). The elements R_1 and C_1, which act as an integrating circuit, provide negative feedback. Assuming $V_d = 0$, the simplified equivalent circuit is shown in Fig. 4.40(d).

Using the relationship of an inverting op-amp circuit in Fig. 4.40(b), V_2 is given by

$$V_2 = -\frac{R_2}{1/sC_2} V_x = -R_2 C_2 s V_s \tag{4.103}$$

Appling Kirchhoff's current law at node A in Figure 4.40(d), we get

$$\frac{1}{R_1}(V_x - V_1) + sC_2 V_x + (V_x - V_2) sC_1 = 0 \tag{4.104}$$

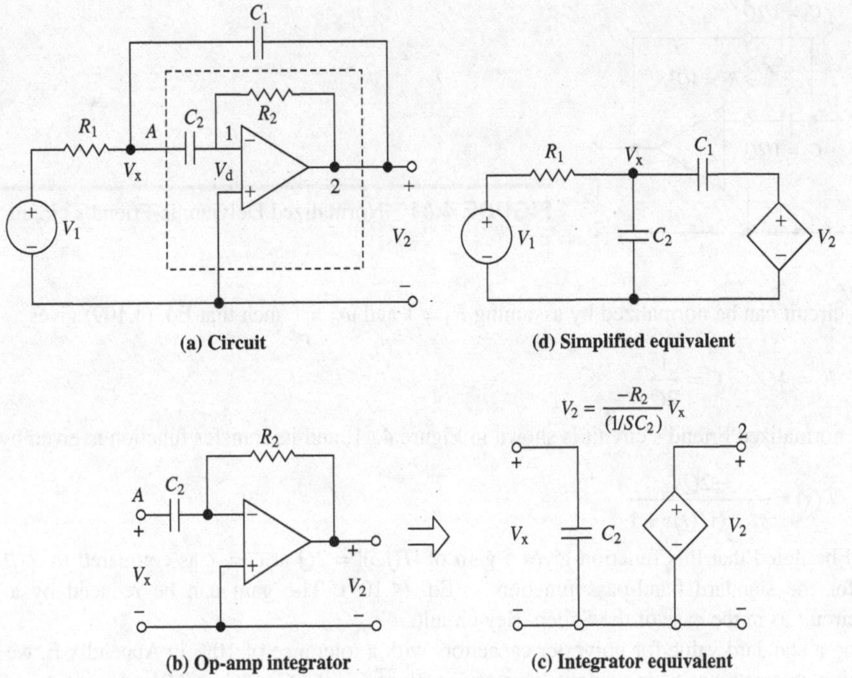

FIGURE 4.40 Delyiannis-Friend circuit

Substitute V_2 from Eq. (4.103) in Eq. (4.104) and solve for the transfer function as

$$T(s) = \frac{V_2}{V_1}(s) = \frac{(-1/R_1C_1)s}{s^2 + (1/R_2)(1/C_1 + 1/C_2)s + 1/R_1C_1R_2C_2} \tag{4.105}$$

Assuming $C_1 = C_2 = C$, Eq. (4.105) can be simplified to give

$$T(s) = \frac{(-1/R_1C)s}{s^2 + (2/R_2C)s + 1/(R_1R_2C^2)} \tag{4.106}$$

The standard form of the band-pass filter response as listed in category (b) in Table 4.3 is given by

$$T(s) = \frac{-(\omega_o/Q)s}{s^2 + (\omega_o/Q)s + \omega_o^2} \tag{4.107}$$

Comparing the coefficients of Eq. (4.106) with those of Eq. (4.107), we get

$$\omega_o = \frac{1}{C\sqrt{R_1R_2}}; \quad Q = \frac{1}{2}\sqrt{\frac{R_2}{R_1}}; \quad \text{BW} = \frac{\omega_o}{Q} = \frac{2}{CR_2} \tag{4.108}$$

Substituting Eq. (4.108) into Eq. (4.105) gives the transfer function of Friend's circuit as

$$T(s) = \frac{-(2\omega_o Q)s}{s^2 + (\omega_o/Q)s + \omega_o^2} \tag{4.109}$$

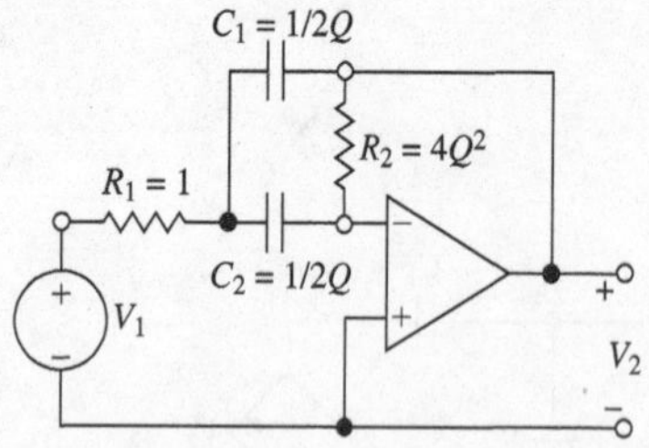

FIGURE 4.41 Normalized Delyiannis-Friend's circuit

The circuit can be normalized by assuming $R_1 = 1$ and $\omega_o = 1$ such that Eq. (4.109) gives

$$R_2 = 4Q^2; \quad C = \frac{1}{2Q} \tag{4.110}$$

The normalized Friend's circuit is shown in Figure 4.41, and its transfer function is given by

$$T(s) = \frac{-2Qs}{s^2 + (1/Q)s + 1} \tag{4.111}$$

It should be noted that this function gives a gain of $|T(j\omega)| = 2Q$ at $\omega = 1$ as compared to $|T(j1)| = 1$ at $\omega = 1$ for the standard band-pass function in Eq. (4.107). The gain can be reduced by a voltage-divider circuit as in the case of the Sallen–Key circuit.

Using a standard value for polyester capacitors with a tolerance of 10% in Appendix E, we can find the circuit elements by selecting a new value of C such as $C_{new} = C_1 = C_2 = 0.01\ \mu F$.

$$k_f = \omega_o; \quad k_m = \frac{C_{old}}{C'_{new} k_f} = \frac{1/2Q}{0.01\mu F \times \omega_o} \tag{4.112a}$$

$$R_1 = k_m; \quad R_2 = 4Q^2 k_m \tag{4.112b}$$

KEY POINTS OF SECTION 4.12

- The Delyiannis-Friend circuit exhibits the characteristic of a band-pass filter.
- This circuit can be used as a module and cascaded to realize the Butterworth band-pass specifications.

4.13 Stagger-Tuned Band-Pass Filter

One or more first-order and second-order pass-band filters are generally connected in cascade to form a stagger filter. The specifications include stop and pass band requirements rather than defining the cut-off at 70.7% of the maximum value as in the case of Butterworth response. A typical band-pass characteristic with brick wall is shown in Fig. 4.42(a). This can be represented by normalized low-pass brick wall with the frequency ratio u as shown in Fig. 4.42(b). The frequency ratio u can be transformed to frequencies that define the pass bands and stop bands shown in Fig. 4.42(c). The frequency transformation from the low pass to the band pass can be expressed by a *Foster reactance function* [3] given by

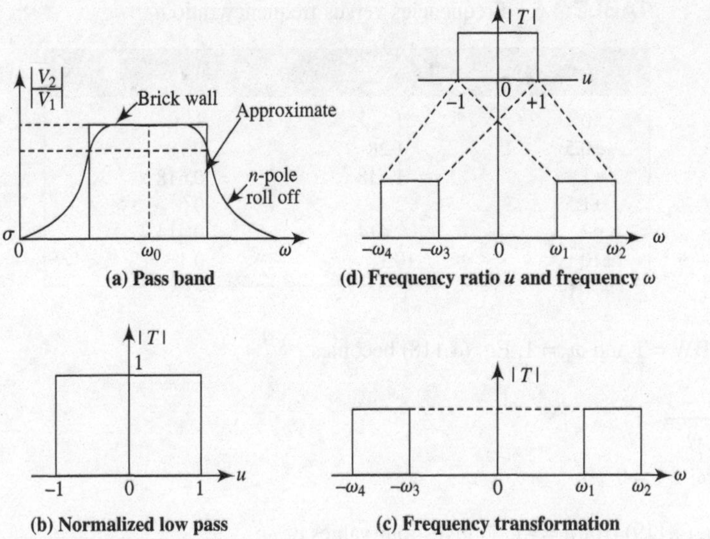

FIGURE 4.42 Brick-wall characteristics to frequency transformation

$$u(\omega) = K\frac{\omega^2 - \omega_o^2}{\omega} \tag{4.113}$$

where K is a constant and ω_o is the brick-wall center frequency. For the brick function defined by $u = 1, -1$, there are four values of frequencies as shown in Fig. 4.42(d). This function has to match the conditions $u(\omega_2) = +1$ and $u(\omega_1) = -1$. Substituting these conditions in Eq. (4.113), we get

$$u(\omega_1) = -1 = K\frac{\omega_1^2 - \omega_o^2}{\omega_1} \tag{4.114}$$

$$u(\omega_2) = +1 = K\frac{\omega_2^2 - \omega_o^2}{\omega_2} \tag{4.115}$$

Solving Eqs. (4.114) and (4.115), we get

$$\omega_o^2 = \omega_1\omega_2 \tag{4.116}$$

$$K = \frac{1}{\omega_2 - \omega_1} = \frac{1}{\text{BW}} \tag{4.117}$$

Thus, ω_o is the geometrical mean of the two frequencies ω_1 and ω_2. K is the reciprocal of the bandwidth BW. The bandwidth is defined as the ends of the brick wall at the half-power or 3-dB point. Substituting Eqs. (4.116) and (4.117) in Eq (4.113), we get the frequency transformation function as

$$u(\omega) = \frac{1}{\text{BW}}\frac{\omega^2 - \omega_1\omega_2}{\omega} \tag{4.118}$$

TABLE 4.5 Frequencies versus frequency ratio u

u	ω_2	ω_1
0	1	1
±0.5	1.28	0.78
±1	1.618	0.618
±1.5	2	0.5
±2	2.414	0.414
±10	10.1	0.1

Assuming BW = 1 and $\omega_o = 1$, Eq. (4.118) becomes

$$u = \frac{\omega^2 - 1}{\omega}$$

$$\text{or} \quad \omega^2 - u\omega + 1 = 0 \tag{4.119}$$

Solving Eq. (4.119) (for $u = -1, 1$) gives four values of ω:

$$\omega_1, \omega_2, \omega_3, \omega_4 = \frac{\pm u \pm \sqrt{u^2 + 4}}{2} \tag{4.120}$$

If ω_1 and ω_2 are the two positive values, then ω_3 and ω_4 are the corresponding negative values. These values are tabulated in Table 4.5. One value moves toward the origin and the other toward the infinity for increasing values of u.

A low-pass pole can be transformed into two equivalent band-pass poles and vice versa as shown in Fig. 4.43 and related by Eq. (4.118). Let S_L be the low-pass variable such that $S_L = a + ju$. Substituting $\omega = s/j$ and $u = S/j$, we can express Eq. (4.118) in Lapalce's domain of s:

$$S_L(s) = \frac{1}{\text{BW}} \frac{s^2 + \omega_{o2}^2}{s} \tag{4.121}$$

This can also be expressed as

$$S_L = \frac{\omega_o}{\text{BW}}\left(\frac{s}{\omega_o} + \frac{\omega_o}{s}\right) = q_c\left(\frac{s}{\omega_o} + \frac{\omega_o}{s}\right) \tag{4.122}$$

where q_c is the quality factor defined by

$$q_c = \frac{\omega_o}{\text{BW}} = \frac{\omega_o}{\omega_2 - \omega_1} \tag{4.123}$$

Assuming BW = 1 and $\omega_o = 1$, Eq. (4.121) becomes

$$S_L = s + \frac{1}{s} \tag{4.124}$$

4.13.1 Transformation of Butterworth Function

Let us apply the frequency transformation to the third-order Butterworth filter with the transfer function as given by

$$T(s) = \frac{1}{s^3 + 2s^2 + 2s + 1} = \frac{1}{(s+1)(s^2+s+1)} \tag{4.125}$$

This consists of a first-order function and a second-order function such that $T(s) = T_1(s)\, T_2(s)$. We can extension the transformations to higher order filters. Substituting the low-pass variable $S = a + ju$ in Eq. (4.124) in $T_1(s)$ and $T_2(s)$, we get

$$T_1(S) = \frac{1}{S+1} = \frac{1}{s + \frac{1}{s} + 1} = \frac{s}{s^2 + s + 1} \tag{4.126}$$

The roots of Eq. (4.126) can be calculated from the following equation:

$$s^2 + s + 1 = 0 \tag{4.127}$$

which can be solved for the roots and the radius of the circle in the s-plane as

$$s_1, s_2 = \sigma \pm j\omega = -0.5 \pm j0.866$$

$$a_1 = \sqrt{\sigma_1^2 + \omega_1^2} = \sqrt{(-0.5)^2 + 0.866^2} = 0.99996$$

Therefore, the first-order function has a pole on the negative real axis that has transformed into two conjugate poles at ±60° with respect to the negative real axis and a zero at the origin as shown in Fig. 4.43(a).

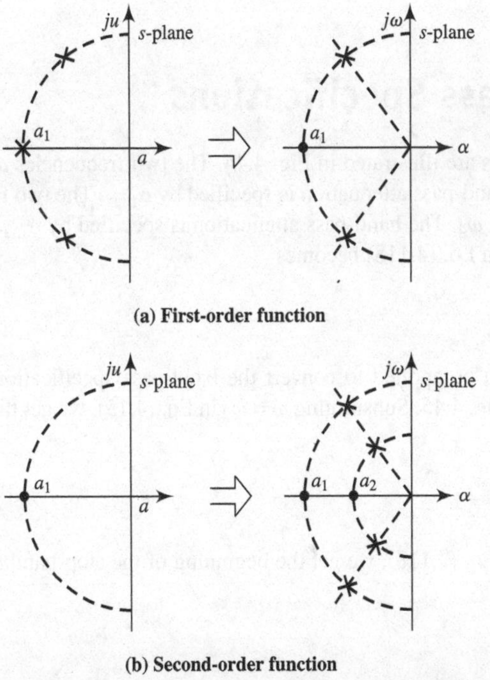

FIGURE 4.43 Locations of poles and zeros

Similarly, the second-transfer function $T_2(S_L)$ becomes

$$T_2(S_L) = \frac{1}{S_L^2 + S_L + 1} = \frac{1}{\left(s + \frac{1}{s}\right)^2 + \left(s + \frac{1}{s}\right) + 1} = \frac{s^2}{s^4 + s^3 + 3s^2 + s + 1} \quad (4.128)$$

The roots of Eq. (4.128) can be calculated from

$$s^4 + s^3 + 3s^2 + s + 1 = 0 \quad (4.129)$$

which can be solved for the roots (by a Mathcad or Mathlab software) and the radiuses of the circles in the s-plane as

$$s_1, s_2 = -0.3516 \pm j1.49853$$

$$s_3, s_4 = -0.1484 \pm j0.6325$$

$$a_1 = \sqrt{\sigma_1^2 + \omega_1^2} = \sqrt{(-0.3516)^2 + 1.49853^2} = 1.53923$$

$$a_3 = \sqrt{\sigma_3^2 + \omega_3^2} = \sqrt{(-0.1484)^2 + 0.6325^2} = 0.64968$$

Using the values of the roots, Eq. (4.128) can be expressed in quadratic terms as given by

$$T_2(S_L) = \frac{s}{(s^2 + 0.7032s + 2.36921)} \times \frac{s}{(s^2 + 0.2968s + 0.42208)} \quad (4.130)$$

Therefore, the second-order function has two poles transform into four poles and two zeros at the origin as shown in Fig. 4.43(b).

4.13.2 Band-Pass Specifications

The band-pass specifications are illustrated in Fig. 4.44. The two frequencies ω_1 and ω_2 define the bandwidth BW = $\omega_2 - \omega_1$. The band-pass attenuation is specified by α_{max}. The two frequencies at which the attenuation is α_{min} are ω_3 and ω_4. The band-pass attenuation is specified by α_{max}. Assuming $\omega_1\omega_2 = \omega_3\omega_4$, the frequency transformation Eq. (4.118) becomes

$$u = -\frac{-\omega^2 + \omega_o^2}{\omega(\omega_2 - \omega_1)} \quad (4.131)$$

This transformation can be applied to convert the band-pass specification into equivalent low-pass specifications as shown in Fig. 4.45. Substituting $\omega = \omega_2$ in Eq. (4.15), we get the end of the pass-band frequency ratio u_p as given by

$$u_p = -\frac{-\omega_2^2 + \omega_o^2}{\omega_2(\omega_2 - \omega_1)} = 1 \quad (4.132)$$

Substituting $\omega = \omega_4$ in Eq. (4.118), we get the beginning of the stop-band frequency ratio u_s as given by

$$u_s = \frac{\omega_4 - \omega_3}{\omega_2 - \omega_1} \quad (4.133)$$

Therefore, the band-pass specifications include the normalized low-pass specifications as given by

$$\alpha_{min}; \quad \alpha_{max}; \quad u_p = 1; \quad u_s = \frac{\omega_4 - \omega_3}{\omega_2 - \omega_1} \quad (4.134)$$

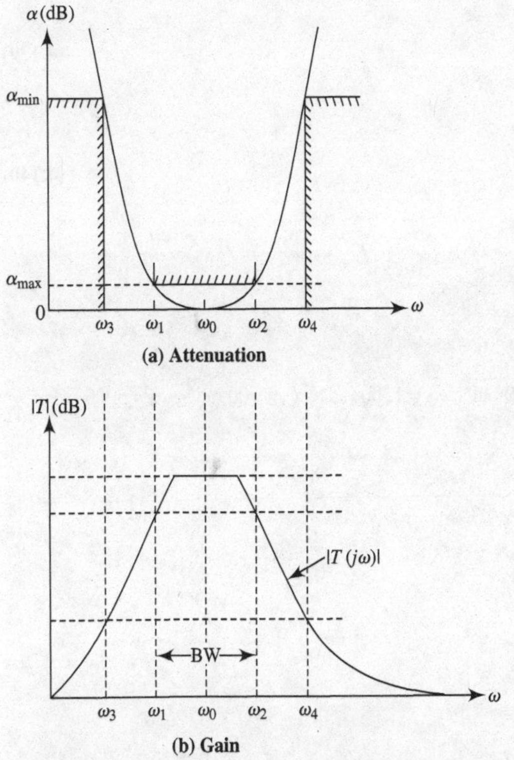

FIGURE 4.44 Band-pass frequency specifications

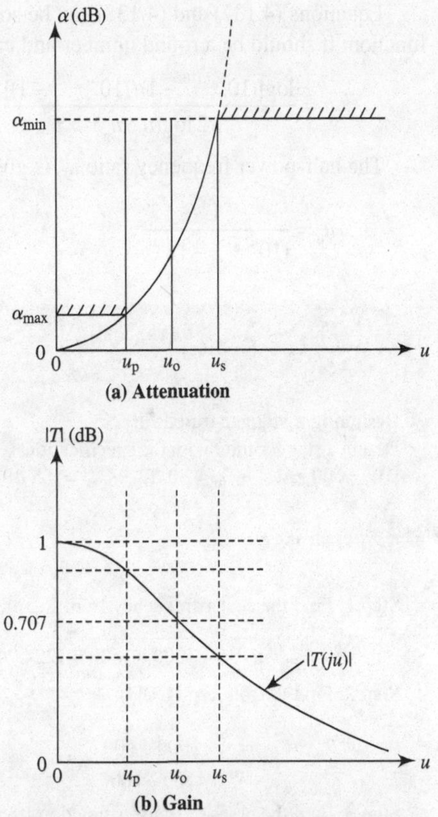

FIGURE 4.45 Low-frequency specifications

The attenuation for the Butterworth response is given by

$$\alpha = -20 \log |T_n(j1)| = -20 \log\left(\frac{1}{\sqrt{1+u^{2n}}}\right) = 10 \log(1+u^{2n}) \quad \text{dB} \tag{4.135}$$

which after dividing by 10 and then taking the antilogarithm gives

$$u = \frac{\omega}{\omega_o} = [10^{\alpha/10} - 1]^{1/2n} \tag{4.136}$$

Since at $u = u_p$, $\alpha = \sigma_{max}$ and at $u = u_s$, $\alpha = \sigma_{min}$, Eq. (4.136) gives

$$u_p = \frac{\omega_p}{\omega_o} = [10^{\alpha_{max}/10} - 1]^{1/2n} \tag{4.137}$$

$$u_s = \frac{\omega_s}{\omega_o} = [10^{\alpha_{min}/10} - 1]^{1/2n} \tag{4.138}$$

Equations (4.137) and (4.138) can be solved for the number n, which is the order of the filter transfer function. It should be a round number and can be found from

$$n = \frac{\log[(10^{\alpha_{min}/10} - 1)/(10^{\alpha_{max}/10} - 1)]}{2\log(u_s/u_p)} \tag{4.139}$$

The half-power frequency ratio u_o is given by

$$u_o = \frac{1}{(10^{\alpha_{max}/10} - 1)^{1/2n}} \tag{4.140}$$

EXAMPLE 4.12

D

Designing a stagger-tuned filter

Design a stagger filter with the specifications: $\omega_1 = 1000$ rad/s, $\omega_2 = 1500$ rad/s, $\omega_3 = 500$ rad/s, $\omega_4 = 3000$ rad/s, BW = 500 rad/s, $\alpha_{min} = 30$ dB, $\alpha_{max} = 0.5$ dB.

SOLUTION

Step 1: Find the center frequency from Eq. (4.116).

$$\omega_o = \sqrt{\omega_1\omega_2} = \sqrt{1000 \times 1500} = 1224.74 \text{ rad/s}$$

Step 2: Find u_s from Eq. (4.133).

$$u_s = \frac{\omega_4 - \omega_3}{\omega_2 - \omega_1} = \frac{3000 - 500}{1500 - 1000} = 5$$

Step 3: Find the order of transfer function from Eq. (4.139).

$$n = \frac{\log\left[\dfrac{10^{\alpha_{min}/10} - 1}{10^{\alpha_{max}/10} - 1}\right]}{2\log(u_s/u_p)} = \frac{\log\left[\dfrac{10^{30/10} - 1}{10^{0.5/10} - 1}\right]}{2\log(3.5)} = 2.79922$$

which gives $n = 3$. Three stages would be connected in cascaded as shown in Fig. 4.46.

Step 4: Find the normalized u_o from Eq. (4.140)

$$u_o = \frac{1}{(10^{\alpha_{max}/10} - 1)^{1/2n}} = \frac{1}{(10^{0.5/10} - 1)^{1/(2\times3)}} = 1.41992 \text{ rad/s}$$

Step 5: Find Q and the frequency ω_o of the first stage from Eq. (4.126).

$$\omega_{on} = 0.99998 \text{ rad/s}; \qquad Q = \omega_{on} = 0.99998$$

$$\omega_o = 1224.72 \text{ rad/s}$$

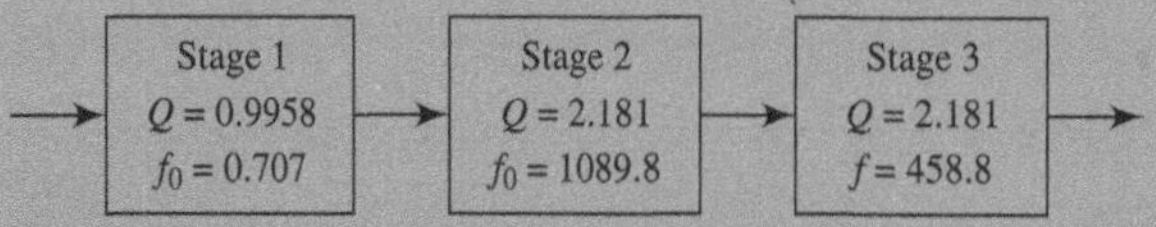

FIGURE 4.46 Three cascaded stages

Step 6: Find Q and the frequency for the second stage from Eq. (4.130).

$$\omega_{on}^2 = 2.36921; \quad \frac{\omega_{on}}{Q} = 0.7032$$

$$\omega_o = 1224.74 \times \sqrt{2.36921} = 1885.12 \text{ rad/s}$$

$$= 1224.74 \times a_1 = 1224.74 \times 1.53923 = 1885.16 \text{ rad/s}$$

$$Q = \frac{\sqrt{2.36921}}{0.7032} = 2.18889$$

Step 7: Find Q and the frequency for the third stage from Eq. (4.130).

$$\omega_{on}^2 = 0.42208; \quad \frac{\omega_o}{Q} = 0.2968$$

$$\omega_o = 1224.74 \times \sqrt{0.42208} = 795.687 \text{ rad/s}$$

$$= 1224.74 \times a_3 = 1224.74 \times 0.64968 = 795.687 \text{ rad/s}$$

$$Q = \frac{\sqrt{0.42208}}{0.2968} = 2.18894$$

Step 8: Use the normalized Friend's circuit in Fig. 4.41 to implement the stages in Fig. 4.47 for the specified values of Q and ω_o.

Step 9: Using a standard value for polyester capacitors with a tolerance of 10% in Table E.10 of Appendix E, we can find the circuit elements by selecting a new value of C such as $C_{new} = C_1 = C_2 = 0.01\ \mu\text{F}$.

$$k_f = \omega_o; \quad k_m = \frac{C_{old}}{C_{new} k_f} = \frac{1/2Q}{0.01\mu\text{F} \times \omega_o} \tag{4.141a}$$

$$R_1 = k_m; \quad R_2 = 4Q^2 k_m \tag{4.141b}$$

For stage 1:

$$k_f = \omega_{o1} = 11.22472 \times 10^3; \quad k_m = \frac{C_{old}}{C_{new} k_f} = \frac{1/2Q_1}{0.01\mu\text{F} \times \omega_{o1}} = 4.08266 \times 10^4$$

$$R_1 = k_m = 4.08266 \times 10^4\ \Omega; \quad R_2 = 4Q^2 k_m = 1.63299 \times 10^4\ \Omega$$

For stage 2:

$$k_f = \omega_{o2} = 1.88516 \times 10^4; \quad k_m = \frac{C_{old}}{C_{new} k_f} = \frac{1/2Q_1}{0.01\mu\text{F} \times \omega_{o2}} = 1.21171 \times 10^4$$

$$R_1 = k_m = 1.21171 \times 10^4\ \text{k}\Omega; \quad R_2 = 4Q^2 k_{m1} = 2.32223 \times 10^5\ \text{k}\Omega$$

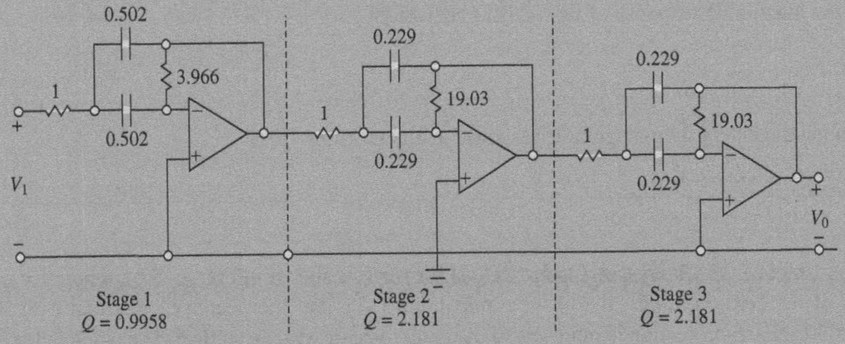

FIGURE 4.47 Normalized three-stage stagger-tuned Friend's circuits

For stage 3:

$$k_f = \omega_{o3} = 795.68726;\quad k_m = \frac{C_{old}}{C_{new}k_f} = \frac{1/2Q_1}{0.01\mu F \times \omega_{o2}} = 2.87075 \times 10^4$$

$$R_1 = k_m = 2.87075 \times 10^4\ \Omega;\quad R_2 = 4Q^2 k_{m1} = 5.50176 \times 10^5\ \Omega$$

NOTE: We should use the values of standard resistances listed in Table E.1 of Appendix E.

Step 10: Calculate the overall gain at $\omega = \omega_o$. We should note that Friend's circuit gives a gain and there will be an overall gain of the stagger filter. If we wish the combined circuit to behave as a filter and not as a filter amplifier, then we must add a compensator to compensate for this gain with voltage-divider attenuator.

KEY POINTS OF SECTION 4.13

- One or more first-order and second-order pass-band filters are generally connected in cascade to form a stagger filter.
- The specifications include stop and pass band requirements rather than defining the cut-off at 70.7% of the maximum value as in the case of Butterworth response.
- The frequency ratio u is transformed to frequencies that define the pass bands and stop bands.

4.14 Chebyshev Filters

The magnitude-squared of the Butterworth response can be described in a generalization form as given by

$$|T_n(j\omega)|^2 = \frac{1}{1 + |F_n(\omega)|^2} \tag{4.142}$$

For a Butterworth response, $F_n(\omega) = (\omega_n)^2$. The Chebyshev response has a function $F_n(\omega)$, which is scaled to fit into a square box as shown in Fig. 4.48(a). There are two units on each side: the limits of the amplitude are ±1 and the limits of ω are ±1. The function F_n is the Lissajous function with x replaced by ω, scaled by the constant ε, which is less than or equal to 1.

Therefore, the magnitude response of Eq. (4.142) is given by

$$|T_n(j\omega)|^2 = \frac{1}{1 + \varepsilon^2 C_n^2(\omega)} \tag{4.143}$$

$C_n(\omega)$, is called the Chebyshev magnitude response and is given by [3]

$$C_n(\omega) = \cos[n(\cos^{-1}\omega)] \qquad \text{for } |\omega| \le 1 \tag{4.144}$$

$$= \cosh[n(\cosh^{-1}\omega)] \quad \text{for } |\omega| > 1 \tag{4.145}$$

The y versus x plot of the Chebyshev function $C_n(x)$ for $n = 6$ is shown in Fig. 4.48(b), where $x = \omega t$. The projections of the zero-crossings, the positive maximum (+1), and the negative minimum (–1) are shown in Fig. 4.48(c). Each projection is represented by a unity vector with an angle θ. The angle θ_i for ith projection is θ_i (= 180° $i/2n$) for i = 0 to n. The difference between two angles is $\Delta\theta = 90°/n$. Connecting the tips of all vectors will be like a unity semicircle over a period 0–180° as shown in Fig. 4.48(c).

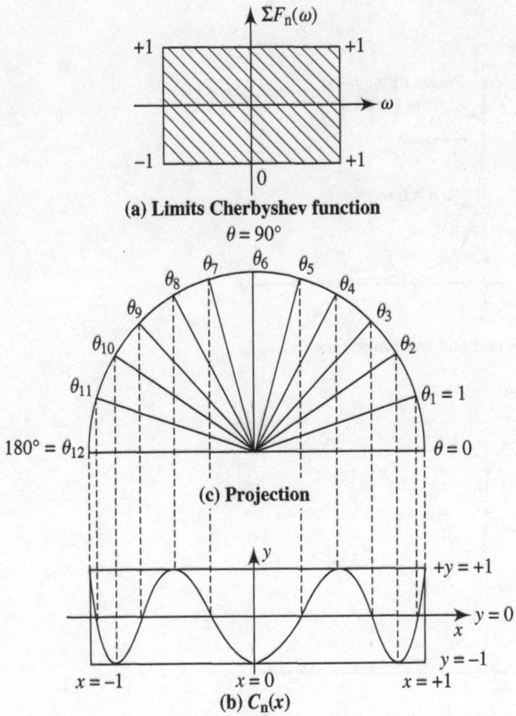

FIGURE 4.48 Plot of Chebyshev response function

4.14.1 Characteristics of Chebyshev Response

The magnitude response can be found from the square root of Eq. (4.143) as given by

$$|T_n(j\omega)| = \frac{1}{\sqrt{1+\varepsilon^2 C_n^2(\omega)}} \tag{4.146}$$

The value of $C_n(\omega)$ should be determined from either Eq. (4.144) or Eq. (4.145) depending on the value of ω. The characteristics of the magnitude response will depend on the properties of $C_n(\omega)$. A typical plot of $|T_n(j\omega)|$ is shown in Fig. 4.49 for $n = 6$ and one value of ε. The plots in Figs. 4.48 and 4.49 for $\omega = -1$ to $\omega = 1$ describe the following characteristics:

1. At $\omega = 0$, the values of $C_n(0)$ and $|T_n(j0)|$, as shown in Fig. 4.49(b), are given by

$$C_n(0) = 0 \quad \text{for } n \text{ odd} \tag{4.147a}$$

$$= 1 \quad \text{for } n \text{ even} \tag{4.147b}$$

$$|T_n(j0)| = 1 \quad \text{for } n \text{ odd} \tag{4.148a}$$

$$= \frac{1}{\sqrt{1+\varepsilon^2}} \quad \text{for } n \text{ even} \tag{4.148b}$$

2. At $\omega = 1$, the values of $C_n(1)$ and $|T_n(j1)|$ are given by

$$C_n(1) = 1 \quad \text{for all } n \tag{4.149}$$

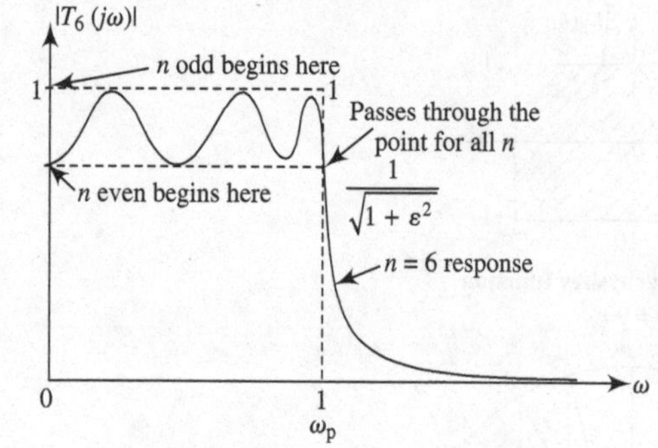

(a) Limits for *n* odd and even numbers

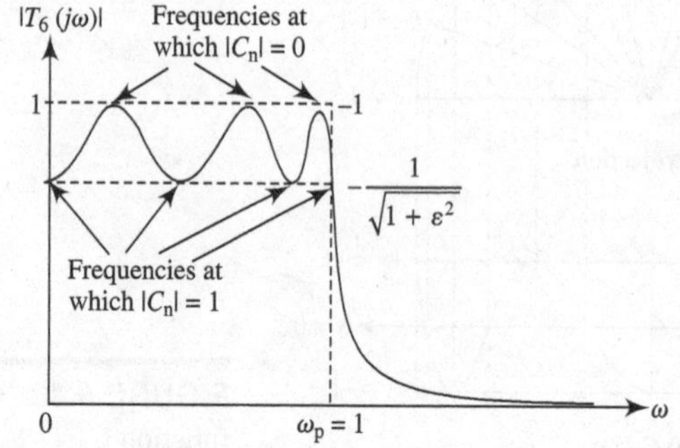

(b) For *n* even numbers

FIGURE 4.49 Magnitude plots of Chebyshev function

$$|T_n(j1)| = \frac{1}{\sqrt{1+\varepsilon^2}} \quad \text{for all } n \tag{4.150}$$

3. The minimum of $|T_n(j\omega)|$ will occur when $C_n^2 = 1$, and the maximum of $|T_n(j\omega)|$ will occur when $C_n^2 = 0$. These are shown in Fig. 4.49(b). During the interval from $\omega = -1$ to $\omega = +1$, there will be n maxima and $(n + 1)$ minima.
4. For the range of frequencies from $\omega = -1$ to $\omega = 1$, there will be n ripples. For example, if the magnitude function is even, there will be n half-ripples or half-cycles (from maximum to minimum or from minimum to maximum).
5. There are ripples in the pass band, and the number of ripples depends on n. Also the value of scaling factor ε for maintaining $\omega_o = 1$ is not constant as shown in Fig. 4.50(a). For the Butterworth response as shown in Fig. 4.50(b), the pass-band remains constant and the value of $\varepsilon = 1/\sqrt{2} = 0.707$. In the Butterworth response, $\omega_o = 1$ identifies the half-power frequency, while in the Chebyshev response, $\omega_o = 1$ identifies the end of the ripple band.

4.14.2 Specifications of Chebyshev Response

A typical attenuation plot of the Chebyshev response is shown in Fig. 4.51 for $n = 7$. This also shows the rippling nature of the Chebyshev response. The pass band extends from $\omega = 0$ to $\omega = 1$. Unlike the

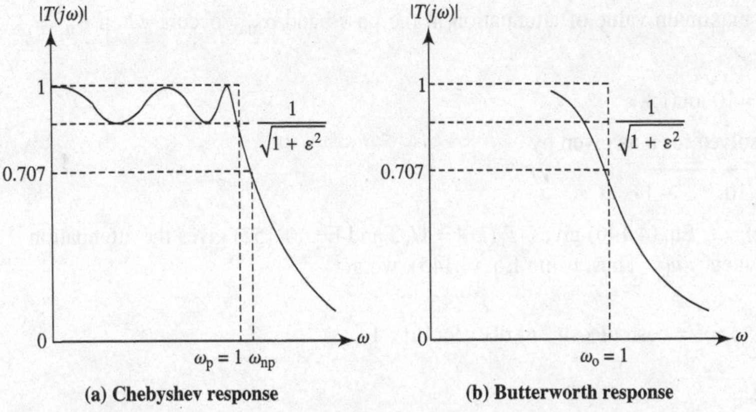

FIGURE 4.50 Typical Chebyshev and Butterworth responses

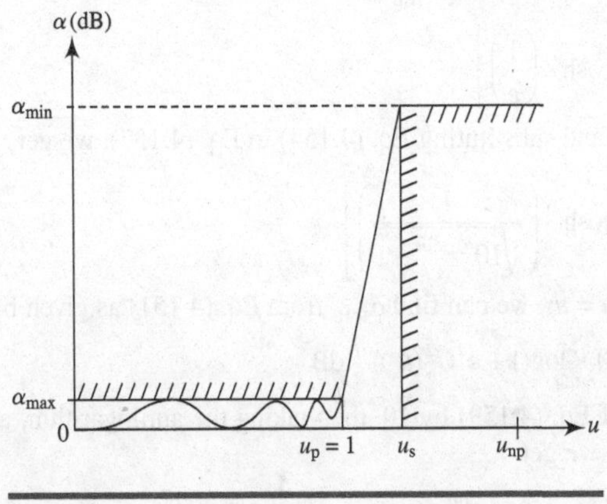

FIGURE 4.51 Attenuation of the Chebyshev response

Butterworth response, the end of the pass band (or ripple band) is always $\omega = 1$. At some higher frequency ω_s, the stop band begins and the attenuation is required to have the minimum value α_{min} in the stop band. Thus, the specifications for a Chebyshev response in normalized pass-band frequency u_p and stop-band frequency u_s are

$$\alpha_{min};\ \alpha_{max};\ u_p = 1;\ u_s$$

The attenuation for the n-order Chebyshev response in the normalized frequency ratio u is given by

$$\alpha_n = -20 \log |T_n(j1)| = -20 \log\left(\frac{1}{\sqrt{1+\varepsilon^2 C_n^2(u)}}\right) = 10 \log(1+\varepsilon^2 C_n^2(u)) \quad \text{dB} \tag{4.151}$$

Substituting $C_n(\omega)$ from Eqs. (4.144) and (4.145), Eq. (4.151) becomes

$$\alpha_n = 10 \log[1+\varepsilon^2 (\cos n \cos^{-1} u)] \text{ dB} \quad \text{for } |u| \le 1 \tag{4.152a}$$

$$= 10 \log[1+\varepsilon^2 (\cosh n \cosh^{-1} u)] \text{ dB} \quad \text{for } |u| > 1 \tag{4.152b}$$

Since the maximum value of attenuation in the pass band α_{max} occurs when $C_n^2 = 1$, Eq. (4.152*a*) gives

$$\alpha_{max} = 10 \log(1 + \varepsilon^2) \quad (4.153)$$

which can be solved for ε as given by

$$\varepsilon = \sqrt{10^{\alpha_{max}/10} - 1} \quad (4.154)$$

When $\varepsilon^2 C_n^2(u) = 1$, Eq. (4.146) gives $|T_n(ju)| = 1/\sqrt{2}$ and Eq. (4.152) gives the attenuation $\alpha = 3.01$ at the half-power frequency u_{hp}. Thus, using Eq. (4.145), we get

$$C_n(u_{hp}) = \frac{1}{\varepsilon} = \cosh n(\cosh^{-1} u) \text{ dB} \quad \text{for } |u| > 1 \quad (4.155)$$

This gives

$$\cosh^{-1} u_{hp} = \frac{1}{n} \cosh^{-1}\left(\frac{1}{\varepsilon}\right) \quad (4.156)$$

This gives the half-power frequency u_{hp} as

$$u_{hp} = \cosh\left[\left(\frac{1}{n}\right)\cosh^{-1}\left(\frac{1}{\varepsilon}\right)\right] \quad (4.157)$$

Since $0 < \varepsilon < u_{hp} > 1$ and substituting Eq. (4.154) in Eq. (4.157), we get

$$u_{hp} = \cosh\left[\left(\frac{1}{n}\right)\cosh^{-1}\left(\frac{1}{\sqrt{10^{\alpha_{max}/10} - 1}}\right)\right] \quad (4.158)$$

Since α_{min} occurs at $u = u_s$, we can find α_{min} from Eq. (4.151) as given by

$$\alpha_{min} = \alpha_n(u_s) = 10 \times \log(1 + \varepsilon^2 C_n^2(u_s)) \text{ dB} \quad (4.159)$$

Dividing both sides of Eq. (4.159) by 10, then taking the antilogarithm, and then subtracting 1 from both sides of the equation, we get

$$\varepsilon^2 \cosh^2(n \cosh^{-1} u_s) = 10^{\alpha_{max}/10} - 1 \quad (4.160)$$

Taking the square root of Eq. (4.160), we get

$$\cosh(n \cosh^{-1} u_s) = \frac{\sqrt{10^{\alpha_{min}/10} - 1}}{\varepsilon} = \sqrt{\frac{10^{\alpha_{min}/10} - 1}{10^{\alpha_{max}/10} - 1}} \quad (4.161)$$

This can be solved for n

$$n = \frac{\cosh^{-1}\left[\sqrt{\frac{10^{\alpha_{min}/10} - 1}{10^{\alpha_{max}/10} - 1}}\right]}{\cosh^{-1} u_s} \quad (4.162)$$

4.14.3 Chebyshev Poles

Replacing ω by *s*/*j*, Eq. (4.143) becomes

$$|T_n(ju)|^2_{u=s/j} = T(s)T(-s) = \frac{1}{1 + \varepsilon^2 C_n^2(s/j)} \quad (4.163)$$

The pole locations are determined by setting the denominator of Eq. (4.163) to zero and then solving the equation, i.e.,

$$1+\varepsilon^2 C_n^2\left(\frac{s}{j}\right)=0 \qquad (4.164)$$

This gives

$$C_n\left(\frac{s}{j}\right)=0\pm j\frac{1}{\varepsilon} \qquad (4.165)$$

Replacing ω by s/j in Eq. (4.144) for $\omega < 1$, we get

$$C_n\left(\frac{s}{j}\right)=\cos\left[n\cos^{-1}\left(\frac{s}{j}\right)\right]=\cos ny \qquad (4.166)$$

where y is a complex number such that

$$\cos^{-1}\left(\frac{s}{j}\right)=y=z+jv \qquad (4.167)$$

Expanding Eq. (4.166) into the real and imaginary parts, we get

$$\cos(ny)=\cos(nz+jnv)=\cos(nz)\cosh(nv)-j\sin(nz)\sinh(nv) \qquad (4.168)$$

Since Eq. (4.168) must be equal to Eq. (4.165), i.e.,

$$\cos(nz)\cosh(nv)-\sin(nz)\sinh(nv)=0\pm j\frac{1}{\varepsilon} \qquad (4.169)$$

equating the real and imaginary parts on both sides, we get

$$\cos(nz)\cosh(nv)=0 \qquad (4.170)$$

$$\sin(nz)\sinh(nv)=\mp\frac{1}{\varepsilon} \qquad (4.171)$$

Since $\cosh(nv)$ can never be zero because its smallest value is 1, it is necessary that $\cos(nz)$ be zero. The values of z for $\cos(nz)=0$ are

$$z_k=\frac{\pi}{2n},\ \frac{3\pi}{2n},\ \frac{5\pi}{2n},\ \frac{7\pi}{2n}\ldots \qquad (4.172a)$$

$$=\frac{\pi}{2n}(2k+1)\quad \text{for } k=0,1,\ldots(2n-1) \qquad (4.172b)$$

For these values of z_k in Eq. (4.172), we get

$$\sin(nz_k)=\pm 1 \qquad (4.173)$$

Substituting Eq. (4.173) in Eq. (4.171), we get the value of v_k as

$$v_k=\pm\frac{1}{n}\sinh^{-1}\left(\frac{1}{\varepsilon}\right)=\pm a \qquad (4.174)$$

For each value of $y_k = \cos^{-1}(s_k/j) = z_k + jv_k$, Eq. (4.166) gives the corresponding value of s_k as given by

$$s_k = j\cos y_k = j\cos\left[\frac{\pi}{2n}(2k+1) + jc\right] \quad \textbf{(4.175)}$$

Assuming $s_k = \sigma_k + j\omega_k$, Eq. (4.175) gives the real and imaginary parts of the poles as follows:

$$\sigma_k = \pm\sinh(a)\sin\left(\frac{2k+1}{2n}\pi\right) \quad \text{for } k = 0, 1, \ldots(2n-1) \quad \textbf{(4.176)}$$

$$u_k = \cosh(a)\cos\left(\frac{2k+1}{2n}\pi\right) \quad \text{for } k = 0, 1, \ldots(2n-1) \quad \textbf{(4.177)}$$

These are the s-plane locations of the poles of $T(s)T(-s)$ and are assigned to the left half plane of the transfer function $T(s)$. Since $\sin^2 x + \cos^2 x = 1$ for any value of x, adding the squared values of σ_k and u_k, we get

$$\left(\frac{\sigma_k}{\sinh a}\right)^2 + \left(\frac{u_k}{\cosh a}\right)^2 = 1 \quad \textbf{(4.178)}$$

which is the equation of an ellipse whose major semi-axis is cosh σ, minor semi-axis is sinh σ, and the foci is $+j1$ or $-j1$ as shown in Fig. 4.52.

The Chebyshev response as shown in Fig. 4.50(a) does decrease more rapidly beyond the pass band. At the cut-off frequency $\omega = \omega_p$ and $\varepsilon = 1$, the Chebyshev response is n times steeper at cut-off than the Butterworth response; that is,

$$\left.\frac{dT(s)}{d\omega}\right|_{\omega=1} (\text{for Chebyshev}) = n\left.\frac{dT(s)}{d\omega}\right|_{\omega=1} (\text{for Butterworth}) \quad \textbf{(4.179)}$$

The choice between a Butterworth and a Chebyshev response will require trade-off between the conflicting requirements of (1) high attenuation in the stop band and a steeper roll-off near the cut-off frequency, and (2) higher values of Q leading to difficulties in achieving circuit realizations and a nonlinear phase characteristic. The filter designer needs to make a choice of these conflicting requirements.

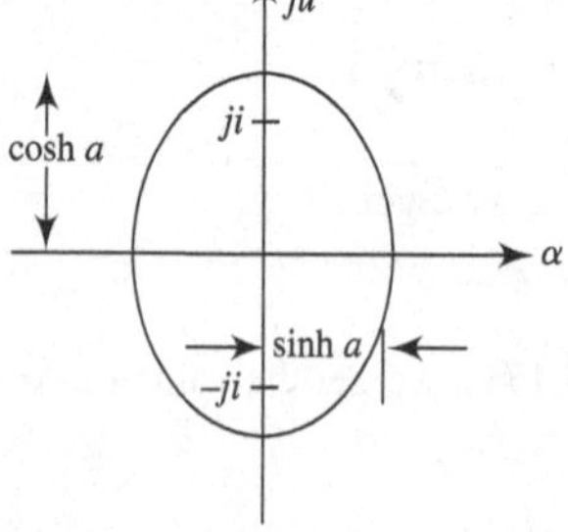

FIGURE 4.52 Ellipse trajectories for the s-plane locations of the poles

EXAMPLE 4.13

D

Designing a low-pass Chebyshev filter

Design a Chebyshev filter with the specifications: pass-band frequency f_p = 10 kHz, stop-band frequency f_s = 24 kHz, α_{min} = 20 dB, and α_{max} = 0.4 dB.

SOLUTION

Step 1: Find the normalized stop-band frequency ratio u_s. For $u_p = 1$, we get

$$u_s = \frac{f_s}{f_p} = \frac{24k}{10k} = 2.4$$

Step 2: Find the order of transfer function from Eq. (4.162).

$$n = \frac{\cosh^{-1}[(10^{\alpha_{min}/10} - 1)/(10^{\alpha_{max}/10} - 1)]}{\cosh^{-1} u_s} = \frac{\cosh^{-1}[(10^{20/10} - 1)/(10^{0.4/10} - 1)]}{\cosh^{-1}(2.4)} = 2.7329$$

which is rounded to $n = 3$. Two stages would be connected in cascade as shown in Fig. 4.53(a).

Step 3: Find the scaling factor ε from Eq. (4.154).

$$\varepsilon = \sqrt{10^{\alpha_{max}/10} - 1} = \sqrt{10^{20/10} - 1} = 0.3106$$

Step 4: Find the 3-dB frequency from Eq. (4.158).

$$u_{hp} = \cosh\left[\left(\frac{1}{n}\right)\cosh^{-1}\left(\frac{1}{\sqrt{10^{\alpha_{max}/10} - 1}}\right)\right] = \cosh\left[\left(\frac{1}{3}\right)\cosh^{-1}\left(\frac{1}{\sqrt{10^{0.4/10} - 1}}\right)\right] = 1.1935$$

Step 5: Find the value of the constant from Eq. (4.174).

$$a = \frac{1}{n}\sinh^{-1}\left(\frac{1}{\varepsilon}\right) = \frac{1}{3}\sinh^{-1}\left(\frac{1}{0.3106}\right) = 0.6286$$

Step 6: Find the value of z from Eq. (4.172).

$$z_{1,3,5} = 1.5708;\quad 3.6652;\quad 5.7596$$

Step 7: Find the real and imaginary parts of the poles from Eqs. (4.176) and (4.177).

$$\sigma_{1,3,5} = \sinh(a)\sin(z_{1,3,5}) = -0.6708;\quad -0.3354;\quad -0.3354$$

$$u_{1,3,5} = \cosh(a)\cos(z_{1,3,5}) = 0;\quad +1.0428;\quad -1.0428$$

Step 8: Find the location of the poles.

$$p_1 = -0.6708$$

$$p_2, p_3 = -0.3354 \pm j1.0428$$

Step 9: Find the frequency ω_o of the first-stage transfer function.

$$\omega_o = -\sigma_1 = 0.6708 \text{ rad/s}$$

Step 10: Find the denominator of the second-stage transfer function.

$$s^2 + (-2\sigma_3)s + \sigma_3^2 + u_3^2 = s^2 + (2 \times 0.3354)s + (0.3354)^2 + 1.0428^2 = s^2 + 0.6708s + 1.1935$$

Step 11: Find Q and the frequency for the second stage.

$$u_o^2 = 1.19351;\ u_o = \sqrt{1.19351} = 1.0954;\quad \frac{u_o}{Q} = 0.6708$$

$$Q = \frac{0.6708}{u_o} = \frac{0.6708}{1.0954} = 1.6331$$

Step 12: Use the normalized Sallen–Key circuit to implement the second stages for the specified values of Q and ω_0. This is shown in Fig. 4.53(b).

Step 13: Calculate the final element using the scaling factors. Using a standard value for carbon resistors with a tolerance of 10% in Table E.1 of Appendix E, we can find the circuit elements by selecting a new value of C such as $C_{new} = C_1 = C_2 = 0.01\ \mu F$.

$$k_f = 2\pi u_o f_p = 2\pi \times 1.0954 \times 10 \times 10^3 = 6.8826 \times 10^4;\quad \text{Let } R = R_1 = R_2 = k_m = 20\ k\Omega$$

$$C = \frac{1}{k_m k_f \sigma_1} = \frac{1}{20 \times 10^3 \times 6.8826 \times 10^4 \times 0.6708} = 1083\ \text{pF}$$

$$C_1 = \frac{2Q}{k_m k_f} = \frac{2 \times 1.6331}{20 \times 10^3 \times 6.8826 \times 10^4} = 2372.8\ \text{pF}$$

$$C_2 = \frac{1/(2Q)}{k_m k_f} = \frac{1/(2 \times 1.6331)}{20 \times 10^3 \times 6.8826 \times 10^4} = 222.242\ \text{pF}$$

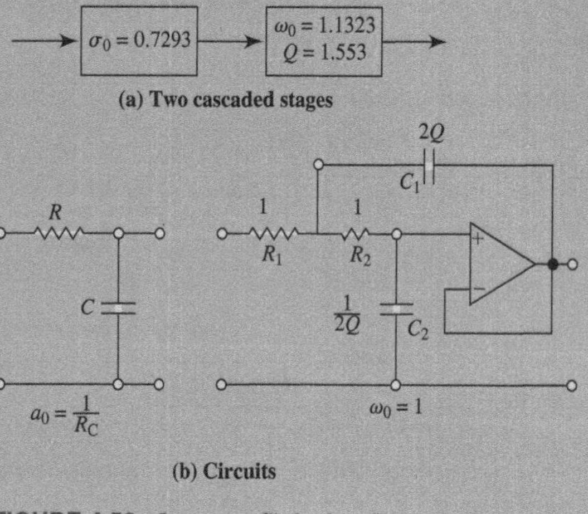

FIGURE 4.53 Low-pass Chebyshev filter

KEY POINT OF SECTION 4.14

- The Chebyshev response has a function $F_n(\omega)$ which is scaled to fit into a square box. There are two units on each side: the limits of the amplitude are ±1 and the limits of ω are ±1.
- The choice between a Butterworth and a Chebyshev response will require trade-off between the conflicting requirements of (1) high attenuation in the stop band and a steeper roll-off near the cut-off frequency, and (2) higher values of Q leading to difficulties in achieving circuit realizations and a nonlinear phase characteristic.

4.15 Delay Filters

Delays occur naturally in the transmission of a signal through medium such as on a coaxial cable or an optical fiber. But, the signal processing requires a defined time delay. A delay filter can be selected and designed to provide a certain amount of delay.

If the frequency of a signal is ω and it is delayed by a time t, then the delay angle ϕ can be expressed as

$$\phi = -\omega t \tag{4.180}$$

The negative sign represents the delay. This gives the rate of change of the delay angle as

$$\frac{d\phi}{d\omega} = -t = -D \tag{4.181}$$

This gives the delay magnitude D as

$$D = t = \frac{d\phi}{d\omega} \tag{4.182}$$

Therefore, there are two types of delays: ϕ is known as the *phase delay* and D is known as the *group delay* or sometimes as *signal delay* or *envelope delay* or simply as *delay*. Figure 4.54 shows the input and output signals related by a delay function.

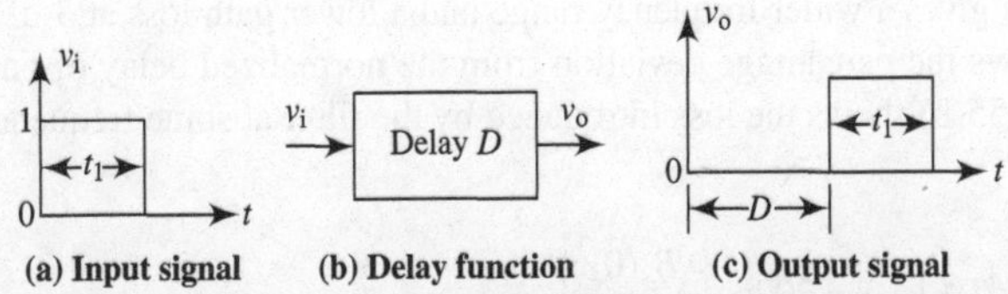

FIGURE 4.54 Input and output signals with a delay

4.15.1 Delay Functions

If $V_i(s)$ is the input voltage and $V_o(s)$ is the output of a delay circuit, then the transfer function with a unity gain and a delay of D can be expressed as

$$\frac{V_i(j\omega)}{V_o(j\omega)} = 1e^{-j\omega D} \tag{4.183}$$

which in Lapalce's domain becomes

$$T(s) = \frac{V_i(s)}{V_o(s)} = e^{-s} = \frac{1}{e^s} \tag{4.184}$$

Equation (4.184) can be expressed as the Bessel–Thomson function of nth order as given by

$$T_n(s) = \frac{B_n(0)}{B_n(s)} \tag{4.185}$$

where $B_n(s)$ is the Bessel polynomial for any value of n and $B_n(0) = 1$. $B_n(s)$ can be found from the recursion formula as given by [3]

$$B_n(s) = (2n-1)B_{n-1} + s^2 B_{n-2} \tag{4.186}$$

For examples

$$B_0(s) = 1$$
$$B_1(s) = s + 1$$
$$B_2(s) = s^2 + 3s + 3$$
$$B_3(s) = s^3 + 6s^2 + 15s + 15 = (s^2 + 3.67782s + 6.45944)(s + 2.32219) \tag{4.187}$$
$$B_4(s) = s^4 + 10s^3 + 45s^2 + 105s + 105$$
$$= (s^2 + 5.79242s + 9.14013)(s^2 + 4.20758s + 11.4878)$$
$$B_5(s) = s^5 + 15s^4 + 105s^3 + 420s^2 + 945s + 945$$
$$= (s^2 + 6.70391s + 14.2725)(s^2 + 4.64934s + 18.15631)(s + 3.64674)$$
$$B_6(s) = s^6 + 21s^5 + 210s^4 + 1260s^3 + 4725s^2 + 10395s + 10395$$
$$= (s^2 + 8.49672s + 18.80113)(s^2 + 7.47142s + 20.85282)(s^2 + 5.03186s + 26.51402)$$
$$B_7(s) = s^7 + 28s^6 + 378s^5 + 3150s^4 + 17325s^3 + 62370s^2 + 135135s + 135135$$
$$B_8(s) = s^8 + 36s^7 + 630s^6 + 6930s^5 + 51975s^4 + 270270s^3 + 945945s^2 + 2027025s + 2027025$$

A higher order function gives a wider frequency range and a lower gain loss at 3-dB.

Figure 4.55(a) shows the percentage deviation from the normalized delay of one second at a specific frequency ω_1. Figure 4.55(b) shows the loss introduced by the filter at some frequency ω_1. Such a loss can be calculated from

$$d(\omega_1) = 20 \log |T_n(j\omega_1)| = 20 \log \left| \frac{B_n(0)}{B_n(j\omega_1)} \right| \tag{4.188}$$

Figure 4.55 illustrates the limitations of the BesselThomson delay filters. The delay decreases with increasing frequency, and the magnitude of the transfer function also decreases with frequency. The departure from the ideal response is a function of the order of the filter n, and better as n increases. The second-order function can be used to synthesize 4th, 6th, and 8th order delays. Similarly, the second- and third-order functions can be use do synthesize 5th and 7th order delays. Using Eq. (4.187), the second-order function is given by

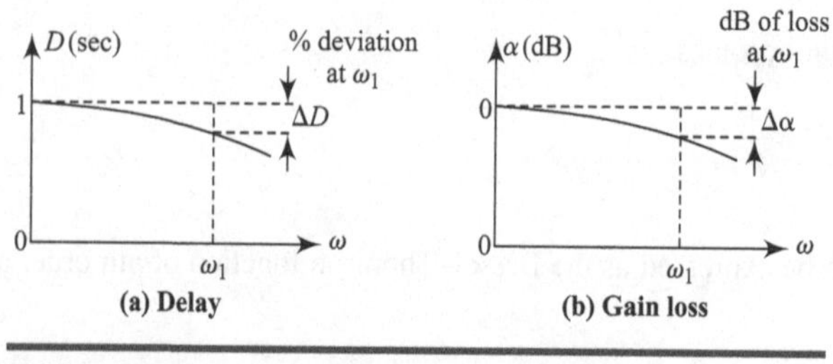

FIGURE 4.55 Normalized delay and gain loss

$$T_2(s) = \frac{a_0}{s^2 + a_1 s + a_0} \tag{4.189}$$

and the corresponding delay is

$$\phi_2 = -\tan^{-1}\left(\frac{a_1\omega}{a_0 - \omega^2}\right) \tag{4.190}$$

Differentiating Eq. (4.189) with respect to ω gives the delay function as

$$D_2(\omega) = \frac{d\phi}{d\omega} = \frac{a_1\omega^2 + a_1 a_0}{(a_1\omega)^2 + (a_0 - \omega^2)^2} \tag{4.191}$$

For $n = 2$, $a_0 = 9$ and $a_1 = 3$. Thus, Eqs. (4.189) and (4.191) give the transfer and delay functions as

$$T_2(s) = \frac{3}{s^2 + 3s + 9} \tag{4.192}$$

$$D_2(\omega) = \frac{d\phi}{d\omega} = \frac{3\omega^2 + 9}{\omega^4 + 3\omega^2 + 9} \tag{4.193}$$

The magnitude of the function $T_2(j\omega)$ for $s = j\omega$ is

$$|T_2(j\omega)| = \frac{3}{\sqrt{(3\omega)^2 + (9 - \omega^2)}} \tag{4.194}$$

Similarly, the third-order transfer and the delay functions are given by

$$T_3(s) = \frac{15}{s^3 + 6s^2 + 15s + 15} \tag{4.195}$$

$$D_3(\omega) = \frac{d\phi}{d\omega} = \frac{6\omega^4 + 45\omega^2 + 225}{\omega^6 + 6\omega^4 + 45\omega^2 + 225} \tag{4.196}$$

The magnitude of the function $T_3(j\omega)$ for $s = j\omega$ is given by

$$|T_3(j\omega)| = \frac{15}{\sqrt{(15 - 6\omega^2)^2 + (15\omega - \omega^3)}} \tag{4.197}$$

4.15.2 Realization of Delay Functions

The second-order function in Eq. (4.192) can be implemented with a low-pass Sallen–Key circuit shown in Fig. 4.56 for a gain of $K = 1$. The actual circuit values are given by

$$\begin{aligned} R &= k_m R_n = R_o R_n \\ C &= \frac{k_d}{k_m k_f} C_n \\ k_f &= \omega_o \\ k_d &= D \end{aligned} \tag{4.198}$$

where R_n and C_n are the normalized values of resistances and capacitances, respectively, as shown in Fig. 4.56. k_m, k_f, and k_d are the scale factors for resistances, frequencies, and delays, respectively, to be selected by the designer.

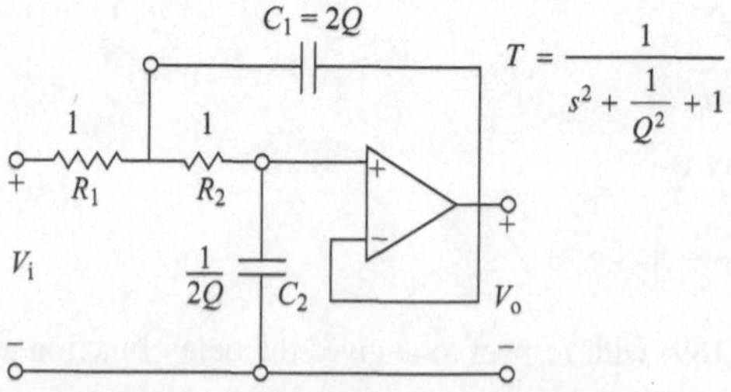

FIGURE 4.56 Sallen–Key delay filter

EXAMPLE 4.14

D

Designing fourth-order delay filter with a specified delay

(a) Design a fourth-order Sallen–Key delay filter to implement the Bessel–Thomson function with a delay of 100 μs. Find the (b) magnitude of the function and (c) delay at the first break frequency ω_{o1}.

SOLUTION

$D = 100$ μs

(a) For $n = 4$, Eqs. (4.185) and (4.187) give the transfer function consisting of the two second-order sections as

$$T_4(s) = \frac{105}{(s^2 + 5.79242s + 9.14013)(s^2 + 4.20758s + 11.4878)}$$

The Q and ω_o specifications for the two sections are

$\omega_{o1} = \sqrt{9.14013} = 3.023$, $Q_1 = \omega_{o1}/5.79242 = 0.522$

$\omega_{o2} = \sqrt{11.4878} = 3.3389$, $Q_2 = \omega_{o2}/4.20758 = 0.806$

These specifications give the normalized values of the Sallen–Key two-stage filter as shown in Fig. 4.56, and the new values can be found from

$$C_{new} = \frac{D}{k_m k_f} C_{old}; \quad R_{new} = k_m C_{old}$$

Elements	Stage I Normalized	Stage I Values	Stage 2 Normalized	Stage 2 Values
C_1	$2Q_1 = 2 \times 0.522 = 1.044$	3.45 nF	$2Q_2 = 2 \times 0.806 = 1.612$	4.765 nF
C_2	$1/2Q_1 = 1/(2 \times 0.522) = 0.958$	3.17 nF	$1/2Q_2 = 1/(2 \times 0.806) = 0.620$	1.83 nF
$R_1 = R_2$	1	10 kΩ	1	10 kΩ
k_f	$\omega_{o1} = 3.023$		$\omega_{o2} = 3.389$	
k_m	10^4		10^4	

(b) The magnitude response at $\omega = \omega_{o1} = 3.023$ rad/s is given by

$$|T_4(\omega)| = \frac{9.14013}{\sqrt{33.64\omega^2 + (9.14 - \omega^2)^2}} \times \frac{11.4878}{\sqrt{17.72\omega^2 + (11.49 - \omega^2)^2}} = 0.4627$$

(c) Applying Eq. (4.191), the total delay is the sum of delays associated with the two sections.

$$D_4(\omega) = \frac{5.8\omega^2 + 53.01}{33.64\omega^2 + (9.14 - \omega^2)^2} + \frac{4.21\omega^2 + 48.37}{17.72\omega^2 + (11.49 - \omega^2)^2} = 0.8634 \text{ s}$$

NOTE: This is the normalized value of the delay, not the actual delay. At $\omega = 0$, $T_4(0) = 0.9998$ and $D_4(0) = 1.0009$ s.

KEY POINT OF SECTION 4.15

- Delays occur naturally in the transmission of a signal through medium such as a coaxial cable or an optical fiber. But, the signal processing requires a defined time delay.

4.16 All-Pass Filters

An all-pass filter passes all frequency components of the input signals without attenuation. However, this filter provides predictable phase shifts for different frequencies of the input signals. Transmission lines (e.g., telephone wires) usually cause phase changes in the signals; all-pass filters are commonly used to compensate for these phase changes. An all-pass filter is also called a *delay equalizer* or a *phase corrector*.

Figure 4.57(a) shows the characteristic of an all-pass filter; the circuit diagram is shown in Fig. 4.57(b). The output voltage in Laplace's domain can be obtained by using the superposition theorem:

$$V_o(s) = -\frac{R_F}{R_1}V_i(s) + \frac{1/sC}{R + 1/sC}\left(1 + \frac{R_F}{R_1}\right)V_i(s) \tag{4.199}$$

If we assume $R_F = R_1$, Eq. (4.199) can be reduced to

$$V_o(s) = -V_i(s) + \frac{2}{1 + sRC}V_i(s)$$

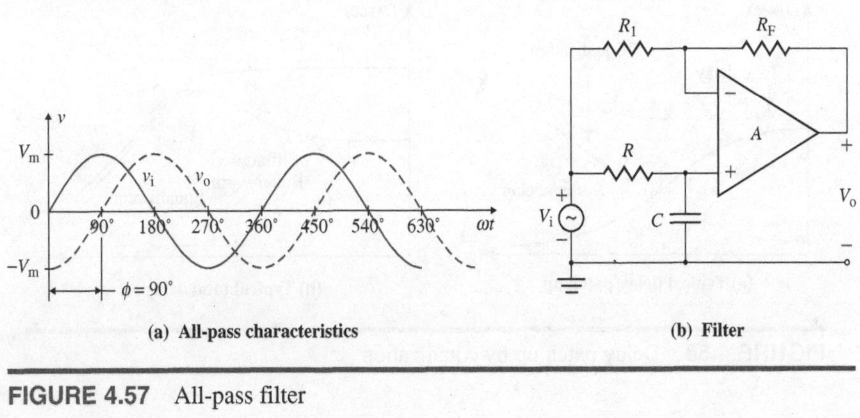

(a) All-pass characteristics (b) Filter

FIGURE 4.57 All-pass filter

which gives the voltage gain as

$$H(s) = \frac{V_o(s)}{V_i(s)} = \frac{1 - sRC}{1 + sRC} \tag{4.200}$$

Substituting $s = j\omega$ into Eq. (4.200) gives the magnitude of the voltage gain as

$$|H(j\omega)| = 1$$

and the phase angle ϕ as

$$\phi = -2\tan^{-1}(\omega RC) = -2\tan^{-1}(2\pi fRC) \tag{4.201}$$

Equation (4.201) indicates that, for fixed values of R and C, the phase angle ϕ can change from 0° to −180° as the frequency f of the input signal is varied from 0 to ∞. For example, if R = 21 kΩ and C = 0.1 μF, we will get ϕ = −64.4° at 60 Hz. If the positions of R and C are interchanged, the phase shift ϕ will be positive. That is, the output signal leads the input signal.

4.16.1 Delay Equalization

A normal filter is required to remove the high-frequency noise while maintaining the integrity of the signal insofar as possible. Filtering also introduces delay which decreases with the frequency. An equalization circuit does the "patch up" of the total delay known as equalization. Equalization is realized by a cascade-connected circuit which provides delay as a function of frequency, but does not introduce attenuation as a function of frequency. This is shown in Fig. 4.58(a).

Therefore, the delay equalization should not affect the desired magnitude response or the attenuation. The phase of the equalization should supplement to the system delay such that the summation of delays is approximately constant over a prescribed frequency range from $\omega = 0$ to $\omega = 0.9$ of normalized frequencies. The total delay of a fifth-order Butterworth filter with a cascaded equalization is illustrated in Fig. 4.58(b).

The equalization circuits are distinguished by polezero patterns as shown in Fig. 4.59. Figure 4.59(a) shows one pole and one zero at equal distances from the origin. Figure 4.59(b) has equal poles and zeros but with opposite real and imaginary parts. Figure 4.59(c) shows the polezero locations with the superposition of the patterns in Fig. 4.59(a) and (b). Figure 4.59(d) shows double pairs of polezero locations with the patterns of Fig. 4.59(b).

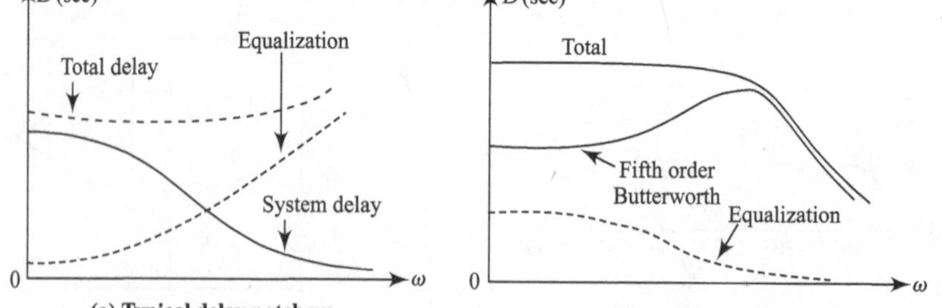

(a) Typical delay patch up

(b) Typical total delay

FIGURE 4.58 Delay patch up by equalization

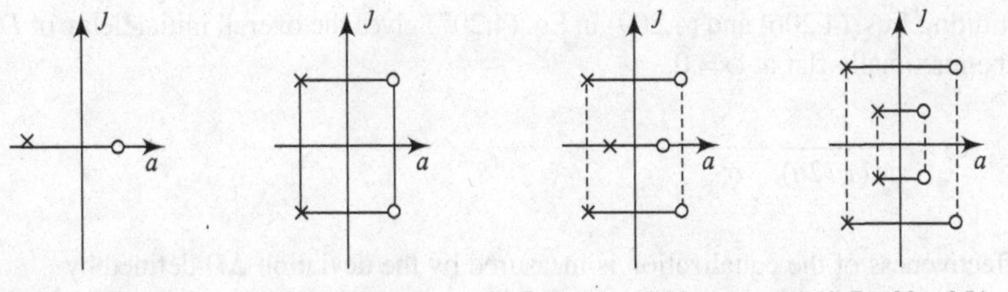

FIGURE 4.59 Pole–zero locations of equalization circuits

4.16.2 First-Order Function and Circuits

Equations (4.200) and (4.201) describe the transfer function and phase delay of a first-order equations. The first- and second-order equalization circuits can be cascaded to produce higher-order equalization. Equation (4.199) can be expressed in general form as given by

$$T_1(s) = K\frac{s-\sigma_1}{s+\sigma_1} \quad (4.202)$$

This function corresponds to the pole–zero pattern of Fig. 4.59(a). The gain and phase delay are given by

$$T_1(j\omega) = K \quad (4.203)$$

$$\phi = -2\tan^{-1}\left(\frac{\omega}{\sigma_1}\right) \quad (4.204)$$

The delay function is given by

$$D_1(\omega) = -\frac{d}{d\omega}\phi(\omega) = \frac{2/\sigma_1}{1+(\omega/\sigma_1)^2} \quad (4.205)$$

Therefore, the delay decreases with increasing frequency starting from the initial value of

$$D_{1(\max)} = D_1(0) = \frac{2}{\sigma_1} \quad (4.206)$$

If the first-oder equalization is cascaded with a Butterworth low-pass filter with a delay of $D_B(\omega)$, the overall or total delay will be the sum of the two delays, that is,

$$D_T(\omega) = D_B(\omega) + D_1(\omega) \quad (4.207)$$

The constraint for $D_1(\omega)$ that must be satisfied in order to ensure the lower-order derivative of $D_T(\omega)$ vanishes at $\omega = 0$ is given by [3]

$$\sigma_1 = \left(2\sin\frac{3\pi}{2n}\right)^{1/3} \quad (4.208)$$

The initial delay of Butterworth low-pass filters is given by

$$D_B(0) = \frac{1}{\sin(\pi/2n)} \quad (4.209)$$

Thus, substituting Eqs. (4.206) and (4.209) in Eq. (4.207) gives the overall initial delay of $D_T(0)$ that is optimized to be maximally flat at $\omega = 0$.

$$D_T(0) = \frac{1}{\sin(\pi/2n)} + \frac{2}{\sigma_1} \tag{4.210}$$

The effectiveness of the equalization is measured by the deviation ΔD defined by

$$\Delta D = D_{\max} - D(0) \tag{4.211}$$

4.16.3 Secord-Order Function

As listed in Table 4.3(e), the second-order function is given by

$$T_2(s) = \frac{s^2 - (\omega_o/Q)s + \omega_o^2}{s^2 + (\omega_o/Q)s + \omega_o^2} \tag{4.212}$$

The phase angle of the transfer function is

$$\phi_2(\omega) = -2\tan^{-1}\left[\frac{\omega_o/Q}{\omega_o^2 - \omega^2}\right] \tag{4.213}$$

Letting $u = \omega/\omega_o$, Eq. (4.213) can be simplified to

$$\phi_2(u) = -2\tan^{-1}\left[\frac{u/Q}{1-u^2}\right] \tag{4.214}$$

Differentiating Eq. (4.214) gives the delay function as

$$D_2(u) = -\frac{d}{d\omega}\phi_2(u) = \frac{2}{\omega_o Q}\frac{1+u^2}{(1-u^2)^2 + (u/Q)^2} \tag{4.215}$$

This gives the initial delay as

$$D_2(0) = \frac{2}{\omega_o Q} \tag{4.216}$$

The maximum value of $D_2(u)$ in Eq. (4.215) can be found by setting the derivative to zero, that is,

$$\frac{d}{du}D_2(u) = \frac{d}{du}\left[\frac{2}{\omega_o Q}\frac{1+u^2}{(1-u^2)^2 + (u/Q)^2}\right] = 0 \tag{4.217}$$

This gives the condition of the maximum value as

$$u^4 + 2u^2 + \left(\frac{1}{Q^2} - 3\right) = 0 \tag{4.218}$$

Solving Eq. (4.218) gives the value of u for $D_{2(\max)}$ as

$$u = \sqrt{-1 + \sqrt{4 - \frac{1}{Q^2}}} \tag{4.219}$$

For $Q > 1$, $u \approx 1$ and $\omega_{\max} = \omega_o$. As u becomes large tending to infinity, $D_2(\infty) \to 0$. Substituting $u = 1$ in Eq. (4.215) gives the maximum delay as

$$D_{2(\max)}(u=1) = \frac{4}{\omega_o Q} \tag{4.220}$$

4.16.4 Secord-Order Circuit

The transfer function in Eq. (4.212) can be realized with the Delyiannis circuit as shown in Fig. 4.60. The voltage at the positive terminal of the op-amp is

$$V_C = \frac{R_4}{R_3 + R_4} V_1 = k\,V_1 \tag{4.221}$$

Applying Kirchhoff's current law at node A gives

$$\left(2Cs + \frac{1}{R_1}\right)V_A - CsV_2 - \left(\frac{1}{R_1} + kCs\right)V_1 = 0 \tag{4.222}$$

Applying Kirchhoff's current law at node B gives

$$-CsV_A - \frac{1}{R_2}V_2 + k\left(Cs + \frac{1}{R_2}\right)V_1 = 0 \tag{4.223}$$

Eliminating V_A and solving for V_2/V_1 give the transfer function as

$$T_2(s) = \frac{V_2}{V_1}(s) = k\frac{s^2 + [2/CR_2 + (1 - 1/k)/CR_1]s + 1/C^2R_1R_2}{s^2 + (2/CR_2)s + 1/C^2R_1R_2} \tag{4.224}$$

For the circuit in Fig. (4.60) to have a transfer function of the form of Eq. (4.212), it is necessary that the coefficient of s in the numerator be the negative of the coefficient of s in the denominator of Eq. (4.224). This requires that the following conditions must the satisfied:

$$\frac{2}{CR_2} + \frac{1 - 1/k}{CR_1} = -\frac{2}{CR_2} \tag{4.225}$$

This equation can be simplified to give

$$\frac{R_2}{R_1} = 4\frac{R_4}{R_3} \quad \text{or} \quad \frac{R_4}{R_3} = \frac{1}{4}\frac{R_2}{R_1} \tag{4.226}$$

The resonant frequency is

$$\omega_o = \frac{1}{C\sqrt{R_1R_2}} \tag{4.227}$$

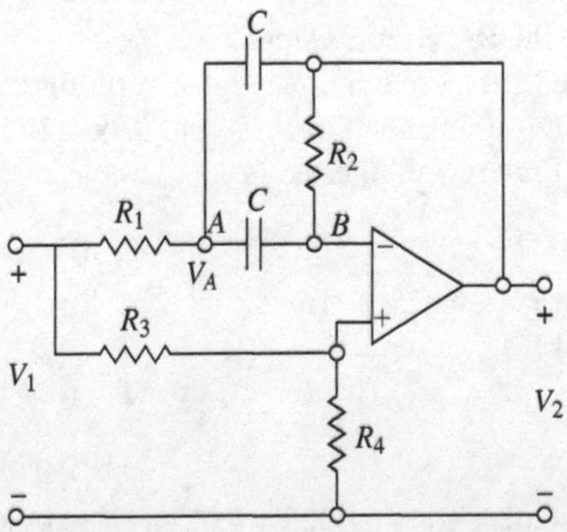

FIGURE 4.60 Delyiannis circuit for the second-order function

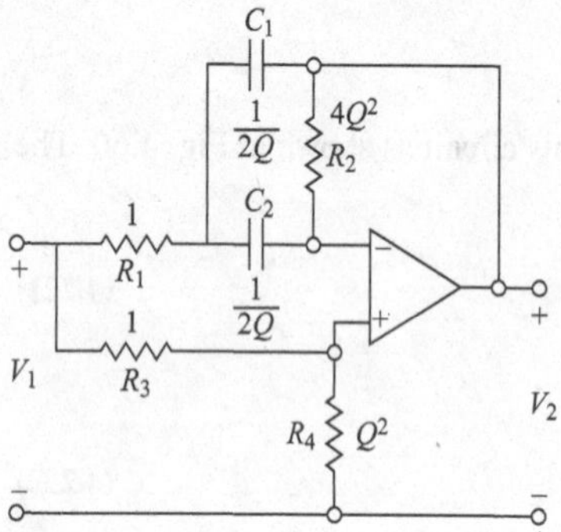

FIGURE 4.61 Frequency normalized second-order all-pass filter

And Q of the circuit is given by

$$Q = \frac{2}{CR_2}\omega_o \tag{4.228}$$

Scaling the frequency $\omega_o = 1$, and selecting $R_1 = 1$ and $R_2 = 4Q_2$, Eq. (4.226) gives

$$\frac{R_4}{R_3} = Q^2 \tag{4.229}$$

Assuming $R_3 = 1$ and $R_4 = Q^2$, $C_1 = C_2 = C = 1/2Q$. The frequency normalized circuit is shown in Fig. 4.61. At $\omega = 0$, Eq. (4.224) gives the initial value of the constant k as a function of Q as

$$T_2(0) = k = \frac{R_4}{R_3 + R_4} = \frac{Q^2}{1+Q^2} \tag{4.230}$$

Therefore, the frequency-scaled transfer function of Fig. 4.61 is given by

$$T_2(s) = \frac{Q^2}{1+Q^2} \times \frac{s^2 - (1/Q)s + 1}{s^2 + (1/Q)s + 1} \tag{4.231}$$

EXAMPLE 4.15

D

Finding the optimized delays and the RC circuit values

A first-order equalization is cascaded with a fourth-order Butterworth filter and optimized to be maximally fit at $\omega = 0$. Determine (a) $D_T(0)$, (b) $D_T(\omega = 0.9998)$, (c) ΔD, and (d) the RC circuit values of the equalization. The transfer function of fifth-order Butterworth filter is

$$T_5(s) = \frac{1}{(1+s)(s^2 + 1.625s + 1)(s^2 + 0.625s + 1)}$$

SOLUTION

$n = 5$, $\omega = 0.9998$, $b_1 = 1.625$, $b_2 = 0.625$, $Q_1 = 1/1.625 = 0.6154$, $Q_2 = 1/0.625 = 1.6$.
From Eq. (4.208),

$$\sigma_1 = \left(2\sin\frac{3\pi}{2n}\right)^{1/3} = \left(2\sin\frac{3\pi}{2\times 4}\right)^{1/3} = 1.2271$$

(a) From Eq. (4.210),

$$D_{\mathrm{T}}(0) = \frac{1}{\sin(\pi/2n)} + \frac{2}{\sigma_1} = \frac{1}{\sin(\pi/2\times 4)} + \frac{2}{1.2271} = 4.243 \text{ s}$$

(b) The phase angle of the fourth-order Butterworth filter is

$$\phi_{\mathrm{B4}}(\omega) = \tan^{-1}\left(\frac{1.625}{1-\omega^2}\right) + \tan^{-1}\left(\frac{0.625}{1-\omega^2}\right) = 3.1415 \text{ rad (at } \omega = 0.9998)$$

Differentiating this equation with respect to ω, we get

$$D_{\mathrm{B4}} = \frac{d}{d\omega}\phi_{B4}(\omega) = \frac{1.625(1+\omega^2)}{(1-\omega^2)^2 + 2.62\omega^2} + \frac{0.625(1+\omega^2)}{(1-\omega^2)^2 + 0.384\omega^2} = 4.4958 \text{ s (at } \omega = 0.9998)$$

From Eq. (4.205), we get

$$D_1(\omega) = \frac{2/\sigma_1}{1+(\omega/\sigma_1)^2} = \frac{2/1.2271}{1+(\omega/1.2271)^2} = 0.9794 \text{ s}$$

Therefore, the overall delay is given by

$$D_{\mathrm{T}}(\omega) = D_{\mathrm{B4}}(\omega) + D_1(\omega) = 4.4958 + 0.9794 = 5.4752 \text{ s}$$

(c) From Eqs. (4.210) and (4.220), we get

$$D_{\mathrm{T(max)}}(u=1) = \frac{4}{\omega Q_1} + \frac{4}{\omega Q_2} + \frac{2}{\sigma_1} = \frac{4}{0.6158} + \frac{4}{1.6} + \frac{2}{1.2271} = 10.63 \text{ s}$$

From Eq. (4.211), we get

$\Delta D = D_{\mathrm{T(max)}} - D_{\mathrm{T}}(0) = 10.63 - 4.243 = 6.493$ s

(d) $\sigma_1 = 1/RC = 1.2271$. Assume $C = 1$ F, $R = 0.8149\ \Omega$. These values can be sealed to different values to match the values of the Butterworth filter.

KEY POINT OF SECTION 4.16

- An all-pass filter does not give any gain attenuation, but it provides predictable phase shifts for different frequencies of the input signals.
- An equalization circuit does the "patch up" of the total delay known as equalization. Equalization is realized by a cascade-connected circuit which provides delay as a function of frequency, but does not introduce attenuation as a function of frequency.

4.17 Sensitivity

The active filters consist of an op-amp with a negative feedback. The feedback reduces the effects of any gain and parametric variations. Let us consider a simple inverter amplifier as shown in Fig. 4.62. Using Eq. (2.5), we can determine the closed-loop transfer function as given by

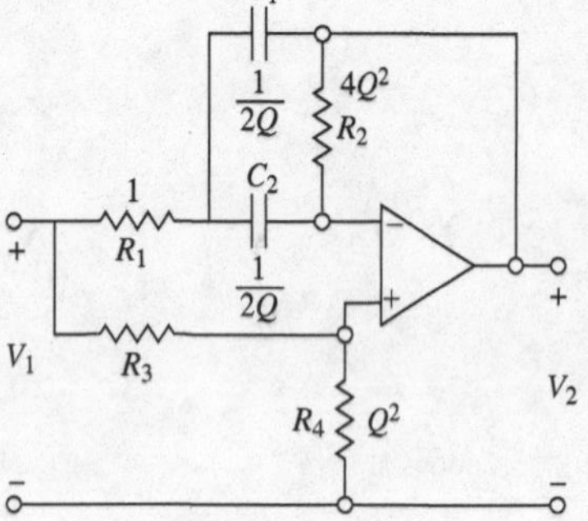

FIGURE 4.62 Inverting op-amp amplifier

$$T = \frac{V_2}{V_1} = \frac{-A}{1+\beta A} \tag{4.232}$$

where $\beta = R_1/(R_1 + R_s)$ is the feedback factor.

We can notice from Eq. (4.232) that T will vary with any variations in A or β. β will also vary with any variations in R_1 or R_2. The sensitivity is defined as the percentage change in variation of any variable with respect the percentage change in another variable. Thus, we can find the sensitivity of T with respect to A from Eq. (2.13) as

$$S_A^T = \frac{\delta T/T}{\delta A/A} = \frac{-1}{1+\beta A} \cong \frac{-1}{\beta A} \tag{4.233}$$

For $\beta A >> 1$, which is generally the case, the sensitivity of T with respect to β becomes very small. Thus, a significant change in A will cause only a small change in T. The sensitivity of T with respect to β can be found from Eq. (2.17) as

$$S_A^T = \frac{\delta T/T}{\delta \beta/\beta} = \frac{\beta A}{1+\beta A} \cong +1 \tag{4.234}$$

For $\beta A >> 1$, $S_\beta{}^T = 1$. Therefore, T is directly sensitive to any change in β. The positive sign in Eq. (4.234) signifies that an increase in β will cause an increase in T. β will also be sensitive to both R_1 and R_2.

4.17.1 Sensitivity Functions

We can generalize the sensitivity function of function T with respect to a variable x which in turn is dependent on variable z by the expressions

$$S_x^T = \frac{\delta T/T}{\delta x/x} \tag{4.235}$$

$$S_z^x = \frac{\delta x/x}{\delta z/z} \tag{4.236}$$

From which we can find the overall density of T with respect to z as given by

$$S_z^T = \frac{\delta x/x}{\delta z/z} \times \frac{\delta T/T}{\delta x/x} = S_z^x S_x^T \tag{4.237}$$

We can rewrite Eq. (4.235) as

$$S_x^T = \frac{x}{T} \times \frac{\delta T}{\delta x} \tag{4.238}$$

Since the derivative of T can be expressed as a logarithmic function, we get

$$\frac{d}{dx}\ln T = \frac{1}{T}\frac{dT}{dx} \tag{4.239}$$

This gives the relation as

$$d(\ln T) = \frac{dT}{T} \tag{4.240}$$

Using the relationship in Eq. (4.240), Eq. (4.235) can also be written as

$$S_x^T = \frac{\delta T/T}{\delta x/x} = \frac{\delta \ln(T)}{\delta \ln(x)} \tag{4.241}$$

There the sensitivity of T with respect to x can be found by using any of the expressions.

$$S_x^T = \frac{\delta T/T}{\delta x/x} = \frac{x}{T} \times \frac{\delta T}{\delta x} = \frac{\delta \ln(T)}{\delta \ln(x)} \tag{4.242}$$

The obvious question arises which one of these expressions is used? T is often a function of ω and Q as in the case of normalized transfer functions, Eq. (4.238) can be used. On the other hand, if the function involves square roots such as the resonant frequency ω_o is the square root of RC elements, Eq. (4.241) can simplify the analysis and calculations.

Case 1: Let us assume that T is a function of x or several values of x, such as x_1, x_2, x_3. The form which encounter most often is given by

$$T = x_1^a x_2^b x_3^c x_4^d \tag{4.243}$$

The natural logarithm of this function is given by

$$\ln T = a \ln x_1 + b \ln x_2 + c \ln x_3 + d \ln x_4 \tag{4.244}$$

Differentiating Eq. (4.244) with respect to x, Eq. (4.242) gives the sensitivities as

$$S_{x_1}^T = \frac{\delta \ln(T)}{\delta \ln(x)} = a; \quad S_{x_2}^T = b; \quad S_{x_3}^T = c; \quad S_{x_4}^T = d \tag{4.245}$$

Case 2: T is a product of two other functions: $T = T_1 T_2$.

$$S_x^T = \frac{\delta \ln(T)}{\delta \ln(x)} = \frac{\delta \ln(T_1 T_2)}{\delta \ln(x)} = \frac{\delta \ln(T_1)}{\delta \ln(x)} + \frac{\delta \ln(T_2)}{\delta \ln(x)} \tag{4.246a}$$

or $$S_x^T = S_x^{T_1} + S_x^{T_2} \tag{4.246b}$$

Case 3: T is a quotient of two other functions: $T = T_1/T_2$.

$$S_x^T = \frac{\delta \ln(T)}{\delta \ln(x)} = \frac{\delta \ln(T_1/T_2)}{\delta \ln(x)} = \frac{\delta \ln(T_1)}{\delta \ln(x)} - \frac{\delta \ln(T_2)}{\delta \ln(x)} \quad \textbf{(4.247a)}$$

or $$S_x^T = S_x^{T_1} - S_x^{T_2} \quad \textbf{(4.247b)}$$

Case 4: T is the inverse function of T_1 such as $T = 1/T_1$.

$$S_x^T = \frac{\delta \ln(T)}{\delta \ln(x)} = \frac{\delta \ln(1/T_1)}{\delta \ln(x)} = -\frac{\delta \ln(T_1)}{\delta \ln(x)} \quad \textbf{(4.248a)}$$

or $$S_x^T = -S_x^{T_1} \quad \textbf{(4.248b)}$$

Case 5: T is the sum of two other functions: $T = T_1 + T_2$

$$S_x^T = S_x^{T_1+T_2} = \frac{T_1 S_x^{T_1} + T_2 S_x^{T_2}}{T_1 + T_2} \quad \textbf{(4.249)}$$

4.17.2 Sensitivity of Transfer Functions

The normalized transfer function $T(s)$ is a function of elements such as ω_o and Q. Using Eq. (4.238), the sensitivity of $T(s)$ with respect to a variable x is given by

$$S_x^{T(s)} = \frac{x}{T(s)} \times \frac{\delta T(s)}{\delta x} \quad \textbf{(4.250)}$$

Assuming $s = j\omega$, the transfer function in the frequency domain will have a magnitude and a phase angle as given by

$$T(j\omega) = |T(j\omega)|\, e^{j\theta(\omega)} \quad \textbf{(4.251)}$$

Therefore, the sensitivity function becomes

$$S_x^{T(j\omega)} = \frac{x}{|T(j\omega)|\, e^{j\theta(\omega)}} \times \frac{\delta}{\delta x}[|T(j\omega)|\, e^{j\theta(\omega)}]$$

$$= \frac{x}{|T(j\omega)|} \times \frac{\delta}{\delta x}|T(j\omega)| + jx\frac{\delta\theta(\omega)}{\delta x} \quad \textbf{(4.252)}$$

This is a complex function. The real part gives the sensitivity of the magnitude and the imaginary part gives the sensitivity of the phase angle, i.e.,

$$S_x^{|T(j\omega)|} = \text{Re}\, S_x^{T(j\omega)} \quad \textbf{(4.253)}$$

$$S_x^{\theta(\omega)} = \frac{1}{\theta(\omega)} \text{Im}\, S_x^{T(j\omega)} \quad \textbf{(4.254)}$$

We are more interested in finding the sensitivity of the magnitude, and Eq. (4.253) can be used for computing the sensitivity function for a magnitude $|T(j\omega)|$.

Case 1: First-order Transfer Function

For the first-odder transfer function,

$$T(s) = \frac{\omega_o}{s + \omega_o} \tag{4.255}$$

The sensitivity function are given by

$$S_Q^{T(s)} = \frac{Q}{T}\frac{\delta T}{\delta Q} = 0 \tag{4.256}$$

$$S_{\omega_o}^{T(s)} = \frac{\omega_o}{T}\frac{\delta T}{\delta Q} = \frac{s}{s + \omega_o} \tag{4.257}$$

At $s = j\omega_o$, the real part is

$$S_{\omega_o}^{|T(j\omega)|} = \text{Re } S_{\omega_o}^{T(j\omega)} = \frac{1}{2} \tag{4.258}$$

Case 2: Low-pass Transfer Function

For the second-order transfer function,

$$T(s) = \frac{\omega_o^2}{s^2 + (\omega_o/Q)s + \omega_o^2} \tag{4.259}$$

The sensitivity function are given by

$$S_Q^{T(s)} = \frac{Q}{T}\frac{\delta T}{\delta Q} = \frac{(\omega_o/Q)s}{s^2 + (\omega_o/Q)s + \omega_o^2} \tag{4.260}$$

$$S_{\omega_o}^{T(s)} = \frac{\omega_o}{T}\frac{\delta T}{\delta \omega_o} = \frac{2s^2 + (\omega_o/Q)s}{s^2 + (\omega_o/Q)s + \omega_o^2} \tag{4.261}$$

For $s = j\omega_o$, the real parts are given by

$$S_Q^{[T(j\omega_o)]} = \text{Re } S_Q^{T(j\omega)} = -1 \tag{4.262}$$

$$S_{\omega_o}^{[T(j\omega_o)]} = \text{Re } S_{\omega_o}^{T(j\omega)} = 1 \tag{4.263}$$

Case 3: Band-pass Transfer Function

For the second-order transfer function,

$$T(s) = \frac{(\omega_o/Q)s}{s^2 + (\omega_o/Q)s + \omega_o^2} \tag{4.264}$$

The sensitivity function are given by

$$S_Q^{T(s)} = \frac{Q}{T}\frac{\delta T}{\delta Q} = \frac{-(s^2 + \omega_o^2)}{s^2 + (\omega_o/Q)s + \omega_o^2} \tag{4.265}$$

$$S_{\omega_o}^{T(s)} = \frac{\omega_o}{T}\frac{\delta T}{\delta \omega_o} = \frac{s^2 - \omega_o^2}{s^2 + (\omega_o/Q)s + \omega_o^2} \tag{4.266}$$

For $s = j\omega_o$, the real parts are given by

$$S_Q^{[T(j\omega_o)]} = \text{Re } S_Q^{T(j\omega)} = 0 \tag{4.267}$$

$$S_{\omega_o}^{[T(j\omega_o)]} = \text{Re } S_{\omega_o}^{T(j\omega)} = 0 \tag{4.268}$$

4.17.3 Sensitivity of Biquad Filter

The transfer function of a biquadratic function as listed in Table 4.2 is

$$T(s) = \frac{K\omega_o^2}{s^2 + (\omega_o/Q)s + \omega_o^2} \tag{4.269}$$

The circuit for the realization of this function is shown in Fig. 4.63. The transfer function of the unity gain amplifier is $V_2/V_4 = -1$ such that

$$\frac{V_4}{V_1} = \frac{V_2}{V_1} \times \frac{V_4}{V_2} = -\frac{V_2}{V_1} \tag{4.270}$$

V_4 is related to V_3 by

$$\frac{V_4}{V_3} = \frac{1}{R_4 C_2 s} \tag{4.271}$$

From the chain rule, which states that the overall transfer function is the product of the transfer functions of the individual circuits, we get

$$\frac{V_3}{V_2} = \frac{V_3}{V_4} \times \frac{V_4}{V_2} \times \frac{V_2}{V_1} \tag{4.272}$$

This gives

$$\frac{V_3}{V_1} = \frac{(-1/R_3C_1)s}{s^2 + (1/R_1C_1)s + 1/R_2R_4C_1C_2} \tag{4.273}$$

This gives the overall transfer function as given by

$$T(s) = \frac{V_2}{V_1} = \frac{V_3}{V_1} \times \frac{V_2}{V_3} = \frac{-1/(R_3R_4C_1C_2)}{s^2 + (1/R_1C_1)s + 1/(R_2R_4C_1C_2)} \tag{4.274}$$

Equating with the co-efficient of Eqs. (4.38), we get ω_o, Q and K as follows:

$$\omega_o = \frac{1}{\sqrt{C_1C_2R_2R_4}}; \quad Q = \sqrt{\frac{C_1R_1^2}{C_2R_2R_4}}; \quad K = \frac{R_2}{R_3} \tag{4.275}$$

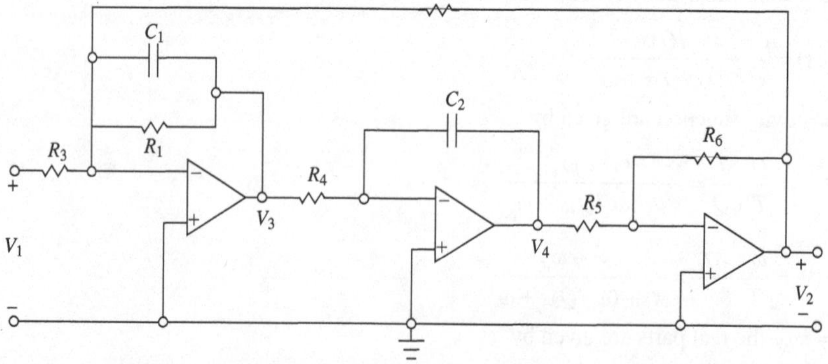

FIGURE 4.63 Biquadratic circuit

We can rewrite ω_o, Q, and K as

$$\omega_o = C_1^{-1/2} C_2^{-1/2} R_2^{-1/2} R_4^{-1/2}; \quad Q = C_1^{1/2} C_2^{-1/2} R_1^{1} R_2^{-1/2} R_4^{-1/2}; \quad K = R_2^{1} R_3^{-1} \tag{4.276}$$

Using Eq. (4.245), we can get the sensitivities of ω_o as

$$S_{C1}^{\omega_o} = -\frac{1}{2}; \quad S_{C2}^{\omega_o} = -\frac{1}{2}; \quad S_{R_2}^{\omega_o} = -\frac{1}{2}; \quad S_{R_4}^{\omega_o} = -\frac{1}{2} \tag{4.277}$$

Therefore, if all of these elements change by +10%, ω_o will change by (– 1/2 – 1/2 – 1/2 – 1/2 = –2) × 10% = –20%. Using Eq. (4.245), we can get the sensitivities of Q as

$$S_{C1}^{Q} = \frac{1}{2}; \quad S_{C2}^{Q} = -\frac{1}{2}; \quad S_{R_2}^{Q} = 1; \quad S_{R_2}^{Q} = -\frac{1}{2}; \quad S_{R_4}^{Q} = -\frac{1}{2} \tag{4.278}$$

Therefore, if all of these elements change by +10%, Q will change by (1/2 – 1/2 + 1 – 1/2 – 1/2 = 0) × 10% = 0%. Using Eq. (4.245), we can get the sensitivities of K as

$$S_{R_2}^{K} = 1; \quad S_{R_3}^{K} = -1 \tag{4.279}$$

Therefore, if all of these elements change by +10%, K will change by (1 – 1 = 0) × 10% = 0%. Equations (4.30) and (4.31) can be applied to calculate the sensitivity of the transfer function with respect to ω_o and Q.

4.17.4 Sensitivity of Delyiannis and Friend Circuit

Equation (4.108) gives ω_o and Q_o of Friend's circuit as shown in Fig. 4.40(a):

$$\omega_o = \frac{1}{C\sqrt{R_1 R_2}}; \quad Q = \frac{1}{2}\sqrt{\frac{R_2}{R_1}}; \quad \text{BW} = \frac{\omega_o}{Q} = \frac{2}{CR_2} \tag{4.280}$$

We can rewrite ω_o and Q as

$$\omega_o = C^{1} R_1^{-1/2} R_2^{-1/2}; \quad Q = \frac{1}{2} R_1^{-1/2} R_2^{1/2} \tag{4.281}$$

Using Eq. (4.245), we can get the sensitivities as

$$S_{C}^{\omega_o} = -1; \quad S_{R_1}^{\omega_o} = -\frac{1}{2}; \quad S_{R_2}^{\omega_o} = -\frac{1}{2}; \quad S_{R_1}^{Q} = -\frac{1}{2}; \quad S_{R_2}^{Q} = \frac{1}{2} \tag{4.282}$$

Therefore, if all of these elements change by +10%, ω_o will change by (–1 – 1/2 – 1/2 = –2) × 10% = –20% and Q will change by (– 1/2 + 1/2 = 0) × 10% = 0%.

Since the bandwidth BW = ω_o/Q, Eq. (4.247*b*) gives

$$S_{x}^{\text{BW}} = S_{x}^{\omega_o} - S_{x}^{Q} \tag{4.283}$$

Substituting values, we get

$$S_{C}^{\text{BW}_o} = -1; \quad S_{R_1}^{\text{BW}} = -\frac{1}{2} - \left(-\frac{1}{2}\right) = 0; \quad S_{R_2}^{\text{BW}} = -\frac{1}{2} - \frac{1}{2} = 1 \tag{4.284}$$

Therefore, if all of these elements change by +10%, BW will change by (–1 – 0 + 1 = 02) × 10% = 0%.

4.17.5 Sensitivity of Sallen–Key Circuit

The Sallen–Key circuit as shown in Fig. 4.13(b) is repeated in Fig. 4.64. Note the symbols are different from those in Fig. 4.13(b). The transfer function can be derived as given by

$$T(s)=\frac{V_2}{V_1}=\frac{K/(R_1R_2C_1C_2)}{s^2+(1/R_1C_1+1/R_2C_1+1/R_2C_2-K/R_2C_2)s+1/(R_1R_2C_1C_2)} \tag{4.285}$$

This shows ω_0, Q, and K to be consistent with the normalized transfer function as given by

$$\omega_o=\frac{1}{\sqrt{C_1C_2R_1R_2}}=C_1^{-1/2}C_2^{-1/2}R_1^{-1/2}R_2^{-1/2} \tag{4.286}$$

$$\frac{\omega_o}{Q}=Y=\frac{1}{R_1C_1}+\frac{1}{R_2C_1}+\frac{1-K}{R_2C_2} \tag{4.287}$$

$$K=1+\frac{R_B}{R_A} \tag{4.288}$$

Using Eq. (4.245), we can get ω_0 sensitivities as

$$S_{R_1}^{\omega_o}=-\frac{1}{2};\quad S_{R_2}^{\omega_o}=-\frac{1}{2};\quad S_{C_1}^{Q}=-\frac{1}{2};\quad S_{C_2}^{Q}=-\frac{1}{2};\quad S_{K}^{\omega_o}=0 \tag{4.289}$$

Since $Q=\omega_0/Y$, we can apply Eq. (4.247*b*) for a variable x as given by

$$S_x^Q=S_x^{\omega_o}-S_x^Y \tag{4.290}$$

Using Eq. (4.238), we get

$$S_x^Y=\frac{x}{Y}\times\frac{\delta Y}{\delta x}=x\frac{Q}{\omega_o}\times\frac{\delta Y}{\delta x} \tag{4.291}$$

Differentiating Eq. (4.287) with respect to R_1 gives

$$\frac{\delta Y}{\delta R_1}=\frac{-1}{R_1^2C_1} \tag{4.292}$$

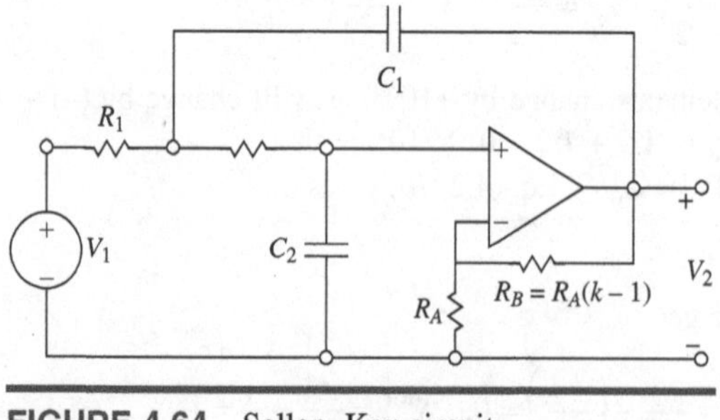

FIGURE 4.64 Sallen–Key circuit

which after substituting into Eq. (4.291) gives

$$S_{R_1}^{Y} = R_1 Q\sqrt{C_1C_2R_1R_2} \times \left(\frac{-1}{R_1^2 C_1}\right) = -Q\sqrt{\frac{C_2R_2}{C_1R_1}} \tag{4.293}$$

Therefore, substituting Eqs. (4.293) and (4.289) into Eq. (4.290), we get

$$S_{R_1}^{Q} = -\frac{1}{2} - Q\sqrt{\frac{C_2R_2}{C_1R_1}} \tag{4.294}$$

$$= -\frac{1}{2} - Q \quad \text{for } C_1 = C_2 = C, R_1 = R_2 = R$$

Differentiating Eq. (4.287) with respect to R_2 gives

$$\frac{\delta Y}{\delta R_2} = \frac{-1}{R_2^2 C_1} + \frac{-(1-K)}{R_2^2 C_2} \tag{4.295}$$

This after substituting into Eq. (4.291) gives

$$S_{R_2}^{Y} = R_2 Q\sqrt{C_1C_2R_1R_2} \times \left(\frac{-1}{R_2^2 C_1} + \frac{-(1-K)}{R_2^2 C_2}\right) = -Q\left[\sqrt{\frac{C_2R_1}{C_1R_2}} + (1-K)\sqrt{\frac{C_1R_1}{C_2R_2}}\right] \tag{4.296}$$

Therefore, substituting Eqs. (4.296) and (4.289) into Eq. (4.290), we get

$$S_{R_2}^{Q} = -\frac{1}{2} - Q\left[\sqrt{\frac{C_2R_1}{C_1R_2}} + (1-K)\sqrt{\frac{C_1R_1}{C_2R_2}}\right] \tag{4.297}$$

$$= -\frac{1}{2} - Q \times (2-K) \quad \text{for } C_1 = C_2 = C, R_1 = R_2 = R$$

The sensitivities of Q with respect to C_1, C_3, and K can be shown to be

$$S_{C_1}^{Q} = -\frac{1}{2} + Q\left[\sqrt{\frac{C_2R_1}{C_1R_2}} + \sqrt{\frac{C_2R_2}{C_1R_1}}\right] \tag{4.298a}$$

$$= -\frac{1}{2} + 2Q \quad \text{for } C_1 = C_2 = C, R_1 = R_2 = R \tag{4.298b}$$

$$S_{C_2}^{Q} = -\frac{1}{2} + (1-K)Q\sqrt{\frac{C_1R_1}{C_2R_2}} \tag{4.299a}$$

$$= -\frac{1}{2} + (1-K)Q \quad \text{for } C_1 = C_2 = C, R_1 = R_2 = R \tag{4.299b}$$

Substituting K from Eq. (4.288) into Eq. (4.287) gives Y as

$$Y = \frac{1}{R_1C_1} + \frac{1}{R_2C_1} + \frac{-R_B/R_A}{R_2C_2} \tag{4.300}$$

The sensitivities of Q with respect to R_A and R_3 can be shown to be

$$S_{R_B}^{Q} = -S_{R_A}^{Q} = (1-K)\sqrt{\frac{C_1R_1}{C_2R_2}} \tag{4.301}$$

Appling Eq. (4.242) to Eq. (4.288), we get

$$S_{R_B}^{K} = -S_{R_A}^{K} = \frac{K-1}{K} = 1 - \frac{1}{K} \tag{4.302}$$

All of these sensitivity equations are derived in terms of circuit *RC* parameters and the gain *K*. It is generally assumed that $C_1 = C_2 = C$ and $R_1 = R_2 = R$ in order to simplify the filter design. As a result the sensitivities depend mostly on the *Q*-factor and *K*-value of the circuit.

NOTE: The hand calculation of the sensitivities scan is time-consuming and laborious. SPICE can calculate the sensitivities with respect to circuit components more effectively [2].

KEY POINTS OF SECTION 4.17

- The sensitivity is defined as the percentage change in variation of any variable with respect the percentage change in another variation of any variable.
- The feedback reduces the effects of any gain and parametric variations. The sensitivity of active filters can be determined by hand calculations.

4.18 Switched-Capacitor Filters

Switched-capacitor filters use on-chip capacitors and MOS switches to simulate resistors. The cutoff frequencies are proportional to and determined by the external clock frequency. In addition, the cutoff or center frequency can be programmed to fall anywhere within an extremely wide range of frequencies—typically more than a 200,000:1 range. Switched-capacitor filters are becoming increasingly popular since they require no external reactive components, capacitors, or inductors. They offer the advantages of low cost, fewer external components, high accuracy, and excellent temperature stability. However, they generate more noise than standard active filters.

4.18.1 Switched-Capacitor Resistors

In all the filters discussed so far, discrete resistors and capacitors were connected to one or more op-amps to obtain the desired cutoff frequencies and voltage gain. Use of discrete resistors is avoided in integrating circuits to reduce chip size; instead, resistor behavior is simulated by using active switches. A resistor is usually simulated by a capacitor and switches. The value of this *simulated resistor* is inversely proportional to the rate at which the switches are opened or closed.

Consider a capacitor with two switches, as shown in Fig. 4.65. The switches are actually MOS transistors that are alternately opened and closed. When S_1 is closed and S_2 is open, the input voltage is applied to the capacitor. Therefore, the total charge on the capacitor is

$$q = V_i C \tag{4.303}$$

When S_1 is open and S_2 is closed, the charge *q* flows to the ground. If the switches are ideal (i.e., they open and close instantaneously and have zero resistance when they close), the capacitor C will charge and

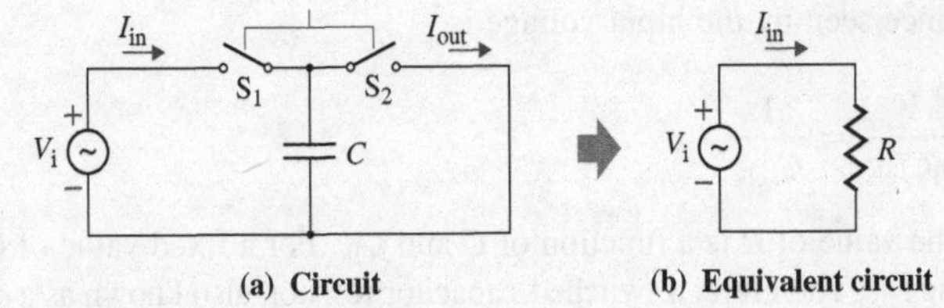

FIGURE 4.65 Switched-capacitor resistor

discharge instantly. The charging current I_{in} and the discharging current I_{out} of the capacitor are shown in Fig. 4.66. If the switches are opened and closed at a faster rate, the current pulses will have the same magnitude but will occur more often. That is, the average current will be higher at a higher switching rate. The average current flowing through the capacitor of Fig. 4.41 is given by

$$I_{av} = \frac{q}{T} = \frac{V_i C}{T} \tag{4.304}$$

$$= V_i C f_{clk}$$

where q = capacitor's charge

T = time between closings of S_1 or closings of S_2, in seconds

$f_{clk} = 1/T$ = clock frequency, in hertz

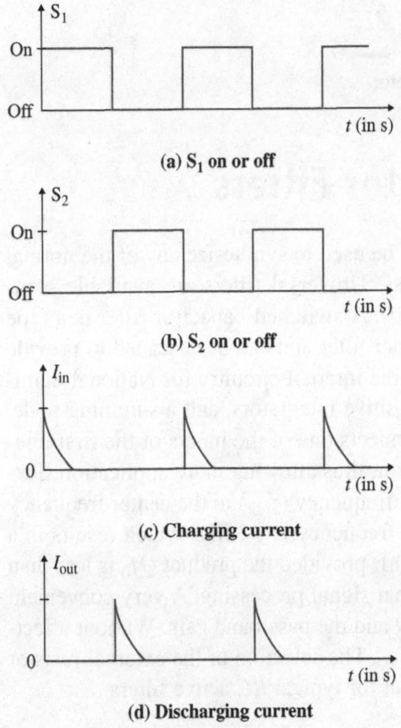

FIGURE 4.66 Current into and out of switched-capacitor resistor

The equivalent resistance seen by the input voltage is

$$R = \frac{V_i}{I_{av}} = \frac{V_i}{V_i C f_{clk}} = \frac{1}{C f_{clk}} \tag{4.305}$$

which indicates that the value of R is a function of C and f_{clk}. For a fixed value of C, the value of R can be adjusted by adjusting f_{clk}. Therefore, a switched-capacitor resistor, also known as a *clock-tunable resistor*, can be built in IC form with a capacitor and two MOS switches. Note that any change in V_i must occur at a rate much slower than f_{clk}, especially when V_i is an AC signal.

4.18.2 Switched-Capacitor Integrators

A simulated resistor can be used as a part of an IC to build a switched-capacitor integrator, as shown in Fig. 4.67. The switches S_1 and S_2 must never be closed at the same time. That means the clock waveform driving the MOS switches must not overlap if the filter is to operate properly.

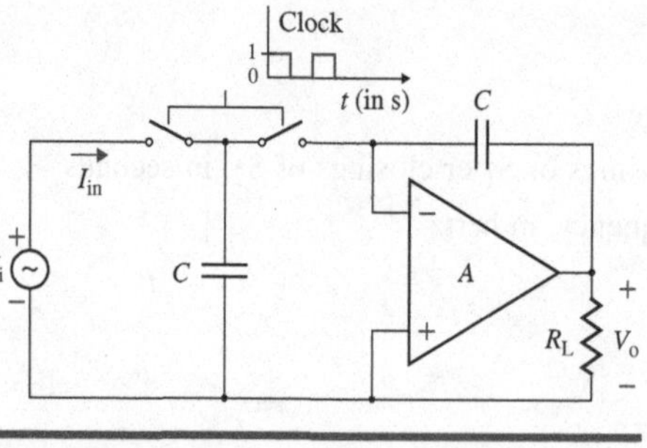

FIGURE 4.67 Switched-capacitor integrator

4.18.3 Universal Switched-Capacitor Filters

A *universal filter* combines many features in an op-amp and can be used to synthesize any of the normal filter types: band-pass, low-pass, high-pass, notch, and all-pass. Universal filters are available commercially (e.g., the type FLT-U2 manufactured by Datel-Intersil). A switched-capacitor filter is a type of universal active filter. It has the characteristics of a second-order filter and can be cascaded to provide very steep attenuation slopes. Figure 4.68 is a block diagram of the internal circuitry for National Semiconductor's MF5. The basic filter consists of an op-amp, two positive integrators, and a summing node. An MOS switch, controlled by a logic voltage on pin 5 (S_A), connects one of the inputs of the first integrator either to the ground or to the output of the second integrator, thus allowing more application flexibility. The MF5 includes a pin (9) that sets the ratio of the clock frequency (f_{clk}) to the center frequency (f_C) at either 50:1 or 100:1. The maximum recommended clock frequency is 1 MHz, which results in a maximum center frequency f_C of 20 kHz at 50:1 or 10 kHz at 100:1, provided the product Qf_o is less than 200 kHz. An extra uncommitted op-amp is available for additional signal processing. A very convenient feature of the MF5 is that f_o can be controlled independently of Q and the pass-band gain. Without affecting the other characteristics, one can tune f_o simply by varying f_{clk}. The selection of the external resistor values is very simple, so the design procedure is much easier than for typical RC active filters.

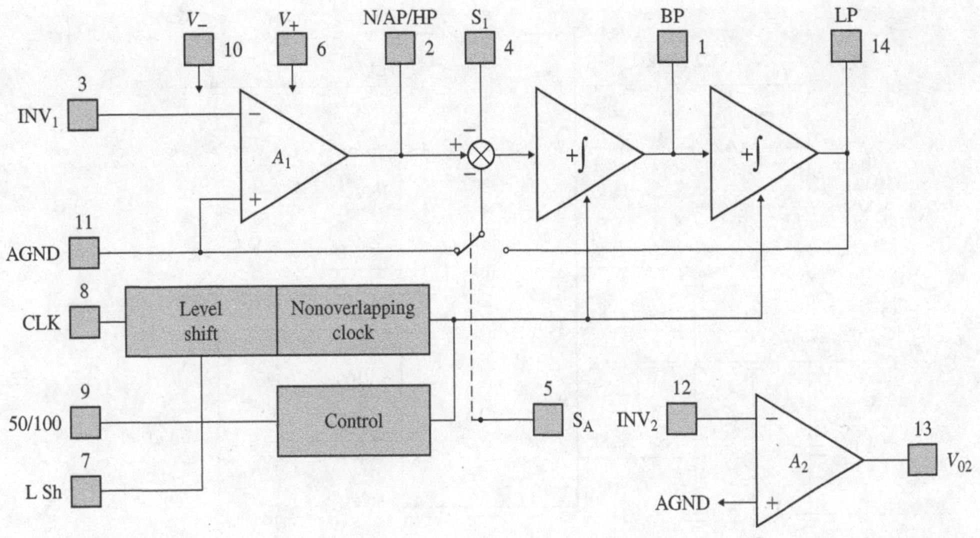

FIGURE 4.68 MF5 universal monolithic switched-capacitor filter (Courtesy of National Semiconductor, Inc.)

EXAMPLE 4.17

D

Designing a second-order Butterworth filter using a universal filter Using the MF5, design a second-order Butterworth low-pass filter with a cutoff frequency of 1 kHz and a pass-band gain of -4. Assume a power supply of ± 5 V and a CMOS clock.

SOLUTION

Step 1. Choose the mode in which the MF5 filter will be operated. Let us choose the simplest one: mode 1, which has low-pass, band-pass, and notch output and inverts the output signal polarity.

Step 2. Determine the values of the external resistors. The MF5 requires three external resistors that determine Q and the gain of the filter. The external resistors are connected as shown in Fig. 4.45. For mode 1, the relationship among Q, K_{LP}, and the external resistors is given by the data sheet (on which only three out of six possible modes are shown) as

$$Q = \frac{f_C}{\mathrm{BW}} = \frac{R_3}{R_2} \tag{4.306}$$

$$K_{LP} = -\frac{R_2}{R_1} \tag{4.307}$$

In this mode, the input impedance of the filter is equal to R_1 since the input signal is applied to INV (pin 3) through R_1. To provide a fairly high input impedance, let $R_1 = 10\ \mathrm{k\Omega}$. From Eq. (4.307), we get

$$R_2 = -K_{LP}R_1 = -(-4) \times 10\ \mathrm{k\Omega} = 40\ \mathrm{k\Omega}$$

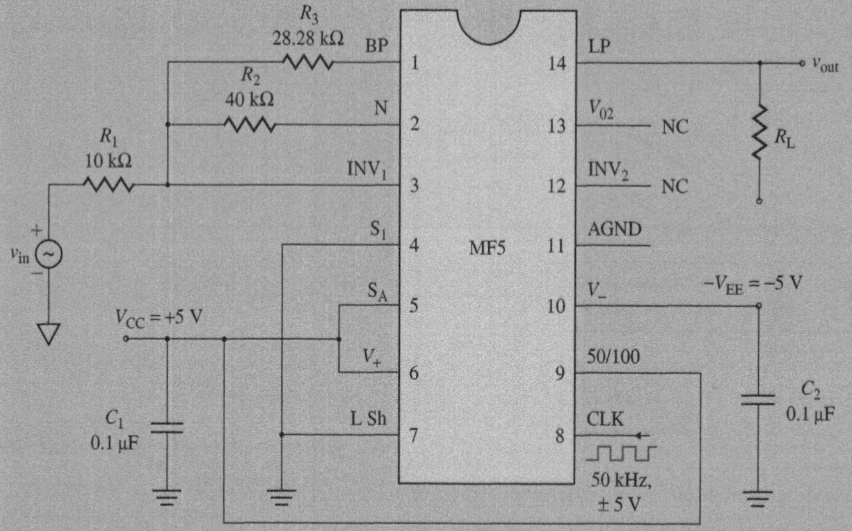

FIGURE 4.45 MF5 configured as a second-order low-pass filter

For a second-order Butterworth low-pass filter, $Q = 0.707$. Therefore, Eq. (4.109) gives

$$R_3 = QR_2 = 0.707 \times 40\text{ k}\Omega = 28.28\text{ k}\Omega$$

Step 3. Choose the power supplies and complete their connections. Since a power supply of ±5 V is required, V_+ (pin 6) is connected to +5 V, V_- (pin 10) is connected to −5 V, and AGND (pin 11) is connected to the ground. To eliminate any ripples, two 0.1-μF capacitors are connected across the power supplies.

Step 4. Choose the clock frequency f_{clk}. The 50/100 (pin 9) must be connected to V_+ (pin 6) for a ratio of 50:1 or to V_- (pin 10) for a ratio of 100:1. Let us choose an f_{clk}-to-f_o ratio of 50:1. That means the 50/100 (pin 9) must be connected to V_+ (pin 6). Since the cutoff frequency is 1 kHz, the external clock frequency is $f_{clk} = 50 \times 1\text{ kHz} = 50\text{ kHz}$.

Step 5. For a CMOS clock, the L Sh (pin 7) should be connected to the ground (pin 11). The low-pass filter S_A (pin 5) is connected to V_+ (pin 6), and S_1 (pin 4) is connected to the ground (pin 11).

The complete circuit for the second-order low-pass filter is shown in Fig. 4.45.

KEY POINT OF SECTION 4.18

- Switched-capacitor filters use on-chip capacitors and MOS switches to simulate resistors. The cutoff frequencies depend on the external clock frequency. In addition, the cutoff or center frequency can be programmed to fall anywhere within an extremely wide range of frequencies.

4.19 Filter Design Guidelines

Designing filters requires selecting the values of R and C that will satisfy two requirements: the bandwidth and the gain. Normally, more than two resistors and capacitors are necessary, and the designer has to assume values for some of them, so there is no unique solution to the design problem. The general guidelines for designing an active filter are as follows:

Step 1. Decide on the design specifications, which may include the cutoff frequencies f_L and f_H, pass-band gain K_{PB}, bandwidth BW, damping factor of $\zeta = 0.707$ for a flat response, $|H(j\omega_o)| = 0.707$, and $|H(j0)| = 1$.

Step 2. Assume a suitable value for the capacitor. The recommended values of C are 1 μF to 5 pF. (Mylar or tantalum capacitors are recommended because they give better performance than other types of capacitors.)

Step 3. Having assumed a value for the capacitor, find the value for the resistor that will satisfy the bandwidth or frequency requirement.

Step 4. If the value of R does not fall within the practical range of 1 kΩ to 500 kΩ, choose a different value of C.

Step 5. Find values for the other resistances that will satisfy the gain requirements and fall in the range of 1 kΩ to 500 kΩ.

Step 6. If necessary, change the filter's cutoff frequency. The procedure for converting the original cutoff frequency f_o to the new cutoff frequency f_n is called *frequency scaling*. It is accomplished by multiplying the value of R or C (but not both) by the ratio of the original frequency f_o to the new cutoff frequency f_n. The new value of R or C can be found from

$$R_n \,(\text{or } C_n) = \frac{\text{Original cutoff frequency } f_o}{\text{New cutoff frequency } f_n} R \,(\text{or } C) \tag{4.308}$$

KEY POINTS OF SECTION 4.19

- Designing filters requires selecting values of R and C to meet the specifications for bandwidth and DC gain. Normally more than two resistors and capacitors are used, and values must be assumed for some of them (normally C). There is no unique solution to a design problem.
- Once a filter has been designed for one cutoff frequency, new values of R and C can be determined by multiplying the value of R or C (but not both) by the ratio of the original frequency f_o to the new cutoff frequency f_n.

Summary

Active filters offer many advantages over passive filters. The many types of active filters—low-pass, high-pass, band-pass, band-reject, and all-pass—are based on the frequency characteristics. A second-order filter has a sharper stop band and is preferable to a first-order filter. An all-pass filter gives a phase shift that is proportional to the input signal frequency.

Universal filters are very popular because of their flexibility in synthesizing frequency characteristics with a very high accuracy. A switched-capacitor filter is a type of universal filter that uses on-chip capacitors and MOS switches to simulate resistors. Its cutoff frequency is proportional to and determined by the external clock frequency.

The Delyiannis–Friend circuit exhibits the characteristic of a band-pass filter. One or more first-order and second-order pass-band filters are generally connected in cascade to form a stagger filter. The Chebyshev response has a function $F_n(\omega)$ which is scaled to fit into a square box. There are two units on each side; the limits of the amplitude are ±1 and the limits of ω are ±1. The choice between a Butterworth and a Chebyshev response will require trade-off between the conflicting requirements.

The delay circuit provides a defined delay for signal processing. An equalization circuit does the "patch up" of the total delay known as equalization. The sensitivity is an important performance parameter and it is defined as the percentage change in variation of any variable with respect to the percentage change in another variation of any variable.

References

1. W. K. Chen, *Passive and Active Filters—Theory and Implementation*. New York: Wiley, 1986.
2. M. H. Rashid, *Introduction to SPICE using* OrCAD *for Circuits and Electronics*. Englewood Cliffs, NJ: Prentice Hall, 2003, Chapter 10.
3. M. E. Van Valkenburg, *Analog Filter Design*. New York: CBS College Publishing, 1982.
4. R. Schaumann and M. E. Van Valkenburg, *Design of Analog Filters*. New York: Oxford University Press, 2001.
5. L. P. Huelsman and P. E. Allen, *Introduction to the Theory and Design of Active Filters*. New York: McGraw-Hill, 1980.
6. A. V. Oppenheim and A. S. Willsky, with S. H. Nawab, *Signals and Systems*. Upper Saddle River, NJ: Prentice Hall, 1997.
7. R. Schaumann, M. S. Ghausi, and K. R. Laker, *Design of Analog Filters—Passive, Active RC, and Switched Capacitor*. Englewood Cliffs, NJ: Prentice Hall, 1990.
8. R. A. Gayakwad, *Op-Amps and Linear Integrated Circuits*. Englewood Cliffs, NJ: Prentice Hall, 1993.
9. G. C. Temes and L. Lapatra, *Introduction to Circuit Synthesis and Design*. New York: McGraw-Hill, 1977.

Review Questions

1. What is an active filter?
2. What are the advantages of active filters over passive ones?
3. What are the types of active filters?
4. What are the passband and the stop band of a filter?
5. What is a cutoff frequency?
6. What is the Butterworth response of a filter?
7. What are the differences between first-order and second-order filters?
8. What is frequency scaling of filters?
9. What is a notch filter?
10. What is a notch-out frequency?
11. What is an all-pass filter?
12. What is a universal filter?
13. What is a switched-capacitor resistor?

14. What is a switched-capacitor filter?
15. What is a clock-tunable resistor?
16. What is a Delyiannis–Friend circuit?
17. What is stagger-tuned band-pass filter?
18. What is a Chebyshev filter?
19. What is a delay filter?
20. What is an equalization circuit?
21. What is sensitivity of an electronic circuit?

Problems

The symbol **D** indicates that a problem is a design problem. The symbol **P** indicates that you can check the solution to a problem using PSpice/SPICE or Electronics Workbench. For PSpice/SPICE simulation, assume op-amps with parameters $R_i = 2\ \text{M}\Omega$, $R_o = 75\ \Omega$, and $A_o = 2 \times 10^5$.

4.4 *First-Order Filters*

4.1 Determine **(a)** the transfer function of the network shown in Fig. P4.1 and **(b)** its poles and zeros.

FIGURE P4.1

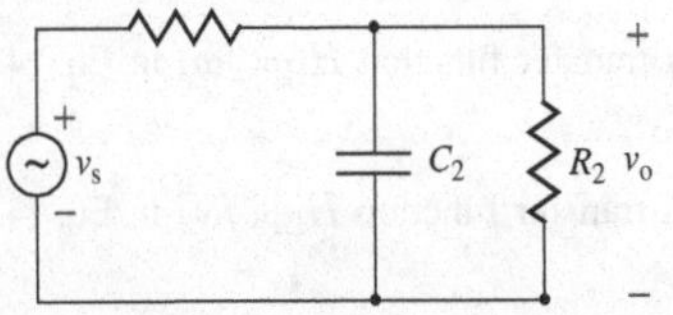

4.2 Determine **(a)** the transfer function of the network shown in Fig. P4.2 and **(b)** its poles and zeros.

FIGURE P4.2

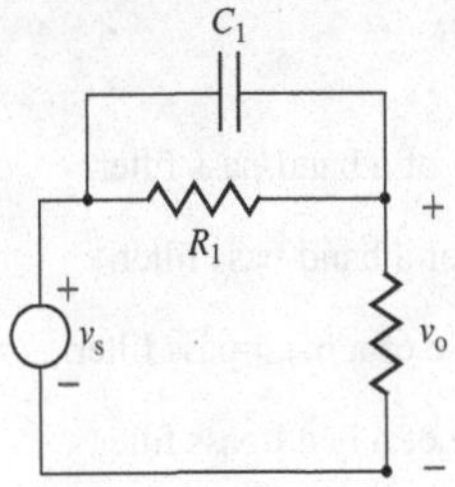

4.3 Determine **(a)** the transfer function of the network shown in Fig. P4.3 and **(b)** its poles and zeros.

FIGURE P4.3

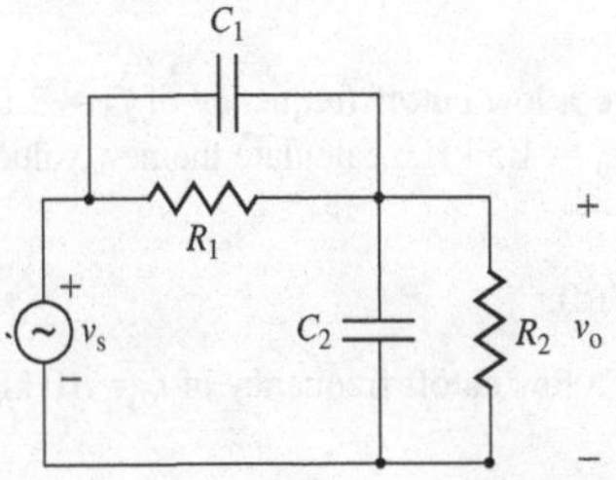

4.4 Determine **(a)** the transfer function of the network shown in Fig. P4.4 and **(b)** its poles and zeros.

FIGURE P4.4

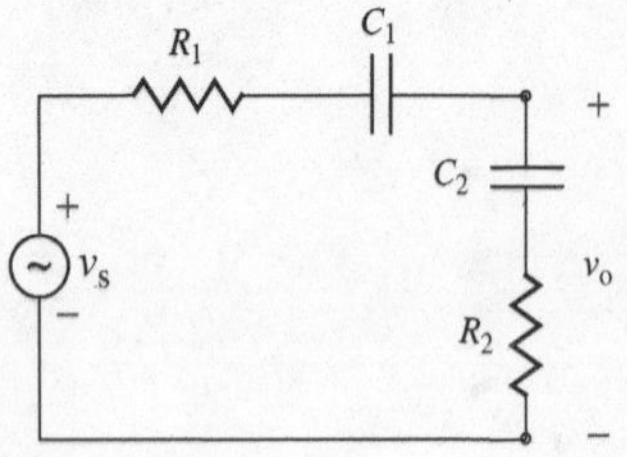

4.5 *The Biquadratic Function*

4.5 Determine **(a)** the pole and zero quality factors Q_p and Q_z, **(b)** the pole and zero resonant frequencies ω_p and ω_z, **(c)** the pole factor β_p, and **(d)** the pole angle ϕ_p. The transfer function has the general form as given by

$$H(s) = \frac{5s^2 + 15s + 100}{s^2 + 20s + 200}$$

4.6 Plot the frequency and phase responses of the low-pass transfer function $H_{LP}(j\omega)$ in Eq. (4.20) versus $u = 0.5$ to 5 for $Q = 0.5, 0.707$, and 1 and $K = 1$.

4.7 Plot the frequency and phase responses of the high-pass transfer function $H_{HP}(j\omega)$ in Eq. (4.21) versus $u = 0.5$ to 5 for $Q = 0.5, 0.707$, and 1 and $K = 1$.

4.8 Plot the frequency and phase responses of the band-pass transfer function $H_{BP}(j\omega)$ in Eq. (4.22) versus $u = 0.5$ to 5 for $Q = 0.5, 0.707$, and 1 and $K = 1$.

4.9 Plot the frequency and phase responses of the band-reject transfer function $H_{BP}(j\omega)$ in Eq. (4.23) versus $u = 0.5$ to 5 for $Q = 0.5, 0.707$, and 1 and $K = 1$.

4.6 *Butterworth Filters*

4.10 Determine the transfer function for fourth-order Butterworth response of a band-pass filter.

4.11 Determine the transfer function for sixth-order Butterworth response of a band-pass filter.

4.12 Determine the transfer function for seventh-order Butterworth response of a band-pass filter.

4.13 Determine the transfer function for eighth-order Butterworth response of a band-pass filter.

4.14 Determine the transfer function for ninth-order Butterworth response of a band-pass filter.

4.8 *Low-Pass Filters*

4.15 Design a first-order low-pass filter as in Fig. 4.10(b) to give a low cutoff frequency of $f_o = 2$ kHz with a pass-band gain of 1. If the desired frequency is changed to $f_n = 1.5$ kHz, calculate the new value of R_n.

4.16 Derive the transfer function $H(s)$ of the network in Fig. 4.13(d).

4.17 Design a second-order low-pass filter as in Fig. 4.14 to give a low cutoff frequency of $f_o = 10$ kHz, a pass-band gain of $K = 5$, and $Q = 0.707$, 1, and ∞.

4.18 Design a second-order Butterworth low-pass filter as in Fig. 4.17(c) to yield $|H(j\omega_o)| = 1$ (or 0 dB), a cut-off frequency of $f_o = 10$ kHz, and $Q = 0.707$.

D P

4.19 Design a second-order Butterworth filter as in Fig. 4.14 to yield $|H(j0)| = 1$ (or 0 dB), a cutoff frequency of $f_o = 10$ kHz, and $Q = 0.707$.

D P

4.20 Design a third-order Butterworth low-pass filter as in Fig. P4.20 to give a high cutoff frequency of $f_o = 10$ kHz and a pass-band gain of 10. The transfer function has the general form

D P

$$H_3(s) = \frac{10\omega_o^3}{s^3 + 2\omega_o s^2 + 2\omega_o^2 s + \omega_o^3}$$

FIGURE P4.20

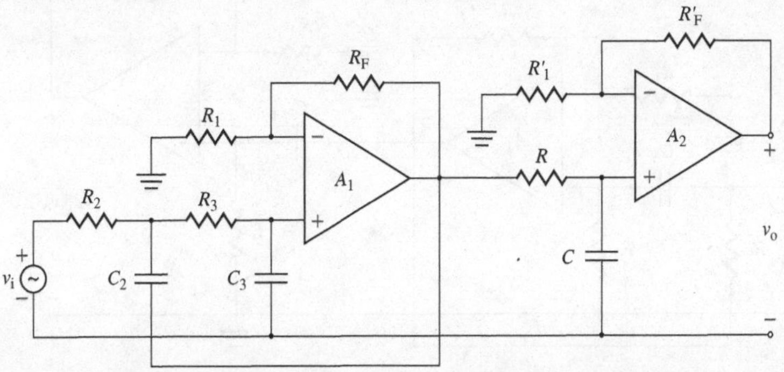

4.21 Design a fourth-order Butterworth low-pass filter as in Fig. P4.21 to give a high cutoff frequency of $f_o = 10$ kHz and a pass-band gain of 25. The transfer function has the general form

D

$$H_4(s) = \frac{25\omega_o^4}{(s^2 + \sqrt{2}\omega_o s + \omega_o^2)^2}$$

FIGURE P4.21

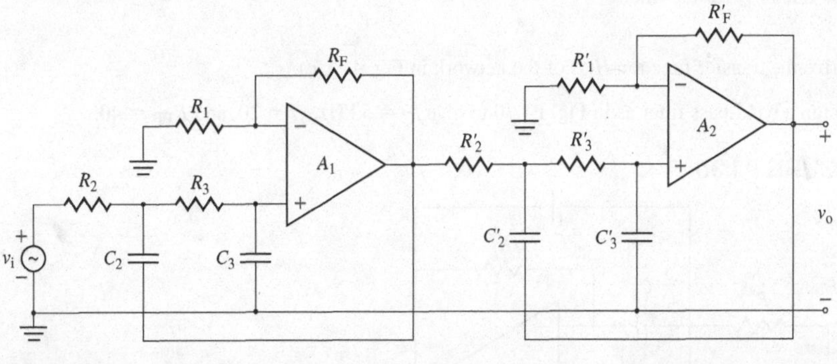

4.9 *High-Pass Filters*

4.22 Design a first-order high-pass filter as in Fig. 4.20(b) to give a low cutoff frequency of $f_o = 400$ Hz and a pass-band gain of $K = 2$. If the desired frequency is changed to $f_n = 1$ kHz, calculate the new value of R_n.

D P

4.23 Design a second-order high-pass filter as in Fig. 4.22 to give a low cutoff frequency of $f_o = 2$ kHz and a pass-band gain of 2. If the desired frequency is changed to $f_n = 3.5$ kHz, calculate the new value of R_n.

D P

4.24 Design a second-order Butterworth high-pass filter as in Fig. 4.25(c) to yield $|H(j\infty)| = 1$ (or 0 dB), a cutoff frequency of $f_o = 10$ kHz, and $Q = 0.707$.

D P

4.25 Design a second-order Butterworth high-pass filter as in Fig. 4.25(c) to yield $|H(j\omega_o)| = 1$ (or 0 dB), a cutoff frequency of $f_o = 10$ kHz, and $Q = 0.707$.

D P

4.26 Design a third-order Butterworth high-pass filter as in Fig. P4.26 to give a low cutoff frequency of $f_o = 10$ kHz and a pass-band gain of 10. The transfer function has the general form

D P

$$H_3(s) = \frac{10s^3}{s^3 + 2\omega_o s^2 + 2\omega_o^2 s + \omega_o^3}$$

FIGURE P4.26

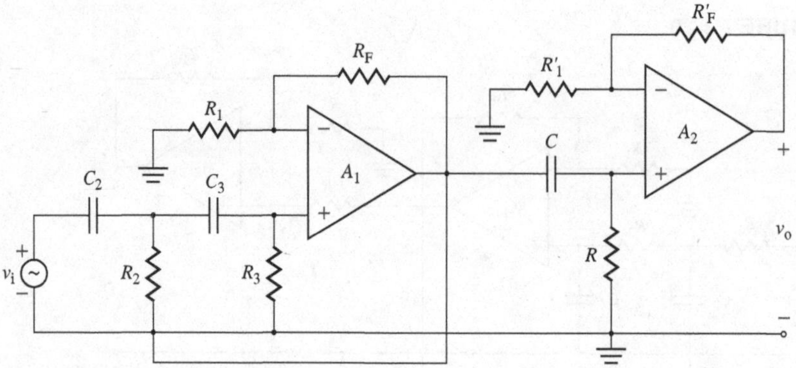

4.10 Band-Pass Filters

4.27 Design a wide-band-pass filter with $f_L = 400$ Hz, $f_H = 2$ kHz, and a pass-band gain of $K_{PB} = 4$. Calculate the value of Q for the filter.

D P

4.28 Design a wide-band-pass filter with $f_L = 1$ kHz, $f_H = 10$ kHz, and a pass-band gain of $K_{PB} = 20$. Calculate the value of Q for the filter.

D P

4.29 Derive the transfer function $H(s)$ of the network in Fig. 4.31(c).

4.30 Design a band-pass filter as in Fig. P4.30 to give $f_C = 5$ kHz, $Q = 20$, and $K_{PB} = 40$.

D P

FIGURE P4.30

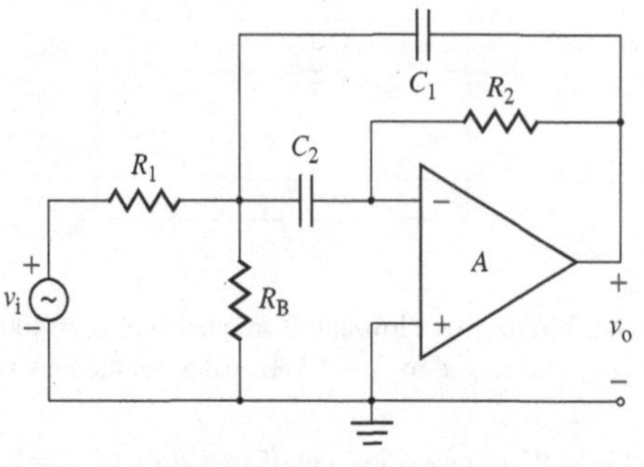

4.31 a. Design a narrow-band-pass filter as in Fig. 4.31(b) such that $f_C = 2$ kHz, $Q = 20$, and $K_{PB} = 10$.

D P b. Calculate the value of R_B required to change the center frequency from 2 kHz to 5.5 kHz.

4.11 *Band-Reject Filters*

4.32 Design a wide-band-reject filter as in Fig. 4.34(c) to give $f_H = 400$ kHz, $f_L = 2$ kHz, and $K_{PB} = 10$. Calculate the value of Q for the filter.
D P

4.33 Design a wide-band-reject filter with a falloff rate of 40 dB/decade to give $f_H = 400$ kHz, $f_L = 2$ kHz, and $K_{PB} = 40$.
D P

4.34 Design an active notch filter as in Fig. 4.37(b) with $f_n = 400$ Hz.
D P

4.35 Derive the transfer function $H(s)$ of the network in Fig. 4.37(b).

4.36 Design an active notch filter as in Fig. P4.36 with $f_n = 400$ Hz and $Q = 5$.
D P

FIGURE P4.36

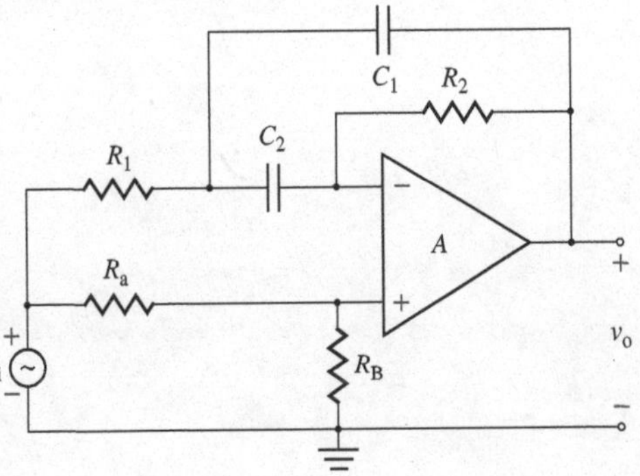

4.16 *All-Pass Filters*

4.37 Design an all-pass filter as in Fig. 4.40(b) so that the phase shift is $\phi = \pm 150°$ at 60 Hz.
D P

4.38 Using the MF5, design a second-order Butterworth low-pass filter with a cutoff frequency of 2 kHz and a pass-band gain of −2. Assume a power supply of ±5 V and a CMOS clock.
D

4.12–4.17: Delyiannis-Friend, Stagger-tuned, and Chebyshev Filters, Delay and Equalization Circuit, and Sensitivity

4.39 Design a stagger filter with the following specifications: $\omega_1 = 1000$ rad/s, $\omega_2 = 1500$ rad/s, $\omega_3 = 500$ rad/s, $\omega_4 = 3000$ rad/s, BW = 500 rad/s, $\alpha_{min} = 20$ dB, $\alpha_{max} = 0.3$ dB.

4.40 Design a stagger filter with the following specifications: $f_p = 10$ kHz, stop-band frequency $f_s = 24$ kHz, $\alpha_{min} = 22$ dB, and $\alpha_{max} = 0.3$ dB.

4.41 Design a Chebyshev filter with the following specifications: pass-band frequency f_p = 10 kHz, stop-band frequency f_s = 24 kHz, α_{min} = 22 dB, and α_{max} = 0.3 dB.

4.42 Design a Chebyshev filter with the following specifications: ω_1 = 1000 rad/s, ω_2 = 1500 rad/s, ω_3 = 500 rad/s, ω_4 = 3000 rad/s, BW = 500 rad/s, α_{min} = 20 dB, α_{max} = 0.3 dB

4.43 (a) Design a fourth-order Sallen–Key delay filter to implement the Bessel–Thomson function with a delay of 200 μs. Find the (b) magnitude of the function and (c) delay at the first break frequency ω_{o1}.

4.44 Design a fourth-order Sallen–Key delay filter to implement the Bessel–Thomson function with a delay of 300 μs. Find the (b) magnitude of the function and (c) delay at the first break frequency ω_{o1}.

4.45 A first-order equalization is cascaded with a fourth-order Butterworth filter and optimized to be maximally fit at $\omega = 0$. Determine (a) $D_T(0)$, (b) $D_T(\omega = 0.9998)$, (c) ΔD, and (c) the RC circuit values of the equalization. The transfer function of fifth-order Butterworth filter is

$$T_5(s) = \frac{1}{(1+s)(s^2+1.725s+1)(s^2+0.725s+1)}$$

CHAPTER 5

OSCILLATORS

Learning Outcomes

After completing this chapter, students should be able to do the following:

- Describe the operating principle of oscillators for generating sinusoidal voltage and the conditions for sustained oscillations.
- Describe the characteristics and types of oscillators.
- Select an oscillator circuit to meet specific frequency requirements.
- Design an oscillator circuit to meet desired frequency specifications.

Symbols and Their Meanings

Symbol	Meaning
A, A_f	Open-loop and closed-loop voltage gains of an amplifier
f_o, ω_o	Oscillation frequencies in hertz and radians per second
g_m, r_π, r_o	Transconductance, input resistance, and output resistance of a transistor
h_{fe}	Hybrid current gain parameter of a BJT transistor
K_p, V_t	MOSFET constant and threshold voltage

Symbol	Meaning
L_p, R_p	Equivalent primary side load inductance and resistance of a transformer-coupled load
$\beta(s)$, $H(s)$	Feedback factor and transfer function of an amplifier
$v_o(t)$	Instantaneous output voltage of an oscillator

5.1 Introduction

We know from Sec. 2.11 that an amplifier with negative feedback will be unstable if the magnitude of the loop gain is greater than or equal to 1 and its phase shift is $\pm 180°$. Under these conditions, the feedback becomes positive and the output of the amplifier oscillates. An oscillator is a circuit that generates a repetitive waveform of fixed amplitude at a fixed frequency without any external input signal. A waveform of this characteristic can be obtained by applying positive feedback in amplifiers. Positive feedback provides enough feedback signal to maintain oscillations. Although oscillations are very undesirable in linear amplifier circuits, oscillators are designed specifically to produce controlled and predictable oscillation. Thus, the strategy for designing oscillators is quite different from that for designing linear amplifiers. Occasionally, oscillators have inputs that are used to control the frequency or to synchronize the oscillations with an external reference. Oscillators are used in many electronic circuits, such as in radios, televisions, computers, and communications equipment.

5.2 Principles of Oscillators

An oscillator is an amplifier with positive feedback. The block diagram of an amplifier with positive feedback, shown in Fig. 5.1(a), suggests the following relationships:

$$v_e = v_i + v_f$$

$$v_o = A v_e$$

$$v_f = \beta v_o$$

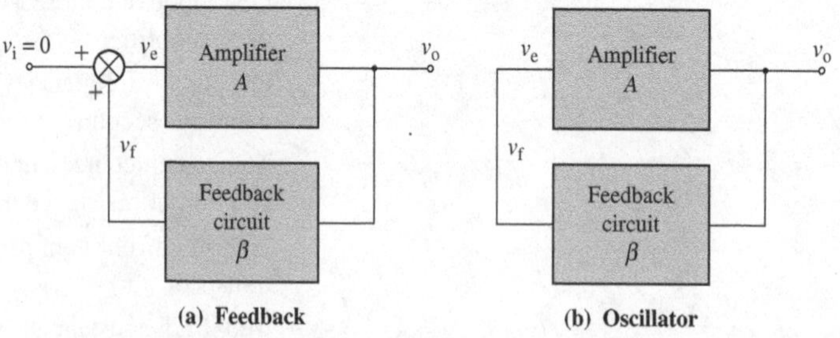

FIGURE 5.1 Oscillator block diagram

Using these relationships, we get the closed-loop voltage gain A_f:

$$A_f = \frac{v_o}{v_i} = \frac{A}{1 - A\beta} \tag{5.1}$$

5.2.1 Conditions for Oscillations

A_f in Eq. (5.1) can be made very large by making $1 - A\beta = 0$. That is, an output of reasonable magnitude can be obtained with a very small-value input signal, tending to zero, as shown in Fig. 5.1(b). Thus, the amplifier will be unstable when $1 - A\beta = 0$, which gives the loop gain as

$$A\beta = 1 \tag{5.2}$$

Expressing Eq. (5.2) in polar form, we get

$$A\beta = 1\angle 0^\circ \quad \text{or} \quad 1\angle 360^\circ \tag{5.3}$$

The above discussion leads to the following design criteria for oscillators:

1. The magnitude of the loop gain $|A\beta|$ must be unity or slightly larger at the desired oscillation frequency.
2. The total phase shift ϕ of the loop gain must be equal to 0° or 360° at the same frequency.
3. The first two conditions must not be satisfied at other frequencies. This condition is normally met by carefully selecting the component values.
4. The first two conditions must continue to be satisfied as parameter values change in response to component tolerance, temperature change, aging, and device replacement. Meeting this criterion often requires special design considerations.

If an amplifier provides a phase shift of 180°, the feedback circuit must provide an additional phase shift of 180° so that the total phase shift around the loop is 360°. The type of waveform generated by an oscillator depends on the types of components used in the circuit; hence the waveform may be sinusoidal, square, or triangular. The frequency of oscillation is determined by the feedback components:

- *RC* components generate a sinusoidal waveform at audio frequencies—that is, in the range from several hertz (Hz) to several kilohertz (kHz).
- *LC* components generate a square wave at radio frequencies—that is, in the range from 100 kHz to 100 MHz.
- Crystals generate a triangular or sawtooth wave over a wide range—that is, in the range from 10 kHz to 10 MHz.

Oscillators can be classified into many types depending on the feedback components, amplifiers, and circuit topologies used [1]. This chapter covers the following types of oscillators: phase-shift oscillators, quadrature oscillators, three-phase oscillators, Wien-bridge oscillators, Colpitts oscillators, Hartley oscillators, crystal oscillators, and active-filter tuned oscillators.

EXAMPLE 5.1

Finding the gain and phase for oscillation A block diagram of an oscillator is shown in Fig. 5.2. Determine the values of the gain A and the phase angle θ that will produce a steady-state oscillation.

SOLUTION

Applying the condition of Eq. (5.3), we know that the loop gain must be

$$A\angle 180^\circ \times 0.01\angle 0^\circ \times 0.5\angle\theta \times 10\angle 0^\circ = 1\angle 360^\circ$$

from which we get

$$A \times 0.01 \times 0.5 \times 10 = 1 \quad \text{or} \quad A = \frac{1}{0.01 \times 0.5 \times 10} = 20$$

and $\quad 180 + 0 + \theta + 0 = 360 \quad \text{or} \quad \theta = 360 - 180 = 180^\circ$

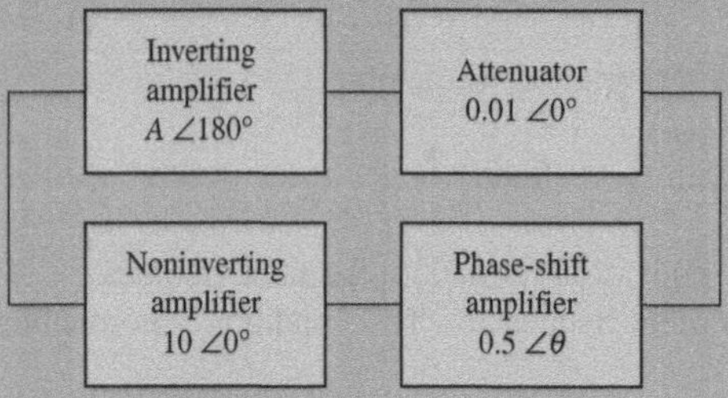

FIGURE 5.2 Oscillator circuit

EXAMPLE 5.2

Finding the frequency and conditions to sustain oscillation The amplifier shown in Fig. 5.3(a) has a voltage gain of $A = 50$, input resistance $R_i = 10\ \text{k}\Omega$, and output resistance $R_o = 200\ \Omega$. Find the resonant frequency ω_o and the values of R and R_3 that will sustain the oscillation.

SOLUTION

The voltage-controlled current source representation is shown in Fig. 5.3(b), where transconductance $G_m = A/R_o$. R_i is in parallel with R, and we let $R_1 = R_i \| R$. Also, R_o is in series with R_3, and we let $R_F = R_o + R_3$. The equivalent circuit is shown in Fig. 5.4, which represents the amplifier as an ideal voltage amplifier. The feedback transfer function β of the feedback circuit is given by

$$\beta(j\omega) = \frac{V_f}{V_o}(j\omega) = \frac{j\omega L \,\|\, (-j/\omega C) \,\|\, R_1}{j\omega L \,\|\, (-j/\omega C) \,\|\, R_1 + R_F}$$

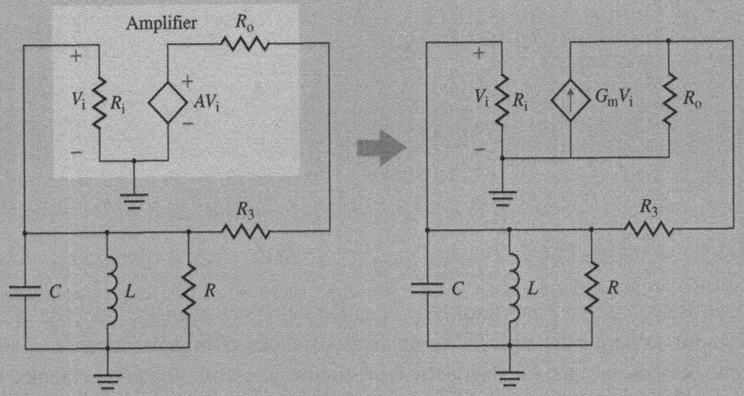

FIGURE 5.3 *LC*-feedback amplifier

which can be simplified to give the loop gain as

$$A\beta(j\omega) = \frac{j\omega LA}{R_F(1 - \omega^2 LC) + j\omega L(1 + R_F/R_1)} \tag{5.4}$$

This will provide a 0° phase shift at the resonant frequency ω_o given by

$$\omega_o = \frac{1}{\sqrt{LC}} \quad \text{(in rad/s)} \tag{5.5}$$

At this frequency, the magnitude of $A\beta(j\omega)$ becomes

$$|A\beta(j\omega)| = \frac{A}{1 + R_F/R_1} \tag{5.6}$$

which must be equal to 1 and gives the condition for oscillation as

$$\frac{R_F}{R_1} = A - 1 \tag{5.7}$$

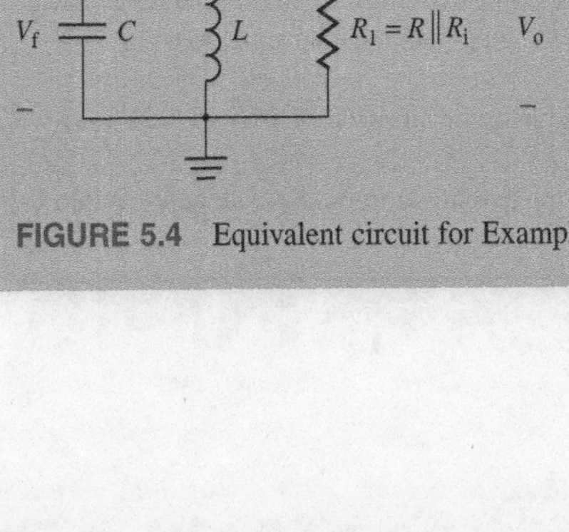

FIGURE 5.4 Equivalent circuit for Example 5.2

That is, for $A = 50$, $R_F/R_1 = 49$. Let $R_1 = R_i \| R = 5\ \text{k}\Omega$; then, for $R_i = 10\ \text{k}\Omega$, $R = 10\ \text{k}\Omega$. Therefore,

$$R_F = R_o + R_3 = 49R_1 = 245\ \text{k}\Omega$$

Thus, for $R_o = 200\ \Omega$, $R_3 = 244.8\ \text{k}\Omega$.

5.2.2 Frequency Stability

The ability of an oscillator to oscillate at an exact frequency is called *frequency stability*. The oscillating frequency is a function of circuit components (e.g., *LC* components) and can change in response to temperature changes, device replacement, or parasitic elements. Good frequency stability can be obtained by making the phase shift a strong function of frequency at resonance. That is, $|d\phi/d\omega|$ (at $\omega = \omega_o$) is made large so that only a slight change in ω is required to correct any phase shift and restore the loop gain to zero phase shift.

The *quality factor* (or *figure of merit*) Q of a circuit also determines the frequency stability. The higher the Q factor, the better the stability will be because the variation in phase shift with frequency near resonance is greater. Crystal oscillators are far more stable than *RC* or *LC* oscillators, especially at higher frequencies. The equivalent electrical circuit of a crystal has a very high Q-value, leading to a high value of $d\phi/d\omega$. *LC* and crystal oscillators are generally used for the generation of high-frequency signals; *RC* oscillators are used mostly for audio-frequency applications.

5.2.3 Amplitude Stability

Like the frequency, the gain of practical amplifiers can change in response to changes in parameters such as temperature, age, and operating point. Therefore, $|A\beta|$ might drop below unity. If the magnitude of $A\beta$ falls below unity, an oscillating circuit ceases oscillating. In practice, an oscillator is designed with a value of $|A\beta|$ that is slightly higher than unity—say, by 5%—at the oscillating frequency. The greater the value of $|A\beta|$, the greater will be the amplitude of the output signal and the amount of its distortion. This distortion will usually lower the gain A to the value required to sustain oscillation.

For good stability, the change in the gain A with a change in the amplitude of output voltage v_o should be made large; an increase in amplitude must result in a decrease in gain. That is, dA/dv_o must be a large negative number. An oscillator is often stabilized by adding nonlinear limiting devices or elements such as diodes.

KEY POINTS OF SECTION 5.2

- To sustain oscillations, the magnitude of the loop gain $|A\beta|$ must be unity or slightly larger at the desired oscillation frequency, and the total phase shift ϕ of the loop gain must be equal to 0° or 360° at the same frequency.
- For good frequency stability, the phase shift must be made a strong function of frequency at resonance; that is, the Q factor should be high.
- For good stability, the change in the gain with a change in the amplitude of output voltage v_o should be made large; an increase in amplitude must result in a decrease in gain.

5.3 Audio-Frequency Oscillators

An audio-frequency (AF) oscillator uses *RC* components for generating a sinusoidal waveform in the frequency range from several hertz (Hz) to several kilohertz (kHz). There are different ways to make an amplifier oscillate. We will consider the following types of AF oscillators: phase-shift oscillator, quadrature oscillator, three-phase oscillator, Wien-bridge oscillator, and ring oscillator. The op-amp Wien bridge is the most common and easy to implement. The MOS oscillators allow CMOS implementation.

5.3.1 Phase-Shift Oscillators

A phase-shift oscillator consists of an inverting amplifier with a positive feedback circuit [1]. The amplifier gives a phase shift of 180°, and the feedback circuit gives another phase shift of 180°, so the total phase shift around the loop is 360°. A phase-shift oscillator consisting of an inverting op-amp amplifier with positive feedback is shown in Fig. 5.5(a). The feedback circuit provides voltage feedback from the output back to the input of the amplifier. Any signal that appears at the inverting terminal is shifted by 180° at the output. Therefore, an additional 180° shift is required for oscillation at a specific frequency f_o in order to give a total phase shift around the loop of 360°. Since the feedback network consists of resistors and capacitors, as shown in Fig. 5.5(b), this type of oscillator is also known as an *RC* oscillator. The transfer function of the feedback network in Laplace's domain of *s* is given by

$$\beta(s) = \frac{V_f(s)}{V_o(s)} = \frac{R^3C^3s^3}{R^3C^3s^3 + 6R^2C^2s^2 + 5RCs + 1} \tag{5.8}$$

(see Prob. 5.3). The closed-loop voltage gain of the op-amp circuit is

$$A(s) = \frac{V_o(s)}{V_f(s)} = -\frac{R_F}{R_1} \tag{5.9}$$

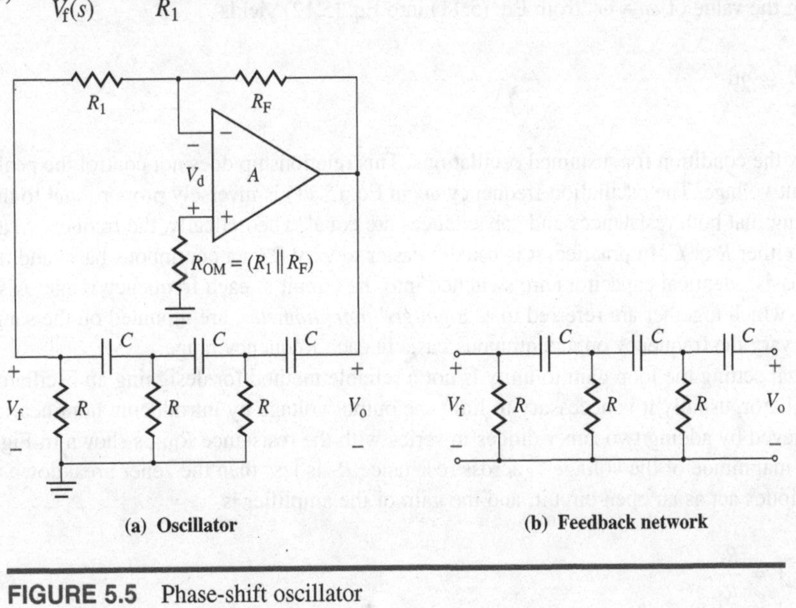

FIGURE 5.5 Phase-shift oscillator

Since $A\beta = 1$ for an oscillator, from Eqs. (5.8) and (5.9), we get

$$-\frac{R_F}{R_1}\left[\frac{R^3C^3s^3}{R^3C^3s^3 + 6R^2C^2s^2 + 5RCs + 1}\right] = 1 \tag{5.10}$$

Substituting $s = j\omega$ into Eq. (5.10) and canceling the elements in the denominator, we get

$$-R_F(-jR^3C^3\omega^3) = R_1(-jR^3C^3\omega^3 - 6R^2C^2\omega^2 + j5RC\omega + 1)$$

Equating the real parts to zero, we get

$$R_1(-6R^2C^2\omega^2 + 1) = 0$$

which gives the oscillation frequency ω_o as

$$\omega_o = \omega = 2\pi f_o = \frac{1}{\sqrt{6}RC} \quad \text{(in rad/s)} \tag{5.11}$$

where f_o is the frequency in hertz. Equating the imaginary parts on both sides yields

$$-R_F(-jR^3C^3\omega^3) = R_1(-jR^3C^3\omega^3 + j5RC\omega)$$

which gives

$$R_F = R_1\left[\frac{5}{R^2C^2\omega^2} - 1\right] \tag{5.12}$$

Substituting the value of $\omega = \omega_o$ from Eq. (5.11) into Eq. (5.12) yields

$$\frac{R_F}{R_1} = 29 \tag{5.13}$$

which gives the condition for sustained oscillations. This relationship does not control the peak amplitude of the output voltage. The oscillation frequency ω_o in Eq. (5.11) is inversely proportional to the *RC* product, assuming that both resistances and capacitances are equal. Theoretically, the frequency can be varied by varying either *R* or *C*. In practice, it is usually easier to vary *R* on a continuous basis and to vary *C* on a discrete basis. Identical capacitors are switched into the circuit at each frequency range. Also, identical resistances, which together are referred to as a *gauged potentiometer*, are mounted on the same shaft and are used to vary the frequency on a continuous basis in each frequency range.

Note that setting the loop gain to unity is not a reliable method for designing an oscillator. To stabilize an oscillator, usually it is necessary to limit the output voltage by introducing nonlinearity. Stability can be achieved by adding two zener diodes in series with the resistance R_B, as shown in Fig. 5.6(a). As long as the magnitude of the voltage v_f across resistance R_1 is less than the zener breakdown voltage V_Z, the zener diodes act as an open circuit, and the gain of the amplifier is

$$|A_1| = \frac{R_F}{R_1} \tag{5.14}$$

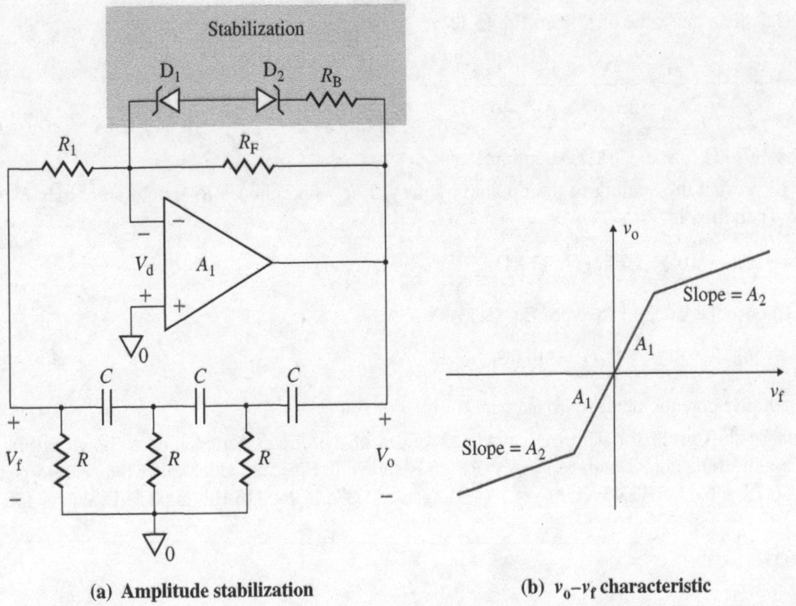

FIGURE 5.6 Stabilization of a phase-shift oscillator

As soon as the magnitude of v_f starts to increase above V_Z, the zener diodes conduct, and the resistor R_B suddenly becomes in parallel with R_F so that the gain is reduced. The new gain becomes

$$|A_2| = \frac{R_F \| R_B}{R_1} \tag{5.15}$$

which is less than $|A_1|$. Furthermore, if the output amplitude starts to decrease, the gain $|A|$ is increased again. The v_o–v_f characteristic of the amplifier is shown in Fig. 5.6(b).

EXAMPLE 5.3

D **Designing a phase-shift oscillator**

(a) Design the phase-shift oscillator shown in Fig. 5.5(a) so that the oscillating frequency is $f_o = 400$ Hz.

(b) Use PSpice/SPICE to plot the transient response of the output voltage $v_o(t)$ in part (a) from 0 to 4 ms. Assume $V_{CC} = V_{EE} = 12$ V.

SOLUTION

(a) The following steps can be used to complete the design.

Step 1. Choose a suitable value of C: Let $C = 0.1$ μF.

Step 2. Calculate the value of R from Eq. (5.11):

$$R = \frac{1}{2\pi\sqrt{6}f_oC} = \frac{1}{2\pi \times \sqrt{6} \times 400\ \text{Hz} \times 0.1\ \mu\text{F}} = 1624\ \Omega$$

Choose $R = 1.7\ \text{k}\Omega$ (use a 2.7-kΩ potentiometer).

Step 3. To prevent the loading of the op-amp by the RC network, choose R_1 much larger than R by making $R_1 \geq 10R$. Therefore, let

$$R_1 = 10R = 10 \times 1.7\ \text{k}\Omega = 17\ \text{k}\Omega$$

Step 4. Choose the value of R_F from Eq. (5.13):

$$R_F = 29R_1 = 29 \times 17\ \text{k}\Omega = 493\ \text{k}\Omega$$

Choose a 500-kΩ potentiometer R_F to account for tolerance.

(b) The phase-shift oscillator with the calculated values of the circuit parameters is shown in Fig. 5.7. The PSpice plot of the output voltage $v_o \equiv$ V(RF:2) is shown in Fig. 5.8, which gives the peak-to-peak voltage $V_{pp} = 6.78 + 6.65 = 13.43$ V at $f_o = 1/(4.3125\ \text{m} - 1.875\ \text{m}) = 410$ Hz (expected value is 400 Hz).

NOTES:

1. If you observe carefully, you will notice that the amplitude of the output voltage is falling slowly and the oscillation will not be sustained for a long time. A nonlinear device is often necessary to stabilize the oscillator.
2. An initial voltage of 1 V has been assigned to the capacitor C in order to start the oscillator, and the UIC (use initial condition) is used in transient analysis. Otherwise, PSpice will first calculate the biasing values and then use those values to find the solutions, and the circuit will not oscillate. In practice, random noise or transients can cause the oscillations to begin, and they are sustained by the feedback of the appropriate signal.

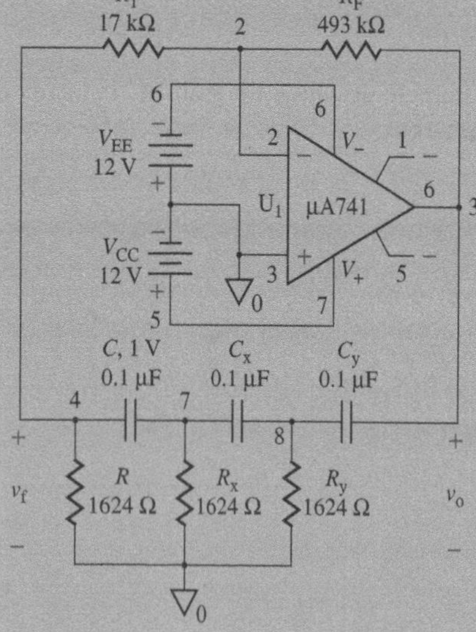

FIGURE 5.7 Phase-shift oscillator for PSpice simulation

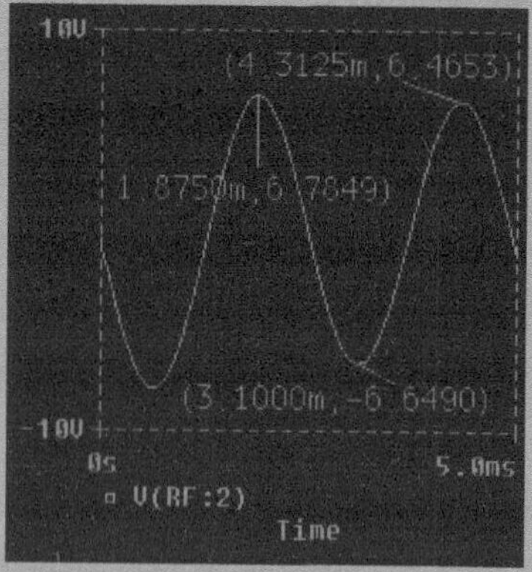

FIGURE 5.8 PSpice plot of output voltage for Example 5.3

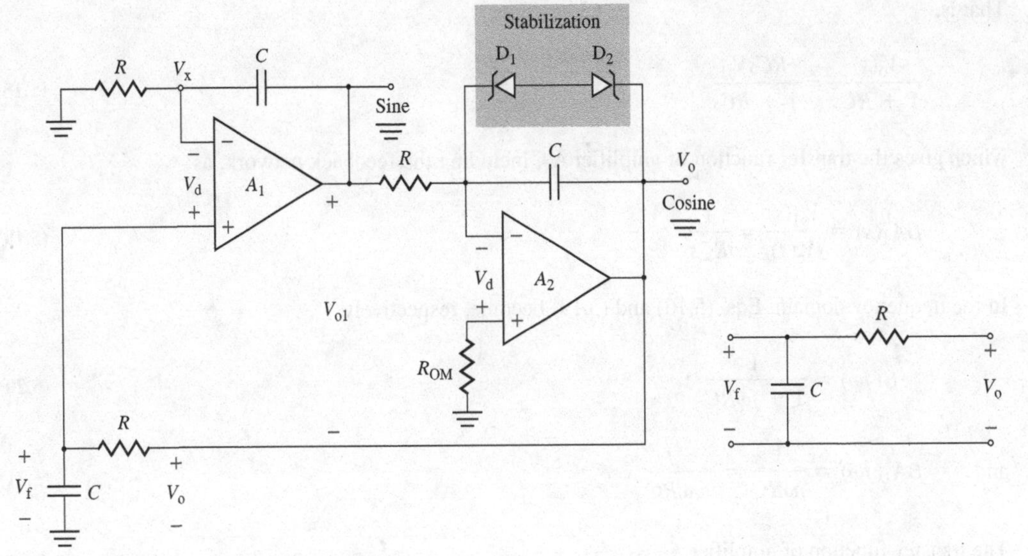

(a) Oscillator (b) Feedback network

FIGURE 5.9 Quadrature oscillator

5.3.2 Quadrature Oscillators

A quadrature oscillator, as shown in Fig. 5.9(a), generates two signals (sine and cosine) that are in quadrature—that is, out of phase by 90°. The actual location of sine and cosine signals is arbitrary. In Fig. 5.9(a), the output of amplifier A_1 is labeled as sine and that of amplifier A_2 as cosine. This oscillator requires a dual op-amp. Amplifier A_2 operates as an inverting integrator and provides a phase shift of $-270°$ (or 90°); amplifier A_1, in combination with the feedback network, operates as a noninverting integrator and provides the remaining $-90°$ (or 270°) to give the total phase shift of 360° that is required to satisfy the condition of oscillation. The transfer function of the feedback network shown in Fig. 5.9(b) in Laplace's domain of s is given by

$$\beta(s) = \frac{V_f(s)}{V_o(s)} = \frac{1/Cs}{R + 1/Cs} = \frac{1}{1 + RCs} \tag{5.16}$$

If V_{o1} is the voltage at the output of amplifier A_1, the voltage V_x at its inverting terminal is given by

$$V_x = \frac{RV_{o1}}{R + 1/Cs} = \frac{RCsV_{o1}}{1 + RCs} \tag{5.17}$$

Since the differential voltage between the op-amp terminals is very small ($V_d \approx 0$), we can write

$$V_f = V_x - V_d \approx V_x$$

That is,

$$\frac{V_o(s)}{1 + RCs} = \frac{RCsV_{o1}}{1 + RCs} \tag{5.18}$$

which gives the transfer function of amplifier A_1, including the feedback network, as

$$\beta A_1(s) = \frac{V_{o1}(s)}{V_o(s)} = \frac{1}{RCs} \tag{5.19}$$

In the frequency domain, Eqs. (5.16) and (5.19) become, respectively,

$$\beta(j\omega) = \frac{1}{1 + j\omega RC} \tag{5.20}$$

and $$\beta A_1(j\omega) = \frac{1}{j\omega RC} = -\frac{j}{\omega RC} \tag{5.21}$$

The transfer function of amplifier A_2 is

$$A_2(j\omega) = \frac{V_o}{V_{o1}}(j\omega) = -\frac{1}{j\omega RC} = \frac{j}{\omega RC}$$

which gives a phase shift of 90°. Thus, $\beta G_1(j\omega)$ must give a phase shift of $-90°$.

To provide a phase shift of $-90°$, $|\beta G_1(j\omega)|$ in Eq. (5.21) must equal unity. That is, $\omega RC = 1$, and the frequency of oscillation is given by

$$f_o = \frac{1}{2\pi RC} \quad \text{(in Hz)} \tag{5.22}$$

At this frequency, the magnitude of $\beta(j\omega)$ in Eq. (5.20) becomes

$$\beta = |\beta(j\omega)| = \left|\frac{1}{1 + j1}\right| = \frac{1}{\sqrt{2}} \tag{5.23}$$

The loop gain becomes

$$\beta A(j\omega) = A_2(j\omega)\beta A_1(j\omega) = \frac{j}{\omega RC} \times \frac{1}{j\omega RC} = \frac{1}{\omega^2 RC} = 1$$

Therefore, the overall closed-loop gain A_v of amplifiers A_1 and A_2 is given by

$$A_f = \frac{1}{\beta} = \sqrt{2} = 1.4142 \tag{5.24}$$

which implies a constant gain of 1.4142. The design of a quadrature oscillator is very simple. For $f_o = 200$ Hz and assuming $C = 0.1$ μF, Eq. (5.22) gives $R = 7958\ \Omega$ (use a 10-kΩ potentiometer). This oscillator can be stabilized by connecting two zener diodes back to back across one of the integrating capacitors, as shown in Fig. 5.9(a) by the shaded area.

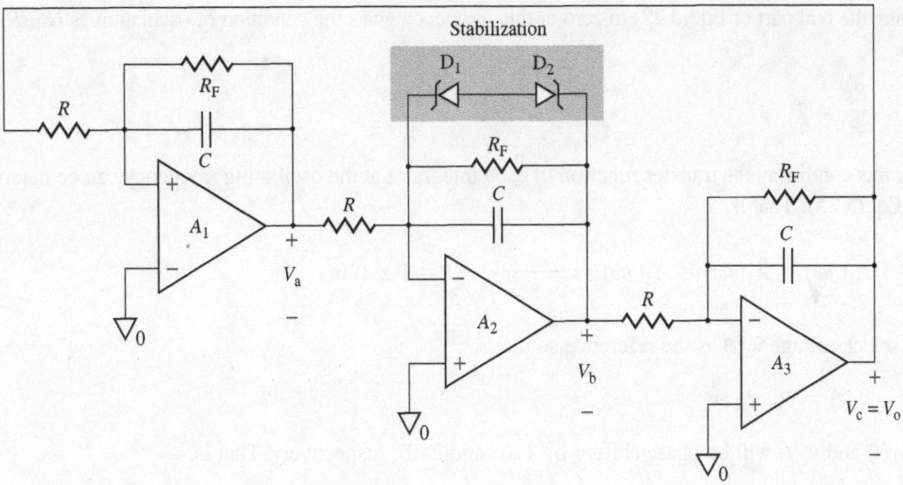

FIGURE 5.10 Three-phase oscillator

5.3.3 Three-Phase Oscillators

A three-phase oscillator generates three sinusoidal voltages of equal magnitude, but displaced by 120° from each other. They have the same form as the voltages in a three-phase power system and are normally used for generating control signals synchronized to the power system. A three-phase oscillator consisting of three "lossy" integrator circuits connected in cascade with unity feedback is shown in Fig. 5.10. The transfer function of each of the integrators is given by

$$A_1(s) = A_2(s) = A_3(s) = -\frac{R_F \| (1/Cs)}{R} = \frac{-R_F/R}{1 + R_F Cs} \tag{5.25}$$

Since $\beta = 1$ for unity feedback, the loop gain is given by

$$\beta A(s) = A_1(s)A_2(s)A_3(s) = \frac{-(R_F/R)^3}{(R_F Cs)^3 + 3(R_F Cs)^2 + 3(R_F Cs) + 1} \tag{5.26}$$

Then the characteristic equation, which is the numerator of $1 - \beta A(s) = 0$, is

$$(R_F Cs)^3 + 3(R_F Cs)^2 + 3(R_F Cs) + 1 + \left(\frac{R_F}{R}\right)^3 = 0 \tag{5.27}$$

Substituting $s = j\omega$ into Eq. (5.27) and then equating the imaginary part to zero, we get the frequency of oscillation ω_o as

$$\omega_o = \frac{\sqrt{3}}{R_F C} \quad \text{(in rad/s)} \tag{5.28}$$

Equating the real part of Eq. (5.27) to zero at this frequency gives the condition of oscillation as $(R_F/R)^3 = 8$; that is,

$$\frac{R_F}{R} = 2 \tag{5.29}$$

Under this condition, the transfer function of each integrator at the oscillating frequency can be determined from Eq. (5.25). That is,

$$A_1(j\omega) = A_2(j\omega) = A_3(j\omega) = -\frac{2}{1 + j\sqrt{3}} = 1\ \angle 120° \tag{5.30}$$

If we select voltage $v_a(t)$ as the reference so that

$$v_a(t) = V_m \sin \omega t \tag{5.31}$$

then $v_b(t)$ and $v_c(t)$ will be phase shifted by 120° and 240°, respectively. That is,

$$v_b(t) = V_m \sin(\omega t + 120°) \tag{5.32}$$

and

$$v_c(t) = V_m \sin(\omega t + 240°) = V_m \sin(\omega t - 120°) \tag{5.33}$$

For $f_o = 60$ Hz and assuming $C = 0.1\ \mu$F, Eq. (5.28) gives

$$R_F = \frac{\sqrt{3}}{2\pi \times 60 \times 0.1\ \mu\text{F}} = 45.94\ \text{k}\Omega \quad \text{(use a 50-k}\Omega\text{ potentiometer)}$$

and Eq. (5.29) gives $R = R_F/2 = 22.97\ \text{k}\Omega$. This oscillator can be stabilized by connecting two zener diodes back to back across one of the integrating capacitors, as shown in Fig. 5.10 by the shaded area.

5.3.4 Wien-Bridge Oscillators

A Wien bridge, which is used for making measurements of unknown resistors or capacitors, is shown in Fig. 5.11(a). The bridge has a series *RC* network in one arm and a parallel *RC* network in the adjoining arm. R_1 and R_F are connected in two other arms. While the bridge is making measurements, either R_1 or R_F acts as a calibrated resistor; the resistance is varied until the null voltage $V_d = 0$ is found. If all component values are known except one, the value of that one can be determined from the following relation:

$$\frac{R_2}{R_3} + \frac{C_2}{C_1} = \frac{R_F}{R_1} \tag{5.34}$$

If an op-amp is inserted into the basic bridge, as shown in Fig. 5.11(b), the bridge is known as a *Wien-bridge oscillator*, provided the elements are adjusted so that $R_2 = R_3 = R$ and $C_1 = C_2 = C$. The op-amp, along with R_1 and R_F, operates as a noninverting amplifier, as shown in Fig. 5.12(a). The Wien-bridge oscillator is one of the most commonly used AF oscillators.

The transfer function of the feedback network shown in Fig. 5.12(b) is given by

$$\beta(s) = \frac{V_f(s)}{V_o(s)} = \frac{RCs}{R^2C^2s^2 + 3RCs + 1} \tag{5.35}$$

(see Prob. 5.9).

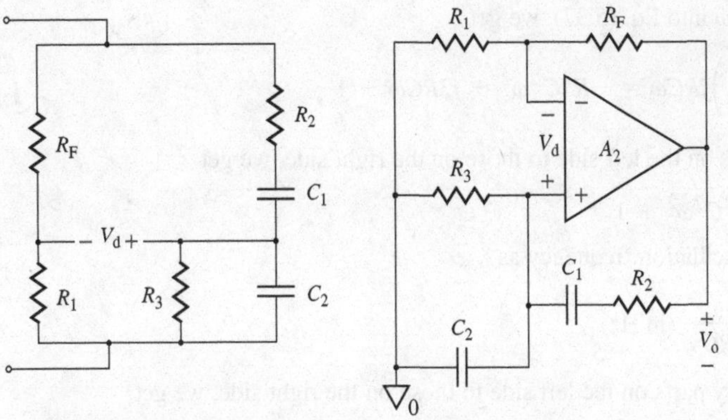

(a) Basic Wien bridge **(b) Wien-bridge oscillator**

FIGURE 5.11 Wien bridge

The closed-loop voltage gain of the noninverting amplifier is given by

$$A(s) = \frac{V_o(s)}{V_f(s)} = 1 + \frac{R_F}{R_1} \tag{5.36}$$

For an oscillator, $A\beta = 1$. Using Eqs. (5.35) and (5.36), we get

$$\left(1 + \frac{R_F}{R_1}\right)\frac{RCs}{R^2C^2s^2 + 3RCs + 1} = 1 \tag{5.37}$$

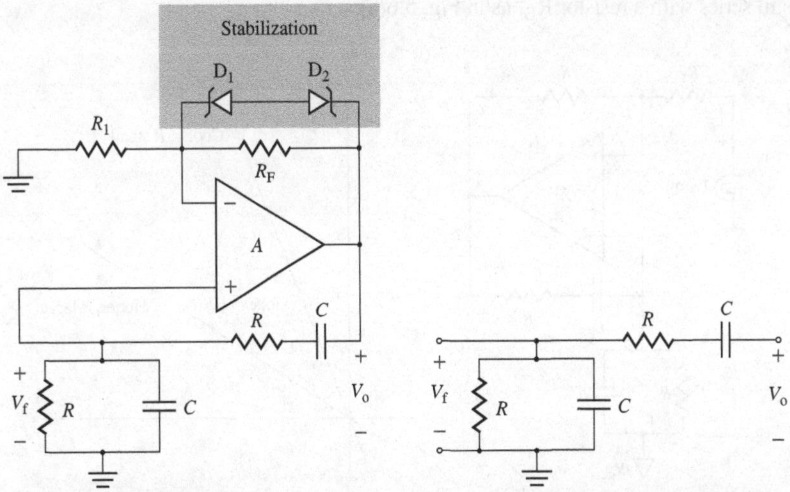

(a) Oscillator **(b) Feedback network**

FIGURE 5.12 Wien-bridge oscillator

Substituting $s = j\omega$ into Eq. (5.37), we get

$$\left(1 + \frac{R_F}{R_1}\right) jRC\omega = -R^2C^2\omega^2 + j3RC\omega + 1$$

Equating real parts on the left side to those on the right side, we get

$$0 = -R^2C^2\omega^2 + 1$$

which gives the oscillation frequency as

$$f_o = \frac{1}{2\pi RC} \quad \text{(in Hz)} \tag{5.38}$$

Equating imaginary parts on the left side to those on the right side, we get

$$\left(1 + \frac{R_F}{R_1}\right) jRC\omega = j3RC\omega$$

which gives the condition for oscillation as

$$1 + \frac{R_F}{R_1} = 3$$

or

$$\frac{R_F}{R_1} = 2 \tag{5.39}$$

For stabilization, a power-sensitive resistor such as a lamp or a thermistor is usually used to adjust dynamically the loop gain of the oscillator. Figure 5.13(a) shows the use of a small incandescent lamp, whose resistance characteristic is shown in Fig. 5.13(b). When the filament of the lamp is cold, the resistance is small and the gain A is large. But when the lamp filament becomes hot, the resistance becomes larger, and the gain A becomes small. This automatic adjustment of the gain causes distortion of the amplifier to be low and stabilizes the oscillator. The nonlinear characteristic can be provided by two back-to-back zener diodes in series with a resistor R_B, as in Fig. 5.6(a).

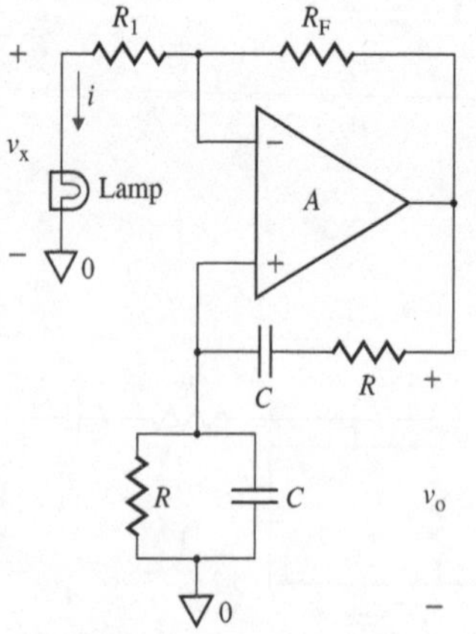

(a) Amplitude stabilization by lamp

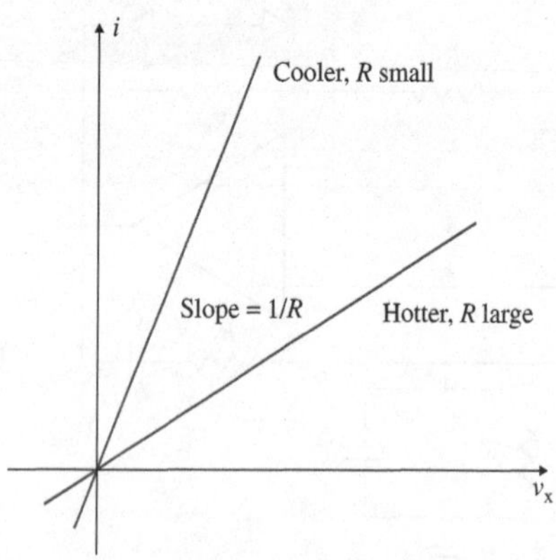

(b) Lamp characteristic

FIGURE 5.13 Stabilization of a Wien-bridge oscillator

EXAMPLE 5.4

D

Designing a Wien-bridge oscillator

(a) Design the Wien-bridge oscillator in Fig. 5.12(a) so that $f_o = 1$ kHz.

(b) Use PSpice/SPICE to plot the transient response of the output voltage $v_o(t)$ in part (a) from 0 to 2 ms. Assume $V_{CC} = V_{EE} = 12$ V.

SOLUTION

(a) The following steps can be used to complete the design.

Step 1. Choose a suitable value of C: Let $C = 0.01\ \mu$F.

Step 2. Calculate the value of R from Eq. (5.38):

$$R = \frac{1}{2\pi f_o C} = \frac{1}{2\pi \times 1\text{ kHz} \times 0.01\ \mu\text{F}} = 15{,}915\ \Omega$$

Choose $R = 16$ kΩ.

Step 3. Choose the value of R_F from Eq. (5.39). Letting $R_1 = 10$ kΩ, we have

$$R_F = 2R_1 = 2 \times 10\text{ k}\Omega = 20\text{ k}\Omega$$

NOTE: The exact numbers of the resistances are used to demonstrate the validity of the calculated values with the PSpice simulation results. In a practical design, the commercially available standard values will be used (i.e., 16 kΩ instead of 15,916 Ω; see Appendix E).

(b) The Wien-bridge oscillator with the desired values is shown in Fig. 5.14. The PSpice plot of the output voltage $v_o \equiv$ V(U1:OUT) is shown in Fig. 5.15, which gives the peak-to-peak voltage $V_{pp} = 5.94 + 5.97 = 11.91$ V at $f_o = 1/(1.269\text{ m} - 0.247\text{ m}) = 978$ Hz (expected value is 1 kHz).

NOTE: An initial voltage of 1 V has been assigned to the capacitor C_p in order to start the oscillator, and the UIC (use initial condition) is used in transient analysis. In practice, random noise or transients can cause the oscillations to begin, and they are sustained by the feedback of the appropriate signal.

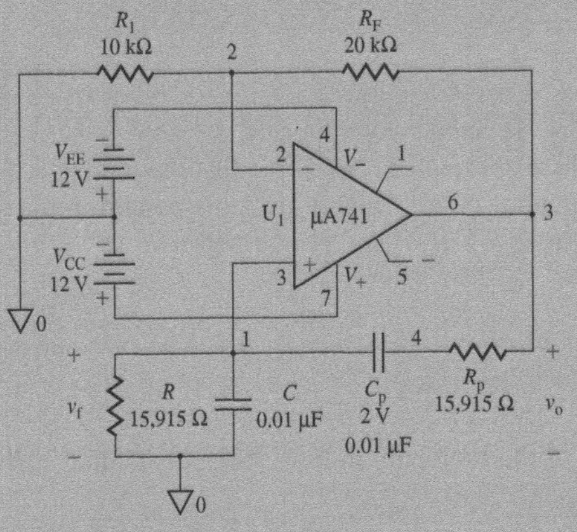

FIGURE 5.14 Wien-bridge oscillator for PSpice simulation

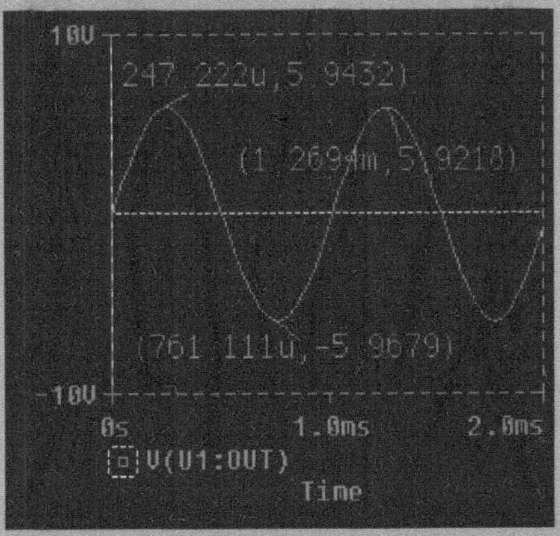

FIGURE 5.15 PSpice plot of output voltage for Example 5.4

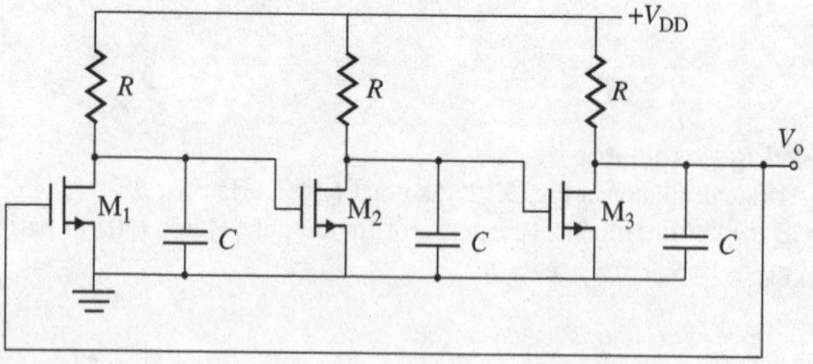

FIGURE 5.16 Three-stage ring oscillator

5.3.5 Ring Oscillators

A ring oscillator consists of a number of gain stages in a loop in order to make the total phase shift of 180°. A three-stage ring oscillator is shown in Fig. 5.16. By neglecting the MOS capacitance, we can find the transfer function of each stage as given by

$$A(s) = -g_m\left(R \parallel \frac{1}{sC}\right) = -\frac{g_m R}{1 + RCs} = -\frac{A_o}{1 + RCs} \tag{5.40}$$

which in frequency domain becomes

$$A(j\omega) = -\frac{A_o}{1 + (j\omega/\omega_n)} \tag{5.41}$$

where $A_o = g_m R$ is the small-signal low-frequency gain and $\omega_n = 1/RC$ is the natural frequency. Since three stages are cascaded, the open-loop transfer function becomes

$$H(j\omega) = -\frac{A_o^3}{1 + (j\omega/\omega_n)^3} \tag{5.42}$$

The circuit oscillates only if the frequency-dependent phase shift equals $-180°$ so that the total phase of $H(j\omega) = 360°$ or $0°$. Therefore, each stage must contribute $\theta = 60°$, which gives the condition for the oscillation frequency ω_o as $\tan^{-1}(\omega_o/\omega_n) = 60°$. This gives

$$\omega_o = \sqrt{3}\,\omega_n = \frac{\sqrt{3}}{RC} \tag{5.43}$$

At the oscillation frequency of $\omega = \omega_o$, the magnitude of the loop gain in Eq. (5.42) must equal 1; that is,

$$|H(j\omega)| = \frac{A_o^3}{[\sqrt{1 + (\omega_o/\omega_n)^2}]^3} = 1 \tag{5.44}$$

which, for $\omega_o = \sqrt{3}\,\omega_n$, gives the minimum value of A_o to sustain the oscillation as

$$A_o = g_m R \geq 2 \tag{5.45}$$

EXAMPLE 5.5

Designing a MOS ring oscillator

(a) Design the ring oscillator in Fig. 5.16 so that $f_o = 1$ kHz. The MOSFET parameters are MOS constant $K_p = 25\ \mu A/V^2$, the threshold voltage $V_t = 1$ V, length $L = 10\ \mu m$, width $W = 20\ \mu m$, and modulation length $\lambda = 0.01$.

(b) Use PSpice [2] to plot the transient response of the output voltage $v_o(t)$ in part (a) from 0 to 2 ms. Assume $V_{DD} = 5$ V.

SOLUTION

(a) Assuming that the MOS is operating in the saturation region, the transconductance of each MOS is $g_m = 2K_p(v_{GS} - V_t)$, which can give the desired value of R to obtain the required gain A_o as given by

$$R = \frac{A_o}{g_m} = \frac{A_o}{2K_p(v_{GS} - V_t)} \tag{5.46}$$

Using KVL in the loop formed by V_{DD}, R, and the gate–source voltage of M_2, we can write

$$V_{DD} = Ri_{D1} + v_{GS1} = \frac{A_o}{2K_p(v_{GS1} - V_t)} \times K_p(v_{GS1} - V_t)^2 + v_{GS1}$$

which, for identical MOSFETs of $v_{GS} = v_{GS1} = v_{GS2}$, gives the gate–source voltage as

$$v_{GS} = \frac{V_{DD} + (A_o/2)V_t}{1 + (A_o/2)} \tag{5.47}$$

Letting $A_o = 4$ (higher than 2), Eq. (5.47) gives the DC biasing gate–source voltage

$$v_{GS} = \frac{5 + (4/2) \times 1}{1 + (4/2)} = 2.333\ \text{V}$$

$$i_D = K_p(v_{GS1} - V_t)^2 = 25\ \mu \times (2.333 - 1)^2 = 44.44\ \mu A$$

$$g_m = 2K_p(v_{GS} - V_t) = 2 \times 25\ \mu \times (2.333 - 1) = 66.67\ \mu A/V$$

$$R = \frac{A_o}{g_m} = \frac{4}{66.67\ \mu A/V} = 60\ k\Omega$$

For $f_o = 1$ kHz, Eq. (5.43) gives

$$C = \frac{\sqrt{3}}{2\pi f_o R} = \frac{\sqrt{3}}{2\pi \times 1\ k \times 60\ k} = 4.59\ nF$$

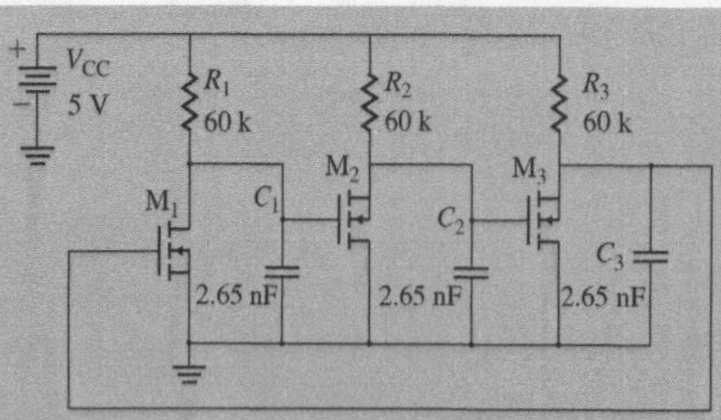

FIGURE 5.17 PSpice schematic for Example 5.5

(b) The PSpice schematic with desired values is shown in Fig. 5.17. The PSpice plot of the output voltage $v_o \equiv$ V(C3:2) is shown in Fig. 5.18, which gives the peak-to-peak voltage of $V_{pp} = (3.7974 - 1.4154) = 2.38$ V at $f_o = 1/(2.6499 \text{ ms} - 1.7135 \text{ ms}) = 1046$ Hz (expected 1 kHz). Note that an initial voltage of 0 has been assigned to the capacitor C_1 in order to start the oscillator, and the UIC (use initial condition) is used in transient analysis. In practice, random noise or transients can cause the oscillations to begin, and they are sustained by the feedback of the appropriate signal.

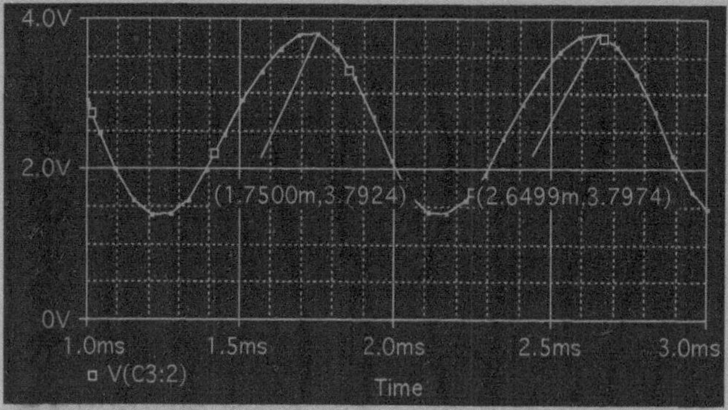

FIGURE 5.18 Output voltage waveform for Example 5.5

KEY POINTS OF SECTION 5.3

- A phase-shift oscillator uses an inverting amplifier and a phase-shifting network to satisfy the requirements of unity loop gain with 0° or 360° phase shift. The oscillation frequency ω_o is inversely proportional to the *RC* product of the feedback network. A nonlinear device is often introduced, however, to stabilize the oscillator.
- A quadrature oscillator uses two op-amp inverting integrators and an *RC* phase shifter. The output could be sine or cosine. The oscillation frequency ω_o is inversely proportional to the *RC* product of the feedback network.
- A three-phase oscillator uses three inverting integrators with a unity feedback loop. The oscillation frequency ω_o is inversely proportional to the *RC* product of the feedback network.
- A Wien-bridge oscillator uses a noninverting amplifier and an *RC* phase-shifting network. The oscillation frequency ω_o is inversely proportional to the *RC* product of the feedback network.

5.4 Radio Frequency Oscillators

A radio frequency (RF) oscillator uses LC components for generating a waveform in the frequency range from 100 kHz to 100 MHz. There are different ways to make amplifiers by using op-amps, BJTs, and MOSFETs; choose the LC components to create conditions for oscillations. We consider the following types of RF oscillators: Colpitts oscillator, Hartley oscillator, and two-stage MOS oscillator. The op-amp Colpitts oscillator is the most common and easy to implement. The MOS oscillators allow CMOS implementation.

5.4.1 Colpitts Oscillators

A Colpitts oscillator is a tuned LC-type oscillator, as shown in Fig. 5.19(a). LC oscillators have the advantage of having relatively small reactive elements. They exhibit higher Q than RC oscillators, but they are difficult to tune over a wide range. For a positive feedback circuit to operate as an oscillator, the loop gain must be zero; that is,

$$1 - A\beta = 0$$

which is really the characteristic equation of the circuit. Therefore, the condition for oscillation can be found from the characteristic equation without deriving the transfer function. Nodal analysis can be applied to find the determinant, which is then set to zero.

The op-amp operates as an inverting amplifier of gain $A = R_F/R_1$. If the amplifier is replaced by its equivalent circuit, Fig. 5.19(a) can be simplified to Fig. 5.19(b). If the voltage source Av_f is replaced by its equivalent current source $g_m v_f$, Fig. 5.19(b) can be reduced to Fig. 5.19(c). Using nodal analysis in Fig. 5.19(c) in Laplace's domain of s, we can write

$$\left[sC_2 + \frac{1}{R_L} + \frac{1}{sL}\right]V_o(s) + \left[g_m - \frac{1}{sL}\right]V_f(s) = 0 \quad \text{(at node B)} \tag{5.48}$$

$$-\frac{1}{sL}V_o(s) + \left[sC_1 + \frac{1}{sL} + \frac{1}{R_1}\right]V_f(s) = 0 \quad \text{(at node A)} \tag{5.49}$$

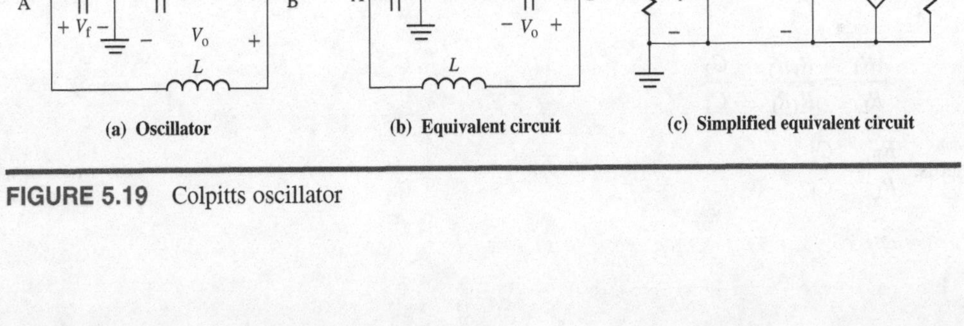

FIGURE 5.19 Colpitts oscillator

To find the condition for oscillation, we set the determinant to zero; that is,

$$\left(sC_2 + \frac{1}{R_L} + \frac{1}{sL}\right)\left(sC_1 + \frac{1}{sL} + \frac{1}{R_1}\right) + \left(g_m - \frac{1}{sL}\right)\frac{1}{sL} = 0$$

which, after simplification, yields

$$s^3C_1C_2LR_1R_L + s^2L(C_1R_1 + C_2R_L) + s(C_1R_1R_L + C_2R_1R_L + L) + (R_1 + R_L + g_mR_1R_L) = 0 \tag{5.50}$$

where $g_m = A/R_L = R_F/(R_1R_L)$. Substituting $s = j\omega$ and equating the imaginary parts to zero, we get

$$-j\omega^3(C_1C_2LR_1R_L) + j\omega(C_1R_1R_L + C_2R_1R_L + L) = 0$$

which gives the frequency of oscillation ω_o as

$$\omega_o = \left[\frac{C_1 + C_2}{C_1C_2L} + \frac{1}{C_1C_2R_1R_L}\right]^{1/2} \quad \text{(in rad/s)} \tag{5.51}$$

Assuming R_L is large, such that $R_1R_L > 1/(C_1C_2)$, Eq. (5.51) can be approximated by

$$\omega_o = \left[\frac{C_1 + C_2}{C_1C_2L}\right]^{1/2} \quad \text{(in rad/s)} \tag{5.52}$$

Similarly, equating the real parts of Eq. (5.50) to zero yields

$$-\omega^2L(C_1R_1 + C_2R_L) + (R_1 + R_L + g_mR_1R_L) = 0$$

which gives

$$\omega^2L(C_1R_1 + C_2R_L) = R_1 + R_L + g_mR_1R_L$$

Substituting the value of $\omega = \omega_o$ from Eq. (5.51) gives

$$L(C_1R_1 + C_2R_L)\left[\frac{C_1 + C_2}{C_1C_2L} + \frac{1}{C_1C_2R_1R_L}\right] = R_1 + R_L + g_mR_1R_L$$

After simplification, the above equation becomes

$$g_mR_1 \approx \frac{C_2}{C_1} + \frac{C_1}{C_2}\frac{R_1}{R_L} + \frac{L}{C_2R_L^2} + \frac{L}{C_1R_1R_L}$$

which, for a large value of R_L, becomes

$$g_mR_1 = \frac{C_2}{C_1} \tag{5.53}$$

or $$\frac{AR_1}{R_L} = \frac{R_FR_1}{R_1R_L} = \frac{C_2}{C_1}$$

That is, $$\frac{R_F}{R_L} = \frac{C_2}{C_1} \tag{5.54}$$

which is independent of R_1 and gives the relationship among R_L, R_F, C_1, and C_2. Equation (5.53) gives the minimum value of g_m (or R_F/R_1R_L) required to sustain the oscillation with a constant amplitude. If g_m is smaller than this value, the oscillation will die exponentially to zero. On the other hand, if g_m is larger than this value, the amplitude will grow exponentially until the nonlinearity of the op-amp limits the amplitude. Therefore, to ensure oscillation, the value of g_m must exceed the minimum value.

In the above analysis, we used a simple op-amp model and neglected the loss in the resistance of the inductor. As a result, we obtained relatively simple expressions for the frequency and the condition to sustain oscillation. If a complex op-amp model including the inductor loss were used, the oscillation frequency would depend (generally only slightly) on other circuit parameters. Usually, the inductor or one of the capacitors is made adjustable so that the frequency can be initially tuned to the desired value.

EXAMPLE 5.6

D **Designing a Colpitts oscillator** Design the Colpitts oscillator of Fig. 5.19(a) so that the oscillating frequency is $f_o = 150$ kHz.

SOLUTION

Step 1. Choose suitable values of C_1 and C_2: Let $C_1 = 0.01\ \mu\text{F}$ and $C_2 = 0.1\ \mu\text{F}$; that is, $C_2/C_1 = 0.1/0.01 = 10$.

Step 2. Calculate the value of L from Eq. (5.52):

$$L = \frac{C_1 + C_2}{4\pi^2 C_1 C_2 f_o^2} = \frac{0.01\ \mu\text{F} + 0.1\ \mu\text{F}}{4\pi^2 \times 0.01\ \mu\text{F} \times 0.1\ \mu\text{F} \times (150\ \text{kHz})^2} = 124\ \mu\text{H}$$

Step 3. Choose the values of R_F and R_L from Eq. (5.54):

$$\frac{R_F}{R_L} = \frac{C_2}{C_1} = \frac{0.1}{0.01} = 10$$

Let $R_L = 100\ \text{k}\Omega$. Therefore, $R_F = 10R_L = 1\ \text{M}\Omega$.

Step 4. Choose a value of A: Let $A = 10 = R_F/R_1$. Then

$$R_1 = \frac{R_F}{A} = \frac{1\ \text{M}\Omega}{10} = 100\ \text{k}\Omega$$

Step 5. Check the values of g_m and f_o:

$$g_m = \frac{A}{R_1} = \frac{10}{100\ \text{k}\Omega} = 0.1\ \text{mA/V}$$

Equation (5.51) gives $\omega_o = 941.86$ krad/s, and $f_o = \omega_o/2\pi = 149.9$ kHz.

EXAMPLE 5.7

Finding the oscillation frequency of a Colpitts BJT-tuned oscillator A Colpitts BJT oscillator is shown in Fig. 5.20. The circuit parameters are $r_\pi = 1.1\ \text{k}\Omega$, $h_{fe} = 100$, $L = 1.5$ mH, $C_1 = 1$ nF, $C_2 = 99$ nF, and $R_L = 10\ \text{k}\Omega$. R_2, R_3, and R_E set the DC-biasing circuit. L, C_1, and C_2 form a parallel resonant circuit. The *RF* choke offers very high impedance at the frequency of oscillation and acts as a constant-current source feeding the *LC* resonant circuit.

(a) Calculate the frequency of oscillation f_o.
(b) Check to make sure the condition for oscillation is satisfied.
(c) Calculate the value of R_2.

SOLUTION

Since the *RF* choke offers a very high impedance, it can be considered as a short circuit for small-signal analysis. Thus, the AC equivalent circuit is as shown in Fig. 5.21(a). Replacing the transistor by its transconductance model (voltage-controlled current source) gives the small-signal AC equivalent circuit shown in Fig. 5.21(b), which is similar to the circuit shown in Fig. 5.21(c). Thus, the analysis of the Colpitts op-amp oscillator in Sec. 5.4.1 is applicable in this case:

$$g_m = \frac{h_{fe}}{r_\pi} = \frac{100}{1.1\ \text{k}\Omega} = 90.91\ \text{mA/V}$$

(a) From Eq. (5.52), the frequency of oscillation is

$$f_o = \frac{1}{2\pi}\left[\frac{C_1 + C_2}{C_1 C_2 L}\right]^{1/2} = \frac{1}{2\pi}\left[\frac{1\ \text{nF} + 99\ \text{nF}}{1\ \text{nF} \times 99\ \text{nF} \times 1.5\ \text{mH}}\right]^{1/2} = 130.6\ \text{kHz}$$

(b) For $R_2 >> r_\pi$, $R_1 = r_\pi \| R_2 \approx r_\pi = h_{fe}/g_m$. Therefore, Eq. (5.53) becomes

$$g_m R_1 \approx g_m r_\pi = \frac{g_m h_{fe}}{g_m} = \frac{C_2}{C_1}$$

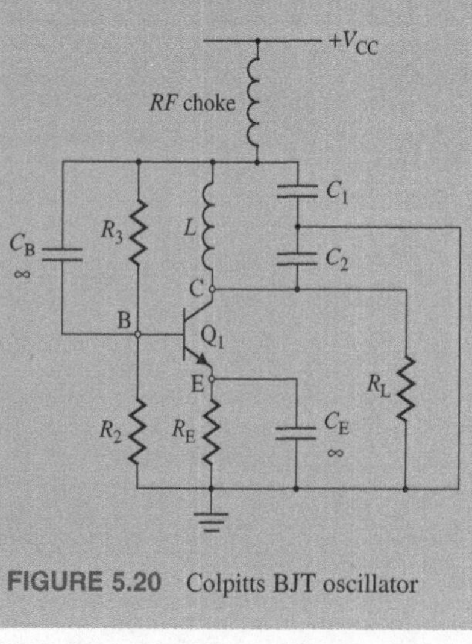

FIGURE 5.20 Colpitts BJT oscillator

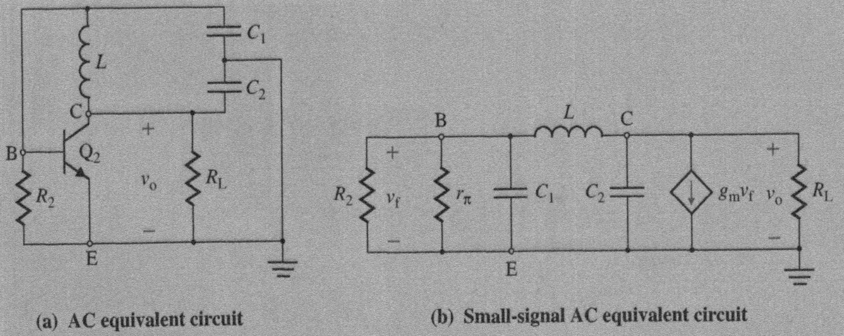

FIGURE 5.21 Equivalent circuits for Example 5.7

which gives

$$h_{fe} = \frac{C_2}{C_1} \tag{5.55}$$

$$= \frac{99}{1} = 99$$

This value is approximately equal to the value of transistor $h_{fe} = 100$. Thus, the condition for oscillation is satisfied.

(c) From Eq. (5.53),

$$R_1 = \frac{C_2}{C_1 g_m} = \frac{99}{g_m} = \frac{99 r_\pi}{h_{fe}} = 99 \times \frac{1.1\,\text{k}\Omega}{100} = 1089\ \Omega$$

Since $R_1 = r_\pi \| R_2 = 1089\ \Omega$,

$$\frac{r_\pi R_2}{r_\pi + R_2} = 1089$$

which gives $R_2 = 108.9\ \text{k}\Omega$.

EXAMPLE 5.8

Finding the oscillation frequency of a Colpitts BJT oscillator A Colpitts BJT oscillator is shown in Fig. 5.22. The circuit parameters are $r_\pi = 2.7\ \text{k}\Omega$, $h_{ie} = 150$, $L = 10\ \mu\text{H}$, $C_1 = 1$ nF, $C_2 = 100$ nF, and $R_L = 10\ \text{k}\Omega$. The BJT acts like a normal amplifier, and the LC resonant signal is fed back to the input base in order to sustain the oscillation. Therefore, the configuration is different from the oscillator circuit in Fig. 5.20.

(a) Find the approximate value of the frequency of oscillation f_o.

(b) Use PSpice/SPICE to plot the transient response of the output voltage $v_o(t)$ from 20 ms to 2 ms.

SOLUTION

(a) From Eq. (5.52), the frequency of oscillation is

$$f_o = \frac{1}{2\pi}\left[\frac{C_1 + C_2}{C_1 C_2 L}\right]^{1/2} = \frac{1}{2\pi}\left[\frac{1\text{ nF} + 100\text{ nF}}{1\text{ nF} \times 100\text{ nF} \times 10\ \mu\text{H}}\right]^{1/2} = 1.6\text{ MHz}$$

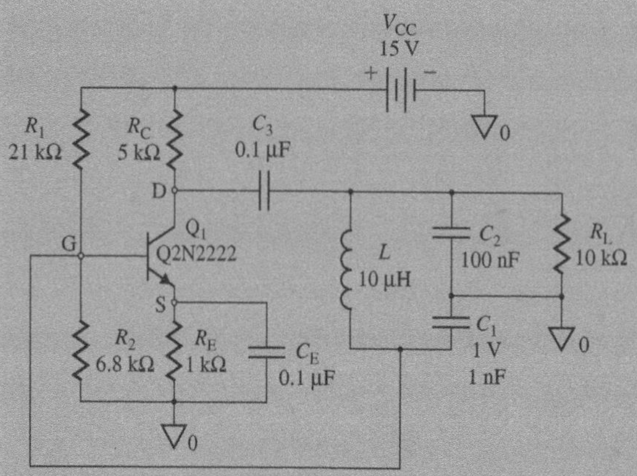

FIGURE 5.22 BJT Colpitts oscillator for PSpice simulation

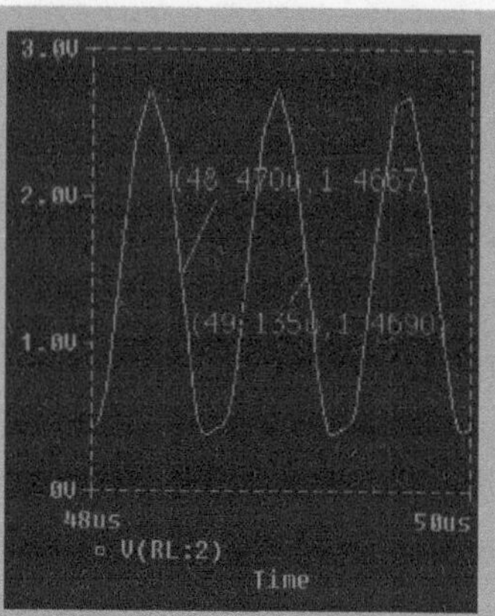

FIGURE 5.23 PSpice plot of output voltage for Example 5.8

(b) The PSpice plot of the output voltage $v_o \equiv$ V(RL:2) is shown in Fig. 5.23, which gives $f_o = 1/(49.149\ \mu s - 48.505\ \mu s) = 1.5$ MHz (expected value is 1.6 MHz).

NOTE: An initial voltage of 1 V has been assigned to the capacitor C_2 in order to start the oscillator, and the UIC (use initial condition) is used in transient analysis.

EXAMPLE 5.9

Finding the oscillation frequency of an *LC*-tuned MOSFET oscillator An *LC*-tuned MOSFET oscillator is shown in Fig. 5.24. Find the values of *L*, *C*, and *n* for an oscillation frequency of $f_o = 150$ kHz. The parameters of the MOSFET are $g_m = 5$ mA/V, $r_d = 25$ kΩ, and $R_G = 10$ kΩ.

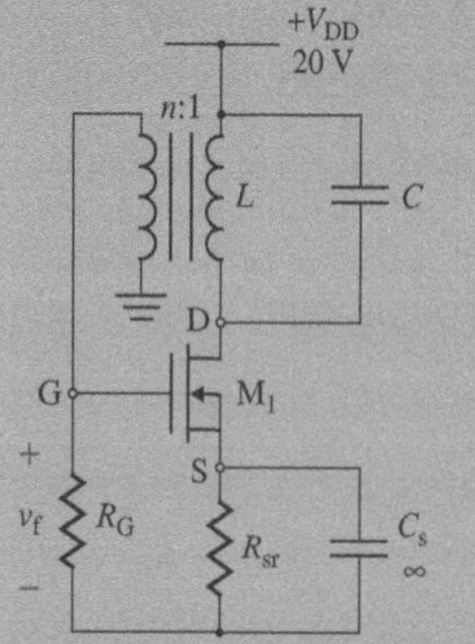

FIGURE 5.24 Colpitts MOSFET oscillator

SOLUTION

This oscillator uses the transformer to feed back the oscillating signal to the gate input, to sustain the oscillation. The small-signal AC equivalent circuit is shown in Fig. 5.25(a). Replacing the MOSFET by its transconductance model gives the small-signal equivalent circuit shown in Fig. 5.25(b), which can be simplified to Fig. 5.25(c). The transfer characteristic of v_o versus v_f in Laplace's domain can be written as

$$\frac{V_o(s)}{g_m V_f(s)} = Z(s) = \frac{1}{1/R_1 + sC + 1/sL}$$

which gives the voltage gain A as

$$A(s) = \frac{V_o(s)}{V_f(s)} = \frac{g_m}{(1/R_1) + sC + (1/sL)} \tag{5.56}$$

$$\text{where} \quad R_1 = r_d \,\|\, \frac{R_G}{n^2} \tag{5.57}$$

Substituting $s = j\omega$ into Eq. (5.56), we get

$$A(j\omega) = \frac{g_m}{(1/R_1) + j(\omega C - 1/\omega L)} \tag{5.58}$$

For oscillation, $A = |A(j\omega)| = 1\angle 0°$. Therefore, the imaginary part of the denominator must equal zero; that is,

$$j\left(\omega C - \frac{1}{\omega L}\right) = 0$$

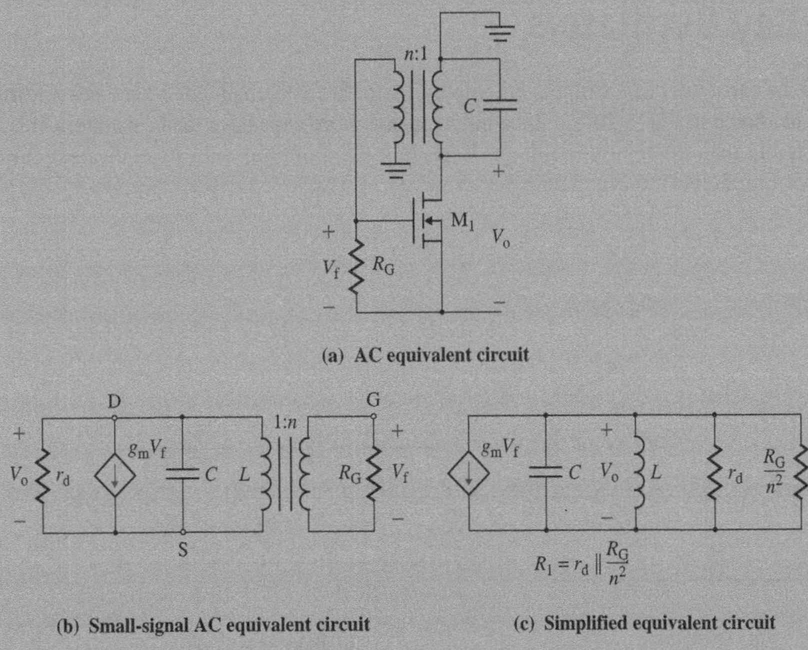

FIGURE 5.25 Equivalent circuits for Example 5.9

which gives the oscillation frequency f_o as

$$f_o = \frac{\omega_o}{2\pi} = \frac{1}{2\pi\sqrt{LC}} \tag{5.59}$$

At this frequency, the gain must be unity; that is,

$$|A(j\omega)| = g_m R_1 = 1 \tag{5.60}$$

Step 1. Choose a suitable value of C: Let $C = 0.01$ μF.

Step 2. Calculate the value of L from Eq. (5.59):

$$L = \frac{1}{4\pi^2 C f_o^2} = \frac{1}{4\pi^2 \times 0.01\ \mu\text{F} \times (150\ \text{kHz})^2} = 112.6\ \mu\text{H}$$

Step 3. Find the value of R_1. From Eq. (5.60),

$$R_1 = \frac{1}{g_m} = \frac{1}{5\ \text{mA/V}} = \frac{1000}{5\ \text{A/V}} = 200\ \Omega$$

Step 4. Using Eq. (5.57), calculate the value of the turns ratio n: Since

$$200 = 25\ \text{k}\Omega \| \left(\frac{10\ \text{k}\Omega}{n^2}\right)$$

$n^2 = 49.6$ and $n = 7.04$. Thus, $n = 7$.

5.4.2 Hartley Oscillators

If the inductor and the capacitors of a Colpitts op-amp oscillator are interchanged, it becomes a Hartley op-amp oscillator, as shown in Fig. 5.26(a). Since inductors are more expensive than capacitors, this oscillator is less desirable than a Colpitts oscillator. Replacing the amplifier with its equivalent current source $g_m V_f$ reduces Fig. 5.26(b) to Fig. 5.26(b).

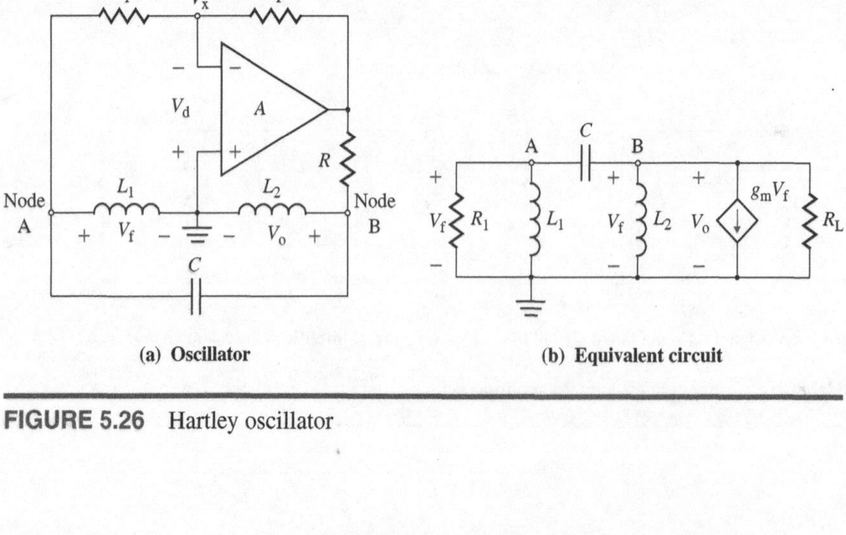

FIGURE 5.26 Hartley oscillator

Using nodal analysis in Fig. 5.26(b), we can write

$$\left[sC + \frac{1}{R_L} + \frac{1}{sL_2}\right]V_o(s) + (g_m - sC)V_f(s) = 0 \quad \text{(at node B)} \tag{5.61}$$

$$-sCV_o(s) + \left[sC + \frac{1}{sL_1} + \frac{1}{R_1}\right]V_f(s) = 0 \quad \text{(at node A)} \tag{5.62}$$

To find the condition for oscillation, we set the determinant to zero; that is,

$$\left(sC + \frac{1}{R_L} + \frac{1}{sL_2}\right)\left(sC + \frac{1}{sL_1} + \frac{1}{R_1}\right) + (g_m - sC)sC = 0$$

which, after simplification, yields

$$s^3CL_1L_2(R_1 + R_L + g_mR_1R_L) + s^2[CR_1R_L(L_1 + L_2) + L_1L_2] + s(L_1R_L + L_2R_1) + R_1R_L = 0 \tag{5.63}$$

where $g_m = A/R_L = R_F/(R_1R_L)$. Substituting $s = j\omega$ and equating the real parts of Eq. (5.63) to zero, we get

$$-\omega^2[CR_1R_L(L_1 + L_2) + L_1L_2] + R_1R_L = 0$$

which gives the frequency of oscillation ω_o as

$$\omega_o = \frac{1}{[C(L_1 + L_2) + L_1L_2/R_1R_L]^{1/2}} \quad \text{(in rad/s)} \tag{5.64}$$

For $C(L_1 + L_2) >> L_1L_2/R_1R_L$, Eq. (5.64) can be approximated by

$$f_o = \frac{1}{2\pi}\left[\frac{1}{C(L_1 + L_2)}\right]^{1/2} \quad \text{(in Hz)} \tag{5.65}$$

Equating the imaginary parts of Eq. (5.63) to zero, we get

$$-j\omega^3CL_1L_2(R_1 + R_L + g_mR_1R_L) + j\omega(L_1R_L + L_2R_1) = 0$$

Substituting $\omega = 2\pi f_o$ into Eq. (5.65), we get

$$\frac{1}{C(L_1 + L_2)} \times CL_1L_2(R_1 + R_L + g_mR_1R_L) = L_1R_L + L_2R_1$$

which, solved for g_mR_1, gives

$$g_mR_1 = \frac{L_1}{L_2} + \frac{R_1L_2}{R_LL_1} \tag{5.66}$$

For a large value of R_L, Eq. (5.66) gives the approximate value of g_m:

$$g_m \approx \frac{L_2}{R_1L_1} \tag{5.67}$$

Equation (5.67) gives the minimum value of g_m required to sustain the oscillation with a constant amplitude. To ensure oscillation, the value of g_m must exceed the minimum value. The capacitor or one of the inductors is usually made adjustable so that the frequency can be initially trimmed to the desired value.

EXAMPLE 5.10

D

Designing a Hartley oscillator

(a) Design the Hartley oscillator shown in Fig. 5.27 so that $f_o = 5$ MHz. Use a depletion NMOS whose parameters are $K_p = 250\ \mu A/V^2$ and $V_p = -2.83$ V. The load resistance is $R = 150\ \Omega$, and the power supply voltage is $V_{DD} = 15$ V. This oscillator uses the capacitor C to feed back the oscillating signal to the gate input, to sustain the oscillation.

(b) Use PSpice to plot the transient response of the output voltage $v_o(t)$ in part (a) from 10 μs to 10.5 μs. Assume $V_{DD} = 15$ V.

SOLUTION

The small-signal AC equivalent circuit is shown in Fig. 5.28, which is similar to Fig. 5.26(b):

$$g_m = 2K_p\,|V_p| = 2 \times 250\ \mu A/V^2 \times |-2.83| = 1.415\ mA/V$$

Step 1. Choose suitable values of L_1 and L_2: Let $L_1 = L_2 = 10\ \mu H$.

Step 2. Calculate the value of C from Eq. (5.65):

$$C = \frac{1}{4\pi^2 f_o^2 (L_1 + L_2)} = \frac{1}{4\pi^2 \times (5\ \text{MHz})^2 (10\ \mu H + 10\ \mu H)} = 50.66\ \text{pF}$$

Step 3. Find the value of effective load resistance R_L. The Q-point of the circuit is $V_{DSQ} = V_{DD} = 15$ V, and $V_{GSQ} = 0$. The value of g_m varies from 3 mA/V to 6.5 mA/V. To ensure oscillation, we choose $g_m = 3$ mA/V. Since R_1 is infinity in Fig. 5.28, Eq. (5.66) can be reduced to

$$g_m = \frac{L_1}{R_1 L_2} + \frac{L_2}{R_L L_1} = \frac{L_2}{R_L L_1}$$

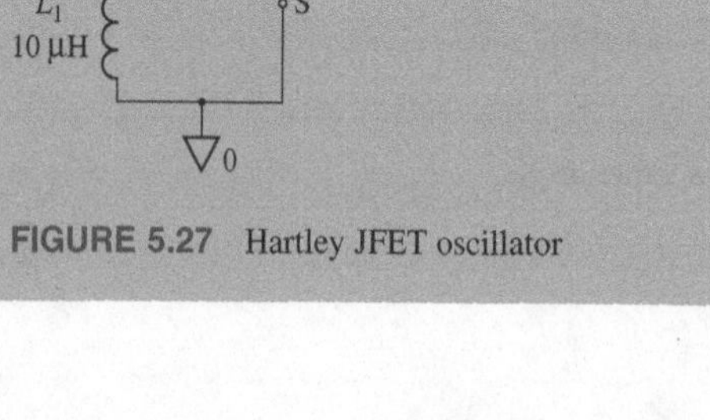

FIGURE 5.27 Hartley JFET oscillator

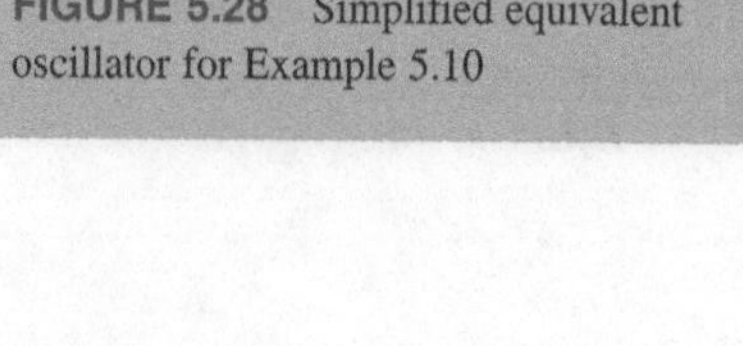

FIGURE 5.28 Simplified equivalent oscillator for Example 5.10

which gives the value of effective load resistance R_L as

$$R_L = \frac{L_2}{g_m L_1} = \frac{10\ \mu\text{H}}{1.415\ \text{mA/V} \times 10\ \mu\text{H}} = 707\ \Omega$$

This is the lowest value of R_L at which sustained oscillation can occur for $g_m = 1.415$ mA/V. Therefore, we select a higher value—say, $R_L = 1.5$ kΩ.

Step 4. Calculate the value of turns ratio n, which is related to the load resistance R and the effective load resistance R_L, by

$$n = \sqrt{\frac{R}{R_L}} = \sqrt{\frac{100\ \Omega}{1.5\ \text{k}\Omega}} = 0.316$$

Step 5. Calculate the inductance of the transformer secondary L_3, which is related to L_2, by

$$L_3 = n^2 L_2 = 0.316^2 \times 10\ \mu\text{H} = 1\ \mu\text{H}$$

(b) The PSpice schematic with the desired values is shown in Fig. 5.29. The PSpice plot of the output voltage $v_o \equiv$ V(R:2) is shown in Fig. 5.30, which gives the peak-to-peak voltage of $V_{pp} = +3.75$ V to $-$ 4.0 V at $f_o = 1/(10.326\ \mu\text{s} - 10.124\ \mu\text{s}) = 4.95$ MHz (expected 5 MHz). Note that an initial voltage of 2 V has been assigned to the capacitor C in order to start the oscillator, and UIC (use initial condition) is used in transient analysis. In practice, random noise or transients can cause the oscillations to begin, and they are sustained by the feedback of the appropriate signal.

NOTE: Increasing the value of R will increase the output voltage and the readers are encouraged to simulate the circuit with different values of R. * symbols in Figs. 5.27 and 5.29 are the dot symbols of the transformer and refer to the direction of the positive induced voltages in the windings.

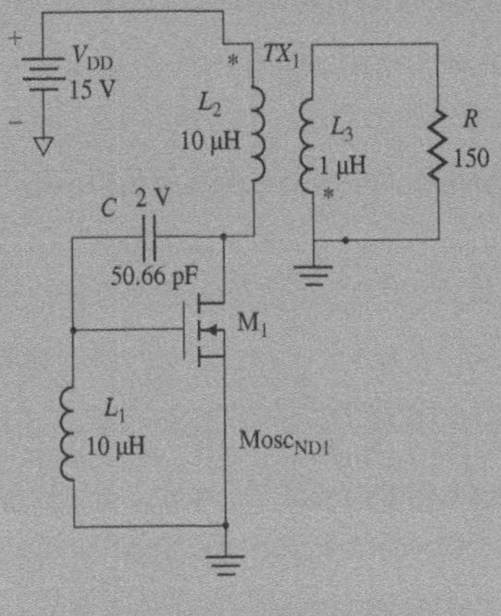

FIGURE 5.29 PSpice schematic for Example 5.10

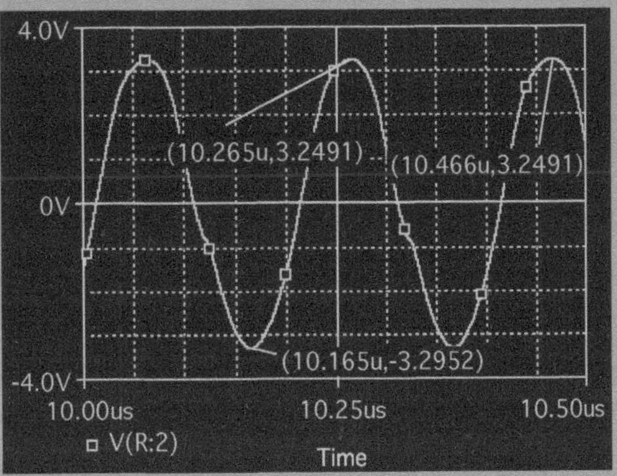

FIGURE 5.30 Output voltage waveform for Example 5.10

5.4.3 Two-Stage MOS Oscillators

In the analyses of the Colpitts and Hartley oscillators, we assumed that inductors are lossless, with no resistances. In practice, inductors suffer from a resistive component that can be modeled as a parallel resistance with an ideal inductor. The parallel equivalent makes the analysis simpler. The series impedance can

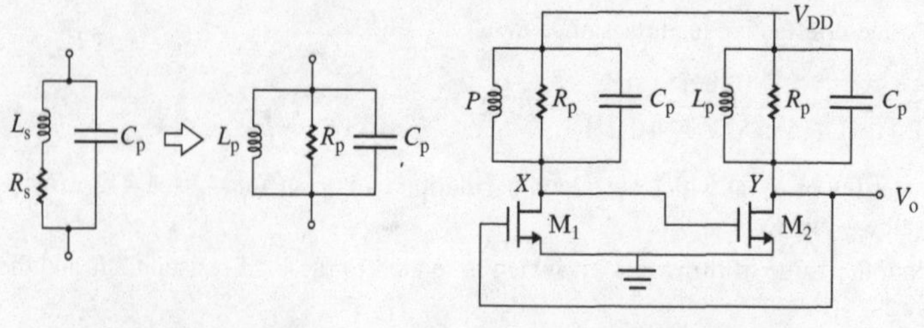

(a) Series-parallel equivalent (b) MOS oscillator

FIGURE 5.31 Two-stage MOS oscillator

be transformed to an equivalent parallel topology as shown in Fig. 5.31(a). A two-stage MOS oscillator [3] is shown in Fig. 5.31(b). The drain of one transistor is connected to the gate of the other transistor, and it is also called the cross-coupled oscillator. The frequency-dependent phase around the loop is zero (0°) because each stage contributes to a zero frequency-dependent phase shift at resonance. The equivalent R_p and L_p can be determined by equating their impedances.

$$j\omega L_s + R_s = \frac{R_p(j\omega L_p)}{R_p + j\omega L_p} \tag{5.68}$$

or
$$(L_sR_p + L_pR_s)j\omega L_s + R_sR_p - \omega^2 L_sL_p = j\omega L_sR_p \tag{5.69}$$

Equating the real and imaginary parts on both sides, we get

$$R_sR_p - \omega^2 L_sL_p = 0 \quad \text{and} \quad L_sR_p + L_pR_s = L_sR_p \tag{5.70}$$

Solving for R_p and L_p, we get

$$L_p = L_s\left(1 + \frac{R_s^2}{L_s^2\omega^2}\right) = L_s\left(1 + \frac{1}{Q^2}\right) \tag{5.71}$$

$$R_p = \frac{\omega^2 L_sL_p}{R_s} \tag{5.72}$$

where $Q = \omega L_s/R_s$ is the quality factor of an inductor. For $Q > 3$, $L_p \simeq L_s$ and $R_p \simeq \omega^2 L_s^2/R_s$. Assuming identical transistors, their transconductances and the output resistances will be equal, $g_m = g_{m1} = g_{m2}$ and $r_o = r_{o1} = r_{o2}$. The open-loop voltage gain is

$$A_o = -g_{m1}(R_p \| r_{o1}) \times -g_{m1}(R_p \| r_{o1}) = g_m^2(R_p \| r_o)^2 \tag{5.73}$$

To satisfy the oscillation condition, it requires a gain of 1 only—that is, $A_o \geq 1$—which is an advantage of this oscillator. The oscillation frequency f_o is given by

$$f_o = \frac{1}{2\pi\sqrt{C_pL_p}} \tag{5.74}$$

The MOS oscillator in Fig. 5.31(b) can be made to operate as a Colpitts oscillator by applying a voltage feedback from the drain to the source terminal of the MOSFET, as shown in Fig.5.32(a). Capacitors C_1 and

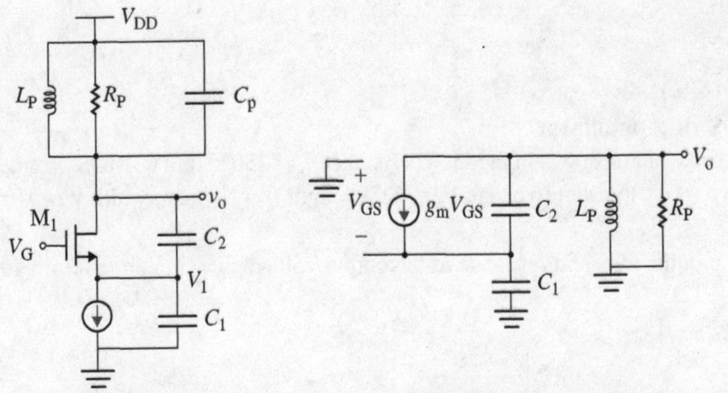

(a) MOS circuit (b) Small-signal AC equivalent

FIGURE 5.32 MOS Colpitts oscillator

C_2 form a potential divide. The equivalent capacitance C_F offered by C_1 and C_2 is in parallel with C_p. To simplify the analysis, we will omit C_p to derive the oscillation conditions, and the small-signal AC equivalent circuit is shown in Fig. 5.32(b). Using the equivalent circuit in Fig. 5.19(c), we can express the equivalent impedance of M_1, C_1, and C_2 as

$$Z_M = \frac{1}{sC_1} + \frac{1}{sC_2} + \frac{1}{sC_1}\frac{1}{sC_2}g_m = \frac{s(C_1 + C_2) + g_m}{s^2C_1C_2} \tag{5.75}$$

Therefore, the equivalent impedance of Z_M, L_p, and R_p is given by

$$\frac{1}{Z_T} = \frac{1}{Z_M} + \frac{1}{sL_p} + \frac{1}{R_p} = \frac{s^2C_1C_2}{s(C_1 + C_2) + g_m} + \frac{1}{sL_p} + \frac{1}{R_p}$$

$$= \frac{s^3C_1C_2L_pR_p + s^2(C_1 + C_2)L_p + s(C_1 + C_2)R_p + sg_mL_p + g_mR_p}{[s(C_1 + C_2) + g_m]sL_pR_p} \tag{5.76}$$

For resonant oscillation, $Z_T(j\omega) = 0$, which gives the following expression:

$$(j\omega)^3C_1C_2L_pR_p + (j\omega)^2(C_1 + C_2)L_p + j\omega(C_1 + C_2)R_p + j\omega g_mL_p + g_mR_p = 0$$

Equating the imaginary part gives the resonant frequency as

$$\omega = \sqrt{\frac{g_mL_p + (C_1 + C_2)R_p}{C_1C_2L_pR_p}} \simeq \sqrt{\frac{(C_1 + C_2)}{C_1C_2L_p}}$$

Equating the real part gives the gain as

$$g_mR_p = \omega^2(C_1 + C_2)L_p$$

which, for $\omega^2 = C_1 + C_2/C_1C_2L_p$, becomes

$$g_mR_p = \frac{(C_1 + C_2)^2}{C_1C_2} \tag{5.77}$$

EXAMPLE 5.11

Designing a MOS ring oscillator

(a) Design the MOS oscillator shown in Fig. 5.31(b) so that $f_o = 250$ kHz. The MOS parameters are MOS constant $K_p = 25\ \mu\text{A/V}^2$, threshold voltage $V_t = 1$ V, length $L = 10\ \mu$m, width $W = 10\ \mu$m, and modulation $\lambda = 0.01$.

(b) Use PSpice to plot the transient response of the output voltage $v_o(t)$ in part (a) from 4000 μs to 4020 μs. Assume $V_{CC} = 12$ V.

SOLUTION

(a) The DC-biasing analysis is similar to the MOS ring oscillator in Fig. 5.16. Choosing a voltage gain of $A_o = 2$, Eq. (5.47) gives the DC biasing gate–source voltage

$$v_{GS} = \frac{5 + (2/2) \times 1}{1 + (2/2)} = 6.5 \text{ V}$$

$$i_D = K_p(v_{GS1} - V_t)^2 = 25\ \mu\text{A/V}^2 \times (6.5 - 1)^2 = 765.2\ \mu\text{A}$$

$$g_m = 2K_p(v_{GS} - V_t) = 2 \times 25\ \mu\text{A/V}^2 \times (6.5 - 1) = 275\ \mu\text{A/V}$$

$$R_p = \frac{\sqrt{A_o}}{g_m} = \frac{\sqrt{2}}{275\ \mu\text{A/V}} = 5.1\text{ k}\Omega$$

Let $C_p = 0.01\ \mu$F. For $f_o = 250$ kHz, Eq. (5.74) gives

$$L_p = \frac{1}{(2\pi f_o)^2 C_p} = \frac{1}{(2\pi \times 250\text{ kHz})^2 \times 0.01\ \mu\text{F}} = 40\ \mu\text{H}$$

(b) The PSpice schematic with the desired values is shown in Fig. 5.33. The PSpice plot of the output voltage $v_o \equiv$ V(M2:d) is shown in Fig. 5.34, which gives the peak-to-peak voltage of $V_{pp} = 23{,}635$ V at $f_o = 1/(4.0049\text{ ms} - 4.0009\text{ ms}) = 250$ kHz (expected 250 kHz). Note that an initial voltage of 1.5 V has been assigned to the capacitor C_{p1} in order to start the oscillator, and UIC (use initial condition) is used in transient analysis. In practice, random noise or transients can cause the oscillations to begin, and they are sustained by the feedback of the appropriate signal.

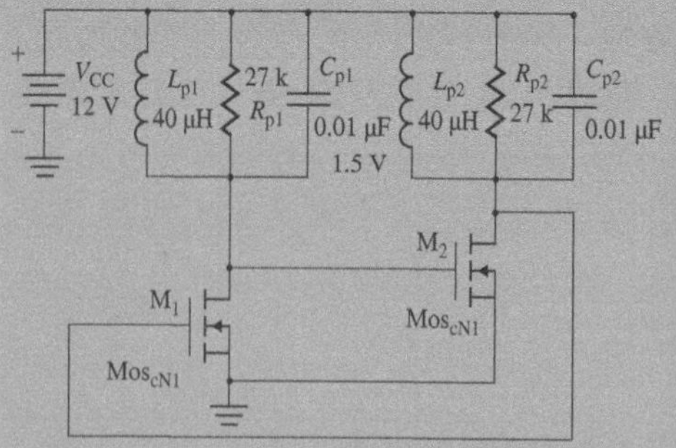

FIGURE 5.33 PSpice schematic for Example 5.11

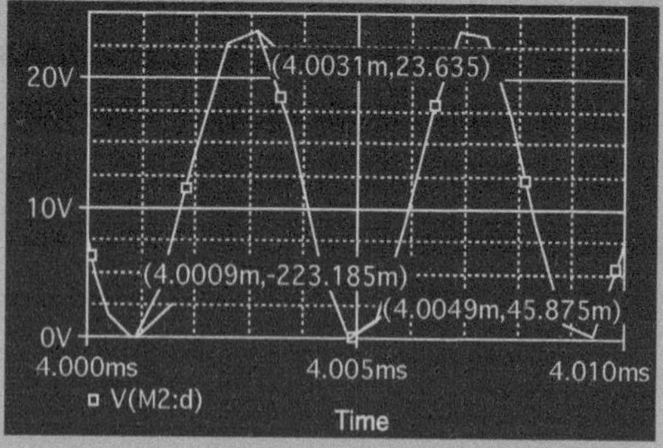

FIGURE 5.34 Output voltage waveform for Example 5.11

KEY POINTS OF SECTION 5.4

- A Colpitts oscillator uses an inverting amplifier and a phase-shifting network consisting of two capacitors and one inductor. The oscillation frequency ω_o is inversely proportional to the LC product.
- A Hartley oscillator is a derivation of a Colpitts oscillator. The phase-shifting network consists of two inductors and a capacitor. The oscillation frequency ω_o is inversely proportional to the LC product.
- In a cross-coupled oscillator, the drain of a MOS transistor is connected to the gate of another transistor. The frequency-dependent phase around the loop is zero (0°) because each stage contributes to zero frequency-dependent phase shift at resonance. A MOS oscillator can be operated as a Colpitts oscillator by applying a voltage feedback from the drain to the source terminal.

5.5 Crystal Oscillators

Because of their excellent frequency stability, quartz crystals are commonly used to control the frequency of oscillation. If the inductor L of the Colpitts oscillator in Fig. 5.19(a) is changed to a crystal, the oscillator is called a *crystal oscillator*. Crystal oscillators are commonly used in digital signal processing. The symbol for a vibrating piezoelectric crystal is shown in Fig. 5.35(a); its circuit model is shown in Fig. 5.35(b), which can be simplified to Fig. 5.35(c). The *quality factor* Q of a crystal can be as high as several hundred thousands. C_p represents the electrostatic capacitance between the two parallel plates of the crystal. L has a large value (as high as hundreds of henries) and is determined from $L \approx 1/C_s\omega_o^2$, where ω_o is the resonant frequency of the crystal. R_s can be as high as a few hundred thousand ohms and is determined from $R_s \approx \omega_o L/Q$, where the quality factor Q is in the range of 10^4 to 10^6. Typical values for a 2-MHz quartz crystal are $Q = 80 \times 10^3$, $C_p/C_s = 350$, $L = 520$ mH, $C_s = 0.0122$ pF, and $R_s = 82\ \Omega$. Table 5.1 shows typical component values for common cuts of quartz oscillator crystals.

Since Q is very high in the typical quartz crystal, we may neglect R_s. The crystal impedance is given by

$$Z(s) = \frac{1}{sC_p + sL + 1/sC_s} = \frac{1}{sC_p}\left[\frac{s^2 + 1/LC_s}{s^2 + (C_p + C_s)/(LC_sC_p)}\right]$$

$$= \frac{1}{sC_p}\left[\frac{s^2 + \omega_s^2}{s^2 + \omega_p^2}\right] \tag{5.78}$$

TABLE 5.1 Common cuts of quartz oscillator crystals (RCA Corp.)

Frequency	32 kHz	280 kHz	525 kHz	2 MHz	10 MHz
Cut	XY bar	DT	DT	AT	AT
R_s	40 kΩ	1820 Ω	1400 Ω	82 Ω	5 Ω
L	4800 H	25.9 H	12.7 H	0.52 H	12 mH
C_s, in pF	0.00491	0.0126	0.00724	0.0122	0.0145
C_p, in pF	2.85	5.62	3.44	4.27	4.35
C_p/C_s	580	450	475	350	300
Q	25,000	25,000	30,000	80,000	150,000

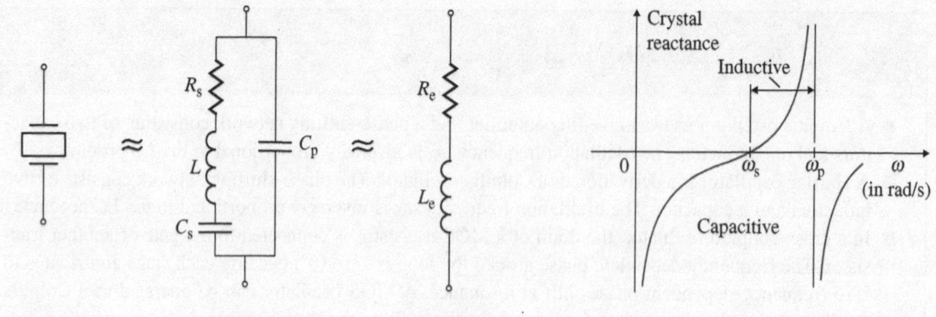

(a) Symbol of piezoelectric crystal (b) Equivalent circuit (c) Simplified equivalent circuit (d) Crystal reactance versus frequency

FIGURE 5.35 Symbol and circuit model of piezoelectric crystal

If we substitute $s = j\omega$, the impedance in Eq. (5.78) becomes

$$Z(j\omega) = -\left(\frac{j}{\omega C_p}\right)\left(\frac{\omega^2 - \omega_s^2}{\omega^2 - \omega_p^2}\right) \tag{5.79}$$

Therefore, the crystal exhibits two resonant frequencies: series resonance at

$$\omega_s = \frac{1}{\sqrt{LC_s}} \tag{5.80}$$

and parallel resonance at

$$\omega_p = \left[\frac{C_s + C_p}{C_s C_p L}\right]^{1/2} \tag{5.81}$$

Note that $\omega_p > \omega_s$. However, since $C_p >> C_s$, the two resonance frequencies are very close. The plot of crystal reactance against frequency in Fig. 5.35(d) illustrates that the crystal exhibits the characteristic of an inductor over the narrow frequency range between ω_s and ω_p.

It is possible to have a variety of crystal oscillators. A Colpitts-derived op-amp crystal oscillator is shown in Fig. 5.36(a); its equivalent circuit is shown in Fig. 5.36(b). This circuit should oscillate at the resonance frequency of the crystal inductance L with the series equivalent of C_s and $C_p + C_1C_2/(C_1 + C_2)$. Since C_s is much smaller than C_p, C_1, or C_2, it will be dominant and the oscillating frequency can be approximately found from

$$\omega_o \approx \frac{1}{\sqrt{LC_s}} \tag{5.82}$$

Using nodal analysis in Fig. 5.36(b), we can write

$$\left[sC_2 + \frac{1}{R_e + sL_e} + \frac{1}{R_L}\right]V_o(s) + \left[g_m - \frac{1}{R_e + sL_e}\right]V_f(s) = 0 \tag{5.83}$$

$$-\frac{1}{R_e + sL_e}V_o(s) + \left[sC_1 + \frac{1}{R_e + sL_e} + \frac{1}{R_1}\right]V_f(s) = 0 \tag{5.84}$$

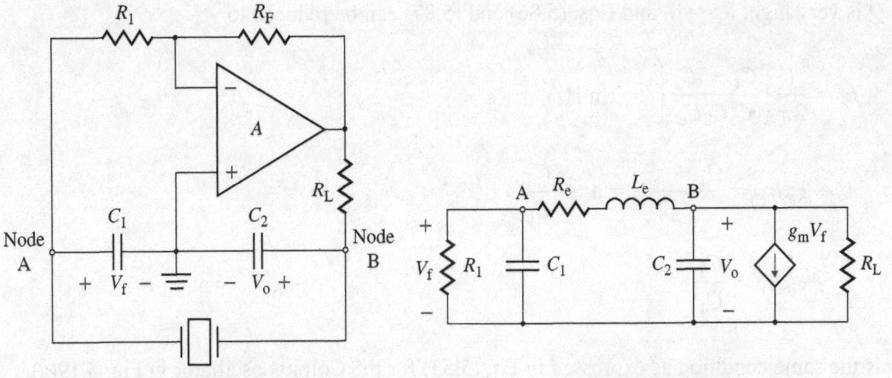

(a) Oscillator

(b) Equivalent circuit

FIGURE 5.36 Crystal oscillator

Assuming R_L is large, tending to infinity, we set the determinant to zero to find the condition for oscillation; that is,

$$\left(sC_2 + \frac{1}{R_e + sL_e}\right)\left(sC_1 + \frac{1}{R_e + sL_e} + \frac{1}{R_1}\right) + \left(g_m - \frac{1}{R_e + sL_e}\right)\frac{1}{R_e + sL_e} = 0$$

which, after simplification, yields

$$s^3C_1C_2L_eR_1 + s^2(C_1C_2R_eR_1 + C_2L_e) + s(C_1R_1 + C_2R_1 + C_2R_e) + 1 + g_mR_1 = 0 \quad (5.85)$$

where $g_m = A/R_L = R_F/(R_1R_L)$. Substituting $s = j\omega$ and equating the imaginary parts of Eq. (5.85) to zero, we get

$$-j\omega^3(C_1C_2L_eR_1) + j\omega(C_1R_1 + C_2R_1 + C_2R_e) = 0$$

which gives the frequency of oscillation f_o as

$$f_o = \frac{1}{2\pi}\left[\frac{C_1R_1 + C_2R_1 + C_2R_e}{C_1C_2L_eR_1}\right]^{1/2} \quad \text{(in Hz)} \quad (5.86)$$

Similarly, equating the real parts of Eq. (5.85) to zero, we get

$$-\omega^2(C_1C_2R_eR_1 + C_2L_e) + 1 + g_mR_1 = 0$$

which gives

$$1 + g_mR_1 = \omega^2(C_1C_2R_eR_1 + C_2L_e)$$

After substitution of the value of $\omega = \omega_o = 2\pi f_o$ from Eq. (5.86), the above equation becomes

$$1 + g_mR_1 = \frac{(C_1R_1 + C_2R_1 + C_2R_e)(C_1C_2R_eR_1 + C_2L_e)}{C_1C_2L_eR_1} \quad (5.87)$$

Since Q is very high, $R_e \approx 0$, and Eqs. (5.86) and (5.87) can be reduced to

$$f_o = \frac{1}{2\pi}\left[\frac{C_1 + C_2}{C_1 C_2 L_e}\right]^{1/2} \quad \text{(in Hz)} \tag{5.88}$$

and
$$1 + g_m R_1 = \frac{C_1 + C_2}{C_1} = 1 + \frac{C_2}{C_1}$$

or
$$g_m R_1 = \frac{C_2}{C_1} \tag{5.89}$$

which is the same condition as expressed in Eq. (5.53) for the Colpitts oscillator in Fig. 5.19(a).

EXAMPLE 5.12

Finding the oscillation frequency of a crystal oscillator The op-amp oscillator in Fig. 5.36(a) uses a 2-MHz crystal and has $C_1 = 0.01$ μF, $C_2 = 0.1$ μF, $R_L = 100$ kΩ, $R_1 = 100$ kΩ, and $R_F = 1$ MΩ.

(a) Find the frequency of oscillation f_o.

(b) Use PSpice/SPICE to verify the frequency in part (a). Assume $V_{CC} = V_{EE} = 15$ V.

SOLUTION

(a) For a 2-MHz crystal, $C_s = 0.0122$ pF, $C_p = 4.27$ pF, $R_s = 82\ \Omega$, and $L = 0.52$ H. Let

$$C_{eqp} = \frac{C_p + C_1 C_2}{C_1 + C_2}$$

$$= \frac{4.27 \text{ pF} + 0.01\ \mu\text{F} \times 0.1\ \mu\text{F}}{0.01\ \mu\text{F} + 0.1\ \mu\text{F}} = 9095 \text{ pF}$$

The effective capacitance C_{eq} is given by C_s in series with $C_{eqp} = C_p + C_1C_2/(C_1 + C_2)$:

$$C_{eq} = \frac{C_s \times C_{eqp}}{C_s + C_{eqp}}$$

$$= \frac{0.0122 \text{ pF} \times 9095 \text{ pF}}{0.0122 \text{ pF} + 9095 \text{ pF}} \approx 0.0122 \text{ pF}$$

Thus, the frequency of oscillation f_o becomes

$$f_o = \frac{1}{2\pi}\left[\frac{1}{C_{eq}L}\right]^{1/2} = \frac{10^6}{2\pi\sqrt{0.0122 \times 0.52}} = 1.998 \text{ MHz}$$

(b) The circuit for PSpice simulation is shown in Fig. 5.37. The PSpice plot of the output voltage across C_s [i.e., $v_o \equiv$ V(Cs:2)] is shown in Fig. 5.38, which gives $f_o = 1/(763.146\text{ n} - 243.137\text{ n}) = 1.923$ MHz, close to the calculated frequency of 1.998 MHz. The amplitude of the output voltage is stable, not falling.

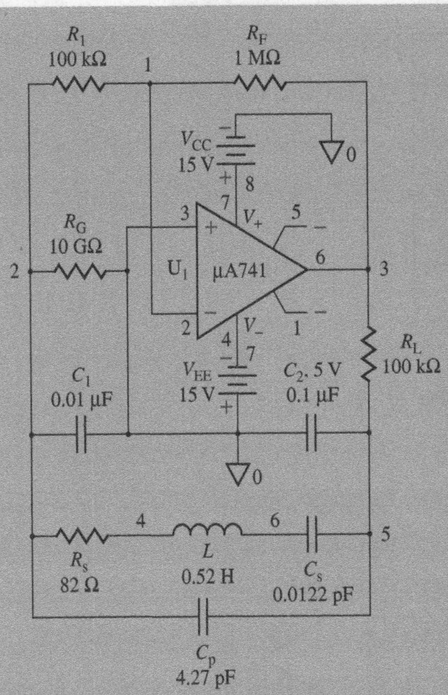

FIGURE 5.37 Crystal oscillator for PSpice simulation

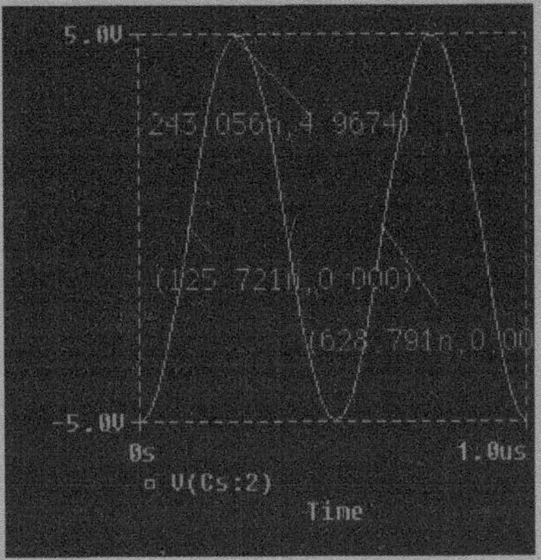

FIGURE 5.38 PSpice plot of output voltage for Example 5.12

NOTE: An initial voltage of 5 V has been assigned to the capacitor C_2 in order to start the oscillator, and the UIC (use initial condition) is used in transient analysis. In practice, random noise or transients can cause the oscillations to begin, and they are sustained by the feedback.

KEY POINT OF SECTION 5.5

- A crystal oscillator has the same circuit topology as a Colpitts oscillator except that it uses a crystal instead of an inductor. A crystal has strong frequency stability.

5.6 Active-Filter Tuned Oscillators

An active band-pass filter with a high Q-value can be operated as an oscillator provided that a positive feedback is applied. This type of oscillator, which consists of a narrow-band filter and a limiter, is illustrated in Fig. 5.39(a). To understand the operation of the circuit, let us assume that the oscillation has already started. The output of the filter v_o is a sine wave whose frequency is the center frequency of the filter f_o. This sine wave is fed to a limiter, which produces a square-wave output v_f of frequency f_o. The peak amplitude of the square wave is determined by the type of limiting devices. The square wave is in turn fed back to the band-pass filter, which filters the harmonics and produces a sinusoidal output v_o at the fundamental frequency f_o. The quality of the sine wave is a direct function of the selectivity (Q-factor) of the band-pass filter. The design of this type of oscillator is very simple, and the oscillator has independent frequency control.

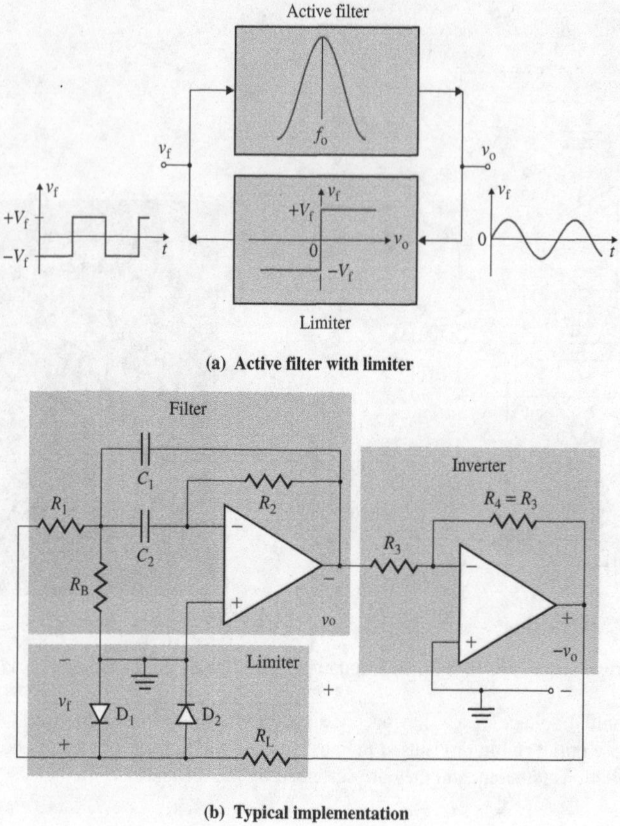

FIGURE 5.39 Active-filter tuned oscillator

A typical practical implementation of an active-filter tuned oscillator using a narrow-band filter is shown in Fig. 5.39(b). An inverter is added to the output of the filter in order to provide positive feedback. A simple diode limiter along with resistance R_L is used to generate a square wave, which is fed back to the input of the filter.

EXAMPLE 5.13

PSpice simulation of an active-filter tuned oscillator Modify the narrow-band filter that was designed in Example 4.10 so that it operates as an oscillator, and use PSpice/SPICE to plot the output voltage.

SOLUTION

The circuit for PSpice simulation is shown in Fig. 5.40. Instead of an inverter, a voltage-controlled voltage source with -1 gain is used to provide the positive feedback required for oscillation.

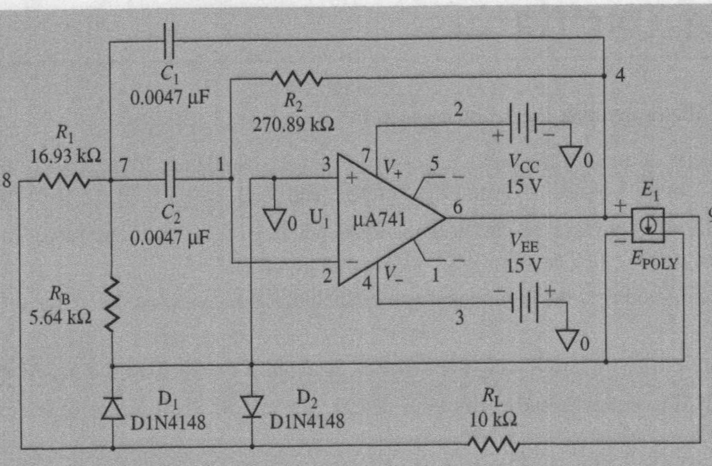

FIGURE 5.40 Active-filter tuned oscillator for PSpice simulation

The PSpice plots of the output voltage $v_o \equiv$ V(E1:3) at the output of the op-amp and the voltage across the diodes $v_{D1} \equiv$ V(RL:1) are shown in Fig. 5.41. We get the oscillation frequency $f_o = 1/(5.913\text{ m} - 4.8548\text{ m}) = 962$ Hz (expected value is 1 kHz), the peak amplitude of the output voltage is ± 5.17 V, and the peak amplitude of the diode limiter is ± 557 mV.

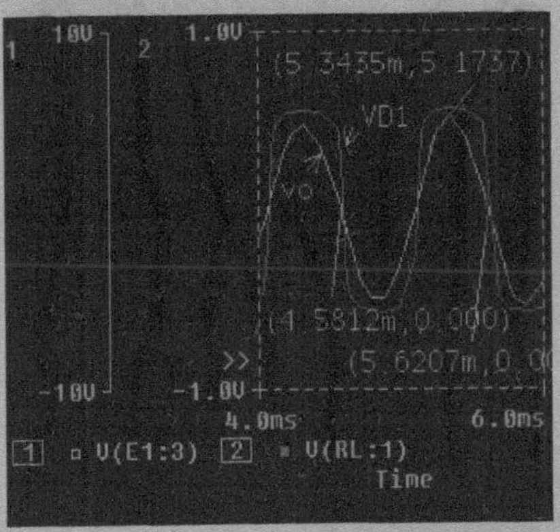

FIGURE 5.41 PSpice plot of output voltage for Example 5.13

KEY POINT OF SECTION 5.6

- An active-filter tuned oscillator uses a narrow-band filter with positive feedback. A limiter is used in the feedback path to provide square-wave input signals to the filter. This type of oscillator has the advantages of independent frequency and amplitude control.

5.7 Design of Oscillators

The design of a sinusoidal oscillator involves the following steps:

Step 1. Identify the specifications of the output stage—for example, oscillation frequency f_o, load resistance R_L, and the DC supply voltages V_{CC} and V_{EE} (or V_{DD} and V_{SS}).

Step 2. Select the type of oscillator and the circuit topology, depending on the oscillation frequency and the types of devices available, such as BJTs, MOSFETs, or op-amps.

Step 3. Analyze the circuit, and find the component values such that the condition $A\beta = 1\ \angle 0°$ or $\angle 360°$ is satisfied.

Step 4. Limit the output voltage by introducing nonlinearity to stabilize the oscillator, if necessary.

Step 5. Use PSpice/SPICE to simulate and verify your design. Use the standard values of components with their tolerances.

EXAMPLE 5.14

Worst-case analysis of the phase-shift oscillator in Example 5.3 Use PSpice/SPICE to find worst-case output voltage and frequency ranges of the phase-shift oscillator in Example 5.3. Use standard component values: $C = 0.1\ \mu\text{F} \pm 10\%$, $R = 1.6\ \text{k}\Omega \pm 5\%$, $R_1 = 17\ \text{k}\Omega \pm 5\%$, and $R_F = 490\ \text{k}\Omega \pm 5\%$.

SOLUTION

To assign tolerances to resistors and capacitors, we will use model RMOD for resistors and CMOD for capacitors. Also, we will add a statement for the worst-case analysis (.WCASE) [2].

The PSpice plots of the worst-case maximum and nominal output voltages $v_o \equiv V(3)$ are shown in Fig. 5.42(a), which give the worst-case peak values of 8.7 V and −9.1 V and an oscillation frequency of 1/2.322 ms = 430 Hz (nominal value is 422 Hz). The plots of the worst-case minimum and nominal output voltages $v_o \equiv V(3)$ are shown in Fig. 5.42(b), which give the worst-case peak values of 8.05 V and −9.45 V and an oscillation frequency of 1/2.246 ms = 445 Hz (nominal value is 422 Hz). Thus, the output frequency can vary from 430 Hz to 445 Hz based on component tolerances.

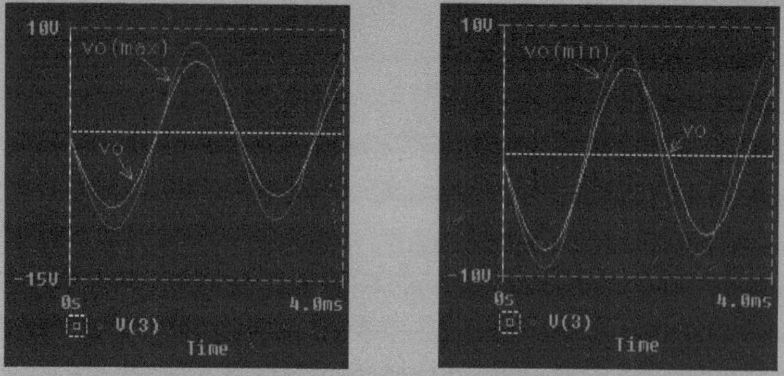

(a) Maximum and nominal values **(b) Minimum and nominal values**

FIGURE 5.42 Worst-case output of the phase-shift oscillator for Example 5.14

Summary

Oscillators use positive feedback and are commonly employed in electronics circuits. There are many types of oscillators. However, *RC*, *LC*, and crystal oscillators are the most commonly used. For a feedback circuit to operate as an oscillator, the magnitude and the phase shift of the loop gain must be unity and 0° (or 360°), respectively. Frequency stability is an important criterion in defining the quality of an oscillator. Crystal oscillators have the highest frequency stability. The frequency of oscillation and the conditions for oscillation can be determined from the transfer function or the determinant of a circuit.

References

1. R. A. Gayakwad, *Op-Amps and Linear Integrated Circuits*. Englewood Cliffs, NJ: Prentice Hall, 1993.
2. M. H. Rashid, *Introduction to PSpice Using OrCAD for Circuits and Electronics*. Upper Saddle River, NJ: Prentice Hall, 2004.
3. B. Razavi, *Design of Analog CMOS Integrated Circuits*. New York: McGraw-Hill, 2001.

Review Questions

1. What is an oscillator?
2. What are the two conditions for oscillation?
3. What are the major types of oscillators?
4. What is an *RC* oscillator?
5. What is an *LC* oscillator?
6. What is a crystal oscillator?
7. What is the frequency stability of oscillators?
8. What is the figure of merit of an oscillator?
9. What is a phase-shift oscillator?
10. What is a Wien-bridge oscillator?
11. What is a quadrature oscillator?
12. What is a Colpitts oscillator?
13. What is a Hartley oscillator?
14. What is a crystal oscillator?
15. What is an active-filter tuned oscillator?

Problems

The symbol **D** indicates that a problem is a design problem. The symbol **P** indicates that you can check the solution to a problem using PSpice/SPICE or Electronics Workbench.

5.2 Principles of Oscillators

5.1 For the circuit in Fig. P5.1, determine the values of A and θ that will produce a steady-state sinusoidal oscillation.

FIGURE P5.1

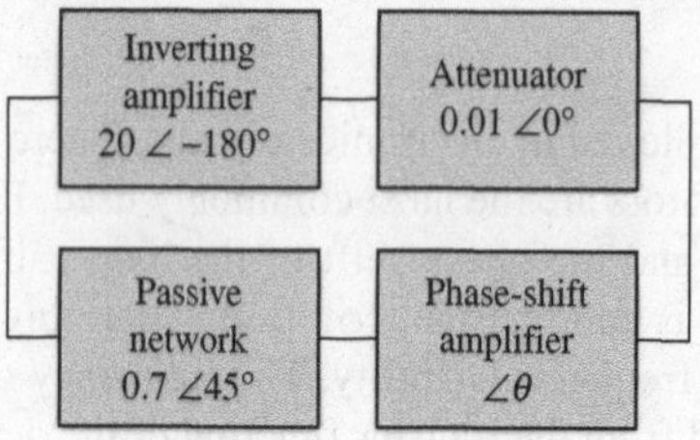

5.2 The amplifier in Fig. 5.3 has a voltage gain of $A = 200$, input resistance $R_i = 50$ kΩ, and output resistance $R_o = 500$ Ω. Find the values of R, R_3, C, and L so that the oscillation frequency is $f_o = 5$ kHz.

D P

5.3 *Audio-Frequency Oscillators*

5.3 Derive the transfer function $\beta(s)$ of the feedback network shown in Fig. 5.5(b).

5.4 Design a phase-shift oscillator as shown in Fig. 5.5(a) so that $f_o = 1$ kHz.

D P

5.5 Find the values of R and C for the phase-shift oscillator in Fig. P5.5 so that the oscillation frequency is $f_o = 5$ kHz.

D

FIGURE P5.5

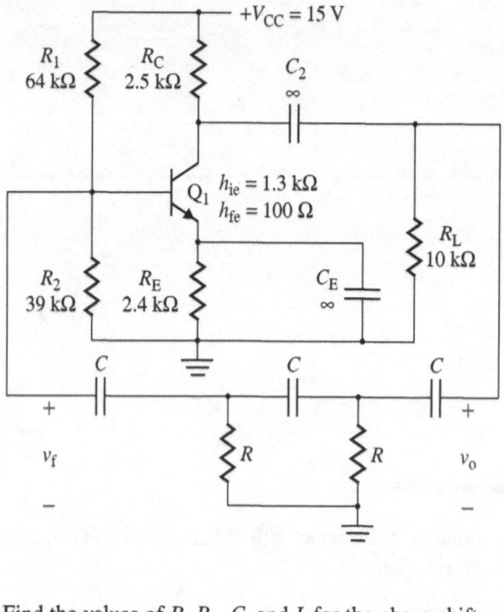

5.6 Find the values of R, R_3, C, and L for the phase-shift oscillator in Fig. P5.6 so that the oscillation frequency is $f_o = 5$ kHz.

D

FIGURE P5.6

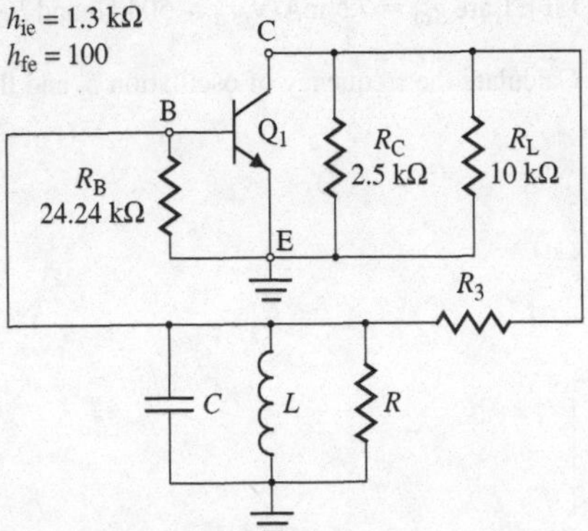

5.7 Design a quadrature oscillator as shown in Fig. 5.9(a) so that f_o = 500 Hz.

D P

5.8 Design a three-phase oscillator as shown in Fig. 5.10 so that f_o = 50 Hz.

D P

5.9 Derive the transfer function $\beta(s)$ of the feedback network in Fig. 5.12(b).

5.10 Design a Wien-bridge oscillator as shown in Fig. 5.12(a) so that f_o = 5 kHz.

D P

5.11 Design the MOS ring oscillator in Fig. 5.16 so that f_o = 2 kHz. The MOSFET parameters are MOS constant K_p = 25 μA/V^2, threshold voltage V_t = 1 V, length L = 10 μm, width W = 20 μm, and modulation length λ = 0.01. Use PSpice to plot the transient response of the output voltage $v_o(t)$ in part (a) from 0 to 1 ms. Assume V_{DD} = 5 V.

D

5.12 Determine the oscillation frequency f_o for the ring oscillator in Fig. 5.16 if R = 40 kΩ and C = 6 nF. The MOSFET parameters are MOS constant K_p = 25 μA/V^2, threshold voltage V_t = 1 V, length L = 10 μm, width W = 20 μm, and modulation length λ = 0.01. Use PSpice to plot the transient response of the output voltage $v_o(t)$ to verify the result. Assume V_{DD} = 5 V.

5.4 *Radio Frequency Oscillators*

5.13 Design a Colpitts oscillator as shown in Fig. 5.19(a) so that the oscillation frequency is f_o = 500 kHz.

D

5.14 A Colpitts BJT oscillator is shown in Fig. 5.20. The circuit parameters are $r_\pi = h_{ie}$ = 500 Ω, h_{fe} = 200, L = 1.5 mH, C_1 = 10 nF, C_2 = 10 nF, and R_L = 5 kΩ. Calculate the frequency of oscillation f_o and the value of R_1 required to sustain the oscillation.

5.15 Design a Colpitts BJT oscillator as shown in Fig. 5.20 so that f_o = 250 kHz. The circuit parameters are r_π = h_{ie} = 500 Ω, h_{fe} = 200, and R_L = 5 kΩ.

D P

5.16 An *LC*-tuned MOSFET oscillator is shown in Fig. 5.24. Find the values of *L*, *C*, and *n* for an oscillation frequency of f_o = 100 kHz. The parameters of the MOSFET are g_m = 7.5 mA/V, r_d = 50 kΩ, and R_G = 20 kΩ.

D

5.17 A Colpitts BJT oscillator is shown in Fig. P5.17. Calculate the frequency of oscillation f_o and the value of R_{E1} required to sustain the oscillation.

FIGURE P5.17

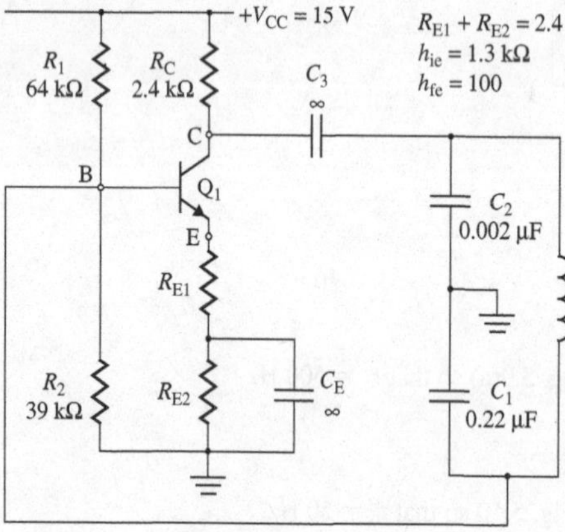

5.18 Design a Colpitts oscillator as shown in Fig. 5.19(a) so that the oscillation frequency is f_o = 5 kHz. Assume $V_{CC} = V_{EE}$ = 12 V.

D P

5.19 The Colpitts oscillator of Fig. 5.19(a) has C_1 = 400 pF, C_2 = 200 pF, and L = 1 mH. Determine the frequency of oscillation f_o and the minimum value of gain $A = R_F/R_1$ needed to sustain the oscillation.

5.20 Determine the frequency of oscillation for the Colpitts MOSFET oscillator in Fig. P5.20(a). The MOSFET can be replaced by its transconductance model, shown in Fig. P5.20(b). The parameters are r_d = 25 kΩ, g_m = 5 mA/V, R_G = 1 MΩ, L = 1.5 mH, C_1 = 10 nF, and C_2 = 10 nF. Calculate the frequency of oscillation and check to make sure the condition for oscillation is satisfied.

FIGURE P5.20

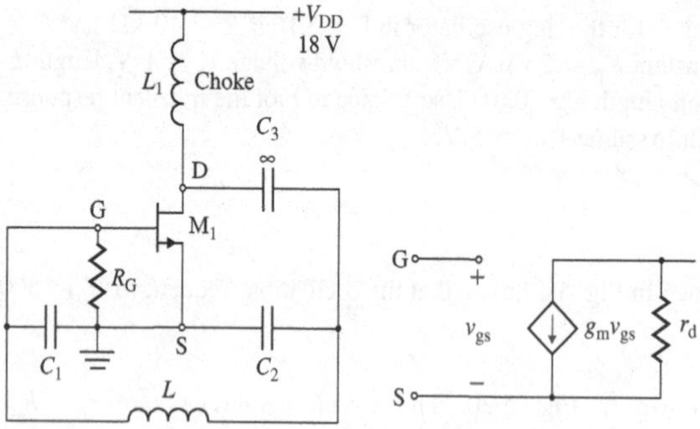

(a) Circuit **(b) JFET model**

5.21 Design a Hartley oscillator as shown in Fig. 5.27 so that f_o = 500 kHz. Use a 2N3821 *n*-channel JFET whose parameters are I_{DSS} = 0.5 mA to 2.5 mA, V_p = −4 V, and g_m = 1.5 mA/V to 4.5 mA/V. The load resistance is R = 50 Ω, and the power supply voltage is V_{DD} = 15 V.

D P

5.22 Design the MOS oscillator in Fig. 5.32(b) so that $f_o = 200$ kHz. The MOS parameters are MOS constant $K_p = 25\ \mu A/V^2$, threshold voltage $V_t = 1$ V, length $L = 10\ \mu m$, width $W = 10\ \mu m$, and modulation $\lambda = 0.01$. Use PSpice to plot the transient response of the output voltage $v_o(t)$ in part (a) from 4000 μs to 4020 μs. Assume $V_{CC} = 12$ V.

5.23 Determine the oscillation frequency f_o for the ring oscillator in Fig. 5.31(a) if $R_p = 3.5\ k\Omega$ and $L_p = 50\ \mu H$. The MOSFET parameters are MOS constant $K_p = 25\ \mu A/V^2$, threshold voltage $V_t = 1$ V, length $L = 10\ \mu m$, width $W = 20\ \mu m$, and modulation length $\lambda = 0.01$. Use PSpice to plot the transient response of the output voltage $v_o(t)$ to verify the result. Assume $V_{DD} = 12$ V.

5.5 *Crystal Oscillators*

5.24 Design a crystal oscillator as shown in Fig. 5.36(a) so that $f_o = 14$ MHz. The crystal parameters are $f_o = 10$ MHz, $Q = 150 \times 10^3$, $C_s = 0.0145$ pF, $C_p = 4.35$ pF, and $L = 12$ mH. Assume a transconductance gain of $|g_m| = 1$ mA/V for the amplifier.

5.6 *Active-Filter Tuned Oscillators*

5.25 The equivalent circuit of a tuned oscillator is shown in Fig. P5.25. Derive the expressions for the oscillation condition and the frequency of oscillation.

FIGURE P5.25

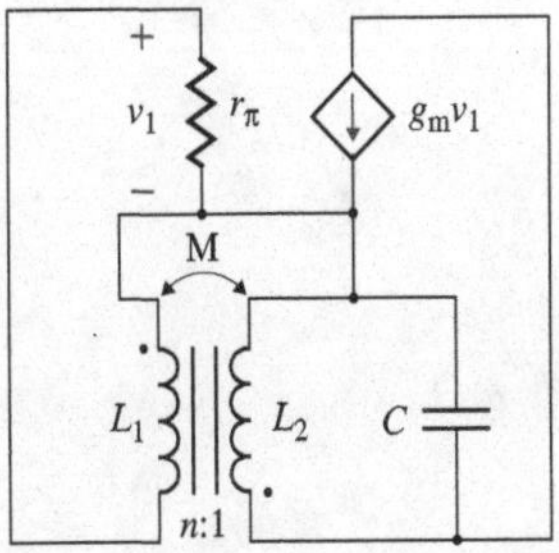

5.26 Design an active-filter tuned oscillator as shown in Fig. 5.39(b) so that $f_o = 10$ kHz. Assume $R_L = 10\ k\Omega$ and $V_{CC} = V_{EE} = 12$ V.

APPENDIX A

INTRODUCTION TO OrCAD

A.1 Introduction

This appendix provides an introduction to the electronic circuit simulation software OrCAD [1, 2]. The objectives of this appendix are the following:

- Becoming familiar with the electronic circuit simulation software OrCAD Capture
- Learning how to draw electrical circuits by OrCAD Capture

After installation instructions presented in Sec. A.2, Sec. A.3 provides an overview of the software package. Then Secs. A.4 to A.8 cover the basic steps in the circuit analysis process:

- Drawing the circuit
- Selecting the type of analysis
- Simulating the circuit
- Displaying the results of the simulation

The remaining sections, Secs. A.9 to A.13, deal with specific operations:

- Copying and capturing schematics
- Varying parameters
- Performing frequency response analysis
- Modeling devices and elements
- Creating netlists
- Adding library files

A.2 Installing the Software

The instructions for installing the OrCAD software are printed on the OrCAD installation disk or the CD-ROM. The software can be downloaded from the Cadence website at http://www.cadence.com/products/orcad/Pages/downloads.aspx

The following steps, which pertain to OrCAD Version 9.2, are applicable to other versions as well:

1. Place the schematics CD-ROM in the CD drive.
2. From Windows, enter the **File Manager** and click the left mouse button on the CD drive.
3. Click on **setup.exe**, **File**, **Run**, and **OK**.
4. Click on **OK** to select the products to install **Capture** and **PSpice**, as shown in Fig. A.1(a).
5. Click on **OK** to select the default **C:OrcadLite**.
6. Click on **Yes** to choose **Program Folder: Orcad Family Release 9.2 Lite Edition**, as shown in Fig. A.1(b).
7. Create the OrCAD Capture Lite and PSpice A/D icons from the Start menu; then go to **Programs**, **Orcad Family**, and **Capture Lite** menu. These icons are shown in Fig. A.2.
8. Click the left mouse button on the Capture Lite icon once, and the window of Capture Lite will open.

▶ **NOTE** Although there are later versions of OrCAD, the 9.2 version has proven to be bug free and works well.

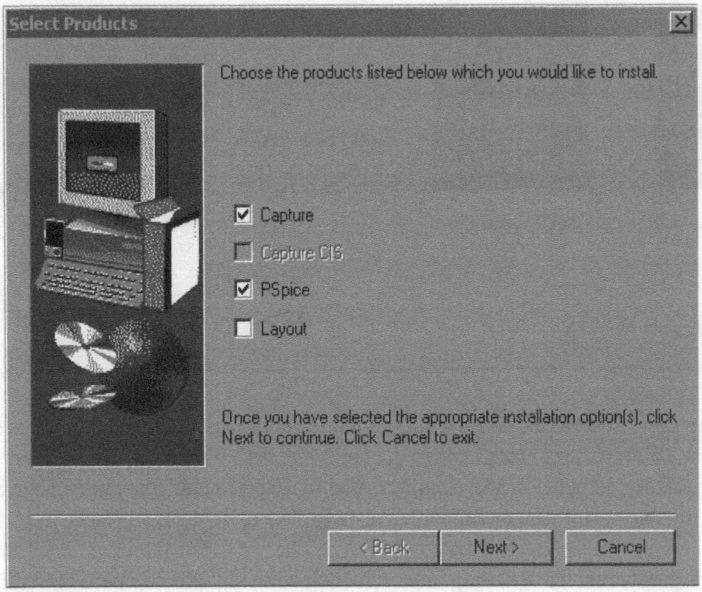

(a) Selecting products

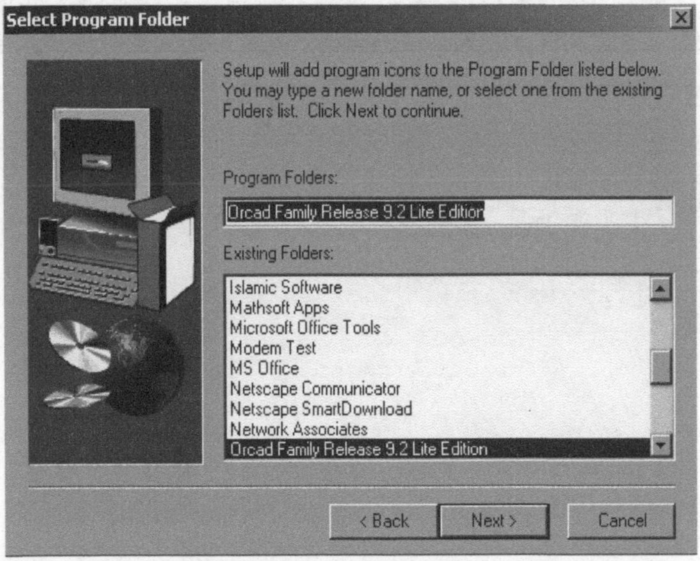

(b) Selecting the program folder

FIGURE A.1 Installation setup for OrCAD Capture

(a) **OrCAD Capture Lite** (b) **PSpice A/D**

FIGURE A.2 Icons for OrCAD Capture Lite and PSpice A/D

A.3 Overview

The OrCAD Capture software package has three major interactive programs: Capture, PSpice A/D, and Probe. *Capture* is a powerful program that lets us build circuits by drawing them within a window on the monitor. *PSpice A/D* lets us analyze the circuit created by Capture and generate voltage and current solutions. *Probe* is a graphic postprocessor and lets us display plots of voltages, currents, impedance, and power.

The general layout of Capture is shown in Fig. A.3. The top menu shows 10 main choices. The right-side menu shows the schematic "drawing" menu for selecting and placing parts. File, Edit, View, Place, and PSpice are most frequently used. For any help, click on the Help menu.

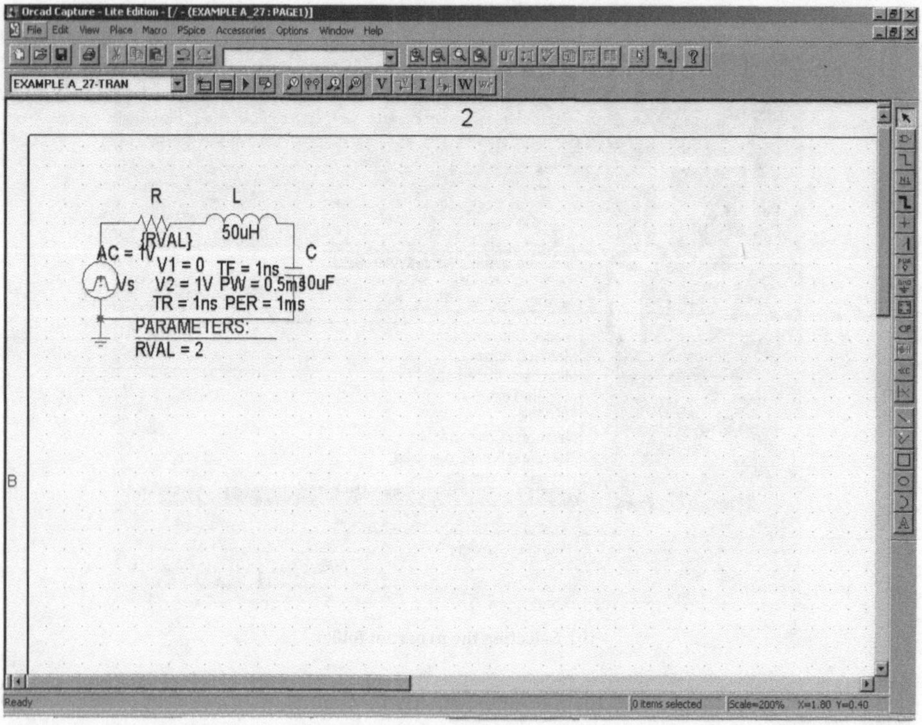

FIGURE A.3 General layout of OrCAD Capture

TABLE A.1 PSpice mouse operations

Button	Action	Function
Left	Single click	Select an item
	Double click	End a mode
	Double click on selected object	Edit a selection
	Single click on selected object and hold	Drag a selection
	Shift + single click	Extend a selection
Right	Single click	Abort the mode
	Double click	Repeat an action

The mouse follows an *object–action* sequence. First select an object and then perform an action, as in the following:

1. Click the mouse on a menu title so that it stays open. Then click on the command that you want.
2. A single click on the left mouse button selects an item.
3. A double click on the left mouse button performs an action such as to end a mode or edit a selection.
4. To drag a selected item, click the left button, hold it down, and move the mouse. Release the left button when placed.
5. To end a mode or edit a selection, click the right mouse button once.
6. To repeat an action, double click the right button.

These mouse operations are summarized in Table A.1.

A.4 The Circuit Analysis Process

As an example, let us draw and analyze the pulse response of an *RLC* circuit as shown in Fig. A.4. The steps to draw and analyze a circuit are as follows:

1. Draw the circuit under Capture.
2. Select the mode of analysis under PSpice.
3. Simulate the circuit under PSpice A/D.
4. Display the results under Probe.

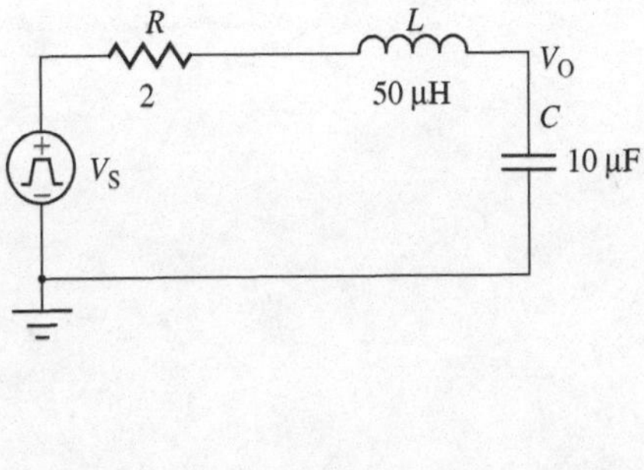

FIGURE A.4 An *RLC* circuit

A.5 Drawing the Circuit

The steps in drawing a circuit are as follows:

Step 1. Begin a new project.
Step 2. Get components from the Get New Part menu and place them on the drawing board.
Step 3. Rotate components as desired.
Step 4. Wire the components together.
Step 5. Label the components and add text as desired.
Step 6. Set attributes of the components.
Step 7. View the schematic.
Step 8. Save the circuit.

A.5.1 Beginning a New Project

Let us begin a new project to draw and analyze; call it Example B-3. Open the File menu, choose **New**, and then select **Project** as shown in Fig. A.5.

The New Project menu opens as shown in Fig. A.6. Give the file name of the new project (e.g., Example B-3), select **Analog or Mixed A/D**, and give the location of this new file (e.g., **C:\Rashid\PT3**). Next select **Create a blank project**, as shown in Fig. A.7.

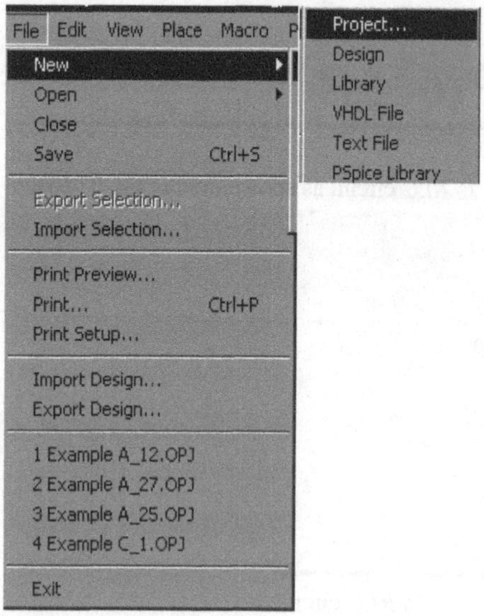

FIGURE A.5 File menu for a new project

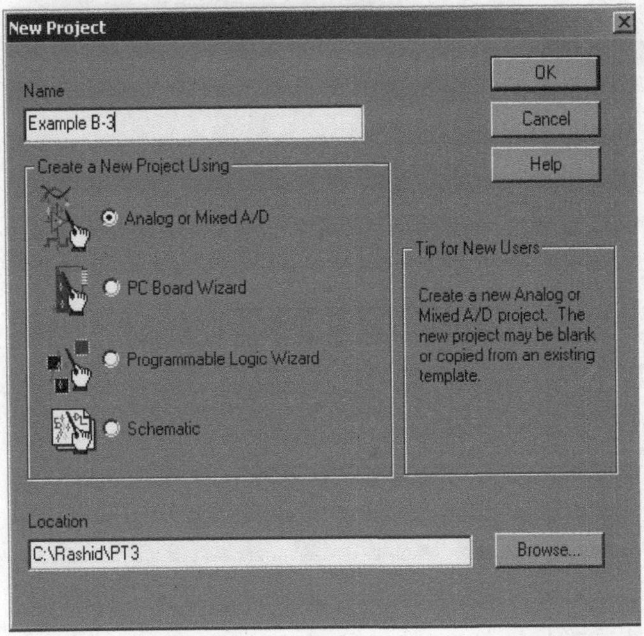

FIGURE A.6 Project menu

A.5.2 Getting and Placing Components

Let us start by placing a pulse source, a resistor, an inductor, a capacitor, and a ground on the drawing board. From the capture editor, you can place parts from the component libraries onto your schematic.

1. Use the Part command in the Place menu shown in Fig. A.8 or from the Place part menu on the right side. Alternately, choose the **Place part** command from the Draw menu on the right side. Choose **Browse** to browse the list of libraries or use the search command to enter the name of a known part, resistor R, as shown in Fig. A.9.
2. Choose **OK** or double click after selecting the part. The chosen part becomes the "current part" and is ready to be placed on your schematic.

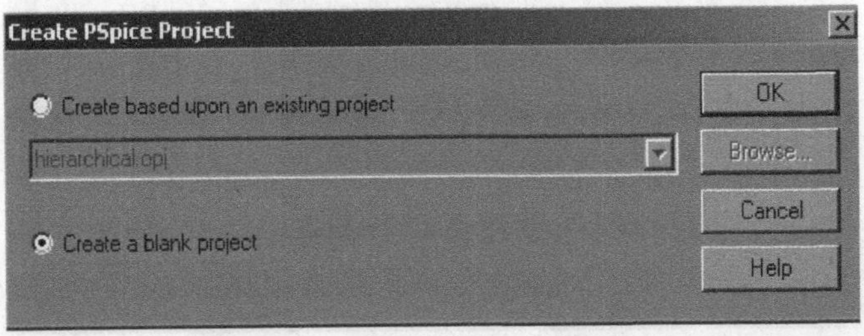

FIGURE A.7 Create a blank project

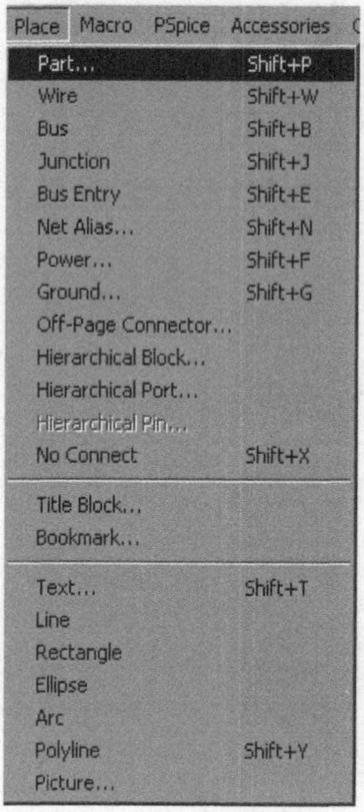

FIGURE A.8 Place menu

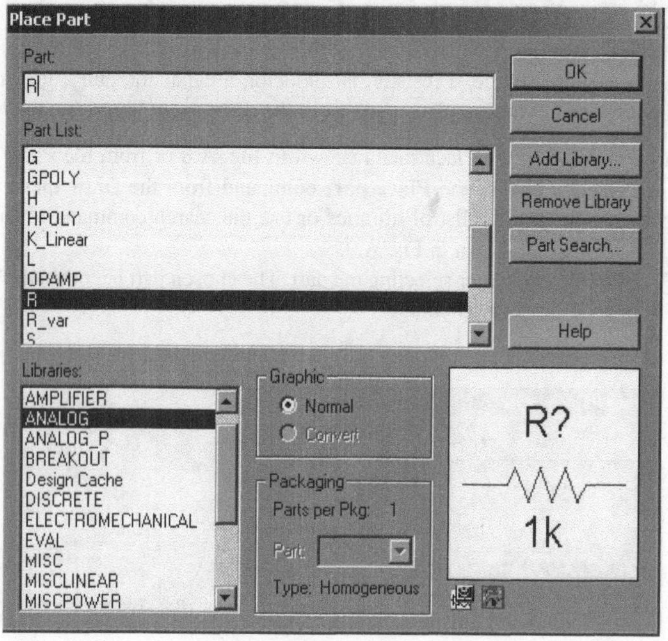

FIGURE A.9 Place part menu

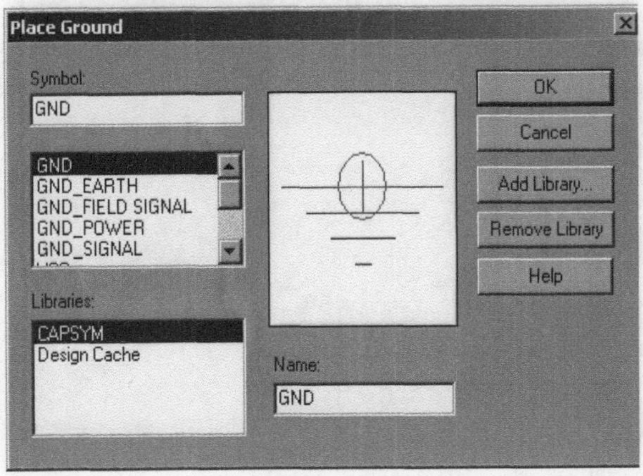

FIGURE A.10 Ground parts for an *RLC* circuit

3. The cursor is replaced by the shape of the part.
4. Click left to place an instance of the part; double click to place the part and end the mode or click right to end the mode without placing the part. To drag, point to an object, press and hold the left mouse button, and move the mouse. When the object is where you want it, release the mouse button.
5. To move a component, point to it, press and hold the left mouse button, and drag it to a new location.
6. To remove a component, select it, and choose **Delete** from the Edit menu.
7. Place a pulse source (VPULSE) from the **source.slb** library, a resistor (R), an inductor (L), and a capacitor (C) from the **analog.slb** library and a ground symbol **(GND)** from the Place menu or Place Ground (zero 0) menu on the right-side menu as shown in Fig. A.10. Arrange them as shown in Fig. A.11.

A.5.3 Rotating Components

Now rotate the capacitor so that it can be wired neatly into the circuit. Each time that you rotate a component, it turns clockwise 90°. The Flip command flips a selected object(s) to produce a mirror image of the object.

1. To rotate the capacitor (or other component), select it and choose **Rotate** from the Edit menu as shown in Fig. A.12. If you select an area of your schematic, the area is rotated around the center of the selection box. This is done as shown in Fig. A.13.
2. To flip a component, select it and choose **Flip** from the Edit menu.

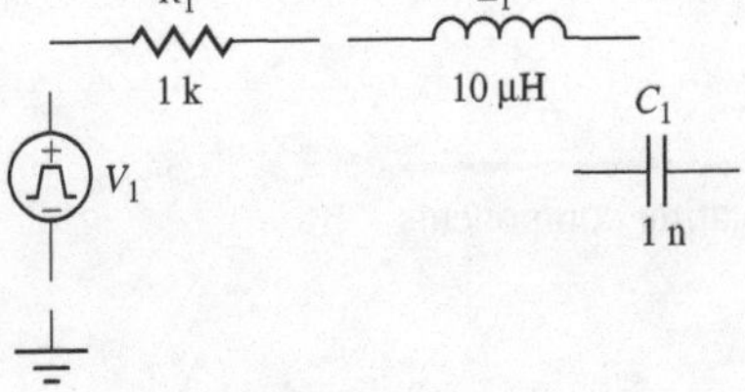

FIGURE A.11 Parts for an *RLC* circuit

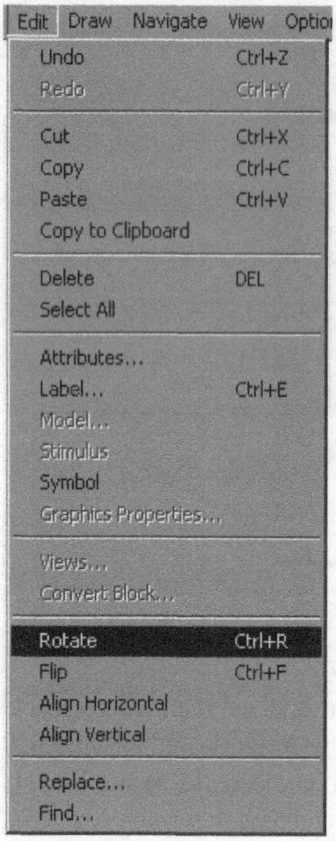

FIGURE A.12 Edit menu

▶ **HELP** If the Rotate command is dimmed, the capacitor is not selected. Try again by pointing to it so that the pointer becomes a hand and then click the left mouse button.

3. To deselect the selected capacitor (or other selected component), click it with the right mouse button or click an empty spot with the left mouse button.
4. To drag or rotate two or more components at once, first select them by drawing a rectangle around the components and then drag or rotate the rectangle.

▶ **HELP** To draw a rectangle around components, point above and beside one of the components that you want to select. Press and hold the left mouse button and drag diagonally until the rest of the components are in the rectangle that appears.

5. To deselect one of the selected components, click it with the right mouse button. To deselect everything on the marked rectangle, click an empty spot with the left mouse button.

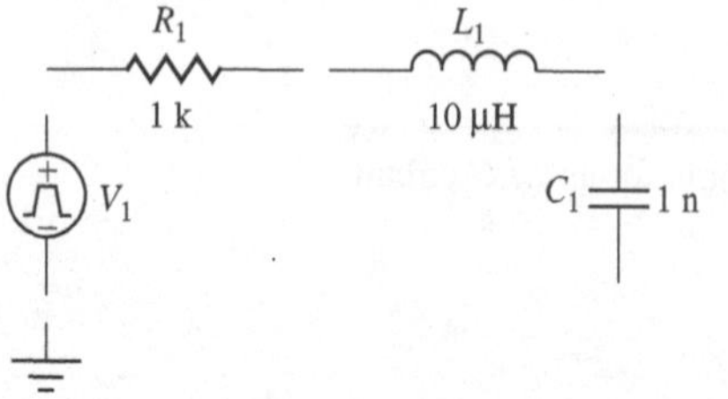

FIGURE A.13 Rotating components

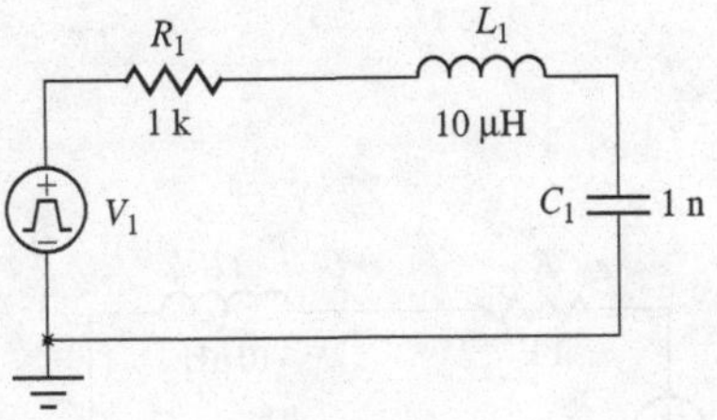

FIGURE A.14 Wiring circuit

A.5.4 Wiring Components

Once components are placed on the drawing board, you need to connect them. You can use the schematic editor to draw wires on your schematic and/or make vertices.

1. To draw a wire, choose **Wire** from the Place menu or Wire menu on the right side.
2. The cursor changes to a "wire" mode and is displayed as crosshairs. Click left to start drawing.
3. Move the mouse in any direction to extend the wire.
4. Click left again to end the wiring of a part. Click left once to start and click left again to stop. Continue until wiring of all parts is completed.
5. Press the **ESC** key or double click left or click right to end the wiring mode (the cursor will change from crosshairs to normal mode).

Now wire the circuit as shown in Fig. A.14.

▶ **TIP** To bring back the last command used (for wire), double click the right mouse button. In the lower right-hand corner, you will see the Wire command after the word *Cmd.*

A.5.5 Labeling Components and Adding Text

We can place a label on selected wires, bus segments, or ports. Wire and bus segments or ports may display multiple labels; however, all labels for a segment will contain the same text. Each component in a circuit can be labeled. We will assign levels R, L, C, V_S, and V_O.

1. To edit a label, select a wire, bus segment, or port for labeling.
2. Double click on the label to bring up the dialog box for **Display Properties**.
3. Enter the text for the label.
4. Click **OK**.
5. Change all levels R, L, C, V_S, and V_O as shown in Fig. A.15.

▶ **TIP** To move the level, left click on it and move to a desired location.

6. Next click on **Text** in the Place menu (see Fig. A.8).

You can place text anywhere on your schematic and size it to suit your needs.

1. To add text to your schematic, choose **Text** from the Place menu or **Place text** on the right-side menu.
2. In the dialog box, type in the desired text.
3. To change the font size, modify the **Font Size** shown in the dialog box to suit your needs.
4. Click **OK**.
5. Move the text to the desired location on the schematic and click left to place the text. Click right to end the mode.

Now place the output voltage, V_O as shown in Fig. A.16.

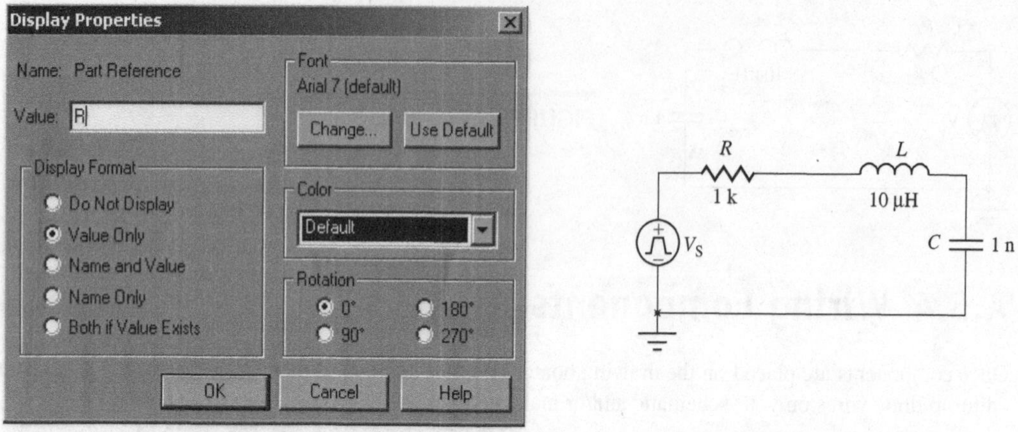

(a) **Display Properties menu** (b) ***RLC* circuit with desired labels**

FIGURE A.15 Labeling components

A.5.6 Setting Attributes

An attribute of a schematic item consists of a name and value pair.

1. To edit the properties of a selected object(s), select the object(s) to edit.
2. Choose **Properties** from the Edit menu or double click on the attribute text to bring up the Display Properties dialog box directly as shown in Fig. A.17.
3. Select an individual attribute: A dialog box appears in which you can enter a new value for the attribute. You can also select an entire part: A dialog box appears showing all attributes that can be edited for that part—for example, **Value Only**.

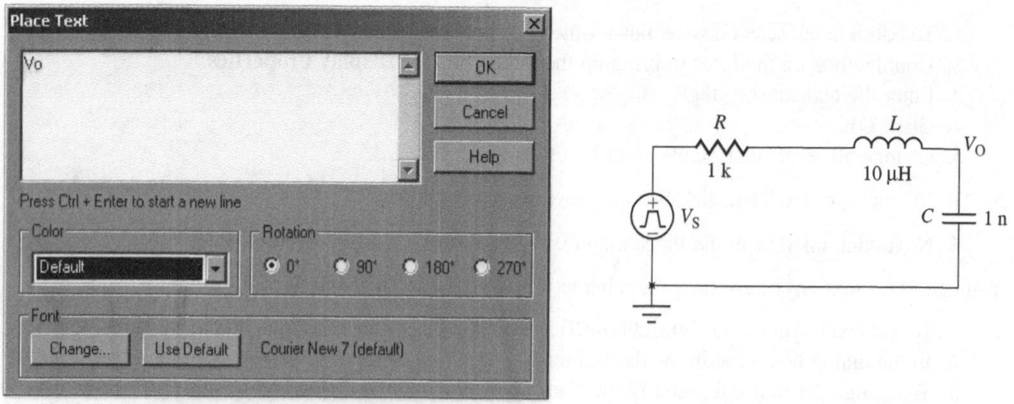

(a) **Place Text dialog box** (b) ***RLC* circuit with desired text**

FIGURE A.16 Adding text

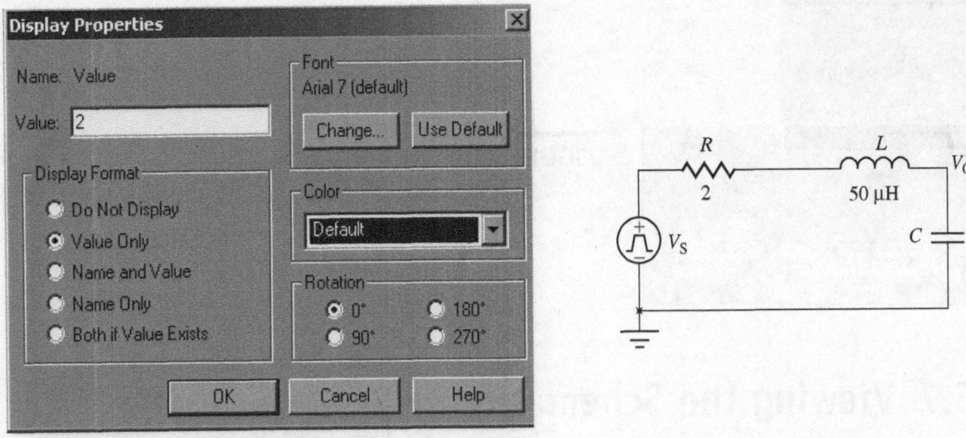

(a) Display Properties dialog box **(b) *RLC* circuit showing attributes**

FIGURE A.17 Setting attributes

4. Set $R = 2\ \Omega$, $L = 50\ \mu H$, and $C = 10\ \mu F$.

▶ **TIP** A quick way to set a component's value is to double click it.

▶ **HELP** Changes made to attributes in the schematic editor occur only on the particular part instance. Changes do not affect the underlying symbol in the library.

5. To change the value of an attribute, select it. The name and value should appear in the properties of the Edit menu.
   a. Change the value in the **Value** edit field and click **OK**.
   b. To delete an attribute, select it from the list and press **Delete**.
   c. To change whether the attribute name and/or value is shown on the schematic, select it and click **OK**.
   d. To add a new attribute, type the name and value in the **Edit** fields and press **OK**. Set the pulse source V_S as shown in Fig. A.18. Click on the attributes and then type V1 = 0, V2 = 1 V, TD (delay time) = 0, TR (rise time) = 1 ns, TF (fall time) = 1 ns, PW (pulse width) = 0.5 ms, PER (period) = 1 ms.
6. Click **OK** to incorporate the change.

▶ **TIP** Protected attributes, marked with an asterisk (*), cannot be changed using the Schematic Editor.

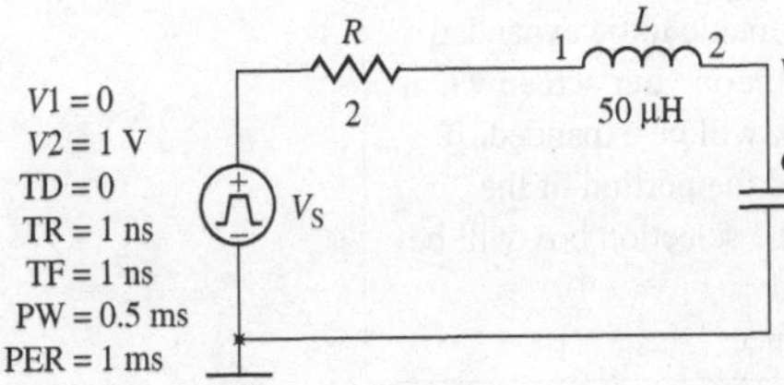

FIGURE A.18 Changing source attributes

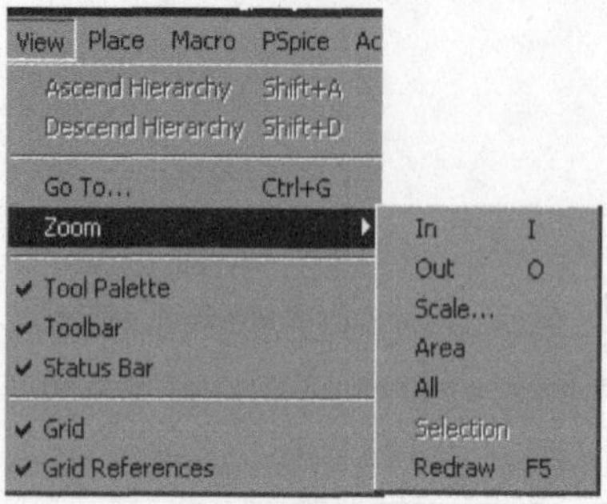

FIGURE A.19 View menu

A.5.7 Viewing the Schematic

You can change the viewing scale on your schematic via the View menu, shown in Fig. A.19. Table A.2 explains the functions of commands in the View menu.

A.5.8 Saving the Circuit File

Your schematic files are automatically saved into the current directory, unless otherwise specified.

- To save your schematic file, choose **Save** from the File menu, as shown in Fig. A.20, to save changes to the current file. You do not need to type a file name extension.

A circuit name must be a valid DOS file name. All schematic files are automatically given the file name extension OPJ. For example, the file name FIGB_3 will be saved as FIG B_3.OPJ.

TABLE A.2 View menu commands

Command	Function
Fit	Resets the viewing scale so that all parts, wires, and text can be seen on the screen.
In	Allows you to view an area on the schematic at closer range (i.e., magnify). After you select this command, a crosshair appears on the screen. Move the crosshair to the area you want to zoom in on.
Out	Changes the viewing scale so that you can view the schematic from a greater distance (i.e., view more of the schematic on the screen). After you select this command, a crosshair appears on the screen. Move the crosshair to define the center of the viewing area.
Area	Allows you to select a rectangular area on the schematic to be expanded to fill the screen. If you already have a selection box on your screen when you choose **Area**, the contents of the selection box will be expanded. If not, drag the mouse to form a selection box around the portion of the schematic you want to expand. The items within the selection box will be expanded to fill the screen.
Entire Page	Allows you to view the entire schematic page at once.

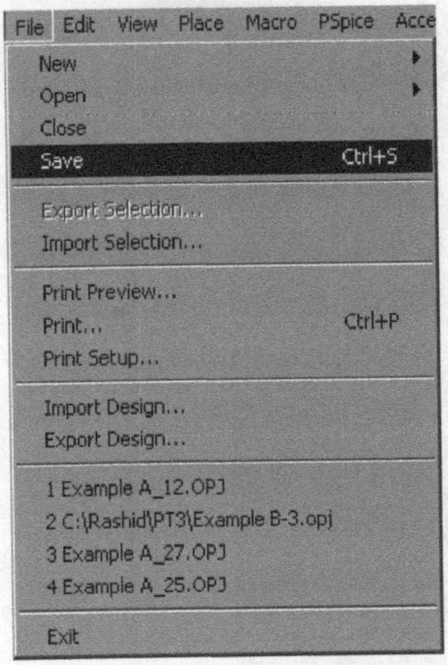

FIGURE A.20 File menu

A.6 Selecting the Type of Analysis

PSpice allows *DC sweep analysis, AC* (frequency response) *sweep analysis*, and *transient analysis*. The Setup command specifies which types of simulation analyses are enabled and allows the user to set up the parameters for selected analyses (described next).

When a signal is first applied to a circuit, there is a short-lived transient state before the circuit settles down to its usual responses. For the sample *RLC* circuit, we will conduct a transient analysis to examine the charging and discharging voltage of the capacitor.

1. Choose **New Simulation Profile** for a new simulation file or **Edit Simulation Profile** for an existing file from the PSpice menu, shown in Fig. A.21, and the Analysis Setup dialog box will open.
2. Enable transient analysis by clicking once in the **Enabled** space so that a check appears in the space, as shown in Fig. A.22.

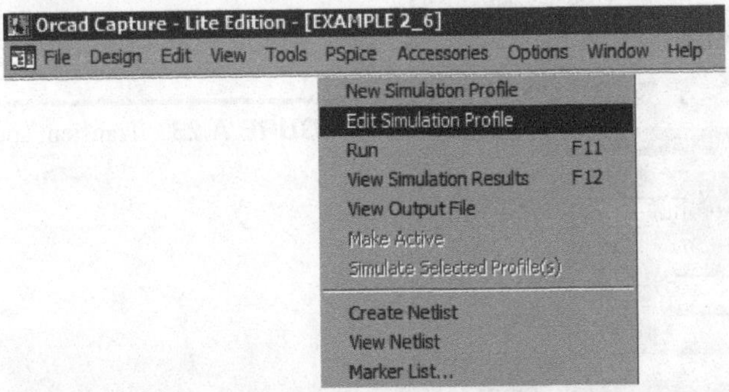

FIGURE A.21 PSpice setup for analysis

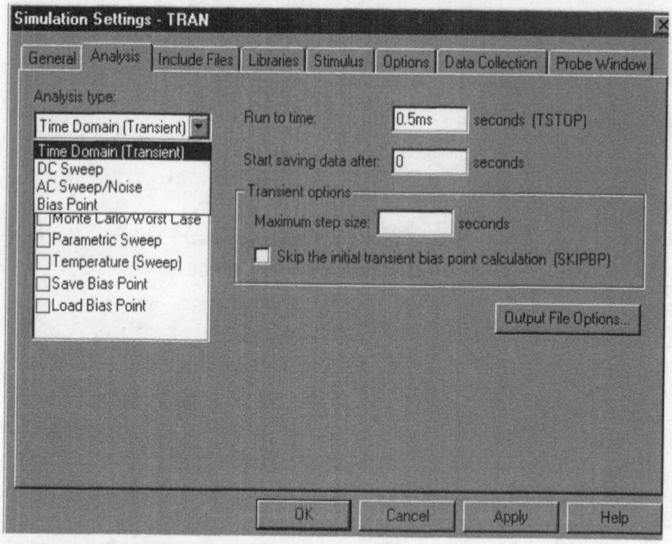

FIGURE A.22 Selecting Analysis type

3. Click on **Transient** to open the Transient specifications dialog box.
4. Type the print particulars, such as a **Print values** of 10 ns and a **Run to time** of 0.5 ms, as shown in Fig. A.23.
5. Click **OK**.

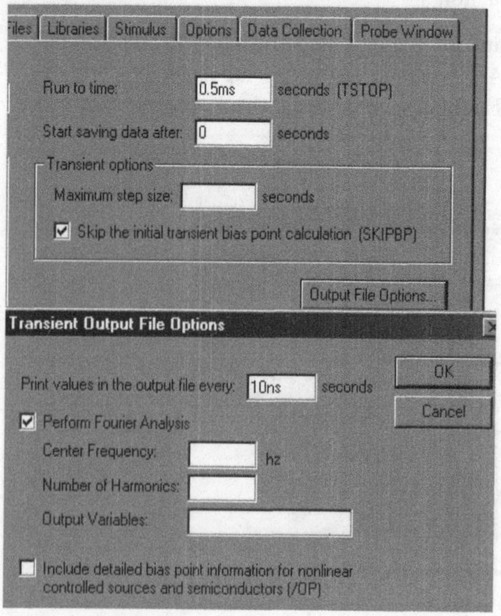

FIGURE A.23 Transient specifications

A.7 Simulation with PSpice

Now we are ready to simulate the circuit with PSpice. To begin the simulation, left click on **Simulate** in the Analysis menu.

During the circuit simulation process, PSpice creates and accesses a number of files. The first file is the *Schematics* file (*.SCH*), generated when a circuit drawn on the screen is saved. When the Schematics file is analyzed, three new files are generated: the *Circuit* file (*.CIR*), the *Netlist* file (*.NET*), and the *Alias* file (*.ALS*). The Circuit file (the master file) contains the *simulation directives* and references to the Netlist, Alias, and Model files. The Netlist file contains a Kirchhoff-like set of equations that lists parts and how they are connected; the equations relate the voltages and currents to the circuit elements through node numbers. The Alias file lists alternative names for circuit nodes. The Model file lists the characteristics and model statement of each component.

▶ **TIP** If there is any problem such as a missing attribute name or value, PSpice will indicate the error, and the simulation will be aborted. You can find the error message in the output file. For example, a message about an error in the file FIG1.1 will appear in FIG1_1.OUT.

When PSpice is run, each simulation directive in the Circuit (master) file specifies the information to be sent to the Output and Data files.

- The *Output* file (*.OUT*) is an ASCII file that holds the "audit trail" for the simulation. It contains a wide variety of information, including the original netlist, all output variables, and various tables.
- The *Data* file (*.DAT*) is sent to Probe, which uses the binary information to generate plots and graphs within the Probe window.

While a simulation is running, you can check the status of the simulation, as shown in Fig. A.24. When the simulation is completed, PSpice will display the message "Transient Analysis finished."

▶ **TIP** *Time step* is the internal simulation step for convergence to give a specified accuracy, whereas *Print step* (which was specified in the Transient specifications dialog box) is for printing or plotting the output variables.

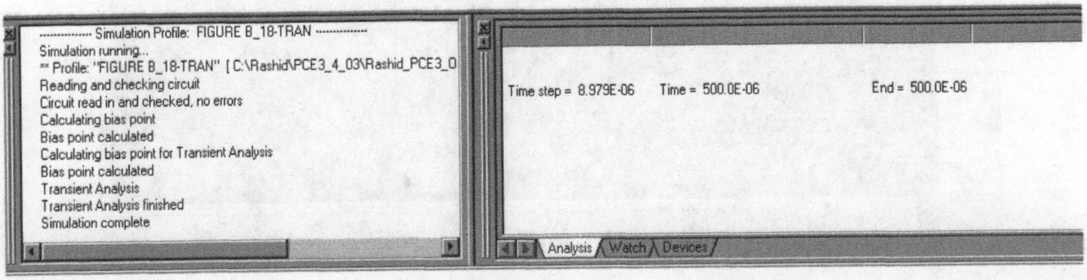

FIGURE A.24 PSpice analysis status box

A.8 Displaying the Results of a Simulation

Probe is a graphics postprocessor that allows the simulation results to be displayed in graphical form. After the calculations are completed, assuming no errors are found, PSpice sends the database it generated (FIG1_1.DAT) to Probe, which displays a graph.

1. To use Probe, choose **Run Probe** from the selecting Analysis type in Fig. A.22 menu (or use the F12 function key). As shown in Fig. A.25, Probe opens with an initial (default) graph in which the x-axis is automatically set to the transient variable, time.

▶ **TIP** To run Probe automatically after a simulation, choose **Automatically Run Probe After Simulation** from **Probe Setup** in the Analysis menu.

2. Choose **Trace** on the Probe menu. Within the large box at the top of the dialog box that appears, you will see the list of default *trace variables* that you can choose from, as shown in Fig. A.26. Before moving to the third step, specify the plot variables(s). For this example, we choose the following variables:
   a. V(R:2): V specifies the variable type (voltage, which is referenced to the ground), R specifies a component (resistor), and 2 specifies one end (node 2) of the component. A 1 indicates what was the left-hand side and a 2 what was the right-hand side when the component was initially placed horizontally. After one counterclockwise rotation, node 2 would be at the top.
   b. V(Vs:+): This is the voltage at the positive terminal of the voltage source V_S.
   c. I(R): This is the current flowing through resistance R from terminal 1 (left) to terminal 2 (right).

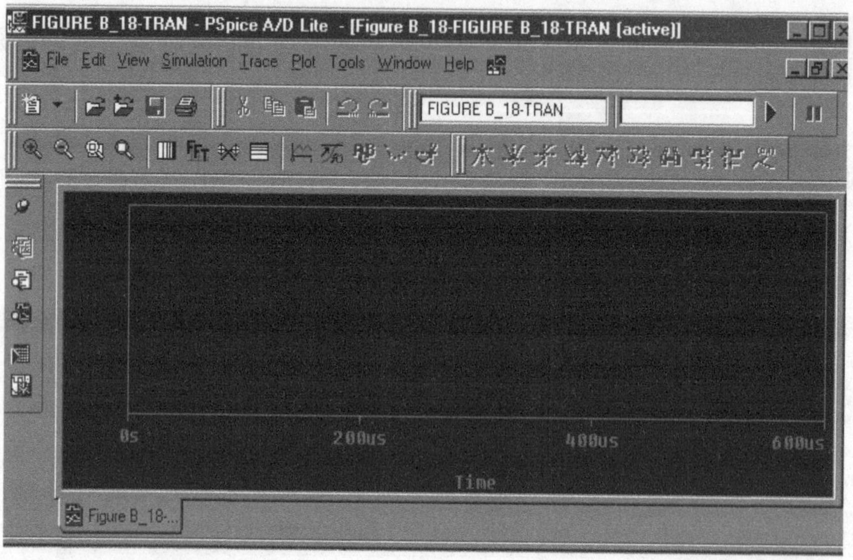

FIGURE A.25 Probe menu

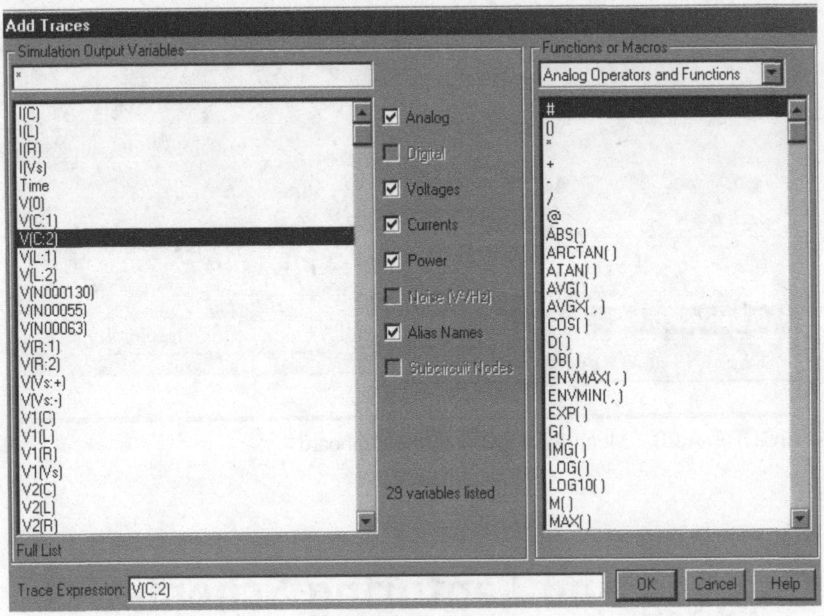

FIGURE A.26 Probe trace variables

3. After specifying the plot variables, type **V(C:2)** in the trace command box. This instructs Probe to display a graph of the voltage at terminal 2 of capacitor C, which is our desired output voltage. Click **OK** and you will see the plot of the output voltage, V(C:2), shown in Fig. A.27.
4. While copying to the clipboard and pasting to another document, the background and foreground colors can be changed as shown in the dialog box in Fig. A.28.

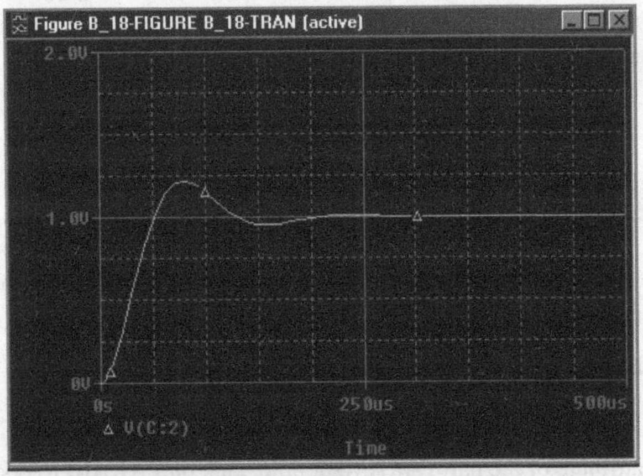

FIGURE A.27 Transient plot of the output voltage

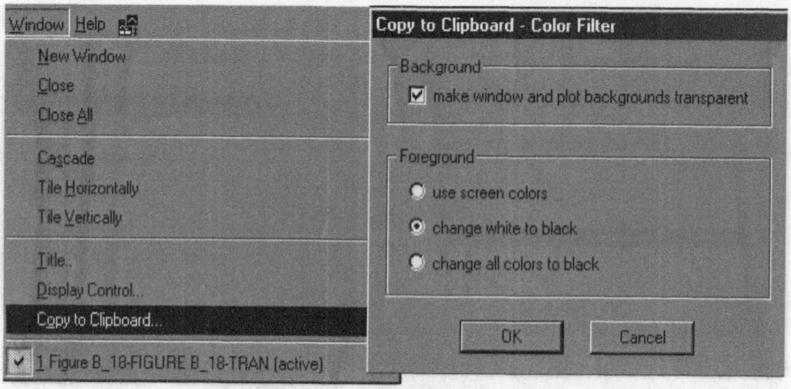

FIGURE A.28 Menu for copying to the clipboard

A.9 Copying and Capturing Schematics

You can use the Copy to Clipboard command from the Edit menu to copy one or more items from your schematic into another Windows program. With the mouse, select a rectangular area on the schematic to copy. Choose **Copy to Clipboard** from the Edit menu. Open the Windows program into which you want to paste the item(s). Use the Paste command from the newly opened Windows program to paste the item into the new file.

The Cut command deletes items from your schematic and places them in the Paste buffer. The Copy command copies items from your schematic to the Paste buffer. The Paste command places the contents of the Paste buffer on the schematic.

To cut or copy one or more items on your schematic, first select the item(s) on your schematic to be cut or copied. Then choose **Cut** or **Copy** from the Edit menu. To paste a cut or copied item onto your schematic, choose **Paste** from the Edit menu. Then place the cursor on the schematic where you'd like the cut or copied item to appear, and left click to place the item; right click to end the mode.

A.10 Varying Parameters

PSpice allows variation of component values or device parameters. We will plot the output voltage of our sample *RLC* circuit for three values: $R = 1\ \Omega$, $2\ \Omega$, and $10\ \Omega$.

1. To begin, get the part PARAM from the special.slb library file.
2. To change the value of R to a variable name such as RVAL, double click on the value of R and type **{RVAL}**.
3. To change the attribute of the part PARAM, double click on **PARAM** to open the Attributes dialog box, shown in Fig. A.29.
4. Choose **NAME1=RVAL** and **VALUE1=2** and click each time on **Save Attr**.

Parameters	
RVAL	10 k
CVAL	10 μF
LVAL	50 μH

(a) Symbol

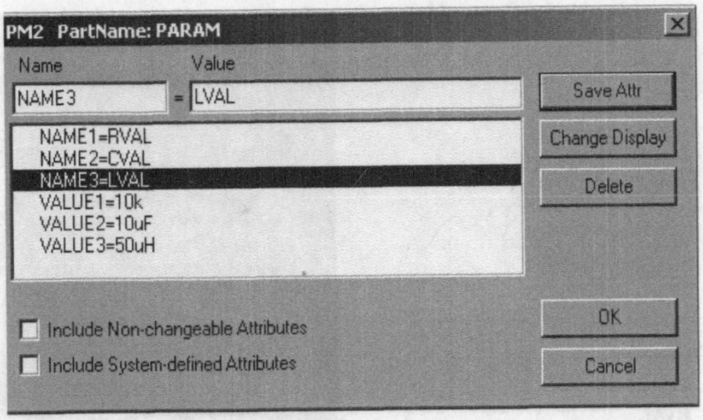

(b) Parameters

FIGURE A.29 Dialog box for PARAM

5. Choose **Setup** from the Analysis menu; then click once in the enabled space to enable parametric analysis.
6. Click on **Parametric** to open the box shown in Fig. A.30.
7. Enter or enable the Parametric specifications as follows: For Sweep variable, choose **Global parameter**; for Parameter name, type in **RVAL**; for Sweep Type, choose **Value list**; and for the values, type in 1 2 10.

▶ **TIP** You can define up to three variables in the same PARAM.

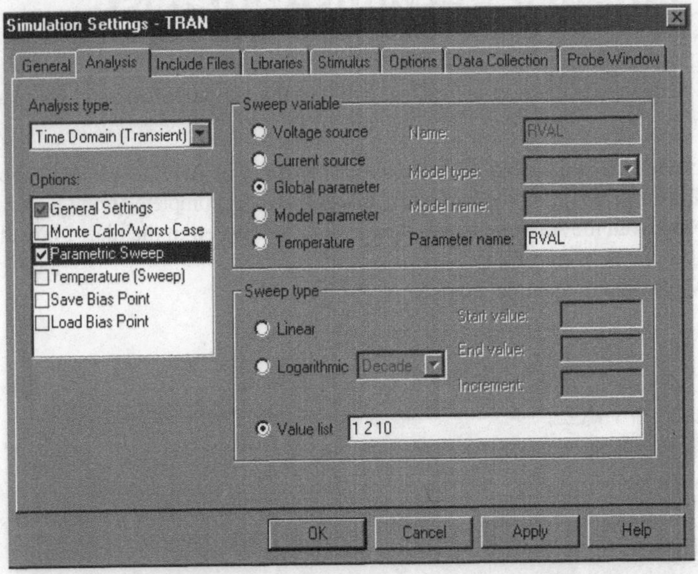

FIGURE A.30 Parametric specifications

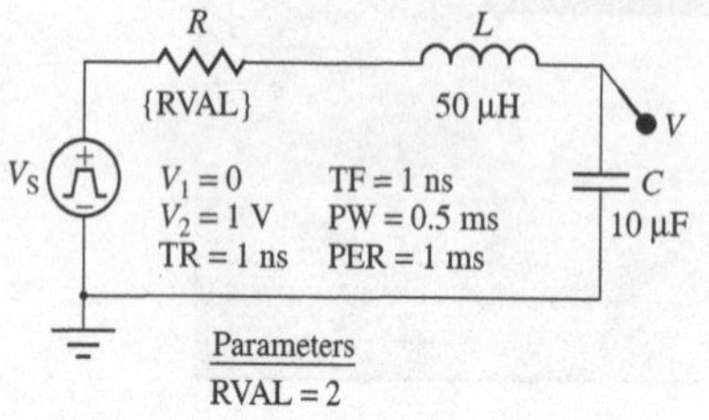

FIGURE A.31 *RLC* circuit with PARAM

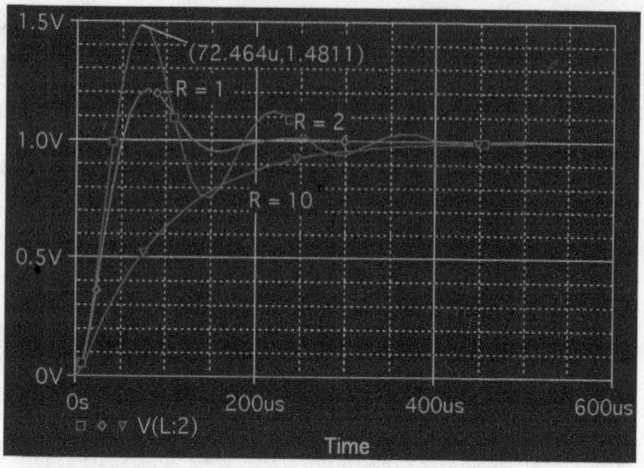

FIGURE A.32 Transient response for $R = 1\ \Omega, 2\ \Omega, 10\ \Omega$

8. Run the simulation for the complete circuit, shown in Fig. A.31. Use **Probe** to plot the output variable, V(C:2), as shown in Fig. A.32.

▶ **TIP** The labels R=1, R=2, and R=10 in Fig. A.32 were typed by choosing (in sequence) **Tools, Label,** and **Text** from the Probe menu. You can copy the Probe plot(s) to other Windows programs by choosing (in sequence) **Tools** and **Copy to Clipboard** from the Probe menu.

A.11 Frequency Response Analysis

As an example of frequency response analysis, we will plot the output voltage and phase angle of the example *RLC* circuit for three values: $R = 1\ \Omega$, $2\ \Omega$, and $10\ \Omega$.

1. To begin, change the attributes of the source V_S, shown in Fig. A.17, to 1 V AC by selecting the part and changing the attributes in the Part Name dialog box. The complete circuit is shown in Fig. A.33.
2. Choose **Setup** from the Analysis menu; then enable **AC Sweep** in the dialog box that appears (see Fig. A.22).
3. Click on **AC Sweep** to open the box shown in Fig. A.34.

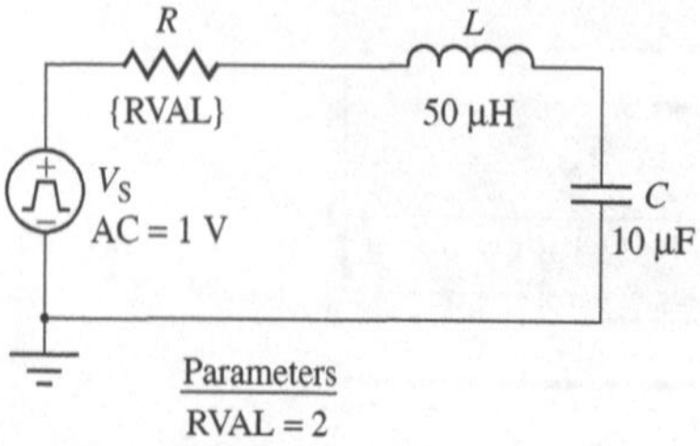

FIGURE A.33 *RLC* circuit for frequency response

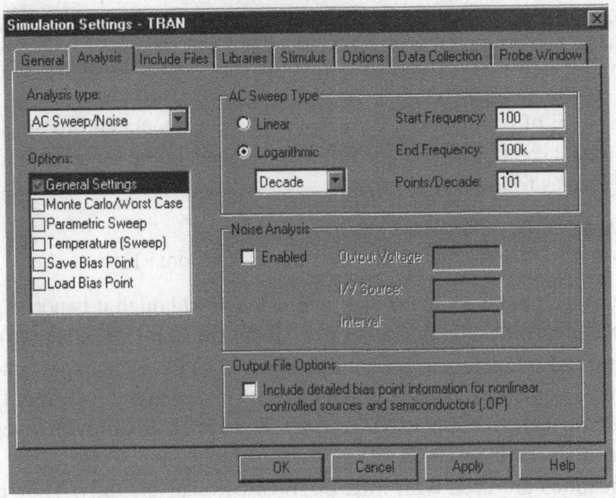

FIGURE A.34 Dialog box for frequency response analysis

4. Enter or enable the **AC Sweep** specifications as follows: For AC **Sweep Type**, choose **Decade**; for Start Frequency, type in 100; for End Frequency, type in 100k; for Points/Decade, type in 101.
5. Run the simulation for the complete circuit, shown in Fig. A.33. Use **Probe** to plot the output voltage, V(C:2), and the phase of the output voltage, VP(C:2), as shown in Fig. A.35.

▶ **TIP** If you add DB, as in VDB(C:2), you will get the output voltage in decibels.

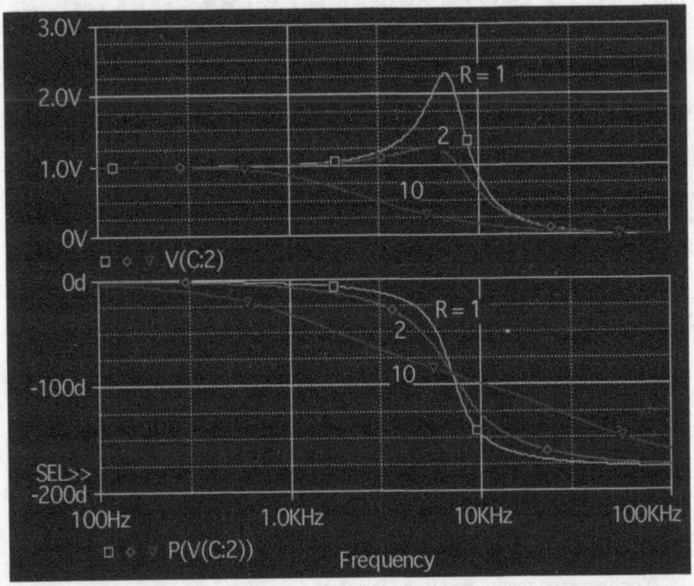

FIGURE A.35 Frequency response for $R = 1\ \Omega$, $2\ \Omega$, and $10\ \Omega$

A.12 Modeling Devices and Elements

A model that specifies a set of parameters for an element can be generated in PSpice using the .MODEL command. The same model can be used by one or more elements in the same circuit. The general form of the model statement [1] is as follows:

```
.MODEL  MNAME  TYPE (P1=A1 P2=A2
+       P3=A3 . . . PN=AN  [<tolerance specifications>])
```

MNAME is the name of the model and must start with a letter. Although it is not required, it is advisable to use the symbol of the element as the first letter (e.g., R for resistor, L for inductor). P1, P2, . . ., PN are the element parameters, and A1, A2, . . ., AN are their values. TYPE is the type name of the element and must be one of the types shown in Table A.3. An element must have the correct type name; that is, a resistor must have the type name RES, not IND or CAP. However, a circuit with several model names can include more than one model of the same type.

Tolerance specifications are used with .MC or .WORSE analysis only. They may be appended to each parameter using the following format:

```
[DEV/<distribution name> <value in % from 0 to 9>]
[LOT/<distribution name> <value in % from 0 to 9>]
```

where <distribution name> is either UNIFORM (in which case uniformly distributed deviations are generated over the range of ±<value>) or GAUSS (in which case deviations with Gaussian distribution are generated over the range ±4 and <value> specifies the ±1 deviation).

Following are some sample model statements:

```
.MODEL RLOAD  RES (R=1 TC1=0.02 TC2=0.005)
.MODEL RLOAD  RES (R=1 DEV/GAUSS 0.5% LOT/UNIFORM 10%)
.MODEL CPASS  CAP (C=1 VC1=0.01 VC2=0.002 TC1=0.02 TC2=0.005)
```

TABLE A.3 Type names of elements

Type Name	Element
RES	Resistor
CAP	Capacitor
D	Diode
IND	Inductor
NPN	*npn* bipolar junction transistor
PNP	*pnp* bipolar junction transistor
NJF	*n*-channel junction FET
PJF	*p*-channel junction FET
NMOS	*n*-channel MOSFET
PMOS	*p*-channel MOSFET
GASFET	*n*-channel GaAs MESFET
VSWITCH	Voltage-controlled switch
ISWITCH	Current-controlled switch
CORE	Nonlinear magnetic core (transformer)

```
.MODEL    LFILTER    IND (L=1 IL1=0.1 IL2=0.002 TC1=0.02 TC2=0.005)
.MODEL    DNOM       D (IS=1E-9)
.MODEL    DLOAD      D (IS=1E-9 DEV 0.5% LOT 10%)
.MODEL    QMOD       NPN (BF=50 IS=1E-9)
```

A.12.1 Resistors

The symbol for a resistor is R. The name of a resistor must start with R, and the model statement takes the following general form:

```
R<name>  N+  N-  RNAME  RVALUE
```

A resistor does not have a polarity, and so the order of the nodes does not matter. However, the current is assumed to flow from the node designated by N+ as the positive node through the resistor to the node designated by N− as the negative node. RNAME is the model name that defines the parameters of the resistor. RVALUE is the nominal value of the resistance.

▶ **NOTE** Some versions of PSpice and SPICE do not make this assumption and thus do not allow you to refer to currents through a resistor. For example, such versions will not allow you to use a notation such as I(RL) to indicate the current through RL.

The model parameters for resistors are shown in Table A.4. If RNAME is omitted, RVALUE is the resistance in ohms; RVALUE can be positive or negative but must *not* be zero. If RNAME is included and TCE is *not* specified, the resistance as a function of temperature is calculated from

$$\text{RES} = \text{RVALUE} * \text{R} * [1 + \text{TC1} * (\text{T} - \text{T0}) + \text{TC2} * (\text{T} - \text{T0})^2]$$

If RNAME is included and TCE is specified, the resistance as a function of temperature is calculated from

$$\text{RES} = \text{RVALUE} * \text{R} * 1.01^{\text{TCE} * (\text{T}-\text{T0})}$$

where T and T0 are the operating temperature and the room temperature, respectively, in degrees Celsius.

Following are some sample resistor statements:

```
RL        5     6     5K
RLOAD     10    13    ARES 1MEG
.MODEL    RMOD        RES (R=1  TC1=0.02  TC2=0.005)
RINPUT    13    17    RRES  2K
.MODEL    ARES        RES (R=1  TCE=1.5)
```

TABLE A.4 Model parameters for resistors

Name	Meaning	Units	Default
R	Resistance multiplier		1
TC1	Linear temperature coefficient	$°C^{-1}$	0
TC2	Quadratic temperature coefficient	$°C^{-2}$	0
TCE	Exponential temperature coefficient	%/°C	0

TABLE A.5 Model parameters for capacitors

Name	Meaning	Units	Default
C	Capacitance multiplier		1
VC1	Linear voltage coefficient	V^{-1}	0
VC2	Quadratic voltage coefficient	V^{-2}	0
TC1	Linear temperature coefficient	$°C^{-1}$	0
TC2	Quadratic temperature coefficient	$°C^{-2}$	0

A.12.2 Capacitors

The symbol for a capacitor is C. The name of a capacitor must start with C, and the model statement takes the following general form:

```
C<name>  N+  N-  CNAME  CVALUE  IC=VO
```

N+ is the positive node, and N− is the negative node. The voltage of node N+ is assumed to be positive with respect to node N−, and the current flows from node N+ through the capacitor to node N−. CNAME is the model name, and CVALUE is the nominal value of the capacitor. IC defines the initial (time zero) voltage of the capacitor, VO.

The model parameters for capacitors are shown in Table A.5. If CNAME is omitted, CVALUE is the capacitance in farads; CVALUE can be positive or negative but must *not* be zero. If CNAME is included, the capacitance as a function of voltage and temperature is calculated from

$$\text{CAP} = \text{CVALUE} * \text{C} * (1 + \text{VC1} * \text{V} + \text{VC2} * \text{V}^2)[1 + \text{TC1} * (\text{T} - \text{T0}) + \text{TC2} * (\text{T} - \text{T0})^2]$$

where T and T0 are the operating temperature and the room temperature, respectively, in degrees Celsius.

Following are some sample capacitor statements:

```
C1        2     6     0.01UF
CLOAD     10    13    10PF   IC=1.5V
CINPUT    14    16    DCAP   5PF
CX        10    25    DCAP   10NF   IC=3.5V
.MODEL    DCAP        CAP (C=1  VC1=0.01  VC2=0.002  TC1=0.02  TC2=0.005)
```

▶ **TIP** The initial conditions (if any) apply only if you specify the UIC (use initial condition) option under the .TRAN command.

A.12.3 Inductors

The symbol for an inductor is L. The name of an inductor must start with L, and the model statement takes the following general form:

```
L<name> N+  N-  LNAME  LVALUE  IC=IO
```

N+ is the positive node, and N− is the negative node. The voltage of N+ is assumed to be positive with respect to node N−, and the current flows from node N+ through the inductor to node N−. LNAME is the model name, and LVALUE is the nominal value of the inductor. IC defines the initial (time zero) current of the inductor IO.

TABLE A.6 Model parameters for inductors

Name	Meaning	Units	Default
L	Inductance multiplier		1
IL1	Linear current coefficient	A^{-1}	0
IL2	Quadratic current coefficient	A^{-2}	0
TC1	Linear temperature coefficient	$°C^{-1}$	0
TC2	Quadratic temperature coefficient	$°C^{-2}$	0

The model parameters for inductors are shown in Table A.6. If LNAME is omitted, LVALUE is the inductance in henries; LVALUE can be positive or negative but must *not* be zero. If LNAME is included, the inductance as a function of current and temperature is calculated from

$$\text{IND} = \text{LVALUE} * \text{L} * (1 + \text{IL1} * \text{I} + \text{IL2} * \text{I}^2)[1 + \text{TC1} * (\text{T} - \text{T0}) + \text{TC2} * (\text{T} - \text{T0})^2]$$

where T and T0 are the operating temperature and the room temperature, respectively, in degrees Celsius.

Following are some sample inductor statements:

```
LE        3      5    5MH
LLOAD     10     14   2UH   IC=0.1MA
LLINE     12     14   LMOD  2MH
LCHOKE    15     29   LMOD  5UH  IC=0.4A
.MODEL    LMOD        IND (L=1  IL1=0.1  IL2=0.002 TC1=0.02  TC2=0.005)
```

▶ **TIP** The initial conditions (if any) apply only if you specify the UIC (use initial condition) option using the .TRAN command.

A.12.4 Diodes

The model statement for a diode has the following general form:

```
.MODEL  DNAME  D (P1=A1 P2=A2 P3=A3 . . PN=AN)
```

DNAME is the name of the model and can begin with any character, but its size is normally limited to eight characters. D is the type symbol for diodes. P1, P2, . . ., PN are the model parameters, and A1, A2, . . ., AN are their values. The model parameters are listed in Table A.7.

A.12.5 Bipolar Transistors

The model statement for *npn* transistors has the following general form:

```
.MODEL QNAME NPN (P1=A1 P2=A2 P3=A3 . . . PN=AN)
```

The general form of the model statement for *pnp* transistors is

```
.MODEL QNAME PNP (P1=A1 P2=A2 P3=A3 . . . PN=AN)
```

QNAME is the name of the BJT model, and NPN and PNP are the type symbols for *npn* and *pnp* transistors, respectively. QNAME can begin with any character, but its size is normally limited to eight characters.

TABLE A.7 Model parameters for diodes

Name	Model Parameter	Units	Default	Typical
IS	Saturation current (I_S)	A	1E-14	1E-14
RS	Parasitic resistance (R_S) (series lead and bulk resistance)	Ω	0	10
N	Emission coefficient (n)		1	1
TT	Transit time	s	0	0.1 ns
CJO	Zero-bias *pn* capacitance (C_{jo})	F	0	2 pF
VJ	Junction potential (V_j)	V	1	0.6
M	Junction grading coefficient		0.5	0.5
EG	Activation energy (0.67 for Shockley and 1.11 for silicon)	eV	1.11	1.11
XTI	IS temperature exponent		3	3
KF	Flicker noise coefficient		0	
AF	Flicker noise exponent		1	
FC	Forward-bias depletion capacitance coefficient		0.5	
BV	Reverse breakdown voltage	V	∞	50
IBV	Reverse breakdown current (reverse current at BV)	A	1E-10	

P1, P2, . . ., PN are the parameters, and A1, A2, . . ., AN are their values. Table A.8 shows the model parameters for BJTs.

A.12.6 JFETs

The model statement for an *n*-channel JFET has the following general form:

```
.MODEL JNAME NJF (P1=A1 P2=A2 P3=A3 . . . PN=AN)
```

The general form of the model statement for a *p*-channel JFET is

```
.MODEL JNAME PJF (P1=A1 P2=A2 P3=A3 . . . PN=AN)
```

JNAME is the name of the model, and NJF and PJF are the type symbols for *n*-channel and *p*-channel JFETs, respectively. P1, P2, . . ., PN are the parameters, and A1, A2, . . ., AN are their values. Table A.9 lists the model parameters for JFETs.

A.12.7 MOSFETs

The symbol for a metal oxide semiconductor field-effect transistor (MOSFET) is M. The name of a MOSFET must start with M, and the model statement takes the following general form:

```
M<name>  ND  NG  NS  NB  MNAME
+        [L=<value>] [W=<value>]
```

where ND, NG, NS, and NB are the drain, gate, source, and bulk (or substrate) nodes, respectively. MNAME is the model name. The positive current is the current flowing into a terminal; that is, the current flows from the drain node through the device to the source node in an *n*-channel MOSFET.

TABLE A.8 Model parameters for BJTs

Name	Model Parameter	Units	Default	Typical
IS	*pn* saturation current	A	1E-16	1E-16
BF	Ideal maximum forward beta		100	100
NF	Forward current emission coefficient		1	1
VAF(VA)	Forward Early voltage	V	∞	100
IKF(IK)	Corner for forward beta high-current roll-off	A	∞	10 MA
NE	Base–emitter leakage emission coefficient		1.5	2
BR	Ideal maximum reverse beta		1	0.1
NR	Reverse current emission coefficient		1	
IKR	Corner for reverse beta high-current roll-off	A	∞	100 MA
RB	Zero-bias (maximum) base resistance (base spreading resistance)	Ω	0	100
RE	Emitter ohmic resistance	Ω	0	1
RC	Collector ohmic resistance (collector lead and bulk resistance)	Ω	0	10
CJE	Base–emitter zero-bias *pn* capacitance (C_{jeo})	F	0	2 pF
VJE(PE)	Base–emitter built-in potential (V_{jbe})	V	0.75	0.7
MJE(ME)	Base–emitter *pn* grading factor		0.33	0.33
CJC	Base–collector zero-bias *pn* capacitance ($C_{\mu o}$)	F	0	pF
VJC(PC)	Base–collector built-in potential (V_{jc})	V	0.75	0.5
MJC(MC)	Base–collector *pn* grading factor		0.33	0.33
CJS(CCS)	Collector–substrate zero-bias *pn* capacitance	F	0	2 pF
MJS(MS)	Collector–substrate *pn* grading factor			0
FC	Forward-bias depletion capacitor coefficient		0.5	
TF	Ideal forward transit time	s	0	0.1 ns
TR	Ideal reverse transit time	s	0	10 ns
EG	Bandgap voltage (barrier height)	eV	1.11	1.11
XTI(PT)	IS temperature effect exponent (temperature coefficient for IS)		3	
KF	Flicker noise coefficient		0	6.6E-16
AF	Flicker noise exponent		1	1

TABLE A.9 Model parameters for JFETs

Name	Model Parameter	Units	Default	Typical
VTO	Threshold voltage	V	−2	−2
BETA	Transconductance coefficient	A/V^2	1E-4	1E-3
LAMBDA	Channel-length modulation	V^{-1}	0	1E-4
RD	Drain ohmic resistance	Ω	0	100
RS	Source ohmic resistance	Ω	0	100
IS	Gate *pn* saturation current	A	1E-14	1E-14
PB	Gate *pn* potential	V	1	0.6
CGD	Gate–drain zero-bias *pn* capacitance (C_{gdo})	F	0	5 pF
CGS	Gate–source zero-bias *pn* capacitance (C_{gso})	F	0	1 pF
FC	Forward-bias depletion capacitance coefficient		0.5	
KF	Flicker noise coefficient		0	
AF	Flicker noise exponent		1	

TABLE A.10 Model parameters for MOSFETs

Name	Model Parameter	Units	Default	Typical
LEVEL	Model type (Shichman-Hodges)		1	
L	Channel length	m	DEFL	
W	Channel width	m	DEFW	
VTO	Zero-bias threshold voltage	V	0	0.1
KP	Transconductance	A/V^2	2E-5	2.5E-5
GAMMA	Bulk threshold parameter	$V^{1/2}$	0	0.35
LAMBDA	Channel-length modulation (LEVEL=1)	V^{-1}	0	0.02
RD	Drain ohmic resistance	Ω	0	10
RS	Source ohmic resistance	Ω	0	10
IS	Bulk *pn* saturation current	A	1E-14	1E-15
JS	Bulk *pn* saturation current/area	A/m^2	0	1E-8
PB	Bulk *pn* potential	V	0.8	0.75
CBD	Bulk-drain zero-bias *pn* capacitance	F	0	5 pF
CBS	Bulk-source zero-bias *pn* capacitance	F	0	2 pF
CJ	Bulk *pn* zero-bias bottom capacitance/length	F/m^2	0	
MJ	Bulk *pn* bottom grading coefficient		0.5	
FC	Bulk *pn* forward-bias capacitance coefficient		0.5	
CGSO	Gate–source overlap capacitance/channel width	F/m	0	
CGDO	Gate–drain overlap capacitance/channel width	F/m	0	
CGBO	Gate-bulk overlap capacitance/channel length	F/m	0	
NSUB	Substrate doping density	$1/cm^3$	0	
TOX	Oxide thickness	m	∞	
UO	Surface mobility	$cm^2/V \cdot s$	600	
KF	Flicker noise coefficient		0	1E-26
AF	Flicker noise exponent		1	1.2

The model statement for an *n*-channel MOSFET has the following general form:

```
.MODEL   MNAME   NMOS (P1=A1 P2=A2 P3=A3 . . . PN=AN)
```

The general form of the model statement for a *p*-channel MOSFET is

```
.MODEL   MNAME   PMOS (P1=A1 P2=A2 P3=A3 . . . PN=AN)
```

NMOS and PMOS are the type symbols for *n*-channel and *p*-channel MOSFETs, respectively. MNAME can begin with any character, but its size is normally limited to eight characters. P1, P2, . . ., PN are the parameters, and A1, A2, . . ., AN are their values. Table A.10 lists the model parameters for MOSFETs.

A.13 Creating Netlists

Once you have drawn a schematic, you can create a netlist to use with other PSpice/SPICE software applications. After drawing your schematic, choose **Create Netlist** from the Analysis menu, shown in Fig. A.36. The netlist for Figs. A.31 and A.33 follows. It can be found in the file EXA-25.NET.

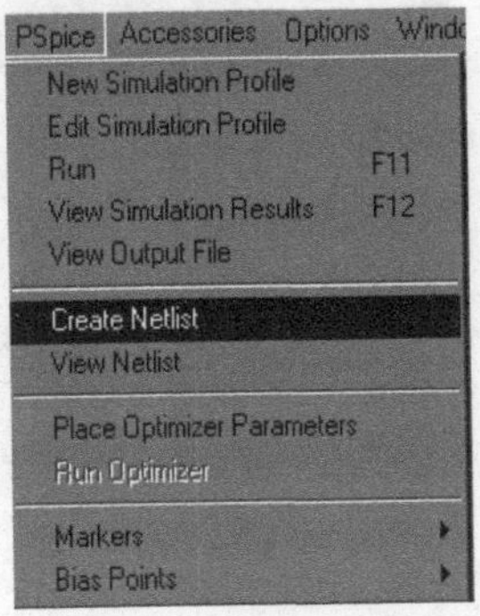

FIGURE A.36 Dialog box for creating netlist

```
* Schematics Netlist
C_C     0 $N_0001  10uF                 ; C is connected between nodes 0 and
                                        ; $N_0001
L_L     $N_0002 $N_0001  50uH           ; L is connected between nodes $N_0002
                                        ; and $N_0001
R_R     $N_0003 $N_0002 {RVAL}          ; R is connected between nodes $N_0003
                                        ; and $N_0002
V_Vs    $N_0003 0  AC  1V               ; Vs is connected between nodes $N_0003
                                        ; and 0
+PULSE 0 1V 0 1ns 1ns 0.5ms 1ms         ; Pulse voltage specifications
```

▶ **TIP** When you open a schematic file (e.g., EX5-1.SCH) for the first time, PSpice may indicate an error with the message "Not Finding Netlist." If this happens, first create the netlist by choosing **Create Netlist** from the Analysis menu, shown in Fig. A.36.

If the commands for transient analysis and AC analysis are included, the netlist for the PSpice circuit file becomes as follows:

```
Frequency Response ; The first line is the title line; PSpice always
                   ; ignores this statement
* Circuit Description
C_C     0 $N_0001  10uF                 ; C is connected between nodes 0 and
                                        ; $N_0001
L_L     $N_0002 $N_0001  50uH           ; L is connected between nodes $N_0002
                                        ; and $N_0001
R_R     $N_0003 $N_0002 {RVAL}          ; R is connected between nodes $N_0003
                                        ; and $N_0002
* Source Descriptions for both ac and pulse sources
V_Vs    $N_0003 0  AC  1V               ; Vs is connected between nodes $N_0003
                                        ; and 0
+ PULSE 0 1V 0 1ns 1ns 0.5ms 1ms        ; Pulse voltage specifications
* Analysis Descriptions for both ac and pulse sources
.AC  DEC 101 100HZ 100KHZ               ; ac analysis from 100 Hz to 100 kHz
                                        ; with decade increments of 101 points
                                        ; per decade
.TRAN  10ns 0.5ms                       ; Transient analysis from 0 to 0.5 ms
                                        ; with 10 ns printing/plotting interval
.PROBE                                  ; Graphics post-processor
.END                                    ; This is the last line and must always
                                        ; be included
```

TABLE A.11 Circuit elements and sources

Circuit Elements and Sources	First Letter	Model Type
Bipolar junction transistor	B	NPN/PNP
Capacitor	C	CAP
Current-controlled current source	F	
Current-controlled switch	W	VSWITCH
Current-controlled voltage source	H	
Diode	D	D
Exponential source	EXP	
GaAs MES field-effect transistor	B	GASFET
Ground	AGND	
Independent current source	I	
Independent DC voltage source	V	VDC
Inductor	L	IND/CORE
Junction field-effect transistor	J	NJF/PJF
MOS field-effect transistor	M	NMOS/PMOS
Mutual inductors (transformer)	K	
Piecewise linear source	PWL	
Polynomial source	POLY(n)	
Pulse voltage source	PULSE	VPULSE
Resistor	R	RES
Single-frequency frequency-modulation source	SFFM	
Sinusoidal voltage source	SIN	VSIN
Transmission line	T	
Voltage-controlled current source	G	
Voltage-controlled switch	S	VSWITCH
Voltage-controlled voltage source	E	EVALUE

The first letter and the model type for each element name are listed in Table A.11. The PSpice/SPICE commands are listed in Table A.12.

A.14 Adding Library Files

The directory of the PSpice circuit simulator (default name C:\MSIMEV61) has a subdirectory LIB, which contains library files of schematics and device models. This library has only a limited number of transistor schematics and models. If you want to use devices or elements with different model parameters, use elements or devices from the symbol library BREAKOUT.SLB, as shown in Fig. A.37. Use Notepad to modify the model statement in the BREAKOUT.LIB file—for example,

```
C:\OrcadLite\Capture\library\PSpice/Breakout.Lib
```

You can either use the library file RASHID.LIB that comes with the book or create your own library file containing the model statement of a device with the same model name as in the PSpice schematic—for example,

```
.MODEL    Q2N2222    Q (BF=100)                ; BJT Q2N2222
.MODEL    D1N4148    D (IS=10E-15 BV=100)      ; Diode D1N4148
```

This way you keep the same model name and use the PSpice schematic symbols, but change the model parameters.

TABLE A.12 PSpice/SPICE commands

Analysis or Function	Command
Absolute value (operator)	ABS
AC/frequency analysis	.AC
DC operating point	.OP
DC sweep	.DC
Difference (operator)	DIF
End of subcircuit	.ENDS
Fourier analysis	.FOUR
Frequency response transfer function	FREQ
Function definition	.FUNC
Gain limit (operator)	GLIMIT
Global nodes	.GLOBAL
Graphics postprocessor	.PROBE
Include file	.INC
Initial conditions	.IC
Library file	.LIB
Model definition	.MODEL
Multiplier (operator)	MULTI
Node setting	.NODESET
Noise analysis	.NOISE
Options	.OPTIONS
Parameter definition	PARAM
Parameter variation	.PARAM
Parametric analysis	.STEP
Plot output	.PLOT
Print output	.PRINT
Probe	.PROBE
Sensitivity analysis	.SENS
Subcircuit call	X_Call
Subcircuit definition	.SUBCKT
Summation (operator)	SUM
Table (operator)	TABLE
Temperature	.TEMP
Transfer function	.TF
Transient analysis	.TRAN
Value	VALUE
Value of voltage-controlled voltage source	EVALUE
Width	.WIDTH

▶ **TIP** If you wish to change the model parameters of any device, use a word processing program in ASCII (DOS) text format (or DOS Editor or Notepad in PSpice) to edit the file EVAL.LIB and change the model parameters.

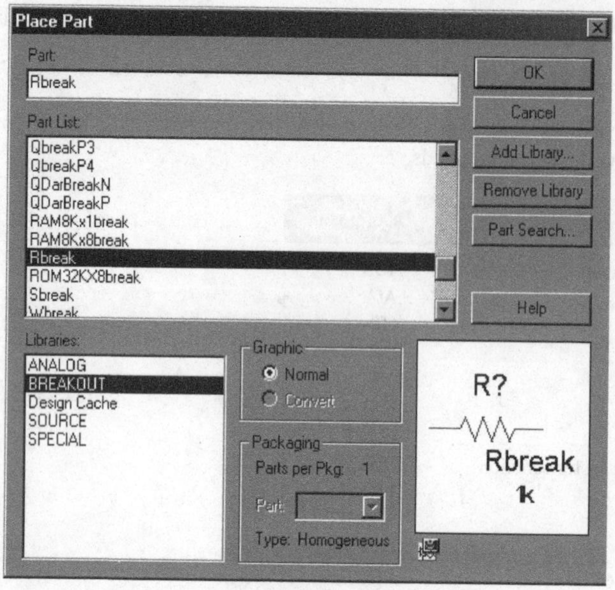

FIGURE A.37 Menu for adding parts

We can add the library file RASHID_MODEL.LIB while running the simulation. First, choose **Library and Include Files** from the Simulation Settings menu, shown in Fig. A.38. Then browse through the library files and select the file RASHID_MODEL.LIB. Click on **Add Library*** and then on **OK**. PSpice will look for device models in this library file while running the simulation. The Include Files dialog box is shown in Fig. A.39.

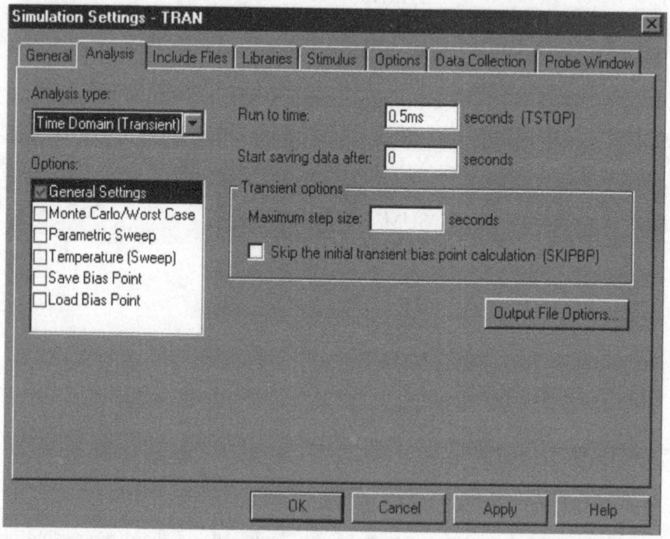

FIGURE A.38 Simulation Settings menu

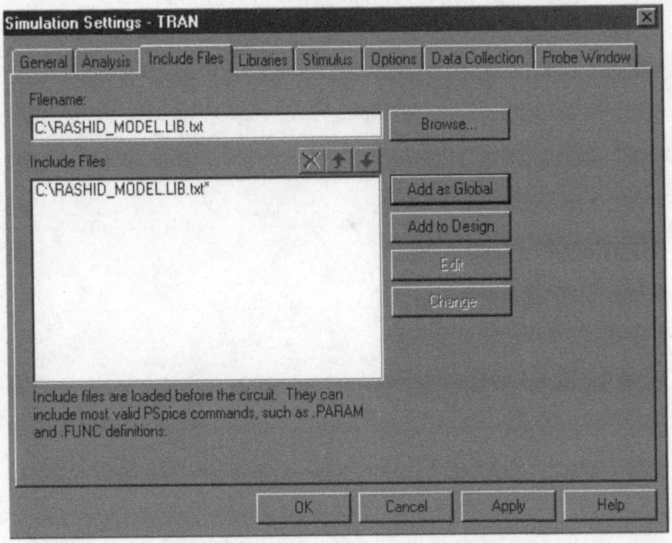

FIGURE A.39 Dialog box for adding library files

References

1. M. H. Rashid, *Introduction to PSpice Using OrCAD for Circuits and Electronics.* Upper Saddle River, NJ: Prentice Hall, 2003.
2. M. E. Herniter, *Schematic Capture with MicroSim PSpice*. Englewood Cliffs, NJ: Prentice Hall, 1996.

APPENDIX B

REVIEW OF BASIC CIRCUITS

B.1 Introduction

Electronic devices that are used as parts of electronic circuits are normally modeled by equivalent circuits. Doing a performance evaluation and design of any electronic circuit requires a knowledge of circuit analysis. This appendix reviews the basic circuit theorems and analysis techniques that are commonly used for electronic circuits.

B.2 Kirchhoff's Current Law

Kirchhoff's current law (KCL) states that the sum of all currents at a node must be zero; that is,

$$\sum I_n = 0$$

where I_n is the current flowing into nth node and $n = 1, 2, 3, \ldots, \infty$.

EXAMPLE B.1

Finding the currents in two parallel resistors For the circuit shown in Fig. B.1, find currents I_1 and I_2.

SOLUTION

Using KCL at node 1 gives

$$I_S - I_1 - I_2 = 0 \quad \text{or} \quad I_S = I_1 + I_2 \tag{B.1}$$

Since the voltage V_S across R_1 is the same as that across R_2, we can write $V_S = R_1 I_1 = R_2 I_2$, which gives $I_2 = R_1 I_1 / R_2$. Substituting for I_2, we get

$$I_S = I_1 + \frac{R_1}{R_2} I_1 = \frac{R_2 + R_1}{R_2} I_1$$

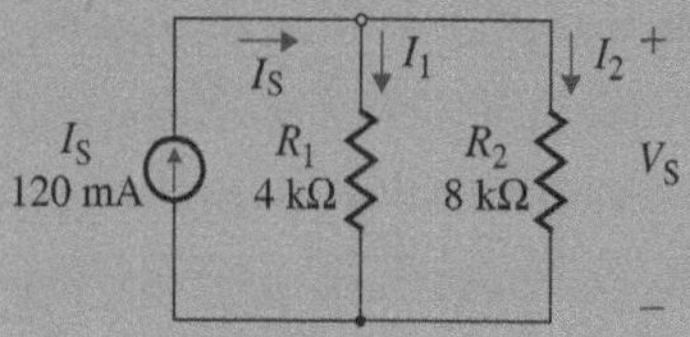

FIGURE B.1 Current distribution in two resistors

which gives the current I_1 through R_1 as

$$I_1 = \frac{R_2}{R_1 + R_2} I_S \tag{B.2}$$

$$= \frac{8\text{ k}\Omega}{4\text{ k}\Omega + 8\text{ k}\Omega} \times 120\text{ mA} = 80\text{ mA}$$

Similarly, substituting $I_1 = R_2I_2/R_1$ in Eq. (B.1) and simplifying, we get the current I_2 through R_2 as

$$I_2 = \frac{R_1}{R_1 + R_2} I_S \tag{B.3}$$

$$= \frac{4\text{ k}\Omega}{4\text{ k}\Omega + 8\text{ k}\Omega} \times 120\text{ mA} = 40\text{ mA}$$

The branch currents I_1 and I_2 can be found by applying Ohm's law as given by

$$V_S = I_S(R_1 \| R_2) = \frac{R_1R_2}{R_1 + R_2} I_S$$

which can be used to find $I_1 = V_S/R_1$ as in Eq. (B.2) and $I_2 = V_S/R_2$ as in Eq. (B.3).

NOTE: Equations (B.2) and (B.3) give the current distribution in two resistors, and the two equations together are often known as the *current divider rule*.

EXAMPLE B.2

Finding the currents in three parallel resistors For the circuit shown in Fig. B.2, find the currents I_1, I_2, and I_3.

SOLUTION

The current source I_S is divided into I_1 through R_1, I_2 through R_2, and I_3 through R_3. Using KCL at node 1 gives

$$I_S - I_1 - I_2 - I_3 = 0 \quad \text{or} \quad I_S = I_1 + I_2 + I_3 \tag{B.4}$$

Since the voltage V_S across R_1 is the same as that across R_2 and R_3, we can write $V_S = R_1I_1 = R_2I_2 = R_3I_3$, which gives $I_2 = R_1I_1/R_2$ and $I_3 = R_1I_1/R_3$. Substituting for I_2 and I_3 in Eq. (B.4), we get

$$I_S = I_1 + \frac{R_1}{R_2}I_1 + \frac{R_1}{R_3}I_1 = \frac{R_1R_2 + R_2R_3 + R_3R_1}{R_2R_3} I_1$$

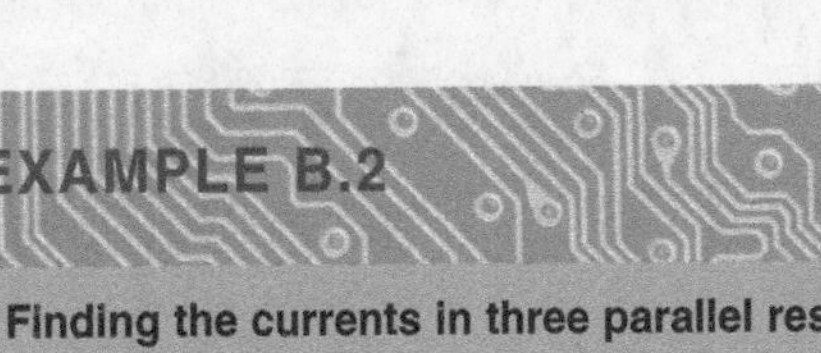

FIGURE B.2 Current distribution in three resistors

which gives the current I_1 through R_1 as

$$I_1 = \frac{1/R_1}{1/R_1 + 1/R_2 + 1/R_3} I_S = \frac{R_2 R_3}{R_1 R_2 + R_2 R_3 + R_3 R_1} I_S \tag{B.5}$$

$$= \frac{4\ \text{k}\Omega \times 6\ \text{k}\Omega}{2\ \text{k}\Omega \times 4\ \text{k}\Omega + 4\ \text{k}\Omega \times 6\ \text{k}\Omega + 6\ \text{k}\Omega \times 2\ \text{k}\Omega} \times 120\ \text{mA} = 65.45\ \text{mA}$$

Substituting $I_1 = R_2 I_2/R_1$ and $I_3 = R_2 I_2/R_3$ in Eq. (B.4) yields the current I_2 through R_2 as

$$I_2 = \frac{1/R_2}{1/R_1 + 1/R_2 + 1/R_3} I_S = \frac{R_1 R_3}{R_1 R_2 + R_2 R_3 + R_3 R_1} I_S \tag{B.6}$$

$$= \frac{2\ \text{k}\Omega \times 6\ \text{k}\Omega}{2\ \text{k}\Omega \times 4\ \text{k}\Omega + 4\ \text{k}\Omega \times 6\ \text{k}\Omega + 6\ \text{k}\Omega \times 2\ \text{k}\Omega} \times 120\ \text{mA} = 32.73\ \text{mA}$$

Similarly, substituting $I_1 = R_3 I_3/R_1$ and $I_2 = R_3 I_3/R_2$ in Eq. (B.4) yields the current I_3 through R_3 as

$$I_3 = \frac{1/R_3}{1/R_1 + 1/R_2 + 1/R_3} I_S = \frac{R_1 R_2}{R_1 R_2 + R_2 R_3 + R_3 R_1} I_S \tag{B.7}$$

$$= \frac{2\ \text{k}\Omega \times 4\ \text{k}\Omega}{2\ \text{k}\Omega \times 4\ \text{k}\Omega + 4\ \text{k}\Omega \times 6\ \text{k}\Omega + 6\ \text{k}\Omega \times 2\ \text{k}\Omega} \times 120\ \text{mA} = 21.82\ \text{mA}$$

The branch currents I_1, I_2, and I_3 can be found by applying Ohm's law as given by

$$V_S = I_S(R_1 \| R_2 \| R_3) = \frac{1}{(1/R_1) + (1/R_2) + (1/R_3)} I_S$$

which can be used to find $I_1 = V_S/R_1$ as in Eq. (B.5), $I_2 = V_S/R_2$ as in Eq. (B.6), and $I_3 = V_S/R_3$ as in Eq. (B.7).

NOTE: If there are *n* parallel resistances, the current I_n through the *n*th resistance R_n can be expressed in general form as

$$I_n = \frac{1/R_n}{(1/R_1) + (1/R_2) + \cdots + (1/R_m)} I_S$$

B.3 Kirchhoff's Voltage Law

Kirchhoff's voltage law (KVL) states that the sum of the voltages around any loop must be zero; that is,

$$\sum V_n = 0$$

where V_n is the voltage across the *n*th segment of a loop and $n = 1, 2, 3, \ldots, \infty$.

EXAMPLE B.3

Finding the voltage distribution in two resistors In Fig. B.3, the voltage V_S is divided into V_1 across R_1 and V_2 across R_2. Find the voltages V_1 and V_2.

SOLUTION

Using KVL around loop I gives

$$V_S - V_1 - V_2 = 0 \quad \text{or} \quad V_S = V_1 + V_2 \tag{B.8}$$

Since the current I_S through R_1 is the same as that through R_2, we can write $V_1 = R_1 I_S$ and $V_2 = R_2 I_S$. Substituting for V_1 and V_2 in Eq. (B.8), we get

$$V_S = V_1 + V_2 = R_1 I_S + R_2 I_S = (R_1 + R_2) I_S$$

which gives the current I_S as

$$I_S = \frac{V_S}{R_1 + R_2}$$

$$= \frac{24\ \text{V}}{4\ \text{k}\Omega + 8\ \text{k}\Omega} = 2\ \text{mA}$$

Therefore, the voltage V_1 across R_1 can be found from

$$V_1 = R_1 I_S = \frac{R_1}{R_1 + R_2} V_S \tag{B.9}$$

$$= \frac{4\ \text{k}\Omega}{4\ \text{k}\Omega + 8\ \text{k}\Omega} \times 24\ \text{V} = 8\ \text{V}$$

Similarly, the voltage V_2 across R_2 can be found from

$$V_2 = R_2 I_S = \frac{R_2}{R_1 + R_2} V_S \tag{B.10}$$

$$= \frac{8\ \text{k}\Omega}{4\ \text{k}\Omega + 8\ \text{k}\Omega} \times 24\ \text{V} = 16\ \text{V}$$

NOTE: The loop current I_S can be found directly by applying Ohm's law, $I_S = V_S/(R_1 + R_2)$. Equations (B.9) and (B.10) give the voltage distribution in two resistors only when the current through R_1 and R_2 is the same. This distribution of voltage is often known as the *voltage* (or *potential*) *divider rule*.

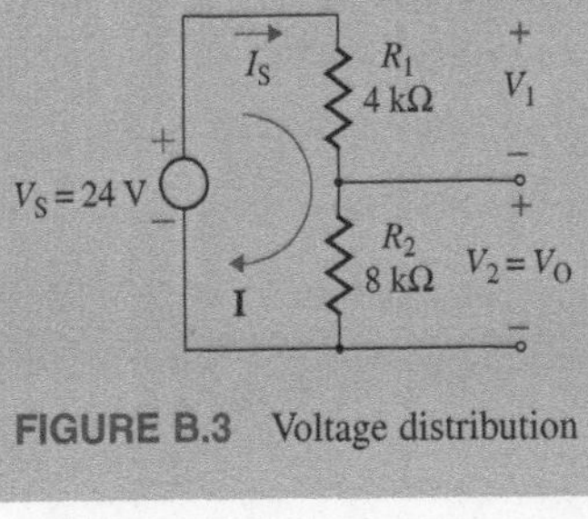

FIGURE B.3 Voltage distribution in two resistors

EXAMPLE B.4

Analyzing a circuit with a current-dependent current source For the circuit shown in Fig. B.4 with a current-dependent current source, find the currents I_B, I_C, and I_E and voltage V_C. Assume $R_{Th} = 15\ k\Omega$, $r_\pi = 1\ k\Omega$, $R_C = 2\ k\Omega$, $R_E = 500\ \Omega$, $\beta_F = 100$, $V_{CC} = 30$ V, and $V_{Th} = 5$ V.

SOLUTION

Using KCL at node 1, we get

$$I_E = I_B + I_C = I_B + \beta_F I_B = (1 + \beta_F)I_B \tag{B.11}$$

Using KVL around loop I, we get

$$V_{Th} = R_{Th}I_B + r_\pi I_B + R_E I_E = R_{Th}I_B + r_\pi I_B + R_E(1 + \beta_F)I_B$$

which gives I_B as

$$I_B = \frac{V_{Th}}{R_{Th} + r_\pi + R_E(1 + \beta_F)} \tag{B.12}$$

$$= \frac{5\ V}{15\ k\Omega + 1\ k\Omega + 500\ \Omega\ \times (1 + 100)} = 75.19\ \mu A$$

Current I_C, which is dependent only on I_B, can be found from

$$I_C = \beta_F I_B = \frac{\beta_F V_{Th}}{R_{Th} + R_\pi + R_E(1 + \beta_F)} \tag{B.13}$$

$$= \frac{100 \times\ 5\ V}{15\ k\Omega + 1\ k\Omega + 500\ \Omega\ \times (1 + 100)} = 7519\ \mu A$$

Then

$$I_E = I_B + I_C = 75.19\ \mu A + 7519\ \mu A = 7594\ \mu A$$

$$V_C = V_{CC} - I_C R_C = 30\ V - 7519\ \mu A \times 2\ k\Omega = 14.96\ V$$

Voltage V_E can be found from

$$V_E = R_E I_E = R_E(I_C + I_B) = R_E(\beta_F I_B + I_B) = R_E I_B(1 + \beta_F)$$

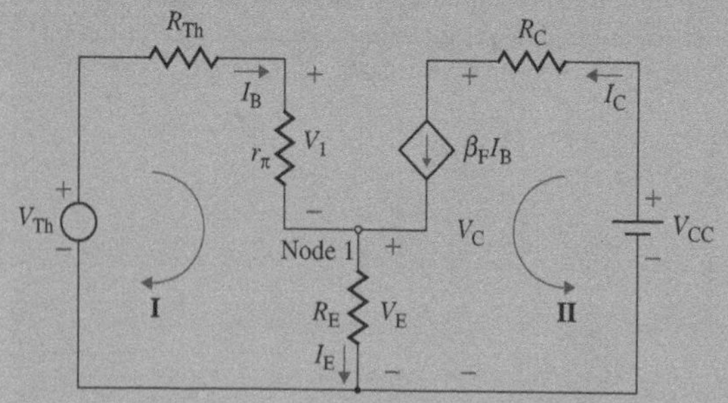

FIGURE B.4 Circuit with current-dependent current source

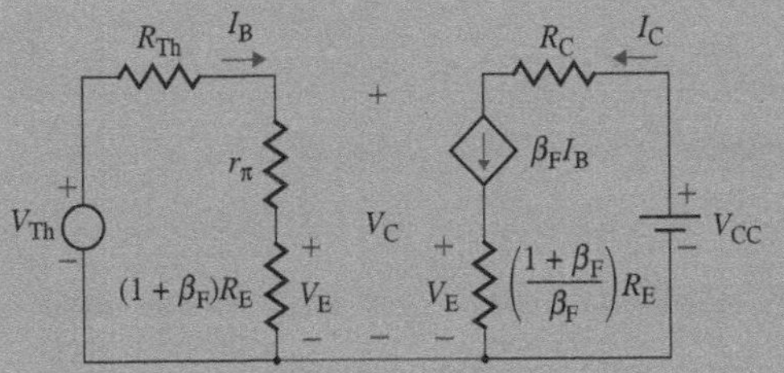

FIGURE B.5 Splitting of current R_E

Since $I_C = \beta_F I_B$, we get

$$V_E = R_E I_B(1 + \beta_F) = R_E I_C[(1 + \beta_F)/\beta_F)]$$

Thus, R_E offers a resistance of $R_E(1 + \beta_F)$ to current I_B in loop I and a resistance of $R_E(1 + \beta_F)/\beta_F$ to current I_C in loop II. Therefore, R_E can be split, or "reflected," into loop I and loop II by adjusting its value such that V_E is preserved on loop I and loop II. This arrangement is shown in Fig. B.5.

NOTE: The current-dependent current source $\beta_F I_B$ is often represented by a voltage-dependent voltage source $g_m V_1$ such that $\beta_F I_B = \beta_F V_1/r_\pi = (\beta_F/r_\pi)V_1$. g_m, which is known as the *transconductance*, is related to β_F by $g_m = \beta_F/r_\pi$. Therefore, by substituting $\beta_F = g_m r_\pi$, we can apply the above equations for a voltage-dependent voltage source that is often used as the small-signal model of bipolar and field-effect transistors.

B.4 Superposition Theorem

The superposition theorem states that the current through or voltage across any element in a linear network is equal to the algebraic sum of the currents or voltages produced independently by each source. To calculate the effect of one source, the other independent sources are removed by short-circuiting the voltage sources and open-circuiting the current sources. Any internal resistance associated with the removed voltage sources must be considered, however.

EXAMPLE B.5

Finding the output voltage using the superposition theorem The circuit in Fig. B.6 has a DC source $V_{DC} = 10$ V, an AC source $v_{ac} = 15 \sin(377\pi t)$, $R_1 = 2$ kΩ, and $R_2 = 3$ kΩ. Use the superposition theorem to determine the instantaneous output voltage v_O.

SOLUTION

$V_{DC} = 10$ V, $v_{ac} = 15 \sin(377\pi t)$, $R_1 = 2$ kΩ, and $R_2 = 3$ kΩ. The DC equivalent circuit with source V_{DC} only is shown in Fig. B.7(a); the output voltage due to V_{DC} is

$$V_{O1} = \frac{R_2}{R_1 + R_2} V_{DC} = \frac{3\text{ k}}{2\text{ k} + 3\text{ k}} \times 10 = 6 \text{ V}$$

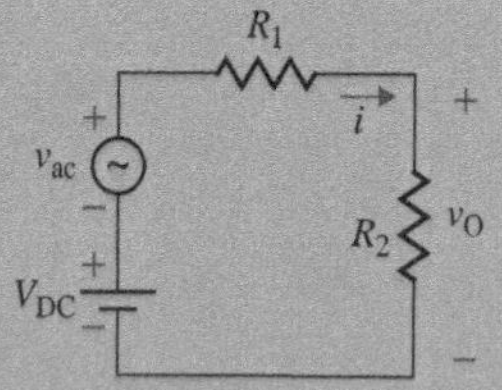

FIGURE B.6 Circuit for Example B.5

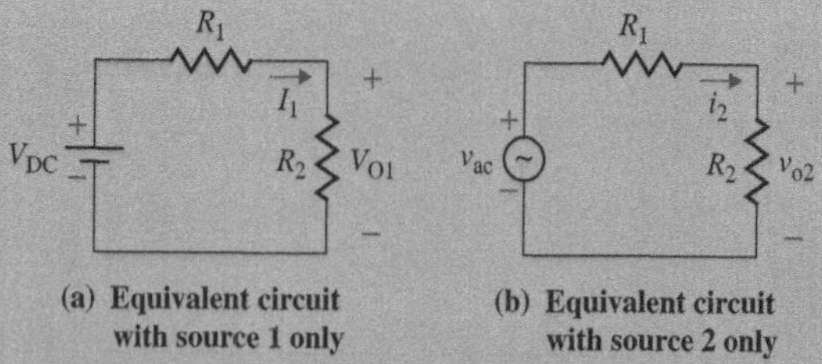

FIGURE B.7 Equivalent circuits for Example B.5

Figure B.7(b) shows the AC equivalent circuit with source v_{ac} only; the output voltage due to v_{ac} is

$$v_{o2} = \frac{R_2}{R_1 + R_2} v_{ac} = \frac{3\text{ k}}{2\text{ k} + 3\text{ k}} \times 15 \sin(377\pi t) = 9 \sin(377\pi t)$$

Therefore, the resultant output voltage v_O can be found by combining the output voltages due to individual sources; that is,

$$v_O = V_{O1} + v_{o2} = 6 + 9 \sin(377\pi t) = 3 \times [2 + 3 \sin(377\pi t)]$$

EXAMPLE B.6

Finding the output voltage using the superposition theorem A circuit with three input voltages V_{S1}, V_{S2}, and V_{S3} is shown in Fig. B.8. Use the superposition theorem to determine the output voltage V_O. Use $R_1 = 2\text{ k}\Omega$, $R_2 = 4\text{ k}\Omega$, $R_3 = 6\text{ k}\Omega$, $V_{S1} = 10$ V, $V_{S2} = 12$ V, and $V_{S3} = 15$ V.

SOLUTION

The equivalent circuit with source V_{S1} only is shown in Fig. B.9(a). Applying the voltage divider rule, we can find the output voltage due to V_{S1} as

$$V_{O1} = \frac{R_2 \| R_3}{R_1 + R_2 \| R_3} V_{S1} = \frac{4\text{ k} \| 6\text{ k}}{2\text{ k} + (4\text{ k} \| 6\text{ k})} \times 10 = 5.45\text{ V}$$

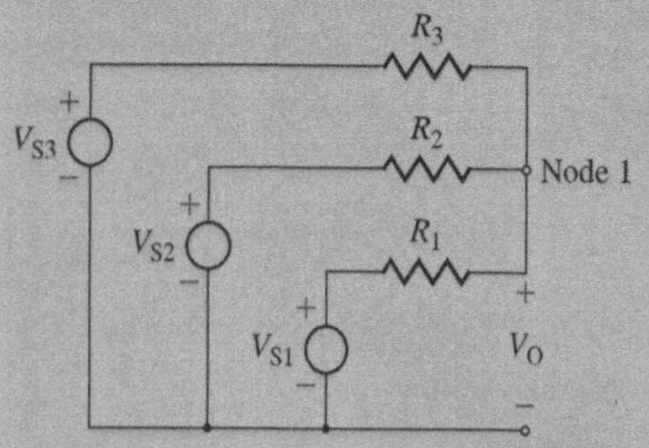

FIGURE B.8 Circuit for Example B.6

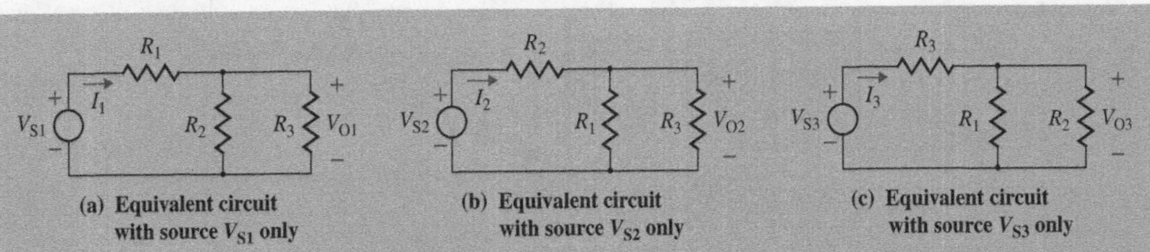

FIGURE B.9 Equivalent circuits for Example B.6

The equivalent circuit with source V_{S2} only is shown in Fig. B.9(b); the output voltage due to V_{S2} is

$$V_{O2} = \frac{R_1 \| R_3}{R_2 + R_1 \| R_3} V_{S2} = \frac{2\text{ k} \| 6\text{ k}}{4\text{ k} + (2\text{ k} \| 6\text{ k})} \times 12 = 3.27\text{ V}$$

The equivalent circuit with source V_{S3} only is shown in Fig. B.9(c); the output voltage due to V_{S3} is

$$V_{O3} = \frac{R_1 \| R_2}{R_3 + R_1 \| R_2} V_{S3} = \frac{2\text{ k} \| 4\text{ k}}{6\text{ k} + (2\text{ k} \| 4\text{ k})} \times 15 = 2.73\text{ V}$$

Therefore, the resultant output voltage V_O can be found by combining the output voltages due to individual sources; that is,

$$V_O = V_{O1} + V_{O2} + V_{O3} = 5.45 + 3.27 + 2.73 = 11.45\text{ V}$$

An alternative approach is to apply KVL at node 1 and look for V_O:

$$V_O = \frac{\text{Currents into node 1 if it were at ground potential}}{\text{Conductances radiating from node 1}}$$

$$= \frac{V_{S1}/R_1 + V_{S2}/R_2 + V_{S3}/R_3}{1/R_1 + 1/R_2 + 1/R_3} = \frac{10/2\text{ k} + 12/4\text{ k} + 15/6\text{ k}}{1/2\text{ k} + 1/4\text{ k} + 1/6\text{ k}} = 11.45\text{ V}$$

B.5 Thevenin's Theorem

Thevenin's theorem states that any two-terminal linear DC (or AC) network can be replaced by an equivalent circuit consisting of a voltage source and a series resistance (or impedance). This theorem is commonly used to find the voltage (or current) of a linear network with one or more sources. It allows us to concentrate on a specific portion of the network by replacing the remaining network by an equivalent circuit. In the case of sinusoidal AC circuits, the reactances are frequency dependent, and Thevenin's equivalent circuit is valid for only one frequency.

Figure B.10(a) shows a general DC network; Thevenin's equivalent circuit is shown in Fig. B.10(b). The steps in determining an equivalent voltage source V_{Th} and an equivalent resistance R_{Th} for Thevenin's equivalent circuit are as follows:

Step 1. Decide on the part of the network for which you desire a Thevenin's representation and mark the terminals, as shown in Fig. B.10(a).

Step 2. Remove that part of the network. In Fig. B.10(a), the load resistance R_L is to be removed.

Step 3. Mark the open-circuit terminals of the remaining network—namely, a and b.

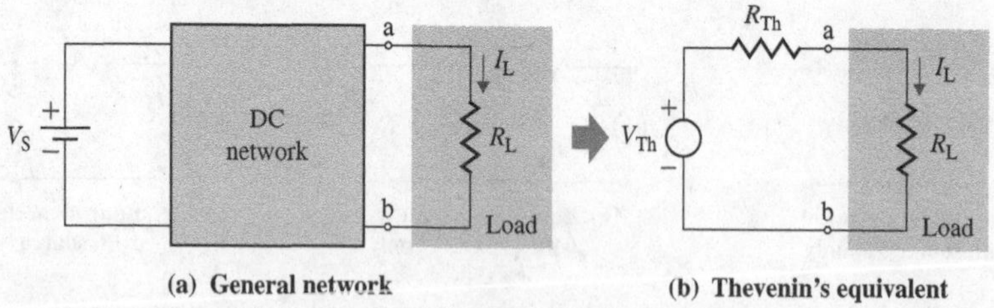

FIGURE B.10 Thevenin's equivalent circuit

Step 4. Determine the open-circuit voltage V_{Th} between terminals a and b.

Step 5. Set all independent sources to zero (voltage sources are replaced by short circuits and current sources by open circuits). Apply a test voltage V_X across terminals a and b. The ratio of V_X to its current I_X gives Thevenin's resistance R_{Th}.

EXAMPLE B.7

Finding Thevenin's equivalent circuit Represent the network shown in Fig. B.11(a) by Thevenin's equivalent, as shown in Fig. B.11(b). Assume $V_{CC} = 12$ V, $R_1 = 15$ kΩ, and $R_2 = 7.5$ kΩ.

SOLUTION

The open-circuit voltage, which is Thevenin's voltage between terminals a and b, can be found from the voltage divider rule in Eq. (B.10); that is,

$$V_{Th} = \frac{R_2}{R_1 + R_2} V_{CC} \tag{B.14}$$

$$= \frac{7.5\text{ k}}{15\text{ k} + 7.5\text{ k}} \times 12 = 4\text{ V}$$

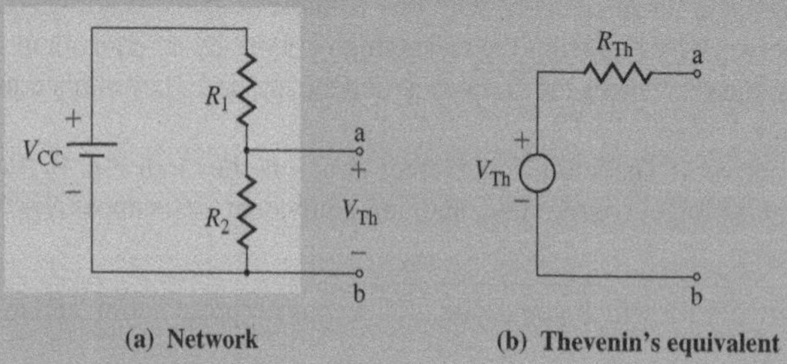

FIGURE B.11 Network for Example B.7

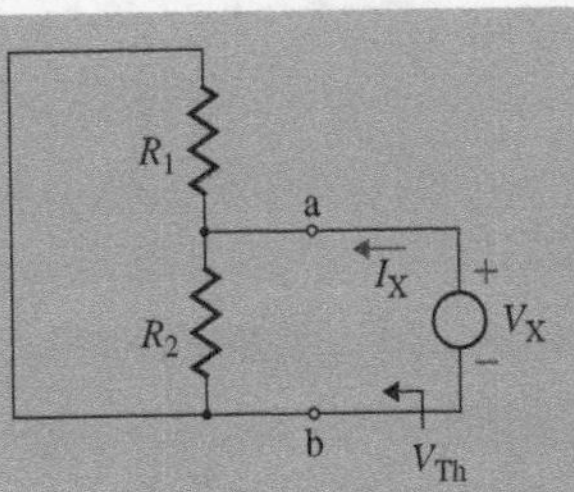

FIGURE B.12 Equivalent circuits

If source V_{CC} is set to zero and test voltage V_X is applied across terminals a and b, the circuit for determining R_{Th} is shown in Fig. B.12. R_{Th} becomes the parallel combination of R_1 and R_2. That is,

$$R_{Th} = \frac{V_X}{I_X} = R_1 \| R_2 \quad \text{(B.15)}$$

$$= 15\text{ k} \| 7.5\text{ k} = 5\text{ k}\Omega$$

EXAMPLE B.8

Finding Thevenin's equivalent circuit Represent the network shown in Fig. B.13(a) by Thevenin's equivalent, as shown in Fig. B.13(b). Assume $V_{CC} = 12$ V, $V_A = 9$ V, $R_1 = 15$ kΩ, and $R_2 = 7.5$ kΩ.

SOLUTION

Since there are two voltage sources V_{CC} and V_A, we will apply the superposition theorem to find V_{Th}. The equivalent circuit with source V_{CC} only is shown in Fig. B.14(a). Applying the voltage divider rule, we can find the output voltage due to V_{CC} only as

$$V_{O1} = \frac{R_2}{R_1 + R_2} V_{CC} = \frac{7.5\text{ k}}{15\text{ k} + 7.5\text{ k}} \times 12 = 4\text{ V}$$

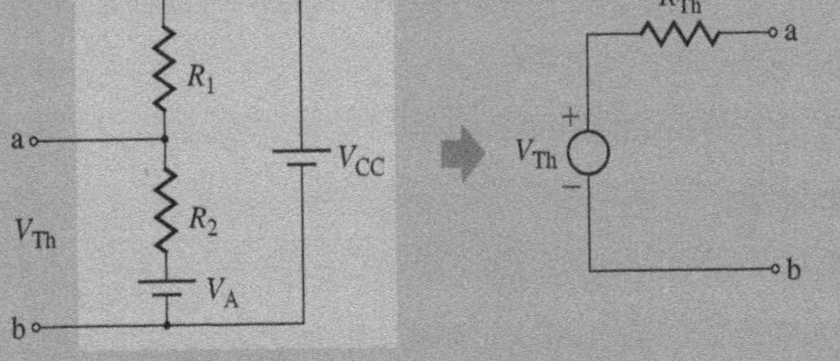

(a) Network (b) Thevenin's equivalent

FIGURE B.13 Network for Example B.8

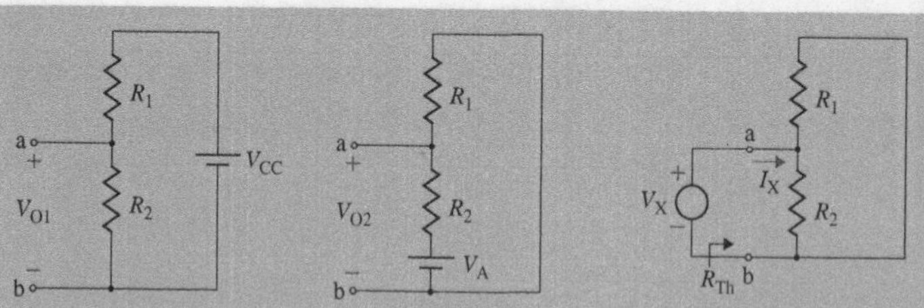

(a) Equivalent circuit with source V_{CC} only (b) Equivalent circuit with source V_A only (c) Finding R_{Th}

FIGURE B.14 Equivalent circuits

The circuit with source V_A only is shown in Fig. B.14(b); the output voltage due to V_A is

$$V_{O2} = \frac{R_1}{R_1 + R_2} V_A = \frac{15\text{ k}}{15\text{ k} + 7.5\text{ k}} \times 9 = 6\text{ V}$$

The resultant output voltage V_O, which equals V_{Th}, can be found by combining the output voltages due to individual sources; that is,

$$\begin{aligned} V_{Th} &= V_O = V_{O1} + V_{O2} \\ &= \frac{R_2}{R_1 + R_2} V_{CC} + \frac{R_1}{R_1 + R_2} V_A \qquad \text{(B.16)} \\ &= 4 + 6 = 10\text{ V} \end{aligned}$$

If sources V_A and V_{CC} are set to zero and test voltage V_X is applied across terminals a and b, the circuit for determining R_{Th} is shown in Fig. B.14(c). R_{Th} becomes the parallel combination of R_1 and R_2; that is,

$$\begin{aligned} R_{Th} &= \frac{V_X}{I_X} = R_1 \| R_2 \qquad \text{(B.17)} \\ &= 15\text{ k} \| 7.5\text{ k} = 5\text{ k}\Omega \end{aligned}$$

It is evident that Eqs. (B.15) and (B.17) are the same.

EXAMPLE B.9

Representing a network by Thevenin's equivalent circuit Represent the network shown in Fig. B.15 by Thevenin's equivalent circuit. The circuit values are $R_i = 1.5\text{ k}\Omega$, $R_C = 25\text{ k}\Omega$, $\beta_F = 50$, $h_r = 3 \times 10^{-4}$, and $V_s = 5\text{ mV}$.

(a) Calculate the parameters of Thevenin's equivalent circuit.

(b) Use PSpice/SPICE to check your results.

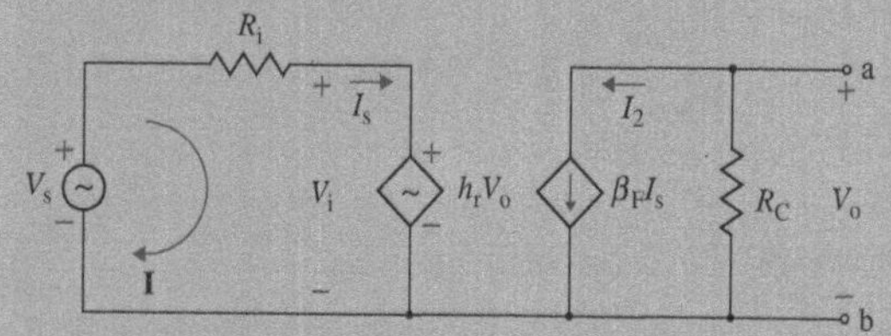

FIGURE B.15 Network for Example B.9

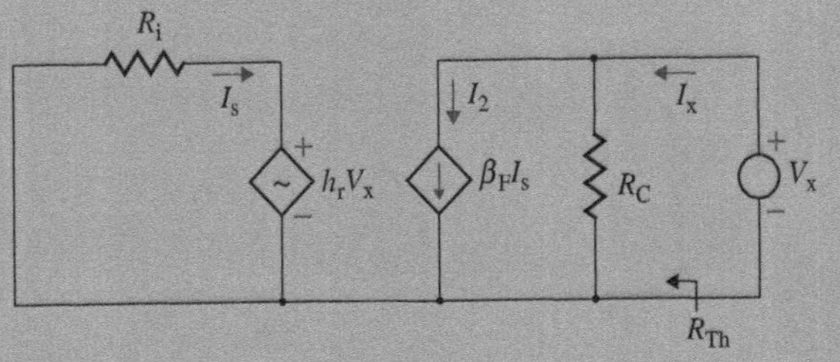

FIGURE B.16 Circuit for determining Thevenin's resistance

SOLUTION

(a) $R_i = 1.5\ \text{k}\Omega$, $R_C = 25\ \text{k}\Omega$, $\beta_F = 50$, $h_r = 3 \times 10^{-4}$, and $V_s = 5$ mV. The output voltage V_o between terminals a and b is

$$V_o = -I_2 R_C = -\beta_F I_s R_C \tag{B.18}$$

The input current I_s can be found from loop I as

$$I_s = \frac{V_s - h_r V_o}{R_i} \tag{B.19}$$

Substituting I_s from Eq. (B.19) into Eq. (B.18) gives the output voltage V_o as

$$V_{Th} = V_o = \frac{-\beta_F R_C}{R_i - \beta_F h_r R_C} V_s \tag{B.20}$$

$$= \frac{-50 \times 25\ \text{k} \times 5\ \text{m}}{1.5\ \text{k} - 50 \times 3 \times 10^{-4} \times 25\ \text{k}} = -5.5556\ \text{V}$$

Thevenin's resistance R_{Th} can be determined from the circuit in Fig. B.16, which is obtained by short-circuiting the independent source V_s. Considering V_x and I_x as the test voltage and current, respectively, we get

$$I_s = -\frac{h_r V_x}{R_i} \tag{B.21}$$

$$I_x = \beta_F I_s + \frac{V_x}{R_C} \tag{B.22}$$

Substituting I_s from Eq. (B.21) into Eq. (B.22) gives

$$I_x = -\frac{\beta_F h_r V_x}{R_i} + \frac{V_x}{R_C} = \frac{R_i - \beta_F h_r R_C}{R_i R_C} V_x$$

which gives Thevenin's resistance R_{Th} as

$$R_{Th} = \frac{V_x}{I_x} = \frac{R_i R_C}{R_i - \beta_F h_r R_C} \tag{B.23}$$

$$= \frac{1.5\ \text{k} \times 25\ \text{k}}{1.5\ \text{k} - 50 \times 3 \times 10^{-4} \times 25\ \text{k}} = 33.33\ \text{k}\Omega$$

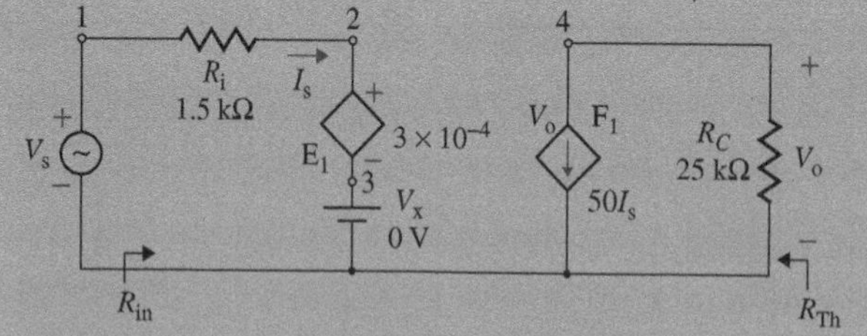

FIGURE B.17 Circuit for PSpice simulation

(b) The circuit for PSpice simulation is shown in Fig. B.17, which is run by OrCAD PSpice A/D. The results of the PSpice simulation are shown below:

```
NODE   VOLTAGE    NODE   VOLTAGE    NODE   VOLTAGE    NODE   VOLTAGE
( 1)   .0050      ( 2)   -.0017     ( 3)   0.0000     ( 4)   -5.5556
```

$V_{Th} = V_o = V(4) = -5.5556$ V

```
****   SMALL-SIGNAL CHARACTERISTICS

V(4)/VS=-1.111E+03

INPUT RESISTANCE AT VS=1.125E+03

OUTPUT RESISTANCE AT V(4)=3.333E+04
```

Gain $A = V_o/V_s = -1111$

$R_{in} = V_s/I_s = 1.125$ kΩ

$R_{Th} = 33.33$ kΩ

B.6 Norton's Theorem

Norton's theorem states that any two-terminal linear DC (or AC) network can be replaced by an equivalent circuit consisting of a current source and a parallel resistance (or impedance). Norton's equivalent circuit can be determined from Thevenin's equivalent circuit; the relationship between them is shown in Fig. B.18. Norton's resistance R_N is identical to Thevenin's resistance R_{Th}, and Norton's current equals the short-circuit current at the terminals of interest.

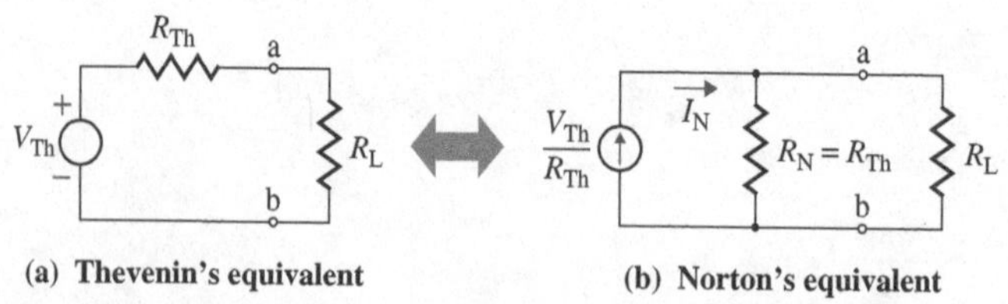

FIGURE B.18 Thevenin's and Norton's equivalent circuits

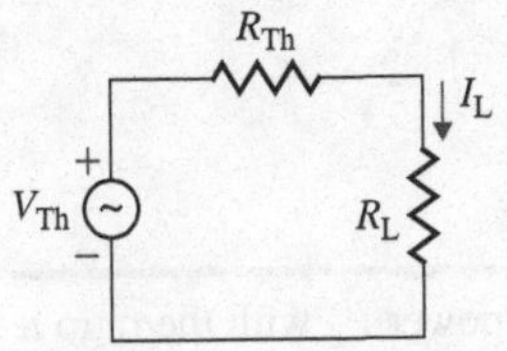

FIGURE B.19 Thevenin's equivalent circuit with a resistive load

B.7 Maximum Power Transfer Theorem

In electronic circuits, it is often necessary to transmit maximum power to the load. Let us consider the circuit in Fig. B.19, which could be Thevenin's equivalent circuit of a network. The power P_L delivered to load resistance R_L can be found from

$$P_L = I_L^2 R_L = \left[\frac{V_{Th}}{R_{Th} + R_L}\right]^2 R_L$$

$$= \frac{V_{Th}^2}{R_{Th}} \times \frac{1}{(1 + R_L/R_{Th})^2} \times \frac{R_L}{R_{Th}} \quad \textbf{(B.24)}$$

For a given circuit, V_{Th} and R_{Th} will be fixed. Therefore, the load power P_L depends on the load resistance R_L. If we set $R_L = uR_{Th}$, Eq. (B.24) becomes

$$P_L = \frac{V_{Th}^2}{R_{Th}} \frac{u}{(1 + u)^2}$$

$$= \frac{u}{(1 + u)^2} P$$

where $P = V_{Th}^2/R_{Th}$. Normalizing P_L with respect to P, we get the normalized power P_n as

$$P_n = \frac{P_L}{P} = \frac{u}{(1 + u)^2} \quad \textbf{(B.25)}$$

Figure B.20 shows the variation of normalized power P_n with u. The power P_n becomes maximum at $u = 1$; that is, $R_{Th} = uR_L = R_L$. The value of R_L for a maximum power transfer can also be determined from the condition $dP_L/dR_L = 0$. From Eq. (B.24),

$$\frac{dP_L}{dR_L} = V_{Th}^2\left[\frac{(R_{Th} + R_L)^2 - 2R_L(R_{Th} + R_L)}{(R_{Th} + R_L)^4}\right] = 0$$

or

$$(R_{Th} + R_L)^2 - 2R_L(R_{Th} + R_L) = 0$$

$$R_L = \pm R_{Th}$$

Since R_{Th} cannot be negative,

$$R_L = R_{Th} \quad \textbf{(B.26)}$$

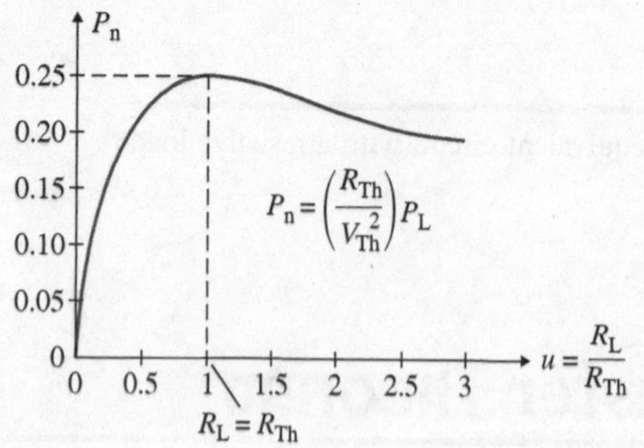

FIGURE B.20 Normalized power P_n with the ratio u

Thus, the maximum power transfer occurs when the load resistance R_L is equal to Thevenin's resistance R_{Th} of the network. For Norton's equivalent circuit of Fig. B.18(b), the maximum power will be delivered to the load when

$$R_N = R_L \tag{B.27}$$

Substituting R_L from Eq. (B.26) into Eq. (B.24) gives the maximum power P_{max} delivered to the load as

$$P_{max} = \frac{V_{Th}^2 R_L}{4R_L^2} = \frac{V_{Th}^2}{4R_L} \tag{B.28}$$

The input power P_{in} supplied by the source V_s is

$$P_{in} = \frac{V_{Th}^2}{R_{Th} + R_L} = \frac{V_{Th}^2}{2R_L} \tag{B.29}$$

Thus, the efficiency η at the maximum power transfer condition is

$$\eta = \frac{P_{max}}{P_{in}} \times 100\% = \frac{V_{Th}^2}{4R_L} \times \frac{2R_L}{V_{Th}^2} \times 100\% = 50\%$$

Therefore, the efficiency is always 50% at the maximum power transfer condition. In electronic circuits, the amount of power being transferred is usually small, and efficiency is often not of primary concern. However, efficiency is of major concern in circuits involving high power—for example, in power systems.

B.8 Transient Response of First-Order Circuits

The transient response gives the instantaneous value of an output voltage (or current) for a specified instantaneous input voltage (or current). The response due to a step-signal input is commonly used in evaluating electronic circuits because such a response allows us to predict the response due to other signals such as a pulse or square-wave input.

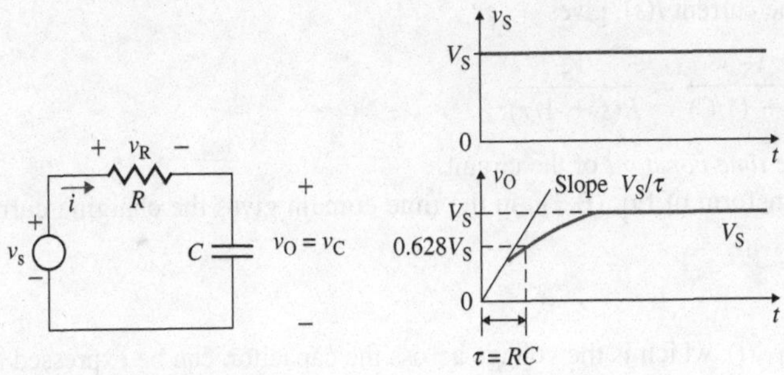

(a) *RC* circuit (b) Step response

FIGURE B.21 Series *RC* circuit

B.8.1 Step Response of Series *RC* Circuits

Consider the series *RC* circuit in Fig. B.21(a) with a step input voltage V_S. The output voltage v_O is taken across capacitor C. For $t \geq 0$, the charging current i of the capacitor can be found from

$$V_S = v_R + v_C = Ri + \frac{1}{C}\int i\,dt + v_C(t = 0) \tag{B.30}$$

with initial capacitor voltage $v_C(t = 0) = 0$.

Using the Laplace transformations in Table B.1, we can transform Eq. (B.30) into Laplace's domain of s as follows:

$$\frac{V_S}{s} = RI(s) + \frac{1}{Cs}I(s)$$

TABLE B.1 Some Laplace transformations

$f(t)$	$F(s)$
1	$\frac{1}{s}$
t	$\frac{1}{s^2}$
$e^{-\alpha t}$	$\frac{1}{s + \alpha}$
$\sin \alpha t$	$\frac{\alpha}{s^2 + \alpha^2}$
$\cos \alpha t$	$\frac{s}{s^2 + \alpha^2}$
$f'(t)$	$sF(s) - F(0)$
$f''(t)$	$s^2F(s) - sF(s) - F'(0)$

which, solved for the current $I(s)$, gives

$$I(s) = \frac{V_S}{sR + (1/C)} = \frac{V_S}{R(s + 1/\tau)} \tag{B.31}$$

where $\tau = RC$ is the *time constant* of the circuit.

The inverse transform of Eq. (B.31) in the time domain gives the charging current:

$$i(t) = \frac{V_S}{R} e^{-t/\tau} \tag{B.32}$$

The output voltage $v_O(t)$, which is the voltage across the capacitor, can be expressed as

$$v_O(t) = \frac{1}{C}\int_0^t i\, dt = \frac{1}{C}\int_0^t \frac{V_S}{R} e^{-t/\tau}\, dt = V_S(1 - e^{-t/\tau}) \tag{B.33}$$

In the steady state (at $t = \infty$), Eq. (B.32) gives

$$i(t = \infty) = 0$$

From Eq. (B.33),

$$v_O(t = \infty) = V_S \tag{B.34}$$

At $t = \tau$, Eq. (B.33) gives

$$v_O(t = \tau) = V_S(1 - e^{-1}) = 0.632V_S \tag{B.35}$$

The initial slope of the tangent to $v_O(t)$ can be found from Eq. (B.33):

$$\left.\frac{dv_O}{dt}\right|_{t=0} = \left.\frac{V_S}{\tau} e^{-t/\tau}\right|_{t=0} = \frac{V_S}{\tau} = \frac{V_S}{RC} \tag{B.36}$$

The transient response of $v_O(t)$, which rises expontentially due to a step input, is shown in Fig. B.21(b).

B.8.2 Step Response of Series *CR* Circuits

In a *CR* circuit, the output voltage is taken across resistance R instead of capacitance C, as shown in Fig. B.22(a). The output voltage v_O, which is the voltage across resistance R, can be found from Eq. (B.32); that is,

$$v_O(t) = Ri(t) = V_S e^{-t/\tau} \tag{B.37}$$

which, in the steady state (at $t = \infty$), gives

$$i(t = \infty) = 0$$

$$v_O(t = \infty) = 0$$

At $t = \tau$, Eq. (B.37) gives

$$v_O(t = \tau) = V_S e^{-1} = 0.368V_S \tag{B.38}$$

From Eq. (B.37), the initial slope of the tangent to $v_O(t)$ is

$$\left.\frac{dv_O}{dt}\right|_{t=0} = -\left.\frac{V_S}{\tau} e^{-t/\tau}\right|_{t=0} = -\frac{V_S}{\tau} = -\frac{V_S}{RC} \tag{B.39}$$

The response of $v_O(t)$, which falls exponentially due to a step voltage input, is shown in Fig. B.22(b).

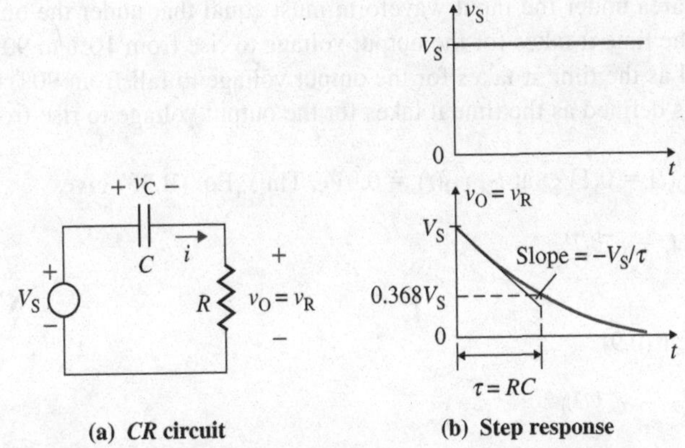

FIGURE B.22 Step response of a series *CR* circuit

B.8.3 Pulse Response of Series *RC* Circuits

An input pulse v_S of duration T, shown in Fig. B.23(a), is applied to the circuit of Fig. B.21(a). The response due to a pulse signal will depend on the ratio of time constant τ to duration T. We consider three cases: $\tau = T$, $\tau << T$, and $\tau >> T$.

In case 1, $\tau = T$, the output voltage $v_O(t)$ has just enough time to reach a near steady-state value of V_S. The capacitor C is charged exponentially to approximately voltage V_S. When the input voltage $v_S(t)$ falls to zero at $t = T$, the output (or capacitor) voltage $v_O(t)$ falls exponentially to zero, as shown

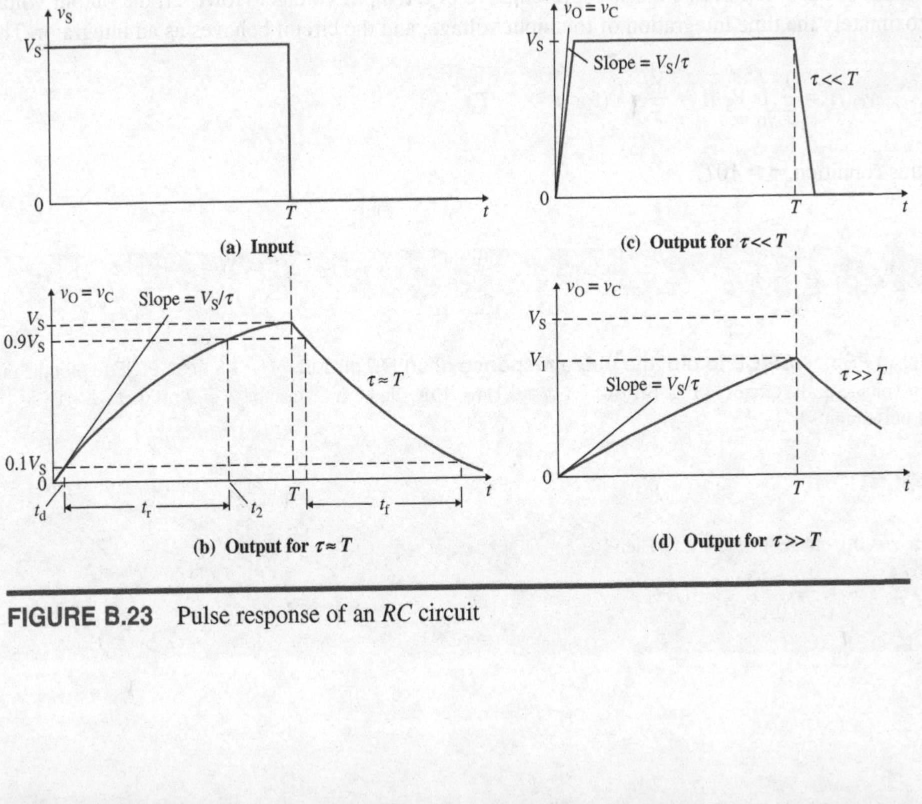

FIGURE B.23 Pulse response of an *RC* circuit

in Fig. B.23(b). The area under the input waveform must equal that under the output waveform. *Rise time* t_r is defined as the time it takes for the output voltage to rise from 10% to 90% of the final value. *Fall time* t_f is defined as the time it takes for the output voltage to fall from 90% to 10% of the initial value. *Delay time* t_d is defined as the time it takes for the output voltage to rise from 0% to 10% of the final value.

At $t = t_1 = t_d$, $v_O(t) = 0.1V_S$; at t_2, $v_O(t) = 0.9V_S$. Thus, Eq. (B.33) gives

$$0.1V_S = V_S(1 - e^{-t_1/\tau})$$
$$e^{-t_1/\tau} = 0.9$$
$$t_1 = -\tau \ln(0.9)$$

and

$$0.9V_S = V_S(1 - e^{-t_2/\tau})$$
$$e^{-t_2/\tau} = 0.1$$
$$t_2 = -\tau \ln(0.1)$$

The rise time t_r, which is equal to the fall time t_f, can be found as follows:

$$t_r = t_f = t_2 - t_1$$
$$= -\tau \ln(0.1) + \tau \ln(0.9) = \tau \ln(9) \approx 2.2\tau \tag{B.40}$$

In case 2, $\tau << T$, t_r and t_f are much smaller than T. The output voltage $v_O(t)$ represents the input signal more closely, as shown in Fig. B.23(c). This condition is generally satisfied by choosing circuit parameters such that $10\tau = T$.

In case 3, $\tau >> T$, there is not enough time for the output voltage $v_O(t)$ to reach the steady-state value of V_S. The output voltage at $t = T$ is V_1, which is much smaller than V_S, as shown in Fig. B.23(d). The output voltage starts to decay exponentially toward zero before reaching its maximum value. Thus, the output voltage will not be a true representative of the input voltage. However, the output voltage is approximately the time integration of the input voltage, and the circuit behaves as an integrator. That is,

$$v_O(t) = \frac{1}{\tau}\int_0^t V_S\,dt = \frac{V_S}{\tau}t \quad (\text{for } \tau >> T)$$

For this condition, $\tau = 10T$.

EXAMPLE B.10

Using PSpice/SPICE to plot the pulse response of an *RC* circuit Use PSpice/SPICE to plot the output voltage of the circuit in Fig. B.21(a) for $\tau = 0.1$ ms, 1 ms, and 5 ms. Assume $T = 2$ ms and $v_S = V_S = 1$ V of pulse input.

SOLUTION

For $\tau = 0.1$ ms, let $C = 0.1$ μF. Then

$$R = \frac{\tau}{C} = \frac{0.1 \text{ ms}}{0.1\ \mu\text{F}} = 1\ \text{k}\Omega$$

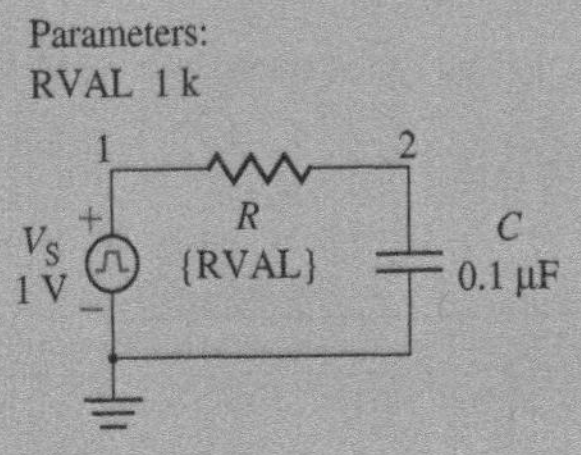

FIGURE B.24 *RC* circuit for PSpice simulation

For $\tau = 1$ ms, let $C = 0.1\ \mu$F. Then

$$R = \frac{\tau}{C} = \frac{1\ \text{ms}}{0.1\ \mu\text{F}} = 10\ \text{k}\Omega$$

For $\tau = 5$ ms, let $C = 0.1\ \mu$F. Then

$$R = \frac{\tau}{C} = \frac{5\ \text{ms}}{0.1\ \mu\text{F}} = 50\ \text{k}\Omega$$

The series *RC* circuit for PSpice simulation is shown in Fig. B.24 with a pulse input voltage. The list of the circuit file, which is run by OrCAD PSpice A/D, is as follows.

```
Example B.10  Pulse Response of Series RC Circuit
.PARAM RVAL=1K
.STEP PARAM RVAL LIST  1K  10K  50K
VS  1  0  PULSE (0V  1V  0  1NS  1NS  2MS  4MS)  ; Pulse input voltage
R  1  2  {RVAL}
C  2  0  0.1UF
.TRAN  0.1MS  4MS                               ; Transient analysis
.PROBE
.END
```

The PSpice plots of the output voltage $v_O(t)$ for three values of the time constant are shown in Fig. B.25. The lower the time constant τ, the faster the output voltage rises and falls.

NOTE: You can use the PSpice Parametric command for variable *R* to vary the time constant.

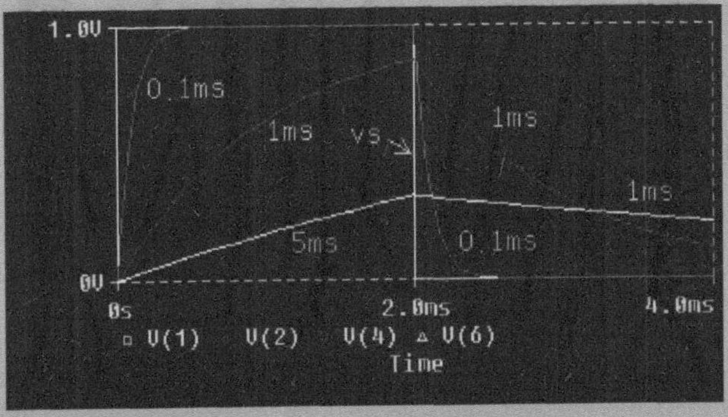

FIGURE B.25 Plots of $v_O(t)$ for Example B.10

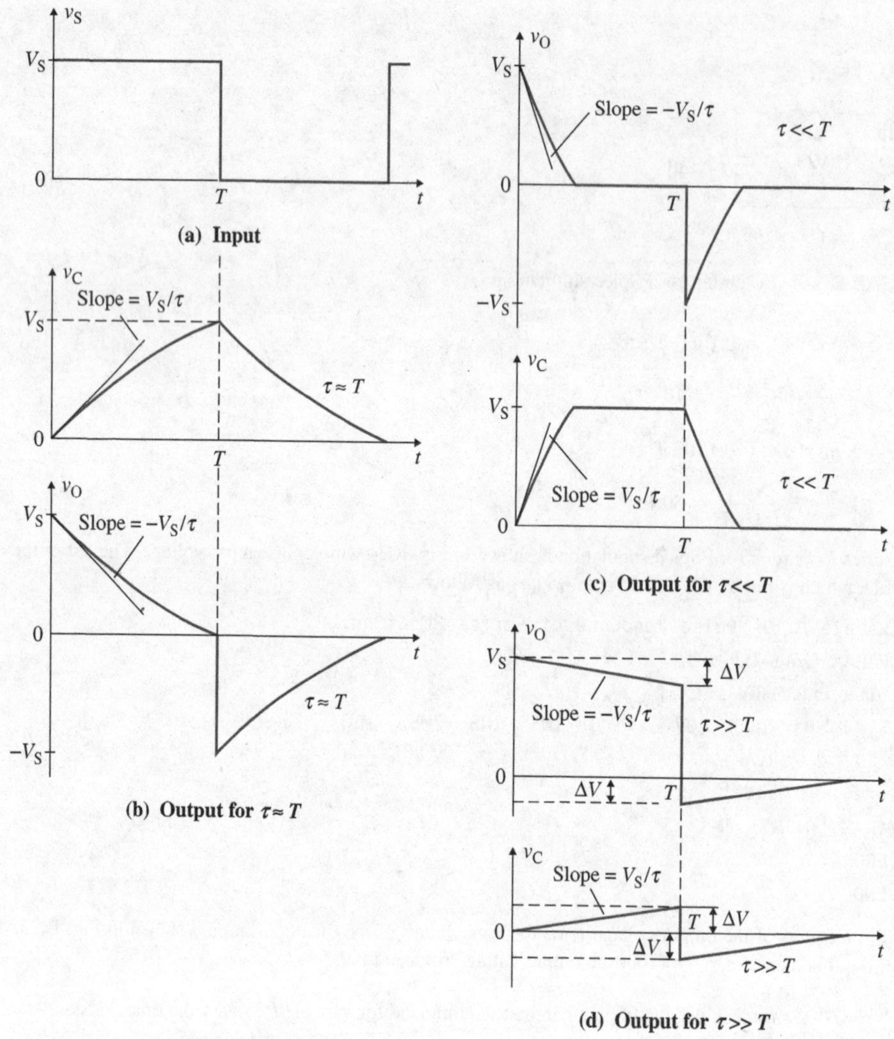

FIGURE B.26 Pulse response of a series CR circuit

B.8.4 Pulse Response of Series *CR* Circuits

An input pulse v_S of duration T, shown in Fig. B.26(a), is applied to the circuit of Fig. B.22(a). The response due to a pulse signal will depend on the ratio of time constant τ to duration T. Let us consider three cases: $\tau = T$, $\tau << T$, and $\tau >> T$.

In case 1, $\tau = T$, the capacitor voltage $v_C(t)$ starts to increase exponentially while the output voltage $v_O(t)$ starts to decay exponentially from V_S. This situation is shown in Fig. B.26(b). At $t = T$, the input signal v_S falls to zero and the capacitor discharges exponentially through the resistance R and the input source v_S. The output voltage v_O decays exponentially from a negative value toward zero.

In case 2, $\tau << T$, the output voltage $v_O(t)$ decays exponentially with a small time constant to zero. During the period $0 \le t \le T$, the capacitor charges exponentially to reach a steady-state value of V_S. For $t > T$, the capacitor discharges exponentially with a small time constant through the resistance R and the input source v_S. The output voltage $v_O(t)$ decays exponentially from a negative value toward zero. The waveforms for $v_O(t)$ and $v_C(t)$ are shown in Fig. B.26(c).

In case 3, $\tau >> T$, output voltage v_O falls only a small amount. The portion of the exponential curve v_O from $t = 0$ to $t = T$ will be almost linear, as shown in Fig. B.26(d). The fall in the output voltage v_O can be found approximately from Fig. B.26(d) as

$$\Delta V = \frac{V_S}{\tau} T \tag{B.41}$$

The *sag* S of the output voltage is defined as

$$S = \frac{\Delta V}{V_S} = \frac{V_S T/\tau}{V_S} = \frac{T}{\tau} = \frac{T}{RC} \tag{B.42}$$

EXAMPLE B.11

Using PSpice/SPICE to plot the pulse response of a *CR* circuit Use PSpice/SPICE to plot the output voltage $v_O(t)$ of the circuit in Fig. B.22(a) for $\tau = 0.1$ ms, 1 ms, and 5 ms. Assume $T = 2$ ms and $v_S = V_S = 1$ V of pulse input.

SOLUTION

For $\tau = 0.1$ ms, let $C = 0.1$ μF. Then

$$R = \frac{\tau}{C} = \frac{0.1 \text{ ms}}{0.1\ \mu\text{F}} = 1 \text{ k}\Omega$$

For $\tau = 1$ ms, let $C = 0.1$ μF. Then

$$R = \frac{\tau}{C} = \frac{1 \text{ ms}}{0.1\ \mu\text{F}} = 10 \text{ k}\Omega$$

For $\tau = 5$ ms, let $C = 0.1$ μF. Then

$$R = \frac{\tau}{C} = \frac{5 \text{ ms}}{0.1\ \mu\text{F}} = 50 \text{ k}\Omega$$

The series *CR* circuit for PSpice simulation is shown in Fig. B.27.

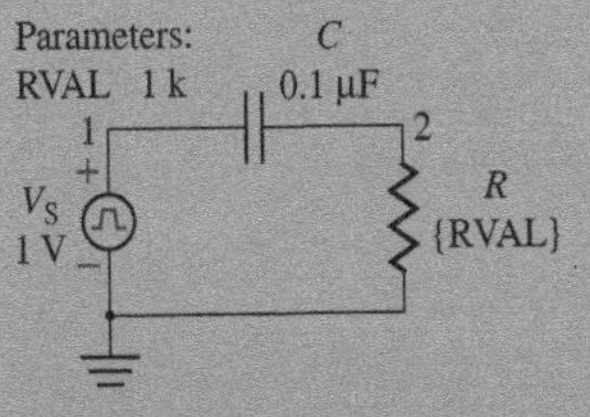

FIGURE B.27 *CR* circuit for PSpice simulation

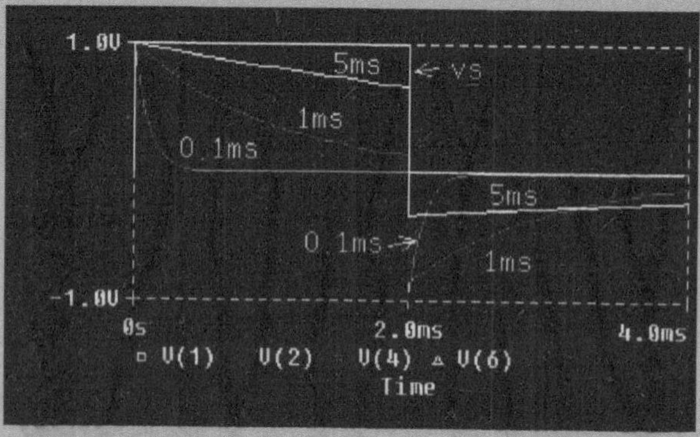

FIGURE B.28 Plots of $v_O(t)$ for Example B.11

The list of the circuit file is as follows.

```
Example B.11  Pulse Response of Series CR Circuit
.PARAM RVAL=1K
.STEP PARAM RVAL LIST  1K  10K  50K
VS  1  0  PULSE (0V  1V  0  1NS  1NS  2MS  4MS)    ; Pulse input voltage
R  2  0  {RVAL}
C  1  2  0.1UF
.TRAN  0.1MS  4MS                                   ; Transient analysis
.PROBE
.END
```

The plots of the output voltage $v_O(t)$ for three values of the time constant are shown in Fig. B.28.

EXAMPLE B.12

Pulse response of a parallel *RC* circuit A constant-current source $i_S = I_S$, shown in Fig. B.29(a), feeds a parallel *RC* circuit with $C = 0.1$ μF and $R = 100$ kΩ, as shown in Fig. B.29(b). The input is a pulse current of duration $T = 0.5$ ms. Determine **(a)** the instantaneous current $i_C(t)$ through capacitance *C*, **(b)** the instantaneous current $i_R(t)$ through resistance *R*, and **(c)** the sag *S* of the capacitor current.

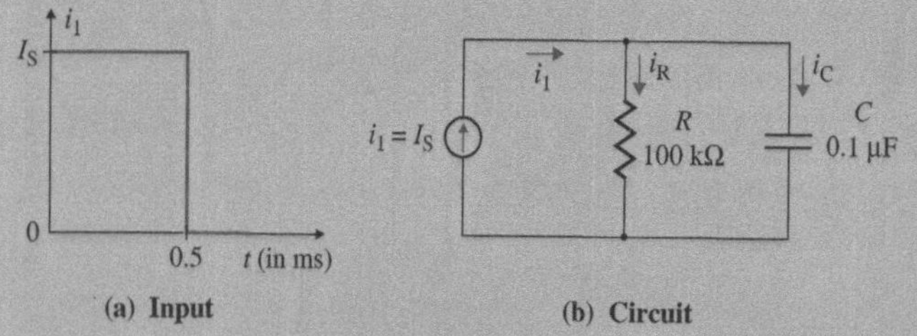

FIGURE B.29 Parallel *RC* circuit with constant current source I_S

SOLUTION

For a pulse signal, the input current in Laplace's domain is $I_S(s) = I_S/s$.

(a) Using the current divider rule, we can find the capacitor current I_C in Laplace's domain:

$$I_C(s) = \frac{R}{R + 1/Cs} I_S(s) = \frac{s}{s + 1/RC} I_S(s) = \frac{s}{s + 1/RC} \times \frac{I_S}{s} = \frac{I_S}{s + 1/RC}$$

$$= I_S \frac{I}{s + 1/RC} \tag{B.43}$$

The inverse transform of $I_C(s)$ in Eq. (B.43) gives

$$i_C(t) = I_S e^{-t/\tau} \tag{B.44}$$

where $\tau = RC$.

(b) The instantaneous current $i_R(t)$ through resistance R is

$$i_R(t) = I_S - i_C(t) = I_S(1 - e^{-t/\tau}) \tag{B.45}$$

(c) $\tau = RC = 100 \times 10^3 \times 0.1 \times 10^{-6} = 10$ ms, and $T = 0.5$ ms. Therefore, $\tau >> T$, and Eq. (B.42) gives

$$S = T/\tau = 0.5/10 = 5\%$$

B.8.5 Step Response of Series *RL* Circuits

A series *RL* circuit with a step-function input is shown in Fig. B.30(a). The output voltage v_O is taken across inductance L. The current i through the inductor can be deduced from

$$V_S = v_L + v_R = L\frac{di}{dt} + Ri \tag{B.46}$$

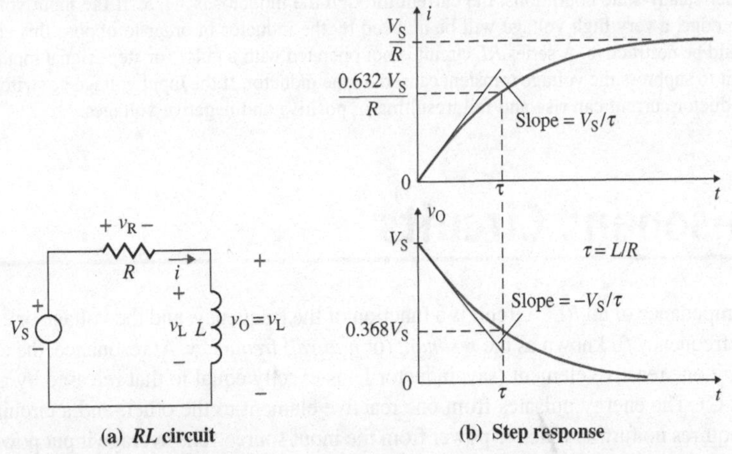

FIGURE B.30 Step response of a series *RL* circuit

with an initial inductor current of zero, $i(t = 0) = 0$. In Laplace's domain of s, Eq. (B.46) becomes

$$\frac{V_S}{s} = LsI(s) + RI(s)$$

which, solved for $I(s)$, gives

$$I(s) = \frac{V_S}{s(sL + R)} = \frac{V_S}{Ls(s + 1/\tau)} = \frac{V_S}{R}\left[\frac{1}{s} - \frac{1}{(s + 1/\tau)}\right] \tag{B.47}$$

where $\tau = L/R$ is the *time constant* of an *RL* circuit. Taking the inverse transform of $I(s)$ in Eq. (B.47) gives the instantaneous current as

$$i(t) = \frac{V_S}{R}(1 - e^{-t/\tau}) \tag{B.48}$$

Using Eq. (B.48), we can find the voltage $v_O(t)$ across the inductance L:

$$v_O(t) = v_L(t) = L\frac{di}{dt} = V_S e^{-t/\tau} \tag{B.49}$$

In the steady state (at $t = \infty$),

$$v_O(t) = 0 \qquad \text{[from Eq. (B.49)]}$$

$$i(t) = \frac{V_S}{R} \qquad \text{[from Eq. (B.48)]}$$

If the output is taken across resistance R, the output voltage $v_O(t)$ becomes

$$v_O(t) = v_R(t) = Ri(t) = V_S(1 - e^{-t/\tau}) \tag{B.50}$$

In the steady state (at $t = \infty$), $v_R(t) = V_S$ and $i(t) = V_S/R$.

▶ **NOTE** Under steady-state conditions, the current through the inductor is V_S/R. If the input voltage v_S is turned off with a sharp edge, a very high voltage will be induced by the inductor in order to oppose this change of current. This voltage could be destructive. A series *RL* circuit is not operated with a pulse (or step) signal input unless there is a protection circuit to suppress the voltage transient caused by the inductor. If the input voltage v_S switches between $+V_S$ and $-V_S$, the inductor current can rise and fall, resulting in positive and negative voltages.

B.9 Resonant Circuits

The effective impedance of an *RLC* circuit is a function of the frequency, and the voltage or current becomes maximum at a frequency f_n, known as the *resonant* (or *natural) frequency*. At resonance, the energy absorbed at any instant by one reactive element (say, inductor L) is exactly equal to that released by another element (say, capacitor C). The energy pulsates from one reactive element to the other, and a circuit with no resistive element requires no further reactive power from the input source. The average input power, which is the power dissipated in the resistive element, becomes maximum at resonance. Resonant circuits are of two types: series resonant circuits and parallel resonant circuits.

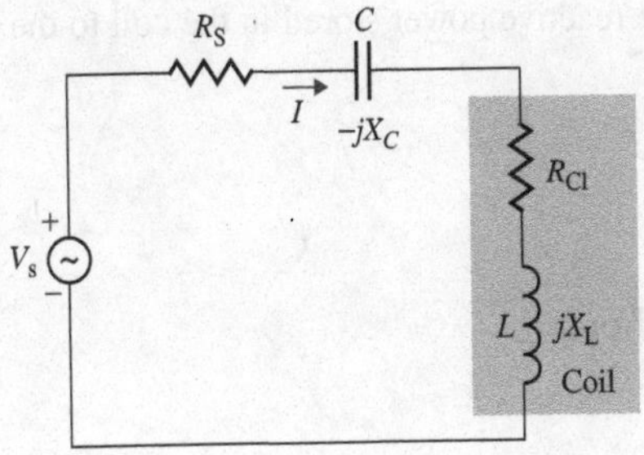

FIGURE B.31 Series resonant *RLC* circuit

B.9.1 Series Resonant Circuits

A series resonant *RLC* circuit is shown in Fig. B.31, where R_{Cl} is the internal resistance of the coil and R_S is the source resistance. If we define $R = R_{Cl} + R_S$, the total series impedance Z of the circuit is given by

$$Z = R + j(X_L - X_C) \tag{B.51}$$

The series resonance will occur at $f = f_n$ when

$$X_L = X_C \tag{B.52}$$

Equation (B.52) can also be stated as

$$\omega L = \frac{1}{\omega C} \quad \text{or} \quad 2\pi f_n L = \frac{1}{(2\pi f_n C)}$$

which gives the series resonant frequency f_n as

$$f_n = \frac{1}{2\pi\sqrt{LC}} \tag{B.53}$$

The impedance Z_n at resonance becomes

$$Z_n = Z = R \tag{B.54}$$

A series resonant circuit is normally defined by a *quality factor* Q_s, which is defined as the ratio of the reactive power stored in either the inductor or the capacitor to the average power dissipated in the resistor at resonance; that is,

$$Q_s = \frac{\text{Reactive power}}{\text{Average power}}$$

$$= \begin{cases} \dfrac{I^2 X_L}{I^2 R} = \dfrac{X_L}{R} = \dfrac{2\pi f_n L}{R} & \text{(for an inductive reactance)} \quad \text{(B.55)} \\[2ex] \dfrac{I^2 X_C}{I^2 R} = \dfrac{X_C}{R} = \dfrac{1}{2\pi f_n C R} & \text{(for a capacitive reactance)} \quad \text{(B.56)} \end{cases}$$

The quality factor Q_{C1} of a coil is defined as the ratio of the reactive power stored in the coil to the power dissipated in the coil resistance R_{C1}; that is,

$$Q_{C1} = \frac{\text{Reactive power}}{\text{Power dissipated}} = \frac{X_L}{R_{C1}}$$

The rms voltage V_L across inductor L at resonance can be found from

$$V_L = \frac{X_L V_s}{Z_n} = \frac{X_L V_s}{R} = Q_s V_s \tag{B.57}$$

The rms voltage V_C across capacitor C at resonance can be found from

$$V_C = \frac{X_C V_s}{Z_n} = \frac{X_L V_s}{R} = Q_s V_s \tag{B.58}$$

In many electronic circuits, the quality factor Q_s is high, in the range of 80 to 400. If $V_s = 30$ V and $Q_s = 80$, for example, then $V_C = V_L = 80 \times 30 = 2400$ V, and all electronic devices in the circuit will be subjected to this high voltage. Thus, a designer must be careful to protect the circuit from a high voltage across the inductor or the capacitor of a resonant circuit.

B.9.2 Parallel Resonant Circuits

A parallel resonant *RLC* circuit is shown in Fig. B.32(a). This circuit is also known as a *tank circuit*. The input signal to a tank circuit is usually a current source. This type of circuit is frequently used with active devices such as transistors, which have the characteristic of a constant current source. By replacing the series *RL* combination of the inductor with a parallel combination, we can obtain the circuit in Fig. B.32(b), for which the admittance Y_{RL} is

$$Y_{RL} = \frac{1}{R_{C1} + jX_L} = \frac{R_{C1}}{R_{C1}^2 + X_L^2} - j\frac{X_L}{R_{C1}^2 + X_L^2} = \frac{1}{R_p} - j\frac{1}{X_p}$$

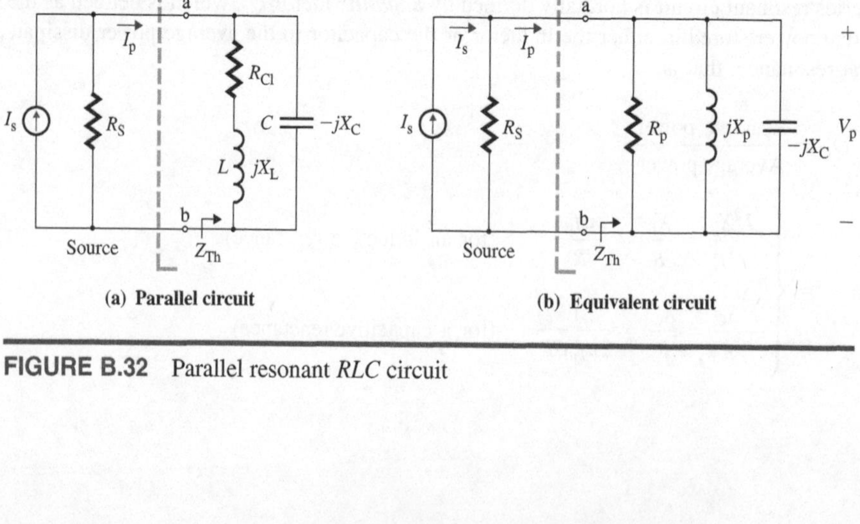

FIGURE B.32 Parallel resonant *RLC* circuit

$$\text{where} \quad R_p = \frac{R_{C1}^2 + X_L^2}{R_{C1}} \tag{B.59}$$

$$X_p = \frac{R_{C1}^2 + X_L^2}{X_L} \tag{B.60}$$

For the resonant condition,

$$X_p = X_C$$

Substituting X_p from Eq. (B.60) into the preceding equation, we get

$$\frac{R_{C1}^2 + X_L^2}{X_L} = X_C$$

$$R_{C1}^2 + X_L^2 = X_C X_L$$

$$X_L^2 = X_C X_L - R_{C1}^2$$

or

$$X_L^2 = \frac{L}{C} - R_{C1}^2$$

$$X_L = \left[\frac{L}{C} - R_{C1}^2\right]^{1/2} \tag{B.61}$$

which gives the parallel resonant frequency f_p as

$$f_p = \frac{1}{2\pi L}\left[\frac{L}{C} - R_{C1}^2\right]^{1/2} = \frac{1}{2\pi\sqrt{LC}}\left[1 - \frac{CR_{C1}^2}{L}\right]^{1/2} \tag{B.62}$$

$$= f_n\left[1 - \frac{CR_{C1}^2}{L}\right]^{1/2} \tag{B.63}$$

Thus, the parallel resonant frequency f_p, which is dependent on the coil resistance R_{Cl}, is less than the series resonant frequency f_n. For the conditions $(CR_{Cl}^2/L) << 1$ or $R_{Cl} << \sqrt{L/C}$, and for $R_{Cl} = 0$, Eq. (B.63) gives

$$f_p = f_n$$

The quality factor Q_p of the parallel resonant RLC circuit can be determined by the ratio of reactive power to the real power at resonance; that is,

$$Q_p = \frac{V_p^2/X_p}{V_p^2/(R_S \parallel R_p)} = \frac{R_S \parallel R_p}{X_p} \tag{B.64}$$

where V_p is the voltage across the parallel branches.

EXAMPLE B.13

Finding the parallel resonant frequency The parameters of the parallel resonant *RLC* circuit in Fig. B.32(a) are $R_{Cl} = 47\ \Omega$, $L = 5$ mH, $C = 50$ pF, $R_S = 20$ kΩ, and the current source $I_s = 6$ mA. Calculate **(a)** the parallel resonant frequency f_p, **(b)** the voltage V_p across the resonant circuit at resonance, **(c)** the quality factor Q_{Cl} of the coil, and **(d)** the quality factor Q_p of the resonant circuit.

SOLUTION

$R_{Cl} = 47\ \Omega$, $L = 5$ mH, $C = 50$ pF, $R_S = 20$ kΩ, and $I_s = 6$ mA.

(a) From Eq. (B.53),

$$f_n = \frac{1}{2\pi \times \sqrt{5 \times 10^{-3} \times 50 \times 10^{-12}}} = 318.3 \text{ kHz}$$

From Eq. (B.63),

$$f_p = 318.3 \times 10^3 \times \left[1 - 50 \times 10^{-12} \times \frac{47^2}{5 \times 10^{-3}}\right]^{1/2} \approx 318.3 \text{ kHz}$$

(b) We know that

$$X_L = 2\pi f_p L = 2\pi \times 318.3 \times 10^3 \times 5 \times 10^{-3} = 9999.7\ \Omega$$

From Eq. (B.59), the effective resistance R_p of the parallel circuit is

$$R_p = \frac{47^2 + 9999.7^2}{47} = 2127.6 \text{ k}\Omega$$

Using the current divider rule, the rms current I_p through the parallel circuit is

$$I_p = \frac{R_S}{R_S + R_p} I_s = \frac{20 \text{ k}\Omega \times 6 \text{ mA}}{20 \text{ k}\Omega + 2127.6 \text{ k}\Omega} = 55.876\ \mu\text{A}$$

$$V_p = I_p R_p = 55.876\ \mu\text{A} \times 2127.6 \text{ k}\Omega = 118.88 \text{ V}$$

(c) We have

$$Q_{Cl} = \frac{X_L}{R_{Cl}} = \frac{9999.7}{47} = 212.8$$

(d) From Eq. (B.60), the effective inductive reactance X_p of the parallel circuit is

$$X_p = \frac{47^2 + 9999.7^2}{9999.7} = 9999.92\ \Omega$$

$$R_S \| R_p = \frac{20 \times 2127.6}{20 + 2127.6} = 19.81 \text{ k}\Omega$$

From Eq. (B.64), $Q_p = 19{,}810/9999.92 = 1.98$.

B.10 Frequency Response of First- and Second-Order Circuits

A sine wave (or sinusoid) is generally used to characterize electronic circuits, such as amplifiers and filters. The frequency response refers to the output characteristic for a sine-wave input. If a sine-wave input voltage

$$v_s(t) = V_m \sin \omega t \tag{B.65}$$

where V_m is the peak input voltage and ω is the frequency of the input voltage in radians per second, is applied to a circuit, the output voltage $v_o(t)$ can have a different amplitude and phase than the input voltage. The output voltage $v_o(t)$ will be of the form

$$v_o(t) = V_p \sin(\omega t - \phi) \tag{B.66}$$

where V_p is the peak output voltage. If f is the input frequency in hertz,

$$\omega = 2\pi f$$

A typical relationship between the input and output voltages of an amplifier is shown in Fig. B.33. An amplifier has in general a voltage gain of $A_v = V_p/V_m > 1$. Thus, the peak output V_p has a higher value than the input voltage V_m.

If $V_s(j\omega)$ and $V_o(j\omega)$ denote the rms values of input voltage and output voltage, respectively, as a function of frequency, the voltage gain $G(j\omega)$ is defined as

$$G(j\omega) = \frac{V_o(j\omega)}{V_s(j\omega)} \quad \text{(dimensionless)} \tag{B.67}$$

$G(j\omega)$ is a complex function with a magnitude and a phase. The magnitude $|G(j\omega)|$ gives the magnitude response, and the phase of $G(j\omega)$ gives the phase response. The magnitude and the phase are generally plotted against the frequency, with a logarithmic scale used for the frequency. The magnitude $|G(j\omega)|$ is normally expressed in decibels (dB):

$$\text{Magnitude in dB} = 20 \log_{10} |G(j\omega)|$$

We will review the frequency responses of the following circuits: first-order low-pass *RC* circuits, first-order high-pass *CR* circuits, second-order series *RLC* circuits, and second-order parallel *RLC* circuits.

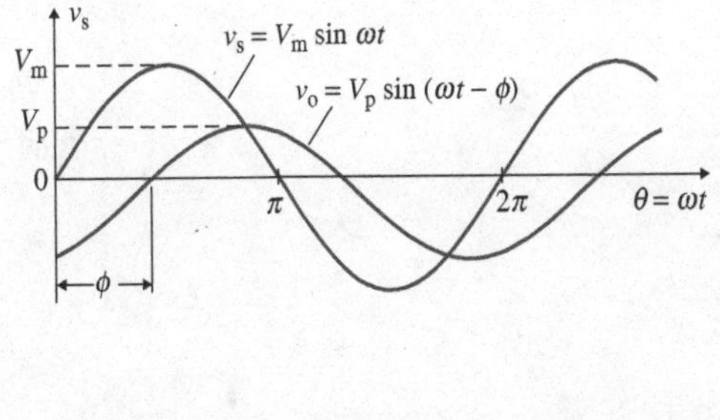

FIGURE B.33 Typical sinusoidal input and output voltages

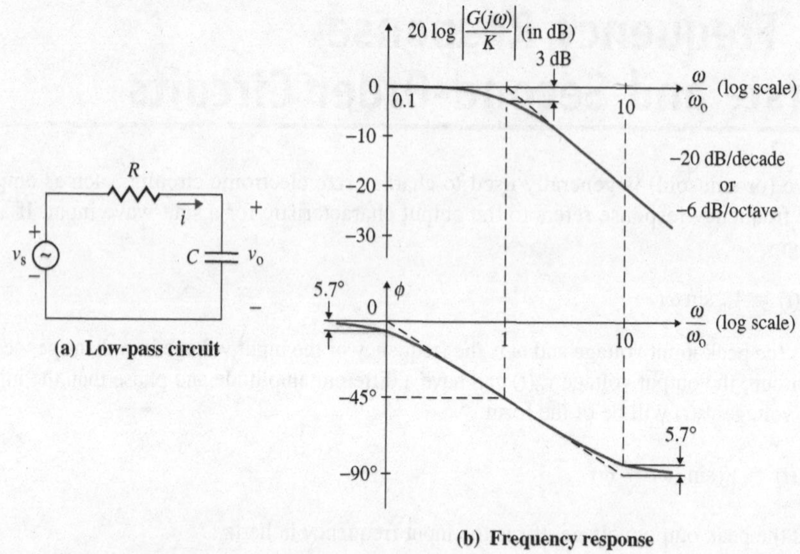

FIGURE B.34 First-order low-pass *RC* circuit

B.10.1 First-Order Low-Pass *RC* Circuits

A typical low-pass *RC* circuit is shown in Fig. B.34(a). The output voltage v_o is taken across capacitance *C*. The capacitor impedance in Laplace's domain is $1/Cs$. Using the voltage divider rule, we can find the voltage gain $G(s)$:

$$G(s) = \frac{V_o(s)}{V_s(s)} = \frac{1/Cs}{R + (1/Cs)} = \frac{1}{1 + sRC} \quad \text{(dimensionless)}$$

In the frequency domain, $s = j\omega$, and

$$G(j\omega) = \frac{1}{1 + j\omega RC} = \frac{1}{1 + j\omega\tau}$$

where $\tau = RC$. Thus, the magnitude $|G(j\omega)|$ of the voltage gain can be found from

$$|G(j\omega)| = \frac{1}{[1 + (\omega\tau)^2]^{1/2}} = \frac{1}{[1 + (\omega/\omega_o)^2]^{1/2}} \tag{B.68}$$

and the phase angle ϕ of $G(j\omega)$ is given by

$$\phi = -\tan^{-1}(\omega\tau) = -\tan^{-1}\left(\frac{\omega}{\omega_o}\right) \tag{B.69}$$

where $\omega_o = 1/RC = 1/\tau$.

For $\omega << \omega_o$,

$$|G(j\omega)| \approx 1$$

$$20 \log_{10}|G(j\omega)| = 0$$

$$\phi = 0$$

Therefore, at low frequency, the magnitude plot is a straight horizontal line at 0 dB. For $\omega >> \omega_o$,

$$|G(j\omega)| \approx \omega_o/\omega$$

$$20 \log_{10}|G(j\omega)| = 20 \log_{10}(\omega_o/\omega)$$

$$\phi = \pi/2$$

For $\omega = \omega_o$,

$$|G(j\omega)| = 1/\sqrt{2}$$

$$20 \log_{10}|G(j\omega)| = 20 \log_{10}(1/\sqrt{2}) = -3 \text{ dB}$$

$$\phi = \pi/4$$

Let us consider a high frequency such that $\omega_1 >> \omega_o$. At $\omega = \omega_1$, the magnitude is $20 \log_{10}(\omega_o/\omega_1)$. At $\omega = 10\omega_1$, the magnitude is $20 \log_{10}(\omega_o/10\omega_1)$. The change in magnitude between $\omega = \omega_1$ and $\omega = 10\omega_1$ becomes

$$20 \log_{10}(\omega_o/10\omega_1) - 20 \log_{10}(\omega_o/\omega_1) = 20 \log_{10}(1/10) = -20 \text{ dB}$$

If the frequency is doubled so that $\omega = 2\omega_1$, the change in magnitude becomes

$$20 \log_{10}(\omega_o/2\omega_1) - 20 \log_{10}(\omega_o/\omega_1) = 20 \log_{10}(1/2) = -6 \text{ dB}$$

The frequency response is shown in Fig. B.34(b). If the frequency doubled, the interval between the two frequencies is called an *octave* on the frequency axis. If the frequency is increased by a factor of 10, the interval between the two frequencies is called a *decade*. Thus, for a decade increase in frequency, the magnitude changes by -20 dB. The magnitude plot is a straight line with a slope of -20 dB/decade or -6 dB/octave. The magnitude curve is therefore defined by two straight-line asymptotes, which meet at the *corner frequency* (or *break frequency*) ω_o. The difference between the actual magnitude curve and the asymptotic curve is largest at the break frequency. The error can be found by finding the gain at $\omega = \omega_o$; that is, $|G(j\omega)| = 1/\sqrt{2}$, and $20 \log_{10}(1/\sqrt{2}) = -3$ dB. This error is symmetrical with respect to the break frequency. The break frequency is also known as the *3-dB frequency*.

The circuit in Fig. B.34(a) passes only the low-frequency signal, and the amplitude falls at higher frequencies. A circuit with this type of response is known as a *low-pass circuit*. The gain function (commonly known as the transfer function) of a low-pass circuit has the general form

$$G(s) = \frac{K}{1 + (s/\omega_o)} \tag{B.70}$$

where K is the magnitude of the gain function at $\omega = 0$ (or the *low-pass gain*). A low-pass circuit exhibits the characteristics of (a) finite output at a very low frequency, tending to zero, and (b) zero output at a very high frequency, tending to infinity.

EXAMPLE B.14

Using PSpice/SPICE to plot the frequency response of a low-pass *RC* circuit Use PSpice/SPICE to plot the frequency response of the low-pass RC circuit in Fig. B.34(a). Assume $V_m = 1$ V (peak AC), $R = 10$ kΩ, and $C = 0.1$ μF. The frequency f varies from 1 Hz to 100 kHz.

SOLUTION

The low-pass RC circuit for PSpice simulation is shown in Fig. B.35. The PSpice plot of the magnitude and phase angle are shown in Fig. B.36, which gives $f_o = 161$ Hz at -3 dB.

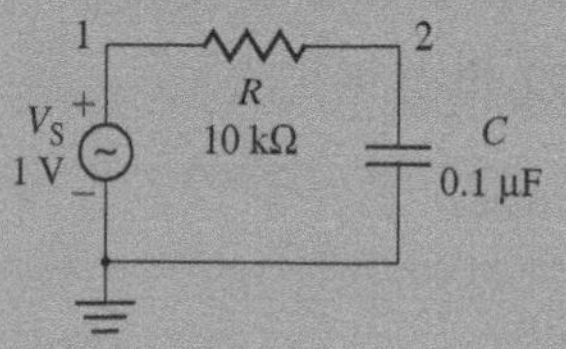

FIGURE B.35 Low-pass RC circuit for PSpice simulation

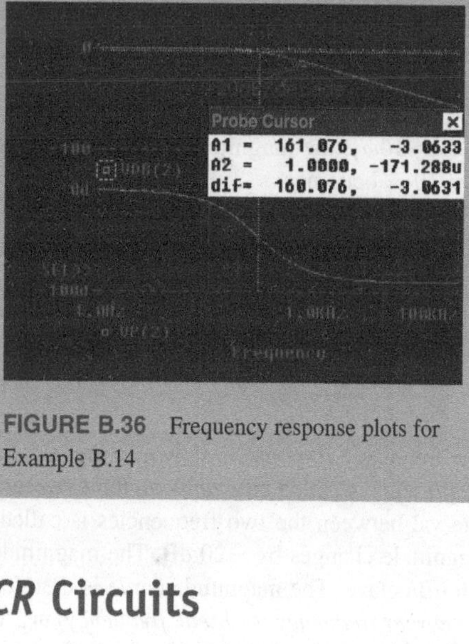

FIGURE B.36 Frequency response plots for Example B.14

B.10.2 First-Order High-Pass *CR* Circuits

A high-pass CR circuit is shown in Fig. B.37(a). The output voltage v_o is taken across resistance R. Using the voltage divider rule, we can find the voltage gain $G(s)$ in Laplace's domain:

$$G(s) = \frac{V_o(s)}{V_s(s)} = \frac{R}{R + (1/Cs)} = \frac{sRC}{1 + sRC} \quad \text{(dimensionless)}$$

In the frequency domain, $s = j\omega$, and

$$G(j\omega) = \frac{j\omega RC}{1 + j\omega RC} = \frac{j\omega\tau}{1 + j\omega\tau}$$

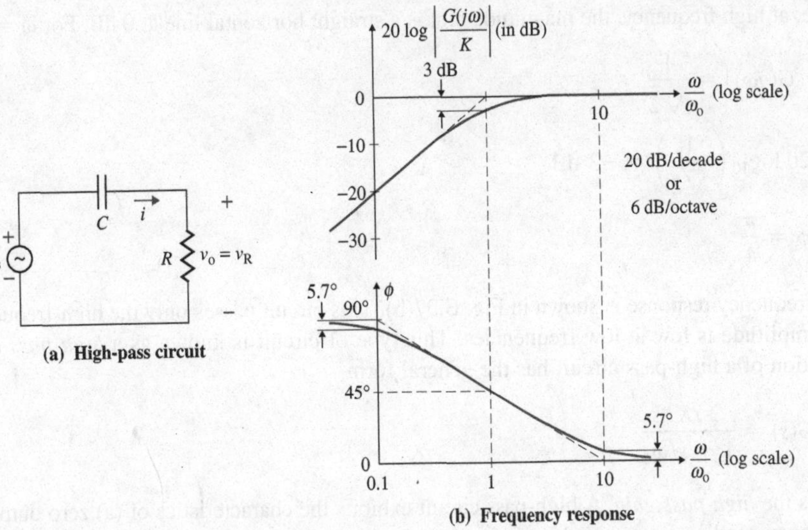

FIGURE B.37 First-order high-pass *CR* circuit

where $\tau = RC$. Thus, the magnitude $|G(j\omega)|$ of the voltage gain can be found from

$$|G(j\omega)| = \frac{\omega\tau}{[1 + (\omega\tau)^2]^{1/2}} = \frac{\omega/\omega_o}{[1 + (\omega/\omega_o)^2]^{1/2}} \tag{B.71}$$

and the phase angle ϕ of $G(j\omega)$ is given by

$$\phi = \frac{\pi}{2} - \tan^{-1}\left(\frac{\omega}{\omega_o}\right) \tag{B.72}$$

where $\omega_o = 1/RC = 1/\tau$.

For $\omega << \omega_o$,

$$|G(j\omega)| \approx \frac{\omega}{\omega_o}$$

$$20 \log_{10}|G(j\omega)| = 20 \log_{10}\left(\frac{\omega}{\omega_o}\right)$$

$$\phi = \frac{\pi}{2}$$

Therefore, for a decade increase in frequency, the magnitude changes by 20 dB. The magnitude plot is a straight line with a slope of +20 dB/decade or +6 dB/octave.

For $\omega >> \omega_o$,

$$|G(j\omega)| = 1$$

$$20 \log_{10}|G(j\omega)| = 0$$

$$\phi \approx 0$$

Therefore, at high frequency, the magnitude plot is a straight horizontal line at 0 dB. For $\omega = \omega_o$,

$$|G(j\omega)| = \frac{1}{\sqrt{2}}$$

$$20 \log_{10}\left(\frac{1}{\sqrt{2}}\right) = -3 \text{ dB}$$

$$\phi = \frac{\pi}{4}$$

The frequency response is shown in Fig. B.37(b). This circuit passes only the high-frequency signal, and the amplitude is low at low frequencies. This type of circuit is known as a *high-pass circuit*. The gain function of a high-pass circuit has the general form

$$G(s) = \frac{sK}{1 + s/\omega_o} \tag{B.73}$$

where K is the *high-pass gain*. A high-pass circuit exhibits the characteristics of (a) zero output at a very low frequency, tending to zero, and (b) finite output at a very high frequency, tending to infinity.

EXAMPLE B.15

Using PSpice/SPICE to plot the frequency response of a high-pass *CR* circuit Use PSpice/SPICE to plot the frequency response of the high-pass *CR* circuit in Fig. B.37(a). Assume $V_m = 1$ V (peak AC), $R = 10$ kΩ, and $C = 0.1$ μF. The frequency f varies from 1 Hz to 100 kHz.

SOLUTION

The high-pass *CR* circuit for PSpice simulation is shown in Fig. B.38. The PSpice plots of the magnitude and phase angle are shown in Fig. B.39, which gives $f_o = 157$ Hz at −3 dB.

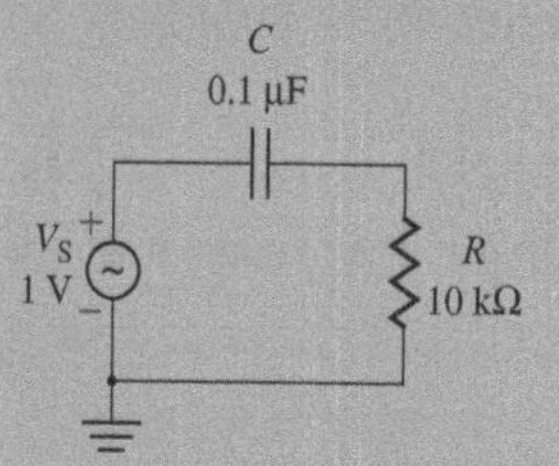

FIGURE B.38 High-pass *CR* circuit for PSpice simulation

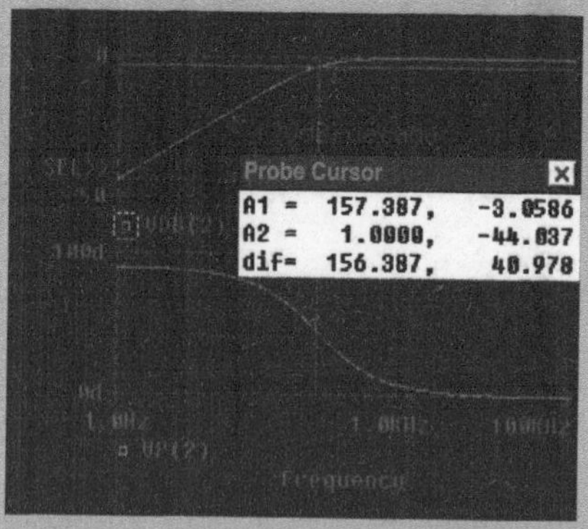

FIGURE B.39 Frequency response plots for Example B.15

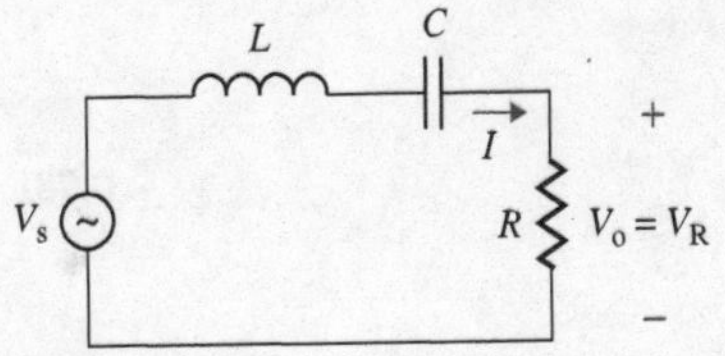

FIGURE B.40 Series *RLC* circuit

B.10.3 Second-Order Series *RLC* Circuits

A series *RLC* circuit is shown in Fig. B.40. The output voltage v_o is taken across resistance *R*. Using the voltage divider rule, we can find the voltage gain (or the transfer function) in Laplace's domain of *s*:

$$G(s) = \frac{V_o(s)}{V_s(s)} = \frac{R}{R + sL + (1/Cs)} = \frac{sR/L}{s^2 + (sR/L) + (1/LC)} \quad \text{(dimensionless)} \tag{B.74}$$

Defining $\omega_n = 1/\sqrt{LC}$ as the *natural frequency* in rad/s and $\alpha = R/(2L)$ as the *damping factor*, we can write Eq. (B.74) as

$$G(s) = \frac{2\alpha s}{s^2 + 2\alpha s + \omega_n^2} \tag{B.75}$$

Let us define

$$\delta = \frac{\alpha}{\omega_n} = \frac{R}{2L}\sqrt{LC} = \frac{R}{2}\sqrt{\frac{C}{L}}$$

as the *damping ratio*. Then Eq. (B.75) becomes

$$G(s) = \frac{2\delta\omega_n s}{s^2 + 2\delta\omega_n s + \omega_n^2} \quad \text{(dimensionless)} \tag{B.76}$$

where $\delta < 1$. (Note that δ is not necessarily less than 1, but has been assumed at less than 1 for this discussion.) In the frequency domain, $s = j\omega$. Thus,

$$G(j\omega) = \frac{2\delta\omega_n j\omega}{(j\omega)^2 + 2\delta\omega_n(j\omega) + \omega_n^2} = \frac{j2\delta\omega/\omega_n}{-(\omega/\omega_n)^2 + (j2\delta\omega/\omega_n) + 1}$$

$$= \frac{j2\delta\omega/\omega_n}{1 + (j2\delta\omega/\omega_n) - (\omega/\omega_n)^2} \tag{B.77}$$

Let us define $u = \omega/\omega_n$ as the *frequency ratio* or the *normalized frequency*. Then Eq. (B.77) can be simplified to

$$G(j\omega) = \frac{j2\delta u}{1 + j2\delta u - u^2}$$

The magnitude $|G(j\omega)|$ can be found from

$$|G(j\omega)| = \frac{2\delta u}{[(1-u^2)^2 + (2\delta u)^2]^{1/2}} \tag{B.78}$$

The phase angle ϕ of $G(j\omega)$ can be found from

$$\phi = \frac{\pi}{2} - \tan^{-1}\left(\frac{2\delta u}{1-u^2}\right) \tag{B.79}$$

For low frequencies $u << 1$,

$$|G(j\omega)| \approx 2\delta u$$

$$20\log_{10}|G(j\omega)| \approx 20\log_{10}(2\delta u)$$

$$\phi \approx \frac{\pi}{2}$$

Therefore, at low frequencies, the magnitude plot is a straight line with a slope of +20 dB/decade or +6 dB/octave. For $u = 1$, $|G(j\omega)| \approx 1$ only if

$$\delta \approx 1$$

$$20\log_{10}|G(j\omega)| = 0 \text{ dB}$$

$$\phi = 0$$

For $u >> 1$,

$$|G(j\omega)| \approx \frac{2\delta u}{u^2} = \frac{2\delta}{u}$$

$$20\log_{10}|G(j\omega)| \approx 20\log_{10}(2\delta) - 20\log_{10}(u) \approx -20\log_{10}(u)$$

$$\phi \approx -\frac{\pi}{2}$$

Therefore, at high frequencies, the magnitude plot is a straight line with a slope of −20 dB/decade or −6 dB/octave. The actual characteristic will differ considerably from the asymptotic lines, and the error will depend on the damping ratio δ. The magnitude and frequency plots of the series *RLC* circuit are shown in Fig. B.41. The magnitude peaks at $u = 1$.

If the output voltage of the series *RLC* circuit falls below 70% of its maximum value, the output is not considered the significant value. The *cutoff frequency* is defined as that value of frequency for which the magnitude of the gain drops to 70.7% of its maximum value $|G(j\omega)|_{max} = 1$. Thus, at cutoff frequencies, Eq. (B.78) gives

$$|G(j\omega)| = \frac{2\delta u}{[(1-u^2)^2 + (2\delta u)^2]^{1/2}} = 0.707 = \frac{1}{\sqrt{2}}$$

or $$\sqrt{2}(2\delta u) = [(1-u^2)^2 + (2\delta u)^2]^{1/2}$$

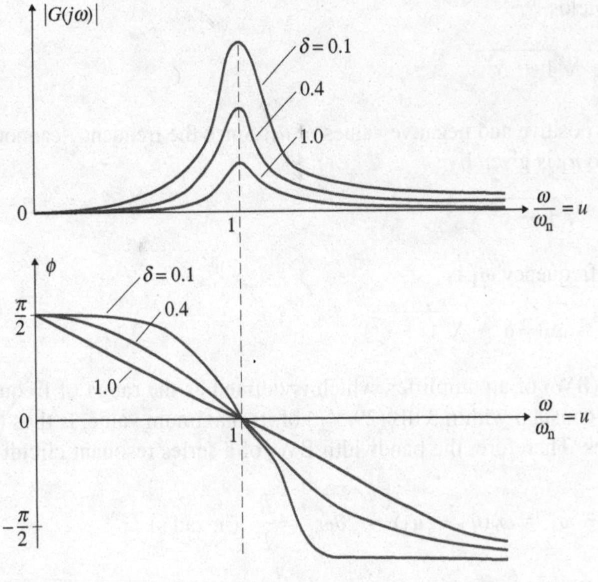

FIGURE B.41 Frequency response of a series *RLC* circuit

Squaring both sides yields

$$2(2\delta u)^2 = (1 - u^2)^2 + (2\delta u)^2$$

or

$$(2\delta u)^2 = (1 - u^2)^2 \tag{B.80}$$

The possible solutions of Eq. (B.80) are

$$2\delta u_1 = 1 - u_1^2$$

$$u_1^2 + 2\delta u_1 - 1 = 0 \tag{B.81}$$

$$2\delta u_2 = -(1 - u_2^2) = u_2^2 - 1$$

$$u_2^2 - 2\delta u_2 - 1 = 0 \tag{B.82}$$

Solving Eq. (B.82) yields

$$u_2 = \delta \pm \sqrt{1 + \delta^2}$$

Since the frequency cannot be negative, the upper cutoff frequency ratio u_2 is given by

$$u_2 = \delta + \sqrt{1 + \delta^2} \tag{B.83}$$

and the upper cutoff frequency ω_2 is

$$\omega_2 = u_2\omega_n = \omega_n(\delta + \sqrt{1 + \delta^2}) \tag{B.84}$$

Solving Eq. (B.81) yields

$$u_1 = -\delta \pm \sqrt{1 + \delta^2}$$

which will give both positive and negative values of u_1. Since the frequency cannot be negative, the lower cutoff frequency ratio u_1 is given by

$$u_1 = -\delta + \sqrt{1 + \delta^2} \tag{B.85}$$

and the lower cutoff frequency ω_1 is

$$\omega_1 = u_1\omega_n = \omega_n(-\delta + \sqrt{1 + \delta^2}) \tag{B.86}$$

The *bandwidth* (BW) of an amplifier, which is defined as the range of frequencies over which the gain remains almost constant within 3 dB (29.3%) of its maximum value, is thus the difference between the cutoff frequencies. Therefore, the bandwidth BW_s of a series resonant circuit can be found from

$$BW_s = \omega_2 - \omega_1 = \omega_n(u_2 - u_1) = 2\delta\omega_n = \frac{R}{L} \quad \text{(in rad/s)} \tag{B.87}$$

$$BW_s = f_2 - f_1 = \frac{1}{2\pi}\frac{R}{L} \quad \text{(in Hz)} \tag{B.88}$$

From Eq. (B.55), $R/L = 2\pi f_n/Q_s$. Thus, Eq. (B.88) can be rewritten as

$$BW_s = \frac{1}{2\pi}\frac{R}{L} = \frac{1}{2\pi}\frac{2\pi f_n}{Q_s} = \frac{f_n}{Q_s} \tag{B.89}$$

which shows that the larger the value of Q_s, the smaller the value of bandwidth BW_s, and vice versa. It can be shown that Eq. (B.89) can be also applied to calculate the bandwidth BW_p of a parallel resonant circuit; that is,

$$BW_p = \frac{f_p}{Q_p} \tag{B.90}$$

where f_p is the parallel resonant frequency in Eq. (B.63) and Q_p is the quality factor of a parallel resonant circuit in Eq. (B.64).

EXAMPLE B.16

Finding the frequency response of a series *RLC* circuit The series *RLC* circuit in Fig. B.40 has $R = 50\ \Omega$, $L = 4$ mH, and $C = 0.15\ \mu$F.

(a) Determine the series resonant frequency f_n, the damping ratio δ, the quality factor Q_s, the cutoff frequencies, and the bandwidth BW_s.

(b) Use PSpice/SPICE to plot the magnitude and phase angle of the output voltage for $R = 50\ \Omega$, $100\ \Omega$, and $200\ \Omega$. The frequency f varies from 100 Hz to 1 MHz. Assume $V_m = 1$ V peak AC.

SOLUTION

(a) $R = 50\ \Omega$, $L = 4$ mH, and $C = 0.15\ \mu$F, so

$$\omega_n = \frac{1}{\sqrt{LC}} = \frac{10^5}{\sqrt{4 \times 1.5}} = 40{,}825 \text{ rad/s}$$

The series resonant frequency is

$$f_n = \frac{\omega_n}{2\pi} = \frac{40{,}825}{2\pi} = 6497.5 \text{ Hz}$$

Since $\alpha = R/(2L) = 50/(2 \times 4 \times 10^{-3}) = 6250$, the damping ratio is

$$\delta = \frac{\alpha}{\omega_n} = \frac{6250}{40{,}825} = 0.1531$$

From Eq. (B.55),

$$Q_s = \omega_n \frac{L}{R} = 40{,}825 \times 4 \times \frac{10^{-3}}{50} = 3.266$$

For the lower cutoff frequency, Eqs. (B.85) and (B.86) give

$$u_1 = -\delta + \sqrt{1 + \delta^2} = -0.1531 + \sqrt{1 + 0.1531^2} = 0.85855$$

$$\omega_1 = u_1 \omega_n = 0.85855 \times 40{,}825 = 35{,}050.4 \text{ rad/s}$$

Thus, $f_1 = 35{,}050.4/2\pi = 5578$ Hz. For the upper cutoff frequency, Eqs. (B.83) and (B.84) give

$$u_2 = \delta + \sqrt{1 + \delta^2} = 0.1531 + \sqrt{1 + 0.1531^2} = 1.16475$$

$$\omega_2 = u_2 \omega_n = 1.16475 \times 40{,}825 = 47{,}551 \text{ rad/s}$$

Thus, $f_2 = 47{,}551/2\pi = 7568$ Hz. From Eq. (B.89), the bandwidth is

$$BW_s = f_2 - f_1 = \frac{f_n}{Q_s} = \frac{6497.5}{3.266} = 1989.4 \text{ Hz}$$

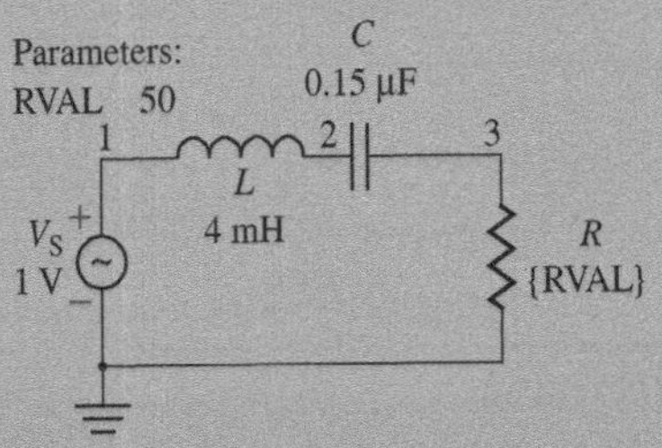

FIGURE B.42 Series *RLC* circuit for PSpice simulation

(b) The series *RLC* circuit for PSpice simulation is shown in Fig. B.42. The list of the circuit file is as follows.

```
Example B.16  Frequency Response of Series RLC Circuit
.PARAM RVAL = 50
.STEP PARAM RVAL LIST  50  100  200
Vm  1  0  AC  1V                     ; AC input of 1 V peak
L  1  2  4MH
C  2  3  0.15UF
R  3  0  {RVAL}
.AC DEC 100 100HZ 1MEGHz             ; AC analysis from f = 100 Hz to 1 MHz
                                     ; with a decade change and 100 points per decade
.PROBE
.END
```

The PSpice plots of the magnitude and phase angle (using EXB-16.SCH) are shown in Fig. B.43. The plot for $R = 50\ \Omega$ gives $f_1 = 5578$ Hz, $f_2 = 7568$ Hz, $f_n = 6457$ Hz, and $BW_s = f_2 - f_1 = 1990$ Hz.

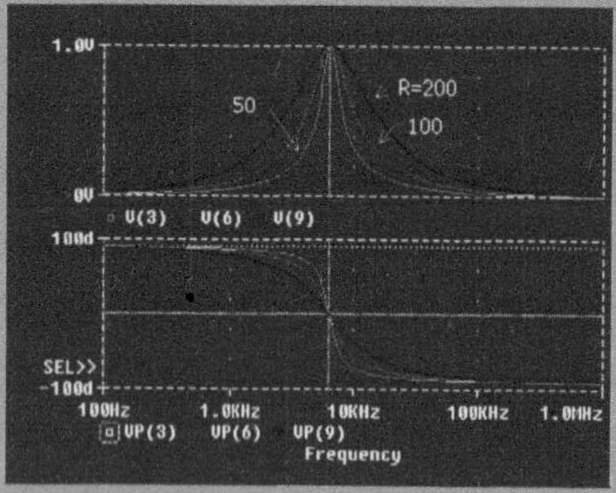

FIGURE B.43 Frequency response plots for Example B.16

B.10.4 Second-Order Parallel *RLC* Circuits

A parallel *RLC* circuit is shown in Fig. B.44. The output voltage v_o is taken across the parallel combination of *R*, *L*, and *C*. The transfer function $G(s) = V_o(s)/I_s(s)$ in Laplace's domain of *s* is the equivalent impedance $Z(s)$.

The function

$$\frac{1}{Z(s)} = \frac{1}{R} + \frac{1}{sL} + sC = \frac{sL + R + s^2LCR}{sRL}$$

$$= \frac{s^2 + (s/RC) + (1/LC)}{s/C} \quad \text{(in siemens, or mhos)}$$

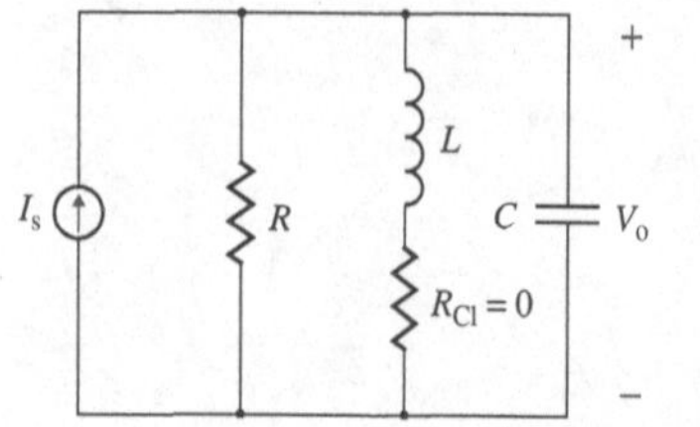

FIGURE B.44 Parallel *RLC* circuit

gives the transfer function $G(s)$ as

$$G(s) = \frac{V_o(s)}{I_s(s)} = Z(s) = \frac{s/C}{s^2 + (s/RC) + (1/LC)}$$

$$= R\frac{s/RC}{s^2 + (s/RC) + (1/LC)} \quad \text{(in ohms)} \tag{B.91}$$

Defining $\omega_n = 1/\sqrt{LC}$ as the *resonant frequency* in rad/s and $\alpha = 1/(2RC)$ as the *damping factor*, we can write Eq. (B.91) as

$$G(s) = R\frac{2\alpha s}{s^2 + 2\alpha s + \omega_n^2} \quad \text{(in ohms)} \tag{B.92}$$

Let us define

$$\delta = \frac{\alpha}{\omega_n} = \frac{1}{2RC} \times \sqrt{LC} = \frac{1}{2R}\sqrt{\frac{L}{C}} \tag{B.93}$$

as the *damping ratio*. Then Eq. (B.92) becomes

$$G(s) = R\frac{\delta\omega_n s}{s^2 + 2\delta\omega_n s + \omega_n^2} \quad \text{(in ohms)} \tag{B.94}$$

where $\delta < 1$. (Note that δ is not necessarily less than 1, but has been assumed less than 1 for this discussion.) The right-hand side of Eq. (B.94) is $R/2$ multiplied by Eq. (B.76). Following the development of Eqs. (B.78) and (B.79), we find the magnitude $|G(j\omega)|$ as

$$|G(j\omega)| = \frac{2\delta uR}{[(1 - u^2)^2 + (2\delta u)^2]^{1/2}} \quad \text{(in ohms)} \tag{B.95}$$

and the phase angle ϕ of $G(j\omega)$ as

$$\phi = \frac{\pi}{2} - \tan^{-1}\left(\frac{2\delta u}{1 - u^2}\right) \tag{B.96}$$

The magnitude and frequency plots of a parallel *RLC* circuit are shown in Fig. B.45. The maximum value $|G(j\omega)|_{max} = |Z(j\omega)|_{max} = 1$. At the cutoff frequencies, the magnitude of the gain drops to 70.7% of its maximum value R. Thus, Eq. (B.95) gives

$$|G(j\omega)| = \frac{2\delta uR}{[(1 - u^2)^2 + (2\delta u)^2]^{1/2}} = 0.707R = \frac{R}{\sqrt{2}}$$

or $\quad \sqrt{2}(2\delta u) = [(1 - u^2)^2 + (2\delta u)^2]^{1/2}$

Squaring both sides yields

$$2(2\delta u)^2 = (1 - u^2)^2 + (2\delta u)^2$$

or $\quad (2\delta u)^2 = (1 - u^2)^2 \quad$ (B.97)

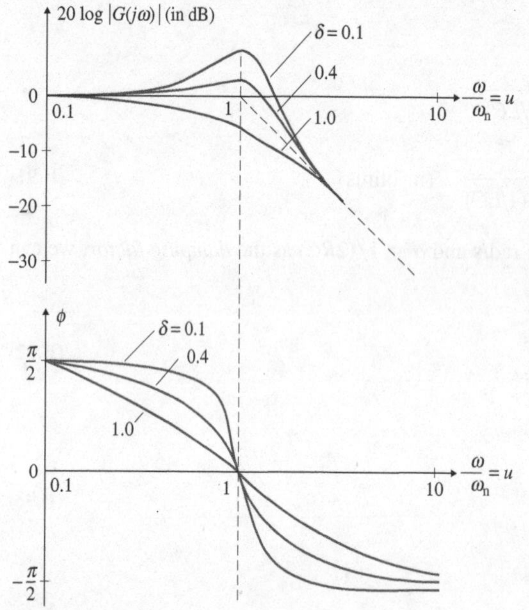

FIGURE B.45 Frequency response of parallel *RLC* circuit

which is the same as Eq. (B.80). Equations (B.81) to (B.86) can be applied to find ω_1 and ω_2. Then the bandwidth BW_p of the parallel resonant circuit can be found from

$$BW_p = \omega_2 - \omega_1 = \omega_n(u_2 - u_1) = 2\delta\omega_n$$

$$= 2\frac{1}{2R}\sqrt{\frac{L}{C}} \times \frac{1}{\sqrt{LC}} = \frac{1}{RC} \quad \text{(in rad/s)} \tag{B.98}$$

▶ **NOTE** For a parallel circuit, $BW_p = 1/RC$ only; for a series circuit, $BW_s = R/L$.

EXAMPLE B.17

Finding the frequency response of a parallel *RLC* circuit The parallel *RLC* circuit in Fig. B.44 has $R = 50\ \Omega$, $L = 4$ mH, and $C = 0.15\ \mu$F.

(a) Determine the parallel resonant frequency f_p, the damping ratio δ, the cutoff frequencies, the bandwidth BW_p, and the quality factor Q_p of the circuit.

(b) Use PSpice/SPICE to plot the magnitude and phase angle of the output voltage for $R = 50\ \Omega$, $100\ \Omega$, and $200\ \Omega$. The frequency f varies from 100 Hz to 100 kHz. Assume $I_m = 1$ A peak AC.

SOLUTION

(a) $R = 50\ \Omega$, $L = 4$ mH, $C = 0.15\ \mu$F, and $I_m = 1$ A peak AC, so

$$\omega_n = \frac{1}{\sqrt{LC}} = \frac{10^5}{\sqrt{4 \times 1.5}} = 40{,}825 \text{ rad/s}$$

The parallel resonant frequency is

$$f_p = \frac{\omega_n}{2\pi} = \frac{40{,}825}{2\pi} = 6497.5 \text{ Hz}$$

Since $\alpha = 1/(2RC) = 1/(2 \times 50 \times 0.15 \times 10^{-6}) = 66.667 \times 10^3$, the damping ratio is

$$\delta = \frac{\alpha}{\omega_n} = 66.667 \times \frac{10^3}{40{,}825} = 1.633$$

For the lower cutoff frequency, Eqs. (B.85) and (B.86) give

$$u_1 = -\delta + \sqrt{1 + \delta^2} = -1.633 + \sqrt{1 + 1.633^2} = 0.28186$$

$$\omega_1 = u_1\omega_n = 0.28186 \times 40{,}825 = 11{,}507 \text{ rad/s}$$

Thus, $f_1 = 11{,}507/2\pi = 1831$ Hz. For the upper cutoff frequency, Eqs. (B.83) and (B.84) give

$$u_2 = \delta + \sqrt{1 + \delta^2} = 1.633 + \sqrt{1 + 1.633^2} = 3.54786$$

$$\omega_2 = u_2\omega_n = 3.54786 \times 40{,}825 = 144{,}841 \text{ rad/s}$$

Thus, $f_2 = 144{,}841/2\pi = 23{,}052$ Hz. From Eq. (B.98), the bandwidth is

$$BW_p = f_2 - f_1 = \frac{1}{RC} = \frac{1}{50 \times 0.15 \times 10^{-6}} = 133{,}333.3 \text{ rad/s, or } 21{,}220 \text{ Hz}$$

Using Eq. (B.90), we can find the quality factor:

$$Q_p = \frac{f_p}{BW_p} = \frac{6497.5}{21{,}220} = 0.3062$$

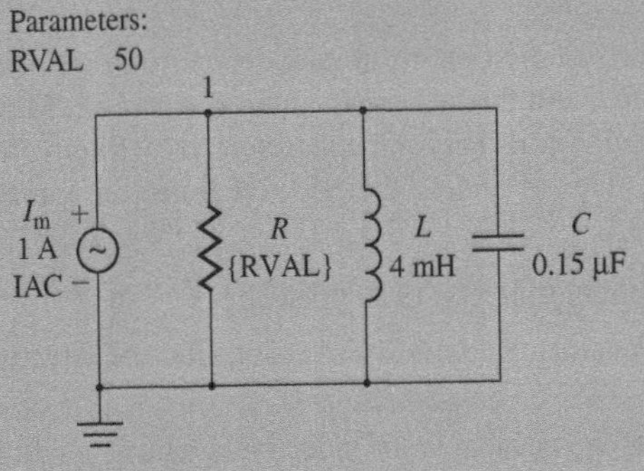

FIGURE B.46 Parallel *RLC* circuit for PSpice simulation

(b) The parallel *RLC* circuit for PSpice simulation is shown in Fig. B.46. The list of the circuit file is as follows.

```
Example B.17 Frequency Response of a Parallel RLC Circuit
.PARAM RVAL = 50
.STEP PARAM RVAL LIST  50  100  200
IM  0  1  AC  1A      ; ac input of 1 V peak
L  1  0  4MH
C  1  0  0.15UF
R  1  0  {RVAL}
.AC  DEC  100  100HZ  1MEGHZ
.PROBE
.END
```

The PSpice plots of the magnitude and phase angle (using EXB-17.SCH) are shown in Fig. B.47. The plot for $R = 50\ \Omega$ gives $f_1 = 1834$ Hz, $f_2 = 22.56$ kHz, $f_p = 6457$ Hz, and $BW_p = f_2 - f_1 = 20{,}726$ Hz.

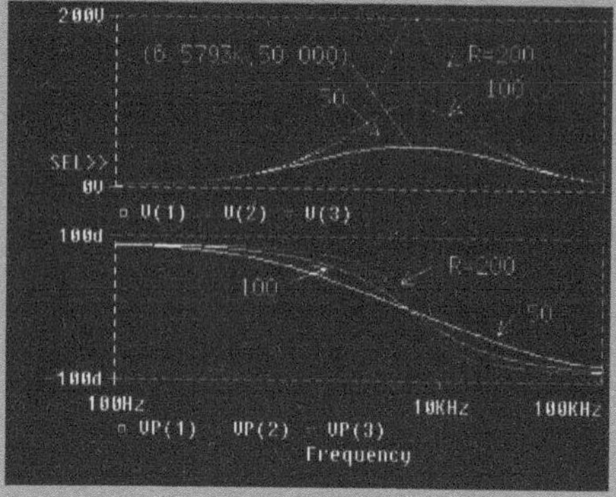

FIGURE B.47 Frequency response plots for Example B.17

B.11 Time Constants of First-Order Circuits

We have seen that the transient and frequency responses of first-order circuits depend on their time constants. The time constant of an *RC* network is $\tau = RC$, and that of an *RL* circuit is $\tau = L/R$. Many circuits have more than two components. The effective time constant can be determined by finding the effective resistance and capacitance of the circuit. The steps in finding the effective time constant are as follows:

Step 1. Set the voltage source(s) to zero and the current source(s) as an open circuit.

Step 2. If there is more than one capacitor (or inductor) and only one resistor, find the effective capacitance or inductance seen by the resistor.

Step 3. If there is more than one resistor and only one capacitive (or inductive) element, find the effective resistance seen by the capacitor (or inductor).

EXAMPLE B.18

Finding the effective time constant The circuit in Fig. B.48 has $R_1 = R_2 = R_3 = 6\ \text{k}\Omega$ and $C = 0.1\ \mu\text{F}$. Determine **(a)** the effective time constant τ, **(b)** the cutoff frequency ω_o, and **(c)** the bandwidth BW.

SOLUTION

If the source is shorted, the effective resistance seen by capacitor C is the parallel combination of R_1, R_2, and R_3. The effective resistance R is given by

$$\frac{1}{R} = \frac{1}{R_1} + \frac{1}{R_2} + \frac{1}{R_3} \quad \text{or} \quad R = \frac{R_1}{3} = \frac{6\ \text{k}}{3} = 2\ \text{k}\Omega$$

(a) The effective time constant is

$$\tau = CR = 2\ \text{k}\Omega \times 0.1\ \mu\text{F} = 0.2\ \text{ms}$$

(b) The cutoff frequency is

$$\omega_o = \frac{1}{\tau} = \frac{1}{0.2\ \text{ms}} = 5000\ \text{rad/s, or } 795.8\ \text{Hz}$$

(c) At $\omega = 0$, capacitor C is open-circuited, and the output voltage has a finite value. At a high frequency, tending to infinity ($\omega = \infty$), capacitor C is short-circuited, and the output voltage becomes zero. This is a low-pass circuit with $f_1 = 0$ and $f_2 = f_o = 795.8$ Hz. Thus, the bandwidth is

$$\text{BW} = f_2 - f_1 = 795.8\ \text{Hz}$$

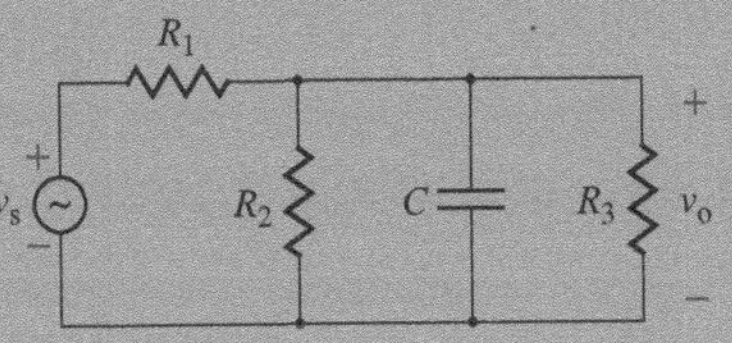

FIGURE B.48 Circuit for Example B.18

EXAMPLE B.19

Finding the effective time constant The circuit in Fig. B.49 has $R_1 = R_2 = R_3 = 10\ \text{k}\Omega$ and $C_1 = 0.1\ \mu\text{F}$. Determine **(a)** the effective time constant τ and **(b)** the cutoff frequency ω_o.

SOLUTION

If the source is shorted, the effective resistance is the sum of R_1 and $(R_2 \| R_3)$; that is,

$$R = R_1 + (R_2 \| R_3) = 10\ \text{k} + 10\ \text{k} \| 10\ \text{k} = 15\ \text{k}\Omega$$

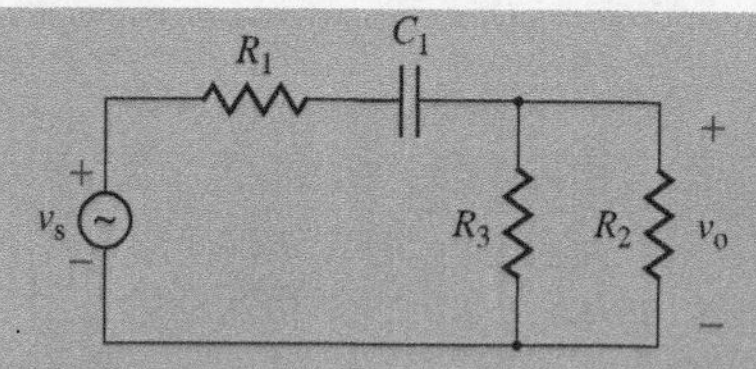

FIGURE B.49 Circuit for Example B.19

(a) The effective time constant is

$$\tau = CR = 15 \text{ k}\Omega \times 0.1 \ \mu\text{F} = 1.5 \text{ ms}$$

(b) The cutoff frequency is

$$\omega_0 = \frac{1}{\tau} = \frac{1}{1.5 \text{ ms}} = 667 \text{ rad/s, or } 106 \text{ Hz}$$

APPENDIX C

LOW-FREQUENCY HYBRID BJT MODEL

A bipolar transistor can be represented by hybrid (h) parameters. If i_b, v_{be}, i_c, and v_{ce} are the small-signal variables of a transistor, as shown in Fig. C.1(a), they are related to the hybrid parameters as follows:

$$v_{bc} = h_{ie}i_b + h_{re}v_{ce} \tag{C.1}$$

$$i_c = h_{fe}i_b + h_{oe}v_{ce} \tag{C.2}$$

where h_{ie} is the *short-circuit input resistance* (or simply the *input resistance*), defined by

$$h_{ie} = \left.\frac{v_{be}}{i_b}\right|_{v_{ce}=0} \quad \text{(in ohms)} \tag{C.3}$$

h_{re} is the *open-circuit reverse voltage ratio* (or the *voltage-feedback ratio*), defined by

$$h_{re} = \left.\frac{v_{be}}{v_{ce}}\right|_{i_b=0} \quad \text{(dimensionless)} \tag{C.4}$$

h_{fe} is the *short-circuit forward-transfer current ratio* (or the *small-signal current gain*), defined by

$$h_{fe} = \left.\frac{i_c}{i_b}\right|_{v_{ce}=0} \quad \text{(dimensionless)} \tag{C.5}$$

and h_{oe} is the *open-circuit output admittance* (or simply the *output admittance*), defined by

$$h_{oe} = \left.\frac{i_c}{v_{ce}}\right|_{i_b=0} \quad \text{(in siemens)} \tag{C.6}$$

The low-frequency hybrid model is shown in Fig. C.1(b). The input has a voltage-controlled voltage source in which the controlling voltage is the output voltage. The output circuit has a current-controlled current source in which the controlling current is the input current. The subscript e on the h parameters indicates that these hybrid parameters are derived for a common-emitter configuration. It is also possible to have common-base and common-collector configurations.

BJT manufacturers specify the common-emitter hybrid parameters. The parameter h_{re}, which takes into account the effect of v_{CE} on i_B, is very small, with a typical value of 0.5×10^{-4}. The parameter h_{oe}, which represents the admittance of the CE junction, is also very small; its value is typically 10^{-6} S. The parameters h_{re} and h_{oe} can often be omitted from the circuit model without significant loss of accuracy, especially for hand calculations.

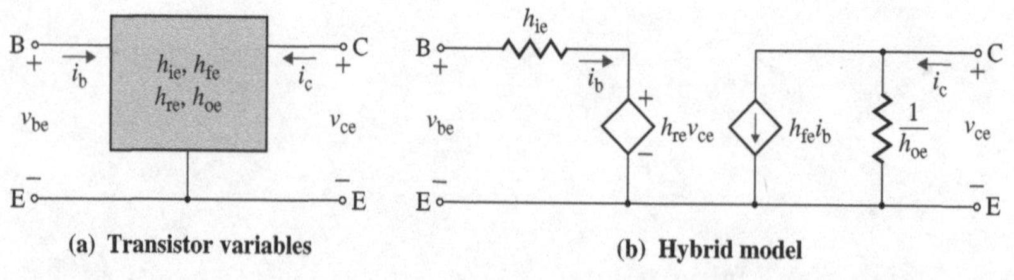

FIGURE C.1 Low-frequency hybrid model

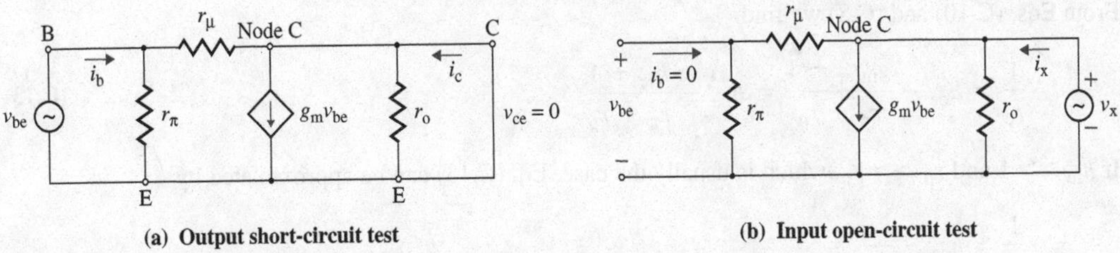

FIGURE C.2 Equivalent circuits for deriving h parameters

The π-model parameters can be related to the h parameters by applying short-circuit and open-circuit tests. The parameters h_{ie} and h_{fe}, which are short-circuit parameters, can be derived in terms of π-model parameters from Fig. C.2(a) by shorting the collector-to-emitter terminals.

$$h_{ie} = \frac{v_{be}}{i_b}\bigg|_{v_{ce}=0} = r_\pi \| r_\mu = \frac{r_\pi r_\mu}{r_\pi + r_\mu} \tag{C.7}$$

$$h_{fe} = \frac{i_c}{i_b}\bigg|_{v_{ce}=0} = \frac{g_m v_{be}}{v_{be}/r_\pi} = g_m r_\pi \tag{C.8}$$

As shown in Fig. C.2(b), h_{oe} and h_{re} can be determined by open-circuiting the base-to-emitter terminals and then applying the voltage divider rule.

$$h_{re} = \frac{v_{be}}{v_x}\bigg|_{i_b=0} = \frac{r_\pi}{r_\pi + r_\mu} \tag{C.9}$$

Summing the currents at the collector node in Fig. C.2(b), we get

$$i_x = \frac{v_x}{r_o} + g_m \frac{r_\pi v_x}{r_\pi + r_\mu} + \frac{v_x}{r_\pi + r_\mu} = v_x\left[\frac{1}{r_o} + \frac{g_m r_\pi + 1}{r_\pi + r_\mu}\right]$$

which gives

$$h_{oe} = \frac{i_x}{v_x} = \frac{1}{r_o} + \frac{g_m r_\pi + 1}{r_\pi + r_\mu} \tag{C.10}$$

Using Eq. (C.8), we get

$$r_\pi = \frac{h_{fe}}{g_m} \tag{C.11}$$

Using Eq. (C.7) and $h_{fe} = g_m r_\pi$ from Eq. (C.8), we get

$$r_\mu = \frac{r_\pi + r_\mu}{r_\pi} h_{ie} = \frac{h_{ie}}{h_{re}} \tag{C.12}$$

From Eqs. (C.10) and (C.8) we find

$$\frac{1}{r_o} = h_{oe} - \frac{g_m r_\pi + 1}{r_\pi + r_\mu} = h_{oe} - \frac{h_{fe} + 1}{r_\pi + r_\mu} \tag{C.13}$$

If $h_{fe} >> 1$ and $r_\mu >> r_\pi$, which is usually the case, Eq. (C.13) can be approximated by

$$\frac{1}{r_o} \approx h_{oe} - \frac{h_{fe}}{r_\mu}$$

which gives

$$r_o = \left[h_{oe} - \frac{h_{fe}}{r_\mu}\right]^{-1} \tag{C.14}$$

$$= \frac{V_A}{I_C}$$

r_μ is very large, and h_{re} is negligibly small. The π model becomes similar to the h model in Fig. C.2(b) such that $r_\pi \equiv h_{ie}$, $\beta_f \equiv h_{fe}$, and $r_o \equiv 1/h_{oe}$. Being open-circuited, r_μ has a high value and can often be neglected.

EXAMPLE C.1

Converting hybrid parameters to π-model parameters The h parameters of a transistor are $h_{ie} = 1\ \text{k}\Omega$, $h_{fe} = 100$, $h_{re} = 0.5 \times 10^{-4}$, and $h_{oe} = 10 \times 10^{-6}$ S. The collector biasing current is $I_C = 11.62$ mA. Calculate the small-signal π-model parameters of the transistor. Assume $V_T = 25.8$ mV at 25°C.

SOLUTION

$$g_m = \frac{I_C}{V_T} = \frac{11.62\ \text{mA}}{25.8\ \text{mV}} = 0.4504\ \text{A/V}$$

From Eq. (C.11),

$$r_\pi = \frac{h_{fe}}{g_m} = \frac{100}{0.4504} = 222\ \Omega$$

From Eq. (C.12),

$$r_\mu = \frac{h_{ie}}{h_{re}} = \frac{1\ \text{k}\Omega}{0.5 \times 10^{-4}} = 20\ \text{M}\Omega$$

From Eq. (C.14),

$$r_o = \left[h_{oe} - \frac{h_{fe}}{r_\mu}\right]^{-1} = \left[10 \times 10^{-6} - \frac{100}{20 \times 10^6}\right]^{-1} = 200\ \text{k}\Omega$$

EXAMPLE C.2

Finding the hybrid-model parameters of the circuit shown in Fig. C.3 The circuit values are $R_1 = 4\ \text{k}\Omega$, $R_2 = 10\ \text{k}\Omega$, and $R_f = 16\ \text{k}\Omega$. Determine the hybrid parameters h_{ie}, h_{re}, h_{fe}, and h_{oe}.

SOLUTION

$R_1 = 4\ \text{k}\Omega$, $R_2 = 6\ \text{k}\Omega$, and $R_f = 20\ \text{k}\Omega$. Using Eq. (C.3) and the equivalent circuit as shown in Fig. C.4(a), h_{ie} can be determined from

$$h_{ie} = \left.\frac{v_S}{i_S}\right|_{v_O=0} = R_1 \| R_f = 4\,\text{k} \| 16\,\text{k} = 3.2\ \text{k}\Omega$$

Using Eq. (C.4) and the equivalent circuit shown in Fig. C.4(b), h_{re} can be determined from

$$h_{re} = \left.\frac{v_S}{v_O}\right|_{i_S=0} = \frac{R_1}{R_1 + R_f} = \frac{4\,\text{k}}{4\,\text{k} + 16\,\text{k}} = 0.2\ \text{V/V}$$

Using Eq. (C.5) and the equivalent circuit shown in Fig. C.4(c), h_{fe} can be determined from

$$h_{re} = \left.\frac{i_O}{i_S}\right|_{v_O=0} = \frac{-R_1}{R_1 + R_f} = \frac{-4\,\text{k}}{4\,\text{k} + 16\,\text{k}} = -0.2\ \text{A/A}$$

Using Eq. (C.6) and the equivalent circuit shown in Fig. C.4(d), h_{oe} can be determined from

$$h_{oe} = \left.\frac{i_O}{v_O}\right|_{i_S=0} = \frac{1}{R_1 \| (R_1 + R_f)} = \frac{1}{4\,\text{k} \| (4\,\text{k} + 16\,\text{k})} = 1.5 \times 10^{-4}\ \mho$$

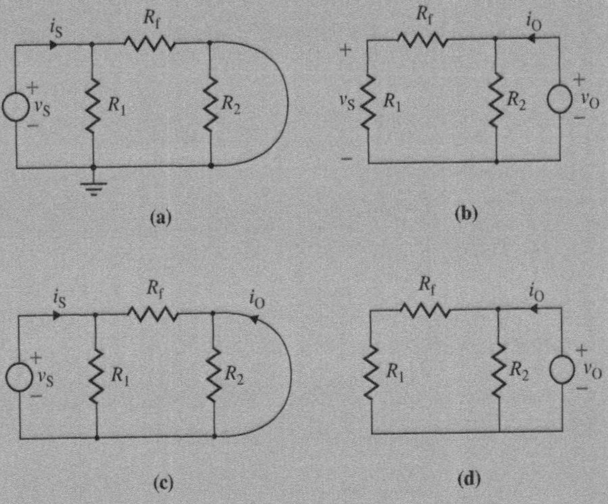

FIGURE C.4 Equivalent circuits for determining hybrid parameters for Example C.2

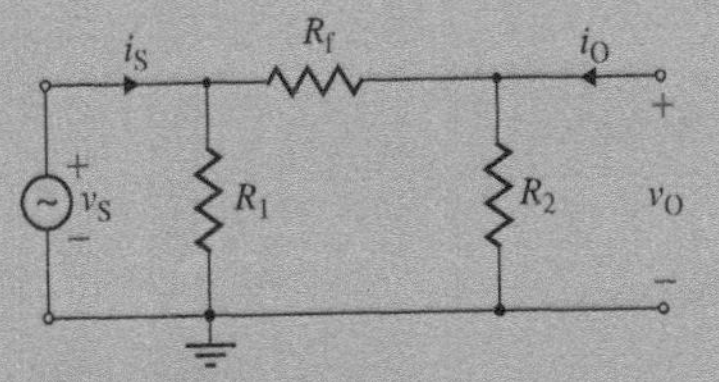

FIGURE C.3 Circuit for Example C.2

APPENDIX D

EBERS–MOLL MODEL OF BIPOLAR JUNCTION TRANSISTORS

A transistor is a nonlinear device that can be modeled using the nonlinear characteristics of diodes. The Ebers–Moll model is a large-signal model commonly used for modeling BJTs. One version of the model is based on the assumption of one forward-biased diode and one reverse-biased diode. This arrangement is shown in Fig. D.1 for an *npn* transistor. This model, referred to as the *injection version* of the Ebers–Moll model, is valid for active, saturation, and cutoff regions. Under normal operation in the active region, one junction of the BJT is forward biased and the other is reverse biased.

The emitter–base and collector–base diodes can be described using the Shockley diode characteristic:

$$I_F = I_{ES}\left(\exp\frac{V_{BE}}{V_T} - 1\right) \tag{D.1}$$

$$I_R = I_{CS}\left(\exp\frac{V_{BC}}{V_T} - 1\right) \tag{D.2}$$

where $V_T = kT/q = 25.8$ mV at 25°C
I_{ES} = reverse saturation current of base–emitter diode
I_{CS} = reverse saturation current of base–collector diode

Both I_{ES} and I_{CS} are temperature dependent. If $V_{BE} > 0$, diode D_F is forward biased and its current I_F causes a corresponding current $\alpha_F I_F$. If $V_{BC} > 0$, diode D_R is reverse biased. The subscripts F and R are used to designate forward and reverse conditions, respectively. Using Kirchhoff's current law (KCL) at the emitter and collector terminals, we can write the emitter current I_E as

$$I_E = -I_F + \alpha_R I_R \tag{D.3}$$

$$= -I_{ES}\left(\exp\frac{V_{BE}}{V_T} - 1\right) + \alpha_R I_{CS}\left(\exp\frac{V_{BC}}{V_T} - 1\right)$$

and the collector current I_C as

$$I_C = \alpha_F I_F - I_R$$

$$= \alpha_F I_{ES}\left(\exp\frac{V_{BE}}{V_T} - 1\right) - I_{CS}\left(\exp\frac{V_{BC}}{V_T} - 1\right) \tag{D.4}$$

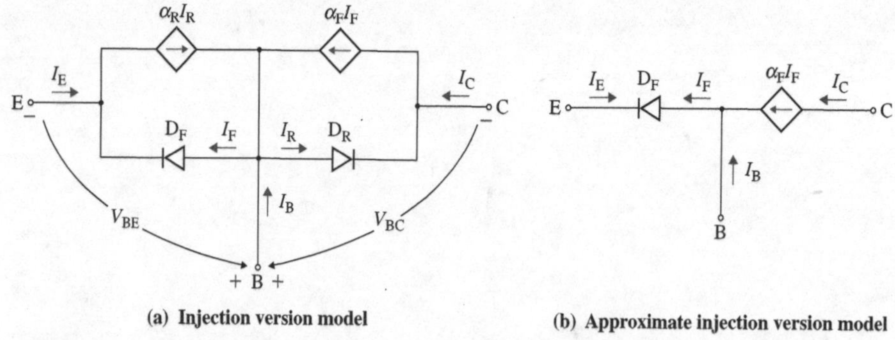

FIGURE D.1 Ebers–Moll injection version model for *npn* transistor

If $V_{BE} = 0$, $\alpha_R I_{CS} = I_S$ represents the reverse saturation leakage current of diode D_R. Similarly, if $V_{BE} = 0$, $\alpha_F I_{ES} = I_S$ represents the reverse saturation leakage current of diode D_F. If we assume ideal diodes, the forward and reverse saturation leakage currents are related by

$$\alpha_R I_{CS} = \alpha_F I_{ES} = I_S \tag{D.5}$$

where I_S is known as the *transistor saturation current*.

The current from the collector to the base with the emitter open-circuited can be found by letting $I_C = I_{CBO}$ and $I_E = 0$. Since the collector–base junction is normally reverse biased, $V_{BC} < 0$ and $|V_{BC}| >> V_T$, $\exp(V_{BC}/V_T) << 1$. With these conditions in Eqs. (D.3) and (D.4), we get

$$0 = -I_{ES}\left(\exp\frac{V_{BE}}{V_T} - 1\right) - \alpha_R I_{CS}$$

$$I_{CBO} = \alpha_F I_{ES}\left(\exp\frac{V_{BE}}{V_T} - 1\right) + I_{CS}$$

Solving these two equations for I_{CBO} yields

$$I_{CBO} = -\alpha_F\alpha_R I_{CS} + I_{CS} = (1 - \alpha_R\alpha_F)I_{CS} = I_{CS} - \alpha_F I_S \tag{D.6}$$

The current from the emitter to the base with the collector open-circuited can be found by letting $I_E = I_{EBO}$ and $I_C = 0$. Since the emitter junction is reverse biased, $V_{BE} < 0$ (i.e., $V_{EB} > 0$) and $|V_{BE}| >> V_T$, $\exp(V_{BE}/V_T) << 1$. Equations (D.3) and (D.4) give

$$I_{EBO} = I_{ES} + \alpha_R I_{CS}\left(\exp\frac{V_{BC}}{V_T} - 1\right)$$

$$0 = -\alpha_F I_{ES} - I_{CS}\left(\exp\frac{V_{BC}}{V_T} - 1\right)$$

Solving these two equations for I_{EBO} gives

$$I_{EBO} = I_{ES} - \alpha_R\alpha_F I_{ES} = (1 - \alpha_R\alpha_F)I_{ES} \tag{D.7}$$

From Eqs. (D.5), (D.6), and (D.7), we get

$$\alpha_F I_{EBO} = I_{ES}(1 - \alpha_R\alpha_F)\alpha_F = \alpha_R I_{CS}(1 - \alpha_R\alpha_F) = \alpha_R I_{CBO} \tag{D.8}$$

Since diode D_F is forward biased and diode D_R is reverse biased, $V_{EB} < V_{BC}$. Thus, I_{EBO} is less than I_{CBO}, and α_F is greater than α_R. In the active region, diode D_R is reverse biased, and $I_R \approx 0$; that is, $I_E = -I_F$, and $I_C = \alpha_F I_F = -\alpha_F I_E$. Thus, Fig. D.1(a) can be approximated by Fig. D.1(b).

The circuit model of Fig. D.1(a) relates the dependent sources to the diode currents. In circuit analysis, it is convenient to express the current source in a form that is controlled by the terminal currents. Eliminating $[\exp(V_{BE}/V_T) - 1]$ from Eqs. (D.3) and (D.4) and then using Eq. (D.6), we get

$$I_C = -\alpha_F I_E - (1 - \alpha_F\alpha_R)I_{CS}\left(\exp\frac{V_{BC}}{V_T} - 1\right)$$

$$= -\alpha_F I_E - I_{CBO}\left(\exp\frac{V_{BC}}{V_T} - 1\right) \tag{D.9}$$

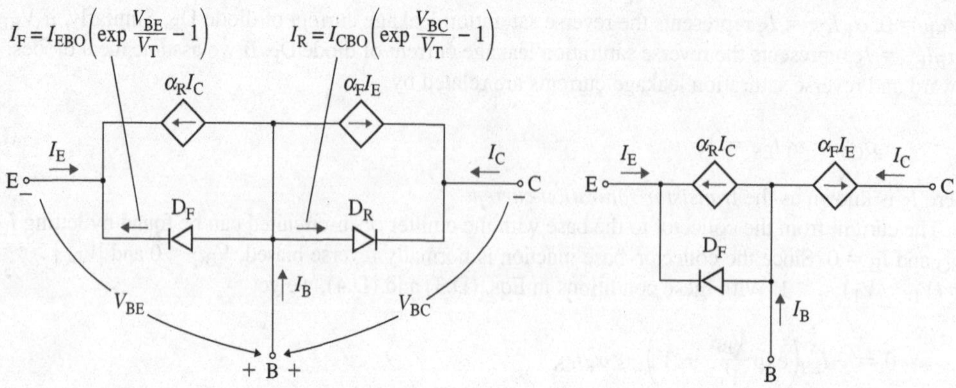

FIGURE D.2 Ebers–Moll transport version model

Similarly, eliminating [exp $(V_{BC}/V_T) - 1$] from Eqs. (D.3) and (D.4) and then using Eq. (D.7) gives

$$I_E = -\alpha_R I_C - (1 - \alpha_R\alpha_F)I_{ES}\left(\exp\frac{V_{BE}}{V_T} - 1\right)$$

$$= -\alpha_R I_C - I_{EBO}\left(\exp\frac{V_{BE}}{V_T} - 1\right) \quad \text{(D.10)}$$

The circuit model corresponding to Eqs. (D.9) and (D.10) is shown in Fig. D.2(a). The current sources are controlled by collector current I_C and emitter current I_E. This model, referred to as the *transport version* of the Ebers–Moll model, is normally used in computer simulations with PSpice/SPICE. In fact, the linear models are the approximate versions of the Ebers–Moll model in Fig. D.2(a).

Assuming that $V_{BC} > 0$ and $I_R \approx 0$, $I_C = -\alpha_F I_E$ and Fig. D.2(a) can be approximated by Fig. D.2(b). If we substitute $I_C = -\alpha_F I_E$, Eq. (D.10) becomes

$$I_E = \alpha_R\alpha_F I_E - I_{EBO}\left(\exp\frac{V_{BE}}{V_T} - 1\right)$$

which relates I_E to α_R, α_F, and V_{BE} by

$$I_E = \frac{I_{EBO}\exp(V_{BE}/V_T - 1)}{1 - \alpha_R\alpha_F} \quad \text{(D.11)}$$

I_{ES} and I_{CS} are also known as the *short-circuit saturation currents*, whereas I_{CBO} and I_{EBO} are known as the *open-circuit saturation currents*. Note that the injection and transport versions of the Ebers–Moll model are interchangeable. Once the parameters of one version are known, the parameters of the other version can be found.

EXAMPLE D.1

Finding the currents for the Ebers–Moll model of an *npn* transistor An *npn* transistor is biased so that $V_{BE} = 0.3$ V and $V_{CE} = 6$ V. If $\alpha_F = 0.99$, $\alpha_R = 0.90$, and $I_{CBO} = 5$ nA, find all the currents for the injection version of the Ebers–Moll model of Fig. D.1(a). Assume thermal voltage $V_T = 25.8$ mV.

SOLUTION

$V_T = 25.8$ mV $= 0.0258$ V. Since V_{BE} is positive, the base–emitter junction is forward biased. The voltage at the collector–base junction is

$$V_{CB} = V_{CE} + V_{EB} = V_{CE} - V_{BE} = 6 - 0.3 = 5.7 \text{ V}$$

or $\quad V_{BC} = -5.7$ V

Since V_{CB} is positive, the collector–base junction is reverse biased, and the transistor operates in the active region. From Eq. (D.8),

$$I_{EBO} = \frac{\alpha_R I_{CBO}}{\alpha_F} = 0.9 \times \frac{5 \text{ nA}}{0.99} = 4.545 \text{ nA}$$

From Eq. (D.6),

$$I_{CS} = \frac{I_{CBO}}{1 - \alpha_R\alpha_F} = \frac{5 \text{ nA}}{1 - 0.9 \times 0.99} = 45.87 \text{ nA}$$

From Eq. (D.7),

$$I_{ES} = \frac{I_{EBO}}{1 - \alpha_R\alpha_F} = \frac{4.545 \text{ nA}}{1 - 0.9 \times 0.99} = 41.697 \text{ nA}$$

From Eq. (D.1), the forward diode current is

$$I_F = 41.697 \times 10^{-9} \times \left(\exp\frac{0.3}{0.0258} - 1\right) \approx 4.678 \text{ mA}$$

From Eq. (D.2), the reverse diode current is

$$I_R = 45.87 \times 10^{-9} \times \left(\exp\frac{-5.7}{0.0258} - 1\right) \approx -45.87 \text{ nA}$$

and $\quad \alpha_R I_R \approx -0.9 \times 45.87 \text{ nA} = -41.28 \text{ nA}$

$$\alpha_F I_F \approx 0.99 \times 4.678 \text{ mA} = 4.63 \text{ mA}$$

$$I_E = -I_F + \alpha_R I_R = -4.678 \times 10^{-3} - 41.28 \times 10^{-6} \approx -4.719 \text{ mA}$$

$$I_C = \alpha_F I_F - I_R = 4.63 \times 10^{-3} + 45.87 \times 10^{-9} \approx 4.63 \text{ mA}$$

$$I_B = -(I_E + I_C) = -(-4.719 \times 10^{-3} + 4.63 \times 10^{-3}) = 89 \text{ } \mu\text{A}$$

EXAMPLE D.2

Finding the currents for the Ebers–Moll model of a *pnp* transistor The model parameters of the *pnp* transistor are $\alpha_R = 0.9$, $\alpha_F = 0.99$, $I_{ES} \approx I_{CS} = 45$ nA, $V_{EB} = 0.4$ V, and $V_{CB} = 0.3$ V. Determine **(a)** the currents for the transport version of the Ebers–Moll model in Fig. D.2(a) and **(b)** $\beta_F = \beta_{forced}$. Assume thermal voltage $V_T = 25.8$ mV.

SOLUTION

For $V_{EB} = 0.4$ V and $V_{CB} = 0.3$ V, the collector–base and emitter–base junctions are forward biased. Therefore, the transistor is operated in the saturation region.

(a) For a *pnp* transistor, all the polarities of voltages and currents are reversed; thus, Eqs. (D.3) and (D.4) become

$$I_E = I_{ES}\left(\exp\frac{V_{EB}}{V_T} - 1\right) - \alpha_R I_{CS}\left(\exp\frac{V_{CB}}{V_T} - 1\right) \tag{D.12}$$

$$= 45 \times 10^{-9}\left(\exp\frac{0.4}{25.8\text{ m}} - 1\right) - 0.9 \times 45 \times 10^{-9}\left(\exp\frac{0.3}{25.8\text{ m}} - 1\right)$$

$$= 243.48\text{ mA} - 4.54\text{ mA} = 238.94\text{ mA}$$

and

$$I_C = -\alpha_F I_{ES}\left(\exp\frac{V_{EB}}{V_T} - 1\right) + I_{CS}\left(\exp\frac{V_{CB}}{V_T} - 1\right) \tag{D.13}$$

$$= -0.99 \times 45 \times 10^{-9}\left(\exp\frac{0.4}{25.8\text{ m}} - 1\right) + 45 \times 10^{-9}\left(\exp\frac{0.3}{25.8\text{ m}} - 1\right)$$

$$= -241.04\text{ mA} + 5.05\text{ mA} = -235.99\text{ mA}$$

Thus, $I_B = -(I_E + I_C) = -(238.94 - 235.99) = -2.95$ mA

(b) Since the transistor is operated in the saturation region, the value of the forward current gain becomes less than that of the active region. The forward current gain is known as the *forced current gain* β_{forced}:

$$\beta_{forced} = \frac{I_C}{I_B} = \frac{235.99\text{ mA}}{2.95\text{ mA}} = 80$$

EXAMPLE D.3

Finding the collector–emitter saturation voltage of a BJT An *npn* transistor is operated in the saturation region, and its parameters are $\alpha_R = 0.9$, $\alpha_F = 0.989$, $\beta_F = 89.91$, and $V_T = 25.8$ mV. Calculate the collector–emitter saturation voltage $V_{CE(sat)}$.

SOLUTION

$$\beta_F = \frac{\alpha_F}{1 - \alpha_F} = \frac{0.989}{1 - 0.989} = 89.91$$

We know that

$$V_{BC} = V_{BE} + V_{EC} = V_{BE} - V_{CE}$$

In the saturation region, $V_{BE} > 0$ and $V_{BC} > 0$; that is,

$$\exp\left(\frac{V_{BE}}{V_T}\right) >> 1 \quad \text{and} \quad \exp\left(\frac{V_{BC}}{V_T}\right) >> 1$$

Using Eqs. (D.3) and (D.5), we get

$$I_E = -I_{ES}e^{V_{BE}/V_T} + a_F I_{ES}e^{V_{BC}/V_T}$$

$$\approx -I_{ES}e^{V_{BE}/V_T} + \alpha_F I_{ES}e^{(V_{BE}-V_{CE})/V_T}$$

$$\approx -I_{ES}e^{V_{BE}/V_T}[1 - \alpha_F e^{-V_{CE}/V_T}]$$

But $I_E + I_C + I_B = 0$, so $-I_E = (I_C + I_B)$; that is,

$$I_C + I_B = I_{ES}e^{V_{BE}/V_T}[1 - \alpha_F e^{-V_{CE}/V_T}] \quad \text{(D.14)}$$

Similarly, using Eqs. (D.4) and (D.5), we get

$$I_C = \alpha_F I_{ES}\left(\exp\frac{V_{BE}}{V_T} - 1\right) - I_{CS}\left(\exp\frac{V_{BC}}{V_T} - 1\right)$$

$$= \alpha_F I_{ES}e^{V_{BE}/V_T}\left[1 - \frac{1}{\alpha_R}e^{-V_{CE}/V_T}\right] \quad \text{(D.15)}$$

Dividing Eq. (D.14) by Eq. (D.15) yields

$$\frac{I_C + I_B}{I_C} = 1 + \frac{I_B}{I_C} = \frac{1 - \alpha_F e^{-V_{CE}/V_T}}{\alpha_F[1 - (e^{-V_{CE}/V_T})\alpha_R]}$$

which, after simplification, becomes

$$V_{CE} = V_T \ln\left\{\frac{\alpha_F[1 + I_C/I_B(1 - \alpha_R)]}{\alpha_R[\alpha_F + I_C/I_B(\alpha_F - 1)]}\right\} \quad \text{(D.16)}$$

In the active region, the base current I_B is related to the collector current I_C by $I_B = I_C/\beta_F$. The saturation region is considered to begin at the point where the forward current gain β_F is 90% of the value in the active region. That is, $\beta_{sat} = \beta_{forced} = 0.9\beta_F$ and $I_C = 0.9I_B\beta_F$. Equation (D.16) gives the collector–emitter saturation voltage as

$$V_{CE(sat)} = V_T \ln\left\{\frac{\alpha_F[1 + 0.9\beta_F(1 - \alpha_R)]}{\alpha_R[\alpha_F + 0.9\beta_F(\alpha_F - 1)]}\right\}$$

Substituting $\beta_F = \alpha_F/(1 - \alpha_F)$ in the denominator gives

$$V_{CE(sat)} = V_T \ln\left[\frac{1 + 0.9\beta_F(1 - \alpha_R)}{\alpha_R(1 - 0.9)}\right] \tag{D.17}$$

For $V_T = 0.0258$ V, $\alpha_R = 0.9$, and $\beta_F = 89.91$, we get

$$V_{CE(sat)} = 0.0258 \times \ln\left[\frac{1 + 0.9 \times 89.91 \times (1 - 0.9)}{0.9 \times (1 - 0.9)}\right] = 0.119 \text{ V}$$

APPENDIX E

PASSIVE COMPONENTS

E.1 Resistors

TABLE E.1 Standard values (in ohms) for metal film resistors (tolerance ±1%) and carbon resistors (tolerance ±5% and ±10%)

1%	5%	10%	1%	5%	10%	1%	5%	10%	1%	5%	10%
10.0	10	10	20.0	20		40.2			61.9	62	
10.2			20.5			41.2			63.4		
10.5			21.0			42.2			64.9		
10.7			21.5			43.2	43		66.5		
11.0	11		22.1	22	22	44.2			68.1	68	68
11.3			22.6			45.3			69.8		
11.5			23.2			46.4			71.5		
11.8	12	12	23.7			47.5	47	47	73.2		
12.1			24.3	24		48.7			75.0	75	
12.4			24.9			49.9			76.8		
12.7			25.5			51.1	51		78.7		
13.0	13		26.1			52.3			80.6		
13.3			26.7	27	27	53.6			82.5	82	82
13.7			27.4			54.9			84.5		
14.0			28.0			56.2	56	56	86.6		
14.3			28.7			57.6			88.7		
14.7			29.4			59.0			90.9	91	
15.0	15	15	30.1	30		60.4			93.1		
15.4			30.9						95.3		
15.8			31.6						97.6		
16.2	16		32.4	33	33						
16.5			33.2								
16.9			34.0								
17.4			34.8								
17.8			35.7	36							
18.2	18	18	36.5								
18.7			37.4								
19.1			38.3								
19.6			39.2	39	39						

▶ **NOTE** The available values for carbon are $1\ \Omega \le R \le 100\ \text{M}\Omega$ and for metal film are $10\ \Omega \le R \le 10\ \text{M}\Omega$. The available values are obtained by multiplying the sequence number by a power of 10 (i.e., 10^{-1}, 10^{0}, 10^{1}, 10^{2}, 10^{3}, and so on).

TABLE E.2 Standard values (in ohms) for wire-wound resistors (tolerance ±5%)

0.008	0.75	7.5	27	62	180	450	1 k	4 k
0.01	1.0	8	30	70	200	470	1.2 k	5 k
0.02	1.5	10	33	75	220	500	1.3 k	10 k
0.03	2.0	12	35	80	250	560	1.5 k	15 k
0.05	2.5	15	40	82	270	600	1.8 k	20 k
0.1	3.0	16	45	100	300	680	2 k	25 k
0.15	3.3	20	47	110	330	700	2.2 k	40 k
0.2	4.0	22	50	120	390	750	2.5 k	50 k
0.26	5.0	22.5	56	150	400	910	3 k	100 k
0.3	6	25	60	160	430		3.5 k	150 k
0.5	7							

TABLE E.3 Power ratings for resistors

Type	Tolerance	Power Rating
Carbon resistors	5% and 10%	$\frac{1}{8}$ W $\frac{1}{4}$ W $\frac{1}{2}$ W 1 W 2 W
Metal film resistors	1%	$\frac{1}{8}$ W $\frac{1}{4}$ W $\frac{1}{2}$ W
Wire-wound resistors	5%	5 W 12 W 25 W 50 W 100 W 225 W

TABLE E.4 Color code for resistors

1st Digit	2nd Digit	Multiplier	Tolerance	Color	1st Digit	2nd Digit	3rd Digit	Multiplier	Tolerance
0	0	$10^0 = 1$	—	Black	0	0	0	$10^0 = 1$	—
1	1	$10^1 = 10$	—	Brown	1	1	1	$10^1 = 10$	±1%
2	2	$10^2 = 100$	—	Red	2	2	2	$10^2 = 100$	—
3	3	$10^3 = 1$ k	—	Orange	3	3	3	$10^3 = 1$ k	—
4	4	$10^4 = 10$ k	—	Yellow	4	4	4	$10^4 = 10$ k	—
5	5	$10^5 = 100$ k	—	Green	5	5	5	$10^5 = 100$ k	—
6	6	$10^6 = 1$ M	—	Blue	6	6	6	$10^6 = 1$ M	—
7	7	$10^7 = 10$ M	—	Violet	7	7	7	$10^7 = 10$ M	—
8	8		—	Gray	8	8	8		—
9	9		—	White	9	9	9		—
—	—		±5%	Gold	—	—	—		—
—	—		±10%	Silver	—	—	—		—
—	—		±20%	No band	—	—	—		—

E.2 Potentiometers

TABLE E.5 Standard values (in ohms) for carbon composition, linear taper potentiometers (tolerance ±10%; power rating 2.25 W)

50	1 k	10 k	100 k	1 M
150	1.5 k	15 k	150 k	1.5 M
200	2 k	20 k	200 k	2 M
250	2.5 k	25 k	250 k	2.5 M
350	3.5 k	35 k	350 k	3.5 M
500	5 k	50 k	500 k	5 M
750	7.5 k	75 k	750 k	

TABLE E.6 Standard values (in ohms) for conductive plastic potentiometers (tolerance ±10%; power rating $\frac{1}{2}$ W)

250			
1 k	10 k	100 k	1 M
2.5 k	25 k	250 k	2.5 M
5 k	50 k	500 k	5 M

TABLE E.7 Standard values (in ohms) for CERMET potentiometers (tolerance ±10%; power rating $\frac{1}{2}$ W)

50				
100	1 k	10 k	100 k	1 M
200	2 k	20 k	200 k	2 M
500	5 k	50 k	500 k	

E.3 Capacitors

TABLE E.8 Standard values for polarized aluminum electrolytic capacitors (tolerance −10% to +50%)

Voltage (V)	Capacitance (μF)	Voltage (V)	Capacitance (μF)	Voltage (V)	Capacitance (μF)
10	22	25	10	50	0.1
	33		22		0.22
	47		33		0.33
	100		47		0.47
	220		100		1.0
	330		220		2.2
	470		330		3.3
	1000		470		4.7
	2200		1000		10
	3300		2200		22
	4700		3300		33
	6800		4700		47
	10,000				100
					220
					330
					470
					1000
					2200

TABLE E.9 Standard values for ceramic disc capacitors (tolerance ±10%)

Voltage (V)	Capacitance (pF)
200	10
	15
	22
	33
	47
	68
	100
	150
	220
	330
	470
	680
	1000
	1500
	2200
	3300
	4700
	6800
	10,000
	15,000

TABLE E.10 Standard values for mylar polyester capacitors (tolerance ±10%)

Voltage (V)	Capacitance (μF)
100	0.001
	0.0015
	0.0022
	0.0033
	0.0047
	0.0068
	0.0082
	0.01
	0.015
	0.022
	0.027
	0.033
	0.039
	0.047
	0.056
	0.068
	0.082
	0.1
	0.12
	0.15
	0.18
	0.22
	0.27
	0.33
	0.39
	0.47
	0.56
	0.68
	0.82
	1

TABLE E.11 Standard values for ceramic variable capacitors (tolerance ±10%)

	Capacitance (pF)	
Voltage (V)	Min	Max
250	1	4.5
	2.5	10
	4	18
	6	35
	7	40
	8	50

APPENDIX F

DESIGN PROBLEMS

Mini Design Projects

F.1 Design a circuit that will sum the sensing signals generated by the probes of an electrocardiogram. The output can be expressed in terms of the input signals (V_A, V_B, V_C, and V_D) as follows:

$$V_O = 1.4V_A + 5V_B + 3V_C + 0.6V_D$$

The accuracy should be better than 2%. The DC power supplies are ±15 V.

F.2 Design a BJT buffer amplifier to give a midfrequency voltage gain of $|A_v| = v_L/v_s \approx 1$ and an input resistance of $R_{in} = v_s/i_s \geq 50\ k\Omega$. The load resistance is $R_L = 10\ k\Omega$. Assume a source resistance of $R_S = 500\ \Omega$ and $V_{CC} = 15$ V.

F.3 Design an NMOS buffer amplifier to give a midfrequency voltage gain of $|A_v| = v_L/v_s \approx 1$ and an input resistance of $R_{in} = v_s/i_s \geq 500\ k\Omega$. The load resistance is $R_L = 10\ k\Omega$. Assume a source resistance of $R_S = 500\ \Omega$ and $V_{DD} = 15$ V.

F.4 Design a BJT buffer amplifier with an active load to give a midfrequency voltage gain of $|A_v| = v_L/v_s \approx 1$ and an output resistance of $R_{in} = v_s/i_s \geq 50\ k\Omega$. The load resistance is $R_L = 10\ k\Omega$. Assume a source resistance of $R_S = 500\ \Omega$ and $V_{CC} = V_{EE} = 15$ V.

Medium Design Projects

F.5 Design a DC power supply to electronic equipment from an AC supply of 120 V (rms) ± 10%, 60 Hz. The load requires ±12 V ± 5% at 0.5 A. Use discrete devices only and design for minimum expense.

F.6 a. Design a BJT amplifier to give a midfrequency voltage gain of $|A_v| = v_L/v_s = 20 \pm 5\%$ at a load resistance of $R_L = 10\ k\Omega$. Assume a source resistance of $R_S = 500\ \Omega$ and $V_{DD} = 15$ V.
b. Modify the design so that the amplifier operates in the frequency range from 10 Hz to 100 kHz.

F.7 a. Design a depletion NMOS amplifier to give a midfrequency voltage gain of $|A_v| = v_L/v_s = 20 \pm 5\%$ at a load resistance of $R_L = 10\ k\Omega$. Assume a source resistance of $R_S = 1.5\ k\Omega$ and $V_{DD} = 15$ V.
b. Modify the design so that the amplifier operates in the frequency range from 5 Hz to 50 kHz.

F.8 a. Design a MOSFET amplifier to give a midfrequency voltage gain of $|A_v| = v_L/v_s = 10 \pm 5\%$ at a load resistance of $R_L = 10\ k\Omega$. Assume a source resistance of $R_S = 1.5\ k\Omega$ and $V_{DD} = 12$ V.
b. Modify the design so that the amplifier operates in the frequency range from 10 Hz to 50 kHz.

F.9 a. Design an NMOS amplifier with an active load to give a midfrequency voltage gain of $|A_v| = v_L/v_s \geq 250 \pm 5\%$ at a load resistance of $R_L = 20\ k\Omega$. Assume a source resistance of $R_S = 1.5\ k\Omega$ and $V_{DD} = 15$ V.
b. Modify the design so that the amplifier operates in the frequency range from 10 Hz to 50 kHz.

Large Design Projects

F.10 The input signal to an amplifier is $v_s = 2$ mV, and the amplifier has a source resistance of $R_S = 1\ k\Omega$.
a. Design a BJT amplifier to give a midfrequency voltage gain A_v ($=v_L/v_s$ with a load resistance $R_L = 10\ k\Omega$) of greater than 650. The input resistance R_{in} of the amplifier should be greater than 70 kΩ, and the output resistance R_{out} should be less than 250 Ω. The DC supply voltage is $V_{CC} = 12$ V.
b. Modify the design so that the amplifier operates in the frequency range from 10 kHz to 80 kHz.
c. Apply feedback and modify the design. The input resistance must be increased by 20 (i.e., $R_{if} \geq 20\ R_{in}$), and the output resistance must be decreased by 20 (i.e., $R_{of} = R_o/20$).
d. Apply feedback and modify the design so that the amplifier oscillates at a frequency of $f_o = 20$ kHz.

F.11 The input signal to an amplifier is $v_s = 2$ mV, and the amplifier has a source resistance of $R_S = 1$ kΩ.

a. Design an NMOS amplifier to give a midfrequency voltage gain A_v ($=v_L/v_s$ with a load resistance $R_L = 10$ kΩ) of greater than 450. The input resistance R_{in} of the amplifier should be greater than 500 kΩ, and the output resistance R_{out} should be less than 250 Ω. The DC supply voltage is $V_{DD} = 12$ V.

b. Modify the design so that the amplifier operates in the frequency range from 20 kHz to 60 kHz.

c. Apply feedback and modify the design. The input resistance must be increased by 10 (i.e., $R_{if} \geq 10\, R_{in}$), and the output resistance must be decreased by 10 (i.e., $R_{of} \leq R_o/10$).

d. Apply feedback and modify the design so that the amplifier oscillates at a frequency of $f_o = 20$ kHz.

F.12 Design a BJT operational amplifier to give a large-signal differential gain of $10^3 \pm 10\%$. The input resistance should be higher than 100 kΩ ± 5%, and the output resistance should be less than 210 Ω ± 5%. The op-amp should have a common-mode rejection ratio of CMRR $= 10^4 \pm 10\%$. The unity-gain bandwidth must be better than $10^4 \pm 10\%$.

F.13 Design a multistage BJT amplifier to meet the following specifications: voltage gain, $|A_v| = v_L/v_s = 600$ ± 5% (with load); input resistance, $R_i = v_s/i_s \geq 25$ kΩ; output resistance, $R_o \leq 300$ Ω; load resistance, $R_L = 25$ kΩ; source resistance, $R_s = 1$ kΩ; DC supply, $V_{CC} = 15$ V; input signal, $v_s = 1$ mV to 2 mV (peak sinusoidal), 1 kHz. The amplifier should operate in the frequency range from 10 Hz to 30 kHz.

F.14 Design a multistage MOS amplifier to meet the following specifications: voltage gain, $|A_v| = v_L/v_s = 400$ ± 5% (with load); input resistance, $R_i = v_s/i_s \geq 25$ kΩ; output resistance, $R_o \leq 300$ Ω; load resistance, $R_L = 25$ kΩ; source resistance, $R_s = 1$ kΩ; DC supply, $V_{DD} = 15$ V; input signal, $v_s = 1$ mV to 2 mV (peak sinusoidal), 1 kHz. The amplifier should operate in the frequency range from 100 Hz to 25 kHz.

F.15 Design a differential amplifier to satisfy the following conditions: a differential voltage gain of $A_o \geq 2500$; common-mode rejection ratio of CMRR ≥ 4500; load of $R_L = 50$ kΩ; coupling capacitor of $C_L = \infty$. Use all BJTs and only one resistance. Use PSpice to verify your design with $v_d = 5$ μV and $v_c = 10$ mV. Supply voltages are $V_{CC} = V_{EE} = 15$ V. The amplifier should operate in the frequency range up to 25 kHz.

F.16 Design a differential amplifier to satisfy the following conditions: a differential voltage gain of $A_o \geq 2500$; common-mode rejection ratio of CMRR ≥ 4500; load of $R_L = 50$ kΩ; coupling capacitor of $C_L = \infty$. Use all MOSFETs. Use PSpice to verify your design with $v_d = 5$ μV and $v_c = 10$ mV. Supply voltages are $V_{DD} = V_{SS} = 15$ V. The amplifier should operate in the frequency range up to 25 kHz.

APPENDIX G

MILLER'S CAPACITOR METHOD

Let us consider the small-signal equivalent of a transistor amplifier as shown in Fig. G.1. C and C_i are the internal capacitances of the transistor. Assuming that the current through capacitor C is very small compared to the voltage dependent current source $g_m V_i$, the output voltage in frequency domain is

$$V_O = -g_m V_i (R_o \| R_L) \tag{G.1}$$

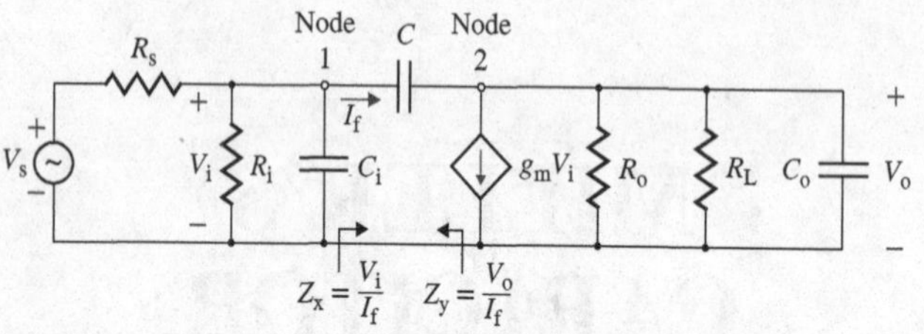

FIGURE G.1 Small signal equivalent of a transistor amplifier

Using Eq. (G.1), the current I_f flowing through C (from the left side to the right side) in frequency domain is given by

$$\begin{aligned} I_f &= j\omega C(V_i - V_o) = j\omega C[V_i + g_m V_i (R_o \| R_L)] \\ &= j\omega C[1 + g_m (R_o \| R_L)]V_i = j\omega C_m V_i \end{aligned} \tag{G.2}$$

where $C_m = C[1 + g_m (R_o \| R_L)]$

Using Eq. (G.1), the current $-I_f$ flowing through C (from the right side to the left side) is given by

$$\begin{aligned} -I_f &= j\omega C(V_o - V_i) = j\omega C\left[V_o + \frac{V_o}{g_m (R_o \| R_L)}\right] \\ &= j\omega C\left[1 + \frac{1}{g_m (R_o \| R_L)}\right]V_o = j\omega C_n V_o \end{aligned} \tag{G.4}$$

$$\text{where } C_n = C\left[1 + \frac{1}{g_m (R_o \| R_L)}\right] \tag{G.5}$$

Thus, capacitor C, which is connected between the input and output terminals of a high-gain amplifier with 180° phase reversal, can be replaced by a shunt capacitor C_m on the input side and another capacitor C_n on the output side. This arrangement is shown in Fig. G.2. The value of C is seen on the input side as a multiplying factor almost equal to the voltage gain $g_m(R_o \| R_L)$. This effect, known as *Miller's effect*, is

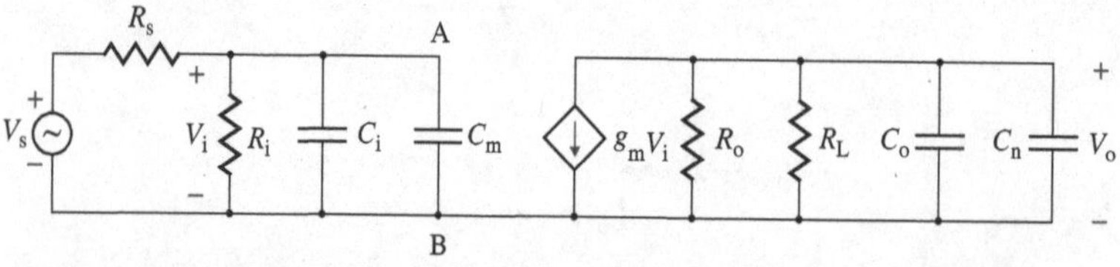

FIGURE G.2 Miller's equivalent high cutoff circuit

dominant in amplifiers with a high voltage gain and a phase reversal. Therefore, the high-frequency poles can be found from

$$f_{H1} = \frac{1}{2\pi(C_i + C_m)(R_s \| R_i)} \tag{G.6}$$

$$f_{H2} = \frac{1}{2\pi(C_o + C_n)(R_o \| R_L)} \tag{G.7}$$

Since $A_v = -g_m(R_o \| R_L)$ is the voltage gain, we can rewrite Eqs. (G.3) and (G.5) as

$$C_m = C(1 - A_v) \tag{G.8}$$

$$C_n = C\left(1 - \frac{1}{A_v}\right) \tag{G.9}$$

Therefore, if the voltage A_v is negative, then $C_m > C$ and there is a capacitance multiplication. Thus, the effective capacitance C_m between nodes 1 and 2 (in Fig. G.1) can be increased by a voltage amplifier. That is, a small capacitance connected between the input and output terminals of a voltage amplifier will have a much larger effective capacitance between the input terminals A and B (in Fig. G.2). C_m is called the *Miller capacitance*. The Miller capacitance plays an important role in designing the high-frequency response of amplifiers and in the design of active filters. For example, f_{H1} in Eq. (G.6), which is contributed mainly by C_m, will be much lower than f_{H2}, and thus f_{H1} will set the dominant high frequency.

G.1 MOSFET Drain Current

The conductance of the channel between the drain and the source can be found from

$$g_{DS} = \frac{1}{r_{DS}} = \frac{W}{L}\mu_n Q_n \tag{G.10}$$

where μ_n is the mobility of the electrons in the reverse-biased (also called the inversion) layer under the oxide layer, Q_n is the magnitude of the reverse-biased layer charge per unit area, W is the channel width, and L is the channel length

Let V_t be the MOSFET threshold voltage. If we consider a small incremental drain-to-source voltage v along the channel, we can find Q_n from the gate oxide capacitance C_{on} and the voltage difference $(v_{GS} - V_t - v)$ as given by

$$Q_n(v) = C_{on}(v_{GS} - V_t - v) \tag{G.11}$$

which can be applied to obtain the drain current as given by

$$i_D = \frac{W}{L}\mu_n C_{on}\int_0^{v_{DS}}(v_{GS} - V_t - v)dv = \frac{W}{L}\mu_n C_{on}\left[\left(v_{GS} - V_t\right)v_{DS} - v_{DS}^2\right] \tag{G.12}$$

which can be expressed in a more general form in terms of external voltages v_{GS} and v_{DS}

$$i_D = K_n[(v_{GS} - V_t)v_{DS} - v_{DS}^2] \tag{G.13}$$

Substituting $v_{DS} = V_{GS} - V_t$ at which the saturation occurs gives the drain current in the saturation region:

$$i_D = K_n(v_{GS} - V_t)^2 \tag{G.14}$$

For a depletion type MOSFET, the threshold voltage V_t is replaced by the pinch-off voltage V_p. Equation (G.14) becomes

$$i_D = K_n(v_{GS} - V_p)^2 \tag{G.15}$$

where $K_n = I_{DSS}/V_p^2$ and I_{DSS} is the maximum steady-state drain current at $v_{GS} = 0$. Equation (G.11) can also be written as

$$i_D = I_{DSS}\left(1 - \frac{v_{GS}}{V_p}\right)^2 \tag{G.16}$$

G.2 Early Effects

The collector current of a BJT can be expressed in general form as

$$i_C = I_{SC}e^{v_{BE}/v_T} \tag{G.17}$$

The slope introduced by the Early effect is almost linear with i_C and v_{CE} characteristics. If the collector current characteristics are extrapolated to zero collector current, the curves intercept the voltage axis at a point known as the *Early voltage*. The Early voltage is positive for an npn transistor and negative for a pnp transistor. From Fig. G.3, we can write the output conductance as

$$g_o = \frac{1}{r_o} = \frac{di_C}{dv_{CE}} = \frac{I_C}{v_{CE} + V_A} \tag{G.18}$$

where r_o is the output resistance of the transistor. If we include the finite slope of i_C versus v_{CE} characteristics due to the Early effect, the collector current in Eq. (G.17) can be modified to

$$i_C = I_{SC}e^{v_{BE}/v_T}\left(\frac{I_C}{v_{CE} + V_A}\right) \tag{G.19}$$

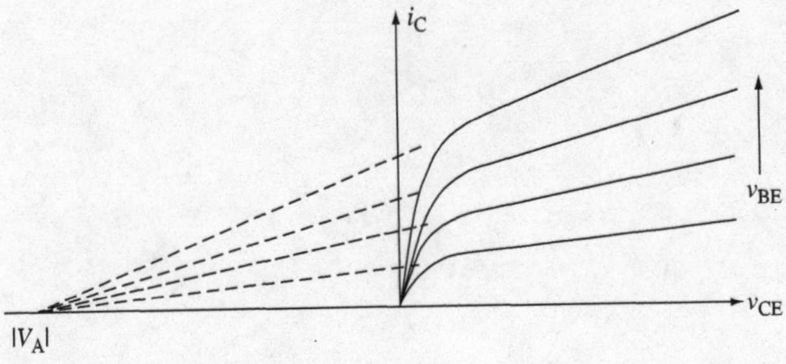

FIGURE G.3 Early voltage

APPENDIX H

HIGH-FREQUENCY TRANSFER FUNCTION METHOD

The high-frequency behavior is determined solely by the internal capacitors of the amplifier. The equivalent circuit for determining the high break frequencies is shown in Fig. H.1. Capacitance C, between the input and output terminals of the amplifier, can be replaced by Miller's equivalent capacitances. Therefore, we can find the frequency response by s-domain analysis or by Miller's capacitor method.

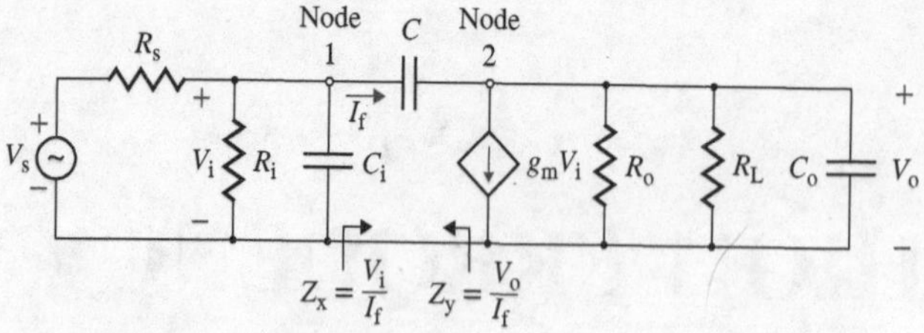

FIGURE H.1 Small signal equivalent high cutoff circuit

As we did for the low break frequencies, we will derive the transfer function for high break frequencies. Applying Kirchhoff's current law at nodes 1 and 2, we get the following equations in Laplace's domain of s:

$$\frac{V_s - V_i}{R_s} = \frac{V_i}{R_i} + V_iC_is + (V_i - V_o)Cs \tag{H.1}$$

$$g_mV_i + \frac{V_o}{R_o} + \frac{V_o}{R_L} + V_oC_os + (V_o - V_i)Cs = 0 \tag{H.2}$$

Equations (H.1) and (H.2) can be solved to give the voltage transfer function:

$$\frac{V_o(s)}{V_s(s)} = \frac{-(g_m - Cs)R_1R_2/R_s}{1 + s[R_1(C_i + C) + R_2(C_o + C) + g_mCR_1R_2] + s^2R_1R_2(C_iC_o + C_iC + C_oC)} \tag{H.3}$$

where $R_1 = (R_s \| R_i)$ and $R_2 = (R_o \| R_L)$. The denominator of Eq. (H.3) has two poles. If p_1 and p_2 are the two poles, the denominator can be written as

$$D(s) = \left(1 + \frac{s}{p_1}\right)\left(1 + \frac{s}{p_2}\right) = 1 + s\left(\frac{1}{p_1} + \frac{1}{p_2}\right) + \frac{s^2}{p_1p_2} \tag{H.4}$$

If the poles are widely separated, which is generally the case, and p_1 is assumed to be the dominant pole, while p_2 is assumed large, then Eq. (H.4) can be approximated by

$$D(s) = 1 + \frac{s}{p_1} + \frac{s^2}{p_1p_2} \tag{H.5}$$

Equating the coefficients of s in Eq. (H.3) to those in Eq. (H.5) yields

$$p_1 = \frac{1}{R_1(C_i + C) + R_2(C_o + C) + g_mCR_1R_2} \tag{H.6}$$

Equating the coefficients of s^2 in Eq. (H.3) to those in Eq. (H.5) yields

$$p_2 = \frac{R_1(C_i + C) + R_2(C_o + C) + g_m C R_1 R_2}{R_1 R_2(C_i C_o + C_i C + C_o C)} \tag{H.7}$$

In practice, the value of C is higher than that of C_i and C_o, and Eqs. (H.6) and (H.7) can be simplified further as follows:

$$p_1 \approx \frac{1}{g_m C R_1 R_2} \tag{H.8}$$

$$p_2 = \frac{g_m C}{C_i C_o + C_i C + C_o C} \tag{H.9}$$

Here p_1 and p_2 correspond to the break frequencies ω_1 and ω_2 in frequency domain.

▶ **NOTES**

1. The dominant pole p_1 decreases as C increases, whereas p_2 increases as C increases. Therefore, increasing C causes the poles to split apart, possibly making p_1 the dominant pole.
2. If $C >> C_i$ and $C >> C_o$, Eq. (H.9) can be approximated as

$$p_2 \approx \frac{g_m C}{C(C_i + C_o)} = \frac{g_m}{C_i + C_o} \tag{H.10}$$

3. If there is no feedback capacitance ($C = 0$), Eq. (H.3) gives the following poles:

$$p_1 = \frac{1}{C_i R_1} \tag{H.11}$$

$$p_2 = \frac{1}{C_o R_2} \tag{H.12}$$

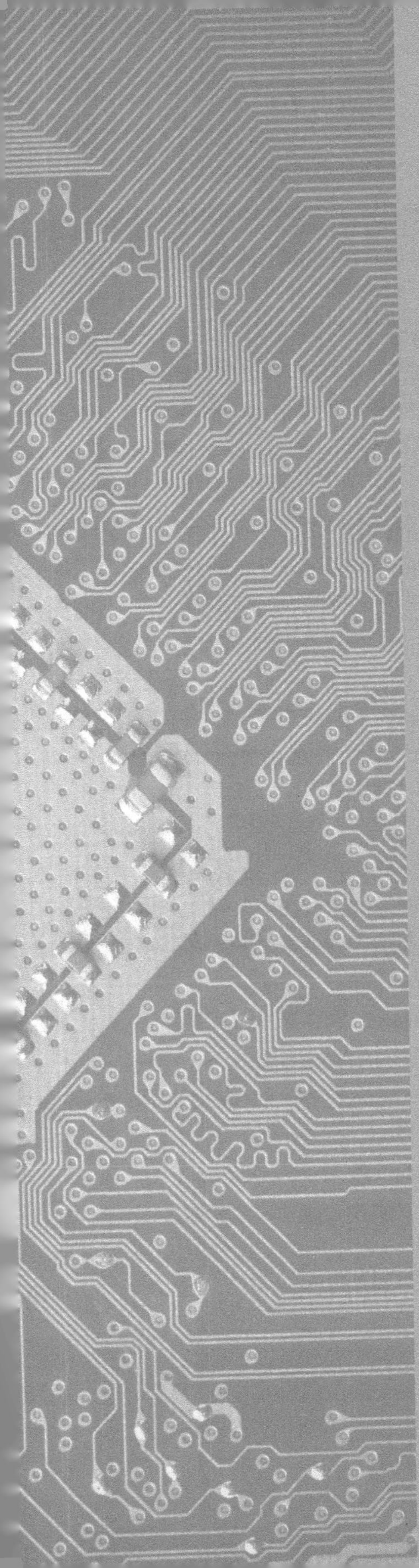

ANSWERS TO SELECTED PROBLEMS

Chapter 1

1.13 $W/L = 39$, $V_t = 5.06$ V

1.19 (a) $I_{D1,2} = 0.168$ mA
(b) $A_d = -1.832$, $A_c = 9.945 \times 10^{-3}$, CMRR = 184.2, $v_O = -54.96$ mV

1.21 (a) 96.46 μA, 103.54 μA

1.22 (a) 0.715 mA (b) 139.8 kΩ
(c) 100 V (d) 1.2909

1.24 (a) 1.36 mA (b) 73.53 kΩ
(c) 100 V (d) 1.282

1.26 (a) 4.152 mA (b) 24 kΩ
(c) 100 V (d) 1.142

1.28 $R_2 = 11.88$ kΩ, $R_o = 55.69$ MΩ

1.30 (a) $R_1 = 2.86$ MΩ (b) $R_o = 760$ MΩ

1.31 1.231 mA, 9.26 MΩ

1.35 For $V_A = 100$ and $\beta_F = 100$, $r_o = 100$ kΩ; for $V_{CC} = 15$ V, $R_1 = 13.6$ kΩ

1.37 (a) 2.955 mA, 2.011 mA (b) −961.5, −0.25, 38.44×10^6, 20.8 Ω, 6.04 MΩ

1.41 769.2, 2.08 MΩ, 8 MΩ, 587.9

1.45 −28.57, 0, 0

1.47 −60.14, 0, 0

1.49 $A_{do} = -961$, $\omega_H = 51.1$ krad/s

Chapter 2

2.1 (a) -9.5×10^{-3} (b) −101.26

2.4 (a) 22.173, 10.4% (b) 9.99, 4%

2.5 9

2.6 (a) 2 (b) 10^6 Hz (c) 2×10^6

2.9 (a) 3.17×10^{11} Ω (b) 0.475 mΩ
(c) 1.25

2.11 (a) 804 kΩ (b) 7.29 Ω (c) 1.31

2.17 5,355,006 Ω, 0.3 Ω, 7.43

2.20 (a) 14.06 kΩ (b) 30.2 Ω
(c) 1427.7

2.24 (a) 28.5 kΩ (c) 3.96×10^{-4} A/V

2.25 (a) 3.15 MΩ (b) 6.3 MΩ
(c) 0.4×10^{-3} A/V

2.28 $R_{if} = 1{,}321{,}326$ Ω, $A_f = 93.6$ mA/V

2.32 (a) 3.5 kΩ (b) 49 Ω (c) −97.67 Ω

2.35 26.26 Ω, 13.72 Ω, −3973 V/A

2.38 (a) 30.5 Ω (b) 342 Ω
(c) 225,077 V/A

2.45 (a) 17,029 Ω (b) 37,072 Ω
(c) 1.88

2.47 (a) 12.86 Ω (b) 5 kΩ
(c) 19.61

2.51 $-\frac{1}{4} + \frac{3}{20}e^{-4t} + \frac{3}{30}e^{6t}$

2.52 $e^{0.5t}[2 \sin 10t]$

2.56 $50/(0.234 - j0.643)$

2.58 44.44 Hz

2.60 (a) 401.9 kHz (b) 127 pF

2.61 10.47 μF

2.64 (a) 3.32 nF, 4.8 kΩ
(b) 79.7 pF, 199.8 kΩ

Chapter 3

3.2 $R_L = 1.43$ kΩ, 29 V

3.6 (a) $v_O = 14.5$ V, $R_1 = 493$ Ω
(b) 525.5 mW (c) 60.4%

3.8 (a) $I_C = 3.75$ A (b) 28.125 W
(c) 50%

3.11 $V_p = 14.8$ V, $P_L = 10.952$ W, 77.5%, $R_1 = 47$ Ω

3.14 (a) 7.5 A, $V_{p(max)} = 9.55$ V
(b) $P_{L(max)} = 56.25$ W (c) 78.55%

3.16 (a) $R_{ref} = 255$ Ω

3.32 76.48 W

Chapter 4

4.1 (b) $p_1 = -\frac{R_1 + R_2}{R_1 R_2 C_2}$

4.4 (b) $p_1 = -\frac{C_1 + C_2}{C_1 C_2 (R_1 + R_2)}$

Zero $z_1 = -\frac{1}{R_2 C_2}$

4.5 **(b)** $\omega_p = 10\sqrt{2}$ rad/s, $\omega_z = 10$ rad/s **(c)** $\beta_p = 10$ **(d)** $\phi_p = 135°$

4.15 $C = 0.05$ μF; $R = 1591\ \Omega$

4.17 $R_1 = R_2 = R_3 = R_4 = R$ and $C_2 = C_3 = C = 0.05$ μF, $f_o = 318\ \Omega$, $R_5 = R_4 = 685\ \Omega$, $Q = 477\ \Omega$

4.18 $C = 0.05$ μF, $R = 318\ \Omega$, $K = 1.586$, $R_F = 186\ \Omega$, $R_a = 357\ \Omega$, $R_b = 2932\ \Omega$

4.19 $R_1 = R_2 = R_3 = R_4 = R$ and $C_2 = C_3 = C = 0.01$ μF, $R = 15{,}916\ \Omega$, $K = 1.586$, $R_F = 9327\ \Omega$

4.20 $K_1 = 2$ and $R_1 = 10$ kΩ $= R_1$, $K_1 = 10$ kΩ, $C = C_2 = C_3 = 0.01$ μF, $R = R_2 = R_3 = 1126\ \Omega$, $K_2 = 40$ kΩ

4.22 $R = 15{,}915\ \Omega$, $R_1 = R_F = 10$ kΩ, $R_n = 6078\ \Omega$

4.24 $R = R_2 = R_3 = R_4$, $C = C_2 = C_3 = 0.01$ μF, and $f_o = 10$ kHz, $R = 1592\ \Omega$, $R_1 = 10$ kΩ, $R_F = 5860\ \Omega$, $C_a = 6.305$ nF, $C_b = 3.695$ nF

4.25 $C_a = 8.9$ nF, $C_b = 1.1$ nF

4.27 $BW = 1600$ Hz, $Q = 0.56$

4.28 $f_H = 1$ kHz and $C = 10$ nF, $R = 15{,}915\ \Omega$, $BW = 9$ kHz, $Q = 0.351$, $R'_1 = 10$ kΩ, $R'_F = 30$ kΩ

4.30 $C_1 = C_2 = C = 4.7$ nF, $R_1 = 3386\ \Omega$, $R_2 = 270{,}902\ \Omega$, $R_B = 178\ \Omega$

4.32 $R_2 = R_3 = R_4 = 10$ kΩ, $Q = 0.56$

4.34 $C = 0.047$ μF, $R_3 = 8466\ \Omega$

4.36 $C_1 = C_2 = C = 0.0047$ μF, $R_1 = 52.9$ kΩ, $R_2 = 846.6$ kHz, $K_{PB} = 8$

4.37 $R_1 = R_F = 10$ kΩ, $C = 0.1$ μF, $R = 98{,}995\ \Omega$

4.38 $R_1 = 10$ kΩ, $R_2 = 20$ kΩ, $Q = 0.707$, $R_3 = 14.14$ kΩ

Chapter 5

5.1 14.29, 135°

5.5 27 Ω

5.7 318 Ω

5.10 For $C = 0.1$ μF, 318 Ω, 20 kΩ

5.14 58.1 kHz, 2.5 Ω

5.16 For $C = 0.01$ μF, 25.3 mH, 133 Ω

5.19 435.8 kHz, $R_L/2R_1$

5.24 0.046 pF

INDEX